Please remember that this is a library book,
and that it belongs only temporarily to each
person who uses it. Be considerate. Do
not write in this, or any, library book.

Fourth Edition

Environmental Science
A GLOBAL CONCERN

William P. Cunningham

University of Minnesota

Barbara Woodworth Saigo

WCB Wm. C. Brown Publishers

Dubuque, IA   Bogotá   Buenos Aires   Caracas   Chicago   Guilford, CT   London
Madrid   Mexico City   Seoul   Singapore   Sydney   Taipei   Tokyo   Toronto

Project Team

Editor *Margaret J. Kemp*

Developmental Editor *Kathleen R. Loewenberg*

Production Editor *Cathy Ford Smith*

Marketing Manager *Thomas C. Lyon*

Designer *K. Wayne Harms*

Art Editor *Brenda A. Ernzen*

Photo Editor *Janice Hancock*

Advertising Coordinator *Heather Wagner*

Permissions Coordinator *Gail I. Wheatley*

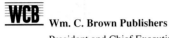 **Wm. C. Brown Publishers**

President and Chief Executive Officer *Beverly Kolz*

Vice President, Director of Editorial *Kevin Kane*

Vice President, Sales and Market Expansion *Virginia S. Moffat*

Vice President, Director of Production *Colleen A. Yonda*

Director of Marketing *Craig S. Marty*

National Sales Manager *Douglas J. DiNardo*

Executive Editor *Michael Lange*

Advertising Manager *Janelle Keeffer*

Production Editorial Manager *Renée Menne*

Publishing Services Manager *Karen J. Slaght*

Royalty/Permissions Manager *Connie Allendorf*

 A Times Mirror Company

Cover Photo © A. Tovy/H. Armstrong Roberts, Inc.

Photo Research by Connie Mueller

Brief Contents

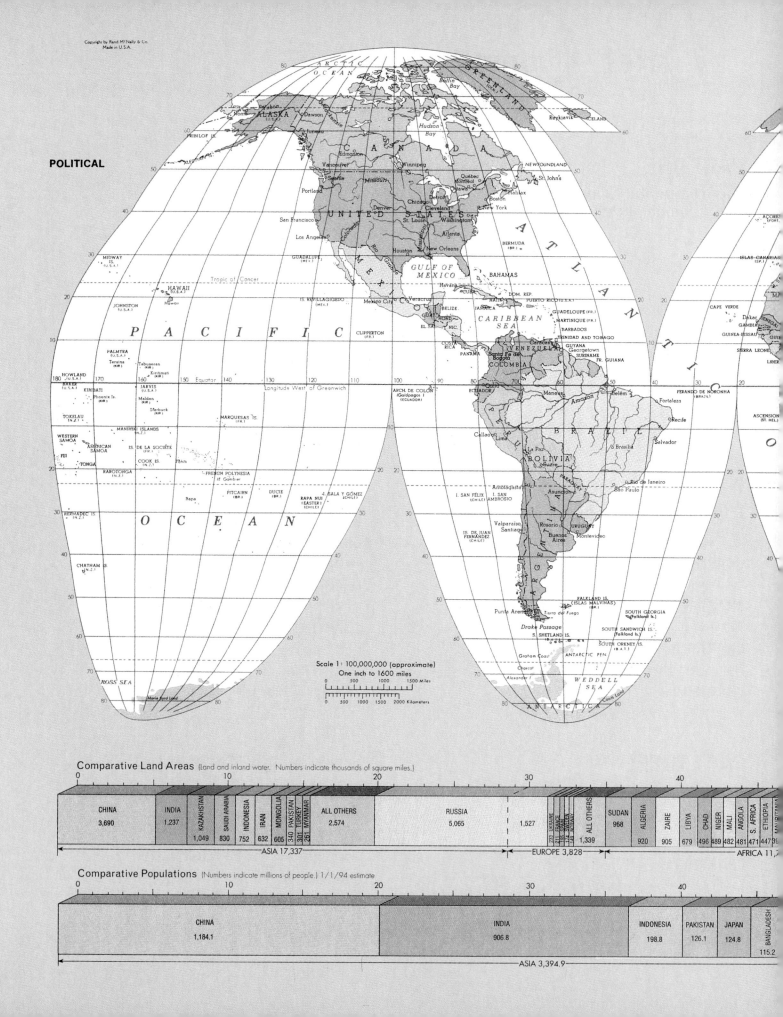

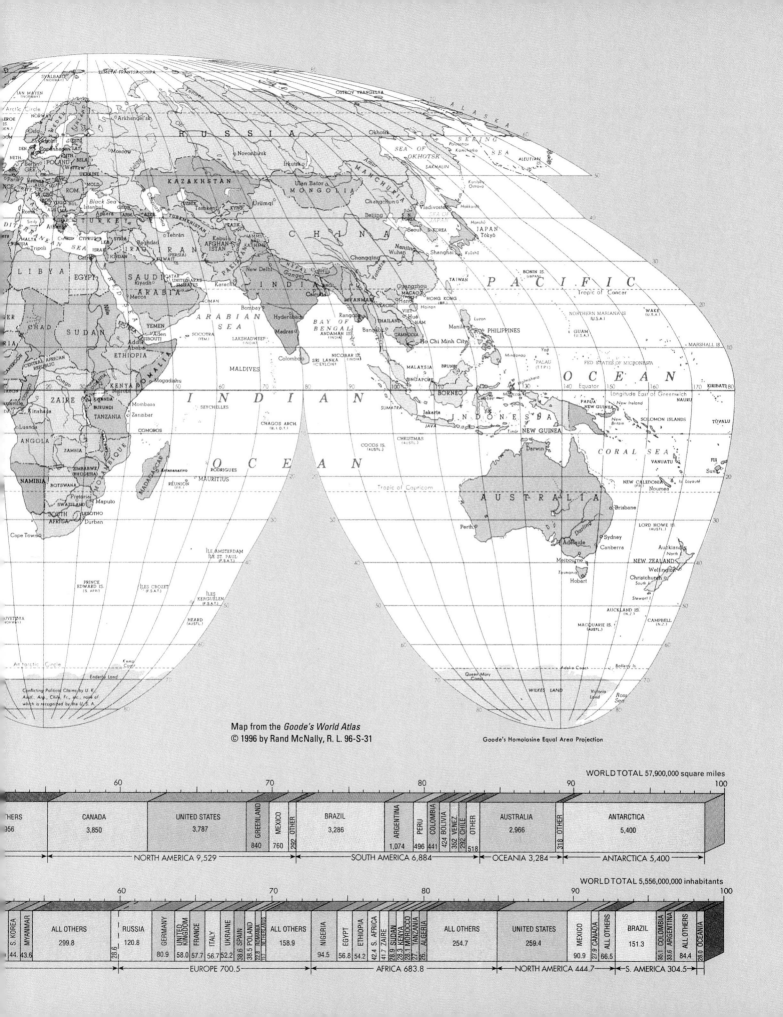

Map from the *Goode's World Atlas*
© 1996 by Rand McNally, R. L. 96-S-31

Goode's Homolosine Equal Area Projection

WORLD TOTAL 57,900,000 square miles

60	70	80	90	100

| HERS 956 | CANADA 3,850 | UNITED STATES 3,787 | GREENLAND 840 | MEXICO 760 | OTHER 292 | BRAZIL 3,286 | ARGENTINA 1,074 | PERU 496 | COLOMBIA 441 | BOLIVIA 424 | VENEZ 352 | CHILE 292 | OTHER 518 | AUSTRALIA 2,966 | OTHER 318 | ANTARCTICA 5,400 |

← NORTH AMERICA 9,529 → ← SOUTH AMERICA 6,884 → ← OCEANIA 3,284 → ← ANTARCTICA 5,400 →

WORLD TOTAL 5,556,000,000 inhabitants

60	70	80	90	100

| S. KOREA 44.4 | MYANMAR 43.6 | ALL OTHERS 299.8 | RUSSIA 120.8 | 28.6 | GERMANY 80.9 | UNITED KINGDOM 58.0 | FRANCE 57.7 | ITALY 56.7 | UKRAINE 52.2 | SPAIN 38.5 | POLAND 38.6 | ROMANIA 22.8 | ALL OTHERS 158.9 | NIGERIA 94.5 | EGYPT 56.8 | ETHIOPIA 54.2 | S. AFRICA 42.4 | ZAIRE 41.7 | SUDAN 28.9 | KENYA 28.3 | MOROCCO 28.1 | TANZANIA 27 | ALGERIA 26 | ALL OTHERS 254.7 | UNITED STATES 259.4 | MEXICO 90.9 | CANADA 27.9 | ALL OTHERS 66.5 | BRAZIL 151.3 | COLOMBIA 35.1 | ARGENTINA 33.6 | ALL OTHERS 84.4 | OCEANIA 28.0 |

← EUROPE 700.5 → ← AFRICA 683.8 → ← NORTH AMERICA 444.7 → ← S. AMERICA 304.5 →

reface

s the Twentieth Century draws to a close, the Earth's inhabitants are faced with both challenges and opportunities. We now see that we are degrading our environment and consuming its resources at unsupportable rates. Biodiversity is disappearing at a pace unequaled since the Age of Dinosaurs ended sixty-five million years ago. Valuable topsoil is washing off farm fields, ancient forests are being chopped down to make patio furniture, rivers and lakes are polluted with industrial wastes and untreated sewage, and air pollution obscures the skies and threatens to alter the global climate. As modern communications and travel bring us into closer contact with every other part of the world, we are forced to recognize that our common environment is shared with all other humans as well as our nonhuman neighbors. Unless we learn to live more sustainably and with less impact on our environment, the prospects for future generations are bleak indeed.

At the same time, we have better tools and knowledge than any previous generation to do something about these crises. Advances in science give us a deeper understanding of natural systems while technology provides the means to reduce pollution and to restore our environment. There are many hopeful signs of progress. Infant mortality is falling, human life expectancy is climbing, and literacy and democracy have advanced in many parts of the world. Population growth rates have decreased dramatically in many developing countries. We can now produce more food and goods with fewer resources then ever before. There is reason to hope that we will be able to continue real human progress while also preserving and renewing the natural world on which we depend.

In 1983, the United Nations established the World Commission on Environment and Development to examine our current situation and to propose ways to reconcile the need to reduce world poverty and improve opportunities with the need to protect our common environment. After public hearings on five continents and nearly three years of deliberations, the Commission published its report, entitled *Our Common Future,* which rejected predictions of inevitable environmental decay, poverty, hardship, and decreasing resources. Instead, it envisioned a new era of development that will sustain and expand the environmental resource base while offering an improved quality of life to everyone. The report suggested that with education, compassion, and hard work, we can bring about the necessary changes to leave a better world for our children than we inherited from our parents. As Margaret Meade once said, "Never doubt that a small group of thoughtful, committed people can change the world; indeed it is the only thing that can."

To bring about these changes, we all need a basic understanding in scientific principles, as well as some insights into the social, political, and economic systems that impact our collective environment. Our purpose in writing this book is to bring together concepts and data from the natural and social sciences in order to provide a foundation for both understanding the problems we face and finding ways to achieve a better future. We hope this book will be an interesting, informative, and inspiring resource toward that end.

Our aim has been to achieve a balanced, realistic view of environmental conditions that neither wallows in "doom and gloom" pessimism, nor glosses over serious problems with unwarranted optimism. We attempt to present relevant information and teach essential skills that will allow students to analyze and discuss challenging issues while forming their own opinions about what should be done. Rather than insist on a rigid orthodoxy, we hope that our readers will become reflective thinkers capable of independent evaluation of critical environmental issues.

Audience

This book is intended for use in a one- or two-semester course in Environmental Science, Human Ecology, or Environmental Studies at the college level. The vocabulary and level of discussion have been kept simple and nontechnical so as to be accessible to lower division students with little or no science background. At the same time, enough data and depth are presented to make this book suitable for upper division classes and as a valuable resource for many students who will keep it in their personal libraries after their formal studies are completed.

Organization

This book is organized into five sections that reflect the major categories of material in environmental science. We have attempted to make each chapter independent and self-contained, however, for use in any order or combination.

Part One presents an overview of environmental science, current global environmental concerns, and a brief history of environmental ethics, resource use, and conservation. Three chapters in this section survey basic ecological principles to provide a foundation for understanding how nature works and how we can live in harmony with nature. **Part Two** reviews principles of population biology and then applies those concepts to human population issues. Fundamental principles of resource economics and environmental health and toxicology round out this section to provide a basis for analyzing what we are doing to our environment and what, in turn, the environment is doing to us.

Part Three describes biological resources that provide practical needs such as food and fiber, as well as intangible benefits such as aesthetics, ecological services, and recreation. We review resource problems and investigate ways that resources are being preserved, extended, and restored. **Part Four** complements Part Three with a discussion of physical resources such as minerals, air, water, and energy supplies. Again, we introduce dilemmas together with possible solutions.

Part Five examines the interactions between society and our environment. It begins with the important issues of solid and toxic wastes. A significant source of many problems discussed in earlier chapters is the rapid urbanization of human populations, especially the growth of megacities in the developing world. To balance this presentation, we illustrate some principles of urban planning as a possible solution to the dangers and degraded environment of modern cities. Finally, we show students some things we can do—individually and collectively—to help build a sustainable future for ourselves and for future generations.

Themes

Global Environment

The global nature of environmental concerns makes it necessary for each of us to be familiar with world geography and to be aware of cultures and conditions in countries far from our own. We have expanded our scope beyond the United States or even North America, therefore, with examples and information from all over the world. This broad view makes environmental science an ideal course for fulfilling the cultural diversity component of liberal education requirements at many institutions.

Holistic Approach

Complex problems demand creative solutions from many disciplines. Understanding our current situation requires knowledge not only from natural sciences such as ecology, geology, and environmental health but also from social sciences such as economics, history, and political science. While we can't represent all these fields completely in a single book, we have tried to weave together information from many fields to give a more complete view of where we are, how we got here, and what our options are for the future.

Critical and Reflective Thinking

Increasingly, we are faced with situations in which experts give us contradictory interpretations of the same data. How can we decide what to believe and what to do in the face of uncertainty and conflicting advice? We need the skills and attitudes necessary to think critically and analytically about the information available to us. In addition to a more extensive presentation of principles of critical thinking in chapter two, nearly every chapter in this edition has a new critical thinking box, most of which were written by Dr. Darby Nelson, and are designed to challenge students to distinguish between fact and value claims, recognize logical fallacies, and make a plan for how to analyze information. Extended questions at the end of each chapter ask students to reflect on and discuss the information presented.

Environmental Justice and Intergenerational Equity

We live in a world of vast inequity in wealth and access to resources. At least 1 billion people live in abject poverty without access to adequate food, shelter, medical care, education, and other necessities required for a healthy, secure life. In fact, most of the 4 billion residents of the less wealthy countries live in what most people who read this book would consider poverty. By contrast, the richest billion—those of us in the industrialized nations—live lives of unparalleled comfort and convenience based on consumption of an inordinate share of the world's resources.

Increasingly, this unbalanced distribution of resources makes the world's poorest citizens both the victims and the agents of environmental degradation. To meet the daily struggle for survival they fell irreplaceable forests, farm erodible hillsides, hunt endangered species, and work in dangerous, degrading conditions. Reducing this inequity by both lowering our own consumption rates and helping others attain less destructive ways of living is an essential requirement for preserving a livable environment. Ethical questions about the world we are leaving to our children also are part of the equity issues that we consider in the "Ethical Considerations" section of every Case Study.

Stewardship

Some environmentalists consider humans irredeemably flawed and conclude that anything we do will end up badly. We take the more optimistic view that while we have made many mistakes, we also have the potential to find solutions. Technology is neither an evil force to be feared nor a magic genie that will save us from errors; rather, it is a tool that must be used carefully. Our power to change and shape our surroundings gives us a responsibility to be caretakers, not only for ourselves and our children, but also for all other creatures.

Sustainable Development

Some authors argue that increasing environmental degradation and critical shortages of nonrenewable resources require immediate, drastic reductions in human populations, levels of consumption, and standards of living. Others argue that continued economic growth—and even population growth in some cases—are the only politically acceptable way to provide the energy, capital, knowledge, and social stability needed for environmental protection. How can we reconcile these two positions? A middle course that we advocate is sustainable development: a real improvement in the quality of life and standard of living for all based on renewable resources, equitable sharing of benefits, and living in harmony with nature. This moderately optimistic outlook is based, at least in part, on a conviction that guilt trips and predictions of disaster are so disempowering and disheartening that they become self-fulfilling prophecies. We hope this book will inspire its readers to become involved—individually and collectively—to work positively to protect and improve our global environment.

New Features in this Edition

Even-Handed Approach

In each edition of this book, we have attempted to present controversial topics in a fair, unbiased manner. While many of the serious environmental problems we face require commitment and decisive action, we recognize that every issue has at least two sides and that we gain little by hectoring or denouncing those who disagree with us. If we truly want students to think for themselves, we have to leave room for those who hold ideas different from our own. Although our individual perspectives are probably obvious in this book, we have tried very hard to acknowledge opposing viewpoints.

Science and Critical Thinking as Ways of Knowing

Because of the importance of scientific literacy and critical thinking in today's world, we have strengthened our coverage of scientific methods and meaning in this edition and have added material on critical and reflective thinking skills and attitudes. The early chapters also include a brief history of conservation and environmentalism, a survey of attitudes toward nature in different groups, and an explicit acknowledgment of the choices between pessimism and optimism in confronting our environmental options.

Updated Ecology

Modernized and strengthened chapters on ecology in this edition reflect the rapidly changing understandings of chance, individualism, and variability in natural systems. We take a less deterministic (Clementsian) view of ecosystems and biological communities in this edition and move toward a stronger Glesonian understanding of community diversity and history. In chapter five ("Biomes, Landscapes, Restoration, and Management"), we reduce the number of community types described in order to consider the effects of human disturbance more fully and to add an innovative section on the exciting new fields of landscape ecology, ecosystem management, and restoration ecology.

Environmental Justice and the Plight of the Poor

Although previous editions of this book contained a strong commitment to social justice, information on linkages between poverty and environmental degradation presented at the United Nations Earth Summit in 1992 have provided new material for this book. Evidence for environmental racism, inequitable access to resources, and foreclosed options for future generations are important considerations in understanding our current situation and what needs to be done.

Personal and Collective Responsibilities and Actions

Although ethics and responsibilities form an underlying theme throughout this book, the final chapter has been rewritten to further emphasize what we can do individually and collectively to improve our environment. We provide information on "green" shopping, responsible consumerism, environmental organizations, environmental education, and environmental careers. We also discuss international NGOs, green government and politics, and examples of successful environmental progress both at home and abroad.

Learning Aids

This book is designed to be useful as a self-education tool for students. To facilitate studying and encourage higher-level thinking, each chapter begins with a set of **objectives** based on major concepts that students should master. Headings within the chapter act as signposts to show relationships and to track the orderly flow of information. Information is also concisely and clearly presented in the many **tables** used in the text. **Key terms,** indicated by boldface type, are defined in context where they are first used and are also listed in the glossary for quick reference. The glossary also contains definitions of terms that students may not recognize.

At the end of each chapter, a **summary** reviews the material just covered and invites students to check their grasp of major concepts. **Questions for review** provide another opportunity for students to test their understanding, while **questions for critical and reflective thinking** are designed to stimulate creative, analytical thinking and to serve as a springboard for in-class or after-class discussions.

Case Studies, In Depth boxes, and **What Do *You* Think?** essays break up the text into manageable segments and give students real-life examples to evaluate. This

edition includes 51 boxed readings and 24 case studies. While users of earlier editions will recognize many of their favorite readings, more than half are new to this text. The case studies are more complete stories than before and most introduce ethical perspectives as well as important environmental issues. Fifteen new What Do *You* Think? boxes, written by Dr. Darby Nelson of Anoka-Ramsey Community College, challenge students to think critically and to formul-

ate their own opinions about issues. The boxes are long enough (about 1000 words, on average) to present real information rather than be simply editorials or trivial envirobriefs.

Profiles of ordinary people in environmental careers have been expanded in this edition to give students models with whom they can identify, which will help them consider how they might pursue a career in an environmental area. While guest editorials and opinion pieces by renowned scientists can be informative, these experts are often too far removed from the life experiences of the typical student to be meaningful. We expect that the profiles presented here will encourage our readers to believe that *they* can make a difference. Although these profiles represent only a few of the many different paths leading to environmental careers, they should serve as a guide for students to

follow in preparing themselves for entry into the work force.

Supplementary Materials

Instructor's Manual

The Instructor's Manual and Test Item File prepared by Mary Ann Cunningham are designed to assist instructors as they plan and prepare for classes using *Environmental Science*. Each chapter begins with an outline of the textbook chapter as well as chapter objectives and a listing of key terms. "Additional Learning Experiences" in each chapter list relevant films, videotapes, and computer programs that will enrich your teaching. The Test Item File offers 50 to 70 questions per chapter, including multiple-choice, fill-in-the-blank, and true-false questions. Critical-thinking questions included with each chapter make excellent essay questions. An important addition to this edition of the Instructor's Manual is a group of questions based on interpretation of graphs and diagrams to test students' grasp of this important, but often misunderstood, set of skills.

Classroom Testing Software

This computerized testing service provides instructions with either a mail-in/call-in testing program or the complete Test Item File on diskette for use with the IBM or Macintosh computer. The software also provides a computerized grade management system for instructors. This program tracks student performance on assignments and examinations. It will compute each student's percentage and corresponding letter grade, as well as the class average. Printouts can be made utilizing both text and graphics.

Student Study Guide

Written by master teacher Darby Nelson, the Student Study Guide contains many innovative features that will help your students master environmental science. Special "Be Alert For" boxes call attention to concepts in each chapter that are especially important or difficult to understand. "Time Out" boxes invite students to stop and reflect on what they have learned and what it all means. This feature fits well with the emphasis on critical thinking in the text. "In Your Own Words" sections in each chapter encourage students to practice writing about basic ideas in each chapter. Fill-in-the-blank questions with answers and references in a separate section guide students in efficient, effective studying.

Electronic Image Bank is a CD-ROM that contains full-color digital versions of the art found in your text. Also included is an easy-to-use program that enables you to quickly move among images, show or hide labels, and create your own multimedia presentation. (ISBN 28671-1)

100 Full-color Acetates of the key line drawings from the text are available free to adopters.

50 Full-color Acetates prepared by the World Resources Institute are available to adopters and provide current data on "Economics," "Population Trends," "Land Use," "Soils," "Agriculture," "Forests," "Biodiversity," "Water Resources," "Energy," and "Atmosphere and Climate," using maps and graphs.

Field and Laboratory Activities in Environment Science by Eldon Enger and Bradley Smith is a flexible guide containing 43 exercises for courses that require hands-on activity. This useful manual offers instruction on basic ecology, energy, pollution, and environmental policy and decision making. (ISBN 15909-4)

You Can Make a Difference by Judith Getis is a short, inexpensive supplement that offers students practical guidelines for recycling, conserving energy, disposing of hazardous wastes, and other environmentally sound practices. It can be shrink-wrapped with the text, at minimal additional cost. (ISBN 13923-9)

Annual Editions: The Environment is a great source for research and discussion, this useful manual provides convenient, affordable access to a wide range of up-to-date articles taken from a variety of the most respected and informative magazines, newspapers, and journals published today. (ISBN 31581-9)

Taking Sides: Clashing Views on Controversial Environmental Issues is a collection of point-counterpoint arguments considering eighteen thought-provoking issues, such as environmental restrictions on property owners, hazardous waste disposal, incentives to combat tropical deforestation, and necessity of efforts to slow global warming. (ISBN 30692-5)

Coming in '97!

Interactive CD-ROM will lend visual excitement to your course while providing students an opportunity to explore a unique and exotic biome. "Discovering Our Environment: A Case Study of the Virgin Islands" features basic environmental principles as portrayed by the daily life and industry in the Virgin Islands. Students will experience the devastation of Hurricane Hugo, explore a coral reef, learn energy conservation, see endangered wildlife, and investigate a slave plantation. The program also features virtual field trips, animations, video and audio segments, quizzing, and interactive maps. The CD-ROM will run on either Mac or Windows programs and will correlate to the fourth edition of Environmental Science: A Global Concern. Ask your sales representative for more information.

Acknowledgments

Many people have contributed in a variety of ways to each edition of this book. We thank, again, friends and colleagues who were acknowledged in the previous editions. Several colleagues have read drafts of chapters and offered helpful suggestions for this editing, as well, for which we are grateful. Mary Cunningham was the initial editor and arbiter of style and taste for the entire manuscript. Mary Ann Cunningham and Chris Sneddon provided research assistance and made many helpful suggestions for revised and new materials. We thank Darby Nelson for his excellent critical thinking boxes and for writing four new profiles of environmental professionals.

We are indebted to all the students and teachers who have sent us helpful suggestions, corrections, and recommendations on how we can improve this book. More than 100 teachers of environmental science reviewed one or more chapters in our third edition and offered a wealth of excellent ideas for how we could improve this edition. We hope that those who read this edition will offer us their advice and insights as well.

Obviously, no individual can have a complete mastery of all the material in a book as complex as this. Those of you who read it undoubtedly know much more about specific areas than we do, and your observations and recommendations would be most sincerely appreciated. Little of the vast range of material in this book represents our own original research. We owe a great debt of gratitude to the multitude of scholars whose work forms the basis of our understanding of environmental science. We stand on the shoulders of giants. If errors persist in spite of our best efforts to root them out, we accept responsibility and ask for your indulgence and assistance in correcting them for the next edition.

We would like to express our appreciation to the entire WCB book team for their encouragement, professionalism, and creativity in producing this book. Kathy Loewenberg was very helpful in the developmental stage and Cathy Smith shepherded the manuscript through the production process. Connie Mueller found superb photographs and Gail Wheatley handled permissions with aplomb. Kevin Kane and Marge Kemp oversaw the whole operation and made this edition possible. To them, and to all the others at WCB who worked on this project, our profound thanks.

Reviewers

We gratefully acknowledge the generous assistance of many reviewers, whose constructive criticism helped us improve this book in each of the successive editions. They include:

Donald D. Adams
State University of New York - Plattsburgh

Max Anderson
University of Wisconsin - Platteville

Joe Arruda
Pittsburg State University

Stephen W. Banks
Louisiana State University Shreveport

Wayne A. Becker
Normandale Community College

Mark C. Belk
Brigham Young University

Bruce Bennett
Community College of Rhode Island

Neil Bernstein
Mount Mercy College

Peter Biddle
North County Community College

Del Blackburn
Clark College

Andrew Blaustein
Oregon State University

Charles Bomar
University of Wisconsin - Stout

Moonyean S. Brower
Armstrong State College

William R. Brown
Prairie View A & M University

Gary Bryner
Brigham Young University

John A. Bumpus
University of Northern Iowa

Warren R. Buss
University of Northern Colorado

Louis H. Cadwell
Providence College

Becky A. Carlson
John Carroll University

Terry Caselnova
Saint Leo College

William R. Chaney
Purdue University

Carl F. Chuey
Youngstown State University

Lu Anne Clark
Lansing Community College

Patricia Clark
Cumberland College

Phillip D. Clem
University of Charleston

Douglas Crawford-Brown
University of North Carolina

Susan L. Dalterio
University of Texas - San Antonio

George A. Damoff
East Texas Baptist University

David Dathe
Alverno College

Elizabeth A. Desy
Southwest State University

Patricia M. Dooris
Saint Leo College

Arthur Driscoll
Westfield State College

John B. Dunning, Jr.
Purdue University

David A. Easterla
Northwest Missouri State University

Debbie Eustis-Grandy
Maine School of Science and Math

David L. Evans
Penn College/PSU

Raymond A. Faber
St. Mary's College of Minnesota

Allen Farrand
Bellevue Community College

Mark Finley
Heartland Community College

Lloyd C. Fitzpatrick
University of North Texas

James R. Fleming
Colby College

Albeno P. Garbin
University of Georgia

Al B. Garlauskas
Kent State University

Janice Gehrke
University of Wisconsin - Stout

Frank S. Gilliam
Marshall University

Brent M. Graves
Northern Michigan University

J. Gray
Otero Junior College

Gian Gupta
University of Maryland - Eastern Shore

Lonnie J. Guralnick
Western Oregon State College

Denny O. Harris
University of Kentucky

Ron Harrison
Mercer University

Edward J. Harvey
California Baptist College

William C. Hauck
Nettleton Junior College

Arthur J. Hawley
University of North Carolina - Chapel Hill

Brian T. Hazlett
Briar Cliff College

Rae Hoisve
Minneapolis Community College

Robert E. Holtz
Concordia College

Huey-Min Hwang
Jackson State University

Dan Ippolito
Anderson University

Mitchell T. Kamlay
Lawrence Technological University

Stephen R. Karr
Carson-Newman College

Penelope M. Koines
University of Maryland - College Park

J. G. Kopachena
Texas A & M University at Commerce

Charles J. Kunert
Concordia University

Hugh Lefcort
Gonzaga University

David A. Lovejoy
Westfield State College

Paul E. Lutz
University of North Carolina at Greensboro

A. Dale Marcy
North Idaho College

John Mathwig
College of Lake County

Ruth E. Mayer
Valencia Community College - Orlando

Mary Lou McReynolds
Hopkinsville Community College

Richard L. Meyer
University of Arkansas

Beth Middleton
Southern Illinois University

Neal D. Mundahl
Winona State University

Andy J. Neill
Joliet Junior College

Darby Nelson
Anoka-Ramsey Community College

Stephen R. Overmann
Southeast Missouri State University

Assad A. Panah
University of Pittsburgh at Bradford

Ken Parejko
University of Wisconsin - Stout

Chris Pennuto
University of Southern Maine

Mark D. Plunkett
Bellevue Community College

Gerald L. Reynolds
University of Central Arkansas

Charles Rhyne
Jackson State University

Gwynne Stoner Rife
The University of Findlay

Sheila J. Roberts
Bowling Green State University

C. Lee Rockett
Bowling Green State University

Terri Ann Rogers
Hawkeye Community College

Neil Sabine
Indiana University East

Jeffrey A. Schneider
SUNY - Oswego

Janet Sherman
Pennsylvania College of Technology

Ravindra P. Sinha
Elizabeth City State University

Philip Smartt
Stephen Austin State University

Kathryn Springsteen
Colby-Sawyer College

Peter G. Sutterlin
Wichita State College

Max R. Terman
Tabor College

Monte Thies
Sam Houston State University

M. Scott Thomson
University of Wisconsin - Parkside

Colin E. Thorn
University of Illinois at Urbana-Champaign

Carl Tobin
Alaska Pacific University

Nels H. Troelstrup, Jr.
South Dakota State University

Stephen C. Trombulak
Middlebury College

Maud M. Walsh
Louisiana State University

Donald E. Wanhala, Sr.
Suomi College

Phillip Watson
Ferris State University

Arlene Westhoven
Ferris State University

H. Warrington Williams
Westminster College

Danielle M. Wirth
Des Moines Area Community College

Richard J. Wright
Valencia Community College

Carol Ecale Zhou
University of California - Davis

Tommy L. Zimmerman
Agricultural Technical Institute Ohio State University

Contents

CHAPTER 4

Biological Communities and Species Interactions

CHAPTER 5

Biomes, Landscapes, Restoration, and Management

PART TWO

Population, Economics, and Environmental Health

CHAPTER 6

Population Dynamics

CHAPTER 7

Human Populations

CHAPTER 8

Environmental Resource Economics

CHAPTER 9

Environmental Health and Toxicology

PART THREE
Food, Land, and Biological Resources

CHAPTER 10
Food, Hunger, and Nutrition

CHAPTER 11
Soil Resources and Sustainable Agriculture

CHAPTER 12
Pest Control

PART FOUR
Physical Resources

CHAPTER 16
The Earth and Its Crustal Resources

CHAPTER 17
Air, Climate, and Weather

CHAPTER 18
Air Pollution

CHAPTER 19

Water Use and Management

CHAPTER 20

Water Pollution

PART FIVE
Society and the Environment

CHAPTER 23
Solid, Toxic, and Hazardous Waste

Rafters enjoy the Merced River in Yosemite National Park.

Part 1

Environmental Science and Ecological Principles

*A*ll history consists of successive excursions from a single starting point, to which man returns again and again to organize yet another search for a durable set of values.

Aldo Leopold

A Lacadon Indian father and son plant seeds in a recently cleared forest opening in the lowland rainforest of the Yucatan Peninsula. In many places indigenous people have lived in harmony with their environment for centuries. Their knowledge of local conditions represents a valuable resource for us all.

CHAPTER 1: Understanding Our Environment

Objectives

After studying this chapter, you should be able to:

- define the term, environment, and identify some important environmental concerns that we face today.

- discuss the history of conservation and the different attitudes toward nature revealed by utilitarian conservation and biocentric preservation.

- briefly describe some major environmental dilemmas and issues that shape our current environmental agenda.

- understand the connection between poverty and environmental degradation, as well as the division between the wealthy, industrialized countries and the poorer, developing countries of the world.

- recognize some of the reasons for feeling both optimistic and pessimistic about our environmental future.

Disappearing Frogs

Among the most beautiful amphibians in the world, golden toads (Bufo periglenes) *are the symbol of the Monteverde Cloud Forest Preserve in Costa Rica (fig. 1.1). Every spring, for as long as anyone can remember, the brilliant, glowing orange-gold males would emerge from their hiding places to congregate by the hundreds in rain pools on the forest floor. As recently as 1987, golden toads were abundant, but then something mysterious began to happen. In 1988, the toads were scarce, and the next year, only a few scattered individuals could be found. Since 1990, not one single toad has been seen.*

What happened to them? No one knows for sure. The forest preserve where they lived appears unchanged, but subtle variations in rainfall or weather patterns may play a role. Air pollution or water-borne contaminants have been suggested as possible culprits. Because their thin skins absorb toxins readily,

Figure 1.1

A golden toad from Costa Rica.

frogs are highly sensitive indicators of environmental quality. They may be early warning indicators of dangerous changes just like the canaries that miners used to take into mines.

Perhaps the most ominous thing about the disappearance of the golden toads is that it is not an isolated event. Around the world, populations of frogs, toads, salamanders, newts, and their amphibian kin have suddenly diminished or vanished entirely from wetlands where they once were common. Denmark, Peru, India, Canada, the United States, and at least a dozen other countries have reported sudden reductions in amphibian numbers in the past 10 or 20 years. Could these sudden disappearances be telling us something important about the state of our global environment?

What is causing this abrupt decline? There is no single obvious factor, but habitat destruction leads the list of suspected causes in many places. Wetlands are drained for agriculture and urban expansion, forests are cleared by lumber operations, and erosion fills streams and ponds with silt. Some other threats include water pollution, acid rain, pesticide runoff, and other sources of toxic chemicals that poison the environment. Game fish and other exotic species introduced by humans eat frogs, eggs, and tadpoles. Millions of frogs are caught every year for the pet, laboratory, and food trade. Climate modification due to increased "greenhouse" gases or increased UV irradiation caused by stratospheric ozone destruction may even be part of the picture.

It's often hard to know whether changing population sizes or distributions are significant because we don't know how many individual animals or different species a particular place would ordinarily have. In some cases, large population fluctuations from year to year in a particular habitat may be perfectly normal. In spite of widespread decreases in many species, some others remain relatively abundant, and a few are even increasing. Still, for the majority of amphibians, something seems to be very wrong.

The good news in this situation is that many dangers to amphibians are things we can do something about. Nature often is resilient and prolific. Remnant populations of many species can still be found. Restoring wetlands and other critical habitat, reducing air and water pollution, and removing exotic predators and competitors all can help restore froggy choruses to their former territories.

The first step in this process is to learn what natural conditions were and what we have done to them. We need to understand how ecological processes work as well as how we can manage human activities to live in harmony with the world around us. In this book, we will survey a variety of global environmental problems and offer some suggestions for remedying them. In many cases—such as the example of disappearing frogs and toads—the problems are complex and only partially understood. Environmental science is a new and evolving field. It will require active participation on your part to think critically about these issues and to work out how you feel personally about them. You are undertaking a journey of discovery that will continue, we hope, long after you finish the course for which you bought this book. ▶

What is Environmental Science?

Humans have always inhabited two worlds. One is the natural world of plants, animals, soils, air, and water that preceded us by billions of years and of which we are a part. The other is the world of social institutions and artifacts that we create for ourselves using science, technology, and political organization. Both worlds are essential to our lives, but integrating them successfully causes enduring tensions.

Where earlier people had limited ability to alter their surroundings, we now have power to extract and consume resources, produce wastes, and modify our world in ways that threaten both our continued existence and that of many organisms with which we share the planet. To ensure a sustainable future for ourselves and future generations, we need to understand something about how our world works, what we are doing to it, and what we can do to protect and improve it.

Environment (from the French *environner:* to encircle or surround) can be defined as (1) the circumstances or conditions that surround an organism or group of organisms, or (2) the complex of social or cultural conditions that affect an individual or community. Since humans inhabit the natural world as well as the "built" or technological, social, and cultural world, all constitute important parts of our environment (fig. 1.2).

Environmental science, then, is the systematic study of our environment and our proper place in it. A relatively new field, environmental science is highly interdisciplinary, integrating natural sciences, social sciences, and humanities in a broad, holistic study of the world around us. In contrast to more theoretical disciplines, environmental science is mission-oriented. That is, it seeks new, valid, generalizable knowledge about the natural world and our impacts on it, but obtaining this information creates a responsibility to get involved in trying to do something about the problems we have created.

As distinguished economist Barbara Ward points out, for an increasing number of environmental issues, the difficulty is not to identify remedies. Remedies are now well understood. The problem is to make them socially, economically, and politically acceptable. Foresters know how to plant trees, but not how to establish conditions under which villagers in developing countries can manage plantations for themselves. Engineers know how to control pollution, but not how to persuade factories to install the necessary equipment. City planners know how to build housing and design safe drinking water systems, but not how to make them affordable for the poorest

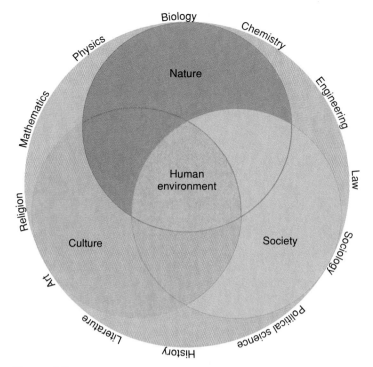

Figure 1.2

The intersections of the natural world with the social and cultural worlds encompass our environment. Many disciplines contribute to environmental science and help us understand how our worlds intertwine as well as our proper role in them.

members of society. The solutions to these problems increasingly involve human social systems as well as natural science.

Criteria for environmental literacy suggested by the National Environmental Education Advancement Project in Wisconsin include: awareness and appreciation of the natural and built environment; knowledge of natural systems and ecological concepts; understanding of current environmental issues; and the ability to use critical-thinking and problem-solving skills on environmental issues. These are good overall goals to keep in mind as you study this book.

Chapter 2 looks more closely at science as a way of knowing, environmental ethics, critical thinking, and other tools that help us analyze and understand the world around us. For the remainder of this chapter, we'll complete our overview with a short history of environmental thought and a survey of some important current issues that face us.

A Brief History of Conservation and Environmentalism

Although many early societies had negative impacts on their surroundings, others lived in relative harmony with nature. In modern times, however, growing human populations and the power of our technology have heightened our concern about what we are doing to our environment. We can divide conservation history and envi-

ronmental activism into at least four distinct stages: (1) pragmatic resource conservation, (2) moral and aesthetic nature preservation, (3) a growing concern about health and ecological damage caused by pollution, and (4) global environmental citizenship. Each era focused on different problems and each suggested a distinctive set of solutions. These stages are not necessarily mutually exclusive, however; parts of each persist today in the environmental movement and one person may embrace them all simultaneously.

Historic Roots of Nature Protection

Recognizing human misuse of nature is not unique to modern times. Plato complained in the fourth century B.C. that Greece once was blessed with fertile soil and clothed with abundant forests of fine trees. After the trees were cut to build houses and ships, however, heavy rains washed the soil into the sea, leaving only a rocky "skeleton of a body wasted by disease" (fig. 1.3). Springs and rivers dried up while farming became all but impossible. Many classical authors regarded the Earth as a living being, vulnerable to aging, illness, and even mortality. Periodic threats about the impending death of nature as a result of human misuse have persisted into our own time. Many of these dire warnings have proven to be premature or greatly exaggerated, but others remain relevant to our own times. As Mostafa K. Tolba, Executive Director of the United Nations Environment Programme has said, "The problems that overwhelm us today are precisely those we failed to solve decades ago."

Some of the earliest scientific studies of environmental damage were carried out in the eighteenth century by French and British colonial administrators who often were trained scientists and who considered responsible environmental stewardship as an

Figure 1.3

Nearly 2500 years ago, Plato lamented land degradation that denuded the hills of Greece. Have we learned from history's lessons?

aesthetic and moral priority, as well as an economic necessity. These early conservationists observed and understood the connection between deforestation, soil erosion, and local climate change. The pioneering British plant physiologist, Stephen Hales, for instance, suggested that conserving green plants preserved rainfall. His ideas were put into practice in 1764 on the Caribbean island of Tobago, where about 20 percent of the land was marked as "reserved in wood for rains."

Pierre Poivre, an early French governor of Mauritius, an island in the Indian Ocean, was appalled at the environmental and social devastation caused by destruction of wildlife (such as the flightless dodo) and the felling of ebony forests on the island by early European settlers. In 1769, Poivre ordered that one quarter of the island was to be preserved in forests, particularly on steep mountain slopes and along waterways. Mauritius remains a model for balancing nature and human needs. Its forest reserves shelter a larger percentage of its original flora and fauna than most other human-occupied islands.

Pragmatic Resource Conservation

Many historians consider the publication of *Man and Nature* in 1864 by geographer George Perkins Marsh as the wellspring of environmental protection in North America. Marsh, who also was a lawyer, politician, and diplomat, traveled widely around the Mediterranean as part of his diplomatic duties in Turkey and Italy. He read widely in the classics (including Plato) and personally observed the damage caused by the excessive grazing by goats and sheep and by the deforesting of steep hillsides. Alarmed by the wanton destruction and profligate waste of resources still occurring on the American frontier in his lifetime, he warned of its ecological consequences. Largely as a result of his book, national forest reserves were established in the United States in 1873 to protect dwindling timber supplies and endangered watersheds.

Among those influenced by Marsh's warnings were President Theodore Roosevelt and his chief conservation advisor, Gifford Pinchot. In 1905, Roosevelt, who was the leader of the populist, progressive movement, moved the Forest Service out of the corruption-filled Interior Department into the Department of Agriculture. Pinchot, who was the first native-born professional forester in North America, became the original head of this new agency. He put resource management on an honest, rational, and scientific basis for the first time in our history. Together with naturalists and activists such as John Muir, William Brewster, and George Bird Grinnell, Roosevelt and Pinchot established the framework of our national forest, park, and wildlife refuge systems, passed game protection laws, and tried to stop some of the most flagrant abuses of the public domain. In 1908, Pinchot organized and chaired the White House Conference on Natural Resources, perhaps the most prestigious and influential environmental meeting ever held in the United States.

The basis of Roosevelt's and Pinchot's policies was pragmatic **utilitarian conservation.** They argued that the forests should be saved "not because they are beautiful or because they shelter wild creatures of the wilderness, but only to provide homes and jobs for people." Resources should be used "for the greatest good, for the greatest number for the longest time." "There has been a fundamental misconception," Pinchot said, "that conservation means nothing but husbanding of resources for future generations. Nothing could be further from the truth. The first principle of conservation is development and use of the natural resources now existing on this continent for the benefit of the people who live here now. There may be just as much waste in neglecting the development and use of certain natural resources as there is in their destruction." This pragmatic approach still can be seen in the multiple use policies of the Forest Service.

Moral and Aesthetic Nature Preservation

John Muir (fig. 1.4), geologist, author, and first president of the Sierra Club, strenuously opposed Pinchot's influence and policies. Muir argued that nature deserves to exist for its own sake, regardless of its usefulness to us. Aesthetic and spiritual values formed the core of his philosophy of nature protection. This outlook has been called **altruistic preservation** because it emphasizes the fundamental right of other organisms to exist and to pursue their own interests. Muir wrote: "The world, we are told, was made for man. A presumption that is totally unsupported by the facts . . . Nature's object in making animals and plants might possibly be first of all the happiness of each one of them. . . . Why ought man to value himself as more than an infinitely small unit of the one great unit of creation?"

Muir, who was an early explorer and interpreter of the Sierra Nevada Mountains in California, fought long and hard for establishment of Yosemite and King's Canyon National Parks. The

Figure 1.4

President Teddy Roosevelt (*left*) and naturalist John Muir pose at Glacier Point in Yosemite National Park during a historical meeting in 1903. How many modern presidents would take time for a camping trip to enjoy and celebrate the natural world?

Figure 1.5

The National Park Service has generally attempted to preserve pristine wilderness—such as this area in Redwood National Park—along the altruistic preservation principles proposed by John Muir.

Figure 1.6

Rachel Carson's book *Silent Spring* was a landmark in modern environmental history. She alerted readers to the dangers of indiscriminate pesticide use.

National Park Service, established in 1916, was first headed by Muir's disciple, Stephen Mather, and has always been oriented toward preservation of nature in its purest state (fig. 1.5). It has often been at odds with Pinchot's utilitarian Forest Service. Environmental ethics is discussed further in chapter 2.

Modern Environmentalism

The undesirable effects of pollution probably have been recognized at least as long as those of forest destruction. In 1273, King Edward I of England threatened to hang anyone burning coal in London because of the acrid smoke it produced. In 1661, the English diarist John Evelyn complained about the noxious air pollution caused by coal fires and factories and suggested that sweet-smelling trees be planted to purify city air. Increasingly dangerous smog attacks in Britain led, in 1880, to formation of a national Fog and Smoke Committee to combat this problem.

The tremendous industrial expansion during and after the Second World War added a new set of concerns to the environmental agenda. *Silent Spring,* written by Rachel Carson (fig. 1.6) and published in 1962, awakened the public to the threats of pollution and toxic chemicals to humans as well as other species. The movement she engendered might be called **environmentalism** because its concerns are extended to include both environmental resources and pollution. Among the pioneers of this movement are activist David Brower and scientist Barry Commoner. Brower, while executive director of the Sierra Club, Friends of the Earth, and, more recently, Earth Island Institute, introduced many of the techniques of modern

environmentalism, including litigation, intervention in regulatory hearings, book and calendar publishing, and using mass media for publicity campaigns. Commoner, who was trained as a molecular biologist, has been a leader in analyzing the links between science, technology, and society. Both activism and research remain hallmarks of the modern environmental movement.

Under the leadership of a number of other brilliant and dedicated activists and scientists, the environmental agenda was expanded in the 1960s and 1970s to include issues such as human population growth, atomic weapons testing and atomic power, fossil fuel extraction and use, recycling, air and water pollution, wilderness protection, and a host of other pressing problems that are addressed in this textbook. Environmentalism has become well established on the public agenda since the first national Earth Day in 1970. A majority of Americans now consider themselves environmentalists, although there is considerable variation in what that term means.

Global Concerns

Increased opportunities to travel, as well as greatly expanded international communications, now enable us to know about daily events in places unknown to our parents or grandparents. We have become, as Marshal McLuhan announced in the 1960s, a global village (In Depth, p. 16). As in a village, we are all interconnected in various ways. Events that occur on the other side of the globe have profound and immediate effects on our lives.

Photographs of the earth from space (fig. 1.7) provide a powerful icon for the fourth wave of ecological concern that might be called **global environmentalism.** These photos remind us how

Figure 1.7
We live in a bountiful and beautiful world. Ours is a unique and irreplaceable planet on whose life-sustaining systems we are totally dependent.

Current Conditions

As you probably already know, many environmental problems now face us. Before surveying them in the following section, we should pause for a moment to consider the extraordinary natural world that we inherited and that we hope to pass on to future generations in as good—perhaps even better—a condition than when we arrived.

A Marvelous Planet

Imagine that you are an astronaut returning to Earth after a long trip to the moon or Mars. What a relief it would be to come back to this beautiful, bountiful planet (fig. 1.8) after experiencing the hostile, desolate environment of outer space. Although there are dangers and difficulties here, we live in a remarkably prolific and hospitable world that is, as far as we know, unique in the universe. Compared to the conditions on other planets in our solar system, temperatures on the earth are mild and relatively constant. Plentiful supplies of clean air, fresh water, and fertile soil are regenerated endlessly and spontaneously by geological and biological cycles (discussed in chapters 3 and 4).

Perhaps the most amazing feature of our planet is the rich diversity of life that exists here. Millions of beautiful and intriguing species populate the earth and help sustain a habitable environment. This vast multitude of life creates complex, interrelated communi-

small, fragile, beautiful, and rare our home planet is. We all share a common environment at this global scale. As our attention shifts from questions of preserving particular landscapes or preventing pollution of a specific watershed or airshed, we begin to worry about the life-support systems of the whole planet.

Minnesota geologist Roger Hooke estimates that current human earth-moving activities now rival those of natural geological forces. In addition, we are changing planetary weather systems and atmospheric chemistry, reducing the natural variety of organisms, and degrading ecosystems in ways that could have devastating effects, both on humans and on all other life forms. Protecting our environment has become an international cause and it will take international cooperation to bring about many necessary changes.

Among the leaders of this worldwide environmental movement have been British economist Barbara Ward, French/American scientist René Dubos, Norwegian Prime Minister Gro Harlem Bruntland, and Canadian diplomat Maurice Strong. All have been central in major international environmental conventions, such as the 1972 U.N. Conference on the Human Environment in Stockholm or the 1992 U.N. "Earth Summit" on Environment and Development in Rio de Janiero. Once again, new issues have become part of the agenda as our field of vision widens. We have begun to appreciate the links between poverty, injustice, oppression, and exploitation of humans and our environment. We will discuss human development in more detail later in this chapter.

Figure 1.8
We are fortunate to live in a beautiful, prolific, agreeable world. It will take care and hard work to keep it this way.

ties where towering trees and huge animals live together with, and depend upon, tiny life-forms such as viruses, bacteria, and fungi. Together, all these organisms make up delightfully diverse, self-sustaining communities, including dense, moist forests, vast sunny savannas, and richly colorful coral reefs. From time to time, we should pause to remember that, in spite of the challenges and complications of life on Earth, we are incredibly lucky to be here. We should ask ourselves: what is our proper place in nature? What *ought* we do and what *can* we do to protect the irreplaceable habitat that produced and supports us? These are some of the central questions of environmental science.

Environmental Dilemmas

While there are many things to appreciate and celebrate about the world in which we live, many pressing environmental problems cry out for our attention. Human populations have grown at alarming rates in this century. Nearly 6 billion people now occupy the earth and we are adding about 100 million more each year. In the next decade, our numbers will increase by nearly as many people as are now alive in China. Most of that growth will be in the poorer countries where resources and services are already strained by present populations.

Some demographers believe that this unprecedented growth rate will slow in the next century and that the population might eventually drop back below its present size. Others warn that the number of humans a century from now could be four or five times our present population if we don't act quickly to bring birth rates into balance with death rates. Whether there are sufficient resources to support 6 billion humans—let alone 25 billion—on a sustainable basis is one of the most important questions we face. How we might stabilize population and what level of resource consumption we and future generations can afford are equally difficult parts of this challenging equation.

Food shortages and famines already are too familiar in many places and may increase in frequency and severity if population growth, soil erosion, and nutrient depletion continue at the same rate in the future as they have in the past (fig. 1.9). We are coming to realize, however, that food security often has more to do with poverty, democracy, and equitable distribution than it does with the amount of food available. Water deficits and contamination of existing water supplies threaten to be critical environmental issues in the future for agricultural production as well as for domestic and industrial uses. Many countries already have serious water shortages and more than one billion people lack access to clean water or adequate sanitation. Violent conflicts over control of natural resources may flare up in many places if we don't learn to live within nature's budget.

How we obtain and use energy is likely to play a crucial role in our environmental future. Fossil fuels (oil, coal, and natural gas) presently supply about 80 percent of the energy used in industrialized countries (fig. 1.10). Supplies of these fuels are diminishing at an alarming rate and problems associated with their acquisition and use—air and water pollution, mining damage, shipping accidents, and political insecurity—may limit where and how we use remaining reserves. Cleaner renewable energy resources—solar

Figure 1.9

Three quarters of the world's poorest nations are in Africa. Millions of people lack adequate food, housing, medical care, clean water, and safety. The human suffering engendered by this poverty is tragic.

power, wind, and biomass—together with conservation, may replace environmentally destructive energy sources if we invest in appropriate technology in the next few years.

As we burn fossil fuels, we release carbon dioxide and other heat-absorbing gases that cause global warming and may bring about sea-level rises and catastrophic climate changes. Acids formed in the air as a result of fossil fuel combustion already have caused extensive damage to building materials and sensitive ecosystems in many places (fig. 1.11). Continued fossil fuel use without pollution control measures could cause even more extensive damage. Chlorinated compounds, such as the chlorofluorocarbons used in refrigeration and air conditioning, also contribute to global warming, as well as damaging the stratospheric ozone that protects us from cancer-causing ultraviolet radiation in sunlight.

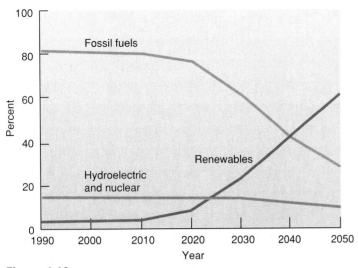

Figure 1.10

Fossil fuel supplies are dwindling at an alarming rate. They could be replaced by renewables such as solar power, wind, and biomass energy that could be more equitably distributed and would be less polluting than our current energy sources.

Source: World Bank estimates; D. Anderson and C. D. Bird, "Carbon Accumulations and Technical Progress—A Simulation Study of Costs," *Oxford Bulletin of Economics & Statistics* 54(1):1–29.

Destruction of tropical forests, coral reefs, wetlands, and other biologically rich landscapes is causing an alarming loss of species and a reduction of biological variety and abundance that could severely limit our future options. Many rare and endangered species are threatened directly or indirectly by human activities (fig. 1.12). In addition to practical values, aesthetic and ethical considerations suggest that we should protect these species and the habitat necessary for their survival.

Toxic air and water pollutants, along with mountains of solid and hazardous wastes, are becoming overwhelming problems in industrialized countries. We produce hundreds of millions of tons of these dangerous materials annually, and much of it is disposed of in dangerous and irresponsible ways. No one wants this noxious stuff dumped in their own backyard, but too often the solution is to export it to someone else's. We may come to a political impasse where our failure to decide where to put our wastes or how to dispose of them safely will close down industries and result in wastes being spread everywhere. The health effects of pollution, toxic wastes, stress, and the other environmental ills of modern society have become a greater threat than infectious diseases for many of us in industrialized countries.

These and other similarly serious problems illustrate the importance of environmental science and environmental education for everyone. What we are doing to our world, and what that may mean for our future and that of our children is of paramount concern as we enter the twenty-first century. We trust that's why you're reading this book.

Signs of Hope

The dismal litany of problems we have just reviewed seems pretty overwhelming, doesn't it? But is there hope that we may find solutions to these problems? We think so. As you will see in subsequent chapters, progress has been made in many areas in controlling air and water pollution and reducing wasteful resource uses. Many cities in North America and Europe are cleaner and less polluted than they were a generation or so ago. Population has stabilized in most industrialized countries and even some very poor countries where social security and democracy have been established. Over the last twenty years, the average number of children born per woman worldwide has decreased from 6.1 to 3.4. This is still above the zero population growth rate of 2.1 children per couple, but it is an encouraging improvement. If this rate of progress continues in the next twenty years as it has in the past twenty, the world population could stabilize early in the twenty-first century.

The incidence of life-threatening infectious diseases has been reduced sharply in most countries during the past century, while the average life-expectancy has nearly doubled. Many new resources have been discovered and more efficient ways of using existing supplies have been invented that allow us to enjoy luxuries and conveniences that would have seemed miraculous only a few generations ago. Although modern life has many stresses and strains, few people would willingly return to conditions that existed 10,000, 1000, or even 100 years ago. Would you?

Still, we can do much more, both individually and collectively, to protect and restore our environment. Being aware of the problems we face is the first step toward finding their solutions. Increased media coverage has brought environmental issues to public attention. More than 80 percent of the Americans polled in public opinion surveys agree that "protecting the environment is so important that requirements and standards cannot be too high and continuing environmental improvements must be made regardless of cost." This growing understanding and concern are themselves hopeful signs. Young people today may be in an unique position to address these issues because, for the first time in history, we now have resources, motivation, and knowledge to do something about our environmental problems. Unfortunately, if we don't act now, we may not have another chance to do so.

North/South: A Divided World

We live in a world of haves and have-nots; a few of us live in increasing luxury while many others lack the basic necessities for a decent, healthy, productive life. The World Bank estimates that more than one billion people—one fifth of the world—live in **acute poverty** in which they lack access to an adequate diet, decent housing, basic sanitation, clean water, education, medical care, and other essentials for a humane existence. Seventy percent of those people are women and children. In fact, four out of five people in the world live in what would be considered poverty in the United States or Canada. The plight of these poor people is not just a hu-

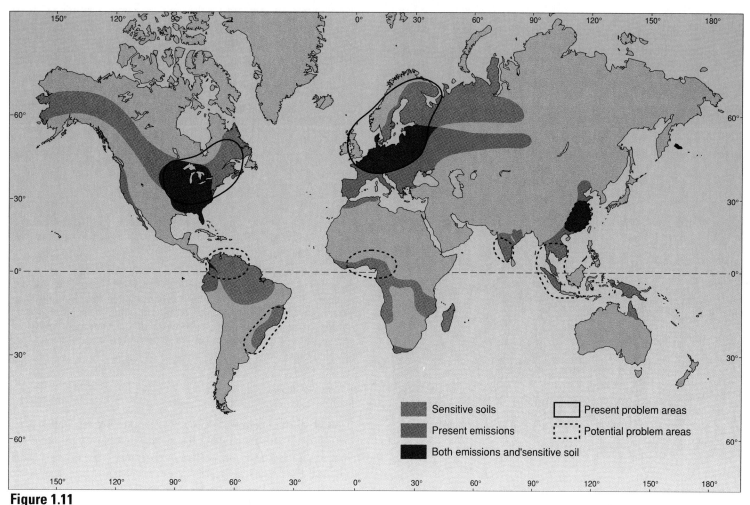

Figure 1.11

This schematic map shows regions that currently have pollution-related acidification problems or where, based on soil types and expected future industrialization, acidification may become severe in the future.

Source: H. Rodhe and R. Herrera, eds., *Acidification in Tropical Countries,* John Wiley & Sons, Chichester, 1988.

manitarian concern. Policymakers are becoming aware that eliminating poverty and protecting our common environment are inextricably interlinked.

The world's poorest people have become both the victims and the agents of environmental degradation. The poorest people are too often forced to meet short-term survival needs at the cost of long-term sustainability. Desperate for croplands to feed themselves and their families, many move into virgin forests or cultivate steep, erosion-prone hillsides where soil nutrients are exhausted after only a few years. Others migrate to the grimy, crowded slums and ramshackle shantytowns that now surround most major cities in the developing world. With no way to dispose of wastes, the residents often foul their environment further and contaminate the air they breathe and the water on which they depend for washing and drinking (fig. 1.13).

The cycle of poverty, illness, and limited opportunities can become a self-sustaining process that passes from one generation to another. People who are malnourished and ill can't work productively to obtain food, shelter, or medicine for themselves or their children, who also are malnourished and ill. About 200 million children—mostly in South and Southeast Asia and some as young as 4 years old—are forced to work under slave-labor conditions weaving carpets, making ceramics and jewelry, or in the sex trade. Growing up in these conditions leads to mental and developmental deficits that condemn these children to perpetuate this cycle.

Faced with immediate survival needs and few options, these unfortunate people often have no choice but to overharvest resources; in doing so, however, they diminish not only their own options, but also those of future generations. And in an increasingly interconnected world, the environments and resource bases damaged by poverty and ignorance are directly linked to those on which we depend. It is in our own self-interest to help everyone find better ways to live.

Figure 1.12

This emperor tamarin is a marmoset from South America. Although we hear a great deal about destruction of tropical *rainforests,* the *dry* tropical forests that once lined the Atlantic coast of South America and the Pacific coast of Central America have been destroyed to a much greater extent. More than 99 percent of the once-extensive dry tropical forests of Latin America have been destroyed, threatening a host of interesting and important species with extinction.

Figure 1.13

While many of us live in luxury, more than a billion people lack access to food, housing, clean water, sanitation, education, medical care, and other essentials for a healthy, productive life. Often the poorest people are both the victims and agents of environmental degradation as they struggle to survive. Helping them meet their needs is not only humane, it is essential to protect our mutual environment.

Rich and Poor Countries

Where do the rich and poor live? About one-fifth of the world's population lives in the twenty richest countries where the average per capita income is above $10,000 (US) per year (fig. 1.14). Most of these countries are in North America or Western Europe, but Japan, Australia, New Zealand, Hong Kong, Singapore, Saudi Arabia, and the United Arab Emirates also fall into this group. Almost every country, however, even the richest, such as the United States and Canada, has poor people. No doubt everyone reading this book knows about homeless people or other individuals who lack resources for a safe, productive life. Physicians for Social Responsibility estimates that 20 million people in the United States are malnourished.

The other four-fifths of the world live in middle- or low-income countries where nearly everyone is poor by our standards. More than 3 billion people live in the poorest nations where the average per capita income is below $580 (US) per year. China and India are the largest of these countries, with a combined population of about 2 billion people. Among the 40 other nations in this category, 31 are in Sub-Saharan Africa. All the other lowest-income nations, except Haiti, are in Asia. Encouragingly, poverty levels in South and East Asia have fallen in recent years and are expected to continue to decline in the future. In Sub-Saharan Africa and Latin America, on the other hand, the percentage of the population living in poverty is rising (fig. 1.15). The destabilizing and impoverishing effects of earlier colonialism continue to play important roles in the ongoing problems of these unfortunate countries.

The ten poorest countries in the world, in 1995, were (in ascending order from very poorest): Mozambique, Tanzania, Ethiopia, Bhutan, Guinea-Bisseau, Malawi, Sierra Leon, Bangladesh, and Madagascar. Each of these countries has annual per capita Gross National Product (GNP) of less than $200 (US) per year. They also have low levels of food security, social welfare, and quality of life as indicated by table 1.1.

By contrast, each of the ten richest countries in the world—Switzerland, Luxembourg, Japan, Finland, Norway, Sweden, Iceland, The United States, Denmark, and Canada (in descending order from richest)—has an annual per capita GNP more than 100 times that of the poorest countries. As you can see in table 1.1, other conditions in the rich countries reflect this wide disparity in wealth.

A Fair Share of Resources?

The affluent lifestyle that many of us in the richer countries enjoy consumes an inordinate share of the world's natural resources and produces a shockingly high proportion of pollutants and wastes. The United States, for instance, with less than five percent of the

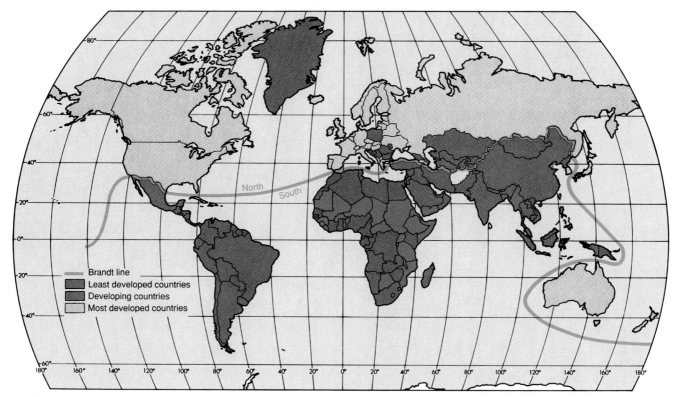

Figure 1.14

The world of underdevelopment. The Brandt line, denoted by the Independent Commission on International Development Issues separates the richer, industrialized nations of Europe, North America, Japan, Australia, and New Zealand from their southern neighbors in the "developing" world. The 42 "least developed" countries, according to the United Nations Development Programme, are mostly in Africa and Asia. Note that since this map was drawn, conditions have worsened in parts of the former USSR and Africa but have often improved elsewhere.

Table 1.1
Average indicators of Quality of Life for the ten richest and poorest nations[1]

Indicator	Poor Countries	Rich Countries
GNP/capita	$176	$22,634
Life expectancy	49 years	77 years
Infant mortality[2]	122	6.4
Child deaths[3]	208	7.9
Percent of calories needed for healthy life	95	130
Grams protein/day	50	95
Safe drinking water	36%	100
Female literacy	20%	NA[4]
Birth rate[5]	45	12.7

Notes: [1]averaged as a group
[2]per 1000 live births
[3]per 1000 children before age 5
[4]not available, but close to 100 percent
[5]per 1000 population

Source: World Resources Institute, *World Resources* 1994–1995.

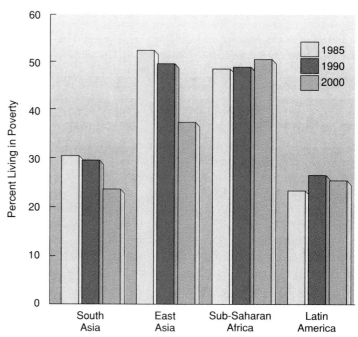

Figure 1.15

People living in poverty as a percentage of regional population. Note that poverty is declining in both South and East Asia but is stable or increasing in sub-Saharan Africa and Latin America.

Source: World Bank, 1992.

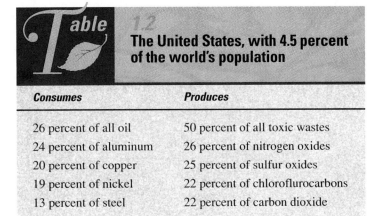

Table 1.2 The United States, with 4.5 percent of the world's population	
Consumes	**Produces**
26 percent of all oil	50 percent of all toxic wastes
24 percent of aluminum	26 percent of nitrogen oxides
20 percent of copper	25 percent of sulfur oxides
19 percent of nickel	22 percent of chloroflurocarbons
13 percent of steel	22 percent of carbon dioxide

Figure 1.16

"And may we continue to be worthy of consuming a disproportionate share of this planet's resources."

Drawing by Lorenz; © 1992 The New Yorker Magazine, Inc.

total population, consumes about one quarter of most commercially traded commodities and produces a quarter to half of most industrial wastes (table 1.2).

To get an average American through the day takes about 450 kg (nearly 1000 lbs) of raw materials, including 18 kg (40 lbs) of fossil fuels, 13 kg (29 lbs) of other minerals, 12 kg (26 lbs) of farm products, 10 kg (22 lbs) of wood and paper, and 450 liters (119 gal) of water. Every year we throw away some 160 million tons of garbage, including 50 million tons of paper, 67 billion cans and bottles, 25 billion styrofoam cups, 18 billion disposable diapers, and 2 billion disposable razors (fig. 1.16).

This profligate resource consumption and waste disposal strains the life-support systems of the Earth on which we all depend. If everyone in the world tried to live at consumption levels approaching ours, the results would be disastrous. Unless we find ways to curb our desires and produce the things we truly need in less destructive ways, the sustainability of human life on our planet is questionable.

North/South Division

Notice in the lists of rich and poor countries discussed earlier that the wealthiest countries tend to be in the north, while the poorest countries tend to be located closer to the equator. It is common to speak of a **North/South division** of wealth and power, even though many poorer nations such as India and China are in the Northern Hemisphere and some relatively rich nations like Australia and New Zealand are in the Southern. Another way of describing the rich nations is that they tend to be highly industrialized and generally are members of the Organization for Economic Development and Cooperation (OECD), made up of Western Europe, North America, Japan, Australia, and New Zealand. This group makes up about 20 percent of the world's population but consumes more than half of most resources. The other, poorer four-fifths of humanity are sometimes called the majority world because they make up the majority of the population.

Since poverty and social welfare are closely linked with environmental degradation, there will be many instances in this book in which we will want to distinguish between conditions in groups of countries. In fact, income, social services, human development, and other indicators of quality of life don't always fit neatly in just two categories. Nations actually spread across a spectrum—or perhaps a three-dimensional matrix—of these indicators, making it difficult to establish inclusive categories. The following categories and corresponding acronyms are among those we will refer to throughout the rest of this book.

Political Economies

Countries are sometimes classified according to their economic system. **First World** describes the industrialized, market-oriented, democracies of Western Europe, North America, Japan, Australia, New Zealand, and their allies. **Second World** originally described the centrally-planned, socialist countries, such as the former Soviet Union and its Eastern European allies. Several Asian socialist countries, such as China, Mongolia, North Korea, and Vietnam, also once belonged in this category but most are rapidly changing to market economies.

In the 1960s, the nonaligned, nonindustrial, ex-colonial nations such as India, Indonesia, Malaysia, Iran, Syria, and many African countries labeled themselves the **Third World** to show their independence from either of the superpower groups. This category was not intended to suggest that its members are third rate or third class. It has come to be used to describe developing countries in general. Some economists suggest that another category, the **Fourth World,** is necessary to classify the poorest nations with neither market economies nor central planning, as well as the indigenous communities within wealthy nations.

The amazing political and economic upheavals of the past decade have thrown all of these categories into a turmoil. Few countries any more are either purely socialist or capitalist. Nearly every government plans centrally and intervenes in its economy to some extent, and nearly every nation has at least some market-orientation. These categories have decreasing significance as old political alliances break down, forcing us to look for other ways to describe nations and peoples.

Human Development

Every year, the United Nations releases a report ranking countries by a **human development index** based primarily on average life expectancy, percentage of literate adults, mean years of schooling, and annual income per capita. Some other indicators factored into this index include infant mortality rates, daily calorie supply, child malnutrition, and access to clean water. The highest possible human development index (HDI) is 1.0; the lowest is 0.0. In 1994, Canada was highest in the world with an HDI of 0.932. As you might expect, there is a close correlation between wealth and human development. All the top 20 nations (all above 0.9) were in North America or Western Europe, except Japan and Israel. The lowest HDI was Guinea at 0.191. In fact, all of the 20 lowest rankings (all below 0.275) were in Africa, except Afghanistan and Bhutan.

Developmental Discrepancies

Of course aggregate numbers such as this hide many important issues. One of these is gender inequities. Men generally fare better than women on almost every socio-economic indicator. For the few countries that keep such data, Japan had the lowest female-to-male wage ratio (51 percent), while Sweden had the highest (90 percent). Similarly, the female-to-male ratio in nonagricultural work varies from a low of 22 percent in Bahrain to 89 percent in Finland. If the HDI is weighed for these gender discrepancies, egalitarian Scandinavian countries like Sweden, Finland, and Norway move into top place.

Race is another variable that determines socio-economic status in many countries. If white South Africa were a separate country, it would rank 24th in the world in human development (just after Spain). Black South Africa, on the other hand, would rank 123rd in the world (just above the Congo). In some countries, regional or ethnic differences create disparities. In Nigeria, the state of Bendel, with a HDI of 0.66 is equivalent to Cuba, while the poor state of Borno, with a HDI of 0.156 is lower than any country in the world.

The United Nations states that these vast discrepancies and the grinding poverty experienced by the poorest of the poor are the greatest threats to political stability, social cohesion, and environmental health on our planet. It warns that a number of nations such as Egypt, South Africa, Nigeria, and Brazil are in danger of joining the list of failed states because of these great disparities. Some suggested strategies for poverty reduction and social justice include:

- basic social services, especially basic education and primary health care;

- agrarian reform for more equitable land distribution;

- credit to tide those without resources through tough times;

- employment to ensure a secure livelihood for all;

- civil rights that enable everyone to participate in planning and management decisions;

- a social safety net to catch those whom markets exclude;

- economic growth that benefits the poor; and

- sustainable resource use that reduces material-intensive and energy-intensive lifestyles and turns, instead, to renewables.

Good News and Bad News

Over the past thirty years, human ingenuity and enterprise have brought about a breathtaking pace of technological innovations and scientific breakthroughs. The world's gross domestic product has increased more than sevenfold during that period, from $3 trillion to $22 trillion. In spite of a doubling of the total population, average real per capita annual income has increased threefold between 1960 and 1990. Unfortunately, not all this increased wealth has been channeled into human development, but there has been significant improvement in this area. In 1960, nearly three-quarters of the world's population lived in abysmal conditions (HDI below 0.5). In 1992, only 34.5 percent were still at this low level of development.

Some of the most successful programs in social improvements have been Malaysia, Republic of Korea, Thailand, and Portugal, all of which moved from low development status to high in just thirty years. Botswana, Tunisia, Syria, and Turkey also increased their HDI by 0.4 or more during this time. Nonetheless, while the general welfare has risen, so has the gap between the rich and poor worldwide. In 1960, the income ratio between the richest 20 percent of the world and the poorest 20 percent was 30 to 1. In 1992, this ratio was 61 to 1.

Sustainable Development

By now, it is clear that security and living standards for the world's poorest people are inextricably linked to environmental protection. One of the most important questions in environmental science is how we can continue improvements in human welfare within the limits of the earth's natural resources. A possible solution to this dilemma is **sustainable development,** a term popularized by *Our Common Future,* the 1987 report of the World Commission on Environment and Development, chaired by Norwegian Prime Minister Gro Harlem Brundtland (and consequently called the Brundtland Commission). In the words of this report, sustainable development means "meeting the needs of the present without compromising the ability of future generations to meet their own needs."

Another way of saying this is that development means improving people's lives. Sustainable development, then, means progress in human well-being that can be extended or prolonged over many generations rather than just a few years. To be truly enduring, the benefits of sustainable development must be available to all humans rather than to just the members of a privileged group.

To many economists, it seems obvious that economic growth is the only way to bring about a long-range transformation to more advanced and productive societies and to provide resources to improve the lot of all people. As former President John F. Kennedy said, "A rising tide lifts all boats." But economic growth is not sufficient in

*I*n Depth: Getting to Know Our Neighbors

Imagine the world as a village of 1000 people: who are our neighbors?

- Ethnically, this global village has 385 East Asians, 209 South Asians, 129 Africans, 91 Europeans, 78 Latin Americans, 52 individuals from the former USSR, 51 from the United States and Canada, and 5 from Oceania (Australia, New Zealand, and the Pacific Islands).

- About half the residents speak one or more of the six major world languages (Mandarin, English, Hindi/Urdu, Spanish, Russian, and Arabic—in that order), while the rest speak some of the other 6000 known languages.

- Children under age 15 make up 32 percent of the village, while those over age 65 make up 6 percent. The number of senior citizens is increasing rapidly while the number of children is falling slowly. At some point in the future, there may be more retirees than workers in the vil-

lage, bankrupting the social security system.

- Every year, 26 children are born in the village and 9 people die. One-third of yearly deaths are children under age 5, mostly from infectious diseases aggravated by malnutrition. One person dies from heart disease, one from cancer, two from chronic or acute lung diseases, and two from violence, accidents, and various infectious diseases.

- The average life expectancy is 63 years for women and 67 for men, but there is considerable variation in longevity. In the richest families, women live to an average age of 78 and men to an average of 76 years, while in the poorest families the life expectancy is less than 43 years for both men and women.

- The average annual income for the village is $3790 (US), but this average also obscures great discrepancies. More than half the village lives in households where

the annual income is less than $580 per person, while the richest 100 citizens (mostly Americans, Canadians, Europeans, and Japanese) enjoy annual incomes over $22,000 each.

- As a consequence of this income gap, the richest 200 villagers own or control 80 percent of the resources and consume 80 percent of all products sold in the market place, while the other 800 people make do with 20 percent of the goods and merchandise available for sale.

- Half of the 620 adults in the village are illiterate. Lacking an education, most of these people work as day laborers or seasonal farm workers. Among the poorer families, girls are half as likely as boys to attend school.

- Women and girls make up slightly more than half the village population. They do two-thirds of all manual labor, receive one-tenth of the wages, and own less than one-

itself to meet all essential needs. As the Brundtland Commission pointed out, political stability, democracy, and equitable economic distribution are needed to ensure that the poor will get a fair share of the benefits of greater wealth in a society.

Can Development Be Truly Sustainable?

Many ecologists regard "sustainable" growth of any sort as impossible in the long run because of the limits imposed by nonrenewable resources and the capacity of the biosphere to absorb our wastes. Using ever-increasing amounts of goods and services to make human life more comfortable, pleasant, or agreeable must inevitably interfere with the survival of other species and, eventually, of humans themselves in a world of fixed resources. But, supporters of sustainable development assure us, both technology and social organization can be managed in ways that meet essential needs and provide long-term—but not infinite—growth within natural limits, if we use ecological knowledge in our planning.

While economic growth makes possible a more comfortable lifestyle, it doesn't automatically result in a cleaner environment. As figure 1.17 shows, people will purchase clean water and sanitation if they can afford to do so. For low-income people, however, more money tends to result in higher air pollution because they can afford to burn more fuel for transportation and heating. Given enough money, people will be able to afford both convenience *and* clean air. Some environmental problems, such as waste generation and carbon dioxide emissions, continue to rise sharply with increasing wealth because their effects are diffuse and delayed. If we are able to sustain economic growth, we will need to develop personal restraint or social institutions to deal with these problems.

Some projects intended to foster development have been environmental, economic, and social disasters. Large-scale hydropower projects, like that in the James Bay region of Quebec or the Brazilian Amazon that were intended to generate valuable electrical power, also displaced indigenous people, destroyed wildlife, and poisoned local ecosystems with acids from decaying vegeta-

hundredth of the property. Seventy percent of the poorest members of the population are women and children.

- About 400 villagers suffer from ill health at any given time. Much of that illness is related to lack of clean water, sanitation, and food. Some 250 of your neighbors don't have clean water to drink or adequate sanitation. About 150 are chronically hungry, lacking the calories and nutrients needed for normal growth and development in children or a healthy, productive life for adults.

- Only 452 people actually live in the village itself; the other 548 live in the surrounding countryside. Sixty percent of the rural families are landless or have too little land to subsist. They make up a majority of those who lack clean water, sanitation, food, housing, and health care.

- Generally the worst pollution problems are borne by those who live in the poorest parts of the village, where air and water pollution, noise, congestion, and toxic wastes are most common. Those who live in the better parts

of town actually enjoy a cleaner, safer environment than their parents did only a few decades ago.

- In the past, people didn't often travel from one part of the village to another. Today, travel is easier and cheaper than ever before. Furthermore, 90 percent of the population has access to

television so that the lower class is exposed to both consumer pressures and news about how the upper class lives (fig. 1).

- How long do you suppose these great discrepancies in opportunity and quality of life will persist? What might we do to reduce them?

Figure 1

Nine out of ten people in the world now have access to television. There may be only one TV per village, as in this scene from Nigeria, but it opens a window to the wider world. Most of what everyone watches is made in the United States an glorifies consumerism and commidification.

tion and heavy metals leached out of flooded soils. Similarly, introduction of "miracle" crop varieties in Asia and huge grazing projects in Africa financed by international lending agencies crowded out wildlife, diminished the diversity of traditional crops, and destroyed markets for small-scale farmers.

Other development projects, however, work more closely with both nature and local social systems. Socially conscious businesses and environmental, nongovernmental organizations sponsor ventures that allow people in developing countries to grow or make high-value products—often using traditional techniques and designs—that can be sold on world markets for good prices (fig. 1.18). Appleseed Fundraising, for example, a company started in 1989 by two Minnesota college students, buys textiles and crafts directly from artisans in Guatemala and sells them to schools, churches, and camps throughout the United States to use for fund-raising. This provides badly needed income for Mayan villagers while also helping charitable organizations here at home.

As the economist John Stuart Mill wrote in 1857, "It is scarcely necessary to remark that a stationary condition of capital and pop-

ulation implies no stationary state of human improvement. There would be just as much scope as ever for all kinds of mental culture and moral and social progress; as much room for improving the art of living and much more likelihood of its being improved when minds cease to be engrossed by the art of getting on." Somehow, in our rush to exploit nature and consume resources, we have forgotten this sage advice.

The 20:20 Compact for Human Development

At the 1995 United Nations Summit for Social Development in Copenhagen, a world social charter was passed that calls on all nations to ensure basic human needs for everyone. Among the goals in this world action plan to vanquish poverty and injustice are:

- universal primary education—for girls as well as for boys;

- adult illiteracy rates to be halved—with the female rate to be no higher than the male one;

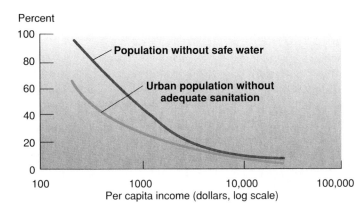

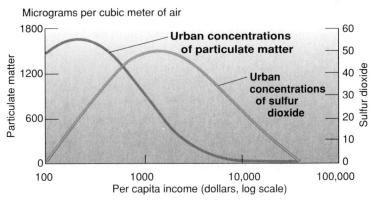

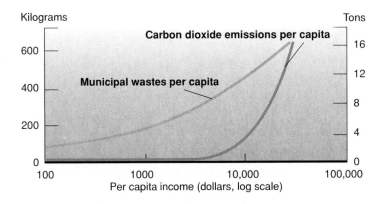

Figure 1.17

Environmental indicators show different patterns as average national income increases. When people have more money they invariably will purchase clean water and better sanitation. Rising income may temporarily produce increased urban air pollution (for example, particulates and sulfur dioxide) as people burn more fuel; eventually, however, people can afford both clean air and the benefits of technology. Some environmental problems such as waste generation and carbon dioxide emissions rise sharply with increasing wealth because of increased demands for goods and services without recognition of adverse environmental consequences.

Sources: Shafik and Bandyopadhyay, background paper; World Bank data.

- elimination of severe malnutrition;

- family planning services for all who wish them;

- safe drinking water and sanitation for all;

- credit for all—to ensure self-employment opportunities.

How much will this cost? A rough estimate provided by the United Nations Development Agency is that $30 to $40 billion a year is needed to meet these targets. This is a large amount but not an impossible one. Developing countries now devote, on average, only $57 billion per year (13 percent of their national budgets) to basic human development. Military spending, in contrast, averages $125 billion per year. If weapons purchases were cut in half, human development could be doubled.

An innovative suggestion called the 20:20 Compact was offered by the World Social Summit. Donor countries now allocate, on average, only 7 percent of their aid to humanitarian concerns. If they were to shift 20 percent of aid to social development, it would provide an additional $12 billion a year. Similarly, if developing countries would earmark at least 20 percent of their budgets to human priority concerns, $88 billion per year would be available. The $100 billion raised by this compact would still be only one-tenth of total world military spending, but it would be three times the minimum needed for the human development agenda (fig. 1.19).

Indigenous People

Often at the absolute bottom of the social strata, whether in rich or poor countries, are the indigenous or native peoples who are generally the least powerful, most neglected groups in the world. Typically descendants of the original inhabitants of an area taken over by more powerful outsiders, they are distinct from their country's dominant language, culture, religion, and racial communities. Of the world's nearly 6000 recognized cultures, 5000 are indigenous ones that account for only about 10 percent of the total world population. In many countries, these indigenous people are repressed by traditional caste systems, discriminatory laws, economics, or prejudice. Unique cultures are disappearing along with biological diversity as natural habitats are destroyed to satisfy industrialized

Figure 1.18

A Mayan woman from Chiapas in southern Mexico weaves on a back-strap loom. A member of a women's weaving cooperative, she sells her work to nonprofit organizations in the United States at much higher prices than she would get at the local market.

Figure 1.19

Every year, military spending equals the total income of half the world's people. The cost of a single large aircraft carrier equals ten years of human development aid given by all the world's industrialized countries. In this post-Cold War era, we ought to be able to declare a "peace dividend" for human development and environmental protection.

world appetites for resources. Traditional ways of life are disrupted further by dominant Western culture sweeping around the globe.

At least half of the world's 6000 distinct languages are dying because they are no longer taught to children. When the last few elders who still speak the language die, so will the culture that was its origin. Lost with those cultures will be a rich repertoire of knowledge about nature and a keen understanding about a particular environment and a way of life (fig. 1.20).

Nonetheless, in many places, the 500 million indigenous people who remain in traditional homelands still possess valuable ecological wisdom and remain the guardians of little-disturbed habitats that are the refuge for rare and endangered species and undamaged ecosystems. Author Alan Durning estimates that indigenous homelands harbor more biodiversity than all the world's nature reserves and that greater understanding of nature is encoded in the languages, customs, and practices of native people than is stored in all the libraries of modern science. Interestingly, just nine countries account

Figure 1.20

Indigenous and tribal cultures, such as these Kung! people of Namibia, have environmental wisdom that may help us understand our proper relationship with nature. Unfortunately, indigenous cultures are disappearing as rapidly as the endangered species and habitats of the world.

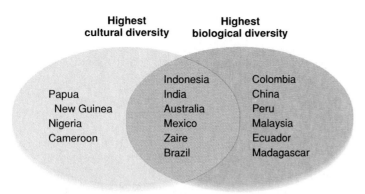

Highest cultural diversity — **Highest biological diversity**

Papua New Guinea
Nigeria
Cameroon

Indonesia
India
Australia
Mexico
Zaire
Brazil

Colombia
China
Peru
Malaysia
Ecuador
Madagascar

Figure 1.21

Cultural and biological diversity often go hand in hand. Six of the countries with the greatest cultural diversity in the world are also on the list of twelve "megadiversity" countries with the highest number of unique biological species.

Sources: Worldwatch Institute, and J. A. McNeely, et al., *Conserving the World's Biological Diversity,* International Union for Conservation of Nature, Gland, Switzerland, 1989.

for 60 percent of all human languages (fig. 1.21). Six of those are also among the "megadiversity" countries that contain exceptional numbers of unique plant and animal species. Conditions that support evolution of many unique species seem to favor development of equally diverse human cultures as well.

Recognizing native land rights and promoting political pluralism is often one of the best ways to safeguard ecological processes and endangered species. As the Kuna Indians of Panama say, "Where there are forests, there are native people, and where there are native people, there are forests." A few countries, such as Papua New Guinea, Fiji, Ecuador, Canada, and Australia acknowledge indigenous title to extensive land areas.

In other countries, unfortunately, the rights of native people are ignored. Indonesia, for instance, claims ownership of nearly three-quarters of its forest lands and all waters and offshore fishing rights, ignoring the interests of indigenous people who have lived in these areas for millennia. Similarly, the Philippine government claims possession of all uncultivated land in its territory, while Cameroon and Tanzania recognize no rights at all for forest-dwelling pygmies who represent one of the world's oldest cultures.

Environmental Perspectives

In the *Christmas Carol* by Charles Dickens, Scrooge questions the Ghost of Christmas Future after seeing the disparity between the rich and poor in London. "Answer me one question," says Scrooge, "are these the shadows of the things that will be or are they shadows of the things that may be only?" We could ask something similar today. Are the problems and dilemmas outlined in this book warnings of what *will* be, or only what *may* be if we fail to take heed and adjust our course of action?

What will be our environmental future and what can we do to shape it? These are, perhaps, the most important questions in environmental science. There are many interpretations of what the data

show and how we should interpret them (What Do You Think?, p. 21). Think about the following worldviews and tactical positions as you read the subsequent chapters in this book.

Pessimism and Outrage

You will find much in this book that justifies pessimism. A number of very serious environmental problems threaten us. Many environmental scientists see our world as one of scarcity and competition in which too many people fight for too few resources. This viewpoint is often called **neo-Malthusian** after Thomas Malthus, who predicted a dismal cycle of overpopulation, misery, vice, and starvation as a result of human fallibility. We will discuss Malthus and his predictions in subsequent chapters as part of our discussion of population growth and resource economics.

This grim view of human nature and resources persuades many environmentalists to issue dire warnings of impending doom unless we make immediate, drastic changes in our way of life. While it is easy to feel moral outrage about the excesses and abuses that have occurred, shock, shame, and fear are generally poor motivation for positive action. How helpful is it in getting you to abandon your bad habits to have someone harp about how guilty and vile you are? Pioneering ecologist Aldo Leopold said: "I have no hope for a conservation based on fear." Furthermore, when the predicted disasters fail to occur, the public may assume that nothing was wrong and that there is no reason to change anything. How many times can we cry wolf before people stop paying attention?

Optimism

Science and technology have provided many benefits to humanity; they also have caused many difficulties, as you will see in the course of this book. **Technological optimists** believe that human ingenuity and enterprise will find cures for all our problems. They see the world as one of abundance and opportunity. Geographer Martin Lewis calls this **"Promethean environmentalism"** after the Greek Titan who stole fire from the Gods and gave it to humans. He fails to mention, however, that Prometheus suffered eternal torment for this sin of pride. Critics describe this optimistic outlook as **"cornucopian fallacy"** (after the mythical horn of plenty) and see it as either wishful thinking or deliberate denial. Optimists argue, however, that they merely expect historic patterns of progress to continue in the future as they have in the past.

While emphasizing only good news is popular and comfortable, it also can mislead the public. Blind faith in technology can be merely an excuse for business as usual. By telling us that everything is just fine, optimists may lull us into complacency and apathy. What do you think is the most appropriate and useful balance between Malthusian and cornucopian worldviews? How can we use environmental information to bring about positive change?

Lessons From The Past

You will see as you read this book that we have tried to be honest about the problems we face but cautiously optimistic about our chances for success. We believe that obstacles can be overcome if

What Do *You* Think?
Environmental Futures: Deciding What to Think

How would you answer if someone asked: *Is it better for the environment if we use paper bags or plastic bags?* or *Will technology ultimately solve our environmental problems?* Questions like these are common in newspapers, magazines, and talk shows, and many people have expressed opinions on them. What about you? What are your positions?

Unless you have researched these matters, perhaps the only reasonable reply is, "I don't know." Such an answer can feel awkward and uncomfortable, however, and there is a natural tendency to respond as though you know, even when you do not.

Many of the issues in environmental studies are among the most significant facing society today. They engage economic and political policies, technology, social values, human health, and the physical and biological environment that nurtures and sustains us. Decisions made, actions taken, or not taken, can have huge consequences for the future.

Unfortunately, the interacting systems are large and complex. Uncertainties and disagreements abound. As we make our collective path through these matters, it is possible to make valid judgments and invalid judgments, sound decisions and unsound decisions. It is not yet clear how successfully we will navigate through the reefs of potential mistakes. What is clear, however, is that our chances of avoiding grievous errors depend on the extent to which we are guided by deliberate, reason-based thought.

Sound thinking is not necessarily automatic. All of us, at times, are prone to careless thinking. The mind tends instinctively to simplify and convert the new to the old and familiar. We are uncomfortable re-examining our fundamental understandings and habits. As a result, our mind tends to see what it expects to see, to accept incomplete evidence that is consistent with previous beliefs and reject credible evidence that is inconsistent with those beliefs. These instincts can lead us to base conclusions on anecdotal or insufficient evidence. We become less responsive to new information than we should be.

Since we may not recognize that a belief rests on insufficient evidence, we give the belief unjustified credence. In fact, we can come to treat the belief not as opinion, but as a logical conclusion, causing further erroneous conclusions, invalid judgments, and unsound decisions.

Another careless thinking habit is the belief that everyone's opinion has credibility, regardless of how poorly it is supported by evidence, and that words and standards of reasoning are arbitrary or relative. It can produce an unwillingness to suspend judgment pending further information.

Critical or reflective thinking, on the other hand, is a more productive thought process. What is critical thinking? In its simplest description, it is deliberate, reason-based thought (fig. 1). Its goal is to avoid basing conclusions on incomplete or irrelevant information and to produce conclusions that are most likely correct given the knowledge one has. The critical thinker evaluates his or her own thinking and the claims of others by posing these kinds of questions: What is the underlying viewpoint of the person making the claim? What kinds of information seem to define an issue and how is this information being interpreted? Are the sources of information credible, unbiased, valid, and reliable? What is being taken for granted? Are the conclusions justified by the factual evidence?

Skepticism and the need for validating information cannot be allowed to paralyze action, however. For many important environmental questions, to wait until all data is in and absolute certainty assured could make any action too late to forestall disaster. Critical thinking involves making inferences and judgments in a timely manner. It prudently weighs the probability of events and the magnitude of the consequences.

Critical thinking involves a number of steps and attitudes that are examined in the next chapter, among the most broadly useful sections presented in this book. The issues in environmental science are dynamic. The goal of this book is to present the most relevant, up-to-date information available, as well as the evaluations and interpretations from the most credible sources. We cannot yet provide you with all the information needed to resolve all the questions, because not all that information is currently known.

To help deepen your understanding of this method of thinking and its application to the issues of environmental science, What Do You Think? boxes such as this are included in most of the chapters to come. You will find the habit of critical thought incredibly empowering.

Figure 1

Critical thinking is deliberate, self-reflective, reason-based thought about approaching our problems and determining our actions.

we face problems candidly, reasonably, and creatively. Furthermore, dwelling excessively on problems and failures has a way of producing self-fulfilling prophecies. If we expect the worst and believe that all efforts to avert doom are useless, the disasters we fear will probably occur. If, on the other hand, we hope for the best and work to build a better world, perhaps we can find ways to improve our lives and protect the environment at the same time.

You will have to decide for yourself as you study environmental science how much hope the future holds. If you feel overwhelmed and discouraged as you read about the many problems in the world, it might help to recall the disasters faced by the people who preceded us. Surely, to many who have experienced the horrors of war, oppression, and terror, the world must have seemed a dreadful place. But they have found the strength to persevere and survive even in horrendous circumstances.

Consider the 14th century, when Europe was wracked by a terrible century-long war in which innocent civilians were slaughtered by both sides (fig. 1.22). The climate changed inexplicably in the "little ice age," so that crops failed across Europe and famine compounded the horrors of war. Claims that nature was dying because of human sins were not unlike some of the apocalyptic warnings we hear today.

Adding to the inexplicable misery of that time, bubonic plague, the "Black Death," swept repeatedly across Europe, killing at least one-third of the population in just its first pass; as many as 70 percent died in some cities. Civilization crumbled as bands of brigands roamed the countryside and people fled in panic. Parents abandoned their children; everything seemed lost. Many interpreted this series of calamities as God's punishment for their sins and fully expected the

Figure 1.22

Many times in the past, human cruelty and destructiveness have seemed about to bring the end of the world. Are we now at a crucial and definitive turning point in history or will our current crises pass away as others have?

end of the world to be imminent. Life went on, however, and things slowly got better. We haven't stopped killing each other yet, but at least those who kill and torture now generally try to hide their deeds. Perhaps progress, although halting and slow, is possible after all.

Summary

Humans always have inhabited two worlds: one of nature and another of human society and technology. Environmental science is the systematic study of the intersection of these worlds. An interdisciplinary field, environmental science draws from many areas of inquiry to help us understand the worlds in which we live and our proper role in them.

The most amazing features of our planet may be the self-sustaining ecological systems that make life possible and the rich diversity of life that is part of, and dependent upon, those ecological processes. In spite of the many problems that beset us, the earth is wonderfully bountiful and beautiful.

Concerns about pollution and land degradation date back at least 2500 years. Clearly, we have pragmatic interests in conserving resources and preserving a habitable environment. There also are ethical reasons to believe that nature has a right to continue to exist for its own sake. Unprecedented population growth, food shortages, scarce energy supplies, air and water pollution, and destruction of habitats and biological resources are all serious threats to our environment and our way of life.

As international travel and communication become easier, we realize that these problems encompass our whole planet and require global cooperation to find solutions. Still, there is good news. Pollution has been reduced and population growth has slowed in many places. Perhaps we can extend these advances to other areas as well.

The twenty percent of us in the world's richest countries consume an inordinate amount of resources and produce a shocking amount of waste and pollution. About one billion people live in acute poverty and lack access to an adequate diet, decent housing, basic sanitation, clean water, education, medical care, and other essentials for a humane existence.

Concern for the poor is more than a humanitarian issue. Faced with immediate survival needs, these desperate people often have no choice but to overharvest resources and reduce long-term sustainability for themselves and their children. Since we share the same environment it is in our own self-interest to help them find better options than they currently have.

There are several ways of describing the economic and developmental status of different countries. The First World is generally industrialized and more highly developed. Many Third World countries have made encouraging progress in improving the quality of life for residents, but much remains to be done.

Indigenous or native peoples are generally among the poorest and most oppressed of any group. Nevertheless, they possess valuable ecological knowledge and remain the guardians of nature in many places. Recognizing the rights of indigenous people and minority communities is an important way to protect natural resources and environmental quality.

There are valid reasons to be pessimistic about our environmental conditions, but we must be careful that dire predictions don't overwhelm us and become self-fulfilling prophecies. Many people find an optimistic outlook a better motivation than fear, but blind faith in technological progress can be simply an excuse for business as usual. Although we still have far to go in protecting our environment, some heartening progress already has been made toward building a just and sustainable world.

▼ Questions for Review

1. Define environment and environmental science.

2. List six environmental dilemmas that we now face and describe how each concerns us.

3. Describe the differences between the North/South or rich/poor or more developed/less developed nations. What do we mean by First, Second, and Third World?

4. Compare some indicators of quality of life between the richest and poorest nations.

5. Why should we be concerned about the plight of the poor? How do they affect us?

6. What benefit to us would there be in protecting the rights of indigenous people?

7. Give some reasons for pessimism and optimism about our environmental future and summarize how you feel personally about the major environmental problems that we face.

8. Do you think that environmental conditions are better now or worse than they were 20 or 100 or 1000 years ago? Why?

▼ Questions for Critical Thinking

1. How could we determine whether the disappearance of frogs has some grand, global significance, or is merely a random, local event?

2. What are the fundamental differences between utilitarian conservation and altruistic preservation? Which do you favor? Why?

3. Do the issues discussed in this chapter as global environmentalism belong in an environmental *science* text? Why would anyone ask this question?

4. Some people argue that we can't afford to be generous, tolerant, fair, or patient. There isn't enough to go around as it is, they say. What questions would you ask such a person?

5. Others claim that we live in a world of bounty. They believe there would be plenty for all if we just shared equitably. What questions would you ask such a person?

6. Around 200 million children are forced into dangerous, degrading slave labor each year. Is it our business what goes on in other countries?

7. What would it take for human development to be really sustainable? What does sustainable mean to you?

8. Are there enough resources in the world for 8 or 10 billion people to live decent, secure, happy, fulfilling lives? What do those terms mean to you? Try to imagine what they mean to others in our global village.

9. What responsibilities do we have to future generations? What have they done for us? Why not use whatever resources we want right now?

10. Do you see any similarities between current conditions and those of the 14th century? Have we made any real progress or do things just stay the same?

▼ Key Terms

acute poverty   10
altruistic preservation   6
cornucopian fallacy   20
environment   4
environmental science   4
environmentalism   7
First World   14
Fourth World   14
global environmentalism   7

human development index   15
neo-Malthusian   20
North/South division   14
Promethean environmentalism   20
Second World   14
sustainable development   15
technological optimists   20
Third World   14
utilitarian conservation   6

▼ Additional Information on the Internet

Center for World Indigenous Studies
    http://www.halcycon.com/FWDP/cwisinfo.html/

Central European Environmental Data Request Facility
    http://pan.cedar.univie.ac.at/

Centre for Earth Observation
    http://ceo-www.jrc.it/

Earth
    http://seds.1pl.arizona.edu/nineplanets/nineplanets/earth.html

Earth Viewer
    http://www.fourmilab.ch/earthview/vplanet.html/

Earthwatch
    http://www.unep.ch/earthw.html/

The Ecology Channel
    http://www.ecology.com/

Infoterra: United Nations Environment Programme
    http://pan.cedar.univie.ac.at/gopher/UNEP/

International Institute for Sustainable Development Linkages
    http://www.mbnet.mb.ca/linkage/

Mission to Planet Earth
    http://www.hq.nasa.gov/office/pao/nasa/mtpe.html/

Planet Earth Home Page
    http://www.nosc.mil/planet_earth/info.html/

Science and Technology
    http://gnn.com/wic/wics/sci.new.html/

United Nations Environment Programme, Geneva Executive Center, Switzerland
    http://www.unep.ch/

Note: Further readings appropriate to this chapter are listed on p. 593.

Spearfishing on the reef of a South Pacific Island. Under what conditions is it ethical to kill other organisms? Does inhabiting a particular place for a long time give us a special relationship with the environment? How would you view this picture if you lived in a culture different from your own?

CHAPTER 2: Tools for Building a Better World

Objectives

After studying this chapter, you should be able to:

- understand some principles of environmental ethics and philosophy.

- compare and contrast how different ethical perspectives shape our view of nature and our role in it.

- explain anthropocentrism, biocentrism, ecocentrism, utilitarianism, and ecofeminism, and what each says about human/nature relationships.

- summarize the methods, applications, and limitations of the scientific method.

- discuss the role of technology in *causing* environmental problems as well as helping us solve them.

- apply the skills of critical thinking to what you read here and elsewhere.

People or Wildlife in South Africa?

South Africa has some of the most successful wildlife refuges in Africa. Although big game populations have been declining elsewhere on the continent, South African parks and refuges have a surplus of rhinos, elephants, and other endangered species. Game wardens have to cull herds periodically to protect their habitat.

Under apartheid, villagers were forcibly removed from national parks and game preserves. Boundaries were patrolled by heavily armed rangers drawn from the ranks of the South African Defense Force. Trespassers were shot on sight, even if they were only gathering firewood or hunting small animals to feed their families. "We had poachers in 1982 and 1983," boasted Anthony Hall-Martin, chief research officer for Pretoria's National Parks Board, "and we killed them off."

More than 560 South African national parks and nature preserves now enclose 3 million hectares (7.4 million acres) of for-

est, grasslands, and open woodlands. They clearly are responsible for saving many rare and endangered species that might otherwise have been lost. Just outside some splendid natural areas, however, former occupants—many never compensated for seizure of their lands—live in squalid shantytowns of mud and tin shacks (fig. 2.1). As apartheid is being dismantled, rural communities are beginning to demand a return of ancestral lands. They claim a rightful place in the landscape and argue they could live in harmony with wildlife.

President Nelson Mandela is sympathetic to their claims, but ecotourism and big game safaris bring much-needed cash to his struggling country. And hunters and tourists like to believe they are in pristine wilderness. They don't like to shoot elephants or watch birds in the midst of cattle herds or domestic fields. Safaris are big business: minimum cost for a three-week hunting trip is

Figure 2.1

Kilptown squatters camp in South Africa. Many former residents of lands seized for national parks now live in squalid shantytowns such as this.

$50,000 per person. Depending on how many animals you want to kill, you can easily spend upwards of $200,000 per person. Little of that money goes to local communities, however.

This case raises some interesting ethical questions. Should protecting wildlife take precedence over human needs? Is it fair to forbid natives to kill animals for food so that rich foreigners can kill them for fun? Is killing other animals justified under any circumstances? Does establishing a national park require removal of all local people and any sign of their occupancy, or can a landscape that includes traditional human habitation be an acceptable representation of nature?

Answers to these questions depend on our ethical perspectives and our attitudes toward nature and our place in it. In fact, ethical questions and dilemmas underlie many of the issues we will study throughout this book. Environmental leaders frequently claim that our environmental problems are evidence, not of technological failures, but of moral breakdowns in society. In this chapter, we present some tools that will help you understand and analyze these controversies as well as many others you will encounter throughout this text. ▶

Environmental Ethics and Philosophy

Ethics is a branch of philosophy concerned with **morals** (the distinction between right and wrong) and **values** (the ultimate worth of actions or things). Ethics evaluates the relationships, rules, principles, or codes that require or forbid certain conduct. Most Western ethicists consider the roots of their field to be the famous questions posed by Socrates and the Greek philosophers 2500 years ago: "What is the good life? How ought we, as moral beings, to behave?"

Environmental ethics asks about the moral relationships between humans and the world around us. Do we have special duties, obligations, or responsibilities to other species or to nature in general? Are there ethical principles that constrain how we use resources or modify our environment? If so, what are the foundations of those constraints and how do they differ from principles governing our relations to other humans? How are our obligations and responsibilities to nature weighed against human values and interests? Do some interests or values supercede others?

Are There Universal Ethical Principles?

The first question considered in ethics is whether *any* moral laws are objectively valid and independent of cultural context, history, or situation. How the question is answered depends largely on the philosophical disposition of the respondent. **Universalists,** such as Plato and Kant, assert that the fundamental principles of ethics are universal, unchanging, and eternal (fig. 2.2). These rules of right and wrong are valid regardless of our interests, attitudes, desires, or pref-

erences. Some believe these rules are revealed by God, while others maintain they can be discovered through reason and knowledge.

Relativists, such as Plato's opponents, the Sophists, claim that moral principles always are relative to a particular person, society, or situation. Although there may be right and wrong—or at least better and worse—things to do, relativists assert that no transcendent, absolute principles apply regardless of circumstances. In this view, ethical values always are contextual.

Nihilists, such as Schopenhauer, on the other hand, claim that the world makes no sense at all. Everything is completely arbitrary, and there is no meaning or purpose in life other than the dark, instinctive, unceasing struggle for existence. According to this view, there is no reason to behave morally. Only power, strength, and sheer survival matter. "Might is right; eat or be eaten." There is no such thing as a "good" life: we live in a world of uncertainty, pain, and despair. Nevertheless, Schopenhauer enjoyed living in a comfortable, civilized society where rules of normative behavior and good conduct prevailed.

Utilitarians hold that an action is right that produces the greatest good for the greatest number of people. This philosophy is usually associated with the English philosopher Jeremy Bentham (1748–1832); but something very similar was suggested by Plato, Socrates, Aristotle, and others. Bentham was an eccentric genius and a hedonist who equated goodness with happiness, and happiness with pleasure. He regarded pleasure as the only thing worth having in its own right. Thus, the good life is one of maximum pleasure. Insofar as people are moral animals, in his view, we should act to produce the greatest pleasure for the greatest number. To do so is good; not to do so is wrong.

Figure 2.2

Plato and Aristotle debate moral philosophy in a painting by Raphael. Plato (*left*) motions upward, indicating a transcendent, universal moral truth, while Aristotle (*right*) motions downward to suggest grounded, situational ethics.

Postmodernism

A relatively new and popular social theory is the **postmodernism** of Jacques Derrida, Jean-Francois Lyotard, Michael Foucault, and others, who argue that no authoritative, definitive expression or conception of reality or truth—and, therefore, ethics—is possible. Originally applied to art and literature, this philosophy holds that no text has a single fixed meaning. Each is created anew with each reading by every individual. We understand the texts only through signs, symbols, and concepts, and each of us interprets those signs and symbols in unique ways. To get a full understanding, then, we need to deconstruct or break down every text into an infinite number of equally valid meanings.

How, you may be wondering, does this apply to environmental ethics? From a postmodernist perspective, nature—or at least our perceptions of it—are social constructions just like texts. If you ask a diverse group of people what nature is, you most likely will get many different answers. To a city dweller, nature might mean a park with a few trees and a couple of squirrels. To a farmer, it means a productive field, while to a backpacker it means a pristine wilderness free of people (except the backpacker). To understand what each of these people mean by nature, then, we have to deconstruct the words and symbols they use to represent it (fig. 2.3).

Interestingly, groups on opposite extremes of the social spectrum sometimes use postmodernist arguments to support their widely diverse agendas. Resource exploiters, for instance, claim that a clearcut forest is just as legitimate as old growth. Social justice advocates, on the other hand, argue that indigenous people have always been a part of certain landscapes and have a right to continue to be there even if their presence threatens some rare or endangered species. What do you think? Is nature whatever we believe it is? Are there ethical absolutes in our relations with nature or is everything relative and subjective?

Values, Rights, and Obligations

For many philosophers, only humans are **moral agents,** beings capable of acting morally or immorally and who can—and should—accept responsibility for their acts. Capacities that enable humans to form moral judgments include moral deliberation, the resolve to carry out decisions, and the responsibility to hold oneself answerable for failing to do what is right.

Of course, not all humans have all these capacities all the time. Children and those who are mentally retarded, mentally ill, or for some other reason, lacking a full use of reason, are not regarded as moral agents. If a child murders someone, we don't hold her/him responsible. Nonetheless, she/he still has rights. Children are considered **moral subjects,** beings who are not moral agents themselves but who have moral interests of their own and can be treated rightly or wrongly by others.

Historically, the idea that weaker members of society deserve equal treatment with those who are stronger was not an universally held opinion. In many societies, women, children, outsiders, serfs, and others were treated as property by those who were more powerful. The Greek philosophers, for example, accepted slavery as

Utilitarianism was modified and made less hedonistic by Bentham's brilliant protégé, John Stuart Mill (1806–1873). Mill believed that pleasures of the intellect are superior to pleasures of the body. He held that the greatest pleasure is to be educated and to act according to enlightened, humanitarian principles. This empirical, intellectual form of utilitarianism inspired Gifford Pinchot and the early conservationists (see chapter 1) who argued that the purpose of conservation is to protect resources for the "greatest good for the greatest number *for the longest time.*"

Although utilitarianism remains widely popular today, it has drawbacks. It can, for instance, be used to justify reprehensible acts. If ten thousand Romans greatly enjoyed watching a few Christians being eaten by lions, did that make it the right thing to do? Does the pleasure of the tormentors outweigh the suffering of victims? Most of us would conclude that it does not. Justice, freedom, morals, and loyalty take precedence over pleasure, or even happiness, although it could be argued that furthering moral ends and right action ought to bring the greatest happiness in the long run.

Figure 2.3

Is nature only a social creation? Can the synthetic world of Disneyland be equated with the natural world?

Figure 2.4

Do other organisms have inherent values and rights? These chickens are treated as if they are merely egg-laying machines. Many people argue that we should treat them more humanely.

natural and necessary. Gradually, we have come to believe that all humans have certain inalienable rights: life, liberty, and the pursuit of happiness, for example. No one can ethically treat another human as a mere object for their own pleasure, gratification, or profit. This gradually widening definition of who we consider ethically significant is called **moral extensionism.**

Do Other Animals Have Rights?

Perhaps the most important question in environmental ethics is whether moral extensionism encompasses nonhumans. Do other species have rights as well? Are they moral agents or at least moral subjects? For many philosophers, the answer is no. Reason and consciousness—or at least a potential thereof—are essential for moral considerability in this view. René Descartes (1596–1650), for instance, claimed that animals are mere automa (machines) and can neither reason nor feel pain (fig. 2.4).

Most pet owners disagree, claiming that while animals may not have the same self-consciousness as humans, they are intelligent and clearly have feelings. As sentient (perceptive) beings, they deserve ethical treatment. But what about nonsentient beings?

Should we extend moral consideration to bugs, rocks, landscapes? Some people think so. Let's see why.

Intrinsic and Instrumental Values

Rather than couch ethics strictly in terms of rights, some philosophers prefer to consider values. Value is a measure of the worth of something. But value can be either inherent or conferred. All humans, we believe, have **inherent value**—an intrinsic or innate worth—simply because they are human. They deserve moral consideration no matter who they are or what they do. Tools, on the other hand, have conferred, or **instrumental value.** They are worth something only because they are valued by someone who matters. If I hurt you without good reason, I owe you an apology. If I borrow your car and smash it into a tree, however, I don't owe the car an apology. I owe *you* the apology—or reimbursement—for ruining your car. The car is valuable only because you want to use it. It doesn't have inherent values or rights of itself.

Do Nonsentient Things Have Inherent Value?

How does this apply to nonhumans? Domestic animals clearly have an instrumental value because they are useful to their owners. But some philosophers would say they also have inherent values and interests. By living, breathing, struggling to stay alive, the animal carries on its own life independent of its usefulness to someone else. Some would draw a line of moral considerability at the limit of sentience. Others argue that it is in the best interest of bugs, worms, and plants to be treated well, even if they aren't aware of it. In this perspective, just being alive gives things inherent value.

Figure 2.5

Mineral King Valley at the southern border of Sequoia National Park was the focus of an important environmental law case in 1969. The Disney Corporation wanted to build a ski resort here, but the Sierra Club sued to protect the valley on behalf of the trees, rocks, and native wildlife.

Some people believe that even nonliving things also have inherent worth. Rocks, rivers, mountains, landscapes, and certainly the earth itself, have value. These things were in existence before we came along and we couldn't recreate them if they are altered or destroyed. In a landmark 1969 court case, the Sierra Club sued the Disney Corporation on behalf of the trees, rocks, and wildlife of Mineral King Valley in the Sierra Nevada Mountains (fig. 2.5) where Disney wanted to build a ski resort. The Sierra Club argued that it represented the interests of beings that could not speak for themselves in court.

A legal brief entitled *Should Trees Have Standing?* written for this case by Christopher D. Stone proposed that organisms as well as ecological systems and processes should have standing (or rights) in court. After all, corporations—such as Disney—are treated as persons and given legal rights even though they are really only figments of our imagination. Why shouldn't nature have similar standing? The case went all the way to the Supreme Court but was overturned on a technicality. In the meantime, Disney lost interest in the project and the ski resort was never built. What do you think? Where would you draw the line of what deserves moral considerability? Are there ethical limits on what we can do to nature?

Figure 2.6

Do humans have a right to use or destroy other species or natural objects in any way we choose? Do we have duties, obligations, or responsibilities toward nature?

Worldviews and Ethical Perspectives

Our beliefs about our proper roles in the world are deeply conditioned by our ethical perspectives (What Do You Think?, p. 32). As historian Lynn White, Jr., said, "What people do about their ecology depends on what they think about themselves in relation to the things around them." In this section, we will look at how some different moral philosophies reflect our attitudes toward nature.

Domination

Throughout history, many cultures have claimed that humans hold a special place in creation. Pride in our power to reshape the world to our liking and a belief that we are superior to other creatures often have been used to justify **domination** of nature (fig. 2.6). In an influential 1967 paper entitled *The Historic Roots of Our Ecological Crisis,* Lynn White, Jr., traced this tradition to the biblical injunction to "Be fruitful, and multiply, and replenish the earth, and subdue it: and have dominion over the fish of the sea, and over the fowl of the air, and over every living thing that moveth upon the earth" (Genesis 1:28). In antiquity, White claimed, "every tree, every spring had its guardian spirit. Before one cut a tree, dammed a brook, or killed an animal, it was important to placate the spirit in charge of that particular situation. By destroying pagan animism, Christianity made it possible to exploit nature in a mood of indifference to feelings of natural objects."

Although many people agree with this analysis, others argue that this passage is translated and interpreted inaccurately. Genesis is really intended, they claim, to teach us to love and nurture creation rather than dominate and exploit it. White himself pointed out that Judeo-Christian culture has a long tradition of caring for nature. He recommended St. Francis of Assisi as an inspiration and patron saint for all environmentalists. Still, many of us clearly behave as if we have a right to use resources and abuse nature as we choose. This

Tools for Building a Better World

What Do *You* Think?
The Role of Worldviews

Should the Endangered Species Act be renewed or abolished? Should oil companies be allowed to search for oil in the Arctic National Wildlife Refuge, called "America's Serengeti"? The national debate on environmental questions is heated. How we feel personally about issues is heavily influenced by the fundamental vision of reality we hold, the collection of values, perceptions, and practices that organizes our lives. This powerful element guiding our thoughts and actions is called our worldview.

Our worldview serves as an invisible guiding hand, directing how we see and interpret events around us and how we employ the critical-thinking skills presented in this chapter. Paradoxically, this powerful molder of how we think and act is all but invisible to us.

Our worldviews are influenced by many factors. Beliefs and practices of our families and society at large, personal experiences, the amount and kind of education we receive and, ultimately, our station in life, all are major determinants of how we think the world works and what our relationship with it should be.

The dominant worldview today has developed over the last several centuries. It emerged from ideas that the universe is a great machine, whose workings can be understood and manipulated to serve human ends, that by applying the tools of science and technology, the materials of earth could become resources to be used to enhance human lives.

Stimulated by the Industrial Revolution, these ideas produced the consumer society, which defines progress as the satisfying of material wants. This worldview is often called Expansionist because it sees endless material growth as necessary for human happiness and possible because of an ever-advancing technology.

Box 2.1
What Is Your Worldview?

Your responses to the ten statements below can provide insight into your own worldview. Cover up the scoring section below the statements with a piece of paper. Then read each statement and decide whether you agree or not.

1. Earth's resources can support unlimited economic growth.
2. It is not possible to protect the environment without having a growing economy.
3. Problems created by past technologies will be solved by future technologies.
4. Perpetual growth is both good and possible.
5. Nature is a storehouse of raw materials to be used to satisfy ever-increasing human needs.
6. Continued material growth is necessary to increase human quality of life.
7. Technological innovations can sustain an ever-increasing human population.
8. Progress means the satisfaction of increasing levels of human wants.
9. Development means essentially the same thing as growth.
10. There is an "away" to throw things to.

Scoring. All ten statements reflect beliefs of the Expansionist Worldview that are disputed by the Ecological Worldview. If you agreed with seven or more of the statements, count yourself among the Expansionists. If you agreed with three or less, you probably hold the Ecological View. Scores from four to six suggest your worldview is mixed.

Across a wide range of issues, differences of opinion represent a clash of underlying worldviews. The reigning Expansionist paradigm is being challenged by a newer view called the Ecological or Finite Earth Worldview. This worldview has coalesced around the work of ecologists, systems analysts, and the biocentric views of nature preservationists. The basic tenets are that since earth is a finite system, no subsystem can exhibit endless growth. Neither populations, resource use, nor production of toxic wastes can grow indefinitely, and to behave as though they can is to threaten the very functioning of the natural, social, and economic systems upon which our future well-being depends.

As you encounter contentious issues in the media and during this course, see if you can connect the policies being advocated with an underlying worldview.

Although the Expansionist mindset prevails in society, the Ecological Worldview has gained increasing acceptance. Most of the contentious environmental issues are a reflection of its challenges to the conventional wisdom of the Expansionists. A dominant worldview is notoriously difficult to replace since it is so wholly integrated into our lives, thoughts, and language. In fact, challenges to our worldview can often seem to challenge who we are and what we stand for.

Must environmental matters be viewed in such an either/or manner? The newly emerging concept of sustainable development appears to incorporate parts of both these worldviews. On what matters might Expansionists and Ecological Worldview adherents agree? Is it possible for a worldview containing elements of both to emerge?

Figure 2.7

For many people a productive, domesticated landscape such as this mosaic of farmland and sugarbush in Vermont presents the ideal perspective. With careful stewardship—including a balance of population density and environmental resources—the land can be stable, harmonious, and fruitful.

Table 2.1 Worldviews and ethical perspectives—A comparison

Philosophy	Intrinsic Value	Instrumental Value	Role of Humans
Anthropocentric	Humans	Nature	Masters
Stewardship	Humans & Nature	Tools	Caretakers
Biocentric	Species	Abiotic nature	One of many
Animal rights	Individuals	Processes	Equals
Ecocentric	Processes	Individuals	Destroyers
Ecofeminist	Relationships	Roles	Caregivers

view of humans as the focus of creation is termed **anthropocentric,** or human-centered.

Stewardship

Many tribal or indigenous people, both hunters and gatherers, as well as traditional agricultural societies have a strong sense of **stewardship** or responsibility to manage and care for a particular place. As custodians of resources, they see their proper role as working together with human and nonhuman forces to sustain life. Humility and reverence are essential in this worldview, where humans are seen as partners in natural processes rather than masters—not outside of nature but part of it.

This attitude also is held by many modern farmers (fig. 2.7) or others in close contact with nature. Authors Rene Dubos and Wendell Berry have written eloquently about the need to nurture and sustain both the rural landscape and culture. In their view, humans can improve the world and make it a better place, both for themselves and for other organisms. As Voltaire said in *Candide,* "This may be the best of all possible worlds, but we must tend our garden."

Biocentrism, Animal Rights, and Ecocentrism

Many environmentalists criticize both stewardship and dominion as being too anthropocentric. They favor, instead, the **biocentric** (life-centered) egalitarianism of John Muir or Aldo Leopold, who claimed that all living organisms have intrinsic values and rights regardless of whether they are useful to us. Leopold wrote, "Of the 22,000 higher plants and animals native to Wisconsin, it is doubtful whether more than 5 percent can be sold, fed, eaten, or otherwise put to economic use. Yet these creatures are members of the land community, and if (as I believe) its stability depends on its integrity, they are entitled to continuance. . . A thing is right when it tends to preserve the integrity, stability, and beauty of the biotic community. It is wrong when it tends otherwise." For many bio-

centrists, biodiversity is the highest ethical value in nature. Species and populations, as the basic units of biodiversity, are the locus of inherent value.

Some animal rights advocates question the importance of species or populations, claiming that each individual organism is of value. They point out that the individual lives, reproduces, and experiences pleasure or pain, not the group. Many ecologists, in contrast, view larger-scale ecological processes such as evolution, adaptation, and the grand biogeochemical cycles as the most important aspects of nature. In this view, which is described as **ecocentric** (ecologically-centered) because it claims moral values and rights for ecological processes and systems, the whole is considered more important than its individual parts. If you kill an individual organism, you deny it a few months or years of life, but if you eliminate an entire species or a whole landscape, you have destroyed something that took millions of years to create.

In the broadest ecocentric view, individuals could be seen to have only instrumental value while abiotic resources, ecological cycles, and the whole earth possess inherent value (table 2.1). Nature doesn't seem to care about individuals. Vastly more offspring are born than can ever survive. Even species come and go. What seems to have longevity in nature are processes like photosynthesis and evolution.

Ecofeminism

Many feminists believe that none of these philosophies is sufficient to solve environmental problems or to tell us how we ought to behave as moral agents. They argue that all these philosophies come out of a patriarchal system based on domination and duality that assigns prestige and importance to some things but not others. In a patriarchal worldview, men are superior to women, minds are better than bodies, and culture is higher than nature. Feminists contend that domination, exploitation, and mistreatment of women, children, minorities, and nature are intimately connected and mutually reinforcing. They reject all "isms" of domination: sexism, racism, classism, heterosexism, and speciesism.

Ecofeminism, a pluralistic, nonhierarchical, relationship-oriented philosophy that suggests how humans could reconceive themselves and their relationships to nature in nondominating ways, is proposed as an alternative to patriarchal systems of domination. It is concerned not so much with rights, obligations, ownership, and responsibilities as with care, appropriate reciprocity, and kinship. This worldview promotes a richly textured understanding or sense of what human life is and how this understanding can shape people's encounters with the natural world. Among ecofeminist leaders are Karen Warren, Vandana Shiva, Carolyn Merchant, Rosemary Ruether, and Ynestra King.

According to ecofeminist philosophy, when people see themselves as related to others and to nature, they will see life as bounty rather than scarcity, as cooperation rather than competition, and as a network of personal relationships rather than isolated egos. Like the postmodernists, ecofeminists reject the view of a single, ahistoric, context-free, neutral observation stance. Instead, they favor multiple understandings, complex relationships, and "embodied objectivity."

In *Healing the Wounds* Ynestra King wrote, "We can use [ecofeminism] as a vantage point for creating a different kind of culture and politics that would integrate intuitive, spiritual, and rational forms of knowledge, embracing both science and magic insofar as they enable us to transform the nature-culture distinction and to envision and create a free, ecological society." Like the Hindu god, Shiva, humans can be both destroyers and creators of nature (fig. 2.8).

Figure 2.8

Shiva Nataraja, Lord of the Dance, dances the world into existence. In another manifestation, Shiva becomes the destroyer of worlds. The modern scientific view bears some resemblance to this image of a world in a constant cycle of creation and destruction.

Environmental Justice

People of color in the United States and around the world are subjected to a disproportionately high level of environmental health risks in their neighborhoods and on their jobs. Minorities, who tend to be poorer and more disadvantaged than other residents, work in the dirtiest jobs where they are exposed to toxic chemicals and other hazards. More often than not they also live in urban ghettos, barrios, reservations, and rural poverty pockets that have shockingly high pollution levels and are increasingly the site of unpopular industrial facilities, such as toxic waste dumps, landfills, smelters, refineries, and incinerators. **Environmental justice** combines civil rights with environmental protection to demand a safe, healthy, life-giving environment for everyone.

Among the evidence of environmental injustice is the fact that 3 out of 5 African-Americans and Hispanics, and nearly half of all Native Americans, Asians, and Pacific Islanders live in communities with one or more uncontrolled toxic waste sites, incinerators, or major landfills. A recent Greenpeace study found that minorities make up twice as large a population share in communities with these locally unwanted land uses (**LULUs**) as in communities without them. And the inequities are growing. Whereas in 1980 the average minority population near a landfill or hazardous waste facility was about 22 percent, in 1994 it was 36 percent.

Race, not class or income, is the strongest determinant of who is exposed to environmental hazards. While poor people in general are more likely to live in polluted neighborhoods than rich people, the discrepancy between the pollution exposure of middle class blacks and middle class whites is even greater than the difference between poorer whites and blacks. Where upper class whites can "vote with their feet" and move out of polluted and dangerous neighborhoods, blacks and other minorities are restricted by color barriers and prejudice (overt or covert) to the less desirable locations (fig. 2.9).

Figure 2.9

Poor people and people of color often live in the most dangerous and least desirable places. Here children play next to a chemical refinery in Texas City, Texas.

Environmental Racism

Racial prejudice is a belief that someone is inferior merely because of their race. Racism is prejudice with power. **Environmental racism** is inequitable distribution of environmental hazards based on race. Evidence of environmental racism can be seen in lead poisoning in children. The Federal Agency for Toxic Substances and Disease Registry considers lead poisoning to be the number one environmental health problem for children in the United States. Some 4 million children—many of whom are African-American, Latino, Native American, or Asian, and most of whom live in inner city areas—have dangerously high lead levels in their bodies. This lead is absorbed from old lead-based house paint, contaminated drinking water from lead pipes or lead solder, and soil polluted by industrial effluents and automobile exhaust. The evidence of racism is that at every income level, whether rich or poor, black children are two to three times more likely than whites to suffer from lead poisoning.

Dumping Across Borders

Because of their quasi-independent status, most Native-American reservations are considered sovereign nations that are not covered by state environmental regulations. Court decisions holding that reservations are specifically exempt from hazardous waste storage and disposal regulations have resulted in a land rush of seductive offers from waste disposal companies to Native-American reservations for onsite waste dumps, incinerators, and landfills. The short-term economic incentives can be overwhelming for communities in which adult unemployment runs between 60 and 80 percent. Uneducated, powerless people often can be tricked or intimidated into signing environmentally and socially disastrous contracts. Nearly every tribe in America has been approached with proposals for some dangerous industry or waste facility.

The practice of targeting poor communities of color in the Third or Fourth World for waste disposal and/or experimentation with risky technologies has been described as **toxic colonialism.** Internationally, the trade in toxic waste has mushroomed in recent years as wealthy countries have become aware of the risks of industrial refuse. Poor, minority communities at home and abroad are being increasingly targeted as places to dump unwanted wastes. Although a treaty regulating international shipping of toxics was signed by 105 nations in 1989, millions of tons of toxic and hazardous materials continue to move—legally or illegally—from the richer countries to the poorer ones every year. This issue is discussed further in chapter 24.

Another form of toxic colonialism is the flight of polluting industries from developed nations and states where control requirements are stringent to less developed areas where regulations are lax and local politicians are easily co-opted. For example, some 1900 *maquiladoras,* or assembly plants operated by American, Japanese, or other foreign companies, are now located along the United States/Mexico border to take advantage of favorable import quotas, low wages, and weak pollution control laws.

Mexican laborers work under appalling conditions in these factories, assembling imported components into consumer goods to be exported to the United States. Although the jobs are demean-

Figure 2.10

Living conditions in the *colonias* or unplanned settlements along the U.S./Mexican border often are appalling. Many U.S. businesses have been attracted to this area by low wages, compliant union bosses, and lax environmental protection.

ing, dangerous, and pay a minimum wage equivalent to 58 cents per hour, large populations have been attracted to squalid shanty towns along the border where industrial effluents poison the air and water (fig. 2.10). A tragic epidemic of birth defects, including a frighteningly high incidence of anacephalic (born without a brain) babies in towns such as Brownsville, Texas, and Matamoras, Mexico, have been linked to pollution, poverty, and occupational hazards caused by this system.

In 1992, an Environmental Justice Act was introduced in the United States Congress to identify areas threatened by the highest levels of toxic chemicals, assess health effects caused by emissions of those chemicals, and ensure that groups or individuals residing within those areas have opportunities and resources to participate in public discussions concerning siting and cleanup of industrial facilities. Perhaps we need something similar worldwide.

Are "Green" Organizations Too White?

Although concern for the environment cuts across race and class lines, environmental activists traditionally have been individuals with above-average education, greater access to economic resources, and a greater sense of personal power than others in society. The agendas of national environmental organizations tend to focus on wilderness and wildlife preservation, wise resource management, outdoor recreation, and other issues that are not high among the concerns of those at the bottom of society who are struggling for immediate survival. Relatively little attention has been paid by the mainline environmental groups to inner city problems. Considerable criticism has been leveled against these groups for their lack of minority staff members and inattention to environmental justice.

Sadly, an unanticipated result of successful lobbying by environmental organizations against polluting industries and LULUs

Figure 2.11

The great Louisiana Toxics March mobilized local residents from the notorious stretch of the Mississippi River between New Orleans and Baton Rouge known as "cancer alley" because of its high concentration of petrochemical industries and environmental illnesses. Environmental justice combines elements of the civil rights movement with environmental protection.

in recent years has been increased vulnerability of minority communities to the siting of unpopular facilities such as municipal landfills, toxic waste dumps, metal smelters, power plants, and incinerators. Too often **NIMBY** (not in my backyard) **protests** result in dumping of pollution in someone else's backyard. And the minorities who live in polluted communities are often subjected to double jeopardy, in that they are exposed to increased environmental risks, while at the same time, they frequently have problems gaining access to health and medical care facilities.

A growing awareness of the inequitable share of environmental hazards borne by communities of color and the long-term risks to their health and survival has stimulated greater activism and organization around environmental protection for these groups (fig. 2.11). The first National People of Color Environmental Leadership Summit was held in 1991 in Washington, DC. More than 650 delegates, participants, and observers, representing all 50 states, Puerto Rico, Mexico, and the Marshall Islands attended this meeting. The stated purpose of the gathering was "to build a unified national voice and an effective agenda for environmental justice." In combining the civil rights movement with environmental protection, the environmental justice movement represents an important new direction in conservation history.

Is Nature Fragile or Resilient?

For many of our earlier ancestors, nature seemed vast, chaotic, and implacable. Wild lands were full of fierce, wild animals, unpredictable storms, and other life-threatening hazards. By comparison, humans seemed frail and endangered. Early writers described the wilderness as a place of danger and evil spirits. The word *panic*

(fear of the nature god Pan) described the hysterical dread that early Greeks felt when forced to travel through wild nature. They believed that if they heard the music of Pan's pipes they would be lured away from the life of reason and civility in the city and doomed to live forever as wild brutes in the forest. This attitude persisted into fairly recent times. As the *Mayflower* approached New England in 1620, Governor William Bradford described the landscape as a "hideous and desolate wilderness . . . a land of spiritual darkness in which demons and wild beasts dwell."

Humans have struggled to control the forces of nature and to create havens of peace, comfort, and safety for themselves and their domestic livestock. *Paradise* is a Persian word describing a walled garden or oasis that offers shelter and respite from the harsh desert (fig. 2.12a). Many cultures had similar ideals of a place of perfect ease, abundance, and solace. The garden metaphor, in which nature is domesticated and cultivated by the human caretaker to produce useful products and aesthetic pleasure, still has a powerful appeal for many of us.

In recent years, however, as technological power to disrupt natural systems has increased, our view of ourselves with respect to nature has changed. We now can change the course of rivers, create vast artificial lakes, turn woodlands and grasslands into deserts, and literally move mountains. "Manmade" earthquakes have been triggered by pumping liquid wastes into deep disposal wells or by the weight of water and sediments accumulated in some artificial lakes. We already have changed regional climates and we now may be modifying the global climate as well.

Increasingly, we see nature as fragmented, threatened, and vulnerable (fig. 2.12b). Critics warn that our actions threaten to irreparably upset a delicate balance that maintains natural functions. But how fragile is that balance? How much can we perturb nature without threatening not only our own existence but that of all the rest of life as well? These may be questions that we don't want to answer through global experiments.

On the other hand, claims of fragility and delicate balances in nature sometimes are overstated. They may be statements of faith rather than fact. In many circumstances, nature is amazingly fecund and resilient. Great Mayan cities in Central America that once housed thousands of people and were surrounded by farms have been almost completely swallowed by the surrounding rainforest (fig. 2.13). Explorers are still finding evidence of this civilization under jungle-covered mounds. Perhaps 1000 years from now, explorers will poke around the weed-covered rubble of our great cities and wonder what happened to us.

Science as a Way of Knowing

Science and technology have become dominant forces in our lives. Many of us have become completely dependent on science and the machines it creates (fig. 2.14) and yet know very little about how they work. As Henry Thoreau said, "Lo! Men have become the tools of their tools." Understanding something about the history and philosophy of technology and the scientific process may help you not only in your current course of study but also in everyday life.

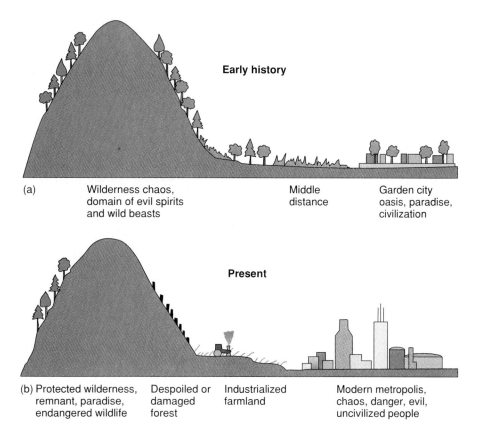

Early history

(a) Wilderness chaos,
domain of evil spirits
and wild beasts

Middle
distance

Garden city
oasis, paradise,
civilization

Present

(b) Protected wilderness,
remnant, paradise,
endangered wildlife

Despoiled or
damaged
forest

Industrialized
farmland

Modern metropolis,
chaos, danger, evil,
uncivilized people

Figure 2.12

Our attitudes toward nature have changed dramatically as cities and domesticated landscapes have expanded and wilderness has diminished. In earlier times (a), the city was seen as a sacred place of comfort, peace, and order, while the wilderness was seen as the domain of chaos, evil, and danger. Now we see the remaining remnants of wild nature as endangered enclaves of paradise while the city represents chaos, evil, and danger (b).

Figure 2.13

Tikal, in what is now Guatemala, was one of the largest Mayan cities 1200 years ago, with a population that may have reached 100,000 people. Today, almost nothing of this great city remains except a few temples rising above the jungle.

Figure 2.14

In modern society, information, education, research, communication, and technology have replaced material goods or natural resources as the main sources of wealth and power.

Tools for Building a Better World

CASE STUDY
Squirrels or Science?

Rising from the hot, dry Sonoran Desert, several small mountain ranges in southeastern Arizona create cool, moist forest refuges for a variety of rare species. Because the desert prevents many species from migrating between ranges, the mountains become islands on which isolated populations evolve along different paths. There are 25 known subspecies of montane red squirrels, for instance. Nearly every range has its own subspecies. One of them, the Mt. Graham Red Squirrel (*Tamiasciurus hudsonicus grahamensis*), has become a pawn in a battle between astronomers and preservationists that illustrates some important differences in our attitudes toward science and nature.

Mountain tops attract astronomers as well as wildlife and these are boom times for astronomy. Technological advances have made it possible to explore questions previously beyond our abilities, but the number of places suitable for high-resolution stargazing is declining rapidly. Urban smog, dust, light, and electromagnetic radiation interfere with observations in many areas. Some of the best locations in the world are in the southern Andes, but travel and logistics at these sites are difficult and expensive.

Mt. Graham, in Arizona's Pinaleno Range, has many attractive features for astronomy (fig. 1). Only about 100 km (60 mi) as the crow flies from Tucson, Mt. Graham is, at 3267 m (10,717 ft), the second highest mountain in the state. It has spectacularly clear air and dark nights. With an international airport nearby and all the modern telecommunications advantages of a modern American city at hand, astronomers can fly in for the weekend or even operate telescopes from their office computers.

Other competing uses for the mountain have made this a battleground, however. An unlikely coalition of animal rights activists, radical environmentalists, hunters, off-road motorcyclists, and traditional Native Americans have fought to keep telescopes off the moun-

Figure 1

Mt. Graham in the Pinaleno Range of southeastern Arizona is the site of highly contested astronomical observatories.

tain. For some, the issue is preserving rare and endangered species and their unique habitat; for others, it is outdoor recreation and personal freedom; for still others, it is a question of despoiling sacred religious sites; and for a few, it is simply a matter of opposing big science and technology on principle.

The institutions proposing new telescopes also make up a surprising alliance. The Smithsonian Institution, the Vatican, the Max Planck Institute from Germany, and the University of Arizona, have all built, or are planning to build, telescopes on Mt. Graham. Several other American universities—including Ohio State, Minnesota, and Texas—expressed an interest but pulled out when students protested.

The Mt. Graham Red Squirrel, which lives only in a small area of spruce-fir forest at the mountain peak—exactly where astronomers want to build telescopes—was declared an endangered subspecies by the Fish and Wildlife Service in 1987. A year later, the United States Congress exempted the project from laws protecting cultural sites or endangered species. Since then, 35 acres of mountain top have been designated for astronomy. Environmentalists still are outraged. Lobbying, legal proceedings, civil disobedience, and eco-sabotage are ongoing.

While beautiful, the mountain is hardly untouched. Logging has gone on there for a century. More than 200,000 people visit every year to hunt, camp, hike, cut Christmas trees, ride dirt bikes, and ski. Some elders of the San Carlos Apache tribe claim that the mountain is a traditional religious site; others dispute that it has any special value. The rarity of finding a cool, moist forest in the midst of the dry rocky desert is probably what endears the mountain to most people.

Some wildlife managers believe that the 200 existing squirrels will do better within the fenced grounds of the astronomical site than in a forest open to hunters and bikers. After all, other squirrels coexist well with humans if they have food and a place to live. Protesters feel that a high-tech facility of this sort is an affront to the wilderness and to free-living wild squirrels.

ETHICAL CONSIDERATIONS

What's your opinion about this situation? Does a telescope complex on the mountain top ruin the landscape? What rights or values would you ascribe to the squirrels? What value does pure science have? Is technology categorically inimical to nature or wilderness? Is nature synonymous with wilderness? Is this a matter of choosing between a unique ecosystem or a world-class scientific facility, or might there be a way to have both?

A Faustian Bargain?

What comes to mind when you think about science and technology? Do they seem mysterious, incomprehensible, and yet powerful? Do you expect science courses to be difficult, disagreeable, and full of obscure details? For many people today, science and the machines it creates seem to be a Faustian bargain (in Goethe's story Dr. Faustus sells his soul to the devil in exchange for power and wealth). One example is the atom bomb, which gives us terrible power to destroy. And yet science, through modern medicine using nuclear equipment, also has saved many lives and eliminated much suffering. Although there have been many misapplications of science and scientific knowledge, we hope to persuade you in this section that science and technology also have many benefits and that scientific methods—especially those of environmental science—can be useful in your everyday life.

What Is Science?

In the best sense, **science** is a systematic, precise, objective way to study the natural world. It is often an exciting and satisfying enterprise that requires creativity, skill, and insight. Science takes many different forms and is done in assorted ways by widely diverse people. When done right, science should be neutral and unbiased. In the wrong hands, it can be perverted and misused, but this is usually the fault of the user, not of the science itself.

Even though you may not be a scientist, you use scientific technique without being aware of it. Suppose your flashlight doesn't work. The problem could be in the batteries, the bulb, or the switch—or all of them could be faulty. How can you distinguish between the possibilities? A series of methodical steps to test each component can be helpful. First, you might try new batteries. If that doesn't help, you might replace the bulb with one that you know works. If neither of these steps solves your problem, perhaps the flashlight has a faulty switch. You could try the original battery and bulb in a different flashlight. By testing the variables one at a time, you should be able to identify the problem.

What you have been following in this example is the **scientific method** for methodical inquiry. Science is a way of knowing about the physical world based on an ordered cycle of observations, methodical investigation, and interpretation of results. The general flow of an experimental scientific study is shown in figure 2.15. We start with observations: in this case, my flashlight doesn't work. From this we formulate a hypothesis, or a provisional explanation: the batteries must be dead. To be useful, our hypothesis must enable us to make predictions that we can test.

Testing Hypotheses

In experimental science, tests of our hypotheses must be controlled in the sense that only one variable is changed at a time. If we change both the bulb and the batteries simultaneously we won't know which is good or bad. We might be discarding perfectly good components. Testing one part with two others that we know work should show which is at fault. Finally, we carry out the experiment, collect data—in this case, look for light—and draw conclusions. Often the results of one experiment gives us information that leads

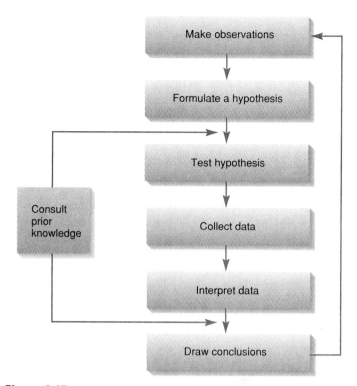

Figure 2.15

Ideally, scientific investigation follows a series of logical, orderly steps to formulate and test hypotheses.

to further hypotheses and additional experiments. If the batteries are OK, then we suspect the bulb is burned out and design a way to test that hypothesis. In every case, prior knowledge and experience help us design experiments and interpret results. Eventually, with evidence from a group of related investigations, we create a theory to explain a set of general principles.

Indirect Scientific Evidence

Not every field of science is accessible to our basic senses—seeing, hearing, touching, smelling, and tasting—nor are all amenable to direct experimentation. We don't regard it as acceptable, for instance, to study the effects of toxins on human health by deliberately poisoning people, no matter how useful that information might be. Although geologists and evolutionary biologists can do limited kinds of experiments, they can hardly build mountain ranges in the laboratory or recreate the millions of years that gave rise to certain species. Scientists in these fields can still generate hypotheses to explain their observations. Their hypotheses are tested indirectly by looking for historic evidence of support or contradiction. Or we use model systems such as laboratory animals as surrogates. While there are difficulties in interpreting these indirect results, they are both scientific and useful.

Proof Is Elusive

A common misconception is that science proves theories. You probably have heard claims that something has been proven "scientifically." In fact, scientific interpretations are always conditional. The

Tools for Building a Better World

evidence from experiments may support a particular hypothesis or provide evidence contrary to it, but we can never be absolutely sure that we have a final proof. A possibility always exists that some new evidence will require a modification of our conclusions. We may even have to discard them and formulate an entirely new set of theories.

Because of these uncertainties, you can often find distinguished scientists who draw opposite conclusions from the same data. This may seem confusing and discouraging. What should we believe when the experts disagree? Learning to evaluate information systematically, keeping an open mind, and remaining alert for new data and alternative interpretations are important lessons from science that can help us understand many aspects of life.

Technology and Progress

For the past two centuries, a central tenet of Western culture has been an almost religious faith in progress: an inevitable march of human betterment. Originally formulated during eighteenth-century enthusiasm over the American and French Revolutions, this theory seemed to be proven by the increase in material wealth and standard of living provided by the Industrial Revolution. But while technology and development brought many benefits, they had a darker side as well (fig. 2.16). Pollution, rapid urbanization, inhumane working conditions for many, and vast disparities in wealth and power between classes still cause social and environmental crises. Whether the root causes of these problems are in technology or human nature, technology clearly allows us to make mistakes faster and on a larger scale than ever before.

A nineteenth-century English backlash against the excesses of industrialization led by Ned Ludd gave rise to the term *Luddites* for opponents of rampant technology. The Luddites smashed power looms and other machines that were threatening the craft guilds, cottage industries, and village networks that sustained traditional rural communities.

In this century, even more dangerous technologies, such as nuclear weapons, biological warfare, and the petrochemical industry, along with the problems caused by earlier technologies, such as biodiversity losses, global climate changes, and destruction of stratospheric ozone, have led many intellectuals and young people to question whether progress is either possible or desirable. **Neo-Luddites** now assert that all large-scale human endeavors eventually fail, that science and technology cause more problems than they solve, and that our only hope is to abandon modern life and go back to a low-tech pastoral or hunting-and-gathering society.

Some neo-Luddites resort to terrorist bombings and sabotage to try to bring down mainstream culture. Others flee to end-of-the-road refuges where they attempt to recreate a simpler, agrarian life. Rural life can be more resource intensive than urban living, however, especially if you demand all the modern conveniences. We probably couldn't all live off the land in remote places without very destructive environmental effects.

Appropriate Technology

As historian Lewis Mumford pointed out, technology consists of more than machines. It includes all the techniques, knowledge, and

Figure 2.16

Science and technology give us power to improve our lives, but they also make it possible to make bigger mistakes faster than ever before, as is the case with these industrial smokestacks shown in the 1890s.

organization that we use to accomplish tasks. When you build a fire, or use a rock as a tool, you are employing some of the oldest and most revolutionary human technologies. Whether our technology is destructive or constructive depends, in part, on the tools themselves, but even more on our worldview about how and why we use them. As sustainable-energy expert Amory Lovins points out, some technologies such as nuclear power, genetic engineering, and nanotechnology might be benign in the hands of a "wise, far-seeing, and incorruptible people." Unfortunately that seems not to describe most of us.

In 1973, British economist E. F. Schumacher published a widely popular book entitled *Small Is Beautiful*. It introduced the concept of **appropriate technology,** which promotes machines and approaches suitable for local conditions and cultures. The appropriate-technology movement attempts to design productive facilities in places where people now live, not in urban areas. It looks for products that are affordably made by simple production methods from local materials for local use. It advocates safe, creative, environmentally sound, emotionally satisfying work conducted in conditions of human dignity and freedom that creates social bonds rather than breaking them down.

Figure 2.17

Appropriate technology, such as this walk-behind cultivator, is designed to suit local needs and conditions in developing countries. Small, simple, and affordable machines that are locally manufactured and repaired can reduce drudgery and improve productivity without reducing the number of jobs available.

Figure 2.18

"There is absolutely no cause for alarm at the nuclear plant!"

Reprint by permission: Tribune Media Services.

Rather than to try to convert local economies and tastes into copies of western culture, appropriate technologists try to work with indigenous people to create sustainable livelihoods suitable for prevailing conditions (fig. 2.17). They hope that appropriate technology can help us avoid future environmental damage and to repair mistakes made in the past. Some advances have been made toward these goals, but the promise and power of the dominant western paradigm are very seductive.

Critical Thinking

Throughout this book you will find many facts, figures, opinions, and theories. Are all of them true? No, probably not. They were the best information available when this text was written, but much in environmental science is in a state of flux. Facts change constantly as does our interpretation of them. Do the ideas presented here give a complete picture of the state of our environment? Unfortunately, they probably don't. No matter how comprehensive our discussion is of this complex, diverse subject, it cannot capture everything worth knowing, nor can it reveal all the possible points of view.

Uncertainty is not limited to environmental science and textbooks. The world around us is changing at a breathtaking pace. Every day we are inundated by a flood of information and misinformation. Competing claims and contradictory ideas battle for our attention. The communication and information revolutions are transforming our lives in ways that our parents could never have imagined. Computers, fax machines, cellular telephones, e-mail, the World Wide Web, direct satellite transmission of hundreds of TV channels, direct mail marketing, and interactive home shopping keep us in touch and provide us with more information and choices than we can possibly manage.

By now, most of us know not to believe everything we read or hear (fig. 2.18). "Tastes great . . . Low, low sale price . . . Vote for me . . . Lose 30 pounds in 3 weeks . . . This product is environmentally friendly . . . My client is innocent . . . Two doctors out of three recommend . . ." More and more of the information we use to buy, elect, advise, judge, or heal has been created not to expand our knowledge but to sell a product or advance a cause.

Still, it would be unfortunate if we become cynical and apathetic due to information overload. It does make a difference what we think and do. But how can we know what to believe when the facts are confusing and experts disagree? Is it simply a matter of what feels good or supports our preconceived notions?

What Is Critical Thinking?

Critical thinking is a set of skills that help us evaluate information and options in a systematic, purposeful, efficient manner. Critical thinking shares many methods and approaches with formal logic but adds some important contextual skills, attitudes, and dispositions. Furthermore, it challenges us to plan methodically and assess the process of thinking as well as the implications of decisions reached. It can help us discover hidden ideas and meanings, develop strategies for evaluating reasons and conclusions in arguments, recognize the differences between facts and values, and avoid jumping to conclusions.

Richard Paul, chair of the National Council for Critical Thinking, breaks this process down into ten steps:

1 What is the *purpose* of my thinking?

2 What precise *question* am I trying to answer?

3 Within what *point of view* am I thinking?

4 What *information* am I using?

5 How am I *interpreting* that information?

6 What *concepts* or ideas are central to my thinking?

7 What *conclusions* am I coming to?

8 What am I taking for granted, what *assumptions* am I making?

9 If I accept the conclusions, what are the *implications*?

10 What would the *consequences* be, if I put my thoughts into action?

Notice that many of these steps are self-reflective and self-correcting. Critical thinking is sometimes called "thinking about thinking." It is an attempt to plan rationally how to analyze a problem, to monitor your progress while you are doing it, and to evaluate how your strategy worked and what you have learned when you have finished. Critical thinking is not critical in the sense of finding fault, but it makes a conscious, active, disciplined effort to be aware of hidden motives and assumptions, to uncover bias, and to recognize the reliability or unreliability of sources.

Developing critical-thinking abilities is an important part of your education and will give you useful skills for life. Students no longer can anticipate the knowledge or proficiencies they will need when they enter the job market because they can no longer predict the careers available in the future or what those fields will require. Even if you know what job you will have, experts predict that about half the information now current in any field will be obsolete in six years. How can you plan for an uncertain future? One of the best things you can do is to learn how to learn. Critical thinking gives you a valuable set of tools to adapt to a rapidly changing world.

What Do I Need to Think Critically?

Certain attitudes, tendencies, and dispositions are essential for well-reasoned analysis. Professor Karen J. Warren of Macalester College suggests the following list:

- *Skepticism and independence.* Question authority. Don't believe everything you hear or read—including this book; even experts are sometimes wrong.

- *Openmindedness and flexibility.* Be willing to consider differing points of view and entertain alternative explanations. Try arguing from a viewpoint different from your own. It will help you identify weaknesses and limitations in your own position.

- *Accuracy and orderliness.* Strive for as much precision as the subject permits or warrants. Deal systematically with parts of a complex whole. Be disciplined in the standards you apply.

- *Persistence and relevance.* Stick to the main point. Don't allow diversions or personal biases to lead you astray. Information may be interesting or even true, but is it relevant?

- *Contextual sensitivity and empathy.* Consider the total situation, relevant context, feelings, level of knowledge, and sophistication of others as you evaluate information. Imagine being in someone else's place to try to understand how they feel.

- *Decisiveness and courage.* Draw conclusions and take a stand when the evidence warrants doing so. Although we often wish

for more definitive information, sometimes a well-reasoned position has to be the basis for action.

- *Humility.* Realize that you may be wrong and that reconsideration may be called for in the future. Be careful about making declarations; you may need to change your mind someday.

Developing these attitudes and skills is not easy or simple. It takes practice. You have to get your mental faculties in shape just as you would train for a sport. Traits such as intellectual integrity, humility, fairness, empathy, and courage are not something to be called up occasionally. They must be cultivated until they become your normal way of thinking.

Applying Critical Thinking

We all use critical thinking at times. Suppose a television commercial tells you that a new breakfast cereal is tasty and good for you. You may be suspicious and ask yourself a few questions. What do they mean by good? Good for whom or what? Does "tasty" simply mean more sugar and salt? Might the sources of this information have other motives in mind besides your health and happiness? Although you may not have been aware of it, you already have been using some of the techniques of critical analysis. Working to expand these skills helps you recognize the ways information and analysis can be distorted, misleading, prejudiced, superficial, unfair, or otherwise defective.

Here are some steps in critical thinking:

1 *Identify and evaluate premises and conclusions in an argument.* What is the basis for the claims made here? What evidence is presented to support these claims and what conclusions are drawn from this evidence? If the premises and evidence are correct, does it follow that the conclusions are necessarily true?

2 *Acknowledge and clarify uncertainties, vagueness, equivocation, and contradictions.* Do the terms used have more than one meaning? If so, are all participants in the argument using the same meanings? Are ambiguity or equivocation deliberate? Can all the claims be true simultaneously?

3 *Distinguish between facts and values.* Are claims made that can be tested? (If so, these are statements of fact and should be able to be verified by gathering evidence.) Are claims made about the worth or lack of worth of something? (If so, these are value statements or opinions and probably cannot be verified objectively.) For example, claims of what we *ought* to do to be moral or righteous or to respect nature are generally value statements.

4 *Recognize and interpret assumptions.* Given the backgrounds and views of the protagonists in this argument, what underlying reasons might there be for the premises, evidence, or conclusions presented? Does anyone have an "axe to grind" or a personal agenda in this issue? What do they think I know, need, want, believe? Is there a subtext based on race, gender, ethnicity, economics, or some belief system that distorts this discussion?

5 *Distinguish the reliability or unreliability of a source.* What makes the experts qualified in this issue? What special knowledge or information do they have? What evidence do they present? How can we determine whether the information offered is accurate, true, or even plausible?

6 *Recognize and understand conceptual frameworks.* What are the basic beliefs, attitudes, and values that this person, group, or society holds? What dominating philosophy or ethics control their outlook and actions (fig. 2.19)? How do these beliefs and values affect the way people view themselves and the world around them? If there are conflicting or contradictory beliefs and values, how can these differences be resolved?

Some Clues for Unpacking an Argument

In logic, an argument is made up of one or more introductory statements (called the premises), and a conclusion that supposedly follows from the premises. Often in ordinary conversation, different kinds of statements are compressed and mixed together, so it is difficult to distinguish between them or to decipher hidden or implied meanings. Social theorists call the process of separating textual components *unpacking.* Applying this type of analysis to an argument can be useful.

An argument's premises are usually claimed to be based on facts; conclusions are usually opinions and values drawn from, or used to interpret, those facts. Words that often introduce a premise include: *as, because, assume that, given that, since, whereas,* and *we all know that. . . .* Words that often indicate a conclusion or statement of opinion or values include: *and so, thus, therefore, it follows that, consequently, the evidence shows,* and *we can conclude that.*

For instance, in the example we used earlier, the television ad might have said: "Since we all need vitamins, and since this cereal contains vitamins, consequently the cereal must be good for you." Which are the premises and which is the conclusion? Does one necessarily follow from the other? Remember, even if the facts in a premise are correct, the conclusions drawn from them may not be. Information may be withheld from the argument such as the fact that the cereal also is loaded with unhealthy amounts of sugar.

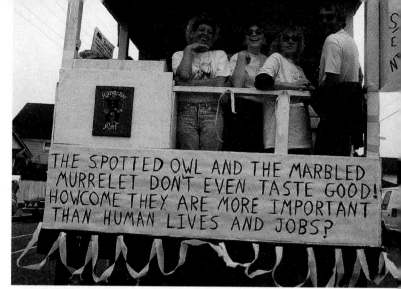

Figure 2.19

How do people's beliefs, attitudes, and values shape the positions they take in controversial environmental issues? Understanding world views and conceptual frameworks can help you evaluate what really motivates your opposition.

As you go through this book, you will have many opportunities to practice these critical thinking skills. Nearly every chapter includes end-of-chapter questions and a boxed reading with clues or examples to help you develop your reasoning abilities. As you read the text, try to distinguish between statements of fact and opinion. Ask yourself if the premises support the conclusions drawn from them. Although we have tried to be fair and even-handed in presenting controversies, we, like everyone, have biases and values—some we may not even recognize—that affect how we present arguments. Watch for examples of areas in which you need to think for yourself and utilize your critical thinking skills.

Summary

Are there universal, eternally valid ethical principles or moral laws? Universalists think so. Nihilists disagree. Relativists believe that everything is contextual. Utilitarians hold that something is good that brings the greatest good to the greatest number. Postmodernists argue that everything is socially constructed. Ecofeminists contend that patriarchal systems of domination and duality cause both environmental degradation and social disfunction. They call for a more pluralistic, nonhierarchical, caring treatment of both nature and other people. Which of these worldviews you hold shapes your views about nature and our place in it?

Anthropocentrists claim that the world was made for our domination, and that only humans have inherent or intrinsic rights and values. Nature, from this perspective is only a source of materials for humans. Many people who live close to the land feel a sense of stewardship or responsibility to care for creation. They see themselves as caretakers rather than dominators. Biocentrists consider all living things to have inherent value. We are merely one of many species. Animal rights advocates place their emphasis on individual animals. Ecocentrics maintain that ecological processes such as evolution, adaptation, and biogeochemical cycles are the most important parts of nature. In their view, individuals don't count for much and humans are mostly a negative influence.

Science is a methodical, meticulous, dispassionate study of the natural world. It takes many different forms and is done in various ways by diverse people, but observation, hypothesis formation, testing, analysis, and re-evaluation of hypotheses in light of new data form the core of scientific methodology. Technology brings us many benefits, but it also creates pollution, consumes resources, despoils nature, and allows us to separate individuals, classes, and nations into those who have and those who do not. It gives us the power to make mistakes faster and on a larger scale than ever before. Then again, appropriate technology may also provide options to avoid environmental damage in the future or to repair mistakes made in the past.

Critical thinking is a systematic, purposeful, open-minded, self-correcting process that employs contextual skills, attitudes, and dispositions to help us decide what to believe and do. It makes a conscious, active, disciplined effort to understand hidden motives and assumptions, uncover bias, and to recognize the reliability of sources. It is sometimes called thinking about thinking. It asks: What am I trying to accomplish? How should I proceed? How will I know when I have reached a conclusion?

▼ Questions for Review

1. Define universalist, nihilist, relativist, utilitarian, postmodern, and ecofeminist ethics.

2. Explain how resource exploiters and social justice advocates use these ethical positions to support their causes.

3. Describe the differences between moral agents, moral subjects, and those who fail to qualify for moral considerability.

4. What are inherent and instrumental values? Who has them? Why?

5. Where is Mineral King Valley and why is it important in environmental history? What happened there?

6. Compare and contrast stewardship, anthropocentric, biocentric, ecocentric, and environmental justice worldviews. Which is closest to your own views?

7. Draw a diagram showing the scientific method and describe, in your own words, what it means.

8. Not every science is experimental. Explain how geologists or evolutionary biologists test their hypotheses.

9. List and explain Richard Paul's ten questions for critical thinking.

10. Itemize six dispositions, tendencies, or attitudes necessary for critical thinking. Elaborate on how they help us think.

▼ Questions for Critical Thinking

1. Review a topic in this chapter—ecofeminism or the question of nature's fragility or resiliency, for example—and arrange the arguments for or against this idea as a series of short statements. Determine whether each is a statement of fact or opinion.

2. How would you verify or disprove fact statements given in question 1? Do the opinions or conclusions reached in this argument *necessarily* follow from the facts given? What alternative proposals could be proposed?

3. Reflect on the preconceptions, values, beliefs, contextual perspective that you bring to the discussion above. Does it coincide with those in the textbook? What different interpretation would your perspective impose on the argument?

4. Try to put yourself in the place of a person from a minority community, an underdeveloped nation, or a Third World country in discussing questions of environmental justice and environmental quality. What preconceptions, values, beliefs, and contextual perspective would you bring to the issue? What would you ask for from the majority society?

5. What is your environmental ethic? What experiences, cultural background, education, religious beliefs help shape your worldview?

6. Try some role-playing with a classmate, friend, or family member. Take the position of a universalist, biocentrist, ecocentrist, postmodernist, or ecofeminist and debate the merits of cutting trees in the National Forest. On what points would you agree and where would you disagree in this issue?

7. This chapter addresses many social issues and theories. Are these areas appropriate for an environmental science textbook? Why or why not? Try answering this question from the perspective of Gifford Pinchot, John Muir, the ecofeminist, or the person of color whom you identified in questions above. Would they agree?

▼ Key Terms

anthropocentric 31
appropriate technology 38
biocentric 31
critical thinking 39
domination 29
ecocentric 31
ecofeminism 32
environmental ethics 26
environmental justice 32
environmental racism 33
inherent value 28
instrumental value 28
LULUs 32
moral agents 27
moral extensionism 28

moral subjects 27
morals 26
neo-Luddites 38
nihilists 26
NIMBY protests 34
postmodernism 27
relativists 28
science 37
scientific method 37
stewardship 31
toxic colonialism 33
universalists 26
utilitarians 26
value 26

▼ Additional Information on the Internet

Association for the Study of Literature and the Environment
http://faraday.clas.virginia.edu/~djp2n/asle.html/

Environmental Ethics
http://www.cep.unt.edu/

CIA World Factbook
http://www.odci.gov/cia/publications/95fact/

The Earth Council
http://terra.ecouncil.ac.cr/

Environmental Philosophers
http://www.envirolink.org/elib/enviroethics/ecophiloindex.html/

John Muir Exhibit
http://ice.ucdavis.edu/John_Muir/

World Wide Web Virtual Library: History of Science, Technology, and Medicine
http://www.asap.unimelb.edu.au/hstm/hstm_ove.html/

Note: Further readings appropriate to this chapter are listed on p. 593.

Interpretive Specialist

Ann Wendland

*A*nn Wendland

Ann Wendland currently works as an Interpretive Specialist for a national forest in Arizona. Her duties include planning displays for a swimming area, an astronomical observatory, an urban pathway, and a ski area.

To begin a project, she talks to a wide variety of people and studies visitor statistics to find out what residents and visitors want or need to know about each of these areas. Then she spends time getting to know the land—looking at its problems and trying to get a sense of place. From all that input, she sorts out the most important stories and finds the best places and ways to tell them. She designs the exhibits, creating signs and writing text for them. Finally, she draws up contracts and sends text and layout to illustrators and fabricators for production.

Ann finds the best aspect of her job to be the diversity of projects that come her way. She loves exploring new places, meeting new people, and learning new things. She enjoys writing and basic design, but most of all, she likes to feel that she is helping people develop a land ethic, "connecting them to the ground they walk on, and protecting a small part of the world."

Are there drawbacks? Yes, of course. Nearly all her work is indoors. While glad to be working to save the land, Ann wishes she could get out to enjoy it more often. due to current budget crises, Ann's job is only temporary and will end after one more year. Neither the Park Service nor the Forest Service can hire permanent employees right now (except for a few special exceptions). Ann advises that you don't look for a government job if you want a relaxed and secure work environment.

How did Ann prepare for this job? Following her interests turned out to be a good route. In the middle of her college years, Ann took time off to live in a small town on the Washington coast. There she volunteered at a science center just so she could look at the octopuses and sea anemones. When a condo and conference center threatened her favorite section of beach, she got active in environmental groups and city planning. Besides being personally fulfilling, these experiences proved to be valuable additions to her resume.

When she "dropped back into college," Ann finished a B.S. in environmental interpretation, with minors in geology and writing. Having a specialized degree, writing and editing skills, a science background, and several interesting internships all added to her credentials. Good academic performance counts: Ann says that good grades boosted her starting salary several thousand dollars. In spite of an excellent resume, however, jobs didn't come easily. After several months of fruitless searching, Ann was helped by a friendly professor who spent a couple of hours on the phone calling friends and contacts to find a job opening for her.

What would Ann suggest to students now in college? "Be flexible, but try to develop a vision or some standards for what you'd like to do. If your friends have jobs you'd like to have, find out how they got them and who their contacts were. And get an internship if you can—they often turn into jobs." The usual way to get a seasonal position is to go to the nearest Park or Forest Service office and look at their vacancy list. The Forest Service hires year-round, but the Park Service has two seasonal application deadlines. Try working for a season or two to get some good experience.

How important is it to live in a beautiful place like the mountains of northern Arizona? For Ann, just getting a job was the first concern, although the mountains and desert also sustain her. She is amazed and discouraged that people who visit her area trash it. If they don't care for and protect the land they drove thousands of miles to see, what *do* they care for? Ann believes that every place has its own beauty that we need to learn to appreciate and care for. "When you get your job, wherever it is," she urges, "believe you're in a special place and make your place beautiful and alive."

A puffin returns to its nest with a bill full of fish for its chick.

CHAPTER 3: Matter, Energy, and Life

Objectives

After studying this chapter, you should be able to:

- describe matter, atoms, and molecules and give simple examples of the role of four major kinds of organic compounds in living cells.

- define *energy* and explain the difference between kinetic and potential energy.

- understand the principles of conservation of matter and energy and appreciate how the laws of thermodynamics affect living systems.

- know how photosynthesis captures energy for life and how cellular respiration releases that energy to do useful work.

- define *species, populations, communities,* and *ecosystems* and understand the ecological significance of these levels of organization.

- discuss food chains, food webs, and trophic levels in biological communities and explain why there are pyramids of energy, biomass, and numbers of individuals in the trophic levels of an ecosystem.

- recognize the unique properties of water and explain why the hydrologic cycle is important to us.

- compare the ways that carbon, nitrogen, sulfur, and phosphorus cycle within ecosystems.

Pigibowin: Poisoning the Wabigoon-English Rivers

In the late 1960s, many native Ojibwa people in the small Canadian villages of White Dog and Grassy Narrows, Ontario, suffered from blurred vision, slurred speech, coordination loss, depression, confusion, and a variety of aches and pains. These symptoms eventually were traced to high mercury levels in fish from the Wabigoon-English River system. For centuries, fish have been a dietary mainstay for native people, but now something was introducing pigibowin (poison in Ojibwa) into their environment.

The mercury had come from a pulp and paper mill in Dryden, Ontario, just upstream from the Indian villages (fig. 3.1). From 1963 to 1970, about 10 metric tons of mercury from a chlorine bleach process was dumped into the river. Mercury is extremely toxic—death can occur from ingesting only a few milligrams (thousandths of a gram) of methyl mercury. Chronic, low-level exposures can cause permanent neurological injuries, birth defects, and kidney damage. Although mercury was present in very low concentrations in river water, it was taken up by mi-

croscopic organisms and concentrated through the food chain until it reached dangerous levels in the fish eaten by humans.

The paper mill stopped dumping mercury in 1970. Mercury contamination in fish and aquatic invertebrates has dropped by 90 percent, but mercury continues to be recycled from sediments through living organisms to humans. In 1990, a six-year-old girl whose grandmother fished regularly was found to have dangerous mercury levels in her blood. Federal and provincial authorities still advise people to limit their fish intake to no more than one meal per week, a hardship for many who can't afford store-boughten food.

In addition to the recycling of mercury already in the ecosystem, heavy metals continue to be added to the river system by long-range wind-borne transport from coal-burning power plants and other industrial facilities, mainly in the United States. Furthermore, naturally occurring mercury in rocks and solids is leached out by acid rain—again caused primarily by fossil-fuel burning in cities far from the Wabigoon-English River.

Figure 3.1

Pulp and paper mills, such as this one in Thunder Bay, Ontario, are major producers of air and water pollution. Twenty years ago, a similar plant in Dryden, Ontario, released mercury into the Wabigoon–English River system that accumulated in fish and caused serious health problems for the native people dependent on the fish as a major food source.

How and why toxic metals—as well as other, more benign, materials—are cycled between the living and nonliving parts of our environment are the domain of **ecology,** *the scientific study of relationships between organisms and their environment. Modern biology covers a wide range of inquiry, ranging from molecules to ecosystems to the entire planet. Ecology examines the life histories, distribution, and behavior of individual species, as well as the structure and function of natural systems at the level of populations, communities, ecosystems, and landscapes (fig. 3.2). The systems approach of ecology encourages us to think holistically and to see interconnections that make whole systems more than just the sum of their individual parts.*

In this chapter, we will survey some fundamental aspects of energy flow and material recycling within ecosystems. We will look at why living organisms need a constant supply of these essential ingredients, how energy and materials are transferred from one organism to another, and how humans benefit from, and also disrupt, those ecological relationships. After reading this chapter, you

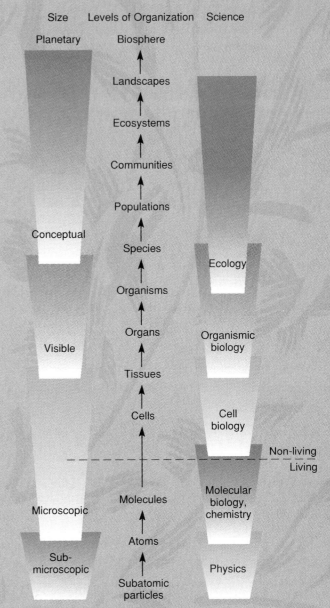

Figure 3.2

Levels of organization and the sciences that study them.

should have a better understanding of how native people could have been exposed to mercury poisoning in their environment and why similar—if less dangerous—risks may apply to all of us. ▶

From Atoms to Cells

In a sense, every organism is a chemical factory that captures matter and energy from its environment and transforms them into structures and processes that make life possible. To understand how these processes work, we need to understand something about the fundamental properties of matter and energy. This section presents a survey of some of these principles.

Atoms, Molecules, and Compounds

Everything that takes up space and has mass is matter. All matter has three interchangeable physical forms, or phases: gas, liquid, and solid. Water, for example, can exist as vapor (gas), fluid (liquid), or ice (solid). Matter also consists of unique chemical forms we call elements, molecules, and compounds. Each of the 111 known elements (65 of which are cycled biogenically) has distinct

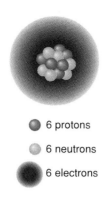

6 protons

6 neutrons

6 electrons

Figure 3.3

As difficult as it may be to imagine when you look at a solid object, all matter is composed of tiny, moving particles, separated by space and held together by energy. It is hard to capture these dynamic relationships in a drawing. This model represents carbon[12], with a nucleus of six protons and six neutrons; the six electrons are represented as a fuzzy cloud of potential locations rather than as individual particles.

chemical characteristics. Among the more common elements in biology are carbon (represented by the symbol C), hydrogen (H), oxygen (O), nitrogen (N), and phosphorus (P).

All elements are composed of discrete units called **atoms,** which are the smallest particles that exhibit the characteristics of the element. Atoms are tiny units of matter composed of positively charged protons, negatively charged electrons, and electrically neutral neutrons. Protons and neutrons, which have approximately the same mass, are clustered in the nucleus in the center of the atom (fig. 3.3). Electrons, which are tiny in comparison to the other units, orbit the nucleus at high speed. Atoms that have an equal number of electrons and protons are electrically neutral. Those that have gained or lost electrons, and therefore are positively or negatively charged, are called **ions.**

Each element has a characteristic number of protons per atom, called its atomic number. The number of neutrons in different atoms of the same element can vary within certain limits. Thus, the atomic mass, which is the sum of the protons and neutrons in each nucleus, also can vary. We call forms of a single element that differ in atomic mass **isotopes.** For example, hydrogen, the lightest element, normally has only one proton (and no neutrons) in its nucleus. A small percentage of hydrogen atoms in nature have one proton and one neutron. We call this isotope deuterium (H^2). An even smaller percentage of natural hydrogen called tritium (H^3) has one proton plus two neutrons.

Tritium is an example of a **radioactive isotope,** an unstable form that spontaneously emits either high-energy electromagnetic radiation or subatomic particles (or both), while gradually changing into another isotope or even a different element. We call this process radioactive decay. Some isotopes decay very rapidly, while others decay very slowly. The half life—the time required for half the atoms in a sample to decay—for tritium is about 13 years. By contrast, some isotopes of iodine have half lives measured in seconds, while plutonium, a waste product of nuclear power reactions, has a half life of 24,000 years. The radioactive emissions from

these atoms may be neutrons, beta particles (high-energy electrons), alpha particles (helium nuclei consisting of two protons and two neutrons), or gamma rays (very short-wavelength radiation).

In a nuclear fission reaction, an unstable isotope is bombarded with free neutron particles, causing it to split into two smaller atoms. This fission releases vast quantities of energy, which provides the power for atomic bombs and nuclear power plants. In a fusion reaction, light atoms of hydrogen or helium are slammed together at such high speeds that they fuse to become a single larger atom. This reaction also releases an enormous amount of energy that is the basis for a hydrogen bomb. Fusion and fission reactions are discussed further in chapter 21.

Atoms can join together to form **molecules.** A **compound** is a molecule containing different kinds of atoms. Water, for example, is a compound composed of two atoms of hydrogen attached to a single oxygen atom, shown by the formula H_2O. In a few cases, two atoms of the same element combine to form a molecule. Hydrogen gas (H_2), molecular oxygen (O_2), and molecular nitrogen (N_2) consist of such diatomic molecules. Most molecules are incredibly small, but some can be relatively large. The genetic information in your cells, for instance, is contained in deoxyribonucleic acid (DNA) molecules, each of which contains billions of atoms and can be seen as long linear chains in very high-power microscopes.

The forces that hold atoms together in molecules are called chemical bonds. Some bonds are very stable, that is, it takes a great deal of energy to create them or break them apart. Other bonds are weak and ephemeral. It's fortunate for us that this range exists. If all bonds were weak, we wouldn't be able to form solid structures such as bones, muscles, or nerves. If, on the other hand, all bonds were of the strongest variety, we wouldn't have any moving parts, and life would be impossible.

Organic Compounds

Organisms use some elements in abundance, others in trace amounts, and others not at all. Certain vital substances are concentrated within cells, while others are actively excluded. Carbon is a particularly important element because chains and rings of carbon atoms form the skeletons of **organic compounds,** the material of which biomolecules, and therefore living organisms, are made. A century ago, scientists thought that organic compounds had some mystical quality and were so complex that they could never be created outside of living cells. We now know this to be false. Many of the millions of known organic compounds can be synthesized fairly easily in the laboratory.

The four major categories of bioorganic compounds are lipids, carbohydrates, proteins, and nucleic acids. Lipids belong to a family of molecules called hydrocarbons because they are composed of chains of carbon atoms, most of which have two hydrogens attached. Some common examples of lipids are fats and oils. They make up an important part of the membranes that surround cells as well as their internal organelles. Carbohydrates, as their name suggests, are composed of carbon, hydrogen, and oxygen $(CH_2O)_n$, where n represents many repeated units arranged in rings or linear chains. Some common examples of carbohydrates are sugars, starch, and cellulose.

Matter, Energy, and Life

Proteins are composed of long chains of subunits called amino acids, which are made primarily of carbon, hydrogen, and nitrogen (along with some other minor elements). Proteins make up much of both the structural and functional components of cells. Enzymes, for instance, are proteins. Nucleic acids are composed of combinations of a sugar molecule, a nitrogen-containing ring structure, and a phosphate bridge that can link many subunits (known as nucleotides) into long molecules such as DNA and RNA. Nucleic acids store the genetic information that directs the processes of life.

Cells: The Fundamental Units of Life

All living organisms are composed of **cells,** minute compartments within which the processes of life are carried out (fig. 3.4). Microscopic organisms such as bacteria, some algae, and protozoa are composed of single cells. Most higher organisms are multicellular, usually with many different cell varieties. Your body, for instance, is composed of several trillion cells of about 200 distinct types. Every cell is surrounded by a thin but dynamic membrane of lipid and protein that receives information about the exterior world and regulates the flow of materials between the cell and its environment. Inside, cells are subdivided into tiny organelles and subcellular particles that provide the machinery for life. Some of these organelles store and release energy. Others manage and distribute

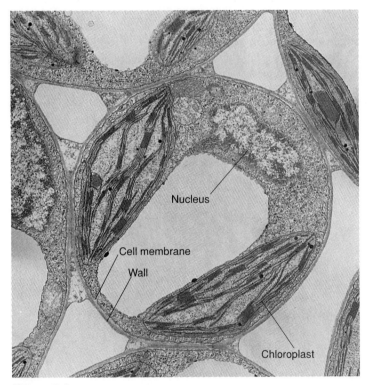

Figure 3.4

A single, spherical plant cell is enclosed within a heavy cellulose cell wall in the center of this high magnification photograph. Parts of five other cells can be seen around it. The bean-shaped structures with multiple layers of internal membranes are chloroplasts. Chlorophyll inside the chloroplasts captures light energy and uses it to carry out photosynthesis, discussed later in this chapter.

information. Still others create the internal structure that gives the cell its shape and allows it to fulfill its role.

All of the chemical reactions required to create these various structures, provide them with energy and materials to carry out their functions, dispose of wastes, and perform other functions of life at the cellular level are carried out by a special class of proteins called **enzymes.** Enzymes are molecular catalysts, that is, they regulate chemical reactions without being used up or inactivated in the process. Think of them as tools: like hammers or wrenches, they do their jobs without being consumed or damaged as they work. There are generally thousands of different kinds of enzymes in every cell, all necessary to carry out the many processes on which life depends. Altogether, the multitude of enzymatic reactions performed by an organism is called its **metabolism.**

Energy and Matter

Energy and matter are essential constituents of both the universe and living organisms. **Matter,** of course, is the material of which things are made. Energy provides the force to hold structures together, tear them apart, and move them from one place to another. In this section we will look at some fundamental characteristics of these components of our world.

Energy Types and Qualities

Energy is the ability to do work such as moving matter over a distance or causing a heat transfer between two objects at different temperatures. Energy can take many different forms. Heat, light, electricity, and chemical energy are examples that we all experience. The energy contained in moving objects is called **kinetic energy.** A rock rolling down a hill, the wind blowing through the trees, water flowing over a dam (fig. 3.5), or electrons speeding around the nucleus of an atom are all examples of kinetic energy. **Potential energy** is stored energy that is latent but available for use. A rock poised at the top of a hill and water stored behind a dam are examples of potential energy. Chemical energy stored in the food that you eat and the gasoline that you put into your car are also examples of potential energy that can be released to do useful work. Energy is measured in **joules** (J). One joule is the energy necessary to lift 1 kilogram 1 meter.

Power is the rate of energy flow from one object or one form to another. **Heat** describes the total kinetic energy of atoms or molecules in a substance not associated with bulk motion of the substance. **Temperature** is a measure of the speed of motion of a typical atom or molecule in a substance. Note that heat and temperature are not the same. A substance can have a low temperature (low average molecular speed) but a high heat content (much mass and many moving molecules or atoms). For example, the average temperature of the ocean is relatively low, but its total heat content is enormous. Conversely, the outer limits of the atmosphere contain gases that have very high kinetic energy (temperature), but there are so few of them per unit volume that their total heat content is very low.

Energy that is diffuse, dispersed, and low in temperature is considered low-quality energy because it is difficult to gather and

Figure 3.5
Water stored behind this dam represents potential energy. Water flowing over the dam has kinetic energy, some of which is converted to heat.

use for productive purposes. The heat stored in the oceans, for instance, is low-quality. Conversely, energy that is intense, concentrated, and high in temperature is high-quality energy because of its usefulness in carrying out work. The intense flames of a very hot fire or high-voltage electrical energy are examples of high-quality forms that are valuable to humans. These distinctions are important, because many of our most common energy sources are low-quality and must be concentrated or transformed into high-quality before they are useful to us.

Conservation of Matter

Under ordinary circumstances, matter is neither created nor destroyed but rather is recycled over and over again. Some of the molecules that make up your body probably contain atoms that once made up the body of a dinosaur and most certainly were part of many smaller prehistoric organisms as chemical elements are used and reused by living organisms. Matter is transformed and combined in different ways, but it doesn't disappear; everything goes somewhere. These statements paraphrase the physical principle of **conservation of matter.**

How does this principle apply to human relationships with the biosphere? Particularly in affluent societies, we use natural resources to produce an incredible amount of "disposable" consumer goods. If everything goes somewhere, where do the things we dispose of go after the garbage truck leaves? As the sheer amount of "disposed-of stuff" increases, we are having greater problems finding places to put it. Ultimately, there is no "away" where we can throw things we don't want any more.

Thermodynamics and Energy Transfers

Organisms use gases, water, and nutrients and then return them to the environment in altered forms as by-products of their metabolic processes. Year after year, century after century, the same atoms find endless reincarnation in new molecules synthesized by suc-

ceeding organisms as they feed, grow, and die. This exchange and continuity are made possible, however, by something that cannot be recycled: energy. *Energy must be supplied from an external source to keep biological processes running.* Energy flows in a one-way path through biological systems and eventually into some low-temperature sink such as outer space. It can be used to accomplish work as it flows through the system, and it can be stored temporarily in the chemical bonds of organic molecules, but eventually it is released and dissipated.

The study of thermodynamics deals with how energy is transferred in natural processes. More specifically, it deals with the rates of flow and the transformation of energy from one form or quality to another. Thermodynamics is a complex, quantitative discipline, but you don't need a great deal of math to understand some of the broad principles that shape our world and our lives.

The **first law of thermodynamics** states that energy is *conserved;* that is, it is neither created nor destroyed under normal conditions. It may be transferred from one place or object to another, but the total amount of energy remains the same. Similarly, energy may be transformed, or changed from one form to another (for example, from the energy in a chemical bond to heat energy), but the total amount is neither diminished nor increased.

The **second law of thermodynamics** states that, with each successive energy transfer or transformation in a system, less energy is available to do work. This is not a contradiction of the first law; the energy is not lost or destroyed, merely degraded or dissipated from a higher-quality form to a lower-quality form. We can think of this process as an energy "expenditure," or the "cost" in terms of useful energy of doing work. The second law recognizes a tendency of all natural systems to go from a state of order (for example, high-quality energy) toward a state of increasing disorder (for example, low-quality energy, such as heat energy). As these systems move from order to disorder, their *entropy,* or amount of disorder, increases, reflecting the loss of energy. Stated simply, the second law of thermodynamics says there is always less useful energy available when you finish a process than there was before you started. Because of this loss, everything in the universe tends to fall apart, slow down, and get more disorganized.

How does the second law of thermodynamics apply to organisms and biological systems? Organisms are highly organized, both structurally and metabolically. Constant care and maintenance is required to keep up this organization, and a constant supply of energy is required to maintain these processes. Every time some energy is used by a cell to do work, some of that energy is dissipated or lost as heat. If cellular energy supplies are interrupted or depleted, the result—sooner or later—is death.

Energy for Life

Ultimately, most organisms depend on the sun for the energy needed to create structures and carry out life processes. A few rare biological communities live in or near hot springs or thermal vents in the ocean where hot, mineral-laden water provides energy-rich chemicals that form the basis for a limited and unique way of life. For most of us, however, the sun is the ultimate energy source. In this

Matter, Energy, and Life

section, we will look at how green plants capture solar energy and use it to create organic molecules that are essential for life.

Solar Energy: Warmth and Light

Our sun is a star, a fiery ball of exploding hydrogen gas. Its thermonuclear reactions emit powerful forms of radiation, including potentially deadly ultraviolet and nuclear radiation (fig. 3.6), yet life here is nurtured by, and dependent upon, this searing, energetic source. Solar energy is essential to life for two main reasons.

First, the sun provides warmth. Most organisms survive within a relatively narrow temperature range. In fact, each species has its own range of temperatures within which it can function normally. At very high temperatures, biomolecules break down or become distorted and nonfunctional. At very low temperatures, the chemical reactions of metabolism occur too slowly to enable organisms to grow and reproduce. Other planets in our solar system are either too hot or too cold to support life as we know it. The earth's water and atmosphere help to moderate, maintain, and distribute the sun's heat.

Second, organisms depend on solar radiation for life-sustaining energy, which is captured by green plants, algae, and some bacteria in a process called **photosynthesis.** Photosynthesis converts radiant energy into useful, high-quality chemical energy in the bonds that hold together organic molecules.

How much potential solar energy is actually used by organisms? The amount of incoming, extraterrestrial solar radiation is enormous, about 1372 watts/m^2 at the top of the atmosphere (1 watt = 1 J per second). However, not all of this radiation reaches the earth's surface. More than half of the incoming sunlight may be reflected or absorbed by atmospheric clouds, dust, and gases. In particular, harmful, short wavelengths are filtered out by gases (such as ozone) in the upper atmosphere; thus, the atmosphere is a valuable shield, protecting life-forms from harmful doses of ultraviolet and other forms of radiation. Even with these energy reductions, however, the sun provides much more energy than biological systems can harness, and more than enough for all our energy needs if technology can enable us to tap it efficiently.

Of the solar radiation that does reach the earth's surface, about 10 percent is ultraviolet, 45 percent is visible, and 45 percent is infrared. Most of that energy is absorbed by land or water or is reflected into space by water, snow, and land surfaces. (Seen from outer space, the earth shines about as brightly as Venus.)

Fortunately for us, some radiation is captured by organisms through photosynthesis. Even then, however, the amount of energy that can be used to build organic molecules is further reduced. Photosynthesis can use only certain wavelengths of solar energy that are within the visible light range of the electromagnetic spectrum. These wavelengths are in the ranges we perceive as red and blue light. Furthermore, half, or more, of the light energy absorbed by leaves is consumed as water evaporates. Consequently, only about 1 to 2 percent of the sunlight falling on plants is captured for photosynthesis. This small percentage represents the energy base for virtually all life in the biosphere!

How Does Photosynthesis Capture Energy?

Photosynthesis occurs in tiny membranous organelles called chloroplasts that reside within plant cells (see fig. 3.4). The most important key to this process is chlorophyll, a unique green molecule that can absorb light energy and use it to create high-energy chemical bonds in compounds that serve as the fuel for all subsequent cellular metabolism. Chlorophyll doesn't do this important job all alone, however. It is assisted by a large group of other lipid, sugar, protein, and nucleic acid molecules. Together these components carry out two interconnected cyclic sets of reactions (fig. 3.7).

Photosynthesis begins with a series of steps called light reactions, because they occur only while light is being received by the chloroplast. During these light reactions, water molecules are split, releasing molecular oxygen. This is the source of all the oxygen in the atmosphere on which all animals, including humans, depend for life. The other products of the light reactions are small, mobile, high-energy molecules called adenosine triphosphate (ATP) and reduced nicotinamide adenine diphosphate (NADPH). These inter-

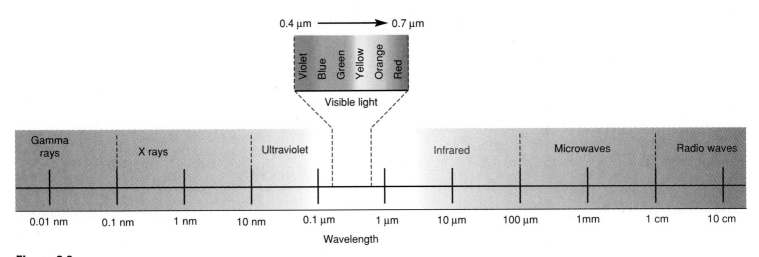

Figure 3.6
The electromagnetic spectrum.

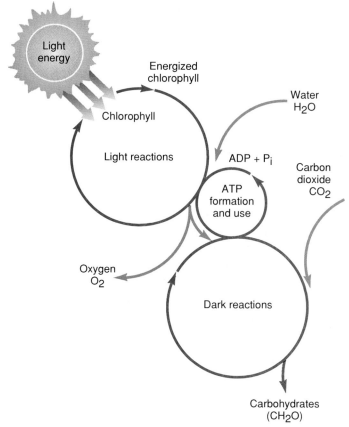

Figure 3.7

Photosynthesis represents the capture of light energy by chlorophyll molecules and the use of that energy to synthesize energy-rich chemical compounds that can be used by cells to drive the processes of life. The first cycle in this complex process—called the light reactions because they only occur while light is present—create ATP and other high-energy compounds that provide power for the next set of reactions—the dark reactions—that carry out carbon fixation to create new carbohydrate molecules. Carbon dioxide and water are consumed in these reactions while oxygen and carbohydrates are produced.

mediates serve as the fuel or energy source for the next set of processes, the dark reactions.

The dark reactions, as their name implies, can occur in the chloroplast after the light has been turned off. The enzymes in this complex use energy from ATP and NADPH to add a carbon atom (from carbon dioxide) to a small sugar molecule.

In most temperate-zone plants, photosynthesis can be summarized in the following equation:

$$6H_2O + 6CO_2 + \text{solar energy} \xrightarrow[\text{chlorophyll}]{} C_6H_{12}O_6 \text{ (sugar)} + 6O_2$$

We read this equation as "water plus carbon dioxide plus energy produces sugar plus oxygen." The reason the equation uses six water and six carbon dioxide molecules is that it takes six carbon atoms to make the sugar product. If you look closely, you will see that all the atoms in the reactants balance with those in the products. This is an example of conservation of matter.

You might wonder how making a simple sugar benefits the plant. The answer is that glucose is an energy-rich compound that serves as the central, primary fuel for all metabolic processes of cells. The energy in its chemical bonds—the ones created by photosynthesis—can be released by other enzymes and used to make ATP, which, in turn, can drive the synthesis of other molecules (lipids, proteins, nucleic acids, or other carbohydrates), or it can drive kinetic processes such as movement of ions across membranes, transmission of messages, changes in cellular shape or structure, or movement of the cell itself in some cases. This process of releasing chemical energy, called **cellular respiration,** involves splitting carbon and hydrogen atoms from the sugar molecule and recombining them with oxygen to re-create carbon dioxide and water. The net chemical reaction, then, is the reverse of photosynthesis:

$$C_6H_{12}O_6 + 6O_2 \rightarrow 6H_2O + 6CO_2 + \text{released energy}$$

Note that in photosynthesis, energy is *captured,* while in respiration, energy is *released.* Similarly, photosynthesis *consumes* water and carbon dioxide to *produce* sugar and oxygen, while respiration does just the opposite. In both sets of reactions, energy is stored temporarily in ATP, which constitutes a kind of energy currency for the cell.

We animals don't have chlorophyll and can't carry out photosynthetic food production. We do have the components for cellular respiration, however. In fact, this is how we get all our energy for life. We eat plants—or other animals that have eaten plants—and break down the organic molecules in our food through cellular respiration to obtain energy (fig. 3.8). In the process, we also consume oxygen and release carbon dioxide, thus completing the cycle of photosynthesis and respiration. Later in this chapter we will see how these feeding relationships work.

From Species to Ecosystems

While cellular and molecular biologists study life processes at the microscopic level, ecologists study interactions at the species, population, biotic community, or ecosystem level. In Latin, *species* literally means *kind.* In biology, **species** refers to all organisms of the same kind that are genetically similar enough to breed in nature and produce live, fertile offspring. There are several qualifications and some important exceptions to this definition of species (especially among plants), but for our purposes this is a useful working definition.

Populations, Communities, and Ecosystems

A **population** consists of all the members of a species living in a given area at the same time. Chapter 6 deals further with population growth and dynamics. All of the populations of organisms living and interacting in a particular area make up a **biological community.** What populations make up the biological community of which you are a part? The population sign marking your city limits announces only the number of humans who live there, disregarding the other populations of animals, plants, fungi, and microorganisms that are part of the biological community within the

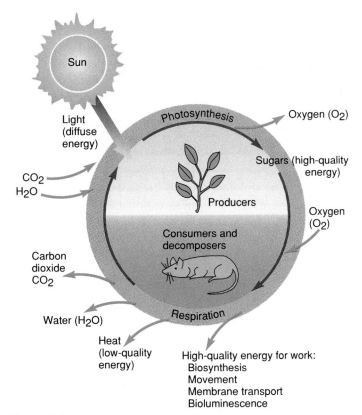

Figure 3.8

Energy exchange in an ecosystem. Plants take water and carbon dioxide from the environment and use the energy from sunlight to convert them into energy-rich sugars and other organic chemicals, releasing oxygen in the process. Consumers and decomposers take up oxygen and break down sugars during cellular respiration to release useful energy for living. In this process, water, carbon dioxide, and low-quality heat are released to the environment.

city's boundaries. Characteristics of biological communities are discussed in more detail in chapter 4.

An ecological system, or **ecosystem,** is composed of a biological community and its physical environment. The environment includes abiotic factors (nonliving components), such as climate, water, minerals, and sunlight, as well as biotic factors, such as organisms, their products (secretions, wastes, and remains), and effects in a given area. Broadening our perspective from the biological community to the community plus its surroundings fosters study of the ways in which energy and materials are obtained, processed, stored, or cycled between components of the ecosystem. This system's approach tends to focus more on roles played by various members of the community rather than on the unique life stories of the individuals themselves. Looking at the organization and functions of a system can give us valuable insights into how it works.

For simplicity's sake, we think of ecosystems as fixed ecological units with distinct boundaries. If you look at a patch of woods surrounded by farm fields, for instance, a relatively sharp line separates the two areas, and conditions such as light levels, wind,

moisture, and shelter are quite different in the woods than in the fields around them. Because of these variations, distinct populations of plants and animals live in each place. By studying each of these areas, we can make important and interesting discoveries about who lives where and why and about how conditions are established and maintained there.

The division between the fields and woods may not be as definite and impenetrable as we might imagine at first glance, however. Air, of course, moves freely from one to another, and the runoff after a rainfall may carry soil, leaf litter, and even live organisms between the areas. Birds may feed in the field during the day but roost in the woods at night, giving them roles in both places. Are they members of the woodland community or the field community? Is the edge of the woodland ecosystem where the last tree grows, or does it extend to every place that has an influence on the woods? Depending on how tenuous the connections you are willing to consider, the whole world might be part of an ecosystem.

As you can see, it may be difficult to draw clear boundaries around communities and ecosystems. To some extent we define these units by what we want to study and how much information we can handle. Thus, an ecosystem might be as large as a whole watershed or as small as a pond or even the surface of your skin. Even though our choices may be somewhat arbitrary, we still can make important discoveries about how organisms interact with each other and with their environment within these units. The woods are, after all, significantly different from the fields around them.

Like the woodland we just considered, most ecosystems are open, in the sense that they exchange materials and organisms with other ecosystems. A stream ecosystem is an extreme example. Water, nutrients, and organisms enter from upstream and are lost downstream. The species and numbers of organisms present may be relatively constant, but they are made up of continually changing individuals. Some ecosystems are relatively closed, in the sense that very little enters or leaves them. A balanced aquarium is a good example of a closed ecosystem. Aquatic plants, animals, and decomposers can balance material cycles in the aquarium if care is taken to balance their populations. Because of the second law of thermodynamics, however, every ecosystem must have a constant inflow of energy and a way to dispose of heat. Thus, at least with regard to energy flow, every ecosystem is open.

Many ecosystems have mechanisms that maintain composition and functions within certain limits. A forest tends to remain a forest for the most part and to have forestlike conditions if it isn't disturbed by outside forces. Some ecologists suggest that ecosystems—or perhaps all life on the earth—may function as superorganisms (What Do *You* Think?, p. 53).

Food Chains, Food Webs, and Trophic Levels

Photosynthesis is the base of the energy economy of all but a few special ecosystems, and ecosystem dynamics are based on how organisms share food resources. In fact, one of the major properties of an ecosystem is its **productivity,** the amount of **biomass** (biological material) produced in a given area during a given period of time. Photosynthesis is described as *primary productivity* because

What Do *You* Think?
Chaos or Stability in Ecosystems?

The questions we ask are shaped by our conceptual frameworks or paradigms, models that serve as general explanations of how things work. You may be surprised to learn that not all ecologists agree about something as central to their science as the basic properties of biological communities and ecosystems.

One paradigm for nature is that of pioneer biogeographer F. E. Clements (1874–1945), who argued that biological communities behave as if they are superorganisms. Species and populations, in his view, work together something like the organs of the body to maintain a stable, well-balanced set of functions and composition. Clements borrowed the physiological term **homeostasis** (from the Greek *homeo*, same, and *stasis*, stationary) to describe the equilibrium achieved by nature when left undisturbed.

Clements also saw a similarity between the growth and development of an individual and the regular developmental stages of an ecosystem from bare ground or fallow field to mature forest (see chapter 4). He claimed that every landscape has a characteristic "climax" community toward which it will develop if free from external interference. The climax community, he believed, represents the maximum state of complexity and stability possible for a given set of environmental conditions.

Many people are attracted by this purposeful, deterministic view of nature. Others, however, believe that assuming intentional design or plan in nature is unscientific. H. A. Gleason, for instance, who was a contemporary of Clements, regarded biological communities as merely chance associations of species able to migrate into and live in a specific place at a particular time. The presence or absence of any one species, Gleason argued, is independent of all others. He suggested that we see ecosystems as stable and uniform only because our lifetimes are so short and our view so limited.

Clements' views were very influential in the early days of ecology, and for many years homeostasis generated by diversity was regarded as the most important principle in nature. More recently, however, the individualistic, probabilistic views of Gleason have attracted more supporters. Nevertheless, both of these viewpoints continue to have proponents.

Recent research on cybernetics and planetary systems has awakened a new interest in the self-maintaining, equilibrating features of biological systems. The Gaia hypothesis, named by biophysicist James Lovelock after the ancient Greek goddess of the earth, originated in speculations about the unique environmental conditions of the earth. In contrast to our neighboring planets, which are either much too hot or much too cold for life as we know it, most of the earth is just right for our existence (some people call Earth the Goldilocks planet).

This fortunate set of circumstances is more than a happy accident, according to the Gaia hypothesis. Rather it is the result of active intervention by the living biota to create and sustain a livable environment. In this view, the whole earth operates as if it were a single superorganism. This homeostasis depends on active feedback processes operated automatically and unconsciously by the biota. Although Lovelock dissociates himself from any mystical or religious implications of his theory, many people believe it suggests design and meaning in the world.

The Gleasonian view of nature as individualistic and unpredictable also has modern supporters. Paleobiogeography (the study of distribution of plants and animals through history based on analysis of fossils and sediments) shows a much greater variability in landscapes than Clements thought possible. What appears to be a stable forest or prairie community may have had dramatically different species composition and ecosystem characteristics over the millennia. From this perspective, patchiness, variability, and randomness of species distribution in communities are more important and characteristic than homogeneity. Constant change seems to be the rule rather than the exception for much of nature.

What triggers these changes? Random, often catastrophic events appear to be major forces in shaping our world. Erratic climate change, irregular volcanic eruptions, giant asteroids crashing into the earth, and the slow but inexorable processes of continental drift, mountain building, and erosion modify landscapes in complex ways. The science of chaos and catastrophe allow us to find patterns—or at least limits—in what had hitherto seemed merely noise. An important revelation from this view is that complex systems can exhibit emergent properties, characteristics that could not have been predicted from a study of individual components.

Both of these contrasting views of nature have interesting implications. From the Gaia hypothesis we might infer that even though nature has self-correcting mechanisms, there may be thresholds of disruption beyond which it may not be able to recover. If nature has a plan for survival, it may not include us. From chaos and catastrophe theory, we realize that there are great uncertainties in nature. If we can't predict what may happen next, we might want to leave a margin of safety in how much we disturb nature.

Perhaps even more useful is the illustration of the importance of conceptual frameworks in shaping our views. If you look for examples of stability and order, you will find them. At the same time, if you look for change, you will find many examples of that as well. Probably neither of these paradigms is exclusively right, but both give useful insights. Which is closest to your worldview? Do you see the world in terms of harmony and stability? Or do you see more evidence of randomness and constant change? What scientific questions and policy implications would you tend to propose if you believed in one paradigm rather than the other?

it is the basis for almost all other growth in an ecosystem. Manufacture of biomass by organisms that eat plants is termed *secondary productivity*. A given ecosystem may have very high total productivity but if decomposers break down organic material as rapidly as it is formed, the *net productivity* will be low.

Think about what you have eaten today and trace it back to its photosynthetic source. If you have eaten an egg, you can trace it back to a chicken, which ate corn. This is an example of a **food chain,** a linked feeding series. Now think about a more complex food chain involving you, a chicken, a corn plant, and a grasshopper. The chicken could eat grasshoppers that had eaten leaves of the corn plant. You also could eat the grasshopper directly—some humans do. Or you could eat corn yourself, making the shortest possible food chain. Humans have several options of where we fit into food chains.

In ecosystems, some consumers feed on a single species, but most consumers have multiple food sources. Similarly, some species are prey to a single kind of predator, but many species in an ecosystem are beset by several types of predators and parasites. In this way, individual food chains become interconnected to form a **food web.** Figure 3.9 shows feeding relationships among some of the larger organisms in a woodland and lake community. If we were to add all the insects, worms, and microscopic organisms that belong in this picture, however, we would have overwhelming complexity. Perhaps you can imagine the challenge ecologists face in trying to quantify and interpret the precise matter and energy transfers that occur in a natural ecosystem!

An organism's feeding status in an ecosystem can be expressed as its **trophic level** (from the Greek *trophe,* food). In our first example, the corn plant is at the **producer** level; it transforms solar energy into chemical energy, producing food molecules. Other organisms in the ecosystem are **consumers** of the chemical energy harnessed by the producers. An organism that eats producers is a primary consumer. An organism that eats primary consumers is a secondary consumer, which may, in turn, be eaten by a tertiary consumer, and so on. Most terrestrial food chains are relatively short (seeds → mouse → owl), but aquatic food chains may be quite long (microscopic algae → copepod → minnow → crayfish → bass → osprey). The length of a food chain also may reflect the physical characteristics of a particular ecosystem. A harsh arctic landscape has a much shorter food chain than a temperate or tropical one (fig. 3.10).

Figure 3.9

Each time an organism feeds, it becomes a link in a food chain. In an ecosystem, food chains become cross-connected when predators feed on more than one kind of prey, thus forming a food web. How many food chains make up the food web shown here? The arrows in this diagram and in figure 3.10 indicate the direction in which matter and energy are transferred through feeding relationships.

Figure 3.10

Harsh environments tend to have shorter food chains than environments with more favorable physical conditions. Compare the arctic food chains depicted here with the longer food chains in the food web in figure 3.9.

Organisms can be identified both by the trophic level at which they feed and by the *kinds* of food they eat (fig. 3.11). **Herbivores** are plant eaters, **carnivores** are flesh eaters, and **omnivores** eat both plant and animal matter. What are humans? We are natural omnivores, by history and by habit. Tooth structure is an important clue to understanding animal food preferences, and humans are no exception. Our teeth are suited for an omnivorous diet, with a combination of cutting and crushing surfaces that are not highly adapted for one specific kind of food, as are the teeth of a wolf (carnivore) or a horse (herbivore).

One of the most important trophic levels is occupied by the many kinds of organisms that remove and recycle the dead bodies and waste products of others. **Scavengers** such as crows, jackals, and vultures clean up dead carcasses of larger animals. **Detritivores** such as ants and beetles consume litter, debris, and dung, while **decomposer** organisms such as fungi and bacteria complete the final breakdown and recycling of organic materials. It could be argued that these microorganisms are second in importance only to producers because without their activity nutrients would remain locked up in the organic compounds of dead organisms and discarded body wastes, rather than being made available to successive generations of organisms.

Ecological Pyramids

If we arrange the organisms in a food chain according to trophic levels, they often form a pyramid with a broad base representing

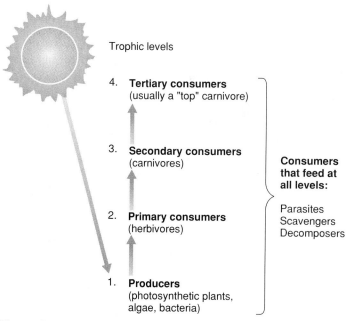

Figure 3.11

Organisms in an ecosystem may be identified by how they obtain food for their life processes (producer, herbivore, carnivore, omnivore, scavenger, decomposer, reducer) or by consumer level (producer; primary, secondary, or tertiary consumer) or by trophic level (1st, 2nd, 3rd, 4th).

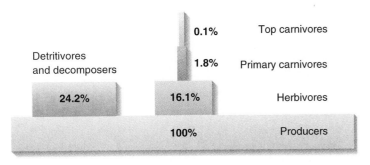

Figure 3.12

A classic example of an energy pyramid from Silver Springs, Florida. The numbers in each bar show the percentage of the energy captured in the primary producer level that is incorporated in the biomass of each succeeding level. Detritivores and decomposers feed at every level but are shown attached to the producer bar because this level provides most of their energy.

Source: Modified from Howard T. Odum, "Trophic Structure and Productivity of Silver Springs, Florida" in *Ecological Monographs*, 27:55–112, 1957, Ecological Society of America.

primary producers and only a few individuals in the highest trophic levels. This pyramid arrangement is especially true if we look at the energy content of an ecosystem (fig. 3.12). True to the second principle of thermodynamics, less food energy is available to the top trophic level than is available to preceding levels. For example, it takes a huge number of plants to support a modest colony of grazers such as prairie dogs. Several colonies of prairie dogs, in turn, might be required to feed a single coyote.

Why is there so much less energy in each successive level in this figure? In the first place, some of the food that organisms eat is undigested and doesn't provide usable energy. Much of the energy that is absorbed is used in the daily processes of living or lost as heat when it is transformed from one form to another and thus isn't stored as biomass that can be eaten.

Furthermore, predators don't operate at 100 percent efficiency. If there were enough foxes to catch all the rabbits available in the summer when the supply is abundant, there would be too many foxes in the middle of the winter when rabbits are scarce. A general rule of thumb is that only about 10 percent of the energy in one consumer level is represented in the next higher level (fig. 3.13). The amount of energy available is often expressed in biomass. For example, it generally takes about 100 kg of clover to make 10 kg of rabbit and 10 kg of rabbit to make 1 kg of fox.

The total number of organisms and the total amount of biomass in each successive trophic level of an ecosystem also may form pyramids (fig. 3.14) similar to those describing energy content. The relationship between biomass and numbers is not as dependable as energy, however. The biomass pyramid, for instance, can be inverted by periodic fluctuations in producer populations (for example, low plant and algal biomass present during winter in temperate aquatic ecosystems). The numbers pyramid also can be inverted. One coyote can support numerous tapeworms for example. Numbers inversion also occurs at the lower trophic levels (for example, one large tree can support thousands of caterpillars).

Material Cycles and Life Processes

To our knowledge, the earth is the only planet in our solar system that provides a suitable environment for life as we know it. Even our nearest planetary neighbors, Mars and Venus, do not meet these requirements. Maintenance of these conditions requires a constant recycling of materials between the biotic (living) and abiotic (non-living) components of ecosystems. The unique properties of water make it indispensable for life (In Depth, p. 60) and give water a central place in the cycling of many elements through ecosystems. Water itself also cycles in a planetwide process that shapes the land, regulates climate, and provides a constant supply of pure, fresh water to terrestrial life (see chapter 19).

The Carbon Cycle

Carbon serves a dual purpose for organisms: (1) it is a structural component of organic molecules, and (2) the energy-holding chemical bonds it forms represent energy "storage." The **carbon cycle** begins with the intake of carbon dioxide (CO_2) by photosynthetic organisms (fig. 3.15). Carbon (and oxygen) atoms are incorporated into sugar molecules during photosynthesis. Carbon dioxide is eventually released during respiration, closing the cycle.

The path followed by an individual carbon atom in this cycle may be quite direct and rapid, depending on how it is used in an organism's body. Imagine for a moment what happens to a simple sugar molecule you swallow in a glass of fruit juice. The sugar molecule is absorbed into your bloodstream where it is made available to your cells for cellular respiration or for making more complex biomolecules. If it is used in respiration, you may exhale the same carbon atom as CO_2 the same day.

Alternatively, that sugar molecule can be used to make larger organic molecules that become part of your cellular structure. The carbon atoms in it could remain a part of your body until it decays after death. Similarly, carbon in the wood of a thousand-year-old tree will be released only when the wood is digested by fungi and bacteria that release carbon dioxide as a byproduct of their respiration.

Can you think of examples where carbon may not be recycled for even longer periods of time, if ever? Coal and oil are the compressed, chemically altered remains of plants or bacteria that lived millions of years ago. Their carbon atoms (and hydrogen, oxygen, nitrogen, sulfur, etc.) are not released until the coal and oil are burned. Enormous amounts of carbon also are locked up as calcium carbonate ($CaCO_3$), used to build shells and skeletons of marine organisms from tiny protozoans to corals. Most of these deposits are at the bottom of the oceans. The world's extensive surface limestone deposits are biologically formed calcium carbonate from ancient oceans, exposed by geological events. The carbon in limestone has been locked away for millennia, which is probably the fate of carbon currently being deposited in ocean sediments. Eventually, even the deep ocean deposits are recycled as they are drawn into deep molten layers and released via volcanic activity. Geologists estimate that every carbon atom on the earth has made about thirty such round trips over the last 4 billion years.

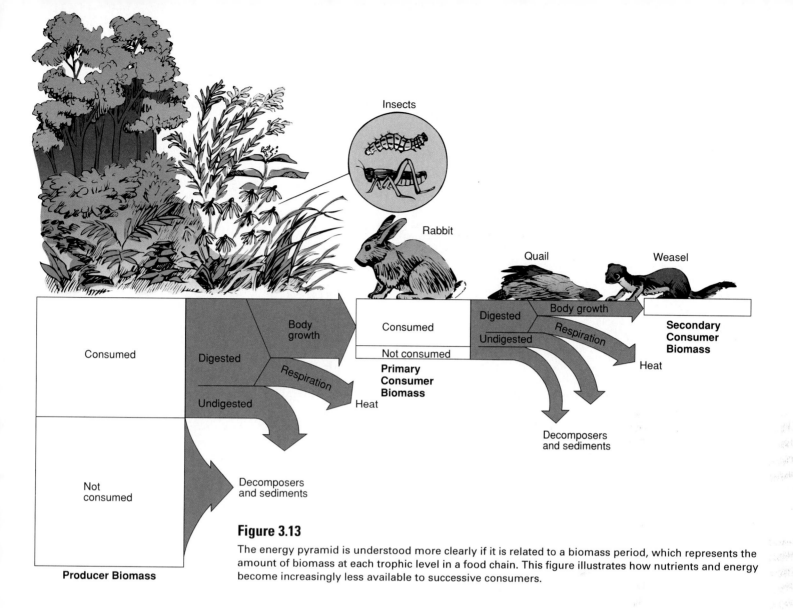

Figure 3.13

The energy pyramid is understood more clearly if it is related to a biomass period, which represents the amount of biomass at each trophic level in a food chain. This figure illustrates how nutrients and energy become increasingly less available to successive consumers.

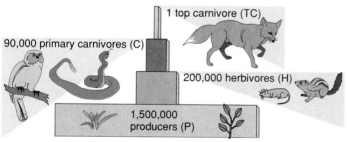

Grassland in summer

Figure 3.14

Usually, smaller organisms are eaten by larger organisms and it takes numerous small organisms to feed one large organism. This statement is demonstrated in studies of food chains in communities and can be represented visually as a numbers pyramid. The classic study represented in this pyramid shows numbers of individuals at each trophic level per 1000 m² of grassland, and reads like this: to support one individual at the top carnivore level, there were 90,000 primary carnivores feeding upon 200,000 herbivores that in turn fed upon 1,500,000 producers.

How does tying up so much carbon in the bodies and byproducts of organisms affect the biosphere? Favorably. It helps balance CO_2 generation and utilization. Carbon dioxide is one of the so-called greenhouse gases because it blocks radiation of heat from the earth's surface, retaining it instead in the atmosphere. This phenomenon is discussed in more detail in chapter 17. Photosynthesis and deposition of $CaCO_3$ remove atmospheric carbon dioxide; therefore, vegetation (especially large forested areas such as the tropical rainforests) and the oceans are very important **carbon sinks** (storage deposits). Cellular respiration and

Matter, Energy, and Life

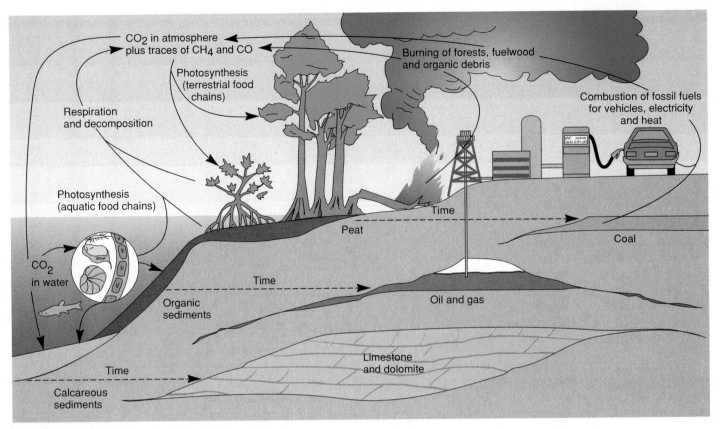

Figure 3.15

Atmospheric carbon dioxide is the "source" of carbon in the carbon cycle. It passes into ecosystems through photosynthesis and is captured in the bodies and products of living organisms. It is released to the atmosphere by weathering, respiration, and combustion. Carbon may be locked up for long periods in both organic (coal, oil, gas) and inorganic (limestone, dolomite) geological formations, which are, therefore, referred to as carbon "sinks."

combustion both release CO_2, so they are referred to as carbon sources of the cycle.

Presently, natural fires and human-created combustion of organic fuels (mainly wood, coal, and petroleum products) release huge quantities of CO_2 at rates that seem to be surpassing the pace of CO_2 removal. Scientific concerns over the linked problems of increased atmospheric CO_2 concentrations, massive deforestation, and reduced productivity of the oceans due to pollution are discussed in later chapters.

The Nitrogen Cycle

Organisms cannot exist without amino acids, peptides, and proteins, all of which are organic molecules containing nitrogen. The nitrogen atoms that form these important molecules are provided by producer organisms. Plants assimilate (take up) inorganic nitrogen from the environment and use it to build their own protein molecules, which are eaten by consumer organisms, digested, and used to build their bodies. This sequence sounds very tidy, but there is one major problem. Even though nitrogen is the most abundant gas (about 78 percent of the atmosphere), plants cannot use N_2, the stable diatomic (2-atom) molecule in the air.

Where and how, then, do green plants get *their* nitrogen? The answer lies in the most complex of the gaseous cycles, the **nitrogen cycle.** Figure 3.16 summarizes the nitrogen cycle, emphasizing its biological aspects, even though some nitrogen is converted to usable form by nonbiological processes. The key step in making nitrogen available is carried out by nitrogen-fixing bacteria (including some blue-green bacteria). These organisms have a highly specialized ability to "fix" nitrogen, meaning they change it to less mobile, more useful forms by combining it with hydrogen to make ammonia (NH_3). It takes a great deal of energy to break the bonds in diatomic nitrogen. Nitrogen-fixing bacteria consume the energy equivalent of 12 grams of glucose to make a single gram of ammonia.

Nitrite-forming bacteria combine the ammonia with oxygen, forming nitrites, which have the ionic form NO_2^-. Another group of bacteria then convert nitrites to nitrates, which have the ionic form NO_3^-, that can be absorbed and used by green plants. After nitrates have been absorbed into plant cells, they are reduced to ammonium (NH_4^+) which is used to build amino acids that become the building blocks for peptides and proteins.

Members of the bean family (legumes) and a few other kinds of plants are especially useful in agriculture because they have nitrogen-fixing bacteria actually living *in* their root tissues (fig. 3.17).

The bacteria are clustered in nodules where they grow in a moist, nutrient-rich environment. This is an example of mutualism, an intimate, mutually beneficial "living-together" relationship between members of different species (see chapter 4). Legumes and their associated bacteria enrich the soil, so interplanting and rotating legumes with crops such as corn that use but cannot replace soil nitrates are beneficial farming practices that take practical advantage of this relationship.

Nitrogen reenters the environment in several ways. The most obvious path is through the death of organisms. Their bodies are decomposed by fungi and bacteria, releasing ammonia and ammonium ions, which then are available for nitrate formation. Organisms don't have to die to donate proteins to the environment, however. Plants shed their leaves, needles, flowers, fruits, and cones; animals shed hair, feathers, skin, exoskeletons, pupal cases, and silk. Animals also produce excrement and urinary wastes that contain nitrogenous compounds. Urinary wastes are especially high in nitrogen because they contain the detoxified wastes of protein metabolism. All of these by-products of living organisms decompose, replenishing soil fertility.

How does nitrogen reenter the atmosphere, completing the cycle? Denitrifying bacteria break down nitrates into N_2 and nitrous oxide (N_2O), gases that return to the atmosphere; thus, it would seem that denitrifying bacteria compete with plant roots for available nitrates. However, denitrification occurs mainly in waterlogged soils that have low oxygen availability and a high amount of decomposable organic matter. These are suitable growing conditions for many wild plant species in swamps and marshes, but not for most cultivated crop species, except for rice, a domesticated wetlands grass.

The Phosphorus Cycle

Minerals become available to organisms after they are released from rocks. Two mineral cycles of particular significance to organisms

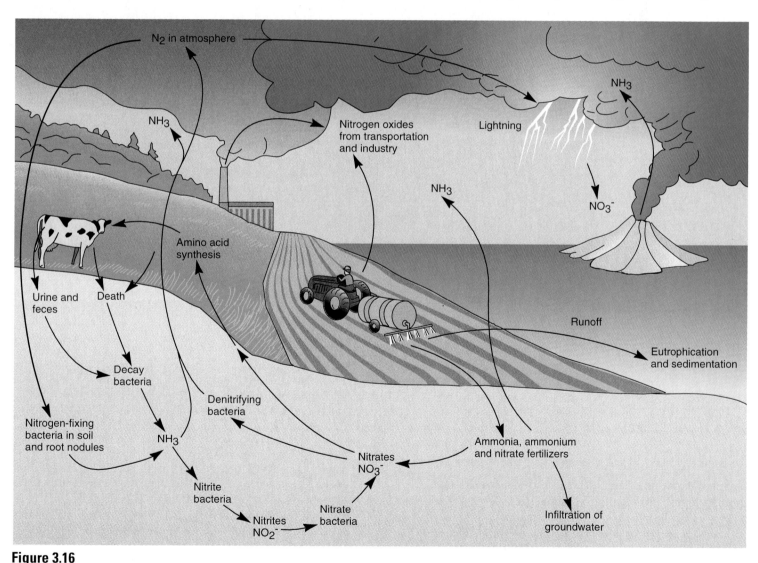

Figure 3.16

Nitrogen is incorporated into ecosystems when plants and bacteria use it to build their own amino acids and is released from ecosystems by bacterial decomposition. Both natural and human interactions with the nitrogen cycle are depicted here.

In Depth: A "Water Planet"

If travelers from other solar systems were to visit our lovely, cool, blue planet, they might call it Aqua rather than Terra because of its outstanding feature: the abundance of streams, rivers, lakes, and oceans of liquid water. The earth is the only place that we know of where water exists in liquid form in any appreciable quantity. Liquid water covers nearly three-fourths of the earth's surface, and during the winter, snow and ice cover a good deal of the rest. Not only is water essential for cell structure and metabolism, but water's unique physical and chemical properties directly impact the earth's surface temperatures, its atmosphere, and the interactions of life-forms with their environments. Water has many unique, almost magical qualities. Without the wonderful properties of water, life on earth would not be possible. Let's look in more detail at some of these properties and how they affect life.

1 Water is the primary component of cells and makes up 60 to 70 percent (on average) of the weight of living organisms. It fills cells, thereby giving form and support to many tissues. More than just being "filler," however, it has vital biological roles. Water is the medium in which all of life's chemical reactions occur, and it is an active participant in many of these reactions.

2 Water is the only inorganic liquid that exists in nature, and it is the solvent in which most substances must be dissolved before cells can absorb, use, or eliminate them. These substances include food molecules, mineral nutrients, gases, hormones and other chemical communicators, and waste by-products of metabolism. When molecules dissolve in water, they often have a tendency to break into positively and negatively charged ions. The resulting electrical attractions and repulsions between ions are an important part of cellular chemistry.

3 Water molecules themselves can ionize, breaking into H^+ (hydrogen ions) and OH^- (hydroxyl ions). These ions help maintain the balance between acids and bases in cells, helping to offset (buffer) fluctuations caused by the release of other ions during metabolism. The term **pH** refers to the relative abundance of H^+ ions in a solution. On a pH scale from 0 to 14, 7 is neutral; values lower than 7 are acidic, and those higher than 7 are basic (fig. 1). A proper pH balance is critical to healthy cellular functioning.

4 Water molecules are cohesive, tending to stick together tenaciously. You have experienced this property if you have ever done a belly flop off a diving board. Because of its cohesiveness, water has the highest surface tension of any common, natural liquid. These forces hold the water molecules at the surface together so strongly that they form a layer strong enough to support small insects (fig. 2). The same cohesion causes capillary action, which is the tendency of water to be drawn into small channels. Without capillary action, movement of water and nutrients into groundwater reservoirs and through living organisms might not be possible.

5 Because water has a high heat capacity, it helps protect us from temperature fluctuations. It can absorb or release large amounts of heat energy before its own temperature changes. For this reason, large bodies of water such as the oceans or the Great Lakes have a moderating effect on their local climates. Without the presence of liquid oceans, the surface temperature of the earth would undergo wide fluctuations between day and night as do the moon and Mars.

6 Water exists as a liquid over a wide temperature range that, for most of the world (at least during summer months), corresponds to the ambient temperature range. For most substances, the freezing point is only a few degrees lower than the boiling point. This means that they exist as either a solid or a gas, but only briefly as a liquid. Organisms synthesize organic compounds such as oils and alcohols that remain liquid at ambient temperatures because they are so valuable to life, but the original and predominant liquid in nature is water.

7 Water is unique in that it expands when it crystallizes. Most substances shrink as they change from liquid to solid. Ice floats because it is less dense than liquid water. When temperatures fall below freezing, the surface layers of lakes, rivers, and

are phosphorus and sulfur. Why do you suppose phosphorus is a primary ingredient in fertilizers? At the cellular level, ATP and other energy-rich, phosphorus-containing compounds are primary participants in energy-transfer reactions, as we have discussed. The amount of available phosphorus in an environment can, therefore, have a dramatic effect on productivity. Abundant phosphorus stimulates lush plant and algal growth, making it a major contributor to water pollution.

The **phosphorus cycle** (fig. 3.18) begins when phosphorus compounds are leached from rocks and minerals over long periods of time. Inorganic phosphorus is taken in by producer organisms, incorporated into organic molecules, and then passed on to con-

Figure 1

The pH scale. The numbers represent the negative logarithm of the hydrogen ion concentration in water.

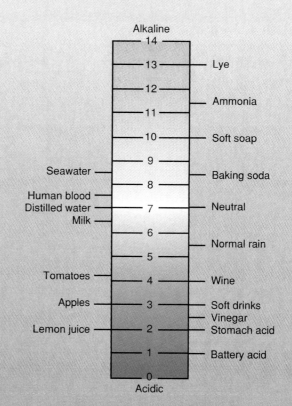

Alkaline

- 14
- 13 — Lye
- 12
- 11 — Ammonia
- 10 — Soft soap
- 9
- Seawater — 8 — Baking soda
- Human blood —
- Distilled water — 7 — Neutral
- Milk —
- 6
- Normal rain
- 5
- Tomatoes — 4 — Wine
- Apples — 3 — Soft drinks
- Vinegar
- Lemon juice — 2 — Stomach acid
- 1 — Battery acid
- 0

Acidic

oceans cool faster and freeze before deeper water. Floating ice then insulates underlying layers, keeping most water bodies liquid throughout the winter in most places. Without this feature, lakes, rivers, and even oceans in high latitudes would freeze solid and never melt.

8 *Water has a high heat of vaporization. Because of the amount of heat it absorbs in changing from a liquid to a vapor state, evaporating water is an effective way for organisms to shed excess heat. Many animals pant or sweat to moisten evaporative cooling surfaces. Why do you feel less comfortable on a hot, humid day than on a hot, dry day? Because the water vapor-laden air inhibits the rate of evaporation from your skin, thereby impairing your ability to shed heat. The heat absorbed when water vaporizes is released when condensation occurs. This accounts for a large transfer of heat from the oceans over the continents as water vapor turns into rain.*

Altogether, these unique properties of water not only shape life at the molecular and cellular level, they also determine many of the features of both the biotic and abiotic components of our world.

Figure 2

Surface tension is demonstrated by the resistance of a water surface to penetration, as when it is walked upon by a water strider.

sumers. It is returned to the environment by decomposition. An important aspect of the phosphorus cycle is the very long time it takes for phosphorus atoms to pass through it. Deep sediments of the oceans are significant phosphorus sinks of extreme longevity. Phosphate ores that now are mined to make detergents and inorganic fertilizers represent exposed ocean sediments that are mil-

lennia old. You could think of our present use of phosphates, which are washed out into the river systems and eventually the oceans, as an accelerated mobilization of phosphorus from source to sink. Aquatic ecosystems often are dramatically affected in the process because excess phosphates can stimulate explosive growth of algae and photosynthetic bacteria populations, upsetting ecosystem

Figure 3.17
The roots of this adzuki bean plant are covered with bumps called nodules. Each nodule is a mass of root tissue containing many bacteria that help to convert nitrogen in the soil to a form the bean plants can assimilate and use to manufacture amino acids.

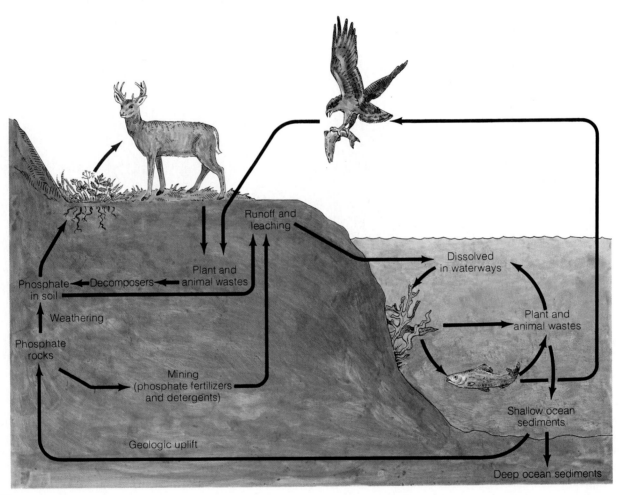

Figure 3.18
The phosphorous cycle carries this essential mineral from its source in rocks and soil, through plants and animals, and back to sediments as wastes and carrion.

stability. Notice also that in this cycle, as in the others, the role of organisms is only one part of a larger picture.

The Sulfur Cycle

Sulfur plays a vital role in organisms, especially as a minor but essential component of proteins. Sulfur compounds are important determinants of the acidity of rainfall, surface water, and soil. In addition, sulfur in particles and tiny air-borne droplets may act as critical regulators of global climate. Most of the earth's sulfur is tied up underground in rocks and minerals such as iron disulfide (pyrite) or calcium sulfate (gypsum). This inorganic sulfur is released into air and water by weathering, emissions from deep seafloor vents, and by volcanic eruptions (fig. 3.19).

The **sulfur cycle** is complicated by the large number of oxidation states the element can assume, including hydrogen sulfide (H_2S), sulfur dioxide (SO_2), sulfate ion (SO_4^{-2}), and sulfuric acid (H_2SO_4), among others. Inorganic processes are responsible for many of these transformations, but living organisms, especially bacteria, also sequester sulfur in biogenic deposits or release it into the environment. Which of the several kinds of sulfur bacteria prevail in any given situation depends on oxygen concentrations, pH, and light levels.

Human activities also release large quantities of sulfur, primarily through burning fossil fuels. Total yearly anthropogenic sulfur emissions rival those of natural processes, and acid rain caused by sulfuric acid produced as a result of fossil fuel use is a serious problem in many areas (see chapter 18). Sulfur dioxide and sulfate

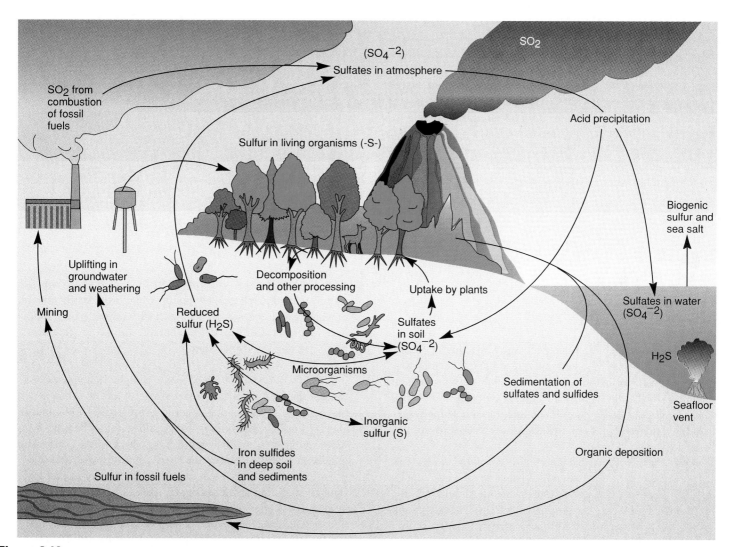

Figure 3.19

Sulfur is present mainly in rocks, soil, and water. It cycles through ecosystems when it is taken in by organisms. Combustion of fossil fuels causes increased levels of atmospheric sulfur compounds, which create problems related to acid precipitation.

Figure 3.20

Dimethylsulfide (DMS), produced by oceanic phytoplankton (microscopic single-celled plants) may act as a governor, or feed-back control, for global climate. As oceans warm, more DMS is produced, causing clouds to block sunlight. Less solar energy means a cooler earth, less active phytoplankton, and, thus, less DMS.

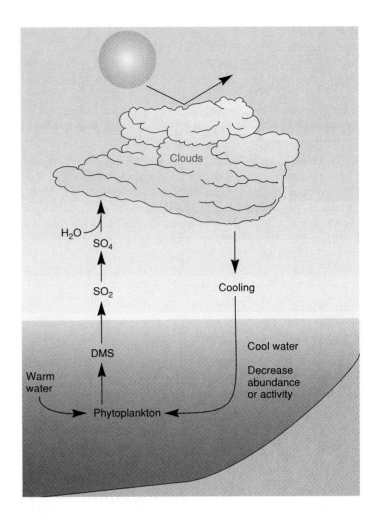

aerosols cause human health problems, damage buildings and vegetation, and reduce visibility. They also absorb UV radiation and create cloud cover that cools cities and may be offsetting greenhouse effects of rising CO_2 concentrations.

Interestingly, biogenic sulfur emissions by oceanic phytoplankton may play a role in global climate regulation. When ocean water is warm, tiny, single-celled organisms release dimethylsulfide (DMS) that is oxidized to SO_2 and then to SO_4 in the atmosphere. Acting as cloud droplet condensation nuclei, these sulfate aerosols increase the earth's albedo (reflectivity) and cool the earth (fig. 3.20). As ocean temperatures drop because less sunlight gets through, phytoplankton activity decreases, DMS production falls, and clouds disappear. Thus DMS, which may account for half of all biogenic sulfur emissions, could be a feedback mechanism that keeps temperature within a suitable range for all life.

Summary

Certain conditions, including the availability of required chemical elements, a steady influx of solar energy, mild surface temperatures, the presence of liquid water, and a suitable atmosphere, are essential for life on earth. Only the earth in our solar system has all these conditions. Life requires and makes possible highly organized exchanges of matter and energy between organisms and their environments.

Matter is the observable material of which the universe is composed. It exists in three interchangeable phases: gas, liquid, and solid. Matter is made up of atoms, which are composed of particles called protons, neutrons, and electrons. The energy that holds atoms together forms the basis for energy transfers in the bodies of living organisms and, therefore, in the biosphere.

A steady influx of solar radiation provides the heat and light energy needed to support life in the biosphere. Water, which covers approximately three-fourths of the earth's surface, is a remarkable substance. Because of its unique characteristics, it stabilizes the biosphere's temperature and provides the medium in which life processes occur. The earth's atmosphere provides gases necessary for life, helps maintain surface temperatures, and filters out dangerous radiation.

Ecosystem dynamics are governed by physical laws, including the law of conservation of matter and the first and second laws of thermodynamics. The recycling of matter is the basis of the cycles of elements that occur in ecosystems. Unlike matter, energy is not cycled. Energy always flows through systems in a one-way process in which some energy is converted from a high-quality, concentrated form to a lower-quality, less useful, dispersed form. We describe this increase in disorder as entropy, a fundamental limit to life. In ecosystems, solar energy enters the system and is converted to chemical energy by the process of photosynthesis. The chemical energy stored in the bonds that hold food molecules together is available for metabolism of organisms.

A species is all the organisms of the same kind that are genetically similar enough to breed in nature and produce live, fertile offspring. The populations of different species that live and interact within a particular area at a given time make up a biological community. An ecosystem is composed of a biological community together with all the biotic and abiotic factors that make up the environment in a defined area. Although ecosystem boundaries may be rather arbitrary, the holistic, systems approach to biology has provided rich insights into who lives where, when, how, and why.

Matter and energy are processed through the trophic levels of an ecosystem via food chains and food webs. At each energy transfer point less energy is available to work because of the laws of thermodynamics, so energy must be supplied to an ecosystem continuously. The relationships between producers and consumers in an ecosystem, often depicted as pyramids, demonstrate this principle. Most of the energy that enters an ecosystem comes, ultimately, from the sun.

The biosphere is a source of large quantities of essential elements. In a given ecosystem, these elements are constantly used and reused by living organisms. Water, carbon, nitrogen, sulfur, and phosphorus, for instance, are recycled in ecosystems through complex biogeochemical cycles.

▼ Questions for Review

1. Define *atom* and *element*. Are these terms interchangeable?

2. Your body contains vast numbers of carbon atoms. How is it possible that some of these carbon atoms may have been part of the body of a prehistoric creature?

3. In the biosphere, matter follows a circular pathway while energy follows a linear pathway. Explain.

4. The oceans store a vast amount of heat, but (except for climate moderation) this huge reservoir of energy is of little use to humans. Explain the difference between high-quality and low-quality energy.

5. Ecosystems require energy to function. Where does this energy come from? Where does it go? How does the flow of energy conform to the laws of thermodynamics?

6. Heat is released during metabolism. How is this heat useful to a cell and to a multicellular organism? How might it be detrimental, especially in a large, complex organism?

7. Photosynthesis and cellular respiration are complementary processes. Explain how they exemplify the laws of conservation of matter and thermodynamics.

8. What do we mean by carbon-fixation or nitrogen-fixation? Why is it important to humans that carbon and nitrogen be "fixed"?

9. The population density of large carnivores is always very small compared to the population density of herbivores occupying the same ecosystem. Explain this in relation to the concept of an ecological pyramid.

10. A species is a specific kind of organism. What general characteristics do individuals of a particular species share? Why is it important for ecologists to differentiate among the various species in a biological community?

▼ Questions for Critical Thinking

1. When we say that there is no "away" where we can throw things we don't want anymore, are we stating a premise or a conclusion? If you believe this is a premise, supply the appropriate conclusion. If you believe it is a conclusion, supply the appropriate premises. Does the argument change if this statement is a premise or a conclusion?

2. Suppose one of your classmates disagrees with the statement above, saying, "Of course there is an 'away.' It's anywhere out of *my* ecosystem." How would you answer?

3. A few years ago laundry detergents commonly contained phosphates for added cleaning power. Can you imagine any disadvantages to adding soluble phosphate to household products?

4. The first law of thermodynamics is sometimes summarized as "you can't get something for nothing." The second law is summarized as "you can't even break even." Explain what these phrases mean. Is it dangerous to oversimplify these important concepts?

5. The ecosystem concept revolutionized ecology by introducing holistic systems thinking as opposed to individualistic life history studies. Why was this a conceptual breakthrough?

6. Why is it important to recognize that ecosystems often are open and that boundaries may be fuzzy? Do these qualifications diminish the importance of the ecosystem study?

7. The holistic or systems approach to biology has sometimes been criticized as "black box" engineering. It allows us to make broad generalizations about what goes into or comes out of a system without knowing the precise details of how the system works. What do you think are the benefits and limitations of this approach?

8. Compare and contrast the views of F. E. Clements and H. A. Gleason concerning the concept of biological communities as superorganisms. How could these eminent biogeographers study the same communities and reach opposite interpretations? What evidence would be necessary to settle this question? Is lack of evidence the problem?

9. The properties of water are so unique and so essential for life as we know it that some people believe it proves that our planet was intentionally designed for our existence. What would an environmental scientist say about this belief?

10. The DMS feedback control of global climate is offered by some people as evidence for the Gaia hypothesis. Why might they take this position?

Key Terms

atoms 47	consumers 54
biological community 51	decomposers 55
biomass 52	detritivores 55
carbon cycle 57	ecology 46
carbon sinks 57	ecosystem 52
carnivores 55	energy 48
cells 48	enzymes 48
cellular respiration 51	first law of thermodynamics 49
compound 47	food chain 54
conservation of matter 49	food web 54

Additional Information on the Internet

ASU Photosynthesis Center
http://www.asu.edu/clas/photosyn/index.html/

The Carbon Cycle
http://seawifs.gsfc.nasa.gov/SEAWIFS/LIVING_OCEAN/carbon_cycle.jpeg/

Cell Biology Chapter Directory
http://esg-www.mit.edu:8001/esgbio/cb/cbdir.html/

Characteristics of Prokaryotes and Eukaryotes
http://esg-www.mit.edu:8001/esgbio/cb/prok_euk.html/

The Chemistry of Carbon
http://www.nyu.edu/pages/mathmol/modules/carbon/carbon1.html

The Hydrological Cycle
http://publish.uwrl.usu.edu/h20cycle1.html#topofh20cycle1page/

7.01 Hypertextbook Chemistry Review
http://esg-www.mit.edu:8001/esgbio/chem/review.html/

Structure and Function of Organelles
http://esg-www.mit.edu:8001/esgbio/cb/org/organelles.html/

UNA's FAQ File
http://www.csc.fi/molbio/una/unapost.index.html/

Water and Ice
http://cwis.nyu.edu/pages/mathmol/modules/water/info_water.html/

What Is a Cell?
http://lenti.med.umn.edu/~mwd/cell_www/cell_intro.html/

Note: Further readings appropriate to this chapter are listed on p. 594.

Environmental Activist

Micaela Martinez

*M*icaela Martinez

Micaela Martinez' concerns about the environment began when she was in high school. Back then, she found herself joining citizen campaigns on a variety of local issues, even helping stage a bike-a-thon to protest a developer's plans to build in habitat crucial to the survival of an endangered butterfly. She went off to college committed to pursuing a career that would enable her to make a contribution to environmental quality. She came to the conclusion that since habitat was so important to organisms' well-being, a major in wildlife biology was the way to make the biggest impact.

She began her college education at Skyline Junior College in a San Francisco suburb. She planned to complete two years there and then transfer to Humboldt State University to complete a major in wildlife management. Her ultimate goal was to work for the California Fish and Game Department. While in college, she continued to volunteer on behalf of assorted environmental causes.

It was during those years that her career goals underwent a sudden and fundamental change. As the result of a national election, a number of pro-environment incumbents lost their offices to candidates with decidedly less environmentally friendly viewpoints. As she watched policy proposals come forth from the new officeholders, she began to feel that effecting environmental change had less to do with biology and habitat and more to do with politics.

Since Micaela was already well on her way to completion of the requirements, she finished her wildlife degree despite the change of heart on her career choice. Soon after graduation she took a job as a canvasser with the California League of Conservation Voters. She participated in door-to-door canvassing, which is used by a number of environmental organizations to inform people about local or regional environmental issues and to identify voters who are concerned about environmental issues, as well as to seek donations.

After gaining canvassing experience with the League of Conservation Voters, Micaela joined the staff of Clean Water Action, a national organization that hires many college students for door-to-door canvassing. She currently serves as supervisor of eight canvass programs around the country, including efforts in California, Pennsylvania, Maryland, and Washington, DC.

Part of Micaela's work is directed at helping the organization set goals and priorities, and the strategies to achieve them. She handles job performance reviews, develops recruiting plans, and makes career day visits to college campuses. She also oversees her organization's efforts in voter identification, phone banks, and get-out-the-vote efforts on election day. Finally, she writes press releases, organizes press conferences, and helps build coalitions with allied organizations and community leaders. Her work generally intensifies during the summer months, and particularly in election years.

Despite not working directly in her area of technical training, Micaela finds her knowledge of the natural sciences helpful in her job. She feels that her scientific training gives her a good background understanding of the issues under debate. She also feels that her training in wildlife biology has helped her understand the mindset of people in both natural resource agencies and industry. She has found this understanding of diverse perspectives very helpful in building coalitions.

Micaela has found her technical writing course of particular value in helping her write succinctly and to present technical information in an understandable form. She wishes she had left college with a better knowledge of human social dynamics, however. Much of her time is spent working with people and she feels that courses in business management or the psychology of team building would have helped.

Her advice to students interested in a career in environmental activism is to learn as much as possible about environmental matters from a scientific or technological perspective. She also feels it is important to volunteer for environmental causes in their college community. Not only does it provide valuable experience, but is also an excellent way to meet lots of people, which may lead to job opportunities as well. She suggests that during summer breaks students gain activism experience by either voluntary or paid positions as canvassers, campaign workers, or legislative aids.

Micaela enjoys working with people drawn to environmental activism. She also gets great satisfaction seeing the results of her work, knowing that she has an impact. She says it's a great feeling going home everyday realizing that one person really can make a difference.

A field of shooting stars and johnny jump ups add spring color to this open woodland community.

CHAPTER 4: Biological Communities and Species Interaction

Objectives

After studying this chapter, you should be able to:

- describe how environmental factors determine which species live in a given ecosystem and where or how they live.

- understand how random genetic variation and natural selection lead to evolution, adaptation, niche specialization, and partitioning of resources in biological communities.

- compare and contrast interspecific predation, competition, symbiosis, commensalism, and mutualism.

- discuss productivity, diversity, complexity, and structure of biological communities and how these characteristics might be connected to resilience and stability.

- explain how ecological succession results in ecosystem development and allows one species to replace another. You should also understand the difference between primary and secondary succession.

- give some examples of exotic species introduced into biological communities and describe the effects such introductions can have on indigenous species.

Sea Otters, Kelp, and Sea Urchins: A Triangular Affair

Few animals look as cute and cuddly as sea otters. Their round, fuzzy faces with dark button eyes and a huge walrus mustache seem to express perpetual curiosity and good nature (fig. 4.1). They are beautiful swimmers, graceful, swift, and supple in the water. Once ranging from Japan across Siberia to Alaska and down the coast of California to the Mexican border, sea otters (Enhydra lutris) *were superbly adapted to their environment. Their thick waterproof fur coat insulates them from cold Arctic waters. Highly skilled hunters, they live mainly on shellfish and sea urchins from the ocean floor. With few natural enemies, otters were one of the top predators in their ecosystem. In Alaska alone, there once may have been two hundred thousand of these beautiful animals.*

The otter's wonderful fur coats almost were their downfall. When European hunters discovered the huge Pacific populations in the eighteenth century, the otters were quickly hunted nearly to extinction. Only a few scattered animals remained in 1911 when Russia, Japan, Canada, and the United States finally agreed to

Figure 4.1

The thick fur coat and marvelous swimming ability of this sea otter make it superbly adapted to the frigid waters of the northern Pacific Ocean where it plays a key role in maintaining giant kelp forests. Almost driven into extinction a century ago by fur hunters, otters are now protected and populations recovering.

stop hunting them. Now protected, otter populations have re-bounded throughout much of their original range. At least 100,000 live in the Aleutian Islands and southeast Alaska. Another 2000 animals can be seen along the California coast between Santa Cruz and San Luis Obispo.

Recently, a transplantation project has been undertaken to reintroduce otters to the Channel Islands west of Los Angeles. Having a second population along the California coast would offer some insurance that the otters won't be wiped out again by a single disaster such as an oil spill. Transplanting otters has not been simple, however. Most of the animals have either returned to their original homes or simply disappeared. Furthermore, commercial shellfishers complain that otters steal the abalone, lobsters, oysters, and crabs on which their business depends.

Others, however, welcome the otters because of the key role they play in the giant kelp forests in which they live (see fig. 4.11).

When otters were removed, sea urchin populations exploded. Urchins eat kelp: too many of them can completely destroy the kelp and the whole biological community to which it offers shelter, food, and structure. Without otters to control the urchins, much of the California coast has become denuded "urchin barrens" with no kelp and few fish or other species. By reintroducing otters, the kelp is recovering and the ecosystem is becoming revitalized.

How do we account for the amazing and complex interrelationships in ecosystems such as this? How do organisms such as sea otters, kelp, and sea urchins become adapted to their particular environment and to each other? How can we sort out the intricate interactions between the various members of biological communities like the oceanic kelp forests? In this chapter, we will look at competition, predation, symbiosis, tolerance limits, and other features of species and communities that affect both the species themselves and the ecosystems of which they are a part.▶

Who Lives Where, and Why?

"Why" questions often are the stimulus for scientific research, but the research itself centers on "how" questions. Why, we wonder, does a particular species live where it does? More to the point, how is it *able* to live there? How does it deal with the physical resources of its environment and are some of its techniques unique? How does it interact with the other species present? And what gives one species an edge over another species in a particular habitat?

In this section we will examine some specific ways organisms are limited by the physical aspects of their environment. We then will discuss how members of a biological community interact, pointing out a few of the difficulties ecologists encounter when they attempt to discern patterns and make generalizations about community interactions and organization.

Critical Factors and Tolerance Limits

Every living organism has limits to the environmental conditions it can endure. Temperatures, moisture levels, nutrient supply, soil and water chemistry, living space, and other environmental factors must be within appropriate levels for life to persist. In 1840, Justus von Liebig proposed that the single factor in shortest supply relative to demand is the critical determinant in the distribution of that species. Ecologist Victor Shelford later expanded this principle of limiting factors by stating that each environmental factor has both minimum and maximum levels, called **tolerance limits,** beyond which a particular species cannot survive (fig. 4.2) or is unable to reproduce. The single factor closest to these survival limits, he postulated, is the critical limiting factor that determines where a particular organism can live.

At one time, ecologists accepted this concept so completely that they called it Liebig's or Shelford's Law and tried to identify unique factors limiting the growth of every population of plants and animals. For many species, however, we find that the interaction of

several factors working together, rather than a single limiting factor, determines biogeographical distribution. If you have ever explored the rocky coasts of New England or the Pacific Northwest, for instance, you probably have noticed that mussels and barnacles endure extremely harsh conditions but generally are sharply limited to an intertidal zone where they grow so thickly that they often completely cover the substrate. No single factor determines this distribution. Instead, a combination of temperature extremes, drying time between tides, salt concentrations, competitors, and food availability limits the number and location of these animals.

For other organisms, there may be a specific *critical factor* that, more than any other, determines the abundance and distribution of that species in a given area. A striking example of cold intolerance as a critical factor is found in the giant saguaro cactus (*Carnegiea gigantea*), which grows in the dry, hot Sonoran desert of southern Arizona and northern Mexico (fig. 4.3). Saguaros are extremely sensitive to low temperatures. A single exceptionally cold winter night with temperatures below freezing for 12 hours or more will kill growing tips on the branches. Young saguaros are more susceptible to frost damage than adults, but seedlings typically become established under the canopy of small desert trees such as mesquite that shield the young cacti from the cold night sky. Unfortunately, the popularity of grilling with mesquite wood has caused extensive harvesting of the nurse trees that once sheltered small saguaros, adversely affecting reproduction of this charismatic species.

Animal species, too, exhibit tolerance limits that often are more critical for the young than for the adults. The desert pupfish (*Cyprinodon*), for instance, occurs in small isolated populations in warm springs in the northern Sonoran desert. Adult pupfish can survive temperatures between 0° C and 42° C (a remarkably high temperature for a fish) and are tolerant to an equally wide range of salt concentrations. Eggs and juvenile fish, however, can only live between 20° C and 36° C and are killed by high salt levels. Reproduction, therefore, is limited to a small part of the range of

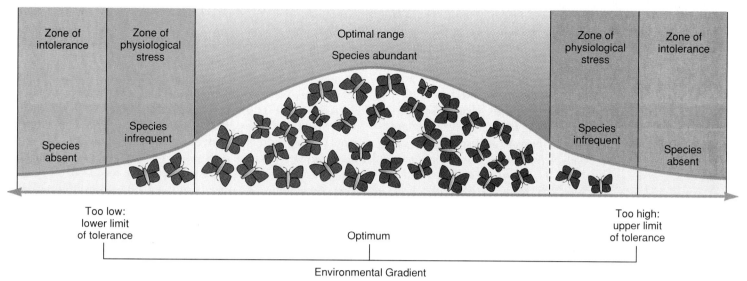

Figure 4.2

The principle of tolerance limits states that for every environmental factor, an organism has both maximum and minimum levels beyond which it cannot survive. The greatest abundance of any species along an environmental gradient is around the optimum level of the critical factor most important for that species. Near the tolerance limits abundance decreases because fewer individuals are able to survive the stresses imposed by limiting factors.

Figure 4.3

Saguaro cacti, symbolic of the Sonoran desert, are an excellent example of distribution controlled by a critical environmental factor. Extremely sensitive to low temperatures, saguaros are found only where minimum temperatures never dip below freezing for more than a few hours at a time.

adult fish, which is often restricted anyway by the size of the small springs and desert seeps in which the species lives.

Sometimes the requirements and tolerances of species are useful indicators of specific environmental characteristics. The presence or absence of such species can tell us something about the community and the ecosystem as a whole. Locoweeds, for example, are small legumes that grow where soil concentrations of selenium are high. Because selenium often is found with uranium deposits, locoweeds have an applied economic value as environmental indicators. Such indicator species also may demonstrate the effects of human activities. Lichens and eastern white pine are less restricted in habitat than locoweeds, but are indicators of air pollution because they are extremely sensitive to sulfur dioxide and acid precipitation. Bull thistle is a weed that grows on disturbed soil but is not eaten by cattle; therefore, an abundant population of bull thistle in a pasture is a good indicator of overgrazing. Similarly, anglers know that trout species require clean, well-oxygenated water, so the presence or absence of trout is an indicator of water quality.

Natural Selection, Adaptation, and Evolution

How is it that mussels have developed the ability to endure pounding waves, daily exposure to drying sun and wind, and seasonal threats of freezing cold or broiling hot temperatures? What enables desert pupfish to tolerate hot, mineral-laden springs? How does the saguaro survive in the harsh temperatures and extreme dryness of the desert? We commonly say that each of these species is "adapted" to its special set of conditions, but what does that mean? In this section, we will examine one of the most important concepts in biology: how species acquire traits that allow them to live in unique ways in particular environments.

We use the term *adapt* in two ways. One is a limited range of *physiological modifications* available to individual organisms. If you keep house plants inside all winter, for example, and then put them out in full sunlight in the spring, they get sunburned. If the damage isn't too severe, your plants will probably grow new leaves with a thicker cuticle and denser pigments that protect them from the sun. They can adapt to some degree, but the change isn't permanent. Another winter inside will make them just as sensitive to the sun as before. Furthermore, the changes they acquire are not passed on to their offspring. Although the potential to adapt is inherited, each generation must develop its own protective epidermis.

The other form of adaptation operates at the population level and is brought about by inheritance of specific genetic traits that allow a species to live in a particular environment. This process is explained by the theory of **evolution,** developed by Charles Darwin and Alfred Wallace. According to this theory, species change gradually through competition for scarce resources and **natural selection,** a process in which those members of a population that are best suited for a particular set of environmental conditions will survive and produce offspring more successfully than their ill-suited competitors.

Natural selection acts on preexisting genetic diversity created by a series of small, random mutations (changes in genetic material) that occur spontaneously in every population. These mutations produce a variety of traits, some of which are more advantageous than others in a given situation. Where resources are limited or environmental conditions place some selective pressure on a population, individuals with those advantageous genes become more abundant in the population, and the species gradually evolves or becomes better suited to that particular environment. Although each change may be very slight, many mutations over a very long time have produced the incredible variety of different life-forms that we observe in nature.

A classic example of natural selection is the changing coloration of European peppered moths, *Biston betularia* (fig. 4.4). Until about 1850, nearly all of these moths had light-colored wings that closely matched the light bark of the trees on which they rest during the day. As industrial pollution darkened the trunks of trees around major urban centers, biologists noticed a dramatic increase in moths with dark coloration. Probably the dark-colored individuals had always been present in the population but were spotted and eaten more frequently by birds when the trees were white. As tree bark darkened, however, the light-colored moths were at a disadvantage. Recently, as factory smoke has been reduced, tree bark is whitening and white moths are becoming more abundant.

The variety of finches observed by Charles Darwin on the Galápagos Islands is another classic example of speciation driven by availability of different environmental opportunities (fig. 4.5). Originally derived from a single seed-eating species that somehow crossed the thousands of kilometers from the mainland, the finches have evolved into a dozen or more distinct species that differ markedly in appearance, food preferences, and habitats they occupy. Fruiteaters have thick parrotlike bills; seedeaters have heavy, crushing bills; insecteaters have thin probing beaks to catch their prey (fig. 4.5). One of the most unusual species is the woodpecker finch, which pecks at tree bark for hidden insects. Lacking the woodpecker's long tongue, however, the finch uses a cactus spine as a tool to extract bugs.

The amazing variety of colors, shapes, and sizes of dogs, cats, rabbits, fish, flowers, vegetables, and other domestic species are evidence of deliberate selective breeding. The various characteristics of these organisms arose through random mutations. We simply kept the ones we liked.

What environmental factors cause selective pressure and influence fertility or survivorship in nature? They include (1) physiological stress due to inappropriate levels of some critical environmental factor, such as moisture, light, temperature, pH, or specific nutrients; (2) predation, including parasitism and disease; (3) competition; and (4) luck. In some cases the organisms that survive environmental catastrophes or find their way to a new habitat where they start a new population may simply be lucky rather than more fit or better suited to subsequent environmental conditions than their less fortunate contemporaries.

Be sure you understand that while selection affects individuals, evolution and adaptation work at the population level. Individuals don't evolve; species do. Each individual is locked in by genetics to a particular way of life. Most plants, animals, or mi-

Figure 4.4

European peppered moths can have either light- or dark-colored wings. Under normal conditions, the lighter moths closely matched the coloration of the birch bark on which they rest during the day. When forests were blackened by industrial pollution in the nineteenth century, the light-colored moths were more visible to predators and the darker moths dominated the population. Since pollution has been reduced, however, and birch bark is once again white, the lighter moths have now become more plentiful.

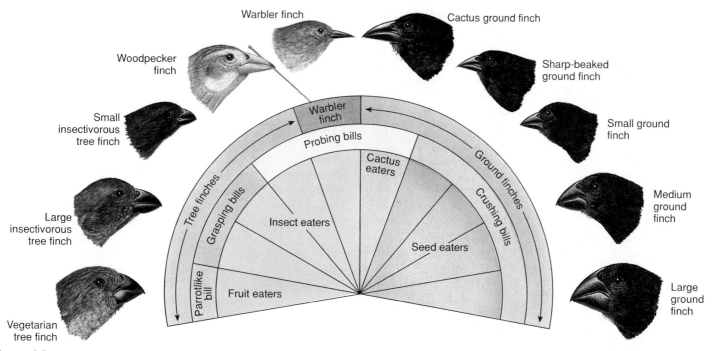

Figure 4.5

Some species of Galápagos Island finches. Although all are descendents of a common ancestor, they now differ markedly in appearance, habitat, and feeding behavior. Ground finches (*lower right*) eat cactus leaves; warbler finches (*upper left*) eat insects; others eat seeds or have mixed diets. The woodpecker finch (*lower left*) pecks tree bark as do woodpeckers, but lacks a long tongue. Instead, it uses cactus spines as tools to extract insects.

crobes have relatively limited ability to modify their physical makeup or behavior to better suit a particular environment. Over time, however, random genetic changes and natural selection can gradually change the whole species.

Given enough time and selective pressure, the members of a population become so different from their ancestors that they may be considered an entirely new species that has replaced the original one. Alternatively, isolation of population subsets by geographical or behavioral factors that prevent exchange of genetic material can result in branching off of new species that coexist with their parental line. Suppose that two populations of the same species become separated by a body of water, a desert, or a mountain range that they cannot cross. Over a very long time—often millions of years—random mutations and different environmental pressures may cause the populations to evolve along such dissimilar paths that they can no longer interbreed successfully even if the opportunity to do so arises. They have now become separate species as in the case of the Galápagos finches. The barriers that divide subpopulations are not always physical. In some cases, behaviors such as when and where members of a population feed, sleep, or mate—or how they communicate—may separate them sufficiently for divergent evolution and speciation to occur even though they occupy the same territory.

Natural selection and adaptation can cause organisms with a similar origin to become very different in appearance and habits over time, but they can also result in unrelated organisms coming to look and act very much alike. We call this latter process *convergent* evolution. The cactus-eating Galápagos finches (fig. 4.5), for example, look and act very much like parrots even though they are genetically very dissimilar. The features that enable parrots to eat fruit successfully work well for these finches also.

A common mistake is to believe that organisms develop certain characteristics because they want or need them. This is incorrect. A duck doesn't have webbed feet because it wants to swim or needs to swim in order to eat; it has webbed feet because some ancestor happened to have a gene for webbed feet that gave it some advantage over other ducks in its particular pond and because those genes were passed on successfully to its offspring. A variety of different genetic types are always present in any population, and natural selection simply favors those best suited for particular conditions. Whether there is a purpose or direction to this process is a theological question rather than a scientific one and is beyond the scope of this book.

The Ecological Niche

Habitat describes the place or set of environmental conditions in which a particular organism lives. An **ecological niche** is a functional description of the role a species plays in a community—how it obtains food, what relationships it has with other species, and the services it provides in its biological community. Some species, such as the panda (fig. 4.6), are specialists and occupy a very narrow

Figure 4.6

The giant panda feeds exclusively on bamboo. Although its teeth and digestive system are those of a carnivore, it is not a good hunter, and has adapted to a vegetarian diet. To support its size, it must eat up to 40 lbs of bamboo leaves and stalks daily. This narrow specialization makes it extremely vulnerable to fluctuations in the bamboo population. In the 1970s, huge acreages of bamboo flowered and died, and many pandas starved.

niche. Others, like raccoons (to which pandas are related), are generalists that eat a wide variety of food and live in a broad range of habitats. Specialists tend to be rarer than generalists and less resilient to disturbance or change.

A few species such as elephants, chimpanzees, and baboons learn how to behave from their social group and can invent new ways of doing things when presented with new opportunities or challenges. Most organisms, however, are limited by genetically determined physical structure and instinctive behavior to follow established niche roles.

Over time, though, niche roles can evolve, just as physical characteristics do. The law of competitive exclusion states that no two populations will occupy the same niche and compete for exactly the same resources in the same habitat for very long. Eventually, one group will gain a larger share of resources while the other will ei-

ther migrate to a new area, become extinct, or change its behavior or physiology in ways that minimize competition. We call this latter process of niche evolution **resource partitioning** (fig. 4.7). It can produce high levels of specialization that allow several species to utilize different parts of the same resource and coexist within a single habitat (fig. 4.8).

Niche specialization also can create behavioral separation that allows subpopulations of a single species to diverge into separate species. Why doesn't this process continue until there is an infinite number of species? The answer is that a given resource can be partitioned only so far. Populations must be maintained at a minimum size to avoid genetic problems and to survive bad times. This puts an upper limit on the number of different niches—and therefore the number of species—that a given community can support.

Perhaps you haven't thought of time as an ecological factor, but niche specialization in a community is a twenty-four hour phenomenon. Swallows and insectivorous bats both catch insects, but some insect species are active during the day and others at night, providing noncompetitive feeding opportunities for day-active swallows and night-active bats.

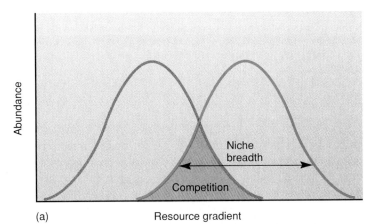

(a)

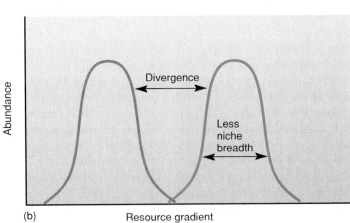

(b)

Figure 4.7

Resource partitioning and niche specialization caused by competition. Where niches of two species overlap along a resource gradient, competition occurs (shaded area in *a*). Individuals occupying this part of the niche are less successful in reproduction so that characteristics of the population diverge to produce more specialization, narrower niche breadth, and less competition between species (*b*).

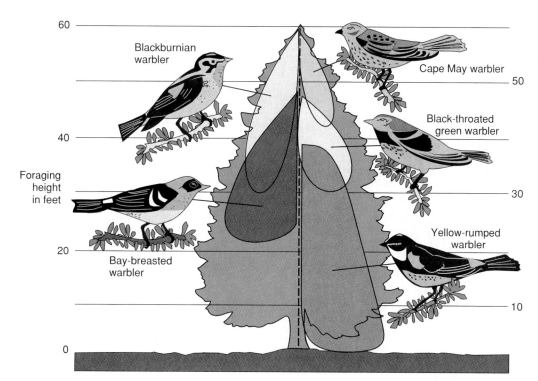

Figure 4.8

Resource partitioning and the concept of the ecological niche are demonstrated by several species of wood warblers that use different strata of the same forest. This is a classic example of the principle of competitive exclusion.

Species Interactions and Community Dynamics

Predation and competition for scarce resources are major factors in evolution and adaptation. Not all biological interactions are competitive, however. Organisms also cooperate with, or at least tolerate, members of their own species as well as individuals of other species in order to survive and reproduce. In this section, we will look more closely at the different interactions within and between species that shape biological communities.

Predation

All organisms need food to live. Producers make their own food, and consumers eat organic matter created by other organisms. In most communities, as we saw in chapter 3, photosynthetic organisms are the producers. Consumers include herbivores, carnivores, omnivores, scavengers, and decomposers. With which of these categories do you associate the term *predator?* Ecologically, the term has a much broader meaning than you might expect. A **predator** in an ecological sense, is an organism that feeds directly upon another living organism, whether or not it kills the prey to do so (fig. 4.9). By this definition herbivores, carnivores, and omnivores that feed on live prey are predators, but scavengers, detritivores, and decomposers that feed on dead things are not. In this broad sense, **parasites** (organisms that feed on a host organism or steal resources from it without killing it) and even **pathogens** (disease-causing organisms) might also be considered predator organisms.

Predation is a potent and complex influence on the population balance of communities involving (1) all stages of the life cycles of predator and prey species, (2) many specialized food-obtaining

Figure 4.9

Insect herbivores are predators as much as are lions and tigers. In fact, insects consume the vast majority of biomass in the world. Complex patterns of predation and defense have often evolved between insect predators and their plant prey.

mechanisms, and (3) specific prey-predator adaptations that either resist or encourage predation (Case Study, p. 76).

Predation throughout the life cycle is very pronounced in marine intertidal animals, for example. Many crustaceans, mollusks, and worms release eggs directly into the water, and the eggs and free-living larval and juvenile stages are part of the floating community, or **plankton** (fig. 4.10). Planktonic animals feed upon each other and are food for successively larger carnivores, including small fish. As prey species mature, their predators change. Barnacle larvae are planktonic and are eaten by fish. Adult barnacles, on the other hand, build hard shells that protect them from fish but can be

CASE STUDY
Where Have All the Songbirds Gone?

Every June, some 2200 amateur ornithologists and bird watchers across the United States and Canada join in an annual bird count called the Breeding Bird Survey. Organized in 1966 by the U.S. Fish and Wildlife Service to follow bird population changes, this survey has discovered some shocking trends. While birds such as robins, starlings, and blackbirds that prosper around humans have increased their number and distribution over the past 30 years, many of our most colorful and melodious forest birds have declined severely. The greatest decreases have been among the true songbirds such as thrushes, orioles, tanagers, catbirds, vireos, buntings, and warblers. These long-distance migrants nest in northern forests but spend the winters in South or Central America or in the Caribbean Islands. Scientists call them neotropical migrants.

In many areas of the eastern United States and Canada, three-quarters or more of the neotropical migrants have declined significantly since the survey was started. Some that once were common have become locally extinct. Grover Archbold Park and Rock Creek Park in Washington, DC, for instance, lost 75 percent of their songbird population and 90 percent of their long-distance migrant species in just 20 years. Nationwide,

Figure 1

Many of our most colorful and melodious woodland songbirds, such as this redstart, are vanishing due to habitat destruction, predation, and nest parasitism.

cerulean warblers, American redstarts (fig. 1), and ovenbirds declined about 50 percent in the single decade of the 1970s. Studies of radar images from National Weather Service stations in Texas and Louisiana suggest that only about half as many birds fly across the Gulf of Mexico each spring now compared to the 1960s. This could mean a loss of about half a billion birds in total.

What causes these devastating losses? Destruction of critical winter habitat is clearly a major issue. Birds often are much more densely crowded in the limited areas available to them during the winter than they are on their summer range. Unfortunately, forests throughout Latin America are being felled at an appalling rate. Central America, for instance, is losing about 1.4 million hectares (2 percent of its forests or an area about the size of Yellowstone National Park) each year. If this trend continues, there will be essentially no intact forest left in much of the region in 50 years.

But loss of tropical forests is not the only threat. Recent studies show that fragmentation of breeding habitat and nesting failures in the United States and Canada may be just as big a problem for woodland songbirds. Many of the most threatened species are adapted to deep woods and need an area of 10 hectares (24.7 acres) or more per pair to breed and raise their young. As our woodlands are broken up by roads, housing developments, and shopping centers, it becomes more and more difficult for these highly specialized birds to find enough contiguous woods to nest successfully.

Predation and nest parasitism also present a growing threat to many bird species. While birds have probably always lost eggs and nestlings to preda-

crushed by limpets and other mollusks. Predators also change their feeding targets. Adult frogs, for instance, are carnivores, but the tadpoles of most species are grazing herbivores. Sorting out the trophic levels in these communities can be very difficult.

Predation is an important factor in evolution. Predators prey most successfully on the slowest, weakest, least fit members of their target population, thus allowing successful traits to become dominant in the prey population, actually benefiting the prey species. Aldo Leopold described this effect in his pioneering studies of predators and deer on the Kaibab Plateau in northern Arizona. When wolves and mountain lions were present, the deer were healthier and in better balance with their environment than after these predators were removed. Without predators, the deer overpopulated the area, exhausted their food supply, and starved to death.

Prey species have evolved many protective or defensive adaptations to avoid predation. In plants, for instance, this often takes the form of thick bark, spines, thorns, or chemical defenses. Animal prey may become very clever at hiding, fleeing, or fighting back against predators. Predators, in turn, evolve mechanisms to overcome the defenses of their prey. This process in which species exert selective pressure on each other is called coevolution.

Keystone Species

A **keystone species** is a species or set of species whose impact on its community or ecosystem is much larger and more influential than would be expected from mere abundance. Originally, keystone species were thought to be top predators, such as

tors, there has been a startling increase in predation in the past 30 years. Raccoons, opossums, crows, bluejays, squirrels, and house cats thrive in human-dominated landscapes. They are protected from larger predators like wolves or owls and find abundant supplies of food and places to hide. Their numbers have increased dramatically, as have their raids on bird nests. A comparison of predation rates in the Great Smoky Mountain National Park and in small rural and suburban woodlands shows how devastating predators can be. In a 1000 hectare study area of mature, unbroken forest in the national park, only one songbird nest in 50 was raided by predators. By contrast, in plots of 10 hectares or less near cities, up to 90 percent of the nests were raided.

Nest parasitism by brown-headed cowbirds is one of the worst threats for woodland songbirds. Originally called buffalo birds, these small blackbirds were adapted to follow migratory bison herds picking up seeds and insects from the droppings. Because they didn't stay in one place long enough to raise a family, they developed the habit of depositing their eggs in the nests of other species, leaving their young to be raised by surrogate parents. The young cowbirds are generally larger and more aggressive than the resident chicks, which generally starve to death because they don't get enough food. Adult cowbirds also find a welcome source of food and shelter around humans. Once fairly uncommon in the United States, there are now about 150 million of these parasites.

A study in southern Wisconsin found that 80 percent of the nests of woodland species were raided by predators and that three-quarters of those that survived were invaded by cowbirds. Another study in the Shawnee National Forest in southern Illinois found that 80 percent of the scarlet tanager nests contained cowbird eggs and that 90 percent of the wood thrush nests were taken over by these parasites. The sobering conclusion of this latter study is that there probably is no longer any place in Illinois where scarlet tanagers and wood thrushes can breed successfully.

What can we do about this situation? First, we can support sustainable development in Third World countries so that people there can enjoy a better standard of living without destroying their forests and natural areas. A number of such projects are discussed elsewhere in this book. Next, we should identify and protect critical habitat at home and abroad on which especially endangered species depend. Buying up inholdings that fragment the forest and preserving corridors that tie together important areas will help. In areas where people already live, we could encourage clustering of houses to protect as much woods as possible. We also might discourage clearing underbrush and trees from yards and parks to leave shelter for the birds.

Could we reduce the number of predators or limit their access to critical breeding areas? Human residents might not like the idea of reintroducing wolves and bears, but they might accept fencing or trapping of small predators. A campaign to keep house cats inside during the breeding season would certainly help.

ETHICAL CONSIDERATIONS

Some wildlife managers already are trapping cowbirds. The Kirtland's warbler is one of the rarest songbirds in the United States. It nests only in young, fire-maintained jackpine forests in Michigan. Controlled burning to maintain habitat for this endangered species was started in the 1960s, but the population continued to decline. Studies showed that 90 percent of the nests were being parasitized by cowbirds. Since 1972, refuge managers have trapped and killed some 7000 cowbirds each year to protect the warblers. In the past two decades, the number of breeding pairs of warblers has risen from about 150 to nearly 400. Would it be possible to do something similar on a nationwide scale? Could we trap and kill 150 million cowbirds? Should we eliminate one species to save another? What do you think?

wolves, whose presence limits the abundance of herbivores and thereby reduces their grazing or browsing on plants. Recently, it has been recognized that less conspicuous species also play essential community roles. Certain tropical figs, for example, bear during seasons when no other fruit is available for frugivores (fruit-eating animals). If these figs were removed, many animals would starve to death during periods of fruit scarcity. With those animals gone, many other plant species that depend on them at other times of year for pollination and seed-dispersal would disappear as well.

Even microorganisms can play vital roles. In some forest ecosystems, mycorrhizae (fungi associated with tree roots) are essential for mineral mobilization and absorption. If the fungi die, so do the trees and many other species that depend on a healthy forest community. Rather than being a single species, mycorrhizae are actually a group of species that together fulfill a keystone function.

Often a number of species are intricately interconnected in biological communities so that it is difficult to tell which is the essential key. Consider the Pacific kelp forests and sea otters that opened this chapter (fig. 4.11). Giant kelp provide shelter for a number of fish and shellfish species and so could be regarded as the key to community structure. Sea urchins, however, feed on the kelp and determine their number and distribution while sea otters regulate urchins and kelp provides a resting place for dozing otters. Which of these species is the most important? Each depends on and affects the others. Perhaps we should think in terms of a "keystone set" of organisms in some ecosystems.

Figure 4.10

Microscopic plants and animals form the basic levels of many aquatic food chains and account for a large percentage of total world biomass. Many oceanic plankton are larval forms that have very different habitats and feeding relationships than their adult forms.

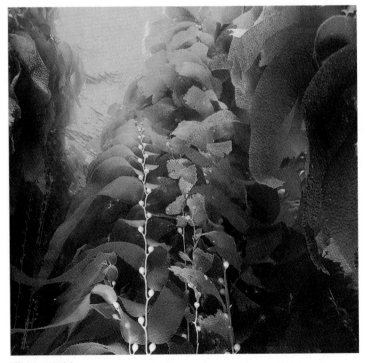

Figure 4.11

Giant kelp is a massive alga that forms dense "forests" off the Pacific coast of California. It is a keystone species in that it provides food, shelter, and structure essential for a whole community. Removal of sea otters by fur hunters allowed sea urchin populations to explode. When the urchins destroyed the kelp, many other species suffered as well.

Competition

Competition is another kind of antagonistic relationship within a community. For what do organisms compete? To answer this question, think again about what all organisms need to survive: energy and matter in usable forms, space, and specific sites for life activities (What Do *You* Think?, p. 79). Plants compete for growing space for root and shoot systems so they can absorb and process sunlight, water, and nutrients. Animals compete for living, nesting, and feeding sites, as well as for food, water, and mates. Competition among members of the same species is called **intraspecific competition,** whereas competition between members of different species is called **interspecific competition.**

If you look closely at a patch of weeds growing on good soil early in the summer, you likely will see several types of interspecific competition. First of all, many weedy species attempt to crowd out their rivals by producing prodigious numbers of seeds. After the seeds germinate, the plants race to grow the tallest, cover the most ground, and get the most sun. You may observe several strategies to do this. For example, vines don't build heavy stems of their own; they simply climb up over their neighbors to get to the light.

We often think of competition among animals as a bloody battle for resources. A famous Victorian description of the struggle for survival was "nature red in tooth and claw." In fact, a better metaphor is a race. Have you ever noticed that birds always eat fruits and berries just before they are ripe enough for us to pick? Having a tolerance for bitter, unripe fruit gives them an advantage in the race for these food resources. Many animals tend to avoid fighting if possible. It's not worth getting injured. Most confrontations are more noise and show than actual fighting.

Intraspecific competition can be especially intense because members of the same species have the same space and nutritional requirements; therefore, they compete directly for these environmental resources. How do plants cope with intraspecific competition? The inability of seedlings to germinate in the shady conditions created by parent plants acts to limit intraspecific competition by favoring the mature, reproductive plants. Many plants have adaptations for dispersing their seeds to other sites by air, water, or animals. Undoubtedly you've seen dandelion seeds on their tiny parachutes (fig. 4.12) or had sticky or burred seeds become attached to your clothing. Some plants secrete leaf or root exudates that inhibit the growth of seedlings near them, including their own and those of other species. This strategy is particularly significant in deserts where water is a limiting factor.

Animals also have developed adaptive responses to intraspecific competition. Two major examples are varied life cycles and territoriality. The life cycles of many invertebrate species have juvenile stages that are very different from the adults in habitat and feeding. Compare a leaf-munching caterpillar to a nectar-sipping adult butterfly or a planktonic crab larva to its bottom-crawling adult form. In these examples, the adults and juveniles of each species do not compete because they occupy different ecological niches.

You may have observed robins chasing other robins during the mating and nesting season. Robins and many other vertebrate species demonstrate **territoriality,** an intense form of intraspecific competition in which organisms define an area surrounding their

What Do *You* Think?
Understanding Competition

Ecology is a relatively young science. Consequently, many ecological processes are incompletely understood. How a community comes to have its particular organization is one area of uncertainty. Some ecologists feel that interactions between organisms are not so important in determining community organization, while others feel that interspecific competition has a major influence on a community.

How can we find out which view is correct? Ecologists employ the scientific method, as described in chapter 2, to better understand community dynamics. This process is mostly refined common sense and its basic elements can be useful in everyday life.

Once ecologists have decided on the concept to be investigated, they look for a specific situation that can either be observed or manipulated to provide relevant information. For example, ecologist Richard Karban was interested in how competition affected a community. He learned that larvae of two insect species, the meadow spittlebug (fig. 1) and the calendula plume moth, both feed and develop on the seaside daisy, a common beach plant on the American west coast. The specific question to be investigated was: Does competition affect these two insect species, therefore impacting community organization?

Competition might reduce survival rate, larval growth, or both. Karban's procedure involved setting up four groups of plants at Bodega Bay, CA : one got both spittlebugs and moths, another got only spittlebugs, another only moths, and a fourth had neither. He compared survival rates of spittlebugs and moths when competitors were present and absent.

Figure 1

Spittlebugs produce mounds of foam under which they hide from predators while feeding on host plants.

There are three important general considerations in designing scientific investigations:

1 *Things need to be organized in such a way that the outcome can clearly be linked to a particular cause. In other words, differences in insect survival rates need to be clearly attributable to competition and not to other factors.*

   Karban accomplished this by making his plant/insect groups as uniform as possible, except for the presence or absence of competitors. He eliminated genetic differences between plants by using plants from the same clone. He was careful to put the same numbers of insects on each plant to eliminate animal density as a factor, and so on.

2 *The data collected must be a reliable representation of the larger situation and not simply the result of chance. This is usually accomplished by replicating the procedure many times. Instead of setting up just a few plants with one or both insects present, Karban set up thirty plants with each treatment. The procedure was repeated a second year. This gave him a cumulative total of 60 plants that had just spittlebugs, 60 plants having just moths, and 60 plants each having both or neither spittlebugs and moths. With such a large number of replications it was highly likely that differences in survival rates were, in fact, the result of competition and not simply chance occurrences.*

3 *Finally, conclusions must be justified by the data. Karban's statistical analysis revealed that spittlebug persistence was nearly 40 percent higher when the plume moths were absent. Plume moth persistence was not significantly affected by spittlebug presence, however.*

His overall conclusion was:

"*Evidence from this and other studies supports the contention that interspecific competition can play an important role in influencing densities of plant-feeding insects.*"

Notice the caution expressed in these words. He did not claim to have *proven* anything. Instead, his study "*supports* the contention." Secondly, he states competition "*can* play an important role," instead of using stronger language. And finally, he restricts these conclusions to plant-feeding insects. Karban carefully avoids drawing conclusions beyond the realm supported by his data.

Based on a healthy skepticism, clarity of language, critical evaluation of relationships and information, and caution in coming to judgment, critical thinking in science has been a very successful tool in enhancing understanding.

Figure 4.12
Many plant species have mechanisms for seed dispersal over long distances, which enables them to reach new areas for growth. The ability to move rapidly into newly disturbed areas gives "weedy" species a competitive advantage in some circumstances.

home site or nesting site and defend it, primarily against other members of their own species. Territoriality helps to allocate the resources of an area by spacing out the members of a population. It also promotes dispersal into adjacent areas by pushing grown offspring outward from the parental territory.

Territory size depends on the size of the species and the resources available. A pair of robins might make do with a suburban yard, but a large carnivore like a tiger may need thousands of square kilometers.

Symbiosis

In contrast to predation and competition, symbiotic interactions between organisms are nonantagonistic. **Symbiosis** is the intimate living together of members of two or more species. **Commensalism** is a type of symbiosis in which one member clearly benefits and the other apparently is neither benefited nor harmed. Cattle often are accompanied by cattle egrets, small white shore birds who catch insects kicked up as the cattle graze through a field. The birds benefit while the cattle seem indifferent. Many of the mosses, bromeliads, and other plants growing on trees in the moist tropics are also considered to be commensals (fig. 4.13). These epiphytes get water from rain and nutrients from leaf litter and dust fall, and often neither help nor hurt the trees on which they grow. In a sense, the robins and sparrows that inhabit suburban yards are commensals with humans.

Lichens are a combination of a fungus and a photosynthetic partner, either an alga or a blue-green bacterium. Their association is a type of symbiosis called **mutualism,** in which both members of the partnership benefit (fig. 4.14). Some ecologists believe that cooperative, mutualistic relationships may be more important in evolution than we have commonly thought. Aggressive interactions often are dangerous and destructive, while cooperation and compromise may have advantages that we tend to overlook. Survival of

Figure 4.13
Plants compete for light and growing space in this Indonesian rainforest. Epiphytes, such as the ferns and bromeliads shown here, find a place to grow in the forest canopy by perching on the limbs of large trees. This may be a commensal relationship if the epiphytes don't hurt their hosts. Sometimes, however, the weight of epiphytes breaks off branches and even topples whole trees.

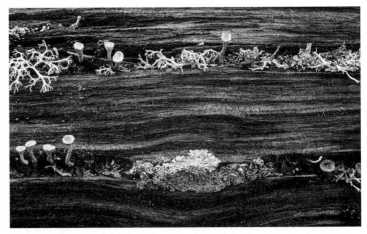

Figure 4.14
Lichens, such as the various species growing on this log, are a classic example of mutualistic symbiosis. A combination of a blue-green bacterium or a green alga with a fungus, lichens have evolved such a close partnership that it is impossible for one member of the pair to survive without the other. Together, however, these organisms create a tough, tenacious combination that endures the harshest environmental conditions.

the fittest often may mean survival of those organisms that can live best with one another.

Parasitism, described earlier as a form of predation, also could be considered a type of symbiosis, where one species benefits and the other is harmed. All of these relationships have a bearing on such ecological issues as resource utilization, niche specialization, diversity, predation, and competition. Symbiotic relationships often enhance the survival of one or both partners.

Symbiotic relationships often entail some degree of coadaptation or coevolution of the partners, shaping—at least in part—their structural and behavioral characteristics. An interesting case of mutualistic coadaptation is seen in Central and South American swollen thorn acacias and their symbiotic ants. Acacia ant colonies live within the swollen thorns on the acacia tree branches and feed on two kinds of food provided by the trees: nectar produced in glands at the leaf bases and special protein-rich structures produced on leaflet tips. The acacias thus provide shelter and food for the ants. Although they spend energy to provide these services, the trees are not physically harmed by ant feeding.

What do the acacias get in return and how does the relationship relate to community dynamics? Ants tend to be aggressive defenders of their home areas, and acacia ants are no exception. They drive off herbivorous insects that attempt to feed on their home acacia, thus reducing predation. They also trim away vegetation that grows around their home tree, thereby reducing competition. This is a fascinating example of how a symbiotic relationship fits into community interactions. It is also an example of coevolution based on mutualism rather than competition or predation.

Mimicry and Camouflage

Many species have evolved amazing structures or behaviors for camouflage or subterfuge. You very likely have seen examples of insects that look exactly like dead leaves or twigs to hide from predators. Scorpion fish, on the other hand, resemble a pile of harmless vegetation so as to lure unsuspecting prey within striking distance (fig. 4.15). Some plants also have evolved in remarkable ways. A tropical orchid has flower structures that look exactly like a female fly. Males attempting to mate unwittingly carry away pollen.

Batesian mimicry, named for British naturalist W. H. Bates, who first described it in 1857, is an adaptation in which an unprotected species evolves in shape and coloration to resemble *related*

Figure 4.15

This highly camouflaged scorpion fish lays in wait for its unsuspecting prey. Natural selection and evolution have created the elaborate disguise seen here.

species that protect themselves with poison or painful stingers. A classic example of Batesian mimicry is the viceroy butterfly, *Limentitis archippus* (fig. 4.16), which closely resembles the bad-tasting monarch, *Danaus plexippus*. Birds learn through trial and experience to avoid monarchs. Thus viceroys, if relatively scarce compared to monarchs, are avoided as well.

Muellerian mimicry, named for the German biologist Fritz Mueller, who first described it in 1878, is evolution of similar coloration and body shapes by several *unrelated* species. For instance, some harmless flies and beetles look very much like dangerous stinging wasps (fig. 4.17). In both Batesian and Müllerian mimicry, the mimic and model must not only look alike but also act alike. To avoid being eaten by predators, the wasp-resembling insects must fly in the same way and must spend most of their time in the same habitat as the wasps they mimic. Sometimes Muellerian mimics all are poisonous or dangerous, but they still gain an advantage from looking alike, thus achieving collective protection.

Figure 4.16

Monarch butterflies (*right*) accumulate poisonous alkaloids from the milkweeds on which they feed. Birds learn to avoid the bad-tasting monarchs and then also avoid the similar-appearing viceroy (*left*) even though they feed on other plants and lack poisonous chemicals. This is an example of Batesian mimicry.

Figure 4.17

Muellerian mimics are unrelated species evolved to resemble one another. The dangerous wasp (*left*) has bold yellow and black bands to warn predators away. The much rarer longhorn beetle (*right*) has no poisonous stinger, but looks and acts like a wasp and thus avoid predators as well.

Community Properties

The processes and principles that we have studied thus far in this chapter—tolerance limits, species interactions, resource partitioning, evolution, and adaptation—play important roles in determining the characteristics of populations and species. In this section we will look at some fundamental properties of biological communities and ecosystems—productivity, diversity, complexity, resilience, stability, and structure—to learn how they are affected by these factors.

Productivity

A community's **productivity** is measured as the rate of biomass production, an indication of the rate of solar energy conversion to chemical energy. Photosynthetic rates are regulated by light levels, temperature, moisture, and nutrient availability. Figure 4.18 shows approximate productivity levels for some major ecosystems. As you can see, tropical forests, coral reefs, and estuaries (bays or inundated river valleys where rivers meet the ocean) have high levels of productivity because they have abundant supplies of all these resources. In deserts, lack of water limits photosynthesis. On the arctic tundra or in high mountains, low temperatures inhibit plant growth. In the open ocean, a lack of nutrients reduces the ability of algae to make use of plentiful sunshine and water.

Some agricultural crops such as corn (maize) and sugar cane grown under ideal conditions in the tropics approach the productivity levels of tropical forests. Because shallow water ecosystems such as coral reefs, salt marshes, tidal mud flats, and other highly productive aquatic communities are relatively rare compared to the vast extent of open oceans—which are effectively biological deserts—marine ecosystems are much less productive *on average* than terrestrial ecosystems. Although oceans cover about 71 percent of the earth's surface, their total productivity is actually less than that of terrestrial areas.

Even in the most photosynthetically active ecosystems, only a small percentage of the available sunlight is captured and used to make energy-rich compounds. Between one-quarter and three-quarters of the light reaching plants is reflected by leaf surfaces. Most of the light absorbed by leaves is converted to heat that is either radiated away or dissipated by evaporation of water. Only 1 to 2 percent of the absorbed energy is used by chloroplasts to synthesize carbohydrates.

In a temperate-climate oak forest, only about half the incident light available on a midsummer day is absorbed by the leaves. Ninety-nine percent of this energy is used to evaporate water. A large oak tree can transpire (evaporate) several thousand liters of water on a warm, dry, sunny day while it makes only a few kilograms of sugars and other energy-rich organic compounds.

Abundance and Diversity

Abundance is an expression of the total number of organisms in a biological community, while **diversity** is a measure of the number of different species, ecological niches, or genetic variation present. The abundance of a particular species often is inversely related to the total diversity of the community. That is, communities with a very large number of species often have only a few members of any given species in a particular area. As a general rule, diversity decreases but abundance within species increases as we go from the equator toward the poles. The arctic has vast numbers of insects such as mosquitoes, for example, but only a few species. The tropics, on the other hand, have vast numbers of species—some of which have incredibly bizarre forms and habits—but often only a few individuals of any particular species in a given area.

Consider bird populations. Greenland is home to 56 species of breeding birds, while Colombia, which is only one-fifth the size of Greenland, has 1395. Why so many species in Colombia and so few in Greenland?

Climate and history are important factors. Greenland has such a harsh climate that the need to survive through the winter or escape to milder climates becomes the single most important critical factor that overwhelms all other considerations and severely limits the ability of species to specialize or differentiate into new forms. Furthermore, because Greenland was covered by glaciers until about 10,000 years ago, there has been little time for new species to develop.

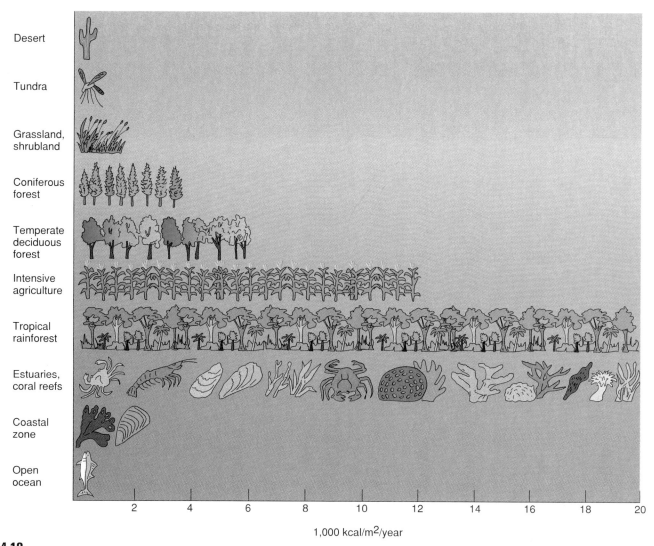

Figure 4.18
Relative gross primary productivity of major world ecosystems.

Many areas in the tropics, by contrast, have relatively abundant rainfall and warm temperatures year-round so that ecosystems there are highly productive. The year-round dependability of food, moisture, and warmth supports a great exuberance of life and allows a high degree of specialization in physical shape and behavior. Coral reefs are similarly stable, productive, and conducive to proliferation of diverse and exotic life-forms. The enormous abundance of brightly colored and fantastically shaped fish, corals, sponges, and arthropods in the reef community is one of the best examples we have of community diversity.

Productivity is related to abundance and diversity, both of which are dependent on the total resource availability in an ecosystem as well as the reliability of resources, the adaptations of the member species, and the interactions between species. You shouldn't assume that all communities and ecosystems are perfectly adapted to their environment. A relatively new community that hasn't had time for niche specialization or a disturbed one where roles such as top predators are missing may not achieve

maximum efficiency of resource use or reach its maximum level of either abundance or diversity.

Complexity and Connectedness

Community complexity and connectedness generally are related to diversity but are important because they help us visualize and understand community functions. **Complexity** in ecological terms refers to the number of species at each trophic level and the number of trophic levels in a community. A diverse community may not be very complex if all its species are clustered in only a few trophic levels and form a relatively simple food chain.

By contrast, a complex, highly interconnected community (fig. 4.19) might have many trophic levels, some of which can be compartmentalized into subdivisions. In tropical rainforests, for instance, the herbivores can be grouped into "guilds" based on the specialized ways they feed on plants. There may be fruit eaters, leaf nibblers, root borers, seed gnawers, and sap suckers, each composed

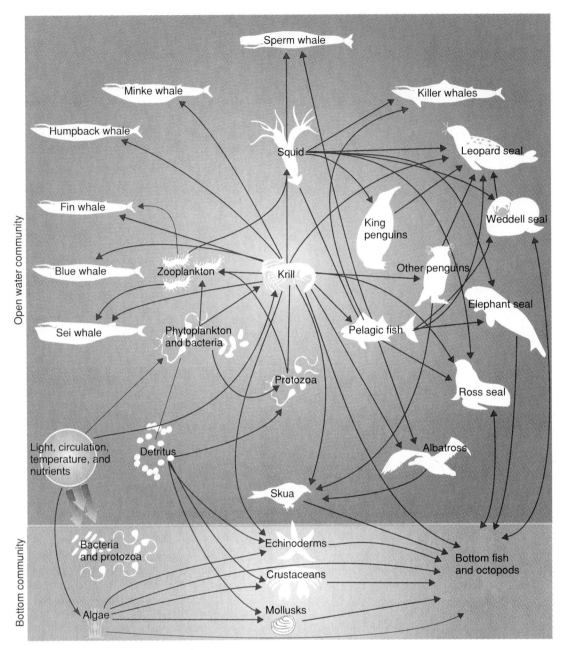

Figure 4.19

A complex and highly interconnected community can have many species at each trophic level and many relationships, as illustrated by this Antarctic marine food web.

of species of very different size, shape, and even biological kingdom, but that feed in related ways. A highly interconnected community such as this can form a very elaborate food web.

Resilience and Stability

Many biological communities tend to remain relatively stable and constant over time. An oak forest tends to remain an oak forest, for example, because the species that make it up have self-perpetuating mechanisms. We can identify three kinds of stability or resiliency in ecosystems: *constancy* (lack of fluctuations in composition or func-

tions), *inertia* (resistance to perturbations), and *renewal* (ability to repair damage after disturbance).

In 1955, Robert McArthur, who was then a graduate student at Yale, proposed that the more complex and interconnected a community is, the more stable and resilient it will be in the face of disturbance. If many different species occupy each trophic level, some can fill in if others are stressed or eliminated by external forces, making the whole community resistant to perturbations and able to recover relatively easily from disruptions. This theory has been controversial, however. Some studies support it, while others do not. For example, Minnesota ecologist David Tilman, in studies of native prairie and recovering farm fields, found that plots with high diversity were better able to withstand and recover from drought than those with only a few species.

Yet, in a highly specialized ecosystem, removal of a few keystone members can eliminate many other associated species. Removing a major tree species from a tropical forest, for example, may destroy pollinators and fruit distributors as well. We might replant the trees, but could we replace the whole web of relationships on which they depend? In this case, diversity has made the forest less resilient rather than more.

Community Structure

Ecological **structure** refers to patterns of spatial distribution of individuals and populations within the community as well as the relation of a particular community to its surroundings. At the local level, even in a relatively homogeneous environment, individuals in a single population can be distributed randomly, clumped together, or in highly regular patterns (fig. 4.20). In randomly arranged populations, individuals live wherever resources are available. Ordered patterns may be determined by the physical environment but are more often the result of biological competition. For example, competition for nesting space in a penguin colony is often fierce. Each nest tends to be just out of reach of the neighbors sitting on their own nests. Constant squabbling produces a highly regular pattern. Similarly, sagebrush releases toxins from roots and fallen leaves that inhibit the growth of competitors and create a circle of bare ground surrounding each bush. As neighbors fill in empty spaces up to the limit of this chemical barrier, a regular spacing results.

Some other species cluster together (fig. 4.20c) for protection, mutual assistance, reproduction, or to gain access to a particular environmental resource. Dense schools of fish, for instance, cluster closely together in the ocean, increasing their chances of detecting and escaping predators (fig. 4.21). Similarly, predators, whether sharks, wolves, or humans, often hunt in packs

Figure 4.21

Fish and birds often flock together in dense bands for protection and mutual feeding.

to catch their prey. A flock of blackbirds descending on a cornfield or a troop of baboons traveling across the African savanna band together both to avoid predators and to find food more efficiently.

Plants, too, can be protected by clustering. A grove of wind-sheared evergreen trees is often found packed tightly together at the crest of a high mountain or along the seashore. They offer mutual protection from the wind not only to each other but also to other creatures that find shelter in or under their branches.

If we look closely at larger and more ecologically varied communities, we often find a mosaic of smaller units or subsets of the whole assemblage. These subunits develop because each species has a preference for specific, localized conditions. Ecologists describe these patterns as **patchiness** or graininess in spatial distribution. Some communities are coarse-grained with large patches; others are so fine-grained as to be almost indistinguishable from random distribution. Even though individual patches may be small and not very distinct from adjacent areas, the specific interactions in each patch may be more important in determining functions of the community than the overall composition.

Distribution in a community can be vertical as well as horizontal. The tropical forest, for instance, has many layers, each with different environmental conditions and combinations of species. While the larger trees are part of each layer, the communities of smaller plants, animals, and microbes that live in each part of this complex community generally are distinct and very different from each other. Similarly, aquatic communities are often stratified into layers based on light penetration in the water, temperature, salinity, pressure, or other factors.

Edges and Boundaries

An important aspect of community structure is the boundary between one habitat and its neighbors. We call these relationships **edge effects.** In many cases, the edge of a patch of habitat is relatively sharp and distinct. In moving from a woodland patch into a grassland or cultivated field, you sense a dramatic change from the

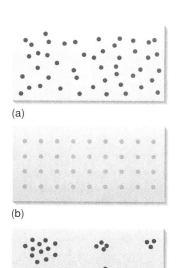

(a)

(b)

(c)

Figure 4.20

Distribution of members of a population in a given space can be random (*a*), ordered (*b*), or clustered (*c*). These patterns are determined both by the physical environment and by biological interactions. They may produce a graininess or patchiness in community structure.

cool, dark, quiet forest interior to the windy, sunny, warmer, open space of the field or pasture. Species composition may change equally suddenly if organisms adapted to one set of conditions can't survive in the other (fig. 4.22).

Ecologists call the boundaries between adjacent communities **ecotones.** A community that is sharply divided from its neighbors is called a closed community. In contrast, communities with gradual or indistinct boundaries over which many species cross are called open communities. Often this distinction is a matter of degree or perception. As we saw earlier in this chapter, birds might feed in fields or grasslands but nest in the forest. As they fly back and forth, the birds interconnect the ecosystems by moving energy and material from one to the other, making both systems relatively open. Furthermore, the forest edge, while clearly different from the open field, may be sunnier and warmer than the forest interior, and may have a different combination of plant and animal species than either field or forest "core."

Depending on how far edge effects extend from the boundary, differently shaped habitat patches may have very dissimilar amounts of interior area (fig. 4.23). In Douglas fir forests of the Pacific Northwest, for example, increased rates of blowdown, decreased humidity, absence of shade-requiring ground cover, and

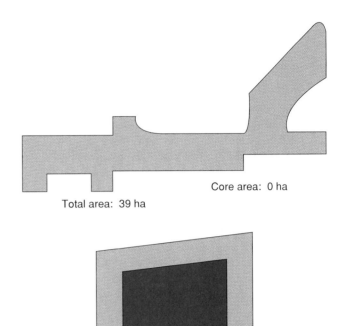

Total area: 39 ha   Core area: 0 ha

Total area: 47 ha   Core area: 20 ha

Figure 4.23

Shape can be as important as size in small preserves. While these areas are close to the same size, no place in the top figure is far enough from the edge to have characteristics of core habitat, while the bottom patch has a significant core.

other edge effects can extend as much as 200 m into a forest. A 40-acre block (200 meters square) surrounded by clear cut would have essentially no true core habitat at all.

Many popular game animals, such as white-tailed deer and pheasants that are adapted to human disturbance, often are most plentiful in boundary zones between different types of habitat. Game managers once were urged to develop as much edge as possible to promote large game populations. Today, however, most wildlife conservationists recognize that the edge effects associated with habitat fragmentation are generally detrimental to biodiversity. Preserving large habitat blocks and linking smaller blocks with migration corridors are among the best ways to protect rare and endangered species (see chapter 15).

Communities in Transition

So far our view of communities has focused on the day-to-day interactions of organisms with their environments, set in a context of survival and selection. In this section, we'll step back to look at some transitional aspects of communities, including where communities meet and how communities change over time.

Ecological Succession

Biological communities have a history in a given landscape. The process by which organisms occupy a site and gradually change environmental conditions so that other species can replace the orig-

Figure 4.22

Ecologists call the sharp edge, or boundary between the woods and sunny glade seen in this picture, an ecotone. Edge effects such as increased sunshine, wind, and temperature, as well as decreased humidity may extend a considerable distance into the forest. When fragmented into many small patches, the forest may become all edge, with no true interior core at all.

Figure 4.24

Primary succession on a terrestrial site is shown in five stages *(left to right),* beginning with rocks that are initially colonized by a pioneer community of lichens and mosses and ending with a climax forest community.

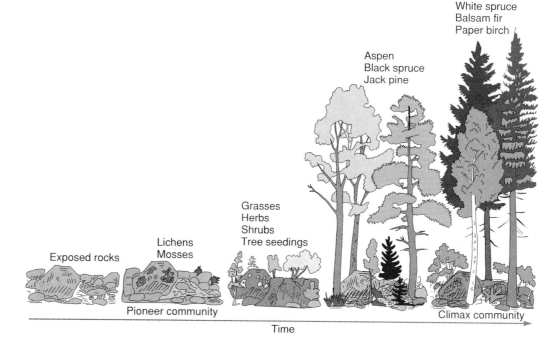

inal inhabitants is called ecological succession or development. **Primary succession** occurs when a community begins to develop on a site previously unoccupied by living organisms, such as an island, a sand or silt bed, a body of water, or a new volcanic flow (fig. 4.24). **Secondary succession** occurs when an existing community is disrupted and a new one subsequently develops at the site. The disruption may be caused by some natural catastrophe, such as fire or flooding, or by a human activity, such as deforestation, plowing, or mining. Both forms of succession usually follow an orderly sequence of stages as organisms modify the environment in ways that allow one species to replace another.

In primary succession on a terrestrial site, the new site first is colonized by a few hardy **pioneer species,** often microbes, mosses, and lichens that can withstand harsh conditions and lack of resources. Their bodies create patches of organic matter in which protists and small animals can live. Organic debris accumulates in pockets and crevices, providing soil in which seeds can become lodged and grow. We call this process of environmental modification by organisms **ecological development.** The community of organisms becomes more diverse and increasingly competitive as development continues and new niche opportunities appear. The pioneer species gradually disappear as the environment changes and new species combinations replace the preceding community. In a global sense, the gradual changes brought about by living organisms have created many of the conditions that make life on earth possible. You could consider evolution to be a very slow, planetwide successional and developmental process.

Figure 4.25 depicts succession in an aquatic ecosystem, showing how a developing community changes its own environment. The amount of open water in the lake or pond gradually decreases as vegetation encroaches from the margins, resulting in gradual community replacement progressing from the edges of the pond toward the center. Succession proceeds from open lake to shallow

pond with highly vegetated edges to marshy area with rooted, emergent vegetation and finally to grassland or forest.

Examples of secondary succession are easy to find. Observe an abandoned farm field or burned-over forest in a temperate climate. The bare soil first is colonized by rapidly growing annual plants (those that grow, flower, and die the same year) that have light, wind-blown seeds and can tolerate full sunlight and exposed soil. They are followed and replaced by perennial plants (those that live for several to many years), including grasses, various nonwoody

Figure 4.25

Succession occurs in a pond as vegetation gradually encroaches from the margins toward the center. Eventually, the pond may fill in completely and become a marsh or woodland.

Biological Communities and Species Interaction

flowering plants, shrubs, and trees. As in primary succession, plant species progressively change the environmental conditions. Biomass accumulates and the site becomes richer, better able to capture and store moisture, more sheltered from wind and climate change, and biologically more complex. Species that cannot survive in a bare, dry, sunny, open area find shelter and food as the field turns to prairie or forest.

Eventually, in either primary or secondary succession, a community develops that seemingly resists further change. Ecologists call this a **climax community** because it appears to be the culmination of the successional process. An analogy is often made between community succession and organism maturation. Beginning with a primitive or juvenile state and going through a complex developmental process, each progresses until a complex, stable, and mature form is reached. It's dangerous to carry this analogy too far, however, because no mechanism is known to regulate communities in the same way that genetics and physiology regulate development of the body.

As mentioned in chapter 3, the concept of succession to a climax community was first championed by the pioneer biogeographer F. E. Clements. He viewed this process as being like a parade or relay, in which species replace each other in predictable groups and in a fixed, regular order, and as being driven almost entirely by climate. This community-unit theory was opposed by Clements's contemporary, H. A. Gleason, who saw community history as a much more individualistic and random process driven by many environmental factors. He argued that temporary associations are formed according to the conditions prevailing at a particular time and the species available to colonize a given area. You might think of the Gleasonian model as a time-lapse movie of a busy railroad station. Passengers come and go; groups form and then dissipate. Patterns and assemblages that seem significant to us may not mean much in the long run.

The process of succession may not be as deterministic as we once thought, yet mature or highly developed ecological communities may tend to be resilient and stable over long periods of time because they can resist or recover from external disturbances. Many are characterized by high species diversity, narrow niche specialization, well-organized community structure, good nutrient conservation and recycling, and a large amount of total organic matter. Community functions, such as productivity and nutrient cycling, tend to be self-stabilizing or self-perpetuating. What once were regarded as "final" climax communities, however, may still be changing. It's probably more accurate to say that the rate of succession is so slow in a climax community that, from the perspective of a single human lifetime, it appears to be stable.

Some landscapes never reach a stable climax in the traditional sense because they are characterized by, and adapted to, periodic disruption. They are called **equilibrium communities** or **disclimax communities.** Grasslands, the chaparral shrubland of California, and some kinds of coniferous forests, for instance, are shaped and maintained by periodic fires that have long been a part of their history. They are, therefore, often referred to as **fire-climax communities** (fig. 4.26). Plants in these communities are adapted to resist fires, reseed quickly after fires, or both. In fact, many of the plant species we recognize as dominants in these communities *require* fire to eliminate competition, to prepare seedbeds for ger-

Figure 4.26

This lodgepole pine forest in Yellowstone National Park was once thought to be a climax forest, but we now know that this forest must be constantly renewed by periodic fire. It is an example of an equilibrium, or disclimax, community.

mination of seedlings, or to open cones or thick seed coats. Without fire, community structure may be quite different.

Introduced Species and Community Change

Succession requires the continual introduction of new community members and the disappearance of previously existing species. New species move in as conditions become suitable; others die or move out as the community changes. New species also can be introduced after a stable community already has become established. Some cannot compete with existing species and fail to become established. Others are able to fit into and become part of the community, defining new ecological niches. If, however, an introduced species preys upon or competes more successfully with one or more populations that are native to the community, the entire nature of the community can be altered.

Human introductions of Eurasian plants and animals to non-Eurasian communities often have been disastrous to native species because of competition or overpredation. Oceanic islands offer classic examples of devastation caused by rats, goats, cats, and pigs liberated from sailing ships. All these animals are prolific, quickly developing large populations. Goats are efficient, nonspecific herbivores; they eat nearly everything vegetational, from grasses and herbs to seedlings and shrubs. In addition, their sharp hooves are hard on plants rooted in thin island soils. Rats and pigs are opportunistic omnivores, eating the eggs and nestlings of seabirds that tend to nest in large, densely packed colonies, and digging up sea turtle eggs. Cats prey upon nestlings of both ground- and tree-nesting birds. Native island species are particularly vulnerable because they have not evolved under circumstances that required them to have defensive adaptations to these predators.

Sometimes we introduce new species in an attempt to solve problems created by previous introductions but end up making the situation worse. In Hawaii and on several Caribbean Islands, for instance, mongooses were imported to help control rats that had escaped from ships and were destroying indigenous birds (fig. 4.27). Since the mongooses were diurnal (active in the day), however, while rats are nocturnal, they tended to ignore each other. Instead, the mongooses also killed native birds and further threatened endangered species. Our lessons from this and similar introductions have a new technological twist. Some of the ethical questions currently surrounding the release of genetically engineered organisms are based on concerns that they are novel organisms, and we might not be able to predict how they will interact with other species in natural ecosystems—let alone how they might respond to natural selective forces. It is argued that we can't predict either their behavior or their evolution.

Figure 4.27

Mongooses were released in Hawaii in an effort to control rats. The mongooses are active during the day, however, while the rats are night creatures so they ignored each other. Instead, the mongooses attacked defenseless native birds and became as great a problem as the rats.

ummary

The principle of limiting factors states that for every physical factor in the environment, there are both maximum and minimum tolerable limits beyond which a given species cannot survive. Often the factor in shortest supply or closest to the tolerance limit for a particular species at a particular time is the critical factor that will determine the abundance and distribution of that species in that ecosystem. Random genetic variations create diversity in a population that gives some individuals advantages in a given set of circumstances. The best suited organisms will survive and reproduce more successfully than the ill-suited ones. Eventually, the genes for these successful characteristics predominate in the population and the species becomes adapted to its environment and to a particular role. This process leads to evolution of a species either through gradual replacement of the original parental type or a splitting of a population into two species.

Habitat describes the place or set of conditions in which an organism lives; niche describes the role an organism plays: how it makes a living. Natural selection often leads to niche specialization and resource partitioning that reduce competition between species. Organisms interact within communities in many ways. Predation—feeding on another organism—involves

pathogens, parasites, and herbivores as well as carnivorous predators. Competition is another kind of antagonistic relationship in which organisms vie for space, food, or other resources. Symbiosis is the intimate living together of two species. Mutualism means that both species benefit; commensalism means that one species benefits while the other is indifferent.

Some fundamental properties of biological communities are productivity, diversity, complexity, resilience, stability, and structure. Productivity is a measure of the rate at which photosynthesis produces biomass made of energy-rich compounds. Tropical rainforests are generally the most productive of all terrestrial communities; coral reefs and estuaries are generally the most productive aquatic communities. Diversity is a measure of the number of different species in a community, while abundance is the total number of individuals. Often the most productive and stable communities are highly diverse and profusely populated, but sometimes a high degree of specialization makes an ecosystem more, rather than less, susceptible to disturbance. Ecological complexity refers to the number of species at each trophic level as well as the total number of trophic levels in a community. Structure concerns the patterns of organization, both spatial and functional, in a community. Often a keystone species, or group of species, plays an unusually important role in determining community structure, composition, or function. All of these characteristics are affected both by physical and chemical factors as well as biological interactions between the organisms that make up the community. Edge effects at the boundaries between different habitat types can be important where landscapes are fragmented into isolated patches.

Ecological succession and development are processes by which organisms alter the environment in ways that allow some species to replace others. Primary succession starts with a previously unoccupied site. Secondary succession occurs on a site that has been disturbed by external forces. Often succession proceeds until a mature, diverse, climax community is established. These mature communities may have self-perpetuating processes that make them resistant to change and resilient to disturbance. Whether diversity always leads to stability, however, is controversial. Communities that are disrupted regularly by fires or other natural disasters sometimes establish dynamic equilibrium or disclimax communities dependent on constant renewal. Introduction of new species by natural processes, such as opening of a land bridge, or through human intervention can upset the natural relationships in a community and cause catastrophic changes for indigenous species.

▼ Questions for Review

1. Explain how tolerance limits to environmental factors determine distribution of a highly specialized species such as the desert pupfish. Compare this to the distribution of a generalist species such as cowbirds or starlings.

2. Productivity, diversity, complexity, resilience, and structure are exhibited to some extent by all communities and ecosystems. Describe how these characteristics apply to the ecosystem in which you live.

3. Describe the general niche occupied by a bird of prey, such as a hawk or an owl. How can hawks and owls exist in the same ecosystem and not adversely affect each other?

4. Define keystone species and explain their importance in community structure and function.

5. All organisms within a biological community interact with each other. The most intense interactions often occur between individuals of the same species. What concept discussed in this chapter can be used to explain this phenomenon?

6. Relationships between predators and prey play an important role in the energy transfers that occur in ecosystems. They also influence the process of natural selection. Explain how predators affect the adaptations of their prey. This relationship also works in reverse. How do prey species affect the adaptations of their predators?

7. Competition for a limited quantity of resources occurs in all ecosystems. This competition can be interspecific or intraspecific. Explain some of the ways an organism might deal with these different types of competition.

8. Each year fires burn large tracts of forestland. Describe the process of succession that occurs after a forest fire destroys an existing biological community. Is the composition of the final successional community likely to be the same as that which existed before the fire? What factors might alter the final outcome of the successional process? Why may periodic fire be beneficial to a community?

9. Explain the concept of climax community. Why does the climax community often exhibit a higher level of stability than that found in other successional stages?

10. Discuss the dangers posed to existing community members when new species are introduced into ecosystems. What type of organism would be most likely to survive and cause problems in a new habitat?

▼ Questions for Critical Thinking

1. Ecologists debate whether biological communities have self-sustaining, self-regulating characteristics or are highly variable, accidental assemblages of individually acting species. What outlook or worldview might lead scientists to favor one or the other of these theories?

2. The concepts of natural selection and evolution are central to how most biologists understand and interpret the world, and yet the theory of evolution is contrary to the beliefs of many religious groups. Why do you think this theory is so important to science and so strongly opposed by others? What evidence would be required to convince opponents of evolution?

3. What is the difference between saying that a duck has webbed feet because it needs them to swim and saying that a duck is able to swim because it has webbed feet?

4. The concept of keystone species is controversial among ecologists because most organisms are highly interdependent. If each of the trophic levels is dependent on all the others, how can we say one is most important? Choose an ecosystem with which you are familiar and ask whether it has a keystone species or keystone set.

5. Some scientists look at the boundary between two biological communities and see a sharp dividing line. Others looking at the same boundary see a gradual transition with much intermixing of species and many interactions between communities. Why such different interpretations of the same landscape?

6. The absence of certain lichens is used as an indicator of air pollution in remote areas such as national parks. How can we be sure that air pollution is really responsible? What evidence would be convincing?

7. We tend to regard generalists or "weedy" species as less interesting and less valuable than rare and highly specialized endemic species. What values or assumptions underlie this attitude?

8. What part of this chapter do you think is most likely to be challenged or modified in the future by new evidence or new interpretations?

▼ Key Terms

abundance 82	edge effects 85
Batesian mimicry 81	equilibrium communities 88
climax community 88	evolution 74
commensalism 80	fire-climax communities 88
complexity 83	habitat 73
disclimax communities 88	interspecific competition 78
diversity 82	intraspecific competition 78
ecological development 87	keystone species 76
ecological niche 73	Muellerian mimicry 81
ecotone 86	mutualism 80

▼ Additional Information on the Internet

Animal and Ecosystem Related Links
http://www.cs.uidaho.edu/~connie/interests-wildlife.html/

Biodiversity and Ecosystems Network
http://straylight.tamu.edu/bene/bene.html/

Enter Evolution: Theory & History
http://UCMP1.berkeley.edu/exhibittext/evolution.html/

Introduction to Evolutionary Biology
http://rumba.ics.uci.edu:8080/faqs/faq-intro-to-biology.html/

List of WWW Sites of Interest to Ecologists
http://biomserv.univ-lyon1.fr/Bota.html/

The Origin of the Species
http://www.wonderland.org/works/Charles-Darwin/origin/

Study Guide for Ecology
http://www.nl.net/~paideia/natsci.html/

The Tree of Life
http://phylogeny.arizona.edu/tree/phylogeny.html/

Collection of World Wide Web Sites of Biological Interest
http://www.abc.hu:80/biosites.html/

Biosciences
http://golgi.harvard.edu:80/biopages.html/

Galaxy Einet Sites for Biology
http://galaxy.einet.net/galaxy/Science/Biology.html/
Tradewave Galaxy is copyright 1993, 1994, 1995, 1996 Enterprise
Integration Network Corporation (Tradewave). All Rights Reserved.
Tradewave is a trademark of Tradewave, Inc.

Habitat Ecology Home Page
http://biome.bio.ns.ca/

Note: Further readings appropriate to this chapter are listed
on p. 594.

Volunteers lay willow branches and brush matting to stabilize dunes and restore beach vegetation.

CHAPTER 5: Biomes, Landscapes, Restoration, and Management

Objectives

After studying this chapter, you should be able to:

- recognize the characteristics of major aquatic and terrestrial biomes and understand the most important factors that determine the distribution of each type.

- describe ways in which humans disrupt or damage each of these ecosystem types.

- summarize the overall patterns of human disturbance of world biomes as well as some specific, important examples of losses obscured by broad aggregate categories.

- explain the principles and practices of landscape ecology and ecosystem management.

- evaluate the pros and cons of restoring, replacing, or substituting ecosystems and resources for those we have damaged.

Integrity, Stability, and Beauty of the Land

In 1935, pioneering wildlife ecologist Aldo Leopold bought 32 ha (80 ac) of worn out, sandy farmland on the banks of the Wisconsin River not far from his home in Madison. Originally intended to be merely a hunting camp, the farm quickly became a year-round retreat from the city, as well as a laboratory in which Leopold could test his theories about conservation, environmental ethics, and ecologically-based land management. A dilapidated chicken shack, the only remaining building from the original farm, was remodeled into a rustic cabin (fig. 5.1). The whole Leopold family participated in tree planting, bird watching, gardening, and exploring nature.

The old farm was not pristine wilderness nor were the Leopolds merely spectators. They regarded themselves as participating citizens of the land community, seeking to restore it to ecological health and beauty. Planting as many as 6000 trees and bushes each spring, they practiced "wild husbandry," using axes and shovels to reverse the abuses of previous owners and to revitalize the land through active management, care, and understanding. "Conservation," Leopold wrote, "is the positive exercise of skill and insight, not merely a negative exercise of abstinence or caution."

While building this relationship with the land—by which he meant all the plants and animals as well as the non-living components of the landscape—Leopold mused on the ethics and meaning of conservation and the proper role of humans in nature. The first part of his much-beloved Sand County Almanac is a collection of essays about experiments and experiences at the farm. All of us, he claimed, should choose a piece of land on which we can practice stewardship and develop a sense of place. It doesn't have to be a beautiful place. In fact, it might be best to adopt a weedy, unwanted patch that needs our love and care. Both we and the land benefit from such connectedness, he maintained.

Figure 5.1
Aldo Leopold's Sand County farm in central Wisconsin served as a refuge from the city and a laboratory to test theories about land conservation, environmental ethics, and ecologically-based land management.

Leopold's essay on "The Land Ethic" is a cornerstone of the conservation movement and one of the most eloquent statements of environmental philosophy in American nature writing. In it, Leopold wrote, "We abuse the land because we regard it as a commodity belonging to us. When we see land as a community to which we belong, we may begin to use it with love and respect. . . A land ethic, then, reflects the existence of an ecological conscience, and this in turn reflects a conviction of individual responsibility for the health of the land. Health is the capacity of the land for self-renewal. Conservation is our effort to understand and preserve this capacity. . . . A thing is right when it preserves the integrity, stability, and beauty of the biotic community. It is wrong when it does otherwise."

The story of Leopold's Sand County farm embodies many of the themes of this chapter. First, we survey some of the major biomes or biological communities, around the world as a measure of what the components of a healthy landscape might be. Then we examine what the relatively new fields of landscape and restoration ecology say about our environment and the roles we might play in it. Finally, we look at the emerging goals of ecosystem management and some of the controversies and questions that have persisted from Aldo Leopold's day to our own about how we can and should care for nature. ▶

Terrestrial Biomes

Many places on the earth share similar climatic, topographic, and soil conditions, and roughly comparable communities have developed in response to analogous conditions in widely separated locations. These broad types of biological communities are called **biomes.** Although, as we pointed out in previous chapters, there can be considerable variation in the individual species that make up biological communities at two similar sites or even at a single site over time, some broad landscape categories exist. Recognizing these categories gives us insights about the general kinds of plants and animals that we might expect to find there and how they may be adapted to their particular environment.

Temperature and precipitation are among the most important determinants in biome distribution. If we know the general temperature range and precipitation level, we can predict what kind of biological community is likely to develop on a particular site if that site is free of disturbance for a sufficient time (fig. 5.2). Biome distribution also is influenced by the prevailing landforms of an area. Mountains, in particular, exert major influences on biological communities.

Figure 5.3 shows the distribution of major terrestrial biomes around the world. Because of its broad scope, this map ignores the many variations present within each major category. Most terrestrial biomes are identified by the dominant plants of their communities (for example, grassland or deciduous forest). The characteristic diversity of animal life and smaller plant forms within each biome is, in turn, influenced both by the physical conditions and the dominant vegetation.

Deserts

Deserts are characterized by low moisture levels and precipitation that is both infrequent and unpredictable from year to year. With little moisture to absorb and store heat, daily and seasonal temperatures can fluctuate widely. Deserts that have less than 2.5 cm (1 in) of measurable precipitation support almost no vegetation. Deserts with 2.5 to 5 cm (1 to 2 in) annual precipitation have sparse vegetation (less than 10 percent of the ground is covered), and plants in this harsh climate need a variety of specializations to conserve water and protect tissues from predation. Seasonal leaf production, water-storage tissues, and thick epidermal layers help reduce water loss. Spines and thorns discourage predators while also providing shade (fig. 5.4).

Warm, dry, descending air (see chapter 17) creates broad desert bands in continental interiors at about 30° north in the American Southwest, North and South Africa, China, and Australia. These descending air currents also help create desert strips along the west coast of South America and Africa that are among the driest regions in the world. Although we think of deserts as hot, barren, and filled with sand dunes, those at high latitudes or high elevations are often cool or even cold and sand dunes are actually rather rare away from coastal areas. Most deserts around the world are gravelly or rocky scrubland, where 5 to 10 cm (2 to 4 in) of annual precipitation supports a sparse but often species-rich community dominated by shrubs or small trees. In rare years when winter rains are adequate, a miraculous profusion of spring ephemerals can carpet the desert with flowers.

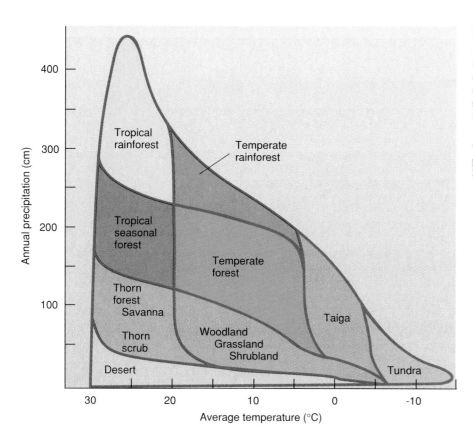

Figure 5.2

Biomes most likely to occur in the absence of human disturbance or other disruptions, according to average annual temperature and precipitation. *Note:* this diagram does not consider soil type, topography, wind speed, or other important environmental factors. Still, it is a useful general guideline for biome location.

Whittaker, R. H., *Communities and Ecosystems*, 2/e, © 1975, p. 167. Adapted with permission of Prentice-Hall, Upper Saddle River, New Jersey.

Animals of the desert have both structural and behavioral adaptations to meet their three most critical needs: food, water, and heat survival. Most desert animals escape the main onslaught of daytime heat by hiding in burrows or rocky shelters from which they emerge only at night. Pocket mice and kangaroo rats (and their Old World counterparts, gerbils) get most of the moisture they need from the seeds and grains they eat. They have many adaptations to conserve water such as producing highly concentrated urine and nearly dry feces that allow them to eliminate body wastes without losing precious moisture.

Deserts may seem formidable, but they are more vulnerable than you might imagine. Desert soils are easily disturbed by human activities and are slow to recover because the harsh desert climate severely reduces the ability of desert communities to recover from damage. Tracks left by army tanks practicing in the California desert during World War II, for instance, are still visible.

Many dry areas have been overgrazed, mainly by domestic livestock. Other areas are being converted to agricultural land in spite of uncertain future availability of water to sustain crops. Often after humans have degraded and abandoned desert areas they remain wastelands unsuitable for native vegetation or wildlife for a very long time.

Grasslands: Prairies and Savannas

The moderately dry continental climates of the Great Plains of central North America, the broad Russian steppes, the African veldt, and the South American pampas support **grasslands,** rich biologi-

cal communities of grasses, seasonal herbaceous flowering plants, and open savannas (fig. 5.5). Seasonal cycles for temperature and precipitation contribute to abundant vegetative growth that both protects and enriches the soils of these prairies and plains, making them among the richest farmlands in the world.

Grasslands have few trees because inadequate rainfall, large daily and seasonal temperature ranges, and frequent grass fires kill woody seedlings. Exceptions are the narrow gallery forests that form corridors along rivers and streams through the grasslands.

In many parts of the world, grasslands are artificially created or maintained by native people using fire. They do this to improve hunting and ease of travel.

In contrast, fire suppression and conversion of the rich prairie soil to farmland have greatly reduced native grasslands elsewhere. The most productive North American prairie lands now grow wheat, corn, oats, sunflowers, flax, and other cultivated crops. Lands that are less suited for crops because of limited water availability are used mainly as rangeland, where overgrazing reduces natural diversity of the community and contributes to soil erosion.

Before humans converted North America's grasslands to croplands and rangelands, they contained herds of wildlife that rivaled those of Africa's Serengeti Plain. Vast herds of bison roamed the plains along with wolves, deer, elk, and pronghorn antelopes. Shorebirds and migratory waterfowl occupied the millions of ponds, potholes, and marshy spots that once dotted our grasslands. Hunting, fencing, wetland drainage, introduction of alien species, and other human modifications to the land have greatly diminished most wildlife populations since the last century.

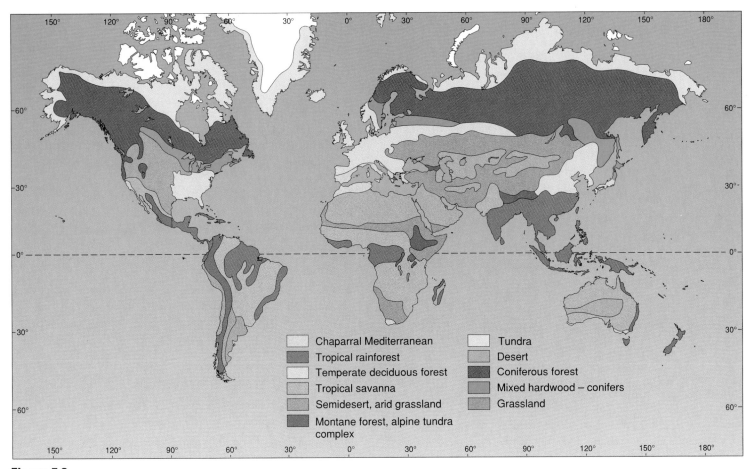

Figure 5.3

This generalized map of world biomes gives an overview of the vegetation types on different landmasses. At this scale, much detail cannot be shown, so such maps should be interpreted with that limitation in mind.

Legend:
- Chaparral Mediterranean
- Tropical rainforest
- Temperate deciduous forest
- Tropical savanna
- Semidesert, arid grassland
- Montane forest, alpine tundra complex
- Tundra
- Desert
- Coniferous forest
- Mixed hardwood – conifers
- Grassland

Figure 5.4

Joshua "trees" in the new Joshua Tree National Park are really large yuccas—members of the lily family. Like other plants in this hot, dry, rocky landscape, they are adapted to conserve water and repel enemies.

Figure 5.5

Grasslands and open woodlands like this one in Kenya develop in moderately dry, highly variable climates. Prairies may seem uniform and monotonous to the untrained eye, but they can be as diverse and productive as forests under optimum climate conditions.

Tundra

Climates in high mountain areas or at far northern or southern latitudes often are too harsh for trees. This treeless landscape, called **tundra,** is characterized by a very short growing season, cold, harsh winters, and the potential for frost any month of the year. Although water may be abundant on the tundra, for much of the year it is locked up in ice or snow and therefore unavailable to plants. As far as plants are concerned, the tundra is a very cold desert.

The *arctic* tundra is a biome of low productivity, low diversity, and low resilience. Winters are long and dark. Only the top several centimeters of the soil thaw out in the summer, and the lower soil is permafrost. This permanently frozen layer prevents snowmelt water from being absorbed into the soil, so the surface soil is waterlogged during the summer. Try to imagine the difficulties encountered by plants in this kind of soil. Most of the year it is completely frozen, and even during the brief growing season, the permafrost is an impenetrable barrier to deep root growth. In addition, the top layer buckles and heaves in response to cycles of freezing and thawing, toppling plants and disrupting root systems.

The *alpine* tundra (fig. 5.6) differs from the arctic tundra in several ways. Plants of the alpine tundra face different challenges than those of the arctic. The thin mountain air permits intense solar bombardment, especially by ultraviolet radiation; thus, many alpine plants have deep pigmentation that shields their inner cells. The glaring summer sun also causes very hot daytime ground temperatures, even though the night temperatures may return to freez-

ing. Alpine soil is windswept and often gravelly or rocky. The sloping terrain causes moisture to drain quickly. Due to this combination of sun, soil, slope, and air currents, drought is a problem—as opposed to the wet conditions in the arctic tundra.

Although the tundra may swarm with life during the brief summer growing season, only a few species are able to survive the harsh winters or to migrate to warmer climates. Dominant tundra plants are dwarf shrubs, sedges, grasses, mosses, and lichens. Its larger life-forms, such as arctic musk ox and caribou, or alpine mountain goats and mountain sheep, must be adapted to survive the harsh climate and sparse food supply. Many animals migrate or hibernate during winter. Flocks of migratory birds nest on the abundant summer arctic wetlands, which also nurture hordes of bloodsucking insects that feed upon the summer flocks and herds (and tourists!).

Damage to the tundra is slow to heal. At present, the greatest threat to this distinctive biome is oil and natural gas wells in the Arctic and mineral excavation in mountain regions. Because plants grow slowly during the brief summer at high altitudes or latitudes, truck ruts and bulldozer tracks on the tundra landscapes may take centuries to heal. Furthermore, some of the most promising sites for oil exploration or mining are summer feeding and breeding grounds for animals such as caribou, grizzly bears, or mountain sheep. Even seemingly minor incursions into this fragile ecosystem can cause great damage.

Conifer Forests

Several distinctive biomes are dominated by **conifer** (cone-bearing) trees. Where moisture is limited by sandy soil, low precipitation, or a short growing season, plants reduced water loss by evolving thin, needlelike evergreen leaves with a thick waxy coating. Although not as efficient in carrying out photosynthesis during summer months as the broad, soft leaves of deciduous trees, conifer needles and scales can survive harsh winters or extended droughts and do accomplish some photosynthesis even under poor conditions. In mountain areas, particularly, fire has been an important and fairly regular factor in maintaining the coniferous forest.

The **boreal forest,** or northern coniferous forest, stretches in a broad band of mixed coniferous and deciduous trees around the world between about 45° and 60° north latitude, depending on altitude and distance from coastlines or major rivers. Among the dominant conifers are pine, hemlock, spruce, cedar, and fir. Some common deciduous trees are birches, aspens, and maples, while mosses and lichens form much of the ground cover. In this moist, cool biome, streams and wetlands abound—especially on recently glaciated landscapes. As a result, there are many lakes, potholes, bogs, and fens (fig. 5.7). Insects that have aquatic stages in their life cycles, such as mosquitoes and biting flies, are particularly abundant, to the consternation of humans as well as wild birds and mammals.

The northernmost edge of the boreal forest is a species-poor black spruce and sphagnum moss (peat moss) woodland—often called by its Russian name, **taiga**—that forms a ragged border with the treeless arctic tundra (fig. 5.8). The harsh climate limits both productivity and resilience of the taiga community. Cold temperatures, very wet soil during the growing season, and acids produced

Figure 5.6
The harsh physical conditions of alpine tundra ecosystems are the result of altitude and slope. The growing season is short, ultraviolet radiation in the thin air is greater than at lower altitudes, and extreme temperature fluctuations are possible even during the summer. Air currents and the thin, rocky soil also contribute to the arid conditions of alpine tundra.

Figure 5.7

Mixed conifer and deciduous forests like this one in northern Minnesota are characteristic of the Great Lakes Region. Although other species assemblages are found elsewhere, temperate cool forests around the world share many common characteristics.

Figure 5.8

This boreal forest in Alaska's Chugach Mountains has small, widely spaced black spruce intermixed with willows and heather on a wet peatland.

by fallen conifer needles and sphagnum inhibit full decay of organic matter. As a result, thick layers of semidecayed organic material, called peat, form. Boreal peat deposits are being explored as energy sources; however, the environmental disturbance caused by peat mining in this boreal community could be severe and long-lasting, perhaps even permanent.

The southern pine forest, another kind of U.S. coniferous forest ecosystem, is characterized by a warm, moist climate and sandy soil and, in the past, was subjected to frequent fires. Now it is managed extensively for timber and such resinous products as turpentine and rosin. The undergrowth includes saw palmetto and various thorny bushes.

The coniferous forests of the Pacific coast represent yet another special set of environmental circumstances. Mild temperatures and abundant precipitation—up to 250 cm (100 in) per year—result in luxuriant plant growth and huge trees such as the California redwood, the largest tree in the world and the largest organism of any kind known to have ever existed. Redwoods formerly were distributed over much of the Washington, Oregon, and California coast, but their distribution has been greatly reduced by logging without regard to sustainable yield or restoration.

In its wettest parts, the coastal forest becomes a **temperate rainforest,** a cool, rainy forest that often is enshrouded in fog. Condensation from the canopy (leaf drip) becomes a major form of precipitation, and annual precipitation exceeds 250 cm (100 in) in some places. Mosses, lichens, and ferns cover tree branches, old stumps, and the forest floor itself. The largest, best-preserved area of temperate rainforest is Olympic National Park in Washington.

Broad-Leaved Deciduous and Evergreen Forests

Forests of broad-leaved trees occur throughout the world where rainfall is plentiful. In temperate regions, the climate supports lush summer plant growth when water is plentiful but requires survival

adaptations for the frozen season. A key adaptation of **deciduous** trees is the ability to produce summer leaves and then shed them at the end of the growing season. This rich and varied biome contains associations of many tree species, including oak, maple, birch, beech, elm, ash, and other hardwoods. These tall trees form a forest canopy over a diverse understory of smaller trees, shrubs, and herbaceous plants, including many annual spring flowers that grow, flower, set seed, and store carbohydrates before they are shaded by the canopy (fig. 5.9). Where the climate is warm year-round, forests are dominated by evergreen trees such as the live oaks and cypresses of the southern United States.

Most of the dense forest that once covered Central Europe was cleared a thousand years ago. When European settlers first came to North America, dense forests of broad-leaved deciduous trees covered most of the eastern half of what is now the United States. Much of that original deciduous forest was harvested for timber a century or more ago. Now, large areas of forest in the eastern U.S. have regrown and are approaching old-growth status, although with a different species composition than the original biome. Vast original deciduous forests remain in eastern Siberia, but they are being harvested at a rapid rate, perhaps the highest deforestation rate in the world. As the forests disappear, so too do Siberian tigers, bears, cranes, and a host of other unique and endangered species.

Tropical Moist Forests

The humid tropical regions of South and Central America, Africa, Southeast Asia, and some of the Pacific Islands support one of the most complex and biologically rich biome types in the world (fig. 5.10). Although there are several kinds of moist tropical forests, they share common attributes of ample rainfall and uniform temperatures. Cool **cloud forests** are found high in the mountains where fog and mist keep vegetation wet all the time. **Tropical rainforests** occur where rainfall is abundant—more than 200 cm (80 in) per year—and temperatures are warm to hot year-round.

Figure 5.9

The deciduous forests of the eastern United States are stratified assemblages of tall trees, understory shrubs, and small, lovely, ground-covering species. A rich fauna is supported by the variety of niche opportunities that are present.

Figure 5.10

Tropical rainforests, like this one in Java, support a luxuriant profusion of species and biomass. Notice the many vines and epiphytes clinging to the buttressed trunk of this giant forest tree.

The soil of both these tropical moist forest types tends to be old, thin, acidic, and nutrient-poor, yet the number of species present can be mind-boggling. For example, the number of insect species found in the canopy of tropical rainforests has been estimated to be in the millions! It is estimated that one-half to two-thirds of all species of terrestrial plants and insects live in tropical forests.

The nutrient cycles of these forests also are unique. Almost all (90 percent) of the nutrients in the system are contained in the bodies of the living organisms. This is a striking contrast to temperate forests, where nutrients are held within the soil and made available for new plant growth. The luxuriant growth in tropical rainforests depends on rapid decomposition and recycling of dead organic material. Leaves and branches that fall to the forest floor decay and are incorporated almost immediately back into living biomass.

When the forest is removed for logging, agriculture, and mineral extraction, the thin soil cannot support continued cropping and cannot resist erosion from the abundant rains. And if the cleared area is too extensive, it cannot be repopulated by the rainforest community. Rapid deforestation is occurring in many tropical areas as people move into the forests to establish farms and ranches, but the land soon loses its fertility.

Tropical Seasonal Forests

Many areas in India, Southeast Asia, Australia, West Africa, the West Indies, and South America have tropical regions characterized by distinct wet and dry seasons instead of uniform heavy rainfall throughout the year, although temperatures are hot year-round. These areas have produced communities of **tropical seasonal forests:** semievergreen or partly deciduous forests tending toward open woodlands and grassy savannas dotted with scattered, drought-resistant tree species.

Tropical dry forests have typically been more attractive than wet forests for human habitation and have suffered greater degradation. Clearing a dry forest with fire is relatively easy during the dry season. Soils of dry forests often have higher nutrient levels and are more agriculturally productive than those of a rainforest. Finally, having fewer insects, parasites, and fungal diseases than a wet forest makes a dry or seasonal forest a healthier place for humans to live. Consequently, these forests are highly endangered in many places. Less than 1 percent of the dry tropical forests of the Pacific coast of Central America or the Atlantic coast of South America, for instance, remain in an undisturbed state.

Aquatic Ecosystems

Water is essential to life in every biological community and is present to varying degrees in all of them. Although freshwater ecosystems and wetlands are often parts of other, larger biomes, they are so important biologically that they deserve individual attention here.

Freshwater and Saline Ecosystems

Freshwater ecosystems include the standing waters of ponds and lakes as well as the flowing waters of rivers and streams. There also are some unique freshwater ecosystems, including underground rivers and subterranean pools. Freshwater ecosystems are as varied as their individual sites because they are influenced not only by characteristics of local climate, soil, and resident communities but also by the surrounding terrestrial ecosystems and anything that happens uphill or upstream from them (fig. 5.11). As with terrestrial ecosystems, the biological communities of freshwater ecosystems are largely determined by the physical characteristics of the environment, except that the surrounding medium is water instead of air.

Aquatic organisms have the same basic needs as terrestrial organisms: carbon dioxide, water, and sunlight for photosynthesis; oxygen for respiration; and food and mineral nutrients for energy, growth, and maintenance. The availability of these necessities is influenced by such site characteristics as (1) substances that are dissolved in the water, such as oxygen, nitrates, phosphates, potassium compounds, and other by-products of agriculture and industry; (2) suspended matter, such as silt and microscopic algae, that affect water clarity and, therefore, light penetration; (3) depth; (4) temperature; (5) rate of flow; (6) bottom characteristics (muddy, sandy, rocky); (7) internal convective currents; and (8) connection to, or isolation from, other aquatic ecosystems.

Vertical stratification is an important aspect of standing water ecosystems, especially in regard to gradients of light, temperature, nutrients, and oxygen. Organisms tend to form distinctive vertical

subcommunities in response to this stratification of physical factors. The plankton subcommunity consists mainly of microscopic plants, animals, and protists (single-celled organisms such as amoebae) that float freely within the water column. Some non-planktonic organisms are specialized to live at the air-water interface (for example, insects such as water striders), and still others are able to swim freely in the open waters (such as fish).

Finally, many animal species are bottom dwellers (such as snails, burrowing worms and insect larvae, bacteria) that make up the **benthos,** or bottom subcommunity. Oxygen levels are lowest in the benthic environment. Anaerobic bacteria can live in the very low-oxygen bottom sediments. Emergent plants such as cattails and rushes are rooted in bottom sediments of the littoral (near shore) zone but spread their leaves above the water surface. They create important structural and functional links between strata of the ecosystem and often are the greatest producers of net primary productivity in the ecosystem.

Deeper lakes are characterized by the presence of a warmer, upper layer that is mixed by the wind (the epilimnion; *epi* = upon) and a colder, deep layer that is not mixed (the hypolimnion; *hypo* = below) (fig. 5.12). The two layers are separated by a distinctive temperature transition zone called the **thermocline,** or metalimnion.

Lakes and ponds have a tendency to undergo succession, which includes changes in the biological community as a response to increases in nutrient levels. Human sources of nutrient input can radically increase the rate at which a lake "ages," as discussed in chapter 20 (eutrophication).

Humans utilize freshwater communities as sources of food and for recreation, waste disposal, cooling of nuclear power plants, and many other industrial uses. Waterways also are major transport lanes for barge and ship traffic, involving channel dredging and contamination from fuel and cargo leaks. Dams, canals, and other diversions have significant effects on freshwater resources. In these and other ways, humans have an enormous impact on the character of individual freshwater ecosystems and their biological communities.

Not all freshwater lakes contain what we would consider "fresh" water. The Caspian Sea, Dead Sea, Great Salt Lake, and numerous

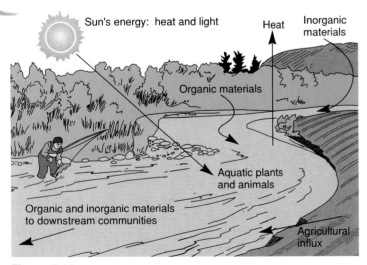

Figure 5.11

The character of freshwater ecosystems is greatly influenced by the immediately surrounding terrestrial ecosystem, and even by ecosystems far upstream or far uphill from a particular site.

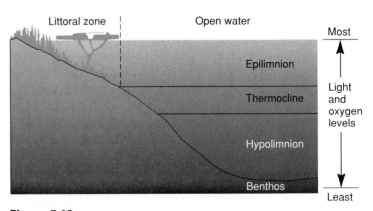

Figure 5.12

The layers of a deep lake are determined mainly by gradients of light, oxygen, and temperature. The epilimnion is affected by surface mixing from wind and thermal convections, while mixing between the hypolimnion and epilimnion is inhibited by a sharp temperature and density difference at the thermocline.

other saline (salty) lakes have been formed by evaporative shrinkage of large bodies of water. Deserts, particularly, may have alkaline or arsenic-containing lakes and potholes. Still other salty and mineral-rich ponds and small lakes are fed by mineral hot springs.

Estuaries and Wetlands: Transitional Communities

Estuaries are bays or semi-enclosed bodies of brackish (salty but less so than seawater) water that form where rivers enter the ocean. Estuaries usually contain rich sediments carried downriver, forming shoals and mudflats that nurture a multitude of aquatic life. Estuaries are sheltered from the most drastic ocean action but do experience tidal ebbs and flows. Daily tides may even cause river levels to rise and fall far inland from the river mouth. The combination of physical factors in estuaries makes them very productive and of high species diversity. They are significant "nurseries" for economically important fish, crustaceans (such as crabs and shrimp), and mollusks (such as clams, cockles, and oysters).

Where the continental shelf is broad and shallow, an extensive fan-shaped sediment deposit called a **delta** may form at the river mouth. Deltas often are channeled by branches of the river, creating extensive coastal wetlands that are part of the larger estuarine zone.

Wetlands of several types form near bodies of water. **Wetlands** are ecosystems in which the land surface is saturated or covered with standing water at least part of the year and vegetation is adapted for growth under saturated conditions. There are special names for specific kinds of wetlands, but we can group them into three major categories: swamps, marshes, and bogs and fens. Defined pragmatically, **swamps** are wetlands with trees (fig. 5.13); **marshes** are wetlands without trees (fig. 5.14); and **bogs** and **fens** are areas that

Figure 5.13

Biological communities, such as the Okefenokee Swamp in Georgia, can be amazingly diverse, complex, and productive. Myriad life-forms coexist in exuberant abundance, interlinked by manifold relationships. Who lives where, and why, are central ecological questions.

Figure 5.14

Coastal wetlands such as this salt marsh are highly productive ecosystems and are vitally important for a wide variety of terrestrial and aquatic species.

may or may not have trees, and have waterlogged soils that tend to accumulate peat. Swamps and marshes tend to be associated with flowing water. Fens are fed by groundwater and surface runoff, whereas bogs are fed solely by precipitation. Swamps and marshes generally have high productivity, while bogs and fens have low productivity.

The water in marshes and swamps usually is shallow enough to allow full penetration of sunlight and seasonal warming. These mild conditions favor great photosynthetic activity, resulting in high productivity at all trophic levels. In short, life is abundant and varied. Wetlands are major breeding, nesting, and migration staging areas for waterfowl and shorebirds.

Wetlands perform major ecosystem services, the importance of which cannot be overstated. As mentioned previously, they support a great diversity of life-forms. What may be less obvious is their role in planetary water relationships. Wetlands act as traps and filters for water that moves through them. Runoff water is slowed as it passes through the shallow, plant-filled areas, reducing flooding. As a result, sediments are deposited in the wetlands instead of traveling into rivers and, eventually, the oceans. In this way, wetlands both clarify surface waters and aid in the accumulation and formation of fertile land. Furthermore, chemical interactions in wetland ecosystems neutralize and detoxify substances in the water. Finally, water in wetlands seeps into the ground, helping to replenish underground water reservoirs called aquifers (see chapter 19).

Wetlands convert naturally to terrestrial communities largely through sedimentation, eutrophication, or stream cutting and draining. Human activities have greatly accelerated these processes in many places, and wetlands are being destroyed or degraded around the world at a disturbing rate (see chapter 15). This destruction is of great concern because it means loss of ecological services to the biosphere as well as loss of essential habitats for a myriad of species.

Shorelines and Barrier Islands

Ocean shorelines, including rocky coasts, sandy beaches, and offshore barrier islands, are particularly rich in life-forms. Rocky shorelines, in particular, support an incredible density and diversity of organisms that grow attached to any solid substrate, including each other. Sandy shorelines, on the other hand, provide homes for organisms that live among the sand grains and in burrows.

Sandy beaches are understandably popular to humans. Cities, resorts, and residences are built on beaches. Grasses and trees that hold the dunes are destroyed, destabilizing the soil system and thus increasing its susceptibility to wind and wave erosion. The constant battle to maintain such real estate is costly. Insurance is very expensive because of the hazards of natural erosion and storms. Building and maintaining sea walls and protecting and replenishing beaches are expensive, continuous processes. Protective structures often have their own effect on shore recontouring and may even increase the rate of natural shoreline loss.

Barrier islands are low, narrow, sandy islands that form offshore from a coastline. In North America, they are particularly characteristic of the Atlantic and Gulf coasts (fig. 5.15). Barrier islands protect inland shores from the onslaught of the surf, especially during severe storms. Because they are so lovely, barrier islands have been tempting targets for real estate development and about 20 percent of the barrier island surface in the United States has been developed. Unfortunately, human occupation of barrier islands often destroys the values that attract us there in the first place. Walking or driving vehicles over dune grass destroys stabilizing vegetation cover and triggers erosion. Cutting roads through dunes and building houses further destabilizes islands. A single, major storm can do massive damage, destroying houses and perhaps washing away most of an island.

Ignoring these environmental realities endangers many rare species and puts human lives and economic investment at risk.

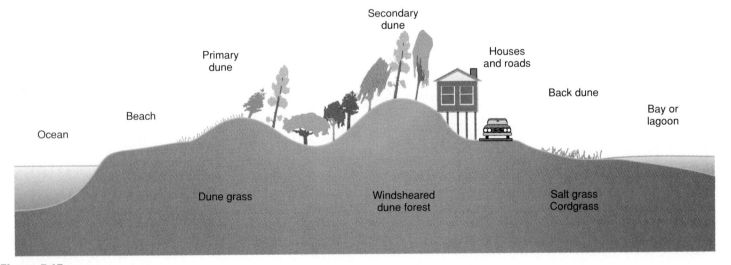

Figure 5.15

A cross section through a barrier island. Fragile dune grass communities on the primary dune must be protected to ensure stability of the dune complex. Roads and houses should be limited to the back dune area, both for safety and to relieve pressure on vulnerable dune vegetation. Only a few narrow footpaths should cross the dunes to the beach where recreational use is permitted.

Although we now have coastal zoning and land use regulations, we still have many environmentally damaging developments in place.

Coral reefs form in clear, warm, tropical seas and are particularly well-developed in the south Pacific (fig. 5.16). They are the accumulated calcareous skeletons of innumerable tiny colonial animals called corals. Each of the interconnected coral animals builds a calcium carbonate chamber on the surface of the accumulated secretions of previous generations of animals. Coral reefs usually form along the edges of shallow, submerged banks or shelves. The depth at which they form is limited by the depth of light penetration, due at least in part to the presence of symbiotic photosynthetic protists in their tissues. Because of the photo-

synthetic relationship, the presence of dead versus living coral growths can be an index of either previous ocean levels or decreased light penetration due to increased turbidity. Coral reef communities rival tropical forest communities in species diversity, numbers of individuals, brilliance of color, and interesting forms of both plants and animals.

Reefs also are among the most endangered biological communities on the earth. Destructive fishing practices (using dynamite or cyanide to stun fish, for instance) and harvesting of coral for building or the pet trade have damaged or destroyed about three-fourths of all reefs in the world. In the next section, we will discuss some other types of human disturbance of natural habitats.

Figure 5.16

Coral reefs harbor some of the most diverse biological communities on the earth, rivaling tropical rainforests in species diversity, and surpassing any biome in the number of phylla (broad taxonomic groups) represented. Surprisingly, almost all the organisms you see here are animals.

Human Disturbance

Humans have become dominant organisms over most of the earth, damaging or disturbing more than half of the world's terrestrial ecosystems to some extent (fig. 5.17). By some estimates, humans preempt about 40 percent of the net terrestrial primary productivity of the biosphere either by consuming it directly, by interfering with its production or use, or by altering the species composition or physical processes of human-dominated ecosystems. Conversion of natural habitat to human uses is the largest single cause of biodiversity losses.

Researchers from the environmental group Conservation International have attempted to map the extent of human disturbance of the natural world (fig. 5.17). The greatest impacts have been in Europe, parts of Asia, North and Central America, and islands such as Madagascar, New Zealand, Java, Sumatra, and those in the Caribbean. Data from this study are shown in table 5.1.

Temperate broad-leaved forests are the most completely human-dominated of any major biome. The climate and soils that support such forests are especially congenial for human occupation. In eastern North America or most of Europe, for example, only remnants of the original forest still persist. Chaparral and thorn scrub grow in areas where summers are hot and dry and winters are mild and moist, often referred to as a Mediterranean climate. This climate often gives rise to high levels of biodiversity. Some of the world's hot spots of endemic species (those restricted to a single area) and biodiversity occur in Mediterranean climate areas such as southern Europe and North Africa, parts of California and Chile, and the Cape region of South Africa. People also like to live in these areas, seriously threatening much of the areas' biodiversity (fig. 5.18).

Temperate grasslands, temperate rainforests, tropical dry forests, and many islands also have been highly disturbed by human activities. If you have traveled through American cornbelt states such as Iowa or Illinois, you have seen how thoroughly former prairies have been converted to farmlands. Intensive cultivation of this land exposes the soil to erosion and fertility losses (see chapter 11). Islands, because of their isolation, often have high numbers of endemic species. Many islands, such as Madagascar, Haiti, and Java have lost more than 99 percent of their original land cover.

Tundra and Arctic deserts are the least disturbed biomes in the world. Harsh climates and unproductive soils make these areas unattractive places to live for most people. Temperate conifer forests also generally are lightly populated and large areas remain in a relatively natural state. However, recent expansion of forest harvesting in Canada and Siberia may threaten the integrity of this biome.

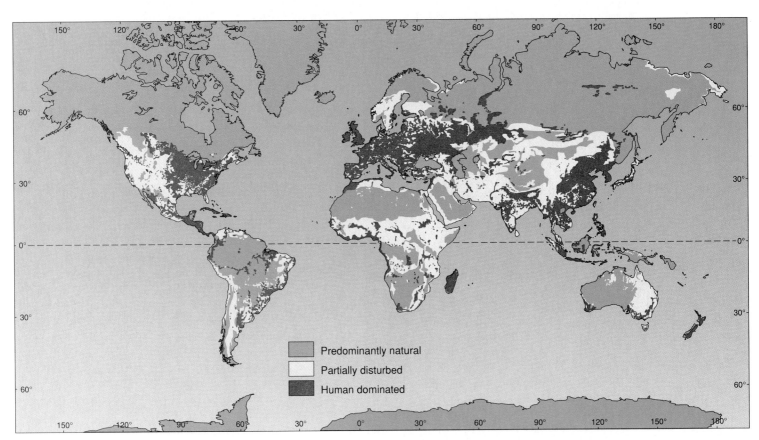

Figure 5.17

A human-disturbance map compares predominantly natural areas with those that are partially disturbed or human dominated.

Source: Data from Lee Hannah and David Lohse, *1993 Annual Report,* Conservation International, Washington, DC.

Table 5.1
Human disturbance

Biome	Total Area (10^6 Km^2)	% Undisturbed Habitat	% Human Dominated
Temperate broadleaved forests	9.5	6.1	81.9
Chaparral and thorn scrub	6.6	6.4	67.8
Temperate grasslands	12.1	27.6	40.4
Temperate rainforests	4.2	33.0	46.1
Tropical dry forests	19.5	30.5	45.9
Mixed mountain systems	12.1	29.3	25.6
Mixed island systems	3.2	46.6	41.8
Cold deserts/semi-deserts	10.9	45.4	8.5
Warm deserts/semi-deserts	29.2	55.8	12.2
Moist tropical forests	11.8	63.2	24.9
Tropical grasslands	4.8	74.0	4.7
Temperate conifer forests	18.8	81.7	11.8
Tundra and arctic desert	20.6	99.3	0.3

Note: where undisturbed and human dominated areas do not add up to 100 percent, the difference represents partially disturbed lands.

Source: Hannah, Lee, et al., 1995, "Human Disturbance and Natural Habitat: A Biome Level Analysis of a Global Data Set," *Biodiversity and Conservation,* vol. 4:128–155.

Figure 5.18

"There was an environment before human beings and there will be an environment after human beings. So what's the big deal?"

© Peter Kohlsaat/Modern Times Syndicate.

Large expanses of tropical moist forests still remain in the Amazon and Congo basins but in other areas of the tropics such as West Africa, Madagascar, Southeast Asia, and the Indo-Malaysian peninsula and archipelago, these forests are disappearing at a rapid rate (see chapter 14).

As mentioned earlier, wetlands have suffered severe losses in many parts of the world. About half of all original wetlands in the United States have been drained, filled, polluted, or otherwise degraded over the past 250 years. In the prairie states, small potholes and seasonally flooded marshes have been drained and converted to croplands on a wide scale. Iowa, for example, is estimated to have lost 99 percent of its presettlement wetlands. Similarly, California has lost 90 percent of the extensive marshes and deltas that once stretched across its central valley. Wooded swamps and floodplain forests in the southern United States have been widely disrupted by logging and conversion to farmland.

Similar wetland disturbances have occurred in other countries as well. In New Zealand, over 90 percent of natural wetlands have been destroyed since European settlement. In Portugal, some 70 percent of freshwater wetlands and 60 percent of estuarine habitats have been converted to agriculture and industrial areas. In Indonesia, almost all the mangrove swamps that once lined the coasts of Java have been destroyed, while in the Philippines and Thailand, more than two-thirds of coastal mangroves have been cut down for firewood or conversion to shrimp and fish ponds.

Landscape Ecology

So far in this chapter, we have considered only very broad ecological regions in which humans act solely as disturbing agents. This is a very Clemensian view (see chapter 3) of nature as static and deterministic. Most modern scientists see nature as much more diverse, dynamic, and complex than is suggested by this perspective. Modern technology, including remote sensing, powerful computers, and geographical information systems (GIS) allow us to study temporal and spatial complexity, heterogeneity, and dynamic fluxes in ecosystems that would previously have been ignored as random noise or unwelcome complications (fig. 5.19).

Landscape ecology is the study of reciprocal effects of spatial pattern on ecological processes. By reciprocal effects, we mean that complex spatial patterns shape, and are shaped by, the dynamic ecological processes that occur in them. In common usage, a landscape is as much as you can see at one time from a high vantage point. By this definition, it is smaller than a biome but generally larger than a single ecosystem. In landscape ecology, a landscape is a geographical unit with a history that shapes the features of the land and organisms in it as well as our reaction to, and interpretation of, the land.

Landscape ecology considers humans important elements of most landscapes, while other ecologists generally have tried to study areas devoid of human influence. Few places on the earth, however, are devoid of human impacts. Human disturbances include hunting, timber harvesting, farming, grazing, pollution, introduction of exotic species, and alteration of atmospheric chemistry and, perhaps, global climate. Restricting our definition of "nature" to a few pristine remnants will lead us either to ignore most of the earth's surface or demand that humans remove themselves entirely from large tracts, something not likely in poor countries that need access to resources.

Patchiness and Heterogeneity

The Amazonian rainforest is a good example of the new perspective provided by landscape ecology. Usually the rainforest is depicted as a vast, homogeneous block of forest that has remained stable and unchanging for hundreds of millions of years. A detailed analysis, however, shows that it actually is a complex mosaic of at least 50 distinct habitat types and species assemblages distributed according to soil type, topography, climate, and site history. These patches are in a constant state of change as big trees fall or fires open up new ground for succession.

Landscape ecologists claim that if we look closely, all landscapes consist of similar mosaics of discrete, bounded patches with different biotic or abiotic composition. Often a predominant or continuous cover type acts as a matrix in which other patch types appear to be embedded. Like individual habitats in the rainforest, these patches are dynamic and change components or functions over time. Among the causes of this patchiness are human and natural disturbances as well as underlying ecological factors such as successional processes.

Landscape heterogeneity can exist across a wide range of scale from burned patches measuring thousands of hectares in Yellowstone National Park to the effects of soil crumb size and insect burrows in a few square centimeters of soil. These spatial patterns exist in marine, freshwater, and wetland environments as well as terrestrial habitats, and thus the term "landscape" can apply to aquatic environments as well as dry land. A basic question in landscape ecology is whether a given phenomenon appears or applies

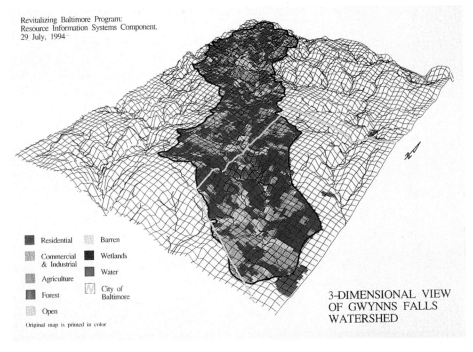

Revitalizing Baltimore Program:
Resource Information Systems Component.
29 July, 1994

Residential Barren
Commercial & Industrial Wetlands
Agriculture Water
Forest City of Baltimore
Open

Original map is printed in color

3-DIMENSIONAL VIEW
OF GWYNNS FALLS
WATERSHED

Figure 5.19

Landscape ecology uses geographic information systems (GIS) to map patch size, type, and configuration as in this 3-D map of the Gwynns Falls watershed near Baltimore, Maryland. This mapping assists planners in analyzing land use patterns.

across many different scales or is restricted to a particular scale. In other words, do patterns emerge, change, or disappear if we look at a landscape up close or from afar?

Landscape Dynamics

Time and space are of special concern in landscape ecology, for that matter, to the rest of us as well. As humorist Garrison Keillor says, "Time exists so everything doesn't happen at once; space exists so everything doesn't happen to us." How does this apply to ecology? A physiological ecologist might want to study the effects of temperature on the rate of photosynthesis in plants. To make the study manageable, she or he might ask, "What is the photosynthetic rate in this particular *plant,* at this *exact* time, at these *specific* temperatures?" A landscape ecologist, in contrast, might ask, "What is the effect of temperature on photosynthesis in this *place,* given its unique combination of history, composition, and characteristics?" Developing a deep relationship with, and understanding of, a particular landscape is sometimes referred to as having a "sense of place."

The boundaries between habitat patches are considered especially significant by landscape ecologists. Edges can induce, inhibit, or regulate movement of materials, energy, or organisms across a landscape. Thus the dynamics between patches may be of greater overall importance than what happens within each patch. As chapter 4 shows, wildlife managers are beginning to be very aware of edge effects as well as the size, shape, and distribution of habitat patches in fragmented landscapes.

This focus on interactions between neighboring communities is quite different from classical ecological focus on the structure and functions of discrete communities, populations, or ecosystems. In its interest in complexity and emergent properties of systems, landscape ecology draws on theories of chaos and complexity from mathematics and physics.

There also are many similarities between landscape ecology and the equally new discipline of conservation biology. Among their shared tenets are: (1) evolutionary change is a central feature of natural systems, (2) non-equilibrium dynamics and uncertainty are more characteristic of nature than are stability and determinism, (3) heterogeneity and diversity are good and ought to be maintained, (4) the context of the surrounding landscape is important, and (5) human needs, desires, abilities, and potential must be considered in efforts to understand and protect nature. All of these points are important issues in the design of nature preserves, parks, and wildlife refuges (chapter 15).

Restoration Ecology

Can the lessons of Aldo Leopold's Sand County farm be applied to disrupted and damaged landscapes elsewhere in the world? Advances in ecological understanding coupled with practical knowledge gained through hands-on field work have given birth to a new discipline of **restoration ecology** that seeks to repair or reconstruct ecosystems damaged by humans or natural forces.

Although land stewardship and husbandry have long been practiced by people in many land-based cultures, the field of ecological restoration suddenly captured public attention in 1988 when a conference called Restoring the Earth, held in Berkeley, California, drew an overflow crowd of more than 800 scientists, policymakers, and activists to share ideas and experiences from restoration projects. There is now a society for, and journal of, restoration ecology. This fledgling science has been given a boost in recent years as courts and regulatory agencies increasingly have required companies and developers to restore or replace illegally damaged wetlands or habitats of endangered species.

Defining Some Terms

Restoration generally means to bring something back to a former condition. Ecological restoration involves active manipulation of nature to re-create species composition and ecosystem processes as close as possible to the state that existed before human disturbance (fig. 5.20). Although there may be many similarities between restoration and conservation, stewardship, or management, the former often entails more direct intervention to achieve a predetermined end than do these other fields.

Rehabilitation refers to attempts to rebuild elements of structure or function in an ecological system without necessarily achieving complete restoration to its original condition. Often rehabilitation means to bring an area back to a useful state for human purposes rather than to a truly natural state. It aims to reverse deterioration of a resource even if it cannot be restored fully.

Remediation is a process of cleaning chemical contaminants from a polluted area by physical or biological methods as a first step towards protecting human and ecosystem health. Incineration, for instance, often is a cost-effected method of cleaning oil-contaminated soils. Volatile organic solvents in aquifers can be

Figure 5.20
Volunteers on a Texas beach stake down discarded Christmas trees to stabilize dunes and shelter beach grasses.

removed by pumping out ground water and aerating it or passing it through absorbent materials.

Living organisms are highly effective cleaning agents for many contaminants. Water hyacinths, for example, have a great capacity for absorbing heavy metals and other toxins from polluted water. Microorganisms (bacteria and fungi) can be found in nature or engineered in the laboratory to destroy many dangerous chlorinated compounds. Chopped-up horseradish roots are very effective in removing phenols from industrial effluents. And common locoweeds can extract selenium from contaminated soils. Sometimes simply adding fertilizer to encourage plant and microbial growth is the best way to clean up surface pollutants.

Reclamation typically is used to describe chemical or physical manipulations carried out in severely degraded sites, such as open-pit mines or large-scale construction. The Surface Mining Control and Reclamation Act, for example, requires mine operators to use reclamation techniques to restore the shape of the land to the original contour and re-vegetate it to minimize impacts on local surface and ground waters.

Historically, reclamation meant irrigation projects that brought wetlands and deserts (considered useless wastelands) into agricultural production. Thus, the Bureau of Reclamation and the Army Corps of Engineers, in the early part of this century dredged, diked, drained, and provided irrigation water to convert millions of acres of wild lands into farmlands. Many of those projects were highly destructive to natural landscapes and we are now using reclamation techniques to restore formerly "reclaimed" land to a more natural state (fig. 5.21).

Re-creation attempts to construct a new biological community on a site so severely disturbed that there is virtually nothing left to restore. The new system may be modeled on what we think was there before human disturbance or it may be something that never existed on that site but that we think suits current conditions. Often developers and government are required to **mitigate** damage caused in one area by re-creating a comparable biological community in another place.

In southern California, for example, part of a saltmarsh near San Diego was destroyed when Interstate 5 was widened where it passed through the Sweetwater Marsh National Wildlife Refuge. In 1984, the courts ordered a mitigation project to replace habitat for the endangered light-footed clapper rail (*Rallus longirostris levipes*). A new wetland was excavated and planted with cordgrass. The grass grew profusely and the marsh appeared to be successful but no rails nested there. It turned out that the sandy soil was nutrient poor and the grass wasn't tall enough to protect nests when the tides rise. Furthermore, after three years, the grass was decimated by an insect outbreak because a key predatory beetle was missing. After ten years of experimentation, the artificial marsh still has no rails.

Many similar experiences make environmentalists suspicious of mitigation projects. "We can't just move ecosystems around like furniture that we put wherever we want," says author Seth Zuckerman. And John Berger, organizer of the first Restoring the Earth Conference, said, "The purpose of restoration ecology is to repair previous damage, not to legitimize further destruction."

Figure 5.21

The naturally meandering Kissimmee River (*right channel*) was straightened and drained by the Army Corps of Engineers (*left*) for flood control thirty years ago. Wetland losses have caused drastic waterfowl reductions and have allowed polluted water to flow into Lake Okeechobee and the Everglades. Now the Corps is attempting to reverse its actions and restore the Kissimmee and its associated wetlands to their original state.

Conflicting Views of Restoration

Controversy has arisen between restorationists and preservationists over both the effectiveness and the ideology of different approaches to protecting nature. Preservationists claim that failures such as the saltmarsh story above show that the best strategy is to avoid destructive projects in the first place. They argue that our faith in science and technology to provide cures for our mistakes is ill-advised. A belief that we can fix up any messes we make just serves as an excuse for business as usual, in their opinion.

Restorationists counter that we are unlikely to preserve more than small areas in pristine parks and nature reserves. While nice, these areas can't solve our global ecological problems, especially in poor countries where immediate survival needs force people to use resources regardless of the impacts on nature. Furthermore, expecting a biological community to remain unchanging is contrary to our modern understanding of ecology. Evolution obviously entails change. If nature is always changing, and that change is one of the essential properties we want to sustain, nature can't simply be locked up in a museum like a piece of art. Restorationists ask, who is to say that the changes we make in restoring or recreating ecosystems is unnatural?

This last point raises questions about the place of humans in nature. Are we members of the biological community or are we separate from it? Should we use our creative energies to try to improve on nature, or should we leave well enough alone? Many ecologists argue that we don't know enough about ecosystems to successfully tinker with them.

Some social scientists, on the other hand, claim that nature is merely a social creation and that we can and should invent ingenious and imaginative new forms of nature to achieve ecological and social goals. Humans have been, are now, and will continue to

Figure 5.22

In 1935, workers from the Civilian Conservation Corps dig up old farm fields in preparation for planting the native grasses and annual plants that eventually became the University of Wisconsin's restored Curtis Prairie.

be, part of the land community for better or worse, they contend. In this view, we have an unavoidable influence on nature that gives us a large degree of responsibility for it. We ought to cultivate, manipulate, use, and even manufacture nature, according to some people. Needless to say, many preservationists regard this as unmitigated arrogance. What do you think?

Tools of Restoration

Some of the earliest examples of restoration have involved labor-intensive horticultural or animal control methods. The Curtis Prairie at the University of Wisconsin in Madison, for instance, was reclaimed by Civilian Conservation Corps workers (fig. 5.22) and student volunteers, starting in 1934. Native species from the area were collected from remnant prairies along railroad rights-of-way and in pioneer cemeteries, and then hand-planted and cultivated on old, abandoned farm fields. Prairie plants initially had difficulty getting established and competing against exotics until it was recognized that periodic fires were essential for maintaining this biological community. The prairie is now flourishing and serves as a seed source for other restoration projects (fig. 5.23).

An even more laborious restoration project in Bermuda involved replanting forests, digging ponds, and recreating a nearly vanished ecosystem to provide habitat for the endangered Bermuda cahow (see Case Study, p. 110). Sometimes the key to rebuilding a community is to remove alien intruders. In the Coachella Valley of California, for example, thousands of exotic salt cedars that dry out riparian habitats and crowd out native vegetation are being removed to protect the endangered Coachella Valley fringe-toed lizard. In Hawaii, feral pigs that root out native plant species and eat native birds are being hunted down and removed. Similarly, on the Galápagos Islands, feral goats and rats are being shot, trapped, and poisoned to protect native species.

A less expensive approach has been adopted in restoring an oak savanna near Chicago. Seeds collected from remnant prairies are simply broadcast rather than being carefully planted. This is not only faster and cheaper, it is more like the natural process of wind-

Figure 5.23

The Curtis Prairie as it looked in 1980. This restored prairie has taught us a great deal about restoration in general and about the dynamics and ecology of prairie ecosystems in particular.

dispersal of seeds. Fire is used to discourage invasions of exotic species (fig. 5.24). This is called successional restoration because it depends on natural succession to determine the outcome.

In the Flint Hills of southern Kansas, the Nature Conservancy has reintroduced both fire and native bison (buffalo) to its 16,000 ha (40,000 ac) tall-grass prairie preserve. The bison disperse seeds and create disturbances necessary for survival of many other native species. And in Costa Rica, an almost vanished dry tropical forest is being rebuilt in the Guanacaste Conservation Area by preventing fires and allowing domestic livestock to act as seed dispersal agents since the native grazing animals have completely disappeared.

CASE STUDY
Restoration of the Bermuda Cahow

The cahow is a seabird endemic (restricted) to Bermuda and adjacent islands off the east coast of North America (fig. 1). A member of the petrel family, related to albatrosses, shearwaters, and other wide-ranging seabirds, cahows once formed dense, noisy colonies that fed on the rich fisheries around the island. When European sailors first landed on Bermuda 400 years ago, cahows were abundant. Like many endemic island species, the ground-nesting cahow had never experienced predation and had no defenses against the pigs, goats, and rats introduced by the first settlers. Overhunting and habitat destruction further decimated the species. By the late 1600s—about the same time that the last dodo was killed on Mauritius—cahows disappeared from Bermuda.

For three centuries, the cahow was assumed to be extinct. In 1951, though, scientists found a few living cahows on some tiny islands in the Bermuda harbor. A protection and recovery program was begun immediately, including establishment of a sanctuary on the 6-hectare (15-acre) Nonsuch Island, which has become an excellent example of environmental restoration.

Nonsuch was a near desert after centuries of abuse, neglect, and habitat destruction. All the native flora and fauna were gone, along with most of its soil. This was a case of re-creating nature rather than merely protecting what was left. Sanctuary superintendent David Wingate, who has devoted his entire professional life to this project, has

Figure 1

The Bermuda cahow or hook-billed petrel was thought to be extinct for nearly three centuries. A breeding population has been reestablished, however, and the restored sanctuary created to protect them is helping many other species as well.

brought about a remarkable transformation of this barren little island. Reestablishing a viable population of cahows has had the added benefit of rebuilding an entire biological community.

The first step in restoration was to reintroduce native vegetation and re-create habitat—more than 5000 native tree and shrub seedlings were planted. Initial progress was slow as trees struggled to get a foothold; once the forest knit itself into a dense thicket that deflected the salt spray and ocean winds, however, the natural community began to reestablish itself. The benefits of in-

digenous species became apparent in 1987 when Hurricane Emily roared across Bermuda. Up to 70 percent of non-native trees were uprooted or snapped off by gale-force winds, littering streets and bringing down power lines. The dense, low-profile, native trees on Nonsuch were barely touched by the winds. Demands soared for hurricane-adapted species to replace those lost along streets and in gardens.

Just providing habitat for the cahows was not enough, however, to restore the population. Each pair lays only one egg per year and only about half survive under ideal conditions. It takes 8 to 10 years for fledglings to mature, giving the species a low reproductive potential. They also compete poorly against the more common long-tailed tropic birds that steal nesting sites and destroy cahow eggs and fledglings. Special underground burrows were built with baffled entrances designed to admit only cahows. Young birds were hand-raised by humans to ensure a proper diet and protection.

By 1990, the cahow population had rebounded to nearly fifty nesting pairs. It is too early to know if this is enough to be stable over the long term, but the progress to date is encouraging. Perhaps more important than rebuilding this single species is that the island has become a living museum of precolonial Bermuda that benefits many species besides its most famous resident. It is a heartening example of what can be done with vision, patience, and some hard work.

Letting Nature Heal Itself

Sometimes all we have to do to reestablish a healthy ecosystem is simply to walk away and let organisms recolonize an area. The demilitarized zone (DMZ) between North Korea and South Korea was totally devastated in 1953 when the Korean War ended. Shattered tree stumps and ruined villages lay strewn across a barren landscape pockmarked by bomb craters and littered with the debris of war. Because neither side was allowed to move back into this political buffer zone, the area has become a wildlife refuge and

a luxuriant oasis for nature in an otherwise densely populated countryside. Scarred slopes have been reclothed with a dense hardwood forest where deer, lynx, and an occasional tiger find shelter. Former rice paddies have reverted to marshland in which waterfowl flourish and the trumpeting cries of the endangered Manchurian cranes are heard once again.

Similarly, parts of the political no-man's-land between the former East and West Berlin became a haven for wildlife during the 40 years that the Berlin Wall separated the two parts of the city. This is a heartening example of the regenerative power of nature.

Figure 5.24
A burn-crew technician sets a back fire to control the prescribed fire that will restore a native prairie.

Now that the city is reunified, the question arises about what to do with this area. Should it be allowed to remain an unkempt urban wilderness or should it be turned into a park?

Authenticity

Authenticity is a contentious issue in restorations of all sorts. Is it necessary, or even desirable, to try to restore a particular place to an exact replica of its original ecosystem? If you could create a pleasing, healthy biological community that resembles what existed in the past, is that close enough, or should you strive for a precise duplicate? Suppose that the original ecosystem was made up of ugly, nasty, diseased plants and animals. Would it be acceptable to try to improve on nature? If you were restoring a northern wetland, would you deliberately reintroduce mosquitoes, black flies, leaches, and ticks? If your wetland was in the South, would you add poisonous water snakes and spiders?

Some ecologists contend that restorations should use only plants and animals that have evolved under the particular environmental conditions of the place to be rehabilitated. According to this viewpoint, you shouldn't introduce organisms—even of the same species—that come from some other place because they may lack unique adaptations that come from evolving in a specific location. Suppose you are restoring a cattail marsh. How far would you go to find cattails to transplant? What if there is a small marsh with cattails right next to the area on which you are working? These plants are probably closely related to the ones that once grew in your marsh. But if you take roots or seeds from the few plants remaining in the neighboring marsh, you may disrupt cattail regeneration there. Is having nearby plants that important, or would it be alright to get stock from further away?

Back To What?

When humans take charge of ecosystems, questions sometimes are raised about what our goals should be. Suppose natural forces, such as hurricanes or fires, disrupt a wilderness area, should we use the principles of restoration ecology to tidy up or improve an area or leave it to natural processes? In some cases where humans have al-

tered an area, there may be more than one historic state to which we could restore it. In The Nature Conservancy's Hassayampa River Preserve near Phoenix, for instance, pollen grains preserved in sediments reveal that 1000 years ago, the area was grassy marsh that was unique in the surrounding desert landscape. Some ecologists would try to rebuild a similar marsh. Corn pollen in the same sediments, however, show that 500 years ago Native Americans began farming the marsh. Which is more important, the natural or early agricultural landscape?

Unfortunately, it may not be possible to return to conditions of either 500 or 1000 years ago since climate changes and evolution may have made the communities existing at that time incompatible with current conditions. This creates a difficult question for restoration ecologists. If change is natural and inevitable, who is to say that present conditions—whatever they are or however they have come about—are bad? How should we distinguish between desirable and undesirable changes? If it is simply a matter of human preference, some people might prefer a golf course or a shopping mall on the site. What would you say to someone who claims that since humans are a part of nature, whatever changes we make to a landscape also are natural?

Creating Artificial Ecosystems

As we learn more about how ecosystems work and about the valuable ecological services they provide, we are coming to realize that we can use some natural principles in human-designed systems. The Arcata, California, marsh and wildlife sanctuary is a famous example of using nature to solve a human problem. In 1974, Arcata was faced with the prospect of spending $56 million to upgrade its municipal sewage treatment plant and reduce dumping of inadequately treated wastewater into Humboldt Bay. City residents and faculty from Humboldt State University devised a low-cost alternative that might serve as a model for many other communities.

An abandoned dump near the existing treatment plant was turned into an artificial wetland (fig. 5.25). Wastewater from conventional treatment processes flows through oxidation ponds where sunlight and air kill pathogenic microbes. The effluent is then filtered through the artificial marsh, where aquatic plants filter out nutrients and zooplankton eat microbes and fish eat the plankton. Finally, the water trickles into a saltwater slough where pelicans and cormorants gather to feed on the plentiful fish and where oysters perform a final filtering step before the "polished" water enters the bay. The marsh has become a favorite bird-watching and recreational site. It also saves millions of dollars and has turned what was formerly an eyesore into a source of civic pride.

Ecosystem Management

As you probably already know, there is a growing sense among ecologists, and the public at large, that our environmental situation is rapidly reaching a crisis point. Many of our management policies, while well-meaning, have made matters worse rather than better. **Ecosystem management** is a relatively new discipline in environmental science that attempts to integrate ecological, economic, and social goals in a unified, systems approach (Case Study, p. 113). It recognizes that we cannot have sustained progress towards social

Figure 5.25

Arcata, California, built an artificial marsh as a low-cost, ecologically based treatment system for sewage effluent.

goals in a deteriorating environment or economy, and vice versa. Each of these domains affects, and is affected by, the others. There are many definitions of ecosystem management. A 1994 study by the U.S. House of Representatives suggested the principles shown in table 5.2.

A Brief History of Ecosystem Management

While the term "ecosystem management" is new, a few visionary ecologists such as Aldo Leopold had the foresight to advocate many specific elements of this science fifty years ago. Leopold's attempts to restore his Sand County farm to ecological health and beauty foreshadowed many of the ideas now widely espoused by ecosystem managers. Another pioneer in this field is environmental pol-

Table 5.2

Principles of ecosystem management

- Managing across whole landscapes, watersheds, or regions over an ecological time scale.
- Considering human needs and promoting sustainable economic development and communities.
- Maintaining biological diversity and essential ecosystem processes.
- Utilizing cooperative institutional arrangements.
- Integrating science and management.
- Generating meaningful stakeholder and public involvement and facilitating collective decision-making.
- Adapting management over time, based on conscious experimentation and routine monitoring.

icy expert Lynton Caldwell, who wrote in 1970 that we should use ecosystems as the basis for public land policy. He understood that to do so "would require that the conventional [political] matrix be unraveled and rewoven in a new pattern." Unfortunately, although land use planning was a high priority of the burgeoning U.S. environmental movement in the 1970s, the forces of inertia and private interest thwarted attempts to manage ecologically.

Most federal and many state natural resource agencies in the United States are attempting to implement ecosystem management as their guiding policy. The U.S. Forest Service, the Bureau of Land Management, and the National Park Service, for instance, all have adopted versions of ecosystem management. This is a marked improvement in many cases over past practices. Traditionally, many of these agencies emphasized commodity production and the commercial or recreational use of natural resources as their first priority. Management objectives often were designed to expedite the development, extraction, and/or production of resources on public lands above all other uses. Wildlife and fish habitats, endangered species, and cultural, scenic, historic, or aesthetic values tended to be viewed mainly as inconveniences or hindrances to economic uses.

Principles and Goals of Ecosystem Management

Conservation biologist E. R. Grumbine, in a seminal article on ecosystem management, pointed out several important differences between this integrative approach and the traditional policies of the past:

1 *Hierarchical context:* a focus on any one level of the biodiversity hierarchy (genes, species, populations, ecosystems, landscapes) is insufficient. Ecosystem managers must see interconnections between all levels.

2 *Ecological boundaries:* Rather than divide administrative units by political boundaries, the watersheds, ecosystems, or other natural units should be managed in an integrated fashion.

3 *Data collection and routine monitoring:* To function correctly, ecosystem management requires ongoing research and data collection so that successes or failures may be recognized and evaluated.

4 *Adaptive management:* Ecosystem management assumes that scientific knowledge is provisional and regards management plans as learning processes or continuous experiments where incorporating the results of previous actions allows managers to remain flexible and adapt to uncertainty (fig. 5.26).

5 *Organizational change:* Implementing ecosystem management requires changes in agency structure and ways of doing business. Accomplishing meaningful changes in entrenched bureaucracies may be the most challenging aspect of this new approach.

6 *Humans in nature:* People cannot be separated from nature. Humans inescapably affect ecological patterns and processes and are in turn affected by them.

7 *Values:* Regardless of the role of scientific knowledge, human values play a dominant role in ecosystem management goals.

CASE STUDY
Yanesha Forestry in Peru

In the 1970s, Peru requested assistance from the U.S. Agency for International Development (USAID) for agricultural development of the Palcazu Valley in the Peruvian Amazon. United States law required an environmental assessment of the project before any aid was made available. A multidisciplinary assessment team warned that the usual pattern of rural development with road building, forest clearing, and colonization by landless peasants from elsewhere in Peru would likely fail because high rainfall (about 7 m, or 275 in, per year) and infertile soil makes the valley unsuitable for farming. Furthermore, an invasion by outsiders would be disastrous for indigenous Arawakan Indians forest dwellers. Instead, the consultants suggested, a locally-run sustainable forestry project could provide long-term economic development, protect biodiversity, and preserve native culture.

One of the first changes needed for this forestry project was recognition of land ownership by native communities in the Palcazu Valley. This was necessary because Indians in Peru typically do not hold title to traditional communal lands. After several years of technical assistance, community education, and political advocacy, land claims of eleven native communities were recognized legally, and an Indian forestry cooperative—the first in South America—was established. Called COFYAL (the Spanish acronym) or the Yanesha Project (after the native people's name for themselves), the cooperative included five native communities and about seventy individual Arawakan Indians.

Rather than clearcut the forest in large blocks, the Yanesha Project is based on strip-cutting. Narrow strips—30 or 40 m wide—are cut through the forest. Oxen drag out logs, resulting in less damage to soil and remaining trees than would be caused by tractors or bulldozers. Strips are never burned or cultivated and are only about twice as wide as the average tree-fall gap, so natural regeneration is rapid. The strips quickly fill with a wide diversity of trees resprouting from stumps or from seeds from nearby trees. Wildlife is disturbed very little by the narrow strips.

COFYAL technicians identify forest suitable for harvesting and locate hauling roads based on tree types, slope, and proximity to streams, wetlands, and other protected areas. Strips are harvested in an alternating pattern (1, 3, 5, then 2, 4, 6, for example) so that it takes 6 to 10 years to complete a cycle and about 30 to 40 years between harvesting on any particular strip (fig. 1). Rather than cut only the biggest trees, which fall on and crush smaller trees, as is done in most timber operations, the Yanesha foresters first cut and remove pole-size stems and then proceed to remove the larger trees.

Everything from the forest is used. A portable sawmill produces lumber from larger timbers, a hydraulic system preserves posts and poles, and scraps are converted to charcoal for cooking and heating. Because lumber is marketed locally, a wider range of species is sold than would be acceptable in national or international markets.

So far the project seems to be meeting its goals admirably: (1) to employ members of the native communities, (2) to manage natural forests for sustained yield and a natural species mix, and (3) to protect the culture of the Yanesha people. By taking forest ecology into account, and managing resources according to local conditions and needs, the COFYAL project is incorporating many principles of ecosystem management.

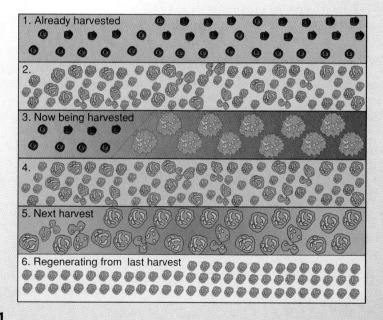

Figure 1

In strip harvesting, a narrow row of trees 30–40 m wide and up to 0.5 km long is clear cut every year or so in an alternating pattern (for example, 1, 3, 5, then 2, 4, 6). Any given strip is allowed to regenerate for 30 to 40 years before being reharvested. Natural regeneration occurs by stump growth or reseeding by nearby trees.

Figure 5.26

Ecosystem management requires a high level of routine monitoring and data collection to assess what works and what doesn't. Adaptive management means that every program is regarded as experimental and subject to change as new information becomes available.

Ecosystem management goals

- Maintain viable populations of native species *in situ*.

- Represent, within protected areas, all native ecosystem types across their natural range of variation.

- Protect essential ecological processes such as nutrient cycles, succession, hydrologic processes, etc.

- Manage over long enough time periods to sustain the evolutionary potential of species and ecosystems.

- Accommodate human use and occupancy within these constraints.

Some commonly reported goals for ecosystem management are presented in table 5.3. In general, the criteria listed by Aldo Leopold, as the basis of environmental ethics—integrity, stability, and beauty of biotic communities—are good benchmarks for ecosystem management.

Critiques of Ecosystem Management

Many of the criticisms of ecosystem management are similar to those mentioned earlier with respect to restoration ecology. Many ecologists argue that we don't understand ecosystems now and we probably never will, given the chaotic, catastrophic, and unpredictable qualities of nature. It is arrogant of us, therefore, to imagine that we can manage nature. Furthermore, in developing countries, political pressures and immediate needs of poor people make it difficult to

enforce environmental protection laws now. This will become even more difficult as populations grow and resources diminish.

Often vision statements and organizational plans for ecosystem management are little more than empty slogans, while nothing really changes. Simply claiming that we will be organized, efficient, and effective doesn't necessarily make us so. Many resource managers worry that ecosystem management encourages us to believe we can have our cake and eat it too. We simply have to accept limits on what we can do and have, they argue. From this perspective we ought to set aside large blocks of untrammeled nature and free ourselves of the illusion that science or technology can provide solutions to our environmental or social problems.

The debate over ecosystem management reflects a deep division that runs through much of environmental science. The cleavage between preservation and management, between conservative and progressive policies, and between pessimism and optimism all are based on different worldviews about what is possible and what is likely to occur. Where do you stand in this spectrum of opinions? Do you think we can—or should even try to—manage ecosystems; restore, reclaim, or rehabilitate landscapes; or create new and imaginative forms of nature? Is it possible for us to do what is necessary to preserve the integrity, stability, and beauty of our environment?

Summary

Major ecosystem types called biomes are characterized by similar climates, soil conditions, and biological communities. Among the major terrestrial biomes are deserts, tundra, grasslands, temperate deciduous forests, temperate coniferous forests, tropical moist forests, and tropical seasonal forests. Aquatic ecosystems include oceans and seas, rivers and lakes, estuaries, marshes, swamps,

bogs, fens, and reefs. Moisture and temperature are generally the most critical determinants for terrestrial biomes. Periodic natural disturbances, such as fires, play a major role in maintaining some biomes.

Humans have disturbed, preempted, or damaged much—perhaps half or more—of all terrestrial biomes and now dominate about 40 percent of all net primary productivity on the land. Some of this disturbance and domination is permanent, but we have opportunities to apply ecological knowledge and practical experience to restoring and repairing ecosystems. In some cases, all we have to do is leave nature alone; in other cases, more active management is required to reestablish functioning ecosystems. Remediation usually means cleaning the soil of pollution, while reclamation is applied to physical repair of severely damaged land. Mitigation means to replace or re-create an ecosystem in restitution for damage caused by a construction project or some other human activity. Whether we can or should try to create new, synthetic forms of nature is controversial.

Landscape ecology is the study of reciprocal effects of spatial pattern on ecological processes. A landscape, in this sense, is a bounded geographical unit that includes both living and non-living components. History is important in that it shapes the land and organisms as well as our reactions to, and interpretations of, the landscape. Humans are important elements of most landscapes and few places on the earth are devoid of human impacts. Restricting our definition of "nature" to a few pristine remnants will lead us either to ignore most of the earth's surface or to demand that humans remove themselves entirely from large tracts, something not very likely.

Patchiness and heterogeneity are characteristic of most landscapes. Movement of organisms and materials across boundaries between habitat patches play an important regulatory role in many ecosystems. Some landscape patterns apply across a wide range of scales while others are restricted to a specific scale.

Ecosystem management involves managing whole landscapes over long timescales, taking human needs into consideration, maintaining biodiversity and ecological processes, utilizing innovative institutional arrangements, integrating science and policy, encouraging public involvement, and adapting management based on conscious experimentation and routine monitoring. Ecosystem management, like sustainable development, recognizes that social goals, environmental quality, and economic health are all inextricably interlinked.

▼ Questions for Review

1. Who was Aldo Leopold and why is he considered important in the history of American conservation?

2. Throughout the central portion of North America is a large biome dominated by grasses. Describe how physical conditions and other factors control this biome.

3. What is taiga and where is it found? Why might logging in taiga be more disruptive than in southern coniferous forests?

4. Why are tropical moist forests often less suited for agriculture and human occupation than tropical deciduous forests?

5. Describe four different kinds of wetlands and explain why they are important sites of biodiversity and biological productivity.

6. Which major biomes have been most heavily disturbed by human activities?

7. Define a landscape. Describe the major ecological and cultural features of the landscape in which you live.

8. Define *restoration, rehabilitation, remediation, reclamation, ecological re-creation,* and *mitigation.* Give an example of each.

9. List and explain either the seven principles or the five goals of ecosystem management.

10. Explain some of the major criticisms of ecosystem management.

▼ Questions for Critical Thinking

1. In which biome do you live? What physical and biological factors are most important in shaping your biological community? How do the present characteristics of your area differ from those 100 or 1000 years ago?

2. Could your biome be returned to something resembling its original condition? What tools (or principles) would you use to do so?

3. A beautiful stand of pine trees in a park in Connecticut was blown down by a windstorm and is now a fire hazard to nearby property as well as being an aesthetic mess. Some people argue that salvage loggers should clean out the area and replant pine trees. Others argue that it was a natural forest and should be left to natural forces. What do you think?

4. Historic records show that indigenous aboriginal people kept the forest described in question 3 as an open, park-like savanna by burning it regularly. Should prescribed fires be a part of the restoration ecology plan?

5. Some environmentalists worry that the science of restoration ecology may give us the arrogant attitude that we can do anything we want now because we can repair the damage later. How would you respond to that concern?

6. What do you think Aldo Leopold meant by integrity, stability, and beauty of the land?

7. Disney World in Florida wants to expand onto a wetland. It has offered to buy and preserve a large nature preserve in a different area to make up for the wetland it is destroying. Is that reasonable? What conditions would make it reasonable or unreasonable?

8. Suppose further that the wetland being destroyed in question 7 and its replacement area both contain several endangered species (but different ones). How would you compare different species against each other? How many plant or insect species would one animal species be worth?

9. Explain why there might be differences in philosophy or worldview between preservationists and restorationists. Which approach do you prefer?

10. Authenticity is a contentious issue in restoration. Is it necessary, or even desirable, to try to create an exact replica of an original ecosystem? Could the changes that humans make be considered part of natural change?

▼ Key Terms

barrier islands 102	coral reefs 103
benthos 100	deciduous 98
biomes 94	delta 101
bogs 101	deserts 94
boreal forest 97	ecosystem management 111
cloud forest 98	estuaries 101
conifer 97	fens 101

▼ Additional Information on the Internet

Bright Edges of the World
http://www.nasm.edu:1995/

Coastal Services Center
http://csc.noaa.gov/

The Coral Health and Monitoring Program
http://coral.aoml.erl.gov/

Sierra Club Critical Ecoregions Program
http://www.sierraclub.org/ecoregions/

International Arid Lands Consortium
http://ag.arizona.edu:80/OALS/IALC/Home.html/

US Fish and Wildlife Service National Wetlands Inventory
http://www.nwi.fws.gov/

The Polar Regions
http://www.stud.unit.no:80/~sveinw/artic/

Society for Ecological Restoration
http://nabalu.flas.ufl.edu/ser/SERhome.html/

Gopher Menu for the Taiga Rescue Network
gopher://gopher.igc.apc.org:70/11/environment/forests/ trn/

US Bureau of Reclamation
http://www.usbr.gov/

Whale Watching Web
http://www.physics.helsinki.fi/whale/

Note: Further readings appropriate to this chapter are listed on p. 595.

A Philippino fisherman mends his nets as his son watches. What social and cultural factors shape their environmental choices?

Part 2

Population, Economics, and Environmental Health

What kind of world do you want to live in? Demand that your teachers teach you what you need to know to build it.

Peter Kropotkin

A herd of wildebeest (gnu) cross the Mara River in their annual migration across the Masri Mara National Park in Kenya.

CHAPTER 6: Population Dynamics

Objectives

After studying this chapter, you should be able to:

- appreciate the potential of exponential growth.

- draw a diagram of J and S curves and explain what they mean.

- explain who Thomas Malthus was and what he said about population growth.

- describe environmental resistance and discuss how it can lead to logistic or stable growth.

- define *fecundity, fertility, birth rates, life expectancy, death rates,* and *survivorship.*

- compare and contrast density-dependent and density-independent population processes.

Wolves and Moose on Isle Royale

Isle Royale National Park occupies the largest island in Lake Superior, the largest freshwater lake in the world. It is a spectacular wilderness setting of high rocky ridges covered by a dense boreal forest (fig. 6.1). Cut off from the mainland of Minnesota and Ontario by 30 km (20 mi) of rough, deep, very cold water, the island is a mostly closed ecosystem that is a unique laboratory for studying large animal population dynamics.

Moose Population Growth and Decline

The original large herbivore of the island was the woodland caribou. Caribou were abundant until about 1900, but disappeared early in this century due to hunting and human-caused changes in the habitat. About the time that caribou became extinct on the island, moose first appeared. We suppose they swam from the mainland or crossed on ice that occasionally forms a bridge to the island. They must have found an ideal situation; the shrubs and

Figure 6.1

A view of Lake Superior with Canada in the background, taken from a high ridge in Isle Royale National Park.

aquatic plants on which they prefer to browse were plentiful, and there were no major predators to limit their population growth.

The number of moose increased slowly at first. In 1915, it is estimated that about two hundred moose were on the island, or one moose for each 2.6 sq km (1 sq mi). Then, in the 1920s, there was a moose **population explosion** on Isle Royale. As figure 6.2 shows, the number jumped from roughly three hundred in 1920 to about five thousand in 1928.

In the summer of 1929, famous wildlife biologist Adolph Murie went to the island to study the moose situation. He reported that all the tender branches on which the moose browse in the winter were eaten back as high as the moose could reach. Much of the summer food (aquatic plants and annuals) was also badly depleted. He predicted that disease and starvation would soon cause a **population crash** extensive die-off. As you can see in figure 6.2, his prediction came true. By 1941, only 171 moose were found on the island—fewer than twenty years before. Clearly, unrestrained growth of the moose population surpassed the limits of the environment and resulted in a catastrophic population decline.

Moose-Wolf Equilibrium?

In the early 1940s, shortly after the moose population on Isle Royale declined so precipitously, wolves appeared on the island in pairs and small groups, presumably having crossed on the ice during previous winters. There were no permanent human inhabitants on the island at the time, so we don't know exactly when the first wolves arrived or how many there were in those early years. By 1957, when the first systematic census was taken, twenty-one wolves were on the island (fig. 6.2).

You might think that the wolves, arriving as they did when the moose were weakened by starvation, could have exterminated their prey completely. Instead, the wolves and moose established a relatively stable balance. Wolf predation seems to have prevented both excess population growth and precipitous decline of moose on the island for many years (fig. 6.3). There was enough vegetation for the moose to eat and stay healthy, and there were enough moose for the wolves to eat. We call the maximum number of individuals of any species that can be supported on a long-term basis by a particular ecosystem its **carrying capacity**. Because of environmental variation and other factors, populations rarely reach or maintain this maximum level, however.

Wolves and Moose in Trouble

In 1988, an alarming decline in the wolf population on Isle Royale was observed. The annual winter survey revealed only twelve wolves in three small packs, down from fifty wolves only six years earlier. Scientists proposed several possible causes for this population crash. It may have been the food supply. Although moose were plentiful, they were young and healthy—perhaps too difficult for the wolves to catch. Diseases might have been introduced by dogs or stray wolves from the mainland. Some scientists believe the problem may be genetic. When a population starts with only a few founding individuals and is highly inbred, as are the

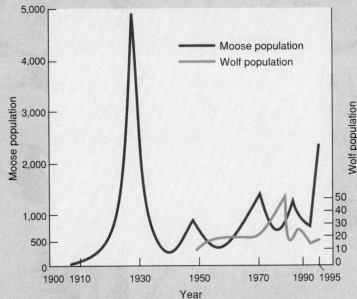

Figure 6.2

Growth of moose and wolf populations on Isle Royale National Park, 1900 to 1985. Moose first appeared on the island shortly after 1900. By 1929, the moose population had grown to about 5000 and had exhausted the easily available food supply. A catastrophic dieback occurred during the 1930s, when 97 percent of the moose died. The appearance of wolves after 1940 at first reduced the moose population but then established a dynamic equilibrium around what appears to be the carrying capacity.

Source: David Mech, *The Wolves of Isle Royale*, 1966, published by the National Park Service.

Figure 6.3

Wolves eating a moose on Isle Royale. The wolves succeed in killing only about one moose in every ten they attack. Usually only the very young, old, and sick moose are vulnerable. A healthy adult can defend itself against a large pack.

Dynamics of Population Growth

What determines whether a population will increase or decrease abruptly, or whether it will maintain a delicate balance with its neighbors? Why do some populations grow to enormous numbers while others do not? Although it is very difficult to determine specific causal mechanisms, in this chapter, we will look at some of the underlying factors that influence population dynamics. We will survey the components that lead to growth or decline of populations, and we will consider the mechanisms that regulate population size, distribution, and growth rate. One of the most important concepts to be gained from the study of population dynamics is that different mechanisms regulate population density, depending on the niche an organism occupies in its ecosystem and the stability of that ecosystem.

To explain the patterns of growth and interspecific interaction exhibited by the wolves and moose of Isle Royale, we need to know something about the principles of population biology. In this section, we will look at growth rates and patterns.

Exponential Growth and Doubling Times

The rapid increase of moose on Isle Royale in the early stages of colonization is an example of **exponential growth,** or growth at a constant *rate* of increase per unit of time. It is called exponential because the rate of increase can be expressed as a constant *fraction,* or exponent, by which the existing population is *multiplied.* This pattern also is called **geometric growth** because the sequence of growth follows a geometric pattern of increase, such as 2, 4, 8, 16, and so on. By contrast, a pattern of growth that increases at a constant *amount* per unit of time is called **arithmetic growth.** The sequence in this case might be 1, 2, 3, 4 or 1, 3, 5, 7. Notice in these examples that a constant amount is *added* to the population (fig. 6.4).

The reason that moose populations (like those of most other organisms) have the ability to grow exponentially, given an appropriate environment, lies in the power of biological reproduction. Each cow moose reaches sexual maturity at age three or four. She then can produce one to three calves each year for the next eight to ten years if she remains healthy, has enough to eat, and survives that long. This means that each female moose can produce as many as thirty calves during her lifetime (although few do). If half of those calves are female, and all survive to have as many offspring as their mother, the moose population will increase four- to tenfold in each ten-year generation. In other words, the population is increasing at a 14 to 35 percent annual growth rate.

As you can see in figure 6.2, the number of moose added to the population at the beginning of the growth curve is rather small. But within a very short time, the numbers begin to increase quickly because a given percentage becomes a much larger amount as the population gets bigger. The growth curve produced by this constant rate of unfettered growth is called a **J curve** because of its shape. At 1 percent per year, a population or a bank account doubles in roughly seventy years (table 6.1). A useful rule of thumb is that if you divide seventy by the annual percentage growth, you will get the approximate doubling time in years. As a result, a moose population growing at 35 percent doubles every two years. You can also apply this rule to calculate doubling time of human populations. Countries growing at 4 percent per year will *double* their populations in 17.5 years. A country growing at a rate of 0.1 percent

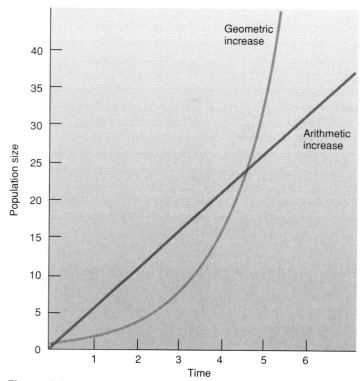

Figure 6.4

Arithmetic and geometric growth curves. Note that the geometric or exponential curve grows more slowly at first, but then accelerates past the arithmetic curve, which grows at a steady incremental pace throughout.

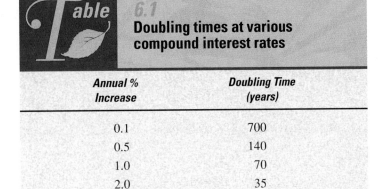

Table 6.1
Doubling times at various compound interest rates

Annual % Increase	Doubling Time (years)
0.1	700
0.5	140
1.0	70
2.0	35
5.0	14
7.5	9
10.0	7
100.0	0.7

Figure 6.5

Reproduction gives many organisms the potential to expand populations explosively. A single female cockroach can produce up to eighty eggs every six months. If everyone survives and reproduces, a kitchen could look like this in only a few generations. This exhibit is in the Smithsonian Institute's National Museum of Natural History.

annually will double its population in seven hundred years. We will use this rule to discuss human population growth in chapter 7.

Biotic Potential

Neither moose nor humans are the fastest reproducing of all organisms. Many species have amazingly high reproductive rates that give them the potential to produce enormous populations very quickly, given unlimited resources and freedom from limiting factors (fig. 6.5). We call the maximum reproductive rate of an organism its **biotic potential.** Table 6.2 shows the potential number of offspring that a single female housefly (*Musca domestica*) and her offspring could produce in a year. The result is astounding. Each female fly lays an average of 120 eggs in each generation. The eggs hatch and mature into sexually active adults and lay their own eggs in fifty-six days. In one year (seven generations), if all its offspring survived long enough to reproduce, a single female could be the ancestor of 5.6 *trillion* flies. If this rate of reproduction continued for ten years, the whole earth would be covered several meters deep with houseflies! Fortunately, this has not happened because of factors that limit the reproductive success of houseflies. This example, however, illustrates the potential for biological populations to increase rapidly.

Population Oscillations and Irruptive Growth

In the real world, there are limits to growth. When a population exceeds the carrying capacity of its environment or some other limiting factor comes into effect, death rates begin to surpass birth rates. The growth curve becomes negative rather than positive, and the population decreases as fast, or faster, than it grew. We call this the population crash or **dieback.** Looking back at figure 6.2, you can see abrupt diebacks of both moose and wolves on Isle Royale at different times.

The extent to which a population exceeds the carrying capacity of its environment is called **overshoot,** and the severity of the

Table 6.2
Biotic potential of houseflies (*Musca domestica*) in one year

Assuming that:

—a female lays 120 eggs per generation

—half of these eggs develop into females

—there are seven generations per year

	Total Population	
Generation	If all females in each generation lay 120 eggs and then die	If all generations survive one year and all females reproduce maximally in each generation
1	120	120
2	7,200	7,320
3	432,000	446,520
4	25,920,000	27,237,720
5	1,555,200,000	1,661,500,920
6	93,312,000,000	101,351,520,120
7	5,598,720,000,000	6,182,442,727,320

Source: Data from E. J. Kormondy, *Concepts of Ecology,* 3d ed., © 1984 Harper & Row Publishers, Inc.

dieback is generally related to the extent of the overshoot. This pattern of population explosion followed by a population crash is

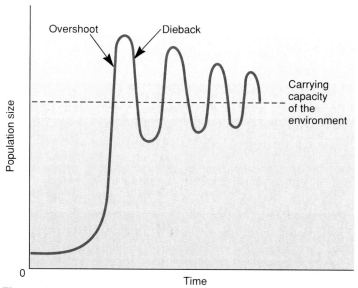

Figure 6.6
Population oscillations. Some species demonstrate a pattern of cyclic overshoot and dieback.

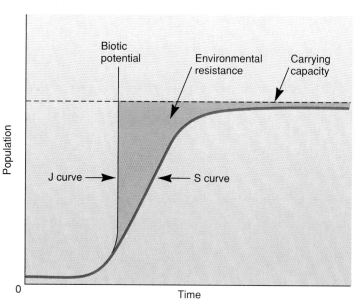

Figure 6.7
J and S population curves. The vertical J represents theoretical unlimited growth. The S represents population growth and stabilization in response to environmental resistance.

called **irruptive or Malthusian growth.** It is named after the eighteenth-century economist Thomas Malthus, who concluded that human populations tend to grow until they exhaust their resources and become subject to famine, disease, or war. Malthus and his theories are discussed further in chapter 7.

Populations may go through repeated oscillating cycles of exponential growth and catastrophic crashes, as shown in figure 6.6. These cycles may be very regular if they depend on a few simple factors, such as the seasonal light- and temperature-dependent bloom of algae in a lake. They also may be very irregular if they depend on complex environmental and biotic relationships that control cycles, such as the population explosions of migratory locusts in the desert or tent caterpillars and spruce budworms in northern forests.

Growth to a Stable Population

Not all biological populations go through these cycles of irruptive population growth and catastrophic decline. The growth rates of many species are regulated by both internal and external factors so that they come into equilibrium with their environmental resources. These species may grow exponentially when resources are unlimited, but their growth slows as they approach the carrying capacity of the environment. This pattern is called **logistic growth,** a mathematical description of its constantly changing rate.

Together, factors that tend to reduce population growth rates are called **environmental resistance.** In later sections of this chapter, we will look in more detail at these factors and how they limit growth and regulate population size. First, we will see how logistic growth compares to the Malthusian growth just discussed and what kinds of organisms we are talking about in these different patterns.

How does the growth curve of a stable population differ from the J curve of an exploding population? Figure 6.7 shows an idealized comparison between biotic potential and sustainable growth.

The J curve on the left in this figure represents the growth without restraint that we just discussed. It rises rapidly toward the maximum biotic potential of the species. The curve to the right represents logistic growth. We call this pattern an **S curve** because of its shape. It is also called a sigmoidal curve (for the Greek letter sigma). The area between these curves is the cumulative effect of environmental resistance. Note that the resistance becomes larger and the rate of logistic growth becomes smaller as the population approaches the carrying capacity of the environment.

Chaotic and Catastrophic Population Dynamics

Real populations under natural conditions rarely follow the linear dynamics and smooth growth curves shown in figures 6.6 and 6.7. Apparently haphazard fluctuations often characterize population dynamics (fig. 6.8). Population ecology has begun to incorporate ideas from nonlinear mathematics to describe these patterns as being chaotic or catastrophic.

Chaotic systems exhibit variability that, while not necessarily random, can be of a complexity whose pattern is not observable over a normal human timescale. In these systems, minute differences in initial conditions can lead to dramatically different outcomes after many iterations.

A famous metaphor for chaotic dynamics is the "butterfly effect," first described by meteorologist Edward Lorenz in 1979. Lorenz observed that seemingly insignificant variations in the beginning parameters of computerized climate models resulted in totally different final results. It is, he said, as if a butterfly flapping its wings in Rio de Janeiro could, through a long series of unpredictable multiplying effects in weather systems, eventually result in a tornado in Texas.

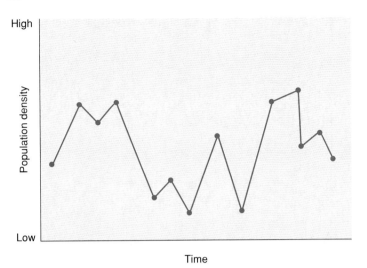

Figure 6.8

Fluctuations in population dynamics of a species that may be chaotic or catastrophic. In real life, it is difficult to distinguish between subtle patterns and simple random variability.

Catastrophic theory was first developed by mathematicians as a purely hypothetical exercise, but some biologists regard it as a good explanation of population dynamics displaying abrupt discontinuities. As conditions change, a **catastrophic system** may jump abruptly from one seemingly steady state to another without any intermediate stages. Balancing a chair on two legs is a good model of catastrophe. You may be stable for a long time with relatively minor corrective input, but, if you pass a certain threshold, suddenly find yourself flat on your back on the floor—and once again, stable. Among populations, the sudden outbursts and diebacks of spruce budworm in the forests of eastern Canada or of migratory locusts in arid regions often are mentioned as examples of catastrophic dynamics.

Is there any practical difference between chaos or catastrophe and simple random behavior? Perhaps so. In a chaotic system, although you can't predict behavior on a short timescale with any certainty, you may be able to identify outer boundary conditions beyond which the system will not go. In a catastrophic system, you can avoid pushing a population close to the edge of an unstable flip point. A good example might be the disastrous population crash of the Canadian and American Atlantic cod fisheries in 1992 (see chapter 10, What Do *You* Think?, p. 213). This calamity was completely unpredicted by management scientists using classical models of population dynamics. Perhaps an understanding of nonlinear behavior may help prevent such disasters in the future.

Strategies of Population Growth

There appear to be evolutionary advantages, as well as disadvantages, in both Malthusian and logistic growth patterns. Although we should avoid implying intention in natural systems where the controlling forces may be entirely mechanistic, it sometimes helps us see these advantages in terms of "strategies" of adaptation and "logic" in different modes of reproduction.

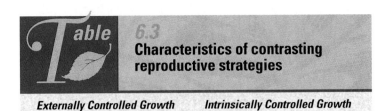

Table 6.3 Characteristics of contrasting reproductive strategies

Externally Controlled Growth	Intrinsically Controlled Growth
1. Short life	1. Long life
2. Rapid growth	2. Slower growth
3. Early maturity	3. Late maturity
4. Many small offspring	4. Fewer large offspring
5. Little parental care or protection	5. High parental care and protection
6. Little investment in individual offspring	6. High investment in individual offspring
7. Adapted to unstable environment	7. Adapted to stable environment
8. Pioneers, colonizers	8. Later stages of succession
9. Niche generalists	9. Niche specialists
10. Prey	10. Predators
11. Regulated mainly by extrinsic factors	11. Regulated mainly by intrinsic factors
12. Low trophic level	12. High trophic level

Malthusian "Strategies"

Organisms with Malthusian growth patterns often tend to occupy low trophic levels in their ecosystems (chapter 3), or to be pioneers in succession (chapter 4). As generalists or opportunists, they move quickly into disturbed environments, grow rapidly, mature early, and produce many offspring. They usually do little to care for their offspring or protect them from predation. They depend on sheer numbers and dispersal mechanisms to ensure that some offspring survive to adulthood (table 6.3). They have little investment in individual offspring, using their energy to produce vast numbers instead.

Many insects, rodents, marine invertebrates, parasites, and annual plants (especially the ones we consider weeds) follow this reproductive strategy. Their numbers generally are limited by predators or other controlling factors in the environment. They reproduce at the maximum rate possible to offset these losses. If the external factors that normally control their populations are inoperative, they tend to rapidly overshoot the carrying capacity of the environment and then die back catastrophically, as we have just seen. Among this group are the weeds, pests, or other species we consider nuisances that reproduce profusely, adapt quickly to environmental change, and survive under a broad range of conditions.

Logistic "Strategies"

While environmental resistance is a factor in controlling population growth in all species, those exhibiting logistic growth tend

What Do *You* Think?
What Is Earth's Carrying Capacity for Humans?

As you will learn in the next chapter, human numbers are doubling about every 40 years. At that rate there would be an astounding 170 quadrillion people 600 years from now. That would put one person on each square meter over the entire land surface of the earth. Carrying capacity reflects the limits imposed on population growth by finite space and finite resources. Earth's resources would never support such ludicrous numbers. So, what is the carrying capacity of the earth for humans? That question cannot be meaningfully addressed without clarifying a basic assumption: the type of lifestyle on which to develop the estimates.

Different lifestyles have different resource requirements. Are people to be vegetarians or will meat be a significant part of the diet? Will the earth's resources be counted upon to provide additional amenities beyond food? The answers to those questions have a profound effect on the numbers of people the earth can sustain. For illustration purposes imagine a miniature planet we could call Terrabase. Consider the following scenarios to see how lifestyle affects carrying capacity.

Scenario 1. **In this scenario, the humans living on Terrabase are vegetarians, and 100 percent of the planet's space is devoted to raising human food. Let's assume that under these conditions 1000 people can be sustained. Carrying capacity equals 1000.**

Scenario 2. **Most people enjoy meat in the diet. Assume that the people of Terrabase do not live as vegetarians, but obtain half their calories by eating herbivores such as beef cattle. This would require a significant amount of plant mass to be fed to animals. But, as you learned in chapter 3, much energy is lost in transfer between trophic levels. It takes about 10 calories of plant food to produce one calorie of beef. After doing the calculations, it turns out that under these conditions, Terrabase will produce enough food to sustain only 180 people. Simply by eating meat, the carrying capacity has been reduced by over 80 percent.**

Scenario 3. **In both scenarios 1 and 2, all of the land has been used exclusively for human food production. But people have consumer needs as well: cars, parking spaces, televisions, washers, dryers, clothing, shopping centers, and much more, all of which require space that would have to be subtracted from that used to produce food. Recreational space is also important to us. We want athletic fields, golf courses, bird sanctuaries, nature preserves, and hunting lands. All of which divert even more land away from food production.**

Assume the residents of Terrabase had the high living standard of industrialized nations, requiring the immense and continuous input of chemicals, energy, paper, and other raw materials, as well as requiring land for waste disposal. How much of Terrabase would be devoted to these uses? If it is 20 percent, the carrying capacity is reduced to 140.

Wild organisms also play ecological roles important to our well-being. Leaving space for wild nature further reduces land available to produce human food.

"What is the earth's carrying capacity for humans?" is not a meaningful question until the cultural context within which people are to live has been clarified.

The magnitude of the cultural impact on carrying capacity is underlined by information provided by Robert Goodland and others. It is estimated that the continuous production of 10 to 15 acres are necessary to sustain just one person at our affluent lifestyle. To support all of the earth's 5.6 billion humans at such a lifestyle would require three times as much productive land as is available. Defined in those terms, the carrying capacity of the earth is around 1.8 billion people, one third the world's current numbers. On the other hand, if carrying capacity is defined as the number of humans the planet could adequately feed without providing other amenities, that number is obviously much larger.

---

to grow more slowly and are more likely to be regulated by intrinsic characteristics than those with Malthusian patterns. These organisms are usually larger, live longer, mature more slowly, produce fewer offspring in each generation, and have fewer natural predators than the species below them in the ecological hierarchy. Some typical examples of this strategy are wolves, elephants, whales, and primates. Each of these species provides more care and protection for its offspring than do "lower" organisms.

Elephants, for instance, are not reproductively mature until they are 18 to 20 years old. During youth and adolescence, a young elephant is part of a complex extended family that cares for it, protects it, and teaches it how to behave. A female elephant normally conceives only once every four or five years after she matures. The gestation period is about eighteen months; thus, an elephant herd doesn't produce many babies in a given year. Since they have few enemies (except humans) and live a long life (often sixty or seventy years), however, this low reproductive rate produces enough elephants to keep the population stable, given appropriate environmental conditions.

An important underlying question to much of the discussion in this book is which of these strategies humans follow. Do we more closely resemble wolves and elephants in our population growth, or does our population growth pattern more closely resemble that of moose and rabbits? Will we overshoot the carrying capacity of our environment (or are we already doing so), or will our population growth come into balance with our resources (What Do You Think?, above)?

Factors that Increase or Decrease Populations

Now that you have seen population dynamics in action, let's focus on what happens *within* populations, which are, after all, made up of individuals. In this section, we will discuss how new members are added to and old members removed from populations. We also will examine the composition of populations in terms of age classes and introduce terminology that will apply in subsequent chapters.

Natality, Fecundity, and Fertility

Natality is the production of new individuals by birth, hatching, germination, or cloning, and is the main source of addition to most biological populations. Natality is usually sensitive to environmental conditions so that successful reproduction is tied strongly to nutritional levels, climate, soil or water conditions, and—in some species—social interactions between members of the species. The maximum rate of reproduction under ideal conditions varies widely among organisms and is a species-specific characteristic. We already have mentioned, for instance, the differences in natality between several different species.

Fecundity is the physical ability to reproduce, while **fertility** is a measure of the actual number of offspring produced. Because of lack of opportunity to mate and successfully produce offspring, many fecund individuals may not contribute to population growth. Human fertility often is determined by personal choice of fecund individuals.

Immigration

Organisms are introduced into new ecosystems by a variety of methods. Seeds, spores, and small animals may be floated on winds or water currents over long distances. This is a major route of colonization for islands, mountain lakes, and other remote locations. Sometimes organisms are carried as hitchhikers in the fur, feathers, or intestines of animals traveling from one place to another. They also may ride on a raft of drifting vegetation. Some animals travel as adults—flying, swimming, or walking—as did the moose and wolves that first colonized Isle Royale. In some ecosystems, a population is maintained only by a constant influx of immigrants. Schools of predatory fish, for example, are important members of ocean ecosystems, but they generally reproduce in shallows, shoals, or estuaries, and their numbers in the open ocean are maintained only by constant recruitment from such nursery areas.

Mortality and Survivorship

An organism is born and eventually it dies; it is mortal. **Mortality,** or death rate, is determined by dividing the number of organisms that die in a certain time period by the number alive at the beginning of the period. Figure 6.9 shows mortality figures for the human population of the United States by age classes and sexes. Note that between age 40 and age 80, the mortality rate of men is higher

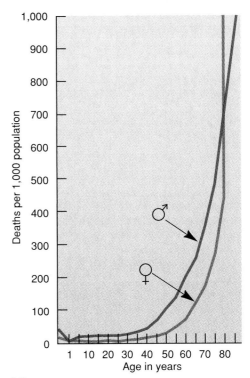

Figure 6.9

Mortality curves by age class in the United States, 1966. Note the high death rates for males in the forty- and seventy-year-old class. It is not yet clear whether this is due to biologic factors or simply lifestyle.

Source: U.S. Bureau of the Census.

than that of women. After age 80, the rate is similar or may slightly favor the remaining men.

Since the number of survivors is more important to a population than is the number that died, mortality is often better expressed in terms of **survivorship** (the percentage of a cohort that survives to a certain age) or **life expectancy** (the probable number of years of survival for an individual of a given age). If more organisms in a population die than are replaced in a given time, the population will decrease. If mortality is low compared to natality, on the other hand, the population will grow. Between these two broad generalizations, many combinations of mortality and natality rates create very different patterns of population growth. To understand how these factors interact, we need to define some more population terms.

Life span is the longest period of life reached by a given type of organism. The process of living entails wear and tear that eventually overwhelm every organism, but maximum age is dictated primarily by physiological aspects of the organism itself. There is an enormous difference in life span between different species. Some microorganisms live their whole life cycles in a matter of hours or minutes. Bristlecone pine trees in the mountains of California, on the other hand, have life spans up to 4600 years.

Most individuals in a population do not live anywhere near the maximum life span for their species. The major factors in early mortality are predation, parasitism, disease, accidents, fighting, and en-

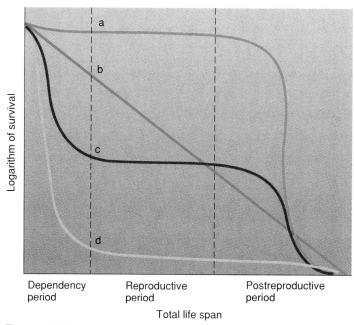

Figure 6.10

Four basic types of survivorship curves for organisms with different life histories. Curve (a) represents organisms, such as humans or whales, which tend to live out the full physiological life span if they survive early growth. Curve (b) represents organisms, such as sea gulls, in which the rate of mortality is fairly constant at all age levels. Curve (c) represents such organisms as white-tailed deer, moose, or robins, which have high mortality rates in early and late life. Curve (d) represents such organisms as clams and redwood trees, which have a high mortality rate early in life but live a full life if they reach adulthood.

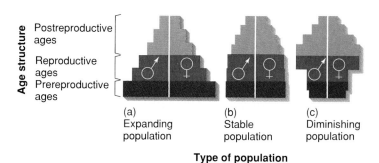

Type of population

Figure 6.11

Typical age structure diagrams for hypothetical populations that are expanding, stable, or decreasing. Different numbers of individuals in each age class create these distinctive shapes. In each diagram, the left half represents males while the right half represents females.

vironmental influences, such as climate and nutrition. Important differences in relative longevity among various types of organisms are reflected in the survivorship curves shown in figure 6.10.

Four general patterns of survivorship can be seen in this figure. Curve (a) is the pattern of organisms that tend to live their full physiological life span if they reach maturity and then have a high mortality rate when they reach old age. This pattern is typical of many large mammals, such as whales, bears, and elephants (when not hunted by humans), as well as humans in developed countries. Interestingly, some very small organisms, including predatory protozoa and rotifers (small, multicellular, freshwater animals) have similar survivorship curves even though their maximum life spans may be hundreds or thousands of times shorter than those of large mammals. In general, curve (a) is the pattern for top consumers in an ecosystem, although many annual plants have a similar survivorship pattern.

Curve (b) represents the survivorship pattern for organisms for which the probability of death is generally unrelated to age. Sea gulls, for instance, die from accidents, poisoning, and other factors that act more or less randomly. Their mortality rate is generally constant with age, and their survivorship curve is a straight line.

Curve (c) is characteristic of many songbirds, rabbits, members of the deer family, and humans in less-developed countries (see chapter 7). They have a high mortality early in life when they

are more susceptible to external factors, such as predation, disease, starvation, or accidents. Adults in the reproductive phase have a high level of survival. Once past reproductive age, they become susceptible again to external factors and the number of survivors falls quite rapidly. The moose on Isle Royale clearly fall into this category.

Curve (d) is typical of organisms at the base of a food chain or those especially susceptible to mortality early in life. Many tree species, fish, clams, crabs, and other invertebrate species produce a very large number of highly vulnerable offspring, few of which survive to maturity. Those individuals that do survive to adulthood, however, have a very high chance of living most of the maximum life span for the species.

Age Structure

An outcome of the interaction between mortality and natality is that growing or declining populations usually will have very different proportions of individuals in various age classes. Figure 6.11 shows three hypothetical population profiles that are distinguished by differences in distribution among prereproductive, reproductive, and postreproductive age classes. Pattern (a) is characteristic of a rapidly expanding population, like the moose of Isle Royale shortly after their arrival on the island. Young animals make up a large proportion of the population. This large number of prereproductive individuals represent **population momentum** because there is potential for rapid increase in natality once the youngsters reach

Population Dynamics

reproductive age. The age structure of human populations in rapidly growing, developing countries generally resembles pattern (a).

When natality comes into balance with mortality, the population enters a stationary phase having an age structure pattern similar to (b). This pattern is characteristic of nations with a stable population size, such as many European countries.

The age structure shown by pattern (c) is characteristic of a diminishing population, where natality has fallen to a level lower than the replacement number. The bulge in the upper age classes represents adults still living but reproducing at a lower rate. When those individuals die, the population will be much smaller.

Emigration

Emigration, the movement of members out of a population, is the second major factor that reduces population size. The dispersal factors that allow organisms to migrate into new areas are important in removing surplus members from the source population. Emigration can even help protect a species. For instance, if the original population is destroyed by some catastrophic change in their environment, their genes still are carried by descendants in other places. Many organisms have very specific mechanisms to facilitate migration of one or more of each generation of offspring.

Factors that Regulate Population Growth

So far, we have seen that differing patterns of natality, mortality, life span, and longevity can produce quite different rates of population growth. The patterns of survivorship and age structure created by these interacting factors not only show us how a population is growing but also can indicate what general role that species plays in its ecosystem. They also reveal a good deal about how that species is likely to respond to disasters or resource bonanzas in its environment. But what factors *regulate* natality, mortality, and the other components of population growth? In this section, we will look at some of the mechanisms that determine how a population grows.

Various factors regulate population growth, primarily by affecting natality or mortality, and can be classified in different ways. They can be *intrinsic* (operating within individual organisms or between organisms in the same species) or *extrinsic* (imposed from outside the population). Factors can also be either **biotic** (caused by living organisms) or **abiotic** (caused by nonliving components of the environment). Finally, the regulatory factors can act in a *density-dependent* manner (effects are stronger or a higher proportion of the population is affected as population density increases) or *density-independent* manner (the effect is the same or a constant proportion of the population is affected regardless of population density).

In general, biotic regulatory factors tend to be density-dependent, while abiotic factors tend to be density-independent. There has been much discussion about which of these factors is most important in regulating population dynamics. In fact, it probably depends on the particular species involved, its tolerance levels, the stage of growth and development of the organisms involved, the specific ecosystem in which they live, and the way combinations of factors interact. In most cases, density-dependent and density-independent factors probably exert simultaneous influences. Depending on whether regulatory factors are regular and predictable or irregular and unpredictable, species will develop different strategies for coping with them.

Density-Independent Factors

In general, the factors that affect natality or mortality independently of population density tend to be abiotic components of the ecosystem. Often weather (conditions at a particular time) or climate (average weather conditions over a longer period) are among the most important of these factors. Extreme cold or even moderate cold at the wrong time of year, high heat, drought, excess rain, severe storms, and geologic hazards—such as volcanic eruptions, landslides, and floods—can have devastating impacts on particular populations.

Abiotic factors can have beneficial effects as well, as anyone who has seen the desert bloom after a rainfall can attest. Fire is a powerful shaper of many biomes. Grasslands, savannas, and some montane and boreal forests often are dominated—even created—by periodic fires. Some species, such as jack pine and Kirtland's warblers, are so adapted to periodic disturbances in the environment that they cannot survive without them.

In a sense, these density-independent factors don't really regulate population *per se,* since regulation implies a homeostatic feedback that increases or decreases as density fluctuates. By definition, these factors operate without regard to the number of organisms involved. They may have such a strong impact on a population, however, that they completely overwhelm the influence of any other factor and determine how many individuals make up a particular population at any given time.

Density-Dependent Factors

Density-dependent mechanisms tend to reduce population size by decreasing natality or increasing mortality as the population size increases. Most of them are the results of interactions *between* populations of a community (especially predation), but some of them are based on interactions *within* a population.

Interspecific Interactions

As we discussed in chapter 4, a predator feeds on—and usually kills—its prey species. While the relationship is one-sided with respect to a particular pair of organisms, the prey species as a whole may benefit from the predation. For instance, the moose that gets eaten by wolves doesn't benefit individually, but the moose *population* is strengthened because the wolves tend to kill old or sick members of the herd. Their predation helps prevent population overshoot, so the remaining moose are stronger and healthier.

Sometimes predator and prey populations oscillate in a sort of synchrony with each other as is shown in figure 6.12. This is a classic study of the number of furs brought into Hudson's Bay Company trading posts in Canada between 1840 and 1930. As you can see, the numbers of Canada lynx fluctuate on about a ten-year cycle that is similar to, but slightly out of phase with, the popula-

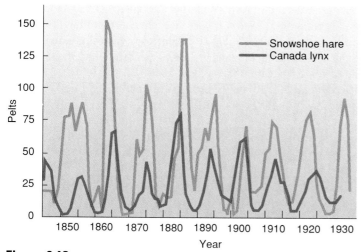

Figure 6.12

Oscillations in the populations of snowshoe hare and lynx in Canada suggest the close interdependency of this prey-predator relationship. This data is based on the number of pelts received by the Hudson Bay Company. Both predator and prey show a ten-year cycle in population growth and decline.

Source: Data from D. A. MacLulich, *Fluctuations in the Numbers of the Varying Hare* (*Lepus americus*). Toronto: University of Toronto Press, 1937, reprinted 1974.

tion peaks of snowshoe hares. When the hare population is high and food is plentiful, lynx reproduction is very successful and lynx populations grow rapidly. Predation diminishes the number of hares available until eventually lynx do not have enough to eat and their population sizes fall. As the predators die off, the prey begin to make a comeback and the cycle starts over again. This predator-prey oscillation is known as the Lotka-Volterra model after the scientists who first described it mathematically.

Parasitism is a form of predation in which the parasite takes nutrients or other benefits at the expense of its host but usually does not kill it. The mosquito or tick that helps itself to a blood meal from your epidermis is a good example. Often, a parasite and its host develop a sort of balance with each other, something like the predator/prey relationship we have just discussed. If the parasite is too virulent and kills its host, food for its offspring will be unavailable. Wherever a parasite and host combination have lived together for a long time, the host is likely to have developed some resistance to its parasite so that enough individuals survive to continue the population. If parasites are introduced to new ecosystems, however, or new host species are introduced into the parasite's ecosystem, the effects can be disastrous. On the plains of Africa, for instance, small biting flies called tsetse flies carry the trypanosome that causes sleeping sickness. The native species of hoofed mammals are mostly resistant to this protozoan parasite, but domestic cattle, which have not been part of the community long enough to adapt, may be killed or become seriously debilitated if they are exposed.

In interspecific competition, two species compete for the same environmental resources in an ecosystem, but this competition rarely involves outright struggles or confrontations. It more often involves a contest in which the species tend to eat faster or grow faster or grab a bigger share of the resources in some other way. As a result, competing populations may work out a balance of resource partitioning (chapter 4), or one population may eclipse the other.

Not all interspecific interactions are harmful to one of the species involved. Mutualism and commensalism, for instance, are interspecific interactions that are beneficial or neutral in terms of population growth (chapter 4).

Intraspecific Interactions

Individuals within a population also compete for resources. When population density is low, resources are likely to be plentiful and the population growth rate will approach the maximum possible for the species, assuming that individuals are not so dispersed that they cannot find mates. As population density approaches the carrying capacity of the environment, however, one or more of the vital resources becomes limiting. The stronger, quicker, more aggressive, more clever, or luckier members get a larger share, while others get less and then are unable to reproduce successfully or survive.

Territoriality is one principal way many animal species control access to environmental resources. The individual, pair, or group that holds the territory will drive off rivals if possible, either by threats, displays of superior features (colors, size, dancing ability), or fighting equipment (teeth, claws, horns, antlers). Members of the opposite sex are attracted to individuals that are able to seize and defend the largest share of the resources. From a selective point of view, these successful individuals presumably represent superior members of the population and the ones best able to produce offspring that will survive.

Stress and Crowding

Stress and crowding also are density-dependent population control factors. When population densities get very high, organisms often exhibit symptoms of what is called **stress shock** or **stress-related diseases.** These terms describe a loose set of physical, psychological, and/or behavioral changes that are thought to result from the stress of too much competition and too close proximity to other members of the same species. There is a considerable controversy about what causes such changes and how important they are in regulating natural populations. The strange behavior and high mortality of arctic lemmings or snowshoe hares during periods of high population density may be a manifestation of stress shock. On the other hand, they could simply be the result of malnutrition, infectious disease, or some other more mundane mechanism at work.

Some of the best evidence for the existence of stress-related disease comes from experiments in which laboratory animals, usually rats or mice, are grown in very high densities with plenty of food and water but very little living space (table 6.4). A variety of symptoms are reported, including reduced fertility, low resistance to infectious diseases, and pathological behavior, such as hypoactivity, hyperactivity, aggression, lack of parental instincts, sexual deviance, and cannibalism. Dominant animals seem to be affected least by crowding, while subordinate animals—the ones presumably subjected to the most stress in intraspecific interactions—seem to be the most severely affected.

Table 6.4

The influence of density on fecundity in the house mouse (*Mus musculus*)

	Sparse	Medium	Dense	Very Dense
Average number/m³	34	118	350	1600
Average percentage pregnant	58.3	49.4	51.0	43.4
Average number per litter	6.2	5.7	5.6	5.1

Source: C. Southwick, "Population Characteristics of House Mice Living in English Corn Ricks: Density Relationships," *Proc. Zool. Soc. Lond.* 131:163–175, 1958, copyright 1958 The Zoological Society of London.

The clinical symptoms reported in these studies include enlargement of the adrenal glands, reduction of the thymus and reproductive glands, and deterioration of the heart, blood vessels, kidney, and liver. The most probable explanation for these changes is that the adrenal gland is the source of epinephrine (adrenalin), the "fight-or-flight" hormone that raises energy levels and stimulates an animal to respond to danger or stress. Ordinarily, the burst of epinephrine released by fear, anger, or excitement is short-lived. When the stimulation is very frequent, however, the adrenal glands become enlarged in response to the continual call for epinephrine, and hormone production is elevated to a constant high level. This puts stress on the heart and circulatory system. In addition to epinephrine, the adrenal glands secrete sex hormones that affect metabolism and behavior. They also secrete corticosteroids that regulate mineral levels, energy metabolism, and the immune systems. Abnormally high levels of these hormones damage the liver and kidneys and can lead to a variety of symptoms including hypertension, arthritis, arteriosclerosis, gastrointestinal ulcers, and reduced resistance to disease. All these factors will tend to increase mortality and decrease fertility.

Are there equivalent stress-related diseases that cause aberrant behavior in human societies? Some people suggest that the high concentrations of social problems in dense urban centers might result from crowding stresses similar to those observed in laboratory animals. They point out that inner cities have much higher rates of disease and social problems, such as mental illness, family dysfunction, alcohol and drug abuse, and criminal behavior, than do other areas. The inner city also has much higher levels of poverty, neglect, undereducation, unemployment, and other social conditions, however, that may be more important than the population density problems found there. While it is tempting to extrapolate conclusions from laboratory experiments to natural populations where a complex of factors are interacting, the conclusions can be misleading. In chapter 24, we discuss the socioeconomic disparities in our urban centers and their effects on human populations.

Summary

Population dynamics play an important role in determining how ecosystems work. Biological organisms generally have the ability to produce enough offspring so populations can grow rapidly when resources are available. Given optimum conditions, populations of many organisms can grow exponentially; that is, they grow at a constant rate of increase so that the population size doubles in some regular interval of time. If conditions appropriate for this kind of growth persist and other factors don't intervene to reduce abundance, exponential growth can produce astronomical numbers of organisms. We describe the rapidly rising curve of an exponentially growing population as a J curve.

Some populations will grow exponentially until they overshoot the carrying capacity of the environment. Mortality rates rise as resources become limited and the population may crash, often dying back as rapidly as it rose. Other species have intrinsic mechanisms that regulate their population growth rate, resulting in an equilibrium at or near the carrying capacity of the environment.

The most important components of population dynamics are natality, fertility, fecundity, life span, longevity, mortality, immigration, and emigration. The sum of all additions to and subtractions from the population determines the net rate of growth. Mortality rates and longevity are often expressed as survivorship rates that reveal much about a species' place in its ecosystem and the kinds of hazards that eliminate members from the population.

The factors that regulate population dynamics can be either intrinsic or extrinsic to the population. They can be caused by biotic or abiotic forces, and they can act on the population in either a density-dependent or density-independent fashion. The most important abiotic regulatory factors are usually climate and weather. The most important biological factors are usually competition (both interspecific and intraspecific), predation, and disease.

Often, organisms develop specific behavioral patterns that reduce conflict and facilitate resource partitioning. In some cases of extreme crowding, it is thought that physiological responses to excessive intraspecific interaction can result in stress-related disease that can lead to aberrant behavior, reduced reproductive success, increased susceptibility to disease, and high rates of mortality. Whether these mechanisms operate in humans at high population densities is a matter of controversy.

▼ Questions for Review

1. Why did moose populations grow so rapidly on Isle Royale in the 1920s?

2. What is the difference between exponential and arithmetic growth?

3. Given a growth rate of 3 percent per year, how long will it take for a population of 100,000 individuals to double? How long will it take to double when the population reaches 10 million?

4. What is environmental resistance? How does it affect populations?

5. What is the difference between fertility and fecundity?

6. Describe the four major types of survivorship patterns and explain what they show about the role of the species in an ecosystem.

7. What are the main interspecific population regulatory interactions? How do they work?

8. What are the suspected causes and effects of stress shock or stress-related disease?

▼ Questions for Critical Thinking

1. Compare the advantages and disadvantages to a species that result from exponential or logistic growth. Why do you think moose have evolved to reproduce as rapidly as possible, while wolves appear to have intrinsic or social mechanisms to limit population growth?

2. We're not sure why wolf populations on Isle Royale are declining. How would you design a study to examine the possible explanations offered in this chapter without compromising the wilderness nature of the park and its wildlife?

3. If wolves are declining due to natural causes, should we intervene or let nature take its course?

4. Are humans subject to environmental resistance in the same sense that other organisms are? How would you decide whether a particular factor that limits human population growth is ecological or social?

5. There obviously are vast differences in birth and death rates, survivorship, and life spans among species. There must be advantages and disadvantages in living longer or reproducing more quickly. Why hasn't evolution selected for the most advantageous combination of characteristics so that all organisms would be more or less alike?

6. Abiotic factors that influence population growth tend to be density-independent, while biotic factors that regulate population growth tend to be density-dependent. Explain.

7. Some people consider stress and crowding studies of laboratory animals highly applicable in understanding human behavior. Other people question the cross-species transferability of these results. What considerations would be important in interpreting these experiments?

8. What implications for human population control might we draw from our knowledge of basic biological population dynamics?

▼ Key Terms

abiotic 128
arithmetic growth 121
biotic 128
biotic potential 122
carrying capacity 120
catastrophic system 124
chaotic systems 123
dieback 122
emigration 128
environmental resistance 123
exponential growth 121
fecundity 126
fertility 126
geometric growth 121
irruptive or Malthusian growth 123

J curve 121
life expectancy 126
life span 126
logistic growth 123
mortality 126
natality 128
overshoot 122
population crash 120
population explosion 120
population momentum 127
S curve 123
stress-related diseases 129
stress shock 129
survivorship 126

▼ Additional Information on the Internet

Centre for Population Biology
    http://forest.bio.ic.ac.uk/cpb/cpb/cpbinto.html/

Landscapes, Cataclysms, and Population Explosions
    http://www.csu.edu.au/edmonton.html/

Population and Demography Information
    http://www.pop.psu.Edu/Demography/Demography.html/

Talk.Origins Archive, Evolution FAQ
    http://rumba.ics.uci.edu:8080/origins/FAQS-evolution.html/

Note: Further readings appropriate to this chapter are listed on p. 595.

A poor fishing family on the steps of their house on Balicasaq Island in the Phillipines. What public and private decisions lead to population growth?

CHAPTER 7: Human Populations

Objectives

After studying this chapter, you should be able to:

- trace the history of human population growth.

- summarize Malthusian and Marxian theories of limits to growth as well as why technological optimists and supporters of social justice oppose these theories.

- explain the process of demographic transition and why it produces a temporary population surge.

- understand how changes in life expectancy, infant mortality, women's literacy, standards of living, and democracy affect population changes.

- evaluate pressures for and against family planning in traditional and modern societies.

- compare modern birth control methods and prepare a personal family planning agenda.

Population Growth

Every second, on average, five children are born somewhere on the earth. In that same second, two other people die. This difference between births and deaths means a net gain of three more humans per second in the world population. If you multiply this out, you will find we are growing at a rate of about 11,000 per hour, 265,000 per day, or almost 100 million more people per year. By 1995, the world population had reached about 5.7 billion, currently making us the most numerous vertebrate species on the planet. For the families to whom these children are born, this may well be a joyous and long-awaited event (fig. 7.1). But is a continuing increase in humans good for the planet in the long run?

Many people worry that overpopulation will cause—or perhaps already is causing—resource depletion and environmental degradation that threaten the ecological life-support systems on which we all depend. These fears often lead to demands for immediate, worldwide birth control programs to reduce fertility rates and to eventually stabilize or even shrink the total number of humans.

Others believe that human ingenuity, technology, and enterprise can extend the world carrying capacity and allow us to overcome any problems we encounter. From this perspective, more people may be beneficial rather than disastrous. A larger population means a larger work force, more geniuses, more ideas about what to do. Along with every new mouth comes a pair of hands. Proponents of this worldview—many of whom happen to be economists—argue that continued economic and technological growth can both feed the world's billions and enrich everyone enough to end the population explosion voluntarily. Not so, counter many ecologists. Growth *is the problem; we must stop both population and economic growth.*

Yet another perspective on this subject derives from social justice concerns. From this worldview, there are enough resources

Figure 7.1

A Mayan family in Guatemala with four of their six living children. Decisions on how many children to have are influenced by many factors, including culture, religion, need for old age security for parents, immediate family finances, household help, child survival rates, and power relationships within the family. Having many children may not be in the best interests of society at large, but may be the only rational choice for individual families.

for everyone. Current shortages are only signs of greed, waste, and oppression. The root cause of environmental degradation, in this view, is inequitable distribution of wealth and power rather than population size. Fostering democracy, empowering women and minorities, and improving the standard of living of the world's poorest people are what are really needed. A narrow focus on population growth only fosters racism and an attitude that blames the poor for their problems while ignoring the deeper social and economic forces at work.

Whether human populations will continue to grow at present rates and what that growth would imply for environmental quality and human life are among the most central and pressing questions in environmental science. In this chapter, we will look at some causes of population growth as well as how populations are measured and described. Family planning and birth control are essential for stabilizing populations. The number of children a couple decides to have and the methods they use to regulate fertility, however, are strongly influenced by culture, religion, politics, and economics, as well as basic biological and medical considerations. We will examine how some of these factors influence human demographics.▶

Human Population History

For most of our history, humans have not been very numerous compared to other species. Studies of hunting and gathering societies suggest that the total world population was probably only a few million people before the invention of agriculture and the domestication of animals around ten thousand years ago. The larger and more secure food supply made available by the agricultural revolution allowed the human population to grow, reaching perhaps 50 million people by 5000 B.C. For thousands of years, the number of humans increased very slowly. Archaeological evidence and historical descriptions suggest that only about 300 million people were living at the time of Christ (table 7.1).

Until the Middle Ages, human populations were held in check by diseases, famines, and wars that made life short and uncertain for most people (fig. 7.2). Furthermore, there is evidence that many early societies regulated their population size through cultural taboos and practices such as infanticide. Among the most destructive of natural population controls were bubonic plagues that periodically swept across Europe between 1348 and 1650. During the worst plague years (between 1348 and 1350), it is estimated that at least one-third of the European population perished. Notice, however, that this did not retard population growth for very long. In 1650, at the end of the last great plague, there were about 600 million people in the world.

As you can see in figure 7.2, human populations began to increase rapidly after 1600 A.D. Many factors contributed to this rapid growth. Increased sailing and navigating skills stimulated

World population growth and doubling times

Date	Population	Doubling Time
5000 B.C.	50 million	?
800 B.C.	100 million	4200 years
200 B.C.	200 million	600 years
1200 A.D.	400 million	1400 years
1700 A.D.	800 million	500 years
1900 A.D.	1,600 million	200 years
1965 A.D.	3,200 million	65 years
1990 A.D.	5,300 million	38 years
2020 A.D. (estimate)	8,230 million	55 years

Source: Population Reference Bureau, Inc., Washington, DC.

commerce and communication between nations. Agricultural developments, better sources of power, and better health care and hygiene also played a role. We are now in an exponential or J curve pattern of growth (What Do You Think?, p. 136). Will this population explosion continue until we overshoot the carrying capacity of our environment—if we have not done so already—and experience

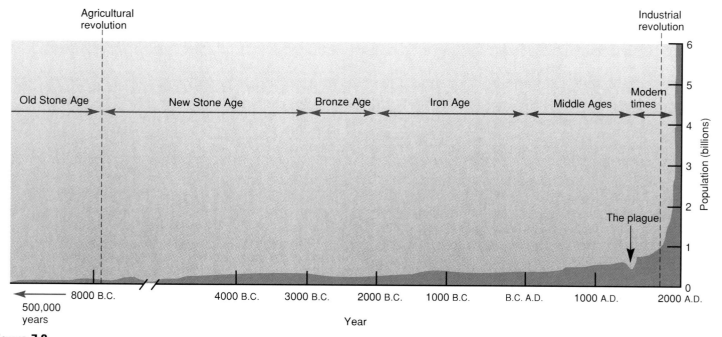

Figure 7.2

Human population levels through history. Since about 1000 A.D., our population curve has assumed a J shape. Are we on the upward slope of a population overshoot? Will we be able to adjust our population growth to an S curve? Or can we just continue the present trend indefinitely?

a catastrophic dieback similar to those described in chapter 6? Or will our populations stabilize at some tolerable level? As you will see later in this chapter, there is some evidence that population growth already is slowing, but whether we will reach equilibrium soon enough and at a size that can be sustained over the long term remains a difficult but important question.

Limits to Growth: Some Opposing Views

As with religion and politics, people have widely differing opinions about population and resources. Some believe that population growth is the ultimate cause of poverty and environmental degradation. Others argue that poverty, environmental degradation, and overpopulation are all merely symptoms of deeper social and political factors. In this section, we will examine some opposing worldviews and their implications.

Malthusian Checks on Population

In 1798, the Rev. Thomas Malthus wrote *An Essay on the Principle of Population as It Affects the Future Improvement of Society, with Remarks on the Speculations of Mr. Godwin, M. Condorcet, and Other Writers* to refute the views of progressives and optimists—including his father—who were inspired by the egalitarian principles of the French Revolution to predict a coming utopia. Malthus argued that human populations tend to increase at an exponential or compound rate while food production either remains stable or increases only slowly. The result, he predicted, is that human pop-

ulations inevitably outstrip their food supply and eventually collapse into starvation, crime, and misery. Malthus's theory might be summarized by the equation in figure 7.3*a*.

According to Malthus, the only ways to stabilize human populations are "positive checks," such as disease or famines that kill people, or "preventative checks," including all the factors that prevent human birth. Among the preventative checks he advocated were "moral restraint," including late marriage and celibacy until a couple can afford to support children. Many social scientists and biologists have been influenced by Malthus. Charles Darwin, for instance, derived his theories about the struggle for scarce resources and survival of the fittest after reading Malthus's essay.

If Malthus's views of the consequences of exponential population growth were dismal, the corollary he drew was even more bleak. He believed that most people are too lazy and immoral to regulate birth rates voluntarily. Consequently, he opposed efforts to feed and assist the poor in England because he feared that more food would simply increase their fertility and thereby perpetuate the problems of starvation and misery.

Not surprisingly, Malthus's ideas provoked a great social and economic debate. Karl Marx was one of his most vehement critics, claiming that Malthus was a "shameless sycophant of the ruling classes." According to Marx, population growth is a symptom rather than a root cause of poverty, resource depletion, pollution, and other social ills. The real causes of these problems, he believed, are exploitation and oppression (fig. 7.3*b*). Marx argued that workers always provide for their own sustenance given access to means of production and a fair share of the fruits of their labor. According to Marxians, the way to slow population growth *and* to alleviate

What Do *You* Think?
Looking for Bias in Graphs

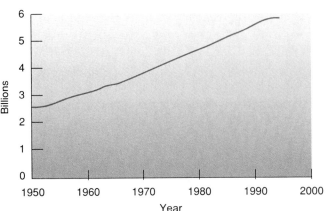

Figure 1

Total world population 1950–94.

Source: Census Bureau.

Graphs, pictorial representations of information, can help others understand what we have to say. They are particularly helpful in conveying patterns of relationship. Why are graphs used so widely and why do they make such powerful impressions on us?

Not only are images expressive, but graphs come equipped with substantiating data. It is one thing to read that "the human population is rising explosively," but it is considerably more compelling to see the J curve pictured together with concrete numbers as evidence. The combination of picture and numbers often conveys powerful impressions beyond the numbers themselves.

An important question addressed in this chapter is: How serious is the worldwide population problem? Does it demand action? The graphs to follow all present information relevant to human population studies. Yet the impressions left with the reader vary considerably depending on the type of data used, and the design of the graph.

Reexamine figure 7.2, which presents an historical perspective on the growth in human numbers. What conclusions do you draw from it? Does it suggest we are experiencing an explosive rise in human numbers? That the current increase is unprecedented in our species history? Does it also imply, perhaps more indirectly, that the earth's population problem is serious, that there is an urgency to respond? In light of concepts of environmental resistance, carrying capacity, and overshoot and dieback, many people would conclude that the current growth, highlighted by the graph, is unsustainable.

Now examine figure 1.

It plots the same variables as figure 7.2, but suggests that population is rising at a modest rate. It does not produce the sense of explosiveness and urgency as figure 7.2. How can the impressions be so different?

Notice that graph 1 covers a much shorter time period. This greatly changes the time interval lengths on the horizontal scale. In figure 7.2, one-millimeter represents about 50 years, but less than one year in graph 1. This changes the line's slope from nearly

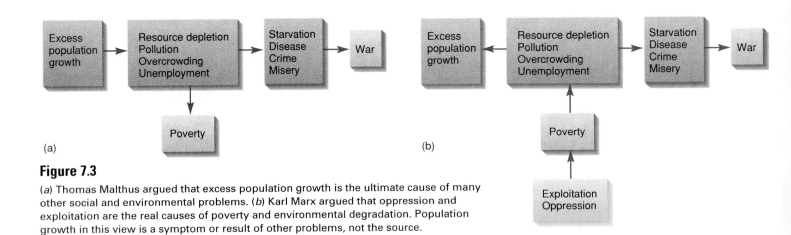

Figure 7.3

(*a*) Thomas Malthus argued that excess population growth is the ultimate cause of many other social and environmental problems. (*b*) Karl Marx argued that oppression and exploitation are the real causes of poverty and environmental degradation. Population growth in this view is a symptom or result of other problems, not the source.

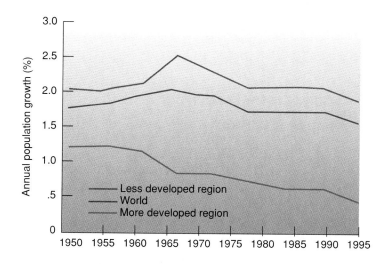

Figure 2

Percent annual population growth 1950–94.

Source: United Nations, *World Population Prospects: The 1994 Revision*
UN Department for Economic and Social Information and Policy
Analysis, 1994, pp. 56–58, 62, 64.

vertical in figure 7.2 to a modest incline in graph 1. Slope impacts the visual impression created by a graph, and therefore on its interpretation. How could you change the time axis in graph 1 to flatten the slope even more?

Next examine figure 2. This graph plots the rate of population growth rather than population size over time.

This graph reveals that the rate has been in decline since about 1970. By itself, does the graph's downward-trending line suggest there really isn't much of a population problem, or at least not much reason for concern?

All three graphs present information relevant to a discussion of worldwide population growth, yet each gives a different impression of the seriousness of the problem. So, how can we analyze graphs and their messages in a thoughtful way? There is no single formula applicable to all situations, certainly, but a few questions are worth keeping in mind.

Is the time frame of reference appropriate or is it too restricted to allow a valid, comprehensive assessment of the issue? Are the unit intervals on the graph appropriately sized? Is the im-

pression created by the graph real or simply an artifact of the graph's format? Does the data presented in a graph present only a partial, perhaps misleading, view of the whole?

Because graphs can create powerful impressions, they need to be interpreted with care.

crime, disease, starvation, misery, and environmental degradation is through social justice.

Malthus and Marx Today

Both Marx and Malthus developed their theories about human population growth in the nineteenth century when understanding of the world, technology, and society were much different than they are now. Still, the questions they raised are relevant today. While the evils of racism, classism, and other forms of exploitation that Marx denounced still beset us, it is also true that at some point available resources must limit the numbers of humans that the earth can sustain (fig. 7.4).

Those who agree with Malthus, that we are approaching—or may already have surpassed—the carrying capacity of the earth are called **neo-Malthusians.** In their view, we should address the issue of surplus population directly by making birth control our highest priority. An extreme version of this worldview is expressed by Cornell University entomologist David Pimentel, who claims that the "optimum human population" would be nearly 5 billion fewer than currently inhabit the planet. He fails to suggest how the 10 percent of us allowed to live would be chosen.

Figure 7.4

Is this our future?

TOLES © 1990 The Buffalo News. Reprinted with permission of *UNIVERSAL PRESS SYNDICATE.* All Rights Reserved.

Human Populations

Neo-Marxians, on the other hand, believe that only eliminating oppression and poverty through technological development and social justice will solve population problems. Perhaps both these viewpoints have some validity. A compromise position is that population growth, poverty, and environmental degradation all are interrelated. No factor exclusively causes any other but each influences and, in turn, is influenced by the others.

Can Technology Make the World More Habitable?

Technological optimists argue that Malthus was wrong in his predictions of famine and disaster two hundred years ago because he failed to account for scientific progress. In fact, food supplies have increased faster than population growth since Malthus' time. There have been terrible famines in the past two centuries, but they were caused more by politics and economics than lack of resources or sheer population size. Whether this progress will continue remains to be seen, but technological advances have increased human carrying capacity more than once in our history.

The burst of growth of which we are a part, was stimulated by the Scientific and Industrial Revolutions. Progress in agricultural productivity, engineering, information technology, commerce, medicine, sanitation, and other achievements of modern life have made it possible to support approximately one thousand times as many people per unit area as was possible ten thousand years ago.

Much of our growth in the past three hundred years has been based on availability of easily acquired natural resources, especially cheap, abundant fossil fuels. Whether we can develop alternative, renewable energy sources in time to avert disaster when current fossil fuels run out is a matter of great concern (chapter 21).

Can More People Be Beneficial?

There can be benefits as well as disadvantages in larger populations. More people mean larger markets, more workers, and efficiencies of scale in mass production of goods. Greater numbers also provide more intelligence and enterprise to overcome problems such as underdevelopment, pollution, and resource limitations. Human ingenuity and intelligence can create new resources through substitution of new materials and new ways of doing things for old materials and old ways. For instance, utility companies are finding it cheaper and more environmentally sound to finance insulation and energy-efficient appliances for their customers rather than build new power plants. The effect of saving energy that was formerly wasted is comparable to creating a new fuel supply.

Economist Julian Simon is one of the most outspoken champions of this rosy view of human history. People, he argues, are the "ultimate resource" and there is no evidence that pollution, crime, unemployment, crowding, the loss of species, or any other resource limitations will worsen with population growth. This outlook is shared by leaders of many developing countries who insist that instead of being obsessed with population growth, we should focus on the inordinate consumption of the world's resources by people in richer countries. What constitutes a resource and which resources

might limit further human population growth are questions we will return to in subsequent chapters in this book. For now, we will move on to discuss how people are counted.

Human Demography

Demography is derived from the Greek words *demos* (people) and *graphos* (to write or to measure). It encompasses vital statistics about people, such as births, deaths, and where they live as well as total population size. In this section, we will survey ways human populations are measured and described, and discuss demographic factors that contribute to population growth.

How Many of Us Are There?

Even in this age of information technology and communication, we do not know exactly how many people there are in the world. Some countries have never even had a census taken, and even those that have been done may not be accurate. Governments may overstate or understate their populations to make their countries appear larger and more important or smaller and more stable than they really are. Individuals, especially if they are homeless, refugees, or illegal aliens, may not want to be counted or identified.

Although there is considerable uncertainty about the exact number of people in some places, most demographers agree that about 5.7 billion people were alive in mid-1995. As we pointed out in chapter 1, only about 20 percent of that population lives in the more-developed or rich countries of the world. Four out of five humans live in the poorer countries of the less-developed world. Demographers estimate that 90 percent of the population growth expected to occur in the next century will take place in the Third World (fig. 7.5).

China has by far the largest population in the world (table 7.2), but India is catching up rapidly because its population control programs have been less successful than China's (Case Study, p. 144).

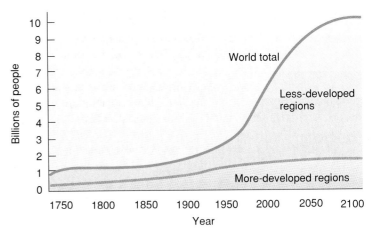

Figure 7.5

Human population growth, 1750–2100, in less-developed and more-developed regions. The human population is now in an exponential growth phase. More than 90 percent of all growth in this century and projected for the next is in the less-developed countries.

Table 7.2
Twelve most populous countries in 2025 (population in millions)

Country	1950	1995	2025	Ratio*
China	554.8	1238.3	1590.8	2.9
India	357.6	931.0	1383.1	3.9
United States	152.3	263.1	295.5	1.9
Indonesia	79.5	201.5	285.9	3.6
Pakistan	39.5	134.9	281.4	7.0
Brazil	53.4	161.4	237.2	4.5
Nigeria	32.9	126.9	216.2	6.6
Bangladesh	41.8	128.3	211.6	5.0
Soviet Union (former)	94.6	149.7	170.7	1.8
Iran	16.9	66.7	159.2	9.4
Mexico	28.0	93.7	143.3	5.3
Japan	83.6	125.9	124.1	1.5

Source: Data from the Population Reference Bureau, Inc., and World Resources Institute.
*Ratio 1950 to 2025.

Notice the interesting changes in relative ranking of countries resulting from demographic changes. The former Soviet Union had the third largest population in the world when it broke apart in 1992. Environmental pollution, civic chaos, and hyperinflation have reduced the standard of living in much of the former Soviet Union so that birth rates have decreased by nearly half and death rates have increased sharply over the past decade. Russia's population is now declining by about 1 million people every year, and Indonesia, Pakistan, Brazil, Nigeria, and Bangladesh are each expected to have more people than Russia in about 30 years. Note that Iran, which had only 16.9 million people in 1950, is expected to have grown almost tenfold by 2025. Sheer numbers do not necessarily equate with power, however. Bangladesh, which is about the size of Iowa, is already overcrowded at 114 million. Doubling the population again in the next 30 years will bring unimaginable calamity, not world power.

Figure 7.6 shows human population distribution around the world. Notice the high densities supported by fertile river valleys of the Nile, Ganges, Yellow, Yangtze, and Rhine Rivers and the well-watered coastal plains of India, China, and Europe. Historic factors such as technology diffusion and geopolitical power also play a role in geographic distribution.

Fertility and Birth Rates

As we pointed out in chapter 6, fecundity is the physical ability to reproduce, while fertility describes the actual production of offspring. Those without children may be fecund but not fertile. The most accessible demographic statistic of fertility is usually the **crude birth rate,** the number of births in a year per thousand persons. It is statistically "crude" in the sense that it is not adjusted for population characteristics such as the number of women in reproductive age.

The **total fertility rate** is the number of children born to an average woman in a population during her entire reproductive life. Upper-class women in seventeenth and eighteenth century England, whose babies were given to wet nurses immediately after birth and who were expected to produce as many children as possible, often had twenty-five or thirty pregnancies. The highest recorded total fertility rates for working-class people is among some Anabaptist agricultural groups in North America who have averaged up to twelve children per woman. In most tribal or traditional societies, food shortages, health problems, and cultural practices limit total fertility to about six or seven children per woman even without modern methods of birth control.

Fertility is usually calculated as births per woman because, in many cases, it is difficult to establish paternity. Some demographers argue, nevertheless, that we should pay more attention to birth rates per male, because in some cultures men have far more children, on average, than do women. In Cameroon, for instance, due to multiple marriages, extramarital affairs, and a high rate of female mortality, men are estimated to have 8.1 children in their lifetime, while women average only 4.8.

The **zero population growth (ZPG)** rate (also called the replacement level of fertility) is the number of births at which people are just replacing themselves. In the more highly developed countries, where infant mortality rates are low, this rate is usually about 2.1 children per couple. It takes slightly more than two children per couple to stabilize the population because some people are infertile, have children who do not survive, or choose not to have children. Total fertility rates have been dropping rapidly in most regions over the past two or three decades (fig. 7.7). The world average fertility has fallen from 6.1 to 3.3 in just 20 years. If this rate of progress were to continue for another decade, we would reach the replacement rate by the end of this century.

Table 7.3 compares some demographic characteristics of the ten countries with the highest total fertility rate to those of the ten with lowest total fertility rates. Cote d'Ivorie had the largest average number of children per woman (7.5) in the world in 1995. Other countries with highest fertility rates (in order) include Uganda, Yemen, Angola, Benin, Mali, Namibia, Ethiopia, Guinea, and Somalia, all of which have more than 7 children per woman. By contrast, Italy, with a fertility rate of 1.3, and Russia, with 1.4, were lowest in the world in 1995. Other countries in this group are Austria, Germany, Greece, Belgium, Denmark, Japan, Netherlands, and Switzerland, all of which were between 1.5 to 1.7 children per woman.

Note that all countries with the highest fertility rates—except Yemen—are in Africa, and most recently have had famines, civil wars, or other socially destabilizing events. Countries with low fertility rates, on the other hand, are generally wealthy, stable, and democratic. A striking exception to this pattern is Russia, where the number of children per woman has fallen from an average of 2.2 in 1990 to 1.4 in 1995. A disastrous economy, an unstable future, poor nutrition and health care, and botched abortions that have left many

Human Populations

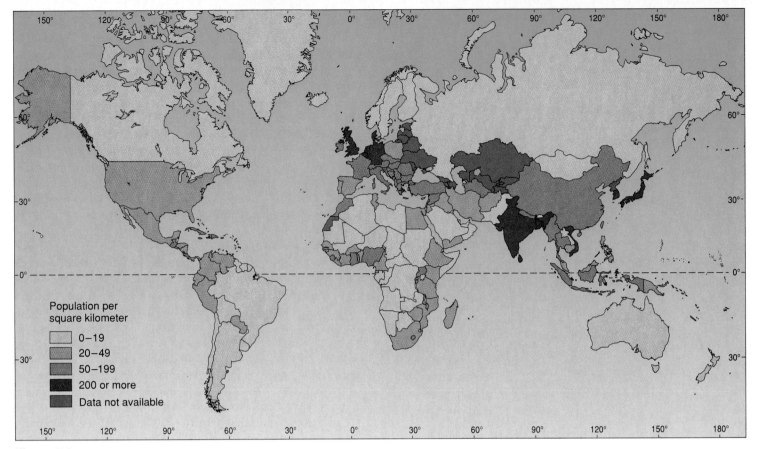

Figure 7.6

World population density. Two-thirds of all humanity lives on about 10 percent of the earth's land surface. Although the world as a whole has about 100 people per square mile (39/km²), Canada and Australia have less than 7 people per square mile on average, compared to more than 2000 people in the same area in Bangladesh and the island of Java in Indonesia.

Source: World Bank, *World Development Report 1992*, Oxford Univeristy Press.

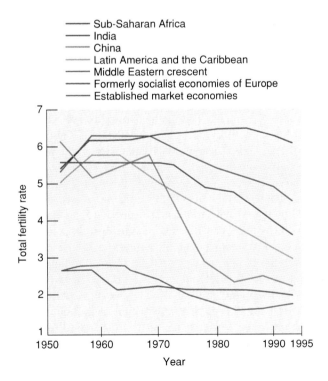

women unwilling or unable to bear children all have contributed to this sudden decline.

In general, high birth rates correlate strongly with poverty, high infant mortality, low female literacy rates, and lack of paid employment for women. Unfortunately, the disparities in social conditions shown in table 7.3 are growing worse rather than better. In the last edition of this book, slow growing countries had about 10 times the income of rapidly growing ones. In 1995, it was 36 to 1. In nearly every category, characteristics of the poorest countries are getting worse, not better.

Mortality and Death Rates

A traveler to a foreign country once asked a local resident, "What's the death rate around here?" "Oh, the same as anywhere," was the reply, "about one per person." In demographics, however, **crude**

Figure 7.7

Total fertility rates by region, 1950–95.
Note: Dotted lines represent projected values.

Source: World Bank, *World Development Report 1993*.

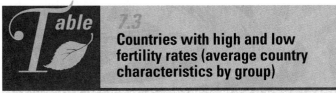

Table 7.3

Countries with high and low fertility rates (average country characteristics by group)

	Ten Highest	Ten Lowest
Total fertility[1]	7.1	1.6
Birth rate[2]	46.6	11.7
Death rate[2]	17.5	10.2
Annual increase[3]	3.2	0.002
Infant mortality[4]	111.9	7.0
Female literacy[3]	15	>95
GNP per capita	$530	$19,304

Source: Data from *World Resources 1994–95*, World Resources Institute.
[1]Children born per woman in a lifetime
[2]Per 1000 people
[3]Percent
[4]Per 1000 live births

death rates (or crude mortality rates) are expressed in terms of the number of deaths per thousand persons in any given year. Countries in Africa where health care and sanitation are limited may have mortality rates of twenty or more per thousand people. Wealthier countries generally have mortality rates around ten per thousand. The number of deaths in a population is sensitive to the age structure of the population. Rapidly growing, developing countries such as Belize or Costa Rica have lower crude death rates (5 per 1000) than do the more-developed, slowly growing countries, such as Denmark (12 per 1000). This is because there are proportionately more youths and fewer elderly people in a rapidly growing country than in a more slowly growing one.

Population Growth Rates

Crude death rate subtracted from crude birth rate gives the **natural increase** of a population. We distinguish natural increase from the **total growth rate,** which includes immigration and emigration, as well as births and deaths. Both of these growth rates are usually expressed as a percent (number per hundred people) rather than per thousand. A useful rule of thumb is that if you divide 70 by the annual percentage growth, you will get the approximate doubling time in years. Qatar, for example, which is growing 4.16 percent per year, is doubling its population every 17.5 years. The United States and Canada, which have natural increase rates of 0.8 percent per year, are doubling in 89 years. Actually, because of immigration, U.S. total growth is about twice as fast as natural increase. Denmark, with a natural increase rate of 0.1 percent, is doubling in about 700 years. Belgium, Germany, Hungary, and Italy have negative growth rates and declining populations. The world growth rate is now 1.7 percent, which means that the population will double in about 41 years if this rate persists.

The most rapid total growth rates in 1995 were Afghanistan (6.7 percent) and Israel (4.7 percent). Afghanistan has many refugees returning from the recent civil war, while Israel has many immigrants fleeing from oppressive conditions in their home countries.

Life Span and Life Expectancy

Life span is the oldest age to which a species is known to survive. Although there are many claims in ancient literature of kings living for one thousand years or more, the oldest age that can be certified by written records is that of Jeanne Louise Calment of Arles, France, who was a spry, healthy 120 years old in 1995. The aging process is still a medical mystery, but it appears that cells in our bodies have a limited ability to repair damage and produce new components. At some point they simply wear out, and we fall victim to disease, degeneration, accidents, or senility.

Life expectancy is the average age that a newborn infant can expect to attain in any given society. It is another way of expressing the average age at death. For most of human history, we believe that life expectancy in most societies has been between thirty-five and forty years. This doesn't mean that no one lived past age forty, but rather that so many deaths at earlier ages (mostly early childhood) balanced out those who managed to live longer. It once was widely believed that differences in life expectancy between ethnic groups were biological, and therefore, difficult to change. We now know that social and environmental, not biological, factors are responsible for most variations in mortality. In the highly developed countries, the average life expectancy has nearly doubled in the past two centuries. The maximum possible life span, however, does not appear to have been changed at all by modern medicine.

Declining mortality, not rising fertility, is the primary cause of most population growth in the past three hundred years. Crude death rates began falling in Western Europe during the late 1700s. Most of this advance in survivorship came long before the advent of modern medicine and is due primarily to better food and better sanitation. As figure 7.8 shows, there is a strong correlation between national wealth and life expectancy.

Figure 7.9 shows changes in life expectancy at birth in four groups of countries over a twenty-five-year period. The greatest gains have been in the lower-income countries, especially in China and East Asia, where life expectancies are approaching those of the developed world. A sobering exception to this general increase is Russia, where life expectancy for adult men plummeted from 70.4 years in 1990 to 59.9 years in 1995. The collapse of the former Soviet Union has resulted in declining standard of living and death rates almost 40 percent higher than birth rates. Russia now has a lower life expectancy than many African or Asian countries.

Some demographers believe that life expectancy is approaching a plateau, while others predict that advances in biology and medicine might extend longevity markedly. If our average age at death approaches one hundred years, as some expect, society will be profoundly affected. In 1970, the median age in the United States was thirty. By 2100, the median age could be over sixty. If workers continue to retire at sixty-five, half of the population could

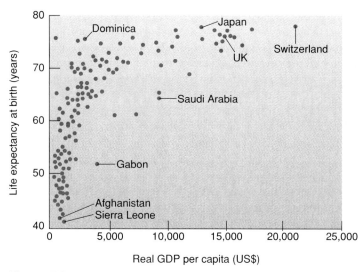

Figure 7.8

The relationship between life expectancy and economic status expressed in GDP per capita. Although there is considerable variation in life expectancy at both high- and low-income levels, in general, the richer you are, the longer you are likely to live.

Source: United Nations Environment Programme, *Human Health 1993*, p. 233.

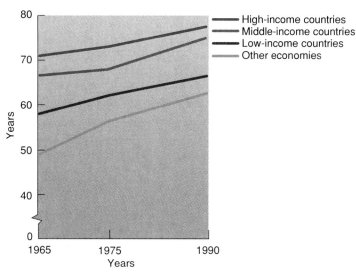

Figure 7.9

Life expectancies have risen sharply in many countries over the past twenty years. The world recession and debt crisis that began in the mid-1970s have slowed previously impressive progress in less-developed countries.

Source: World Book, *World Development Report 1992*, Oxford University Press, 1992.

be unemployed, and retirees might be facing thirty-five or forty years of retirement. We may need to find new ways to structure and finance our lives.

Living Longer: Demographic Implications

A population that is growing rapidly by natural increase has more young people than does a stationary population. One way to show these differences is to graph age classes in a histogram as shown in figure 7.10. In Mexico, which is growing at a rate of 2.5 percent per year, 42 percent of the population is in the prereproductive category (below age fifteen). Even if total fertility rates were to fall abruptly, the total number of births, and population size, would continue to grow for some years as these young people enter reproductive age. This phenomenon is called population momentum.

A population that has recently entered a lower growth rate pattern, such as the United States, will have a bulge in the age classes for the last high-birth-rate generation. A country that has had a stable population for many years, such as Sweden, will have approximately the same numbers in all age classes. Notice that there are more females than males in the older age group in the United States because of differences in longevity between the sexes. The United States has a high percentage of retired people because of long life expectancy.

Countries with a high percentage of children (such as Mexico) or a high percentage of old people (such as Sweden) have a problem with their **dependency ratio,** or the number of nonworking compared to working individuals in a population. Mexico has a high number of children to be supported by each working person. In Sweden, although there are fewer children, a large number of retired persons must be supported by a small working population. This changing age structure and shifting dependency ratio is occurring worldwide (fig. 7.11). There is considerable worry that not enough workers will be available to support retirement systems in countries where population growth has slowed sharply and life expectancies have increased. Retirement might have to be postponed or eliminated altogether.

Emigration and Immigration

Humans are highly mobile, so emigration and immigration play a larger role in human population dynamics than they do in those of many species. Currently, about 800,000 people immigrate legally to the United States each year, but many more enter illegally. Western Europe receives about 1 million applications each year for asylum from economic chaos and wars in former socialist states and the Middle East. Hal Kane, of the Worldwatch Institute estimated in 1995 that at least 33 million people left their countries for political or economic reasons, while another 27 million fled their homes but remained internal refugees in their own countries.

Immigration is a controversial issue in most wealthy countries. "Guest workers" often perform heavy, dangerous, or disagreeable work that citizens are unwilling to do. Many migrants and alien workers are of a different racial or ethnic background than the majority in their new home. They generally are paid low wages and given substandard housing, poor working conditions, and few rights. Local residents often complain that immigrants take away jobs, overload social services, and ignore established rules of behavior or social values.

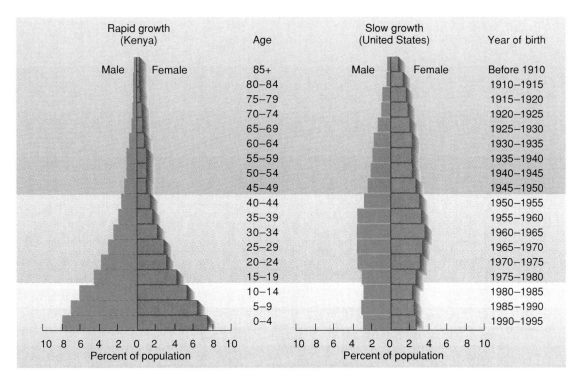

Figure 7.10

Population by age and sex in rapidly growing and slowly growing countries. Middle regions represent individuals of reproductive ages. Note the high proportion of children in the rapidly growing population and the high proportion of elderly in the slow-growing population.

Sources: Data from U.S. Bureau of the Census, the United Nations, and Population Reference Bureau.

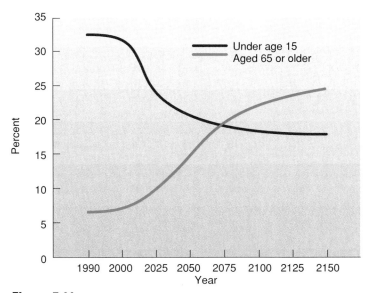

Figure 7.11

Changing age structure of world population. In the twenty-first century, children under 15 years of age will make up a smaller percentage of world population, while people over 65 years old will make up a rapidly rising share of the population.

Source: United Nations Population Division, *Long-Range World Population Projections: Two Centuries of Population Growth, 1950–2150*, United Nations, New York.

Refugees seeking jobs and asylum in the richer countries have been subjected to increasingly violent racist and xenophobic at-

tacks in many countries. Those who are out of work and alienated from society find these people an easy target for their anger. Ironically, those who are themselves immigrants or descendants of immigrants are sometimes the most adamant about closing doors to future immigration. Proponents of an open door policy claim that immigrants make a valuable contribution to society, bringing new energy, ingenuity, cultural diversity, and vigor.

Peaceful internal relocations also are demographically important. In the United States, for instance, a mass migration of rural southern people into northern industrialized cities in the 1930s and 1940s had profound effects on both the areas they left and the new places they settled. A similar shift of labor and business from the Rust Belt of the Great Lakes and mid-Atlantic States to the Sun Belt of the South and West in the 1960s and 1970s left many older industrialized cities empty, poverty-stricken wastelands.

Some countries encourage, or even force, internal mass migrations as part of a geopolitical demographic policy. In the 1970s, Indonesia embarked on an ambitious "transmigration" plan to move 65 million people from the overcrowded islands of Java and Bali to relatively unpopulated regions of Sumatra, Borneo, and New Guinea. Attempts to turn rainforest into farmland had disastrous environmental and social effects, however, and this plan was greatly scaled back. In the 1990s, China announced intentions to move 100 million people into a pristine region along the Amur River adjoining Siberia. This is a region of great biodiversity, home to endangered Siberian tigers, cranes, and other rare species. An influx of millions of people could be a major biological catastrophe.

Human Populations

CASE STUDY
China's One-Child Family Program

The total number of people in China is difficult to comprehend. In 1995, the population was estimated to be 1238 million, about 22 percent of all the people in the world. More than four-fifths of China's population are rural peasants. The per capita income is about $360 per year, placing it among the less-developed countries in the world, and yet it appears to be achieving a demographic transition to a stable population under conditions that experts had predicted were impossible (fig. 1).

For many centuries, China followed a repeating cycle of disasters, famines, and political upheavals. It was widely believed that China was so large and unmanageable that this cycle could never be broken. In the 1950s, as a result of the misguided Great Leap Forward Program, some 20 million people died of starvation. Some experts predicted that China would never be able to feed itself. Today, however, the country provides an adequate, if spartan, nutritional level for all its people. Medical care, housing, education, and social security have also improved markedly. Between 1950 and 1980, the death rate dropped from 20 to 8 per thousand, and average life expectancy increased from 47 to 70 years. By comparison, the United States has a death rate of 9 per thousand and a life expectancy of 74 years.

Pro-family traditions and public policies caused explosive population growth following the 1949 Socialist Revolution, especially in the postfamine recovery in the early 1960s (fig. 2). After Mao Tse-tung's death in 1976, however, new leaders reversed his pro-growth

Figure 1

China's one-child-per-family policy has resulted in a dramatic decline in birth rates. Many people object, however, to the means employed to enforce this policy. Others claim it has created a generation of "little emperors," spoiled, single children.

policies and sought to bring population growth under control. Premier Deng Xiaoping saw rampant population growth as the main obstacle to an im-

proved standard of living. He also worried that modernization of agriculture would make hand labor obsolete, displacing as much as half of the rural population and creating a monumental unemployment problem.

Chinese demographers believe that scarcity of resources (only 11 percent of the land is arable) will limit the population that China can support to a maximum of 1.2 billion people, or about 40 million fewer than the present population. For the past decade, therefore, China has carried on an extensive campaign to persuade couples to have only one child. Although there have been problems, the program has made dramatic progress. Between 1968 and 1988, the crude birth rate fell from about 40 per thousand to 20 per thousand. The Chinese birth rate is now only a little higher than that of the United States, which has more than sixty times the per capita income of China. In Shanghai, Beijing, and other urban centers, 80 to 90 percent of all births are first children.

Communes and production team officials pressure couples to follow birth control regulations through a combination of social pressure, incentives, and disincentives that many of us would find repressive. Birth control specialists in the production team monitor all married women, and couples must be granted permission to be part of the quota of births allowed in any year. Free contraceptives and medical advice are provided. All methods of birth control are used, including pills, condoms, and abortion, but major reliance is placed on the intrauterine device (IUD), in spite of serious questions about its health effects.

Population Growth: Opposing Factors

A number of social and economic pressures affect decisions about family size, which in turn affects the population at large. In this section we will examine both positive and negative pressures on reproduction.

Pronatalist Pressures

Factors that increase people's desires to have babies are called **pronatalist pressures.** Raising a family may be the most enjoyable and rewarding part of many people's lives. Children can be a source of pleasure, pride, and comfort. They may be the only source of support for elderly parents in countries without a social security system. Where infant mortality rates are high, couples may

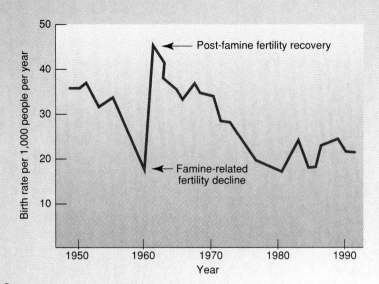

Figure 2

China's crude birth rate, 1950–1992.

Source: Data from H. Yuan Tien, "China: Demographic Billionaire," *Population Bulletin,* Vol. 38, No. 2, (April 1983): table 2; and H. Yuan Tien, "The New Census of China," *Population Today,* Vol. 19., No. 1, (January 1991): table 2.

There are widespread reports that the single child policy is creating a generation of "little emperors." The Chinese call this the 4-2-1 syndrome: four doting grandparents and two overstressed parents lavishing all their attention on one spoiled child. Another chilling outcome is the disappearance of girl babies. The normal ratio at birth is about 105 boys to 100 girls. Because boys have more illnesses and accidents, by the time they reach school age, the ratio is usually even. In China, however, the ratio is now 118 boys to 100 girls at birth and is even higher than that at school age. Parents want to be sure that their only child is a boy to carry on the family name and to care for them in old age. Many use ultrasound scans to determine the sex of their baby *in utero* and have an abortion if it is a girl. Others abandon girl babies at birth or withhold food and medicine. Given China's vast size, this means that about 2 million girls per year are missing.

Although there are many reservations about the means used to enforce China's one-child program and some of its social outcomes, there are some positive aspects nevertheless. It gives us encouraging evidence that a demographic transition can occur without concomitant industrialization and a shift to a high standard of living.

Couples who sign a pledge to have only one child are guaranteed free delivery of that child when they are granted permission to give birth. The child receives preference in education and job placement as it grows up. The parents get better housing, longer vacations, and an extra month's pay each year. Penalties for unsanctioned pregnancies include official reprimands, pay cuts, and public censure. Because each work unit or village is anxious to make a good showing in meeting its birth quotas, zealous local officials may use coercive techniques to demand abortions in cases of unapproved pregnancies and to control the behavior of individuals.

In contrast to most countries where family planning is only effective in urban upper- or middle-class families with a high level of wealth and education, birth control in China is nearly as effective among the masses as in the more affluent classes. There has been resistance to this plan, however, in rural areas and among ethnic minorities. Allowing peasant families to farm private plots has created a demand for large families to work the fields. In many areas, couples have been permitted to have a second or even third child.

ETHICAL CONSIDERATIONS

What do you think about this situation? Is population growth a sufficiently alarming problem to justify the draconian measures employed in China? Should other developing countries be encouraged to adopt similar policies? At what point do individual rights outweigh societal needs?

need to have many children to be sure that at least a few will survive to take care of them when they are old. Where there is little opportunity for upward mobility, children give status in society, express parental creativity, and provide a sense of continuity and accomplishment otherwise missing from life. Often children are valuable to the family not only for future income, but even more as a source of current income and help with household chores. In much of the developing world, children as young as six tend domestic animals and younger siblings, fetch water, gather firewood, and help grow crops or sell things in the marketplace. Parental desire for children rather than an unmet need for contraceptives may be the most important factor in population growth in many cases.

Society also has a need to replace members who die or become incapacitated. This need often is codified in cultural or religious values that encourage bearing and raising children. In some societies, families with few or no children are looked upon with pity or contempt. The idea of deliberately controlling fertility may be shocking, even taboo. Women who are pregnant or have small children are given special status and protection. Boys frequently are more valued than girls because they carry on the family name and are expected to support their parents in old age. Couples may have more children than they really want in an attempt to produce a son.

Male pride often is linked to having as many children as possible. In Niger and Cameroon, for example, men, on average, want 12.6 and 11.2 children, respectively. Women in these countries consider the ideal family size to be only about one half that desired by their husbands. Even though a woman might desire fewer children, however, she may have few choices and little control over her own fertility. In many societies, a woman has no status outside of her role as wife and mother. Without children, she has no source of support.

Figure 7.12 shows a model for the variables determining fertility. Three primary factors interact in this model: the biological supply of children determined by fecundity and infant mortality, the demand for children determined by economics and social values, and the regulation of fertility determined by knowledge, attitudes, and access to birth control. The combined interaction of these factors and their intervening variables determine fertility.

Birth Reduction Pressures

In more highly developed countries, many pressures tend to reduce fertility. Higher education and personal freedom for women often result in decisions to limit childbearing. The desire to have children is offset by a desire for other goods and activities that compete with childbearing and childrearing for time and money. When women have opportunities to earn a salary, they are less likely to stay home and have many children. Not only are the challenge and variety of a career attractive to many women, but the money that they can earn outside the home becomes an important part of the family budget. Thus, education and socioeconomic status are usually inversely related to fertility in richer countries. In developing countries, however, fertility is likely to increase as educational levels and socioeconomic status rise. With higher income, families are better able to afford the children they want; more money means that women are likely to be healthier, and therefore better able to conceive and carry a child to term.

In less-developed countries where feeding and clothing children can be a minimal expense, adding one more child to a family usually doesn't cost much. By contrast, raising a child in the United States can cost hundreds of thousands of dollars by the time the child is through school and is independent. Under these circumstances, par-

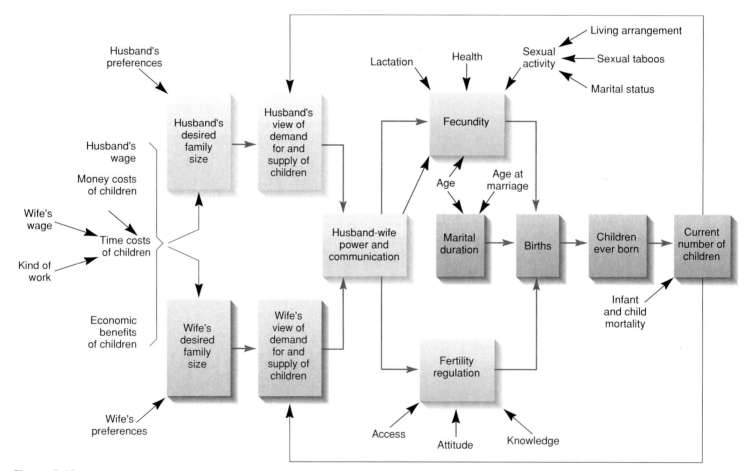

Figure 7.12

A model for the variables determining fertility. Three major factors interact in this model: the biological supply of children, the demand for children, and the regulation of fertility. Other intervening factors have either positive or negative effects on this interaction.

Population, Economics, and Environmental Health   **146**

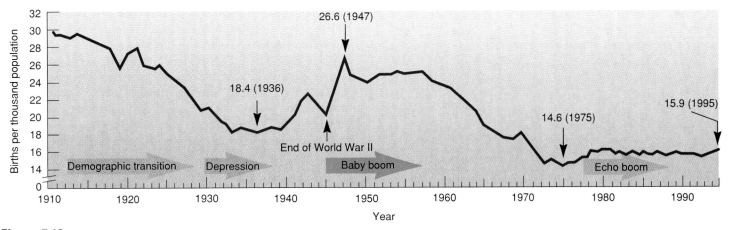

Figure 7.13

Birth rates in the United States, 1910–1995. The falling birth rate from 1910 to 1929 represents a demographic transition from an agricultural to an industrial society. Note that this decline occurred before the start of the Great Depression. The baby boom following the Second World War lasted from 1945 to 1957. A much smaller "echo boom" occurred around 1980 when the baby boomers started to reproduce, but it produced far fewer births than anticipated.

Sources: Data from Population Reference Bureau, Inc., and U.S. Bureau of the Census.

ents are more likely to choose to have one or two children on whom they can concentrate their time, energy, and financial resources.

Figure 7.13 shows U.S. birth rates between 1910 and 1995. As you can see, birth rates have fallen and risen in a complex pattern. The period between 1910 and 1930 was a time of industrialization and urbanization. Women were getting more education than ever before and entering the workforce. The Great Depression in the 1930s made it economically difficult for families to have children, and birth rates were low. The birth rate increased at the beginning of World War II (as it often does in wartime). For reasons that are unclear, a higher percentage of boys are usually born during war years.

At the end of the war, there was a "baby boom" as couples were reunited and new families started. This high birth rate persisted through the times of prosperity and optimism of the 1950s, but began to fall in the 1960s. Part of this decline was caused by the small number of babies born in the 1930s. This meant fewer young adults to give birth in the 1960s. Part was due to changed perceptions of the ideal family size. Whereas in the 1950s women typically wanted four children or more, in the 1970s the norm dropped to one or two (or no) children. A small "echo boom" occurred in the 1980s as people born in the 1960s began to have babies, but changing economics and attitudes seem to have permanently altered our view of ideal family size in the United States.

Birth Dearth?

Most European countries now have birth rates below replacement rates, and Italy, Russia, Austria, Germany, Greece, and Spain are experiencing negative rates of natural population increase. In Asia, Japan, Singapore, Hong Kong, and Taiwan are also facing a "child shock" as fertility rates have fallen well below the replacement level of 2.1 children per couple. There are concerns in all these countries about falling military strength (lack of soldiers), economic power (lack of workers), and declining social systems (not

enough workers and taxpayers) if low birth rates persist or are not balanced by immigration.

Economist Ben Wattenberg warns that this "birth dearth" might seriously erode the powers of Western democracies in world affairs. He points out that Europe and North America accounted for 22 percent of the world's population in 1950. By the 1980s, this number had fallen to 15 percent, and by the year 2030, Europe and North America probably will make up only 9 percent of the world's population. Germany, Hungary, Denmark, and Russia now offer incentives to encourage women to bear children. Japan offers financial support to new parents, and Singapore provides a dating service to encourage marriages among the upper classes as a way of increasing population.

On the other hand, since Europeans and North Americans consume so many more resources per capita than most other people in the world, a reduction in the population of these countries will do more to spare the environment than would a reduction in population almost anywhere else.

Demographic Transition

In 1945, demographer Frank Notestein pointed out that there is a typical pattern of falling death rates and birth rates due to improved living conditions that usually accompanies economic development. He called this pattern the **demographic transition** between high birth and death rates to lower birth and death rates. Population histories of Sweden, Japan, and England and Wales are shown in figure 7.14.

Development and Population

The early stages of each graph in figure 7.14 represent the conditions in a premodern society. Food shortages, malnutrition, lack of sanitation and medicine, accidents, and other hazards keep death

Human Populations

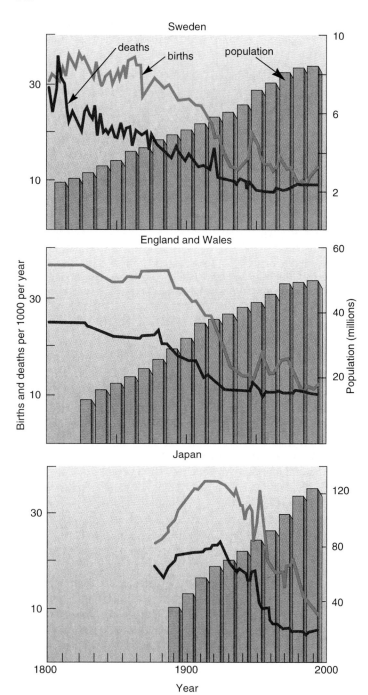

Figure 7.14

Demographic transitions in industrializing countries. Death rates (*bottom lines*) generally drop first as quality of life improves. Birth rates may actually increase as more women achieve wanted fertility and more children survive. Eventually, birth and death rates come into equilibrium. During the transition, the difference between births and deaths (*lines*) produces rapid population growth (*bars*).

rates around 30 per 1000 people. Birth rates are correspondingly high to keep population densities relatively constant.

In the nineteenth century in Europe and in the twentieth century in Japan, better jobs and more income provided by industrialization improved the standard of living, causing death rates to fall—often very rapidly. Birth rates may actually rise at first (note the curve for Japan) but eventually fall as people see that societal improvements are permanent. It takes some years, however, to change attitudes and values. Note that populations grow rapidly during this phase as the rate of natural increase (birth rate minus death rate) rises. Depending on how long it takes to complete this phase, the population may go through one or more rounds of doubling before coming into balance again.

The right-hand side of each curve represents conditions in developed countries where the transition is complete and both birth rates and death rates are low, often one-third or less than those in the predevelopment era. The population comes into a new equilibrium in this phase, but at a much larger size than before. Most of the countries of northern and western Europe went through a demographic transition in the nineteenth or early twentieth century similar to the curves shown here.

Many of the most rapidly growing countries in the world such as Kenya, Yemen, Libya, and Jordan now are in the middle phase of this demographic transition. Their death rates have fallen close to the rates of the fully developed countries, but birth rates have not fallen correspondingly. In fact, both their birth rates and total population are higher than those in most European countries when industrialization began three hundred years ago. The large disparity between birth and death rates means that many developing countries now are growing at 3 to 4 percent per year. This high rate of growth in the Third World could boost total world population to 10 billion or more before the end of the next century. Perhaps the most important questions in this whole chapter are "Why are birth rates not yet falling in these countries?" and "What can be done about it?"

An Optimistic View

Some demographers claim that a demographic transition already is in progress in most developing nations. Problems in taking censuses and a normal lag between falling death and birth rates may hide this for a time, but the world population should stabilize sometime in the next century. Some evidence supports this view. As we mentioned earlier in this chapter, total fertility rates have dropped nearly 40 percent in the past 20 years from 6.1 to 3.3 children per woman worldwide.

Some countries have had remarkable success in population control. In Thailand, for instance, total fertility dropped 50 percent in only 12 years: from 4.6 children per woman in 1975 to 2.3 children in 1987. Indonesia and Colombia both experienced similar fertility declines between 1970 and 1990. Morocco, Dominican Republic, Jamaica, Peru, and Mexico all have seen fertility rates fall between 30 percent and 40 percent over the past 20 years. The following factors contribute to stabilizing populations:

- Growing prosperity and social reforms that accompany development reduce the need and desire for large families in most countries.

- Technology is available to bring advances to the developing world much more rapidly than was the case a century ago, and the rate of technology transfer is much faster than it was when Europe and North America were developing.

- Less-developed countries have historic patterns to follow. They can benefit from our mistakes and chart a course to stability more quickly than they might otherwise do.

- Modern communications (especially television) have caused a revolution of rising expectations that act as a stimulus to spur change and development.

A Pessimistic View

Economist Lester Brown of the Worldwatch Institute takes a more pessimistic view. He warns that many of the poorer countries of the world appear to be caught in a "demographic trap" that prevents them from escaping from the middle phase of the demographic transition. Their populations are now growing so rapidly that human demands exceed the sustainable yield of local forests, grasslands, croplands, or water resources. The resulting resource shortages, environmental deterioration, economic decline, and political instability may prevent these countries from ever completing modernization. Their populations may continue to grow until catastrophe intervenes.

Many people argue that the only way to break out of the demographic trap is to immediately and drastically reduce population growth by whatever means are necessary. They argue strongly for birth control education and bold national policies to encourage lower birth rates. Some agree with Malthus that helping the poor will simply increase their reproductive success and further threaten the resources on which we all depend. Author Garret Hardin described this view as lifeboat ethics. "Each rich nation," he said, "amounts to a lifeboat full of comparatively rich people. The poor of the world are in other much more crowded lifeboats. Continuously, so to speak, the poor fall out of their lifeboats and swim for a while, hoping to be admitted to a rich lifeboat, or in some other way to benefit from the goodies on board. . . . We cannot risk the safety of all the passengers by helping others in need. What happens if you share space in a lifeboat? The boat is swamped and everyone drowns. Complete justice, complete catastrophe."

A Social Justice View

A third view is that **social justice** (a fair share of social benefits for everyone) is the real key to successful demographic transitions. The world has enough resources for everyone, but inequitable social and economic systems cause maldistributions of those resources. Hunger, poverty, violence, environmental degradation, and overpopulation are symptoms of a lack of social justice rather than a lack of resources. Although overpopulation exacerbates other problems, a narrow focus on this factor alone encourages racism and hatred of the poor. A solution for all these problems is to establish fair systems, not to blame the victims. Small nations and minorities often regard calls for population control as a form of genocide. Figure 7.15 expresses the opinion of many people in less-developed countries about the relationship between resources and population.

An important part of this view is that many of the rich countries are, or were, colonial powers, while the poor, rapidly growing countries were colonies. The wealth that paid for progress and security for developed countries was often extracted from colonies, which now suffer from exhausted resources, exploding populations, and chaotic political systems. Some of the world's poorest countries such as India, Ethiopia, Mozambique, and Haiti had rich resources and adequate food supplies before they were impoverished by colonialism. Those of us who now enjoy abundance may need to help the poorer countries not only as a matter of justice but because we all share the same environment.

An Ecojustice View

In addition to considering the rights of fellow humans, we should also consider those of other species. Rather than ask what is the maximum number of humans that the world can possibly support, perhaps we should think about the needs of other creatures. As we convert natural landscapes into agricultural or industrial areas, species are crowded out that may have just as much right to exist as we do. Perhaps we should seek the optimum number of people at which we can provide a fair and decent life for all humans while causing the minimum impact on nonhuman neighbors.

Infant Mortality and Women's Rights

Survival of children is one of the most critical factors in stabilizing population. When infant and child mortality rates are high, as they are in much of the developing world, parents tend to have high numbers of children to ensure that some will survive to adulthood. There has never been a sustained drop in birth rates that was not first preceded by a sustained drop in infant and child mortality. One of the most important distinctions in our demographically divided world is the high infant mortality rates in the less-developed countries. Better nutrition, improved health care, simple oral rehydration therapy, and immunization against infectious diseases (chapter 9) have brought about dramatic reductions in child mortality rates (fig. 7.16), which have been accompanied in most regions by falling birth rates. It has been estimated that saving 5 million children each year from easily preventable communicable diseases would avoid 20 or 30 million extra births.

Increasing family income does not always translate into better welfare for children since men in many cultures control most financial assets. Often the best way to improve child survival is to ensure the rights of mothers. Adult female literacy, for instance, as well as land reform, political rights, opportunities to earn an independent income, and improved health status of women often are better indicators of family welfare than rising GNP (fig. 7.17).

Human Populations

Family Planning and Fertility Control

Family planning allows couples to determine the number and spacing of their children. It doesn't necessarily mean fewer children—people may use family planning to have the maximum number of children possible—but it does imply that the parents will control their reproductive lives and make rational, conscious decisions about how many children they will have and when those children will be born, rather than leaving it to chance. As the desire for smaller families becomes more common, birth control becomes an essential part of family planning in most cases. In this context, **birth control** usually means any method used to reduce births, including celibacy, delayed marriage, contraception, methods that prevent implantation of embryos, and induced abortions.

Traditional Fertility Control

Evidence suggests that people in every culture and every historic period have used a variety of techniques to control population size.

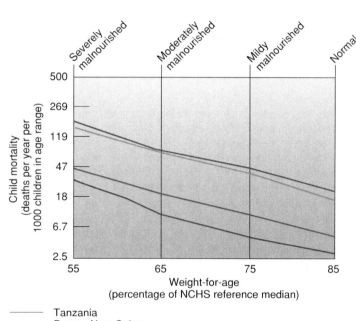

- Tanzania
- Papua, New Guinea
- Matlab, Bangladesh
- Punjab, India

Figure 7.16

Stunted growth due to malnutrition accounts for 25–50% of infant mortality (deaths before age two) each year. A child in Tanzania who weighs only 55 percent of the weight for age norm is 10 times more likely to die before age two than a child at 85 percent of normal weight. Another way to read this graph is that about 200 Tanzanian children out of every 1000 who are 45% underweight can be expected to die in a given year, while only 2 per 1000 of those that are only 15 percent underweight will die.

Source: World Bank, 1993.

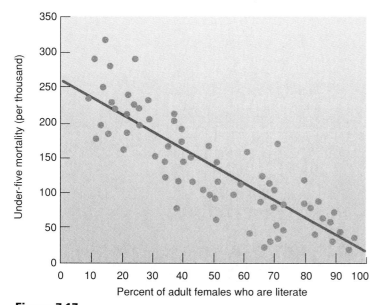

Figure 7.17

Mortality of children under 5 years of age compared to the percentage of adult females in a country who are literate. Note the rapid decline of child mortality as women's literacy increases. Each dot represents one country.

Sources: Food and Agriculture Organization of the United Nations, United Nations Children's Fund, United Nations Development Programme, and United Nations Population Division.

Studies of hunting and gathering people, such as the Kung! or San of the Kalahari Desert in southwest Africa, indicate that our early ancestors had stable population densities, not because they killed each other or starved to death regularly, but because they controlled fertility.

For instance, San women breast-feed children for three or four years. When calories are limited, lactation depletes body fat stores and suppresses ovulation. Coupled with taboos against intercourse while breast-feeding, this is an effective way of spacing children. Other ancient techniques to control population size include celibacy, polygamy, folk medicines, abortion, and infanticide. We may find some or all of these techniques unpleasant or morally unacceptable, but we shouldn't assume that other people are too ignorant or too primitive to make decisions about fertility.

Current Birth Control Methods

Modern medicine gives us many more options for controlling fertility than were available to our ancestors. Some of these techniques are safer, easier, or more pleasant to use (fig. 7.18). The major categories of birth control techniques include (1) avoidance of sex during fertile periods [celibacy; using changes in body temperature or cervical mucus color and viscosity to judge when ovulation will occur]; (2) mechanical barriers that prevent contact between sperm and egg [condoms, spermicides, diaphragm, cervical

cap, and vaginal sponge]; (3) surgical methods that prevent release of sperm or egg [sterilization: tubal ligation or use of the Filshie clip in females, vasectomy in males] (fig. 7.19); (4) chemicals that prevent maturation or release of sperm or eggs or implantation of the embryo in the uterus [the pill: estrogen + progesterone, progesterone alone for females; gossypol for males]; (5) physical barriers to implantation [IUD]; and (6) abortion.

None of these methods is perfect and none suits every contraceptive need. Many require careful, conscientious use. Which choice is best for you depends on your life situation and your plans for the future (table 7.4).

New Developments in Birth Control

In 1991, Norplant, the trade name for flexible, matchstick-sized, silicon-rubber implants containing a slow-release analog of progesterone, was approved for use in the United States. The implants are inserted under the skin where they will release hormones for up to five years (fig. 7.18c). In 1992, Depo-Provera, an injectable progesterone analog, was approved by the Food and Drug Administration after many years of research and controversy. The injections are given four times a year. Both implants and injections are as effective as daily oral contraceptives (about 1 percent failure rate) but eliminate the need to keep track of and take daily pills. They also can be used without knowledge of one's partner, who may oppose birth control. Both injections and implants cause problems for some women who experience increased vaginal bleeding or absence of menstrual periods, and some increased breast cancer and osteoporosis (bone thinning) have been associated with their use. Concerns have been raised about removal of implants as well.

A condom for women called Reality has recently been introduced (fig. 7.18d). It consists of two flexible plastic rings connected by a strong, clear polyurethane sheath. One ring fits over the cervix much like a diaphragm, while the other remains outside the vagina. Each condom costs about two dollars and is designed to be discarded after a single use. The six-month failure rate is about 12 percent, which means that 12 percent of women using only this device will get pregnant in six months. An important advantage of condoms is the protection they offer against sexually transmitted diseases. A female condom gives women control over their own reproduction, but some women may find it difficult or unpleasant to use.

In 1994, the nonprofit Population Council in New York City began clinical tests in the United States of the French drug, RU486 (mifepristone or mifegyne), which blocks the effects of progesterone in maintaining the lining of the uterine wall. It is usually administered together with misoprostol (a prostaglandin analog), which causes uterine contraction and expulsion of the fetus. More than 120,000 women have already received RU486 in clinics in France, Britain, and Sweden. Although for some women, RU486 is preferable to surgical abortion, neither procedure is to be taken lightly. Both require medical supervision and entail cramps, bleeding, and some risk. Interestingly, RU486 also appears to have

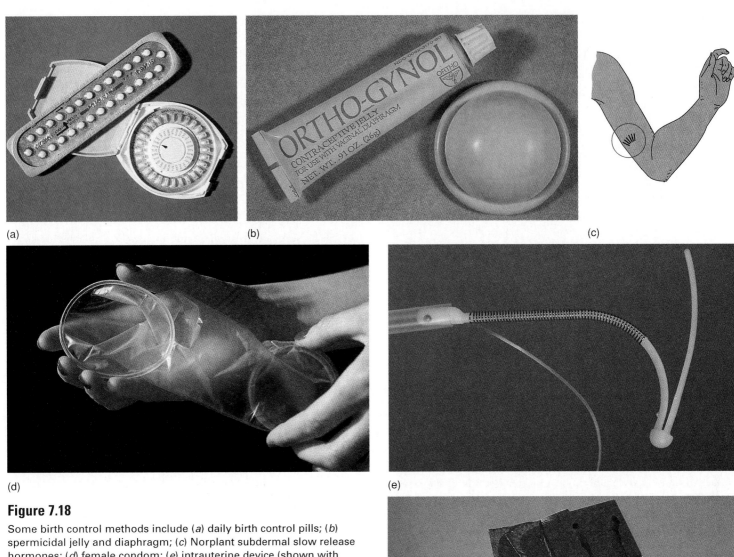

Figure 7.18

Some birth control methods include (*a*) daily birth control pills; (*b*) spermicidal jelly and diaphragm; (*c*) Norplant subdermal slow release hormones; (*d*) female condom; (*e*) intrauterine device (shown with applicator); (*f*) male condom.

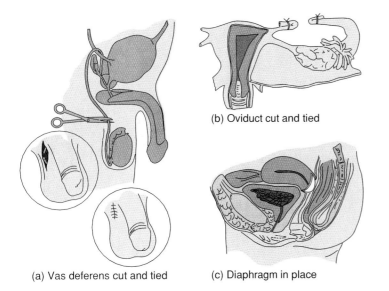

(a) Vas deferens cut and tied

(b) Oviduct cut and tied

(c) Diaphragm in place

Figure 7.19

Surgical birth control procedures include vasectomy for men and tubal ligation for women. (*a*) Vasectomy requires only a small incision under local anesthesia to cut and tie the sperm-conducting vas deferens. (*b*) Tubal ligation in women is a more difficult operation and generally requires general anesthesia. Neither procedure affects sex life or sexual characteristics, but neither is easily reversible. (*c*) The cervical cap, or diaphragm, is an example of a mechanical barrier. It covers the entrance to the cervix and prevents sperm from entering the uterus. While not as dependable as surgical methods, mechanical barriers are easily reversible.

Table 7.4

Personal fertility

What are the options for personal reproductive decisions? More alternatives are available now than were a generation ago, but more concerns about health, sexuality, and parenthood face us also. How can an individual know what to do? Each of us must examine our own values, situation, and choices to design a personal fertility plan.

First, examine your own life and your hopes for the future. How old are you? Are you sexually active now? Will you be in the future? How many children—if any—will you want and when would you like them to be born? Timing and number of births can critically affect education, jobs, travel, and other life aspirations. How many children will you be able to support and educate, given projections for future costs and income? As global citizens, we also should ask how many children it is responsible for us to bring into the world, given the impact we have.

Next, examine the alternatives available to meet your goals. Remember that you don't need to commit yourself to a single option for life. Among the choices are:

Abstinence: One hundred percent effective in birth control and prevention of sexually transmitted diseases. Under some circumstances, this may be the best and safest course of action. As a long-term strategy, however, celibacy may inhibit intimate and fulfilling relationships for many people.

Oral contraceptives (pills): Ninety-nine percent effective. Easy to use and readily reversible. Not recommended for women over age 35 who smoke because of the risk of strokes and other cardiovascular problems. Pills with low estrogen levels or progesterone alone can reduce these risks. Some increase in breast cancer is associated with prolonged use of synthetic hormones, but they seem to give some protection from endometrial and ovarian cancer. Overall, cancer risks seem slightly lower with birth control pills than without.

Implantable or injectable progesterone analogs: Ninety-nine percent effective and highly reversible. Doesn't require daily attention or approval of partner. May be suitable for women who cannot take oral contraceptives. Some women experience dizziness, fatigue, headaches, vaginal bleeding, or absent periods with these methods. May increase risks of osteoporosis. About the same breast cancer risk and other cancer benefits as oral pills.

Barrier methods: The condom, diaphragm, cervical cap, and vaginal sponge are generally 85 to 95 percent effective. They help prevent sexually transmitted diseases, but don't bet your life on them. They require some planning ahead but not the daily attention of the pill. Any of these may be a good option for someone who cannot use hormone analogs or has sex only occasionally.

Spermicidal creams, jellies, douches: Often not very effective by themselves in either birth control or disease prevention. May help, however, when used together with barrier methods.

Intrauterine device (IUD): Ninety-four percent effective. Doesn't require daily attention or alter natural hormone balance. May increase frequency or severity of sexually transmitted diseases as well as vaginal bleeding, cramps, and other side effects. Generally recommended only for mature women in monogamous relationships.

Natural methods: Careful record keeping, temperature measurements, and examination of cervical mucus can be 70 to 90 percent effective in birth control. This method meets ethical or religious scruples but may be difficult and demanding to follow. Prohibits sex on some days.

Vasectomy, tubal ligation, Filshie clip: These surgical methods are more than 99 percent effective in birth control but are not easily reversed. These may be good options for those who are sure they don't want any more children.

As you can see, no birth control method is perfect for everyone. All have some advantages and disadvantages. You should be aware that the failure rates under "perfect" conditions often are very different from the "typical" use. Spermicides, for example, are nearly 100 percent effective in clinical trials, but about one-third of women depending on this method alone get pregnant because of misuse. Almost all techniques require deliberate attention and planning. By thinking carefully and communicating clearly about values and desires, however, couples can make responsible decisions about their fertility and sexual behavior.

---

promise in treating breast cancer, brain cancer, diabetes, and hypertension. Marketing in the United States is still years away.

Recently, a combination of two drugs already on the market for other uses have been shown to be as safe and effective as RU486. Methotrexate and misoprostol are administered a week apart to induce abortion. Equally opposed by abortion foes, these drugs could make the debate over whether to make RU486 available in the United States a moot point.

Some other new contraceptive methods appear to have great promise. Vaginal rings containing slow-release progesterone analogs are being studied. These have all the advantages of Depo-Provera or Norplant but should be much cheaper and easier to use since they can be inserted and removed by the user. Development of simple, inexpensive, do-it-yourself tests for levels of estrogen and progesterone in urine may make the rhythm method easier to follow and more reliable for women who cannot or would rather not use other methods of birth control. Some antipregnancy vaccines (immunization against chorionic gonadotropin—a hormone required to maintain the uterine lining) and antisperm vaccines are being tested that would use the immune system to prevent fertilization or embryonic implantation, but they are years away from the market. The sperm suppressant gossypol is being tested as a male contraceptive, but side effects and male reluctance to take responsibility for reproduction inhibit its introduction. Finally, potent new spermicides are being tested, as are new methods of using them.

The Future of Human Populations

How many people will be in the world a century from now? Most demographers believe that world population will stabilize sometime during the next century. The total number of humans, when we reach that equilibrium, probably will be between 6 billion and 16 billion people, depending on the success of family planning programs and the multitude of other factors affecting human populations. In 1992, the United Nations released new population forecasts, the first in a decade (fig. 7.20). The optimistic (low) projection shows that world population might drop below 5 billion about a century from now. The medium projection, which previously had been expected to peak at 10 billion, was increased to about 13 billion. The most pessimistic (high) projection was increased from 14 billion to more than 25 billion.

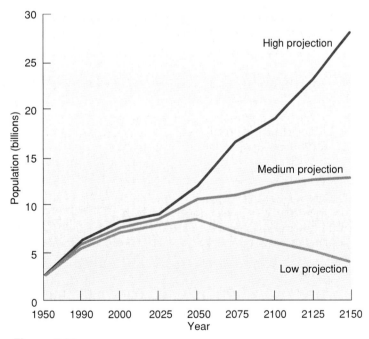

Figure 7.20

Population projections to the year 2150. The low projection assumes rapid expansion and acceptance of family planning. The medium projection assumes progress at current rates. The high projection is the worst-case scenario.

Source: United Nations, 1992.

Which of these scenarios will we follow? As you have seen in this chapter, population growth is a complex subject. To accomplish a stabilization or reduction of human populations will require substantial changes from business as usual. One sign of hope is that the United States has resumed payments to the United Nations Family Planning Fund withheld by the Reagan administration because some of the 135 countries that receive UN aid include abortion as part of population control programs. Population issues were conspicuously absent from the agenda of the UN Conference on Environment and Development in Rio de Janeiro (the Earth Summit) in 1992 because of opposition from religious groups and developing countries. At a special UN Conference on Population held in Cairo, Egypt, in 1994, religion again was a complicating factor. This conference did, however, come out strongly in favor of the rights of women and children.

Another encouraging sign is that worldwide contraceptive use has increased sharply in recent years. About 380 million couples in the Third World used contraceptives in 1990, up nearly tenfold from 1960, but another 300 million couples say they want but do not have access to family planning. Contraceptive use varies widely by region (fig. 7.21), with high levels in Latin America and East Asia but relatively low use in much of Africa. Note the differences both in the percentage of couples using birth control in different regions and in the methods used. The IUD and male sterilization are common in Asia but rarely used in Africa or Latin America. What economic or cultural factors do you see in these patterns?

The World Health Organization estimates that nearly 1 million conceptions occur daily around the world as a result of some 100 million sex acts. At least half of those conceptions are unplanned or unwanted. Still, birth rates already have begun to fall in East Asia and Latin America (fig. 7.22). Similar progress is expected in South Asia in a few years. Only Africa will probably continue to grow in the 21st century. Coordinated, effective family planning requires governmental action, ranging from policies to assistance.

Deep societal changes are often required to make family planning programs successful. Among the most important of these are (1) improved social, educational, and economic status for women (birth control and women's rights are often interdependent); (2) improved status for children (fewer children are born as parents come to regard them as valued individuals rather than possessions); (3) acceptance of calculated choice as a valid element in life in general and in fertility in particular (belief that we have no control over our lives discourages a sense of responsibility); (4) social security and political stability that give people the means and the confidence to plan for the future; (5) knowledge, availability, and use of effective and acceptable means of birth control.

Concerted efforts to bring about these changes can be effective. Twenty years of economic development and work by voluntary family planning groups in Zimbabwe, for example, have lowered total fertility rates from 8.0 to 5.5 children per woman on average. Surveys showed that desired family sizes have fallen nearly by half (9.0 to 4.6) and that nearly all women and 80 percent of men in Zimbabwe use contraceptives. If similar progress can be sustained elsewhere, human populations may be restrained after all.

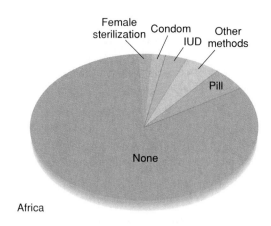

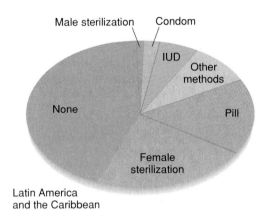

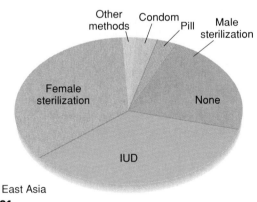

Figure 7.21

Variations in contraceptive use in different regions. Preferences for one form of contraception over another reflect both cultural attitudes and government programs.

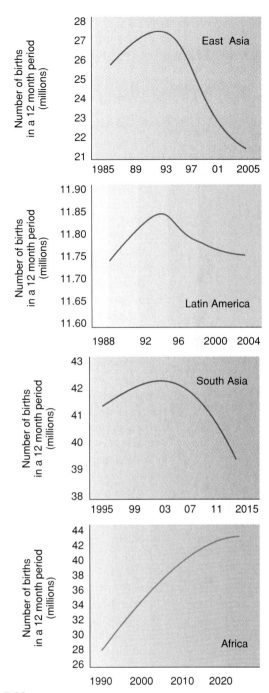

Figure 7.22

In East Asia and Latin America, the number of births per year already has begun to decline. South Asia is expected to began a period of sharply reduced births by about 2000. Only Africa is projected to have a continued increase in birth rates.

Source: *State of the World's Children*, UNESCO.

Human Populations

Summary

Human populations have grown at an unprecedented rate over the past three centuries. By 1993, the world population stood at 5.5 billion people. If the current growth rate of 1.7 percent per year persists, the population will double in 41 years. Most of that growth will occur in the less-developed countries of Asia, Africa, and Latin America. There is a serious concern that the number of humans in the world and our impact on the environment will overload the life-support systems of the earth.

The crude birth rate is the number of births in a year divided by the average population. A more accurate measure of growth is the general fertility rate, which takes into account the age structure and fecundity of the population. The crude birth rate minus the crude death rate gives the rate of natural increase. When this rate reaches a level at which people are just replacing themselves, zero population growth is achieved.

In the more highly developed countries of the world, growth has slowed or even reversed in recent years so that without immigration from other areas, populations would be declining. The change from high birth and death rates that accompanies industrialization is called a demographic transition. Many developing countries have already begun this transition. Death rates have fallen, but birth rates remain high. Some demographers believe that as infant mortality drops and economic development progresses so that people in these countries can be sure of a secure future, they will complete the transition to a stable population. Others fear that excessive population growth and limited resources will catch many of the poorer countries in a demographic trap that could prevent them from ever achieving a stable population or a high standard of living.

While larger populations bring many problems, they also may be a valuable resource of energy, intelligence, and enterprise that will make it possible to overcome resource limitation problems. A social justice view argues that a more equitable distribution of wealth might reduce both excess population growth and environmental degradation.

We have many more options now for controlling fertility than were available to our ancestors. Some techniques are safer than those available earlier; many are easier and more pleasant to use. Sometimes it takes deep changes in a culture to make family planning programs successful. Among these changes are improved social, educational, and economic status for women; higher values on individual children; accepting responsibility for our own lives; social security and political stability that give people the means and confidence to plan for the future; and knowledge, availability, and use of effective and acceptable means of birth control.

▼ Questions for Review

1. At what point in history did the world population pass its *first* billion? What factors restricted population before that time, and what factors contributed to growth after that point?

2. How might growing populations be beneficial in solving development problems?

3. Why do some economists consider human resources more important than natural resources in determining the future of a country?

4. Where will most population growth occur in the next century? What conditions contribute to rapid population growth in some countries?

5. Define *crude birth rate, total fertility rate, crude death rate,* and *zero population growth.*

6. What is the difference between life expectancy and longevity?

7. What is dependency ratio, and how might it affect the United States in the future?

8. What pressures or interests make people want or not want to have babies?

9. Describe the conditions that lead to a demographic transition.

10. Describe the major choices in modern birth control.

▼ Questions for Critical Thinking

1. What do you think is the optimum human population? The maximum human population? Are the numbers different? If so, why?

2. Some people argue that technology can provide solutions for environmental problems; others believe that a "technological fix" will make our problems worse. What personal experiences or worldviews do you think might underlie these positions?

3. Karl Marx called Thomas Malthus a "shameless sycophant of the ruling classes." Why would the landed gentry of the eighteenth century be concerned about population growth of the lower classes? Are there comparable class struggles today?

4. Try to imagine yourself in the position of a person your age in a Third World country. What family planning choices and pressures would you face? How would you choose among your options?

5. Some demographers claim that population growth has already begun to slow; others dispute this claim. How would you evaluate the competing claims of these two camps? Is this an issue of uncertain facts or differing beliefs? What sources of evidence would you accept as valid?

6. What role do race, ethnicity, and culture play in our immigration and population policies? How can we distinguish between prejudice and selfishness on one hand and valid concerns about limits to growth on the other?

▼ Key Terms

birth control 150	natural increase 141
crude birth rate 139	neo-Malthusians 137
crude death rates 140	pronatalist pressures 144
demographic transition 147	social justice 149
demography 138	total fertility rate 139
dependency ratio 142	total growth rate 141
family planning 149	zero population growth (ZPG) 139

▼ Additional Information on the Internet

Carolina Population Center
http://www.cpc.unc.edu/

Center for Demography and Ecology
http://elaine.ssc.wisc.edu/cde/

Demographic and Health Surveys
http://www.macroint.com/dhs/

Population for Countries of the World 1994
gopher://gopher.undp.org/00/ungophers/popin/wdtrends/pop1994/

US Census Bureau
http://www.census.gov/

World Population Clock
http://sunsite.unc.edu/lunarbin/worldpop/

World Wide Web Virtual Library: Demography and Population Studies
http://coombs.anu.edu.au/ResFacilities/DemographyPage.html/

Zero Population Growth
http://www.zpg.org/zpg/

Note: Further readings appropriate to this chapter are listed on p. 596.

Dobi Wallahs (professional launderers) carry on their trade on the banks of the Yumana River in Agra, India.

CHAPTER 8: Environmental Resource Economics

Objectives

After studying this chapter, you should be able to:

- define *natural* and *human resources* and distinguish between economic resource categories.

- describe frontier, industrial, and postindustrial economies and discuss how each uses and affects its environment.

- diagram the relationship between supply and demand at different stages of economic or technological development.

- understand how technology can mitigate scarcity and increase the carrying capacity of our environment.

- appreciate the limits to growth and the features of a steady-state system.

- explain internal and external costs, market approaches to pollution control, and cost/benefit ratios.

- relate and explain the goals and strategies for sustainable development.

Development at What Cost?

Thirty years ago, the island nation of Taiwan was among the world's poorer countries with an annual net income of about $400 per person. Rapid industrialization has brought remarkable economic growth averaging nearly 10 percent per year over the past three decades. By 1996, Taiwan is expected to have a per capita income of more than $14,000 per year, making it one of the 20 richest countries in the world. Many Taiwanese now have a material lifestyle comparable to the average American or European.

This spectacular growth has not been without costs, however. Air pollution in Taipei, the capital city, exceeds health standards 55 days each year, and the multitude of new automobiles, motorcycles, and trucks causes horrendous traffic jams nearly all

Figure 8.1

This view of Taipei, the capital of Taiwan, shows some of the air and water pollution, congestion, and chaotic urbanization produced by industrialization and rapid economic growth. How do we place a value on natural resources, beauty, tranquility, or good health? Resource economics analyzes the costs and benefits of our alternative choices.

the time (fig. 8.1). Cancer has become the leading cause of death in Taiwan, perhaps because of the noxious fumes emitted by the numerous petrochemical plants and plastic factories.

In spite of these drawbacks, however, many developing countries look to Taiwan and its successful neighbors, such as Hong Kong, Singapore, Japan, and South Korea, as models for how they, too, might grow to be wealthy. Those of us who already enjoy a high standard of living can hardly criticize others who crave some of the same benefits. We worry, however, about how many people—and at what level of material consumption—the world's natural resources and ecosystems can support. What will happen if everyone tries to attain the same kinds of industrialization and consumption that we now have?

Some economists argue that technological development and substitution of one resource for another can indefinitely expand the carrying capacity of the environment for human populations. Many ecologists warn, however, that we are approaching or may already have exceeded the capacity of our environment to supply essential resources, absorb wastes, and maintain the web of life on which we ultimately depend. Reconciling these contrasting worldviews and finding optimum levels of population and resource use is one of the most important issues in environmental science.

In this chapter, we will survey some of the principles of natural resource economics and look at how economists assess the problem of sustainable development. Interestingly, ecology and economy are derived from the same root words and concerns. Oikos (ecos) is the old Greek word for household. Economics is the nomos or counting of the household goods and services. Ecology is the logos or logic of how the household works. In both disciplines, household is extended to include the whole world. ▶

Economic Context

Money and politics are the languages of most policy planners and decision makers. They ask, "How much will it cost?" and "What are the benefits?" Economists try to answer those questions. Basically, economics deals with resource allocation or trade-offs, either on the "micro" scale of buying and selling by individual persons and businesses, or on the "macro" scale of national policy and world economic systems. It is a description of how valuable goods and services are to us as we make decisions about how to use time, energy, creativity, or physical resources. If resources were unlimited, there would be no need to choose between alternatives and no need for a system of economics.

Economic Trade-offs

In the real world, however, we usually face decisions about trade-offs between the goods and services we desire and the resources to produce them. Economists ask, what shall we produce, for whom, or for what purpose? Furthermore, in what manner, and when shall we produce these goods and services? What level of pollution and environmental or social disruption is acceptable to obtain the things we want? For those of us in the affluent countries of the world, a new question becomes important as most of our basic necessities are satisfied. We should ask ourselves, how much is enough? How much luxury, convenience, or simple acquisition of "stuff" do we need? Would we—and the world—be better off if we were satisfied with less?

A classic expression of trade-offs in economics is the question "What shall we produce: bicycles or bullets?" This assumes some reciprocal relationship between different productive sectors of society such as that shown in figure 8.2: If we have more bicycles, we will have less bullets, and vice versa. The curve describing the maximum output at a given stage of technological or capital development is called the **production frontier.** A point inside this curve, such as point U in this figure, represents resources and production capacity that are not being fully employed in the most ef-

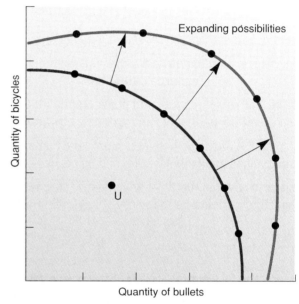

Figure 8.2

Production possibilities: bullets or bicycles? An economy can produce one or the other, or some balanced combination of the two, at a given level of technology represented by the points along the lower curve in this figure. Any point inside the curve, such as U, represents an inefficient use of productive capacities. Technology can expand the production possibility frontier, represented by the outermost curve, so that the economy can produce more of both bullets and bicycles or any other competing goods and services.

ficient manner—at least in economic terms. There may, however, be very good societal or ecological reasons to not fully use all resources at a given time.

Economic Growth and Development

Economic growth is an increase in the total wealth of a nation regardless of the population. If the population grows faster than the economy, there may be real growth but the share per person will decline. **Economic development,** on the other hand, is a growth in the average real income (adjusted for inflation) *per person*. It does not guarantee that everyone is better off, however. Many of us have more income and more "stuff" but less contentment than our parents. Furthermore, development often involves the introduction of labor-saving machines that create greater profits but displace workers. As a result, the average per capita income may rise, but the actual standard of living for much of the population may decline.

The uppermost curve in figure 8.2 shows how growth in an economic system expands the production frontier. With better access to resources, improved technological efficiency, more workers, or more investment in production facilities (all considered to be forms of capital), we can produce more of both bicycles and bullets (or whatever else people want). In most economic systems, continued growth is thought to be essential to maintain full employment and prevent class conflict that arises from inequitable distribution. Recently, however, many of us have become concerned that incessant growth of both human populations and productive systems soon will exhaust natural resources and surpass the capacity of natural systems to withstand disruption, either by effluents and wastes or by harvesting more stock than can be replaced.

Steady-State Economics

Many ecologists and a few unorthodox economists call for a transition to a **steady-state economy** characterized by low birth and death rates, use of renewable energy sources, recycling of materials, and emphasis on durability, efficiency, and stability rather than high throughput of materials. Perhaps the first economist to advocate this new way of looking at the world was John Stuart Mill, whose *Principles of Political Economy* (1848) contained many useful insights. More recently, the eccentric Romanian economist Nicholas Georgescu-Roegen spent his career exploring the limitations imposed on economics and society by the laws of thermodynamics.

Kenneth Boulding's 1966 essay "The Economics of the Coming Spaceship Earth" and Herman E. Daly's 1991 book *Steady-State Economics* both build on the concept that continual growth cannot be sustained. We will discuss the implications of steady-state economic theory further after looking at some of the questions of resource scarcity that make it seem necessary.

Resources and Reserves

What resources are economic systems set up to manage and allocate? How can we determine the available amount of a specific resource, given a particular economic system and technology? These are vital questions in the field of environmental studies because much of our concern about population growth hinges on a continual supply of resources.

Defining Resources

Simply defined, a resource is any useful information, material, or service. Within this broad generalization, we can differentiate between

Figure 8.3
The geological resources of the earth's crust, such as oil from this well, are nonrenewable. Often the limit to our use of these resources is not so much the absolute amount available but the energy cost required to extract the resources and the environmental consequences of doing so.

natural resources (goods and services supplied by our environment, including sinks for wastes) and **human resources** (human wisdom, experience, skill, labor, and enterprise). It is also useful to distinguish between exhaustible, renewable, and intangible resources.

In general, **exhaustible resources** are the earth's geologic endowment: the minerals, nonmineral resources, fossil fuels, and other materials present in fixed amounts in the environment (fig. 8.3). In theory, these exhaustible resources place a strict upper limit on the number of humans and the amount of industrial activity our environment can support. Predictions that we are in imminent danger of running out of one or another of these exhaustible resources are abundant. In practice, however, the available supplies of many commodities such as metals can be effectively expanded by more efficient use, recycling, substitution of one material for another, or better extraction from dilute or dispersed supplies.

Renewable resources include sunlight—our ultimate source of energy—and the biological organisms and biogeochemical cycles powered by solar energy. In contrast to minerals or fossil fuels, biological organisms are self-renewing (fig. 8.4). With careful management, we can harvest surplus plants and animals indefinitely without reducing the available supply. Unfortunately, our stewardship of these resources is often less than ideal. Most species have thresholds of population size, habitat, or other critical factors below which populations can suddenly crash. Once vast populations of species such as passenger pigeons, American bison (buffalo), and Atlantic cod, for instance, were exhausted by overharvesting in only a few years (see chapter 13). Ironically, with human mismanagement these renewable resources may be more ephemeral and limited than fixed geological resources.

Abstract, or **intangible resources,** including open space, beauty, serenity, genius, information, diversity, and satisfaction, are also important to us (fig. 8.5). Strangely, these resources can be

Figure 8.4

Biological resources are renewable in that they replace themselves by reproduction, but if overused or misused, populations die. When a whole species is lost, it cannot be re-created. It is permanently lost as a component of its ecosystem and as a resource to humans. These northern fur seals were almost totally exterminated by overhunting in the nineteenth century.

both infinite *and* exhaustible. There is no upper limit to the amount of beauty, knowledge, or love that can exist in the world, yet they can be easily destroyed. A single piece of trash can ruin a beautiful vista or a single mean-spirited remark can spoil an otherwise perfect day. On the other hand, unlike tangible resources that usually are reduced by use or sharing, intangible resources often are increased by use and multiplied by sharing. These nonmaterial resources can be important economically. Information management and tourism—both based on intangible resources—have become two of the largest and most powerful industries in the world.

Economic Categories

Although we have defined a resource as anything useful, we should distinguish between economic usefulness and the total abundance of a material or service that is available. Vast supplies of potentially important materials are present in the earth's crust, for instance, but they are useful to us only if they can be recovered in reasonable amounts with available technology and with acceptable environmental and economic costs. Within the aggregate total of any natural resource, we can distinguish categories on the basis of economic and technological feasibility, as well as the resource location and quantity (fig. 8.6).

You've probably read about fossil fuel resources or proven reserves of oil and natural gas in newspapers or magazines. What do these terms mean? *Proven reserves* of a resource are those that have been thoroughly mapped and are economical to recover at current prices with available technology. *Known resources* are those that have been located but are not completely mapped. Their recovery or development may not be economical now, but they are likely to become economical in the foreseeable future. *Undiscovered resources* are only speculative or inferred from similarities between known deposits or conditions and unexplored ones. There are also

Figure 8.5

Many of the intangible resources of nature enrich the quality of our lives. Our enjoyment of these resources may be enhanced by sharing them with others. On the other hand, too many other people, or people engaged in conflicting activities, can destroy the very values we seek.

unconceived resources—sometimes called unknown-unknowns—that may be economic but have not even been thought of yet. *Recoverable resources* are accessible with current technology, but are not likely to be economic in the foreseeable future, whereas *nonrecoverable* resources are so diffuse or remote that they are not *ever* likely to be technologically accessible.

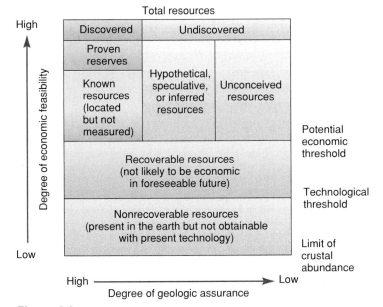

High — Low

Degree of economic feasibility

Total resources

Discovered	Undiscovered

| Proven reserves | | |
| Known resources (located but not measured) | Hypothetical, speculative, or inferred resources | Unconceived resources |

Recoverable resources (not likely to be economic in foreseeable future)

Nonrecoverable resources (present in the earth but not obtainable with present technology)

Potential economic threshold

Technological threshold

Limit of crustal abundance

High — Low

Degree of geologic assurance

Figure 8.6

Categories of natural resources. These categories, based on degree of geological assurance and economic feasibility of recovery, were intended to describe mineral resources. With some modification, these categories can be applied to many other nonrenewable resources.

There is a vast difference in amount between the first and last of these categories. The U.S. Geological Survey estimates, for instance, that only 0.01 percent (one hundredth of one percent) of the minerals in the upper one kilometer of the earth's crust will ever be economically recoverable. That means that ten thousand times as much is not economically available, and billions or even trillions of times as much deeper down is neither economically nor technically feasible to recover. Chapters 16 and 21 give some practical examples of proven, known, and ultimately recoverable fuel resources.

Communal Property Resources

In 1968, biologist Garrett Hardin wrote a widely quoted article entitled "The Tragedy of the Commons" in which he argued that any commonly held resource inevitably is degraded or destroyed because the narrow self-interest of individuals tends to outweigh community values. Hardin offered as a metaphor the common woodlands and pastures held by most New England villages during colonial times. In deciding how many cattle to put on the commons, Hardin supposed, each villager would attempt to maximize his or her personal gain. Adding one more cow to the commons could mean a substantially increased income for an individual farmer. The damage done by overgrazing, however, is shared among all the farmers. The only solution, according to Hardin, is either to give coercive power to the government or to privatize the resource.

Hardin intended this dilemma, known in economics as the "free-rider" problem, to describe resource distribution in general and

to warn specifically about the effects of human overpopulation. Other authors have used his metaphor to explain such diverse problems as African famines, firewood crises in developing countries, acid rain, fisheries declines, and urban crime. Some of these authors claim that the inexorable **tragedy of the commons** means that environmental problems can never be solved through cooperation and that only privatization or a ruthless exercise of coercive government powers can overcome our tendency for selfishness and greed.

In fact, however, Hardin was describing an **open access system** in which there are no rules to manage resource use. Empirical evidence shows that many communal resources have been successfully managed for centuries by cooperative arrangements among the users. Some examples include Native American management of wild rice beds and hunting grounds; Swiss village-owned mountain forests and pastures; Maine lobster fisheries; communal irrigation systems in Spain, Java, and Laos; and near-shore ocean fisheries nearly everywhere in the world. A large body of literature in economics and social sciences describes how these cooperative systems work.

Among the features shared by **communal resource management systems** are: (1) community members have lived on the land or used the resource for a long time and anticipate that their children and grandchildren will as well, thus they have a strong interest in sustaining the resource and maintaining bonds with their neighbors; (2) the resource has clearly defined boundaries; (3) the community group size is known and enforced; (4) the resource is relatively scarce and highly variable so that the community is forced to be interdependent; (5) management strategies appropriate for local conditions have evolved over time and are collectively enforced, that is, those affected by the rules have a say in them; (6) the resource and its use are actively monitored, discouraging anyone from cheating or taking too much; (7) conflict resolution mechanisms reduce discord; and (8) incentives to encourage compliance with the rules and sanctions for noncompliance keep members of the community in line.

Rather than being the only workable solution to problems of common pool resources, privatization and increasing governmental controls often prove to be disastrous. In countries where small villages have owned and operated local jointly-held forests and fishing grounds for generations, nationalization and commodification of resources generally have led to rapid destruction of both society and the ecosystem. While communal systems once enforced restraint over harvesting, privatization encourages narrow self-interest and allows outsiders to take advantage of the weakest members of the community.

A tragic example is the forced privatization of Indian reservations in the United States early in this century. Failing to recognize or value local knowledge systems and forcing local people to participate in a market economy allowed outsiders to disenfranchise native people and resulted in disastrous resource management policies. Whether we accept Hardin's pessimistic view of human nature and resource scarcity or believe that communal systems can be sustained and promoted, our worldview will have important implications in how we choose to manage our environment.

Population, Technology, and Resource Scarcity

In the economist's view, the supply of a particular natural resource available for human use is not determined so much by the absolute amount present on or in the earth as by economic, social, and technological factors. In this section, we will examine how population, technology, and resource scarcity interact in supply-and-demand relationships.

Supply, Price, and Demand Relationships

Supply depends on (1) which raw materials can supply a service using present technology, (2) the availability of those materials in various quantities, (3) the costs of extracting, shipping, and processing them, (4) competition for those materials by other uses and processes, (5) feasibility and cost of recycling already used material, and (6) social and institutional arrangements in force.

In a market system, most of the considerations previously mentioned are expressed in terms of market price for a good or service—the amount it sells for. The available quantity of a resource or opportunity usually increases as the price rises. For example, in 1978, the Congressional Office of Technology Assessment estimated that at $11 per barrel, some 21 billion barrels of oil were available in the United States. If the price were to double (as it soon did) to $22 per barrel, the supply also would double to 42 billion barrels. This increase was not because new oil was being created but because it becomes worthwhile to drill into lower quality and more remote oil fields as prices rise. If the price were to go even higher, furthermore, substitute fuels such as oil shales and tar sands that are not now economical to extract might become competitive with oil. The effect would be as if a whole new resource had been created.

In economic terms, the relationship between available supply of a commodity or service and its price are described by supply/demand curves. **Demand** is the amount of a product that consumers are willing and able to buy at various possible prices, assuming they are free to express their preferences. **Supply** is the quantity of that product being offered for sale at various prices, other things being equal. The inverse relationship between supply and demand is shown in the supply/demand graph in figure 8.7. As the price rises, the supply increases and the demand falls. The reverse holds as the price decreases.

In a mature market of willing and informed buyers and sellers, supply and demand should come into a **market equilibrium,** represented in figure 8.7 by the intersection of the two curves. Ideally, if all parties in the market act independently, competition should result in high efficiency and the best possible products at the lowest possible price. Adam Smith, in his 1776 economic classic *The Wealth of Nations,* described this as an "invisible hand" that leads buyers and sellers who intend only their own gain to promote the public good more effectively than they know or intend. Not all buying and selling decisions follow classic supply and demand curves, however. Some choices depend on other factors. When sellers increase the quantity available faster than prices increase, or if buy-

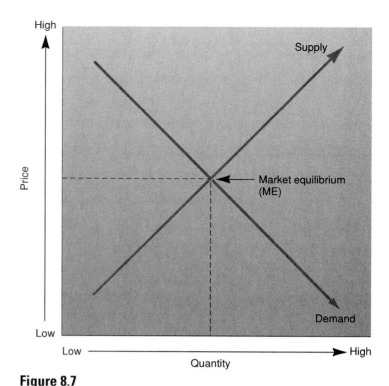

Figure 8.7

Classic supply/demand curves. When price is low, supply is low and demand is high. As prices rise, supply increases but demand falls. Market equilibrium is the price at which supply and demand are equal.

ers increase their purchases more rapidly than prices are falling, we say that the product has **price elasticity.**

Market Efficiencies and Technological Development

In a frontier economy, procedures for gaining access to resources and turning them into useful goods and services tend to be primitive and inefficient. As markets develop, however, experience accumulates in obtaining and working with a particular resource. Specialization and experimentation lead to discovery of new, more efficient technology, making it possible to produce larger quantities of goods at lower prices. The supply curve shifts and the market moves to a new equilibrium point, as is shown in figure 8.8. At each successive stage in this development process, a larger quantity of product is available at a lower price. The effect is that the standard of living increases—at least in economic terms.

Population Effects

Growing populations can offset advances in science and technology. As the number of workers increases, a point may be reached at which there are not enough jobs to employ everyone efficiently. As a result, the productivity per person will decline and wages will fall. This predicament is intensified by the pressure on resources created by more mouths to feed and more bodies to clothe. As more

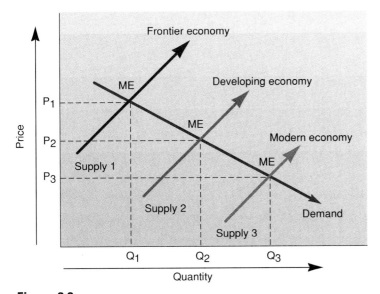

Figure 8.8

Supply and demand curves at three different stages of economic development. At each stage there is a market equilibrium point at which supply and demand are in balance. As the economy becomes more efficient, the equilibrium shifts so there is a larger quantity available at a lower price than before. (P = price, Q = quantity, ME = market equilibrium)

people use more resources, we must look to less accessible or desirable supplies. The prices of raw materials increase, as do the prices of goods and services provided by those resources; thus, the cost of living goes up and the standard of living declines. This "iron law of **diminishing returns**" led Thomas Malthus (chapter 7) to predict that unrestrained population growth would inevitably cause the standard of living to decrease to a subsistence level where poverty, vice, and starvation would make life permanently drab and miserable. This dreary prophecy has led economics to be called "the dismal science."

Growing populations also place a strain on economic development by diverting the capital necessary for growth. In a rapidly growing country, a large proportion of the population is made up of children who require social overhead expenditures, such as new housing, schools, and roads, that contribute little to development. Creating new jobs needed to employ a growing population can trap capital in conventional industries, lessening investments in new technology that might provide a real improvement in the standard of living. This diversion of investment capital is called the **population hurdle.**

On the other hand, growing populations also can create markets that encourage specialization, innovation, and capital investment that result in efficiency. They can bring young, energetic, and better trained workers into the workforce and make changes possible in traditional ways of doing things. Some demographers argue that while growing populations cause problems, they also result in more human ingenuity, energy, and cooperation to solve those problems. Where are we now in the process of economic development and population growth? Are we on a curve of diminishing re-

turns, or are we benefiting from economy of scale in terms of human populations and environmental problems?

Factors that Mitigate Scarcity

Human social systems can adapt to **resource scarcity** (a shortage or deficit in some resource) in a number of ways. Some economists point out that scarcity provides the catalyst for innovation and change (fig. 8.9). As materials become more expensive and difficult to obtain, it becomes cost-efficient to try to discover new supplies or to use the ones we have more carefully; thus, we may be better off in the long run because of these developments.

Several factors can alleviate the effects of scarcity:

- Technological inventions can increase efficiency of extraction, processing, use, and recovery of materials.

- Substitution of new materials or commodities for scarce ones can extend existing supplies or create new ones. For instance, substitution of aluminum for copper, concrete for structural steel, grain for meat, and synthetic fibers for natural ones all remove certain limits to growth.

- Trade makes remote supplies of resources available and may also bring unintended benefits in information exchange and cultural awakening.

- Discovery of new reserves through better exploration techniques, more investment, and looking in new areas becomes rewarding as supplies become limited and prices rise.

Figure 8.9

Scarcity/development cycle. Paradoxically, resource use and depletion of reserves can stimulate research and development, the substitution of new materials, and the effective creation of new resources.

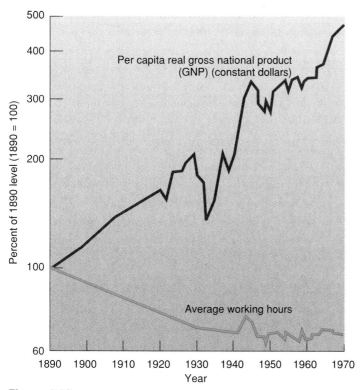

Figure 8.10

Technological improvements, increased capital investments, and more highly trained labor have raised production faster than population growth in the United States. Between 1890 and 1970, real per capita GNP rose fourfold, even though average working hours fell nearly 40 percent. In other words, our increased productivity has given us both more output and more leisure.

Source: U.S. Department of Commerce.

- Recycling becomes feasible and accepted as resources become more valuable. Recycling now provides about 37 percent of the iron and lead, 20 percent of the copper, 10 percent of the aluminum, and 60 percent of the antimony that we consume each year in the United States.

Increasing Environmental Carrying Capacity

Economist Julian Simon says that in spite of recurring fears of natural resource scarcity, the mitigating factors listed above have made every commodity cheaper in real terms as far back as we can find records. In fact, responding to the growing scarcity of resources actually enables us to increase the carrying capacity of the environment for humans. Figure 8.10 shows the change in real GNP (as a percent of that in 1890 adjusted for inflation) during the period of industrial growth and transformation to a postindustrial society. There has been about a 500 percent increase in real per capita GNP during this century even though average working hours have declined, population has tripled, and the easily accessible local resources largely have been used up.

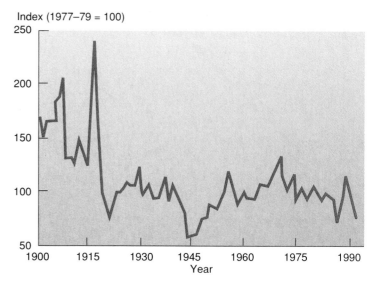

Figure 8.11

Long-run world prices for nonferrous metals (aluminum, copper, tin, and zinc) in the twentieth century, adjusted for inflation. Real prices fell rapidly in the beginning of the century due to technological advances and then remained relatively constant except for wars and temporary shortages or surpluses. There is no evidence that exhaustion of limited supplies is driving up prices.

Source: World Bank data.

Will this economic progress be sustained, however? Ecologist Paul Ehrlich contends that increasing levels of population and consumption will inevitably lead to scarcity and rising prices as more of us try to share less and less. In 1980, Ehrlich made a wager with Simon. They bet on a package of five metals—chrome, copper, nickel, tin, and tungsten—priced at $1000 in 1980. If the 1990 combined prices, corrected for inflation, were higher than $1000, Simon would pay the difference. If prices had fallen, Ehrlich would pay. In 1990, Ehrlich sent Simon a check for $576.07; prices of these five metals had fallen 47.6 percent. In fact, prices for most metals have fallen substantially over the past century (fig. 8.11). Would you care to bet on whether this pattern will hold for the next century?

Most countries have not shared in the rapid increase of wealth enjoyed by the United States in the past century. Figure 8.12 shows GNP and population by region. As you can see, North America and Europe have 65 percent of the world GNP, but only 18 percent of the world population. Asia, Latin America, and Africa have 74 percent of the population, but only 16 percent of the GNP. This may be a very real example of the population hurdle discussed earlier.

Limits to Growth

We return again and again in environmental studies to the underlying question of whether continued population growth, economic growth, or both, would be good or bad. At what point will the number of people or the extent of economic activities that impinge on our environment bring disaster, not only to humans, but to the

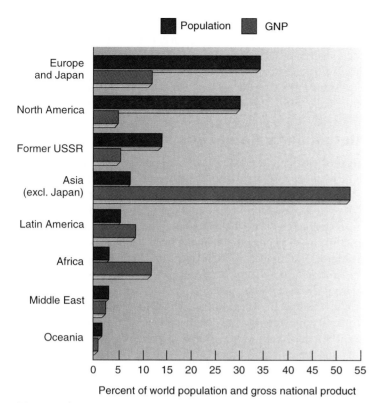

Figure 8.12

Percent of world population and gross national product (GNP) by region in 1990. The more-developed countries (MDCs) had 65 percent of the world GNP, but only 18 percent of the world's population. The less-developed countries (LDCs) had 74 percent of the world's population, but only 16 percent of the total GNP.

Source: Data from Population Reference Bureau.

whole life-supporting system of the biosphere? On the other hand, how will we improve the standard of living in less-developed countries and clean up damage already done to the environment if growth stops? The crux of this argument is whether resources are finite or can be effectively expanded through human ingenuity and enterprise. Let's look at what some economic models predict might be the result of further economic and population growth.

Computer Models of Resource Use

In the early 1970s, an influential study of resource limitations was funded by the Club of Rome, a gathering of wealthy business owners and politicians. The study was undertaken by a team of scientists from the Massachusetts Institute of Technology headed by Donnela Meadows. The results of this study were published in the 1972 book *Limits to Growth*. Several computer models of world economy were used to examine various scenarios of resource depletion, growing population, pollution, and industrial output. Given the Malthusian assumptions built into these early models, catastrophic social and environmental collapse seemed inescapable. If you have x number of cans of soup and you use y cans per day, the supply will last x/y

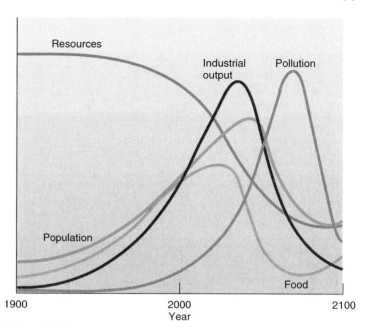

Figure 8.13

A run of one of the world models in *Limits to Growth*. This model assumes business-as-usual for as long as possible until Malthusian limits cause industrial society to crash. Notice that pollution continues to increase well after industrial output, food supplies, and population have all plummeted.

days. If you start with $2x$ cans and consume at the same rate, they last $2x/y$ days, but you still run out sooner or later.

Figure 8.13 is an example of a run of one of the world models given a doubling of present resources but no change in public policies or attitudes. Population, resources, industrial output, and food supplies all crash precipitously sometime about the middle of the next century. Pollution continues to rise after everything else collapses, presumably because of high rates of death and destruction. Note the strong similarity between these curves and "boom and bust" cycles observed when natural populations exceed environmental carrying capacity (chapter 6).

Some authors have criticized *Limits to Growth* because the original computer models used by the Meadows group discounted factors that might mitigate the effects of scarcity. More recently, in *Beyond the Limits*, the Meadows group has modified its model to include technological progress, pollution abatement, population stabilization, and new public policies that work for a sustainable future. If we adopt these changes sooner rather than later, the computer shows an outcome like that in figure 8.14, in which all factors stabilize at an improved standard of living for everyone. Neither of these computer models shows what *will* happen, only what some possible outcomes *might* be, depending on what we do.

Why Not Conserve Resources?

Even if large supplies of resources are available, or the possibility of technological advances to mitigate scarcity exists, wouldn't it be better to reduce our use of natural resources so they will last as long

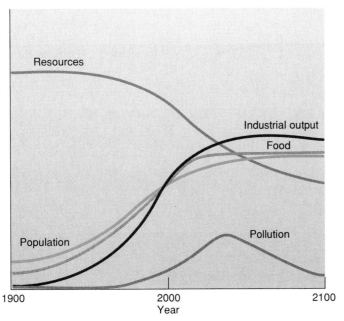

Figure 8.14

A run of the world model from *Beyond the Limits*. This model assumes that population and consumption are curbed, new technologies are introduced, and sustainable environmental policies are introduced immediately rather than after resources are exhausted.

as possible? Will anything be lost if we're frugal now and leave more to be used by future generations? This makes sense if the assumption is correct that resources are limited and cannot be replenished, or that their extraction inevitably degrades the environment. Using them more slowly and sharing them with fewer people will give each of us a larger share, will be kinder to the environment, and will make our supply last longer.

Most economists, however, look at resources as means to an end, rather than having value in themselves. Resources have to be used to be of value. If you bury your savings in a jar in the backyard, it will last a long time but may not be worth much when you dig it up. If you invest it productively, you will have much more in the future than you do now. Furthermore, a window of opportunity for investment may be open now but not later.

Resource Economics

How do we determine the value (or price) of environmental goods and services? Some of the most crucial environmental factors that may shape our future are not represented by monetary values in the marketplace. Certain resource allocation decisions are political or social. Other resources are simply ignored. Groundwater, sunlight, clean air, biological diversity, and other assets that we all share in common often are treated as public goods (benefits) that anyone can use freely. Our economic system typically has not charged for using the absorptive capacity of the environment to dispose of wastes despite ample evidence that this capacity can be exhausted.

In theory, these resources are self-renewing, but as we will show in other chapters of this book, many vital environmental as-

sets are threatened by human activities. If we damage basic life-support systems of the biosphere, we cannot simply substitute another material or service for the ones that have become limited. The crux of this question is how we should manage resources in a market system. Let's look now at how our economic system handles internal and external costs and intergenerational justice.

Internal and External Costs

Internal costs are the expenses (monetary or otherwise) that are borne by those who use a resource. Often, internal costs are limited to the direct out-of-pocket expenses involved with gaining access to the resource and turning it into a useful product or service.

External costs are the expenses (monetary or otherwise) that are borne by someone other than the individuals or groups who use a resource. External costs often are related to public goods and services derived from nature. Some examples of external costs are the environmental or human health effects of using air or water to dispose of wastes. Since these effects usually are diffuse and difficult to quantify, they do not show on the ledgers of the responsible parties. They are likely to be ignored in private decisions about the costs and benefits of a purchase or a project. One way to use the market system to optimize resource use is to make sure that those who reap the benefits of resource use also bear all the external costs. This is referred to as **internalizing costs.**

A controversial provision of the 1990 revision of the Clean Air Act allows companies to market emission quotas as a way of reducing pollution in the most efficient and least costly way possible (What Do *You* Think?, next page). This would have the effect of internalizing external costs but is regarded by its opponents as merely a license to pollute.

Intergenerational Justice and Discount Rates

"A bird in the hand is worth two in the bush." All of us are familiar with sayings that suggest it is better to have something now than in the distant future. **Discount rates** are the economist's way of introducing a time factor in accounting. It is a recognition that a ton of steel delivered today is worth more than the same ton delivered a year from now; the difference is an extra year's worth of use of products made from the steel. In theory, the discount rate should be equal to the interest rate on borrowing money.

The choice of discount rates to apply to future benefits becomes increasingly problematic with intangible resources or long time frames. How much will a barrel of oil or a four-thousand-year-old redwood tree be worth a couple of centuries from now? Maybe there will be substitutes for oil by then so it won't be highly valued. On the other hand, that oil or tree may be priceless. This valuation is complicated by the fact that we are making decisions not only for ourselves but also for future generations.

Although having access to clean groundwater or biological diversity one hundred years from now isn't worth much to us—assuming that we will be long gone by that time—those resources might be quite valuable to our descendants. Future citizens will be affected by the choices we make today, but they don't have a vote. Our decisions about how to use resources raise difficult questions

What Do *You* Think?
Market-Based Incentives for Environmental Protection

What is the most efficient and economical way to eliminate pollution (fig. 1)? Some people argue that we should simply say to polluters, "Stop it! You can't dump garbage into the air or water anymore." Although this approach has a certain moral appeal, it may tend to force all businesses to adopt uniform standards and methods of pollution control regardless of cost or effectiveness. This approach also can lead to an adversarial climate in which resources are used in litigation rather than pollution control.

Furthermore, the "command and control" approach tends to freeze technology by eliminating incentives for continued research and development. Industry is discouraged—even prohibited—from trying new technologies or alternative production methods. These problems can be overcome, many

Figure 1

Allowing firms to buy and sell pollution allowances or "rights to pollute" may be the most efficient way to reduce pollution on a regional level. It could have unfortunate local impacts, however.

economists believe, by using market mechanisms to reduce pollution rather than rigid rules and regulations. Since there may be a one hundred-fold variation in the cost of eliminating a specific pollutant from different sources due to the age of equipment in use, environmental factors, and other considerations, market-based incentives such as pollution charges or tradable permits can be more cost-effective and flexible than simply saying "Thou shalt not."

Pollution charges are fees assessed per unit of pollution based on the "polluter pays" principle. This approach encourages businesses to do as much pollution control as possible: The more pollution you eliminate, the more you save. Charges could be the same for a given pollution type regardless of source or location, or they could be adjusted to reflect relative amounts of damage or other social considerations. Fees might have to be set quite high to discourage some types of pollution, however, and could exaggerate inequities between the rich, who can afford to pollute, and the poor, who suffer the consequences.

Five types of pollution charges are now being considered or are already in place for some industries: (1) effluent charges based on the quantity of discharge, (2) user charges based on the cost of public treatment facilities such as sewage treatment plants that clean up effluents, (3) production charges based on potential damage caused by a product, (4) administrative charges based on government monitoring services, and (5) differential taxes to encourage "green" products.

Tradable permits are based on an assumption of thresholds below which some types of pollution are acceptable. If we can agree on those acceptable levels, industries can be given permits to emit a fair share of pollution. Any company below its limit can sell or lease the excess amount to another company. In theory, this should allow us to reach pollution control goals in the most cost-effective manner. Companies that can lower effluents most cheaply will do so because they can make money by selling credits to others. If what we want is a given result—clean air for instance—and we don't care how that goal is attained, then this may be a good approach.

Like pollution charges, permits tend to encourage some innovation and

technological improvements. The more efficiently pollutants are removed, the more money you can make by selling the excess. While this should make pollution control cheaper, it doesn't create incentives to lower pollution below established targets. Where charges require a large bureaucracy to handle paperwork, measure effluents, and collect fees, however, permits tend to be handled more in the private sector.

Both permits and effluent charges are being considered or are already being used to regulate several types of pollution. In 1993, marketing of sulfur dioxide permits for power plants began as a way of reducing acid precipitation. The goal is to reduce sulfur emissions by 10 million tons per year. Suppose a new plant can remove considerably more sulfur than it is required to by law at a cost of only $100 per ton. An older plant might have to spend $200 per ton to meet air pollution standards. Owners of the older plant might offer to pay $150 per ton for an allowance from the newer plant. Each makes $50 and the target reduction is still met. The overall societal goal of reducing sulfur emissions will be met, but the local residents near the older power plant still have to put up with high levels of air pollution.

A carbon tax on fossil fuels has been proposed as a way of reducing greenhouse gases. A tax is considered more effective than emission permits or effluent standards in this case because the large number of carbon dioxide sources makes emission measurements and enforcement difficult. It is estimated that a $100 per ton tax on fossil fuels could reduce carbon dioxide emissions by 36 percent in the United States by the year 2000 and raise $120 billion per year for pollution control.

How would you weigh the choice between a "just say no" approach and a "tax polluters until they stop" strategy if you were a legislator or administrative official? Market mechanisms may make economic sense, but how can we make sure that their penalties don't fall most heavily on the poor? Is a pollution allowance merely a license to pollute, or is it a rational way to allocate resources? What criteria would you use to choose between these very different approaches to pollution control?

about justice between generations. How shall we weigh their interests in the future against ours right now?

Which discount rate should be used for future benefits and costs is often a crucial question in large public projects such as dams and airports. Proponents may choose to use discount rates that make a venture seem attractive, while opponents prefer rates that show the investment to be questionable. These questions are especially difficult in comparing future environmental costs to immediate financial returns. Environmental activists often need to understand economic nuances when they are fighting environmentally destructive schemes.

Cost/Benefit Ratios

One way to evaluate the outcomes of large-scale public projects is to analyze the costs and benefits that accrue from them in a **cost/benefit analysis.** This process assumes that values can be assigned

to present and future resources, given proper criteria and procedures. It is one of the main conceptual frameworks of resource economics. This process is controversial, however, because it deals with vague and uncertain values and compares costs and benefits that are as different as apples and oranges. Yet we continue doing these analyses because we don't have better ways to allocate resources.

Figure 8.15 presents a flowchart for preparing a cost/benefit analysis. As you can see, several different tributary paths come together to determine the final outcome of this process. The easiest parts of the equation to quantify are the direct costs and benefits to the developer or investor who has proposed the project; that is, the out-of-pocket expenses and the immediate profits that will result from this investment. These direct monetary costs and benefits are usually the most concrete and accurate components in the analysis. It is important that they not outweigh other factors more difficult to ascertain but of equal importance.

Figure 8.15

A flowchart for a cost/benefit analysis.

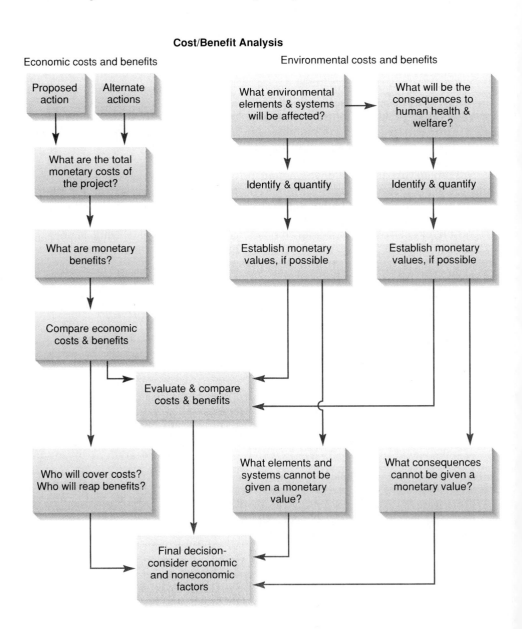

Cost/Benefit Analysis

The other branch of the flowchart involves analysis of more diffuse, nonmonetary factors such as environmental quality, ecosystem stability, human health impacts, historic importance of the area to be affected, scenic and recreational values, and potential future uses. These are difficult values to quantify. It is even more difficult to express them in monetary terms. How much are beauty or tranquility worth? What are the benefits of ethical behavior? How much would you pay for good health?

Some costs and benefits simply cannot be expressed in monetary terms. These invaluable (in a positive sense) factors bypass the mathematical stages of comparison and are considered—we hope—in the final decision-making process, which is more political than mechanical. Also factored in at this stage are distributional considerations, that is, who will bear the costs of the project and who will reap the benefits? If these are two different groups of people, as they usually are, questions of justice arise that must be resolved, perhaps in some other venue.

Criticisms and complications of this process include the following:

- *Absence of standards.* Each person assesses costs and benefits by his or her own criteria, often leading to conflicting conclusions about the comparative values of a project. It has been suggested that an agency or influential group set specifications for how factors should be evaluated.

- *Inadequate attention to alternatives.* To really understand the true costs of a project, including possible loss of benefits from other uses, it is essential to evaluate alternative uses for a resource and alternative ways to provide the same services. These steps are often slighted.

- *Assigning monetary values to intangibles and diffuse or future costs and benefits.* Some critics of this process claim that we should not even try. They believe that attempting to express all values in monetary terms suggests that only monetary gains and losses are important. This can lead to the "slippery slope" argument that everything has a price and that any behavior is acceptable as long as we can pay for it.

- *Acknowledging the degree of effectiveness and certainty of alternatives.* Sometimes speculative or even hypothetical results are given specific numerical values and treated as if they were hard facts. We should use risk-assessment techniques to evaluate and compare uncertainties in the process.

- *Justification of the status quo.* Agencies may make decisions to go ahead with a project for political reasons and then manipulate the data to correspond to preconceived conclusions.

Figure 8.16 shows an example of a cost/benefit analysis for reducing particulate air pollution (soot) in Poland. As you can see, removing the highest 40 percent of particulates is highly cost effective. As you approach 70 percent particulate removal, however, the costs may exceed benefits. This same study, on the other hand, showed negligible benefits compared to high costs for sulfur removal, perhaps for some of the reasons detailed above.

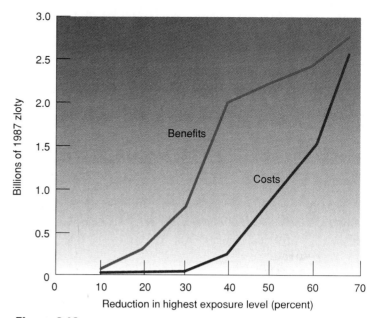

Figure 8.16

Total estimated benefits and costs of reducing particulate air pollution in Tarnobrzeg, Poland. Removing the highest 40 to 50 percent of particulates costs relatively little compared to the large benefits that would be gained. Removing 70 percent of particulates may cost more than the benefits are worth.

Source: J. Cofala, T. Lis, and H. Balandynowicz, *Cost-Benefit Analysis of Regional Air Pollution Control: Case Study for Tarnobrzeg,* 1991, Polish Academy of Sciences, Warsaw.

Distributing Intangible Resources

Can the market work in a socially responsible way in setting priorities for use and protection of intangible assets? Sometimes it does, but there are many instances where it does not. It is important to recognize when we can and cannot depend on market forces to manage natural resource economics. Market systems are notoriously poor at handling distributional questions of equity or justice. Most economists recognize that we need to regulate markets to achieve social goals that market mechanisms alone cannot accomplish. Consider the following points:

- In theory, environmental services, such as pollution absorption, are available to everyone and are not bought and sold. There is, therefore, no market for them and no price-setting mechanism.

- In making decisions about whether it would be more profitable to use a resource now or save it for the future, private interests almost always choose to use it quickly because they benefit directly from immediate use, whereas the benefits from future use may be spread among many users.

- Private discount rates for future benefits do not consider social objectives and, therefore, do not reflect complete values of *in situ* resources.

- The costs for destruction of a resource are often external, that is, they accrue to individuals other than those who benefit from its use.

Since cost/benefit ratios are often the deciding factors in public works projects and setting priorities regarding natural resources, students of environmental science should understand how these values are determined. First, we have to decide who benefits and who pays. Then, we have to set prices on the goods and services provided by the environment. This may be fairly straightforward for some benefits, but it is very difficult when considering intangible assets such as opportunity costs or existence values.

For instance, there is a wild and scenic river not far from my home. Even though I don't go there often, it is important to me that the opportunity to do so exists. How much is this opportunity worth? A cost/benefit study might assume that, since the owner of a large cabin cruiser spends much more per day on the river than I do in my homemade canoe, I don't value the opportunity as highly. (I don't agree.) It is even more difficult to determine existence values. You might like to know that some grizzly bears still live in Alaska and that great whales still swim in the ocean, even though you may never see these animals. How much is this knowledge worth? Does a diffuse value to many of us outweigh a very specific economic gain to the hunter who wants to kill these animals?

Natural Resource Accounting

When economists measure a nation's output to determine **gross national product (GNP),** the sum total of all goods and services produced in a national economy, they typically subtract capital depreciation, that is, the wear and tear on machines, vehicles, and buildings used in production. They rarely, however, account for gains or losses in less tangible capital such as stocks of natural resources or environmental services. Robert Repetto of the World Resources Institute, for instance, calculated that soil erosion in Indonesia reduces the value of crop production about 40 percent per year. In total, natural resource depletion reduces the Indonesian GNP by nearly 20 percent. If these losses are taken into account, what appears by conventional accounting to be a rapidly growing economy is shown to actually be losing ground (no pun intended).

International Development

No single institution has more influence on the financing and policies of developing countries than the World Bank. Of some $25 billion loaned each year for Third World projects by multinational development banks, about two-thirds comes from the World Bank. For every dollar invested by the World Bank, two dollars are attracted from other sources. If you want to have an impact on what is happening in the developing countries, it is imperative to understand how this huge enterprise works.

The World Bank was founded in 1945 to provide aid to war-torn Europe and Japan. In the 1950s, its emphasis shifted to development aid for Third World countries. This aid was justified on humanitarian grounds, but providing markets and political support for Western capitalism was an important by-product.

The Bank is jointly owned by 150 countries, but one-third of its support comes from the United States. The Bank president has always been an American. The bulk of its $66.8 billion capital comes from private investors who buy instruments (bonds and debentures) on the open market. Loan applications are first screened by the professional staff and then voted on by a council of all member countries. No loan has ever been turned down once it arrives at the full council. So far, one hundred countries have borrowed some $140 billion from the World Bank for a wide variety of development projects.

Many World Bank projects have been environmentally destructive and highly controversial. In Botswana, for example, $18 million was provided to increase beef production for export by 20 percent, despite already severe overgrazing on fragile grasslands. The project failed, as did two previous beef production projects in the same area. In Ethiopia, rich floodplains in the Awash River Valley were flooded to provide electric power and irrigation water for cash export crops. More than 150,000 subsistence farmers were displaced and food production was seriously reduced. In India, the sacred Narmada River is being transformed by thirty large dams and 135 medium-sized ones financed by the World Bank. About 1.5 million hill people and farmers will be displaced as a result.

These are only a few of the projects that have aroused concern and protest. They typify the environmental and social costs of many development programs. Loans from the World Bank tend to favor large-scale agricultural, transportation, and energy projects, which make up more than half of all funds the organization distributes. In part, this is because both the World Bank and debtor nations find it easier to manage one big project than many small ones. Furthermore, both the World Bank and the governments borrowing money prefer large, impressive, modern projects to show how the money is being spent. Small cooperatives or cottage industries might do more for the people than a big dam, but they don't look as impressive.

In a response to criticisms of World Bank policies, the U.S. Congress now insists that all loans for international development be reviewed for environmental and social effects before being approved. It asks for assurance that each project (a) uses renewable resources and does not exceed the regenerative capacity of the environment, (b) does not cause severe or irreversible environmental deterioration, and (c) does not displace indigenous people.

As recently as 1986, only six people of the nearly seven thousand World Bank staff members were assigned to do ecological assessments on $17 billion in development projects. Pressure from environmental groups has forced the World Bank to add about a hundred ecologists and environmental specialists to the staff. Critics charge that many of these people are merely recycled economists and that business is going on as usual. However, loans have been canceled for road building that would lead to tropical forest destruction, and loans have been made for environmental restoration projects in Brazil. The World Bank continues to be criticized, at least in part, because it's easy to aim at a large target. A fundamental question remains whether technological and economic growth can bring a better standard of living for everyone without causing unacceptable environmental disruption.

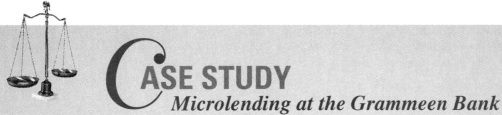

CASE STUDY
Microlending at the Grammeen Bank

Development projects in poorer nations have generally been financed through loans from international development banks. The megaprojects funded by these banks are highly political and often show little sensitivity to local culture or environment. Furthermore, they fail to help the informal sector (street vendors, household industries, or unorganized service workers) even though these people may account for more than half of all economic activity in many less-developed countries.

The greatest barrier to productive self-employment for poor people often is a lack of access to capital. Since they generally have few assets to use for collateral and no credit record, the poor can't go to traditional banks for loans to buy tools and materials to start a small business. An inspiring new approach to banking called microlending has been shown to be both successful and profitable, however, in making loans to the poor. This promises to be a model for grassroots economic development.

Pioneered by economist Muhammad Yunus of Bangladesh, the Grammeen (village) Bank provides credit directly to the poorest people, those who have no collateral and no steady source of income. Started in 1976, the Grammeen Bank now has nearly 1000 local offices and some 1.2 million customers, 90 percent of whom are women who could never have borrowed money from an ordinary bank. Their loans are small, averaging only $67. This is enough, however, to buy a used sewing machine, a bicycle, a loom, a cow, some garden tools—a start in providing needed family income (fig. 1).

Compared to a national average of only 30 percent repayment rate on loans, the recovery rate on Grammeen accounts is an astonishing 98 percent. The key to this success is peer lending. Borrowers are organized into five-member peer groups that act both as mutual aid societies and collection agencies. Payments must be made in regular weekly installments. If one member of the group defaults, the others must repay the loan. Having some dignity, respect, and independence encourages responsibility and self-reliance.

Microlending and self-help programs are now being used around the world. More than a hundred organizations in the United States currently assist microenterprises by providing loans, grants, or training. The Women's Self-Employment Project in Chicago, for instance, is empowering and teaching skills to single mothers in public housing projects. The repayment rate on its loans is nearly 100 percent. Similarly, "tribal circle" banks on Native American reservations successfully finance microscale economic development and group-building projects. In 1992, the Small Business Administration earmarked $15 million for microloan projects. Interestingly, the richest country in the world may have learned a valuable economic lesson from one of the poorest.

Figure 1

A loan from the Grammeen Bank enables this Bangladeshi woman and her son to set up a broom-making business. Collective management of such projects provides mutual support and ensures loan repayment.

An alternative to the huge, often destructive, loans make by the World Bank are the small-scale, community-based programs such as Grammeen Bank of Bangladesh (Case Study, previous page).

International Trade

International issues further complicate questions of resource management. Much of the vast discrepancy between richer and poorer nations is related to economic and political history, as well as to current international trade relations. The banking and trading systems that control credit, currency exchange, shipping rates, and commodity prices were set up by the richer and more powerful nations in their own self-interest. The General Agreement on Tariffs and Trade (GATT), for example, negotiated primarily between the largest industrial nations, regulates 90 percent of all international trade.

These systems tend to keep the less-developed countries in a perpetual role of resource suppliers to the more-developed countries. The producers of raw materials, such as mineral ores or agricultural products, get very little of the income generated from international trade (fig. 8.17). Furthermore, they suffer both from low-commodity prices relative to manufactured goods and from wild "yo-yo" swings in prices that destabilize their economies and make it impossible to either plan for the future or to accumulate capital for further development.

In 1974, the United Nations General Assembly adopted Resolution Number 3281 calling for a "new international economic order" in which "every State has and shall freely exercise full permanent sovereignty, including possession, use and disposal over all its wealth, natural resources and economic activities." This idea of a new relationship between the richer nations of the North and the poorer nations of the South was developed further by the United Nations Commission on North-South Issues chaired in 1980 by Willy Brandt, former mayor of West Berlin, and at the Rio Earth Summit. As you might suppose, most rich nations have been less enthusiastic than the poor ones about changing present economic relationships.

Jobs and the Environment

For years business leaders and politicians have portrayed environmental protection and jobs as mutually exclusive. Pollution control, protecting natural areas and endangered species, and limiting the use of nonrenewable resources will, they claim, strangle the economy and throw people out of work. A new brand of environmental economists dispute this claim, however. Their studies show that only 0.1 percent of all large-scale layoffs in the United States are due to environmental regulations (fig. 8.18). In fact, environmental protection is not only necessary for a healthy economic system, it actually creates jobs and stimulates business.

Recycling, for instance, takes much more labor than extracting virgin raw materials. This doesn't necessarily mean that recycled goods are more expensive than those from virgin resources. We're simply substituting labor in the recycling center for energy and huge machines used to extract new materials in remote places.

Japan, already a leader in efficiency and environmental technology, has recognized the multibillion dollar economic potential of "green business." The Japanese government is investing $4 billion (U.S.) per year on research and development that targets seven areas ranging from utilitarian projects such as biodegradable plastics and heat-pump refrigerants to exotic schemes such as carbon-dioxide-fixing algae and hydrogen-producing microbes (In Depth, next page).

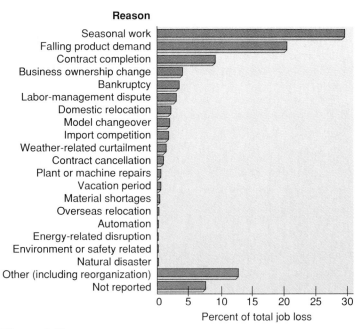

Figure 8.18

Although opponents of environmental regulation often claim that protecting the environment costs jobs, studies by economist E. S. Goodstein show that only 0.1 percent of all large-scale layoffs in the United States were the result of environmental laws.

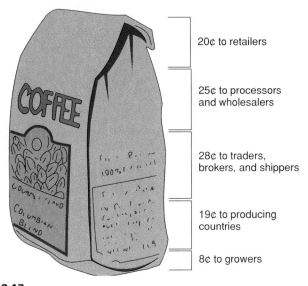

Figure 8.17

What do we really pay for when we purchase a dollar's worth of coffee?

20¢ to retailers

25¢ to processors and wholesalers

28¢ to traders, brokers, and shippers

19¢ to producing countries

8¢ to growers

*I*n Depth: Green Business

What responsibilities do businesses have to protect the environment beyond the legal liabilities spelled out in the law? None whatsoever, according to economist Milton Friedman. In fact, Friedman argues it would be unethical for corporate leaders to consider anything other than maximizing profits in their business decisions. Managers are hired by owners solely for the purpose of making profits and increasing the value of the company. To spend time or resources doing anything else is a betrayal of their duty. In this view, if society considers social justice or environmental protection to be desirable goals, then politicians should pass laws and apply them equally to all businesses. For any single business to voluntarily and unilaterally internalize external costs puts it at a competitive disadvantage against rivals that do not do so.

Some highly successful corporations, however, take a very different view of business ethics. Operating in a socially responsible manner consistent with the principles of sustainable development and environmental protection can be good for employee morale, public relations, and the bottom line simultaneously. Public-spirited companies such as the Body Shop, Patagonia, Aveda, Johnson and Johnson, and Rhino Records consistently earn high marks from community and environmental groups. Conserving resources, reducing pollution, and treating employees and customers fairly may cost a little more initially but can save money and build a loyal customer base in the long run.

The simplest way for businesses to become good citizens and desirable neighbors is to practice the "golden rule:" Treat employees, customers, vendors, shareholders and the general public as you would like them to treat you. Respect human rights and behave as if the place you do business is your home. If every business operated as if every action, decision, and policy were to be disclosed fully on the front page of tomorrow's newspaper, we might have greater accountability.

In 1989, following the wreck of the oil tanker *Exxon Valdez* in Alaska, a group of business executives, environmentalists, and community leaders drew up standards for corporate responsibility and environmental citizenship. These principles, now known as the CERES principles, are:

- *Biosphere protection:* Minimize the release of any pollutant that may damage the air, water, or Earth, including those that contribute to the greenhouse effect, depletion of the ozone layer, acid rain, and smog.

- *Sustainable natural resource use:* Conserve renewable natural resources, such as water, soils, and forests. Wildlife habitat, open spaces, and wilderness will be protected and biodiversity will be preserved.

- *Reduction and disposal of waste:* Minimize waste, especially hazardous waste, and recycle whenever possible. All waste will be disposed of safely and responsibly.

- *Wise energy use:* Make every effort to use environmentally safe and sustainable energy sources and invest in energy efficiency and conservation.

- *Risk reduction:* Minimize environmental and health and safety risks to employees and local communities by employing safe technologies and preparing for emergencies.

- *Marketing of safe products and services:* Sell products or services that minimize adverse environmental impacts and that are safe for consumer use.

- *Damage compensation:* Take responsibility for any harm caused to the environment through cleanup and compensation.

- *Disclosure:* Disclose to employees and community any incidents that cause environmental harm or pose health or safety risks.

- *Environmental directors and managers:* At least one member of the board of directors will be qualified to represent environmental interests. Among management positions there will be a senior executive position in environmental affairs.

- *Assessment and annual audit:* Conduct annual self-evaluation of progress in implementing these principles and make results of independent environmental audits available to the public.

By the year 2000, Japan expects to be selling $12 billion worth of equipment and services per year worldwide. They are already marketing advanced waste incinerators, pollution control equipment, alternative energy sources, and water-treatment systems.

Unfortunately, the United States has been resisting international pollution control conventions rather than recognizing the potential for economic growth *and* environmental protection in the field of green business.

Sustainability: The Challenge

As chapter 1 points out, some 1.5 billion people, or about one-fourth of the world population, are so poor that they cannot, on their own, meet the basic needs for food, shelter, clothing, or medical care. Many scholars and social activists believe that poverty is at the core of many of the world's most serious problems—hunger, child deaths, migrations, insurrections, and environmental degradation. One way to alleviate poverty is to redistribute existing wealth, but those who have plenty often are unwilling to share with those who don't. Another approach is to foster economic growth so there can be a bigger share for everyone. But does the world have enough natural resources—and can the environment withstand—continued economic growth? Or are there alternative ways to bring about human development—a real increase in the standard of living and the quality of life—for everyone without impoverishing our environment in the process? Finding a sustainable way of life, socially, economically, and environmentally, is one of the greatest challenges we face. The World Bank projects that if current trends continue, economic output in developing countries will rise by 4 to 5 percent per year in the next 40 years. The economies of industrialized countries are expected to grow more slowly but could still triple over that period. Altogether, the total world output could be quadruple what it is today.

This economic growth could provide funds to clean up the environmental damage caused by earlier, more primitive technologies and misguided resource uses. It is estimated to cost $350 billion per year to control population growth, develop renewable energy sources, stop soil erosion, protect ecosystems, and provide a decent standard of living for the world's poor. This is a great deal of money, but is small compared to the $1 trillion per year spent on wars and military equipment.

But can the earth sustain that much growth? Many environmental scientists warn that human activities are overwhelming the basic life-support systems of the biosphere. They call for a steady-state economic system that will minimize our impact on the environment.

A moderate position between the extremes of no growth versus unlimited growth is **sustainable development** based on the use of renewable resources in harmony with ecological systems (fig. 8.19). Perhaps the best statement of this principle is that of the World Commission on Environment and Development (see chapter 1), which defined sustainable development in *Our Common Future* as meeting the needs of the present without compromising the ability of future generations to meet their own needs. Some goals of sustainable development are:

- A demographic transition to a stable world population of low birth and death rates.

- An energy transition to high efficiency in production and use, coupled with increasing reliance on renewable resources.

- A resource transition to reliance on nature's "income" without depleting its "capital."

- An economic transition to sustainable development and a broader sharing of the benefits of development.

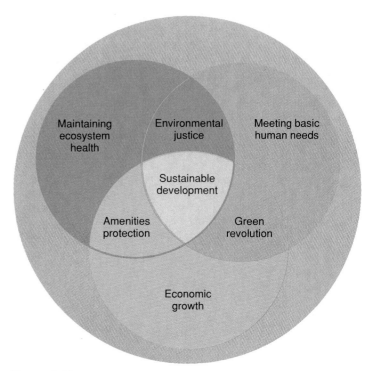

Figure 8.19

A model for integrating ecosystem health, human needs, and sustainable economic growth.

(Modified from Raymond Grizzle and Christopher Barrett, personal communication.)

- A political transition to global negotiation grounded in complementary interests between North and South, East and West.

- An ethical or spiritual transition to attitudes that do not separate us from nature or each other.

Some strategies for attaining these goals are listed in table 8.1.

Will we accomplish the transition to an economy based on sustainable development, intergenerational justice, equitable distribution of resources, and harmony with nature? We will succeed only if we bring the wisdom and knowledge of both ecology *and* economics to bear on our problems. The future will be what we make it.

I know not why it should be a matter of congratulation that persons who are already richer than anyone needs to be, should have doubled their means of consuming things which give little or no pleasure except as representative of wealth. . . . It is only in the backward countries of the world that increased production is still an important object: in those most advanced, what is economically needed is a better distribution, of which one indispensable means is a stricter restraint on population.

—John Stuart Mill
Principles of Political Economy

Table 8.1
Strategies for sustainable development

1. *Greater attention to the problems of the poor people of the world.* In many cases, environmental damage is caused by people who have no other alternatives in their struggle for existence. Modest improvements in their income, health status, political freedom, and access to education, capital, and technology could have major impacts on society and the environment.

2. *Local input in planning and managing resource development.* Local people often have valuable ecological knowledge that is overlooked by planners. Giving them a better share of proceeds from resource development provides an incentive for them to protect and conserve resources.

3. *Proper resource pricing.* Internalizing external costs shows the real trade-offs in resource development and gives resource users the incentive to minimize *all* costs, not just those that directly affect themselves.

4. *Wiser management.* Dramatic and far-reaching improvements in resource availability can be obtained by reducing or eliminating wasteful or unnecessary resource uses. This can be accomplished either by pricing mechanisms or by changes in regulations and institutional mechanisms. Sometimes it is faster and more effective to pay for new efficient equipment (insulation, new furnaces for the poor) than to depend on market forces or laws.

5. *Better management capability.* We need technical personnel, information, and legal and administrative systems to plan and guide resource use so that market forces or legal mechanisms will work as they should to protect and sustain vital resources.

6. *Less throughput.* By emphasizing durable goods, recycling and reuse of materials, efficiency, and less consumption, we can lower our use of resource "capital" and reduce our impact on the environment.

7. *Redistribution of wealth and power.* We can no longer justify inequality as necessary for savings, investment, and growth. As economist Herman Daly points out, "sustainable development will make fewer demands on our environmental resources, but greater demands on our moral resources."

8. *Nondestructive resource uses.* By focusing on activities that use intangible resources, such as information, creativity, communications, leisure, and art, we can have a life rich in values but with minimal impact on environmental resources.

Used by permission from Robert Repetto, *World Enough and Time:Successful Strategies for Resource Management,* p. 18. Copyright © 1986 Yale University Press.

Summary

In this chapter, we have reviewed some economic theories of the effects of natural resource scarcity. A resource is defined as any useful material or service. This includes tangible, physical assets and intangible services of the environment. Resources are defined by their economic and technological feasibility, as well as their location and physical size. We distinguished between proven, known, undiscovered, recoverable, and nonrecoverable resources, and between economically important and technically accessible resources.

Two main mechanisms determine who shall benefit from natural resources. In the case of private goods, we depend on the marketplace to set prices through the interplay of supply and demand. For public goods, where costs and benefits are widely spread and difficult to evaluate in a market price, we use the political process to reflect social values and to make a fair distribution of resources. For a number of reasons, the market may fail to act in an optimum way to set priorities in conservation or utilization of natural resources. Among the most important of these reasons are inadequate reflection of external costs, the public good, and future values in the price system.

Questions about the scarcity of resources and their effects on economic development are important in determining what kind of society we have. In communal property resource systems, self-governing, locally-based, cooperative, community management has successfully sustained natural resources, for centuries in some cases. Some of the properties of these communal property systems might be applicable to larger scale steady-state economic systems, one in which there is sustainable development that brings a real increase in quality of life for everyone.

It is important for us to decide, individually as well as collectively, how we should use our resources and how we can best reach the goal of a just, sustainable society. A new world order and new economic relationships between the richer and poorer nations may be necessary, as is a system for natural resource accounting that considers the value of resource stocks and ecological services along with standard monetary exchanges.

▼ Questions for Review

1. Define a resource and distinguish between tangible and intangible resources.

2. What is the difference between economic growth and economic development?

3. List four economic categories of resources and describe the differences among them.

4. Describe the relationship between supply and demand.

5. What causes diminishing returns in natural resource use? How does population growth affect this phenomenon?

6. Describe how cost/benefit ratios are determined and how they are used in natural resource management.

7. Distinguish between open-access and communal property resources.

8. Describe how research and technology can increase resource supplies.

9. Why does the marketplace sometimes fail to optimally allocate natural resource values?

10. What would be some of the characteristics of a sustainable economic system?

▼ Questions for Critical Thinking

1. If you could retroactively stabilize economic growth or population growth at some point in the past, what point would you choose? What assumptions or values shape your choice?

2. When the ecologist warns that we are using up irreplaceable natural resources and the economist rejoins that ingenuity and enterprise will find substitutes for most resources, are they talking about the same things? What underlying premises and definitions shape these arguments?

3. How can intangible resources be infinite and exhaustible at the same time? Isn't this a contradiction in terms? Can you find other similar paradoxes in this chapter?

4. What is the difference between hypothetical and unconceived (sometimes called unknown-unknown) resources? How can we plan for resources that we haven't even thought of yet? Are there costs in assuming that there are no unknown-unknowns?

5. What would be the effect on the developing countries of the world if all nations were to change to a steady-state economic system? How could we achieve a just distribution of resource benefits while still protecting environmental quality and future resource use?

6. Resource use policies bring up questions of intergenerational justice. Suppose you were asked, "What has posterity ever done for me?" How would you answer?

7. If you were doing a cost/benefit study, how would you assign a value to the opportunity for good health or the existence of rare and endangered species in faraway places? Is there a danger or cost in simply throwing up our hands and saying some things are immeasurable and priceless and therefore off-limits to discussion?

8. What does it really mean to say that sustainable development meets the needs of the present without compromising the ability of future generations to meet their own needs? Is this possible? What is meant here by needs?

▼ Key Terms

communal resource management
   system   163
cost/benefit analysis   170
demand   164
diminishing returns   165
discount rates   168
economic development   161
economic growth   160
exhaustible resources   161
external costs   168
gross national product
   (GNP)   172
human resources   161
intangible resources   161
internal costs   168

internalizing costs   168
market equilibrium   164
natural resources   161
open access system   163
pollution charges   169
population hurdle   165
price elasticity   164
production frontier   160
renewable resources   161
resource scarcity   165
steady-state economy   161
supply   164
sustainable development   176
tradable permits   169
tragedy of the commons   163

▼ Additional Information on the Internet

Agricultural Economics Virtual Library
   http://www.ttu.edu:80/~aecovl/

Association of Environmental and Resource Economists
   http://www.uky.edu/BusinessEconomics/aere.html/

Center for Economic and Social Studies for the Environment
   http://www.ulb.ac.be:80/ceese/

Ecol-Econ
   http://csf.colorado.edu/ecolecon/index.html/

The Financial Sector and the Environment
   http://www.unep.ch/finance.html/

Gopher Menu for the United Nations Environmental Programme:
   Environmental Economics
   gopher://pan.cedar.univie.ac.at:70/11/UNEP/EnvEcon/

Institute for Ecological Economics
   http://kabir.umd.edu/miiee/miiee.html/

International Society for Ecological Economics
   http://kabir.umd.edu/ISEE/ISEEhome.html/

Trade and the Environment
   http://www.unep.ch/trade.html/

The World Bank
   http://www.worldbank.org/

Note: Further readings appropriate to this chapter are listed on p. 596.

Environmental Advocate

*M*arion Taylor

As is true for a growing number of people, Marion Taylor has discovered that life experiences can have a major influence on career plans, changing them in unpredictable ways. Marion is employed as the Director of Environmental Affairs for the Federation of Ontario Naturalists, a membership-based organization headquartered in Toronto. She finds her work both stimulating and meaningful, but quite different from what she expected to be doing, given her formal college training.

Marion graduated from the University of Toronto with majors in English and French. She taught for a few years and then took time out to have a family. She developed an interest in nature and after a broken marriage, began to do volunteer work for environmental causes, including work with the Federation of Ontario Naturalists. When a position in the organization opened up, she applied for it. Although she lacked technical training in the natural sciences, her focus on language in college had provided her with equally important communication and critical thinking skills.

Besides her experience as a member and volunteer, which had familiarized Marion with the work of the organization, she feels her writing ability was a critical factor in her successful employment. Like many others, Marion discovered that job experience, whether gained as part of an internship program or as an unpaid volunteer, can often put you a step ahead of other applicants.

An important part of her work is to advocate for sound environmental management across a broad spectrum of natural resource issues throughout the province of Ontario. One of her projects was to influence development of a new Crown Forest Sustainability Act, intended to guide forest management policy for the extensive public lands in Ontario. She, with other activists, successfully championed the incorporation of the use of sustainability indicators into the Act. These provisions will help insure that wildlife and water quality, as well as biological diversity issues, are considered along with timber harvest in forest management plans.

Marion also worked with the Ministry of Natural Resources on boundary and policy issues for the million-hectare Wabakimi Provincial Park northwest of Lake Nipigon in western Ontario. She helped establish ecologic park boundaries, focusing attention on wilderness recreational values and on preserving a viable unit of boreal forest to serve as an undisturbed benchmark against which the effects of various forest management practices could be measured.

She has also worked with local conservation authorities in southern Ontario on flood control and other river issues, and she serves as a citizen member on a committee of government, community, and industry representatives trying to resolve conflicts between gravel extraction and water, wetland, and community values. In these roles Marions negotiate with government agents, as well as industry representatives, who often arrive with conflicting points of view. While technical backgrounds can be helpful in these settings, to be truly effective requires strong interpersonal skills.

Another aspects of Marion's job includes writing a column for the organization's magazine, attending numerous meetings, occasionally testifying at public hearings, and spending lots of time on the telephone. Budget allowing, she travels to advise and encourage various grassroots citizen efforts.

For students interested in preparing themselves for advocacy, Marion advises getting as broad a base of expertise as possible. Although she wishes she had formal training in computers and the biological sciences, Marion feels the most essential element of preparation is to develop critical thinking. She feels strongly that success in her job requires practice of skepticism, that is, seeking out the factual basis for claims and positions taken by others. Marion finds that writings and positions taken by either side of any environmental issue are often unreasonable, unfair, and sometimes, outright silly. She finds it very important to be able to identify, clarify, and, where necessary, question assumptions. Effective advocacy and resolution of many of the issues with which she works requires skills to clarify issues and language, to make the complex understandable without distortion. She is emphatic that students will be well-served by consciously working at being clear, concise, and fair in thought and language.

Marion also advises students to develop good interpersonal skills. The ability to listen and to understand the conceptual frameworks of other people are very important. Her experience illustrates that skills and attributes other than simple technical knowledge can be important in successfully establishing an environmental career.

A clam digger works the mud flats across from the Eureka, CA, pulp mill. Are these clams safe to eat? What toxic compounds and infectious organisms may they have taken up from their environment?

CHAPTER 9: Environmental Health and Toxicology

Objectives

After studying this chapter, you should be able to:

- define *health* and *disease* in terms of some major environmental factors that affect humans.

- identify some major infectious organisms and hazardous agents that cause environmental diseases.

- distinguish between toxic and hazardous chemicals and between chronic and acute exposures and responses.

- compare the relative toxicity of some natural and synthetic compounds as well as report on how such ratings are determined and what they mean.

- evaluate the major environmental risks we face and how risk assessment and risk acceptability are determined.

Poisoning Bhopal

Just after midnight on December 3, 1984, a thick, acrid, gas cloud rolled through the quiet streets of Bhopal, India. In the still night air, the poisonous fog crept along the ground and quietly seeped into houses where families lay asleep on mats. People awakened coughing, gasping for air, and rubbing their burning eyes. As they emerged from their houses, they joined a panicked crowd surging through the narrow streets trying to escape the toxic cloud. Some never made it beyond their doorstep. Others collapsed in the street and died where they lay. Hospitals overflowed with terrified, suffering victims, many of whom were children and older people.

The noxious gas blanketing the city was methylisocyanate (MIC), a component of the pesticide Temik, which was being made at the Union Carbide plant in Bhopal. Water had gotten into a tank containing about 40 tons of MIC and set off a chemical reaction resulting in an explosive eruption of the toxic cloud. Control panels that should have detected rising temperatures and pressures had been shut down for repairs. Safety equipment that was to neutralize or incinerate the escaping gas failed. Workers blamed management for cutting corners and creating unsafe conditions. Management blamed the staff and claimed that water must have been put in the tank by a "disgruntled worker."

Morning revealed a horrifying sight. Human bodies, along with those of dogs, cats, cows, and birds, littered the streets. Whole families perished. Hardest hit was the crowded shantytown of Jayprakash Nagar, which lay just outside the Union Carbide fence. The official figure was 1754 people dead and 200,000 injured. Eyewitnesses and medical staff claimed the real number may have been as high as 15,000 killed and 300,000—one third of the total city population—injured. A UNICEF study of the number

Figure 9.1

Victims blinded by poisonous gases from the Union Carbide plant in Bhopal wait for medical aid. The sources and effects of toxic substances in our environment are important aspects of environmental science.

of death shrouds and amounts of cremation wood sold in the weeks following this disaster estimated 10,000 immediate deaths.

MIC is a powerful irritant, causing burning and swelling of moist tissues such as eyes, mouth, nasal passages, and lungs. Many people died of heart attacks triggered by lack of oxygen. Some had so much fluid in their lungs that they simply drowned. Long-term effects for those who survived the immediate assault include permanent blindness, emphysema, asthma, birth defects, and reproductive failures (fig. 9.1). Union Carbide has made payments of 10,000 rupees (US $330) to government-certified victims, but eleven years later, hundreds of thousands of claims remain unsettled.

Fortunately, most of us will never be exposed to such high concentrations of toxic chemicals as were the residents of Bhopal. But there are many dangerous substances—both natural and human-made—in our environment. What are these toxic and hazardous materials? How do they move through our environment and how are we exposed to them? Perhaps most important, what do they do to us? Understanding and evaluating health risks is one of the most important topics in environmental science. In this chapter, we will survey some principles of toxicology and environmental health that will help answer these questions. ▶

Types of Environmental Health Hazards

What is health? The World Health Organization defines **health** as a state of complete physical, mental, and social well-being, not merely the absence of disease or infirmity. By that definition, we all are ill to some extent. Likewise, we all can improve our health to live happier, longer, more productive, and more satisfying lives if we pay attention to what we do.

What is a disease? A **disease** is a deleterious change in the body's condition in response to an environmental factor that could be nutritional, chemical, biological, or psychological. Diet and nutrition, infectious agents, toxic chemicals, physical factors, and psychological stress all play roles in **morbidity** (illness) and mortality (death). To understand how these factors affect us, let's look at some of the major categories of environmental health hazards.

Infectious Organisms

For most people in the world, the greatest environmental health threat continues to be, as it has always been, pathogenic (disease-causing) organisms. Although much of our attention in the more-developed countries is focused on lifestyle and toxic chemicals, we should be aware of the biological hazards that pose, by far, the largest threat to a majority of people in the world. Nearly three-quarters of all deaths in industrialized countries are due to cardio-vascular diseases (heart attacks, strokes) or cancer, but in developing countries, infectious agents and parasitic diseases cause about half of all deaths.

Some 12.9 million children under 5 years of age die of infections and parasites each year in developing countries. Altogether, infants and children make up about 80 percent of the deaths from these diseases in developing countries. Whereas under-five mortality strikes only 8 to 10 children per 1000 live births in Canada, the United States, and most of Western Europe, under-five mortality rates for the poorest countries are typically twenty to thirty times higher. The developing world accounts for 98 percent of all deaths of children under five years and 99 percent of all maternal deaths.

Respiratory diseases (pneumonia, tuberculosis, influenza, and pertussis, or whooping cough) probably cause more deaths worldwide than any other group of infectious diseases (table 9.1). Gastrointestinal infections (diarrhea, dysentery, and cholera) caused by bacteria or protozoans (fig. 9.2) run a close second in terms of both new cases each year and total mortality. Note that the estimates in table 9.1 for new cases of infectious diseases add up to nearly 4 billion per year. Although not shown in this table, at least another billion people suffer from nutritional diseases such as anemia, goiter, cretinism, and calorie, protein, or vitamin deficiencies (see chapter 10). How can we explain such high numbers? Many people have more than one disease at a time, and some have more than one case of the same disease in a single year. Still, the World Health Organization estimates that some 2 billion people—about 35 percent of the world population— suffer from some illness at any given time.

Malnutrition and infectious diseases create a vicious cycle. Poor nutrition makes people more susceptible to infection, and infections, in turn, often result in diarrhea and vomiting that make it more difficult to obtain, absorb, and retain food. Improved sanita-

Table 9.1 Estimates of some major infectious and parasitic diseases

Disease	New Cases Each Year	Yearly Deaths
Respiratory diseases*	1 billion	7 million
Diarrhea	1 billion	3 million
Malaria	500 million	2 million
Measles	200 million	1.2 million
AIDS	2 million	1 million
Tetanus	1 million	600,000
Polio	2 million	200,000
Worms and flukes	1 billion	200,000

Source: World Health Organization, 1995.
*Respiratory diseases include pneumonia, tuberculosis, influenza, and pertussis (whooping cough).

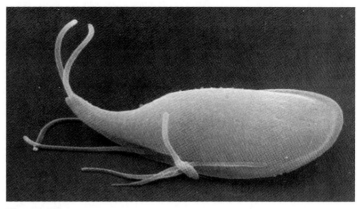

Figure 9.2

Giardia, a parasitic intestinal protozoan, is reported to be the largest single cause of diarrhea in the United States. It is spread from human feces through food and water. Even pristine wilderness areas have *Giardia* outbreaks due to careless campers.

tion and better food could prevent most, if not all, gastrointestinal infections. Simple oral rehydration therapy (ORT), in which patients are given an inexpensive mixture of sugar and salts in water, is highly effective in treating diarrhea and costs only a few cents per patient (In Depth, p. 184).

Tuberculosis and other respiratory diseases (influenza, whooping cough, and pneumonia) are the leading cause of death in many subtropical countries, especially Latin America. In Guatemala, the mortality rate among infants due to influenza and pneumonia is estimated to be 1000 per 100,000 children. This is 125 times the rate in Canada. Until 1900, tuberculosis was the major cause of death in the United States, but this dread disease was largely eliminated by good sanitation and inexpensive inoculations. Now, however, "super-strains" of tuberculosis bacteria that are resistant to multiple antibiotics are appearing in many countries. We may have to go back to quarantines and isolation of patients.

Malaria is an infection of red blood cells by a parasitic protozoan (*Plasmodium* sp.). It occurs in some 100 countries and is spreading rapidly. It may be the leading source of mortality in the world for a single disease and is especially common in the moist tropical countries of Africa, where the *Anopheles* mosquitoes that spread the disease thrive. At least 100 million people have malaria at any given time, and there are around 500 million new cases each year. Once the malaria parasite becomes established in the blood, the disease can recur every few months for many years. Insecticide use greatly reduced malaria incidence (as much as 90 percent in India and Sri Lanka) in the 1950s and 1960s, but pesticide-resistant mosquito populations have developed, allowing the disease to reappear—in some cases at higher levels than before.

Parasitic nematodes (hookworms and other roundworms) and flatworms (flukes and tapeworms) are very common in less-developed countries where sanitation is primitive. People rarely die directly from the infections, but they can be extremely debilitated. Flukes often invade internal organs such as the liver and lungs and interfere with vital functions. Intestinal worms, such as tapeworms and some roundworms, are especially serious in persons who are already malnourished.

Schistosomiasis is a disease caused by waterborne blood flukes. About 200 million people are infected worldwide, and about 200,000 deaths a year are associated with complications of this disease. The adult flukes live in blood vessels of the human digestive tract where they cause dysentery, anemia, general weakness, and greatly reduced resistance to other infections. They have a complex life cycle involving an aquatic snail as an intermediate host. Rice paddy farming, the most common type of agriculture in many tropical countries, creates a nearly perfect environment for transmission of these flukes because human feces are used for fertilizer, the water is shallow and warm—just right for the snails—and people spend hours wading in the water tending rice plants. Large irrigation projects, such as those made possible by building the Aswan Dam in Egypt, bring about increased crop yields but also increase problems with schistosomiasis.

Onchocerciasis (river blindness) is caused by tiny nematodes transmitted by the bite of black flies. Masses of dead worms accumulate in the eyeball, destroying vision. This disease affects 18 million people in the world and permanently blinds 500,000 each year. In some African villages, nearly every adult over thirty is blind from onchocerciasis (fig. 9.3). The World Health Organization undertook a massive pesticide spraying campaign in Africa during the early 1980s that reduced the incidence of river blindness in many areas, but the flies have become pesticide resistant, and many nontarget species were destroyed.

Some other parasitic worms also cause dreadful diseases. Filariasis (one form of which is elephantiasis) is transmitted by mosquitoes. The worms block the lymphatic system, causing large fluid accumulations and swellings in various body areas (see fig. 12.8). Guinea worms (*Dracunculus*) live as larvae inside small aquatic

*I*n Depth: Taking Care of Our Children

Every year in the developing countries of the world, about 12.9 million children under the age of 5 die of common infectious or parasitic diseases (fig. 1). Most of these children could be saved by simple, inexpensive, preventative medicine. Many public health officials argue that it is as immoral and unethical to allow children to die of these easily preventable diseases as it would be to allow them to starve to death or to be murdered. In the 1980s, the United Nations announced a world-wide campaign to prevent unnecessary child deaths, based on four principles designated by the acronym GOBI.

G is for growth monitoring. A healthy child is a growing child. Underweight children are much more susceptible to infectious diseases, retardation, and other medical problems than children who are better nourished.

O is for oral rehydration therapy (ORT). About one-quarter of all deaths under 5 years of age are caused by diarrheal diseases. A simple solution of salts, glucose or rice powder, and boiled water given orally is almost miraculously effective in preventing death from these diseases. The cost of treatment is only a few cents per child.

B is for breast-feeding. Babies who are breast-fed get natural immunity to diseases from antibodies in their mothers' milk, but infant formula companies often try to persuade mothers that bottle-feeding is more "modern" than breast-feeding. "Baby-friendly" hospitals now encourage mothers to breast-feed their newborns.

I is for universal immunization against the six largest, preventable, communicable diseases of the world: measles, tetanus, tuberculosis, polio, diphtheria, and whooping cough. In 1975, less than 10 percent of the developing world's children had been immunized against these diseases. By 1990, this number had risen to 80 percent. Although we

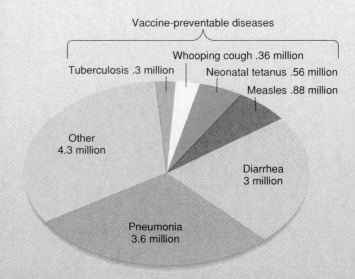

Figure 1

Leading causes of deaths in children under 5 years of age in developing countries. Over 60 percent of the 12.9 million child deaths per year are caused by pneumonia, diarrhea, vaccine-preventable infections, or some other disease that could be prevented at a cost of only a few dollars per child per year.

have not yet reached our goal of full immunization for all children, 3 million lives are being saved every year by this program.

Other encouraging signs of progress include the complete eradication of polio in Latin America and the Caribbean in 1994. India and China once accounted for three-quarters of all polio, but there is a realistic hope of eliminating this disease in all of Asia by 2000. Neonatal tetanus, too, was nearly gone everywhere in the world in 1995.

The United Nations Children's Fund (UNICEF) estimates that within a decade it should be possible to bring an end to the interrelated evils of child malnutrition, preventable disease, and illiteracy. Doing so would cost about $25 billion per year—less than Europeans spend annually on cigarettes or Americans spend on beer.

In addition to being an issue of humanity and compassion, reducing child mortality may be one of the best ways to stabilize world population. There has never been a reduc-

tion in birth rates that was not preceded by a reduction in infant mortality. When parents are confident that their children will survive, many will have only the number of children they actually want, rather than "compensating" for likely deaths by extra births. In Bangladesh, where ORT was discovered, a children's health campaign in the slums of Dacca has reduced infant mortality rates 21 percent in just ten years. In that same period, the use of birth control increased 45 percent and birth rates decreased 21 percent.

The United Nations Children's Fund estimates that if all developing countries had been able to achieve similar birth and death rates, there would be nearly 22 million fewer births each year.

Can we ethically justify allowing millions of children to die each year of easily-preventable diseases? How much would it be worth to you personally to save the life of just one of these children?

Figure 9.3

A child leads a blind adult in West Africa. River blindness, caused by tiny nematodes that are transmitted by biting flies, affects 18 million people. In some African villages, nearly every adult over age thirty is blind due to this disease.

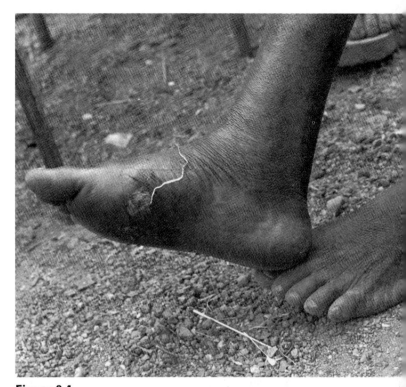

Figure 9.4

Guinea worms burrow under the skin of human hosts, where they live for a year while growing up to 1 m (3 ft) in length. When mature, the worms emerge at night to lay their eggs then retreat back into a festering wound. There is hope that better sanitation and medical care can completely eradicate this terrible parasite by the year 2000.

crustaceans. They infect people who drink unfiltered water containing the crustaceans. Adult worms can be up to 1 m (3 ft) long. After a year of growth, the worm emerges through the skin to lay its eggs (fig. 9.4). This painful ordeal may last several weeks, and the open sores it creates become infected, causing further illness.

The drugs amocarzine and ivermectin have effectively controlled filariasis and a number of other parasitic infections in humans and domestic livestock. Better sanitation and improved drinking water systems are bringing benefits as well. The World Health Organization predicts that river blindness, guinea worms, and perhaps other parasitic infections, could be wiped out by the year 2000 with continued public education and early treatment.

Trachoma is another very widespread eye disease. It is a contagious inflammation of the inner eyelid, tear glands, and cornea caused by a bacterium. This disease is found where sanitation is poor. If not treated, it can cause blindness. Several hundred million people, mostly children, suffer from trachoma.

Although sexually transmitted diseases don't kill as many people currently as some other diseases, they may become more serious problems in the future. The HIV virus, which causes AIDS (acquired immune deficiency syndrome), attacks the immune system with lethal results. HIV is spreading rapidly in Asia and Africa and may infect 26 million people and cause nearly 2 million deaths per year by the end of this century. Gonorrhea and syphilis bacteria have developed multiple antibiotic resistance and also are spreading rapidly.

You might suppose that all these terrible diseases will reduce or reverse population growth rates. If you look at the history of population growth in figure 7.2, however, you will notice that even pandemics such as the bubonic plague (Black Death), which killed about one-third of the population of Europe in the fourteenth century, didn't slow population growth for very long. The great influenza contagion of 1919, which killed at least 20 million people worldwide, doesn't even show up as a ripple on the curve. Fortunately there are more humane and more effective ways than disease to solve population problems.

Toxic Chemicals

As the case of Bhopal illustrates, toxic industrial chemicals in the environment are becoming a source of increasing concern everywhere. Between 1980 and 1990, 15 gas releases in the United States exceeded Bhopal in quantity and toxicity. None were in such densely populated areas as Bhopal, and none caused nearly as much death and disease, but the potential is always present. In this section, we will look at some of the natural as well as synthetic toxic materials in our environment and why they are of concern.

Chemical agents are divided into two broad categories: those that are hazardous and those that are toxic. **Hazardous** means dangerous. This category includes flammables, explosives, irritants, sensitizers, acids, and caustics. Many chemicals that are hazardous in high concentrations are relatively harmless when dilute. **Toxins** are poisonous. This means they react with specific cellular components to kill cells. Because of this specificity, they often are harmful even in dilute concentrations. Toxins can be either general poisons that kill many kinds of cells, or they can be extremely specific in their target and mode of action. Ricin, for instance, is a protein found in castor beans and one of the most toxic organic compounds known. Three hundred picograms (trillionths of a gram) injected intravenously is enough to kill an average mouse. A single molecule can kill a cell. This is about two hundred times more lethal than dioxin. Table 9.2 shows some of the toxic chemicals and elements of greatest concern to the EPA. This group includes heavy metals, inorganic chemicals, and both natural and synthetic organic compounds.

Irritants are corrosives (strong acids), caustics (alkaline reagents), and other substances that damage biological tissues on contact. Some examples are sulfuric and nitric acids, ammonia, sodium hydroxide, toxic metal fumes (such as beryllium or nickel), ozone, chlorine, sulfur or nitrogen oxides, formaldehyde, benzene hexachloride, and dioxin. Skin diseases caused by irritants (dermatoses) are the most common occupational diseases.

Respiratory fibrotic agents are a special class of irritants that damage the lungs, causing scar tissue formation that lowers respiratory capacity. This group includes both chemical reagents and particulate materials. Some conditions are common enough to be given specific names: silicosis (caused by silica dust), black lung (caused by coal dust), brown lung (caused by cotton fibers), asbestosis (caused by asbestos fibers), and farmer's lung (caused by moldy hay), among others. Some of these health problems are simply obstructive diseases in which the lungs fill with residue and tissue that interfere with breathing. Some also lead to cancer.

Asphyxiants are chemicals that exclude oxygen or actively interfere with oxygen uptake and distribution. Pure nitrogen, methane, and carbon dioxide are examples of passive asphyxiants. Under normal circumstances they are relatively inert, but they can be deadly when they fill enclosed spaces like mines, caves, or farm silos. By contrast, active asphyxiants react chemically with blood or lung tissue to prevent oxygen uptake. Some examples are carbon monoxide, hydrogen cyanide, hydrogen sulfide, and aniline. These chemicals are toxic even in low concentrations and their effects tend to be relatively irreversible.

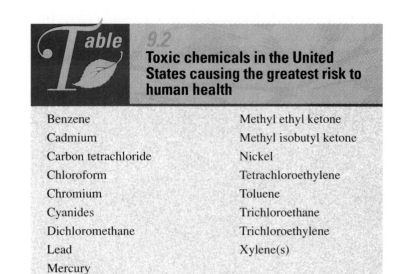

Table 9.2

Toxic chemicals in the United States causing the greatest risk to human health

Benzene	Methyl ethyl ketone
Cadmium	Methyl isobutyl ketone
Carbon tetrachloride	Nickel
Chloroform	Tetrachloroethylene
Chromium	Toluene
Cyanides	Trichloroethane
Dichloromethane	Trichloroethylene
Lead	Xylene(s)
Mercury	

Source: Environmental Protection Agency.

Allergens are substances that activate the immune system. Some allergens act directly as **antigens;** that is, they are recognized as foreign by white blood cells and stimulate the production of specific antibodies. Other allergens act indirectly by binding to other materials and changing their structure or chemistry so they become antigenic and cause an immune response.

Formaldehyde is a good example of a widely used synthetic chemical that is a powerful sensitizer. It is both directly and indirectly allergenic. Some people who are exposed to formaldehyde in plastics, wood products, insulation, glue, fabrics, and a variety of other products become hypersensitive not only to formaldehyde itself but also to many other materials in their environment, sometimes called the "sick house" syndrome. These individuals may have to go to great lengths to protect themselves from these allergenic substances.

Immune system depressants are pollutants that seem to suppress the immune system rather than activate it. Little is known about how this occurs or which chemicals are responsible. Immune system failure is thought to have played a role, however, in widespread deaths of seals in the North Atlantic and of dolphins in the Mediterranean in recent years. These dead animals generally contain high levels of pesticide residues, polychlorinated biphenyls (PCBs), and other contaminants, that may disrupt normal endocrine hormone functions (see Critical Thinking, chapter 12) and make them susceptible to a variety of opportunistic infections. Similarly, some humans with "sick house" syndrome or other environmental illnesses seem to have defective immune responses. Demonstrating a clear cause-and-effect relationship in these cases usually is difficult, however. Exactly which pollutants affect which people often remains unclear.

Neurotoxins are a special class of metabolic poisons that specifically attack nerve cells (neurons). The nervous system is so important in regulating body activities that disruption of its activities is especially fast-acting and devastating. Different types of neurotoxins act in different ways. Anesthetics (ether, chloroform, halothane, etc.), chlorinated hydrocarbons (DDT, Dieldrin, Aldrin), and heavy metals (lead, mercury) disrupt the ion transport across cell membranes necessary for nerve action. Organophosphates (Malathion, Parathion) and carbamates (Sevin, Zeneb, Maneb) inhibit acetylcholinesterase, an enzyme that regulates nerve signal transmission between nerve cells and the tissues or organs they innervate (for example, muscle). Most neurotoxins are both extremely toxic and fast-acting.

Mutagens are agents, such as chemicals and radiation, that damage or alter genetic material (DNA) in cells. This can lead to birth defects if the damage occurs during embryonic or fetal growth. Later in life, genetic damage may trigger neoplastic (tumor) growth. When damage occurs in reproductive cells, the results can be passed on to future generations. Cells have repair mechanisms to detect and restore damaged genetic material, but some changes may be hidden, and the repair process itself can be flawed. It is generally accepted that there is no "safe" threshold for exposure to mutagens. Any exposure has some possibility of causing damage.

Teratogens are chemicals or other factors that specifically cause abnormalities during embryonic growth and development. Some compounds that are not otherwise harmful can cause tragic problems in these sensitive stages of life. One of the most well-known examples of teratogenesis is that of the widely used sedative thalidomide. In the 1960s, thalidomide (marketed under the trade name Cantergan) was the most widely used sleeping pill in Europe. It seemed to have no unwanted effects and was sold without prescription. When used by pregnant women, however, it caused abnormal fetal development resulting in phocomelia (meaning seal-like limbs), in which there is a hand or foot, but no arm or leg (fig. 9.5). There is evidence that taking a single thalidomide pill in the first weeks of pregnancy is sufficient to cause these tragic birth defects. Altogether, at least 12,000 children were affected before this drug was withdrawn from the market. Fortunately, thalidomide was not approved for sale in the United States because the Food and Drug Administration was not satisfied with the laboratory tests of its safety.

Ironically, thalidomide has positive as well as negative features. The drug has been found to be effective in treating leprosy and is being tested against AIDS, cancer, retinal degeneration, and tissue rejection in organ transplants. Tragically, these beneficial applications continue to have a dark side. In Brazil, where thalidomide has been used widely to treat leprosy, some doctors failed to warn patients about the dangers of becoming pregnant while on the drug. Other people, hearing about miraculous cures with thalidomide, obtained it from unlicensed laboratories without knowing about its side effects. In 1994, more than 50 cases of thalidomide-related birth defects were reported in Brazil.

Perhaps the most prevalent teratogen in the world is alcohol. Drinking during pregnancy can lead to **fetal alcohol syndrome**—a cluster of symptoms including craniofacial abnormalities, developmental delays, behavioral problems, and mental defects that last throughout a child's life. Even one alcoholic drink a day during pregnancy has been associated with decreased birth weight.

Carcinogens are substances that cause **cancer,** invasive, out-of-control cell growth that results in malignant tumors. Cancer

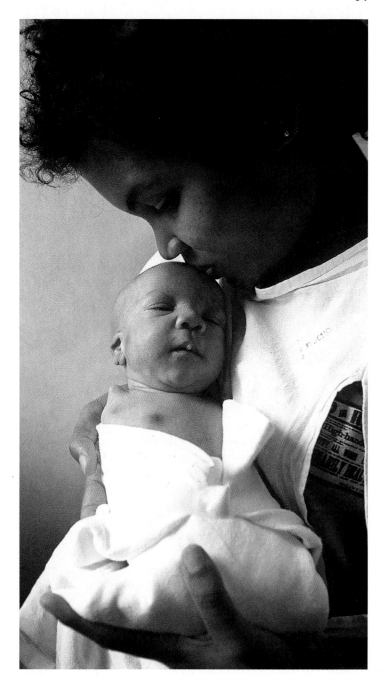

Figure 9.5

Development of this baby's arms and legs was blocked when its mother took the sedative thalidomide early in her pregnancy. Although the drug has been banned in Europe and North America for twenty years, it is still used to treat leprosy in some tropical countries. Unfortunately, some of this potent teratogen is used by pregnant women who are unaware of its tragic side effects.

rates have been rising in most industrialized countries during the 20th century, and cancer is now the second leading cause of death in the United States, killing 515,000 people in 1991. According to the American Cancer Society, 1 in 2 males and 1 in 3 females in the United States will have some form of cancer in their lifetime. Some authors blame this cancer increase on toxic synthetic chemicals in

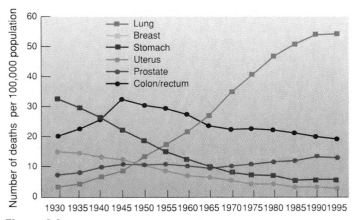

Figure 9.6
Age-adjusted cancer death rates in the United States. Although the total incidence of cancer is up in recent years, better treatment means that more people survive. When adjusted for an aging population, mortality for most major cancers has been stable or falling. One exception is lung cancer, 90 percent of which is attributable to increased smoking, especially among women.

our environment and diet. Others argue that it is attributable mainly to lifestyle (smoking, sunbathing, alcohol) or simply living longer.

If the number of deaths from cancer is adjusted for age, the only major types that have become more prevalent in recent years are prostate cancer in men and lung cancer in women (fig. 9.6). If we look at total incidence rather than death rates, the picture changes considerably. Recently, the number of new cases of breast, testis, and skin cancer, for example, have increased significantly, but so has the success in treating these diseases. Better diagnosis and treatment, rather than lower incidence, is probably the reason that mortality for cancers such as stomach, uterus, and colon have decreased over the past thirty years.

Natural and Synthetic Toxins

There has been so much bad news lately about the dangers of industrial chemicals that some people assume that all human-made compounds are poisonous while all natural materials must be benign and innocuous. In fact, many natural chemicals are just as dangerous as synthetic ones. Since most plants and many animal species can't escape from predators or defend themselves by fighting back, many of them have evolved a kind of chemical warfare, secreting or storing in their tissues a vast arsenal of irritants, toxins, metabolic disrupters, and other chemicals that discourage competitors and predators. Some of the chemical defenses employed by organisms are very sophisticated and specific. Both plants and animals make chemicals similar to—or even identical with—neurotransmitters, hormones, or regulatory molecules of predators or potential enemies. Our cells don't distinguish whether these chemicals are natural or synthetic.

You may have heard the argument that we evolved together with natural toxins whereas synthetic chemicals are so new that we haven't had time to adapt to them. But, in fact, natural chemicals such as arsenic and cyanide have been around longer than we have but are still toxic to us. Organic compounds like solanine and ox-

alic acid are common in many of the foods we eat but we haven't developed an immunity to them.

Toxicologist Bruce Ames claims that there are 10,000 times as many natural pesticides in our diets as synthetic ones. He argues that our fear of synthetic chemicals may divert our attention from more important issues. He finds natural compounds in crops such as potatoes, tomatoes, coffee, celery, and mushrooms, for instance, that are more carcinogenic than some commercial products. Ironically, plants attacked by insects may synthesize natural toxins that are more dangerous than the residues left from protective treatment with synthetic pesticides. Similarly, treatment of crops with fungicides (many of which are carcinogenic) may prevent growth of molds that are even more carcinogenic. Simply because food is raised "organically" may not necessarily make it safer than food raised by current commercial practices.

Other environmental health specialists, however, argue that we should not underrate the dangers of toxic synthetic chemicals. Natural chemicals in our diet are significantly different when mixed with fiber and a multitude of other substances than they may appear as pure chemicals in laboratory tests. Broccoli, for example, clearly contains carcinogens, but reduces the risk of cancer when added to the diet of laboratory animals because it also contains anticancer factors.

Physical Agents, Trauma, and Stress

Physical agents, such as radiation, also are serious environmental health hazards. Radiation associated with nuclear power is discussed in chapter 21. Although the data are not totally conclusive, ubiquitous low-frequency electromagnetic fields associated with power lines and ordinary household appliances may pose a health risk (What Do *You* Think?, next page). Noise, another important physical health threat, is discussed in chapter 24.

Trauma, injury caused by accidents or violence, has surely always been a life-threatening environmental factor for humans. The incidence of trauma-related deaths per 1000 persons is very similar in developed and less-developed countries. Accidents and violence cause far fewer deaths in the United States than infectious diseases do, even though media coverage might make it seem otherwise. Still, trauma is the principal cause of death for people between the ages of 1 and 38 in the United States, and it is the leading cause of years of life lost before age 65 in all industrialized countries. Every year there are about 100,000 premature deaths from accidents and violence in the United States and twice as many cases of permanent disability. More than half of these deaths and injuries are caused by motor vehicle collisions. About 90 percent of the accidents involve private automobiles, trucks, or motorcycles, and at least half are caused by drivers under the influence of alcohol or other drugs.

Stress and lifestyle once were considered factors that might cause unhappiness but were not life-threatening. We now know, however, that stress is clearly related to physical diseases, such as heart attack, stroke, and atherosclerosis, which are the leading causes of nontrauma-related death in the United States and most other industrialized countries. Since stress is a component of our cultural environment, it deserves a place in a discussion of environmental health.

What Do *You* Think?
Electromagnetic Fields and Your Health

Many forms of technology seem scary and mysterious, but few seem as insidious as potential dangers from invisible, unfelt electric and magnetic fields associated with our use of electricity. These fields are generated by power lines, household appliances, video display terminals, or any other device in which electricity flows through a wire or a motor. Although the data are vague and often contradictory, there appears to be some increased risk of cancers, miscarriages, birth defects, and perhaps Alzheimer's disease associated with exposure to these fields. Epidemiological studies generally implicate only the magnetic fields in human health risks, but most studies use the term electromagnetic field (EMF) because of the difficulty in separating electric and magnetic effects.

The first published report of adverse health effects from EMF was a 1966 study of electrical switchyard workers in the former Soviet Union who experienced a variety of symptoms including headaches, fatigue, and reduced fertility. A more alarming study published in 1979 reported that children living near power lines in Denver, Colorado, had two or three times higher rates of childhood leukemia than matched controls. Like the Soviet study, this report was greeted with skepticism because no direct measurement of exposure was available. Instead, researchers estimated field strength based on distance between homes and power lines. Exposure to other possible sources of cancer could not be determined.

These studies have stimulated further research. Some studies have failed to find any association between EMF and cancer, while others point to a cause-effect relationship. Analysis of childhood cancer in Los Angeles, for instance, found links between leukemia and the use of electric hair dryers and black-and-white televisions. Canadian research showed evidence for miscarriages, brain tumors, and birth defects among children whose mothers worked at video display terminals or used electric blankets while pregnant. The statistical significance of these studies is generally weak and they often fail to show clear dose/response relationships expected for a direct and unequivocal link.

In 1992, however, Swedish studies reported that children who live near power lines have an increased risk of leukemia. This research does show a statistically significant dose/response relationship between field strength and cancer incidence. Similarly, studies in the United States and Canada of workers in electric utilities and telephone companies found two to three times the risk for leukemia, brain, and breast cancer among workers exposed to high magnetic fields.

What does this evidence suggest for the average person? First of all, homes and schools should be at least one kilometer away from high-voltage power lines (fig. 1). Electric distribution lines that bring power into homes create much less powerful fields but should still be shielded and routed away from the parts of houses where people spend the most time. An electric blanket generates only minute fields but it lies right on top of you for many hours each night. It might be advisable to use a quilt instead, especially if you are pregnant. People who watch TV or work at a video display terminal (computer screen) for many hours each day should back up at least one meter (three feet) from the screen. Children, especially, should be discouraged from sitting close to TV screens.

Bedside appliances such as electric clocks, telephone answering machines, or anything with an electric motor that runs continuously should be placed at least a meter away from your head. Even better, why not place them

Figure 1

Homes and schools close to high-voltage transmission lines may expose children to dangerous levels of electromagnetic radiation.

across the room? Other electric appliances such as hair dryers, curling irons, electric shavers, can openers, microwave ovens, etc., should be used as briefly as possible and at the greatest distance from your person as is feasible. Don't stand right in front of the microwave door watching your food cook. Consider using a towel to dry your hair or a non-motorized razor to shave.

We also should keep relative risks in mind. If it is true that cancer risks are doubled by exposure to EMFs, remember that smoking increases cancer risks twenty times. Riding in an automobile, being overweight, eating a high-fat diet, engaging in unsafe sex, excessive drinking, risky jobs, radon in your home, and stress are all probably much greater threats to your health than EMF. Still, prudent avoidance makes sense; if you can reduce your exposure to EMF at little cost, why not do it?

In medical terms, **stress** refers to physical, chemical, or emotional factors that place a strain on an organism for which there is inadequate adaptation. This can result in physical responses that contribute to disease. Adverse stress responses are not unique to humans. Plants show signs of environmental stress, as do most animals. When laboratory or zoo animals are kept in cages next to especially aggressive members of the same species, they often show many signs of anxiety and stress. Even though they are separated by glass walls so that no physical contact is possible, the stressed animals often die prematurely of cardiovascular diseases. Gastrointestinal disturbances, such as ulcers, are common human responses to stress. Stress also contributes to susceptibility to infectious diseases, as many students realize at exam time.

Diet

Diet also has an important effect on health. For instance, there is a strong correlation between cardiovascular disease and the amount of salt and animal fat in one's diet. Fat intake seems to correlate with breast cancer in a number of nations (fig. 9.7), but long-term studies of 100,000 nurses in the United States suggest that the association is weaker or more complicated than previously thought. Highly processed foods, fat, and smoke-cured, high-nitrate meats also are associated with cancer.

Fruits, vegetables, whole grains, complex carbohydrates, and dietary fiber (plant cell walls), on the other hand, often have beneficial health effects. Certain dietary components, such as pectins; vitamins A, C, and E; substances produced in cruciferous vegetables (cabbage, broccoli, cauliflower, brussels sprouts);

and selenium, which we get from plants, seem to have anticancer effects.

Eating too much food is a significant dietary health factor in developed countries and among the well-to-do everywhere. At least one-fourth of all Americans are considered overweight. Cutting back on the number of calories consumed reduces the strain on bones, muscles, and other organs and has additional beneficial effects, including reduction of cardiovascular disease, diabetes, and—perhaps—cancer.

In some areas of the world, people seem to live exceptionally long lives. The Abkhasian people in the Caucasus Mountains of Soviet Georgia, the Hunzans in the mountains of Pakistan, and the Vilcabama villagers in Ecuador, for instance, appear to be among the longest-lived people in the world. Many claim to be 120 to 140 years old, although it is difficult to substantiate when they were born. Still, they do appear to live longer and to be more physically active later in life than do most of us.

These people have several factors in common that probably contribute to longevity. They live at moderately high elevations where the climate is cool, dry, and sunny. They lead active, vigorous lives in low-pressure, nonindustrialized settings. Life in their small villages is uniform and predictable. Stress levels are low. All ages work together in the fields and in the home, doing practical, physical work that directly benefits their community. The elderly are respected and live a useful, active life. The whole family shares in decision making, recreation, and religion. Their diet is usually simple, with low fat and salt, high fiber content, little meat, and lots of fruits and vegetables. They enjoy clean air and pure water. Their culture and uncomplicated lifestyles reduce conflict and anxiety.

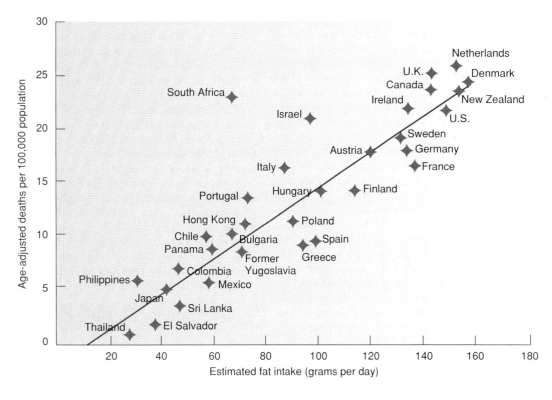

Figure 9.7

There is a strong linear correlation between dietary fat intake and deaths from breast cancer. The exact cause-and-effect relationship is unknown.

Source: Leonard A. Cohen, "Diet and Cancer," *Scientific American,* November 1987, copyright © 1987 by Scientific American, Inc.

Table 9.3
National health recommendations and diet goals

Eat only enough calories to meet body needs (fewer if overweight).

Eat less fat and cholesterol.

Eat less salt.

Eat less sugar.

Eat more whole grains, cereals, fruits, and vegetables.

Eat more fish, poultry, beans, and peas.

Eat less red meat.

Eat less additives and processed foods.

Source: The Surgeon General's Report, *Healthy People,* 1980.

Table 9.4
Factors in environmental toxicity

Factors Related to the Toxic Agent

1. Chemical composition and reactivity
2. Physical characteristics (such as solubility, state)
3. Presence of impurities or contaminants
4. Stability and storage characteristics of toxic agent
5. Availability of vehicle (such as solvent) to carry agent
6. Movement of agent through environment and into cells

Factors Related to Exposure

1. Dose (concentration and volume of exposure)
2. Route, rate, and site of exposure
3. Duration and frequency of exposure
4. Time of exposure (time of day, season, year)

Factors Related to Organism

1. Resistance to uptake, storage, or cell permeability of agent
2. Ability to metabolize, inactivate, sequester, or eliminate agent
3. Tendency to activate or alter nontoxic substances so they become toxic
4. Concurrent infections or physical or chemical stress
5. Species and genetic characteristics of organism
6. Nutritional status of subject
7. Age, sex, body weight, immunological status, and maturity

We might benefit from incorporating some aspects of their lives into our own (table 9.3).

Movement, Distribution, and Fate of Toxins

There are many sources of toxic and hazardous chemicals in the environment and many factors related to each chemical itself, its route or method of exposure, and its persistence in the environment, as well as characteristics of the target organism (table 9.4), that determine the danger of the chemical. We can think of an ecosystem as a set of interacting compartments between which a chemical moves, based on its molecular size, solubility, stability, and reactivity (fig. 9.8). The routes used by chemicals to enter our bodies also play important roles in determining toxicity (fig. 9.9). In this section, we will consider some of these characteristics and how they affect environmental health.

Solubility

Solubility is one of the most important characteristics in determining how, where, and when a toxic material will move through the environment or through the body to its site of action. Chemicals can be divided into two major groups: those that dissolve more readily in water and those that dissolve more readily in oil. Water-soluble compounds move rapidly and widely through the environment because water is ubiquitous. They also tend to have ready access to most cells in the body because aqueous solutions bathe all our cells. Molecules that are oil- or fat-soluble (usually organic molecules) generally need a carrier to move through the environment, into, and within, the body. Once inside the body, however, oil-soluble toxins penetrate readily into tissues and cells because the membranes that enclose cells are themselves made of similar oil-soluble chemicals. Once they get inside cells, oil-soluble materials are likely to be accumulated and stored in lipid deposits where they may be protected from metabolic breakdown and persist for many years.

Bioaccumulation and Biomagnification

Cells have mechanisms for **bioaccumulation,** the selective absorption and storage of a great variety of molecules. This allows them to accumulate nutrients and essential minerals, but at the same time, they also may absorb and store harmful substances through these same mechanisms. Toxins that are rather dilute in the environment can reach dangerous levels inside cells and tissues through this process of bioaccumulation.

The effects of toxins also are magnified in the environment through food chains. **Biomagnification** occurs when the toxic burden of a large number of organisms at a lower trophic level is accumulated and concentrated by a predator in a higher trophic level. Phytoplankton and bacteria in aquatic ecosystems, for instance, take up heavy metals or toxic organic molecules from water or sediments (fig. 9.10). Their predators—zooplankton and small

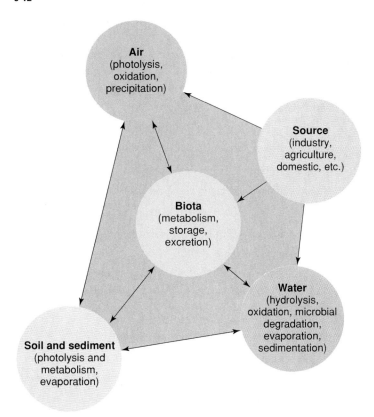

Figure 9.8

Movement and fate of chemicals in the environment. Processes that modify, remove, or sequester compounds are shown in parentheses.

Figure 9.10

Bioaccumulation and biomagnification. Organisms lower on the food chain take up and store toxins from the environment. They are eaten by larger predators, who are eaten, in turn, by even larger predators. The highest members of the food chain can accumulate very high levels of the toxin.

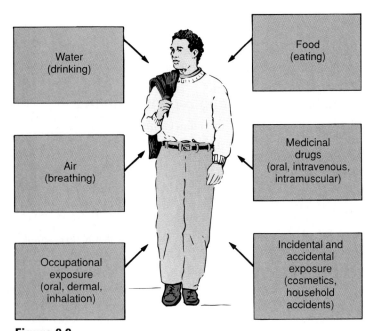

Figure 9.9

Routes of exposure to toxic and hazardous environmental factors.

fish—collect and retain the toxins from many prey organisms, building up higher concentrations of toxins. The top carnivores in the food chain—game fish, fish-eating birds, and humans—can accumulate such high toxin levels that they suffer adverse health effects (chapter 20). One of the first known examples of bioaccumulation and biomagnification was DDT, which accumulated through food chains so that by the 1960s it was shown to be interfering with reproduction of peregrine falcons, brown pelicans, and other predatory birds at the top of their food chains (chapter 12).

Persistence

Some chemical compounds are very unstable and degrade rapidly under most environmental conditions so that their concentrations decline quickly after release. Some of the modern herbicides, for instance, quickly lose their toxicity. Other substances are more persistent and last for long times. Some of the most useful chemicals such as chlorofluorocarbons, plastics, chlorinated hydrocarbons, and asbestos, are valuable because they are resistant to degradation. This stability also causes problems because these materials persist in the environment and have unexpected effects far from the sites of their original use. DDT, for instance, is a useful pesticide because it breaks down very slowly and doesn't have to be reapplied very often. Its toxic effects may spread to unintended victims, however, and it may be stored for long periods of time in organisms that lack mechanisms to destroy it.

Chemical Interactions

Some materials produce *antagonistic* reactions. That is, they interfere with the effects or stimulate the breakdown of other chemicals. For instance, vitamins E and A can reduce the response to some carcinogens. Other materials are *additive* when they occur together in exposures. Rats exposed to both lead and arsenic show twice the toxicity of only one of these elements. Perhaps the greatest concern is *synergistic* effects. Synergism is an interaction in which one substance exacerbates the effects of another. For example, occupational asbestos exposure increases lung cancer rates 20-fold. Smoking increases lung cancer rates by the same amount. Asbestos workers who also smoke, however, have a 400-fold increase in cancer rates. How many other toxic chemicals are we exposed to that are below threshold limits individually but combine to give toxic results?

Mechanisms for Minimizing Toxic Effects

A fundamental concept in toxicology is that every material can be poisonous under some conditions, but most chemicals have some safe level or threshold below which their effects are undetectable or insignificant. Each of us consumes lethal doses of many chemicals over the course of a lifetime. One hundred cups of strong coffee, for instance, contain a lethal dose of caffeine. Similarly, one hundred aspirin tablets, or ten kilograms (22 lbs) of spinach or rhubarb, or a liter of alcohol would be deadly if consumed all at once. Taken in small doses, however, most toxins can be broken down or excreted before they do much harm. Furthermore, damage they cause can be repaired. Sometimes, however, mechanisms that protect us from one type of toxin or at one stage in the life cycle become deleterious with another substance or in another stage of development. Let's look at how these processes help protect us from harmful substances as well as how they can go awry.

Metabolic Degradation and Excretion

Most organisms have enzymes that process waste products and environmental poisons to reduce their toxicity. In mammals, most of these enzymes are located in the liver, the primary site of detoxification of both natural wastes and introduced poisons. Sometimes, however, these reactions work to our disadvantage. Compounds, such as benzepyrene, for example, that are not toxic in their original form are processed by these same enzymes into cancer-causing carcinogens. Why would we have a system that makes a chemical more dangerous? Evolution and natural selection are expressed through reproductive success or failure. Defense mechanisms that protect us from toxins and hazards early in life are "selected for" by evolution. Factors or conditions that affect postreproductive

ages (like cancer or premature senility) usually don't affect reproductive success or exert "selective pressure."

We also reduce the effects of waste products and environmental toxins by eliminating them from our body through excretion. Volatile molecules, such as carbon dioxide, hydrogen cyanide, and ketones are excreted via breathing. Some excess salts and other substances are excreted in sweat. Primarily, however, excretion is a function of the kidneys, which can eliminate significant amounts of soluble materials through urine formation. Accumulation of toxins in the urine can damage this vital system, however, and the kidneys and bladder often are subjected to harmful levels of toxic compounds. In the same way, the stomach, intestine, and colon often suffer damage from materials concentrated in the digestive system and may be afflicted by diseases and tumors.

Repair Mechanisms

In the same way that individual cells have enzymes to repair damage to DNA and protein at the molecular level, tissues and organs that are exposed regularly to physical wear-and-tear or to toxic or hazardous materials often have mechanisms for damage repair. Our skin and the epithelial linings of the gastrointestinal tract, blood vessels, lungs, and urogenital system have high cellular reproduction rates to replace injured cells. With each reproduction cycle, however, there is a chance that some cells will lose normal growth controls and run amok, creating a tumor. Thus any agent, such as smoking or drinking, that irritates tissues is likely to be carcinogenic. And tissues with high cell-replacement rates are among the most likely to develop cancers.

Measuring Toxicity

In 1540, the German scientist Paracelsus said "the dose makes the poison," by which he meant that almost everything is toxic at some level. This remains the most basic principle of toxicology. Sodium chloride (table salt), for instance, is essential for human life in small doses. If you were forced to eat a kilogram of salt all at once, however, it would make you very sick. A similar amount injected into your bloodstream would be lethal. How a material is delivered—at what rate, through which route of entry, and in what medium—plays a vitally important role in determining toxicity.

This does not mean that all toxins are identical, however. Some are so poisonous that a single drop on your skin can kill you. Others require massive amounts injected directly into the blood to be lethal. Measuring and comparing the toxicity of various materials is difficult because not only do species differ in sensitivity, but individuals within a species respond differently to a given exposure. In this section, we will look at methods of toxicity testing and at how results are analyzed and reported.

Animal Testing

The most commonly used and widely accepted toxicity test is to expose a population of laboratory animals to measured doses of a specific substance under controlled conditions. This procedure is expensive, time-consuming, and often painful and debilitating to

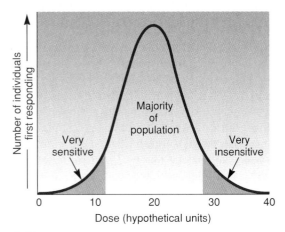

Figure 9.11

Probable variations in sensitivity to a toxin within a population. Some members of a population may be very sensitive to a given toxin, while others are much less sensitive. The majority of the population falls somewhere between the two extremes.

the animals being tested. It commonly takes hundreds—or even thousands—of animals, several years of hard work, and hundreds of thousands of dollars to thoroughly test the effects of a toxin at very low doses. More humane toxicity tests using computer simulation of model reactions, cell cultures, and other substitutes for whole living animals are being developed. However, conventional large-scale animal testing is the method in which we have the most confidence and on which most public policies about pollution and environmental or occupational health hazards are based.

In addition to humanitarian concerns, there are several problems in laboratory animal testing that trouble both toxicologists and policymakers. One problem is differences in sensitivity to a toxin of the members of a specific population. Figure 9.11 shows a typical dose/response curve for exposure to a hypothetical toxin. Some individuals are very sensitive to the toxin, while others are insensitive. Most, however, fall in a middle category forming a bell-shaped curve. The question for regulators and politicians is whether we should set pollution levels that will protect everyone, including the most sensitive people, or only aim to protect the average person. It might cost billions of extra dollars to protect a very small number of individuals at the extreme end of the curve. Is that a good use of resources?

Dose/response curves are not always symmetrical, making it difficult to compare toxicity of unlike chemicals or different species of organisms. A convenient way to describe toxicity of a chemical is to determine the dose to which 50 percent of the test population is sensitive. In the case of a lethal dose (LD), this is called the **LD50** (fig. 9.12).

Unrelated species can react very differently to the same toxin, not only because body sizes vary but also because of differences in physiology and metabolism. Even closely related species can have very dissimilar reactions to a particular toxin. Hamsters, for instance, are nearly 5000 times less sensitive to some dioxins than are guinea pigs. Of 226 chemicals found to be carcinogenic in either rats or mice, 95 caused cancer in one species but not the other. These variations make it difficult to estimate the risks for humans

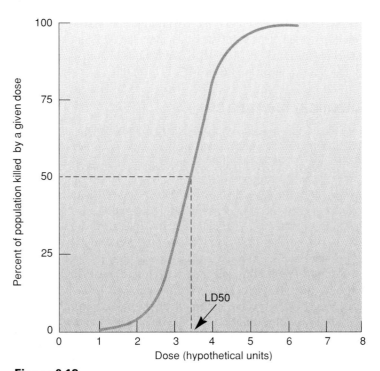

Figure 9.12

Cumulative population response to increasing doses of a toxin. The LD50 is the dose that is lethal to half the population.

since we don't consider it ethical to perform controlled experiments in which we deliberately expose people to toxins.

Toxicity Ratings

It is useful to group materials according to their relative toxicity. A moderate toxin takes about one gram per kilogram of body weight (about two ounces for an average human) to make a lethal dose. Very toxic materials take about one-tenth that amount, while extremely toxic substances take one-hundredth as much (only a few drops) to kill most people. Supertoxic chemicals are extremely potent; for some, a few micrograms (millionths of a gram—an amount invisible to the naked eye) make a lethal dose. These materials are not all synthetic. One of the most toxic chemicals known, for instance, is ricin, a protein found in castor bean seeds. It is so toxic that 0.3 billionths of a gram given intravenously will generally kill a mouse. If aspirin were this toxic, a single tablet, divided evenly, could kill 1 million people. Table 9.5 shows some representative LD50s for a variety of exposure routes and test species. This is not meant to be an exhaustive table but rather a sampling to show the range of variation and some of the various routes of administration.

Many carcinogens, mutagens, and teratogens are dangerous at levels far below their direct toxic effect because abnormal cell growth exerts a kind of biological amplification. A single cell, perhaps altered by a single molecular event, can multiply into millions of tumor cells or an entire organism. Just as there are different levels of direct toxicity, however, there are different degrees of carcinogenicity, mutagenicity, and teratogenicity. Methanesulfonic acid, for

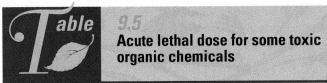

Table 9.5 — Acute lethal dose for some toxic organic chemicals

Chemical	Exposure	LD50
Ricin (castor bean)	Ivn-mus	3 ng/kg
	Orl-rat	100 mg/kg
Botulism toxin	Ipr-mus	160 ng/kg
Dioxin (tetrachlorodioxin)	Orl-gpg	600 ng/kg
	Orl-hmstr	3 mg/kg
Muscarine (mushroom poison)	Ivn-mus	250 µg/kg
Parathion (insecticide)	Ipr-rat	1.5 mg/kg
Aflatoxin (fungal toxin)	Orl-mky	1.75 mg/kg
Nicotine	Ivn-cat	2 mg/kg
	Orl-rat	53 mg/kg
DDT (dichlorodiphenyl-trichloroethane)	Orl-hum	50 mg/kg
Toxaphene	Orl-rat	60 mg/kg
2,4-D (dichlorophenoxyacetic acid)	Orl-hum	80 mg/kg

Source: *Registry of Toxic Effects of Chemical Substances,* National Institute for Occupational Safety and Health, 1985.
orl = oral, ivn = intravenous, ipr = intraperitoneal, mus = mouse, mky = monkey, hmstr = hamster, hum = human, gpg = guinea pig
nanogram (ng) = 1×10^{-9} gm
microgram (µg) = 1×10^{-6} gm
milligram (mg) = 1×10^{-3} gm

instance, is highly carcinogenic, while the sweetener saccharin is a suspected carcinogen whose effects may be vanishingly small.

Acute versus Chronic Doses and Effects

Most of the toxic effects that we have discussed so far have been **acute effects.** That is, they are caused by a single exposure to the toxin and result in an immediate health crisis of some sort. Often, if the individual experiencing an acute reaction survives this immediate crisis, the effects are reversible. **Chronic effects,** on the other hand, are long lasting, perhaps even permanent. A chronic effect can result from a single dose of a very toxic substance, or it can be the result of a continuous or repeated sublethal exposure.

We also describe long-lasting *exposures* as chronic, although their effects may or may not persist after the toxin is removed. It usually is difficult to assess the specific health risks of chronic exposures because other factors, such as aging or normal diseases, act simultaneously with the factor under study. It often requires very large populations of experimental animals to obtain statistically significant results for low-level chronic exposures. Toxicologists talk about "megarat" experiments in which it might take a million rats to determine the health risks of some supertoxic chemicals at very low doses. Such an experiment would be terribly expensive

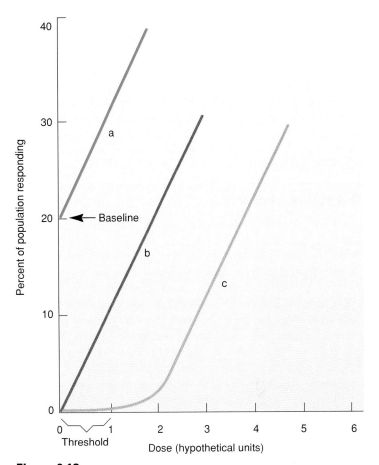

Figure 9.13

Three possible dose-response curves at low doses. (*a*) Some individuals respond, even at zero dose, indicating that some other factor must be involved. (*b*) Response is linear down to the lowest possible dose. (*c*) Threshold must be passed before any response is seen.

for even a single chemical, let alone for the thousands of chemicals and factors suspected of being dangerous.

An alternative to enormous studies involving millions of animals is to give massive amounts—usually the maximum tolerable dose—of a toxin being studied to a smaller number of individuals and then to extrapolate what the effects of lower doses might have been. This is a controversial approach because it is not clear that responses to toxins are linear or uniform across a wide range of doses.

Figure 9.13 shows three possible results from low doses of a toxin. Curve (*a*) shows a baseline level of response in the population, even at zero dose of the toxin. This suggests that some other factor in the environment also causes this response. Curve (*b*) shows a straight-line relationship from the highest doses to zero exposure. Many carcinogens and mutagens show this kind of response. Any exposure to such agents, no matter how small, carries some risks. Curve (*c*) shows a threshold for the response where some minimal dose is necessary before any effect can be observed. This generally suggests the presence of some defense mechanism that prevents the toxin from reaching its target in an active form or repairs the damage that it causes. Low levels of exposure to the toxin in question may have no deleterious effects, and it might not be necessary to try to keep exposures to zero.

Which, if any, environmental health hazards have thresholds is one of the most important questions in environmental science. The Delaney Clause to the U.S. Food and Drug Act forbids the addition of *any* amount of a known carcinogen to foods or drugs. This is based on the assumption that any exposure to these substances will cause some increased risk of cancer. This may not be true in every case. Holding exposures to absolute zero may be impossible and unnecessary. Determining which ones pose an acceptable risk, on the other hand, is not easy. We will discuss risk analysis further in the next section.

Detection Limits

You may have seen or heard dire warnings about toxic materials detected in samples of air, water, or food. A typical headline announced recently that twenty-three pesticides were found in sixteen food samples. What does that mean? The implication seems to be that any amount of dangerous materials is unacceptable and that counting the numbers of compounds detected is a reliable way to establish danger. We have seen, however, that the dose makes the poison. It matters not only what is there, but how much, where it is located, how accessible it is, and who is exposed. At some level, the mere presence of a substance is insignificant.

Toxins and pollutants may seem to be more widespread now than in the past, and this is surely a valid perception for many substances. The daily reports we hear of new materials found in new places, however, are also due, in part, to our more sensitive measuring techniques. Twenty years ago, parts per million were generally the limits of detection for most chemicals. Anything below that amount was often reported as zero or absent rather than more accurately as undetected. A decade ago, new machines and techniques were developed to measure parts per billion. Suddenly, chemicals were found where none had been suspected. Now we can detect parts per trillion or even parts per quadrillion in some cases. Increasingly sophisticated measuring capabilities may lead us to believe that toxic materials have become more prevalent. In fact, our environment may be no more dangerous; we are just better at finding trace amounts.

Risk Assessment and Acceptance

Even if we know with some certainty how toxic a specific chemical is in laboratory tests, it still is difficult to determine **risk** (the probability of harm times the probability of exposure) if that chemical is released into the environment. As you already have seen, many factors complicate the movement and fate of chemicals both around us and within our bodies. Furthermore, public perception of relative dangers from environmental hazards can be skewed so that some risks seem much more important than others.

Assessing Risks

A number of factors influence how we perceive relative risks associated with different situations.

- People with social, political, or economic interests—including environmentalists— tend to downplay certain risks and emphasize others that suit their own agendas. We do this individually as well, building up the dangers of things that don't benefit us while diminishing or ignoring the negative aspects of activities we enjoy or profit from.

- Most people have difficulty understanding and believing probabilities. We feel that there must be patterns and connections in events, even though statistical theory says otherwise. If the coin turned up heads last time, we feel certain that it will turn up tails next time. In the same way, it is difficult to understand the meaning of a 1-in-10,000 risk of being poisoned by a chemical.

- Our personal experiences often are misleading. When we have not personally experienced a bad outcome, we feel it is more rare and unlikely to occur than it actually may be. Furthermore, the anxieties generated by life's gambles make us want to deny uncertainty and to misjudge many risks.

- We have an exaggerated view of our own abilities to control our fate. We generally consider ourselves above average drivers, safer than most when using appliances or power tools, and less likely than others to suffer medical problems, such as heart attacks. People often feel they can avoid hazards because they are wiser or luckier than others.

- News media give us a biased perspective on the frequency of certain kinds of health hazards, overreporting some accidents or diseases while downplaying or underreporting others. Sensational, gory, or especially frightful causes of death like murders, plane crashes, fires, or terrible accidents occupy a disproportionate amount of attention in the public media. Heart diseases, cancer, and stroke kill nearly fifteen times as many people in the United States as do accidents and seventy-five times as many people as do homicides, but the emphasis placed by the media on accidents and homicides is nearly inversely proportional to their relative frequency compared to either cardiovascular disease or cancer. This gives us an inaccurate picture of the real risks to which we are exposed.

- We tend to have an irrational fear or distrust of certain technologies or activities that leads us to overestimate their dangers. Nuclear power, for instance, is viewed as very risky, while coal-burning power plants seem to be familiar and relatively benign; in fact, coal mining, shipping, and combustion cause an estimated 10,000 deaths each year in the United States, compared to none known so far for nuclear power generation. An old, familiar technology seems safer and more acceptable than does a new, unknown one.

Accepting Risks

How much risk is acceptable? How much is it worth to minimize and avoid exposure to certain risks? Most people will tolerate a higher probability of occurrence of an event if the harm caused by that event is low. Conversely, harm of greater severity is acceptable only at low levels of frequency. A 1-in-10,000 chance of being killed might be of more concern to you than a 1-in-100 chance of being injured. For most people, a 1-in-100,000 chance of dying from some event or some factor is a threshold for changing what we do. That is, if the chance of death is less than 1 in 100,000, we are not likely to be worried enough to change our ways. If the risk is greater, we will probably do something about it. The Environmental Protection Agency generally assumes that a risk of 1 in 1 million is acceptable for most environmental hazards. Critics of this policy ask, acceptable to whom?

For activities that we enjoy or find profitable, we are often willing to accept far greater risks than this general threshold. Conversely, for risks that benefit someone else we demand far higher protection. For instance, your chance of dying in a motor vehicle accident in any given year is about 1 in 5000, but that doesn't deter many people from riding in automobiles. Your chances of dying from lung cancer if you smoke one pack of cigarettes per day is about 1 in 1000. By comparison, the risk from drinking water with the EPA limit of trichloroethylene is about 2 in 1 billion. Strangely, many people demand water with zero levels of trichloroethylene, while continuing to smoke cigarettes.

Table 9.6 lists some activities estimated to increase your chances of dying in any given year by 1 in 1 million. These are statistical averages, of course, and there clearly are differences in where one lives or how one rides a bicycle that affect the danger level of these activities. Still, it is interesting how we readily accept some risks while shunning others.

Our perception of relative risks is strongly affected by whether risks are known or unknown, whether we feel in control of the outcome, and how dreadful the results are. Risks that are unknown or unpredictable and results that are particularly gruesome or disgusting seem far worse than those that are familiar and socially acceptable. Figure 9.14 shows the relative acceptability of a variety of technologies and activities plotted on a two-dimensional matrix of factors such as controllability, familiarity, and dread. The relative undesirability of these risks is indicated by the location of the circle that marks its position. Note that factors in the upper right quadrant tend to be much more feared than those in the lower left quadrant, even though the actual numbers of deaths or disease from automobile accidents, smoking, etc. are thousands of times higher than those from pesticides, nuclear energy, or genetic engineering.

Establishing Public Policy

A problem in setting environmental standards is that we are dealing with many sources of harm to which we are exposed simultaneously or sequentially. It is difficult to separate the effects of all these different hazards and to evaluate their risks accurately, especially when the exposures are near the threshold of measurement and response. In spite of often vague and contradictory data, public policymakers must make decisions.

The case of the sweetener saccharin is a good example of the complexities and uncertainties of risk assessment in public health. Studies in the 1970s at the University of Wisconsin and the Canadian Health Protection Branch suggested a link between saccharin and bladder cancer in male rats. Critics of these studies pointed out that humans would have to drink eight hundred cans of diet soda *per day* to get a saccharin dose equivalent to that given to the rats. Furthermore, they argued that people are not just very large rats. Scientists countered that while the doses given to rats was high, the results can be extrapolated to doses actually encoun-

Table 9.6

Activities estimated to increase your chances of dying in any given year by 1 in 1 million

Activity	Resulting Death Risk
Smoking 1.4 cigarettes	Cancer, heart disease
Drinking 0.5 liter of wine	Cirrhosis of the liver
Spending 1 hour in a coal mine	Black lung disease
Living 2 days in New York or Boston	Air pollution
Traveling 6 minutes by canoe	Accident
Traveling 10 miles by bicycle	Accident
Traveling 150 miles by car	Accident
Flying 1,000 miles by jet	Accident
Flying 6,000 miles by jet	Cancer caused by cosmic radiation
Living 2 months in Denver	Cancer caused by cosmic radiation
Living 2 months in a stone or brick building	Cancer caused by natural radioactivity
One chest X ray	Cancer caused by radiation
Living 2 months with a cigarette smoker	Cancer, heart disease
Eating 40 tablespoons of peanut butter	Cancer from aflatoxin
Living 5 years at the site boundary of a typical nuclear power plant	Cancer caused by radiation from routine leaks
Living 50 years 5 miles from a nuclear power plant	Cancer caused by accidental radiation release
Eating 100 charcoal-broiled steaks	Cancer from benzopyrene

From William Allman, "Staying Alive in the Twentieth Century," *Science 85*, 5(6): 31, October 1985. Used by permission of the author.

tered by humans. Although the Food and Drug Act forbids the addition of any substance that causes cancer in any amount in any animal, Congress has repeatedly exempted saccharin from being banned from food because of the uncertainty about its risks.

Experiments testing the toxicity of saccharin in rats merely give a range of probable toxicities in humans. The lower end of this range indicates that only 1 person in the United States would die from using saccharin every 1000 years. That is clearly inconsequential. The higher estimate, however, indicates that 3640 people would die each year from the same exposure. Is that too high a cost for the benefits of having saccharin available to people who must restrict sugar intake? How does the cancer risk compare to the dangers of obesity, cardiovascular disease, and other problems caused by eating too much sugar? What other alternatives might there be to saccharin?

Figure 9.14

Relative perception of various risks. Symbol indicates relative perception of risk; location on graph indicates contribution of different factors in apprehension of risk. Each factor is derived from combinations of characteristics.

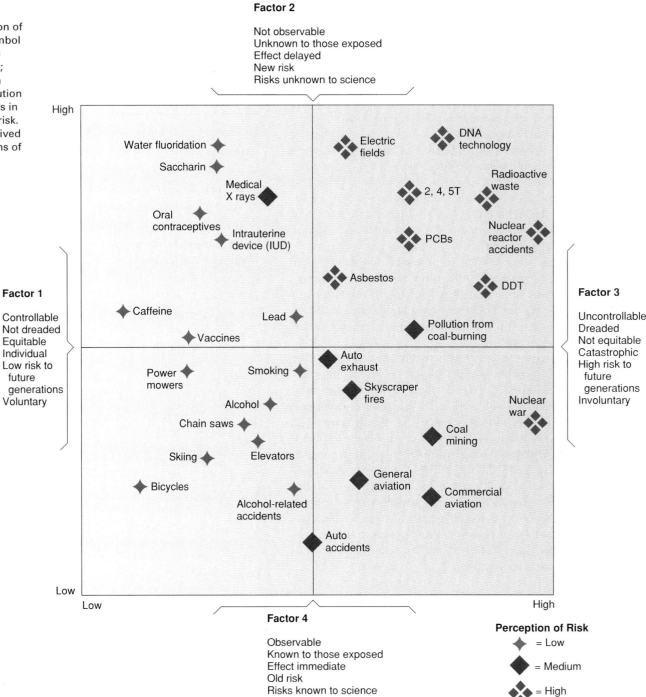

Factor 2

Not observable
Unknown to those exposed
Effect delayed
New risk
Risks unknown to science

Factor 1

Controllable
Not dreaded
Equitable
Individual
Low risk to
  future
  generations
Voluntary

Factor 3

Uncontrollable
Dreaded
Not equitable
Catastrophic
High risk to
  future
  generations
Involuntary

Factor 4

Observable
Known to those exposed
Effect immediate
Old risk
Risks known to science

Perception of Risk
= Low
= Medium
= High

A popular but more expensive alternative (aspartame, derived from the amino acid aspartic acid), which bears the trade name Nutrasweet, also is controversial because of uncertainties about its safety.

In 1989, a near panic swept the United States after the Natural Resources Defense Council issued a report claiming that children were being exposed to dangerous chemicals in fruits and vegetables. The worst of these was believed to be Alar (daminozoide), used to treat apples to promote even ripening and reduce surface blemishes. The apple industry claimed that you would have to eat 28,000 pounds of treated apples every day for 70 years to approach the exposures that caused cancer in laboratory animals. Pesticide opponents, however, claimed that the risks were much greater, especially for children.

In setting standards for environmental toxins, we need to consider (1) combined effects of exposure to many different sources of damage, (2) different sensitivities of members of the population, and (3) effects of chronic as well as acute exposures. Some people argue that pollution levels should be set at the highest amount that does *not* cause measurable effects. Others demand that pollution be reduced to zero if possible, or as low as is technologically feasible. It may not be reasonable to demand that we be protected from every potentially

Figure 9.15

"Do you want to stop reading those ingredients while we're trying to eat?"

Reprinted with permission of the *Star-Tribune*, Minneapolis-St. Paul.

harmful contaminant in our environment, no matter how small the risk. As we have seen, our bodies have mechanisms that enable us to avoid or repair many kinds of damage so that most of us can withstand some minimal level of exposure without harm (fig. 9.15).

On the other hand, each challenge to our cells by toxic substances represents stress on our bodies. Although each individual stress may not be life-threatening, the cumulative effects of all the environmental stresses, both natural and human-caused, to which we are exposed may seriously shorten or restrict our lives. Furthermore, some individuals in any population are more susceptible to those stresses than others. Should we set pollution standards so that no one is adversely affected, even the most sensitive individuals, or should the acceptable level of risk be based on the average member of the population?

Finally, policy decisions about hazardous and toxic materials also need to be based on information about how such materials affect the plants, animals, and other organisms that define and main-

Table 9.7 Relative risks to human welfare

Relatively High-Risk Problems

Habitat alteration and destruction

Species extinction and loss of biological diversity

Stratospheric ozone depletion

Global climate change

Relatively Medium-Risk Problems

Herbicides/pesticides

Toxics and pollutants in surface waters

Acid deposition

Airborne toxics

Relatively Low-Risk Problems

Oil spills

Groundwater pollution

Radionuclides

Thermal pollution

Source: Environmental Protection Agency.

tain our environment. In some cases, pollution can harm or destroy whole ecosystems with devastating effects on the life-supporting cycles on which we depend. In other cases, only the most sensitive species are threatened. In 1991, the Environmental Protection Agency issued a revised assessment of relative risks to human welfare (table 9.7). This ranking reflects a concern that our exclusive focus on reducing pollution to protect human health has neglected risks to natural ecological systems. While there have been many benefits from a case-by-case approach in which we evaluate the health risks of individual chemicals, we have often missed broader ecological problems that may be of greater ultimate importance.

Summary

Health is a state of physical, mental, and social well-being, not merely the absence of disease or infirmity. The cause or development of nearly every human disease is at least partly related to environmental factors. For most people in the world, the greatest health threat in the environment is now, as always, from pathogenic organisms. Bacteria, viruses, protozoans,

parasitic worms, and other infectious agents probably kill more people each year than any other cause of death.

Stress, diet, and lifestyle also are important health factors. Our social or cultural environment may be as important as our physical environment in determining the state of our health. People in some areas in the world five exceptionally long and healthful lives. We might be able to learn from them how to do so as well.

Estimating the potential health risk from exposure to specific environmental factors is difficult because information on the precise dose, length and method of exposure, and possible interactions between the chemical in question and other potential toxins to which the population may have been exposed is often lacking. In addition, individuals have different levels of sensitivity and response to a particular toxin and are further affected by general health condition, age, and sex.

The distribution and fate of materials in the environment depend on their physical characteristics and the processes that transport, alter, destroy, or immobilize them. Uptake of toxins into organisms can result in accumulation in tissues and transfer from one organism to another.

Estimates of health risks for large, diverse populations exposed to very low doses of extremely toxic materials are inexact because of biological variation, experimental error, and the necessity of extrapolating from results with small numbers of laboratory animals. In the end, we are left with unanswered questions. Which are the most dangerous environmental factors that we face? How can we evaluate the hazards of all the natural and synthetic chemicals that now exist? What risks are acceptable? We have not yet solved these problems or answered all the questions raised in this chapter, but it is important that these issues be discussed and considered seriously.

▼ Questions for Review

1. What is the difference between toxic and hazardous? Give some examples of materials in each category.

2. What are some of the most serious infectious diseases in the world? How are they transmitted?

3. How do stress, diet, and lifestyle affect environmental health? What diseases are most clearly related to these factors?

4. How do the physical and chemical characteristics of materials affect their movement, persistence, distribution, and fate in the environment?

5. Define LD50. Why is it more accurate than simply reporting toxic dose?

6. What is the difference between acute and chronic toxicity?

7. Define *carcinogenic, mutagenic, teratogenic,* and *neurotoxic.*

8. What are irritants, sensitizers, allergens, caustics, acids, and fibrotic agents?

9. How do organisms reduce or avoid the damaging effects of environmental hazards?

10. What are the relative risks of smoking, driving a car, and drinking water with the maximum permissible levels of trichloroethylene? Are these relatively equal risks?

▼ Questions for Critical Thinking

1. What consequences (positive or negative) do you think might result from defining health as a state of complete physical, mental, and social well-being? Who might favor or oppose such a definition?

2. How would you feel or act if your child were dying of diarrhea? Why do we spend more money on AIDS or cancer research than childhood illnesses?

3. Some people seem to have a poison paranoia about synthetic chemicals. Why do we tend to assume that natural chemicals are benign while industrial chemicals are evil?

4. Analyze the claim that we are exposed to thousands of times more natural carcinogens in our diet than industrial ones. Is this a good reason to ignore pollution?

5. Describe what is shown in figure 9.7. Find the two countries with similar fat intake that have the greatest difference in deaths from breast cancer. How would you explain this difference in light of negative results from studies done in the United States?

6. What are the premises in the discussion of assessing risk? Could conflicting conclusions be drawn from the facts presented in this section? What is your perception of risk from your environment?

7. Table 9.6 equates activities such as smoking 1.4 cigarettes, having one chest X ray, and riding 10 miles on a bicycle. How was this equation derived? Do you agree with it? Do some items on this list require further clarification?

8. Who were the stakeholders in the saccharin controversy, and what were their interests or biases? Was Congress justified in refusing to ban saccharin? Should soft drink cans have warning labels similar to those on cigarettes?

9. Should pollution levels be set to protect the average person in the population or the most sensitive? Why not have zero exposure to all hazards?

10. What level of risk is acceptable to you? Are there some things for which you would accept more risk than others?

▼ Key Terms

acute effects 195	health 182
allergens 186	irritants 186
antigens 186	LD50 194
asphyxiants 186	morbidity 182
bioaccumulation 191	mutagens 187
biomagnification 191	neurotoxins 186
cancer 187	respiratory fibrotic agents 186
carcinogens 187	risk 196
chronic effects 195	stress 190
disease 182	teratogens 187
fetal alcohol syndrome 187	toxins 186
hazardous 186	trauma 188

▼ Additional Information on the Internet

Agency for Toxic Substance and Disease Registry
    http://atsdr1.atsdr.cdc.gov:8080/atsdrhome.html/

EnviroText Retrieval
    http://tamora.cs.umass.edu/info/envirotext/

Extension Toxicology Network
http://sulaco.oes.orst.edu:70/1/ext/extoxnet/

Gopher Menu for the Material Safety Data Sheets Listing
gopher://atlas.chem.utah.edu:70/11/MSDS

Gopher Menu for the US Consumer Product Safety Commission
gopher://cpsc.gov/

Gopher Menu of Toxics
gopher://ecosys.drdr.Virginia.EDU/11/library/gen/toxics/

HazDat Database
http://atsdr1.atsdr.cdc.gov:8080/hazdat.html/

International Registry of Potentially Toxic Chemicals
http://www.unep.ch/irptc.html/

National Association of Physicians for the Environment
http://intr.net/napenet/

National Center for Environmental Health
http://www.cdc.gov/nceh.htm/

National Library of Medicine
http://www.nlm.nih.gov/

US Occupational Health and Safety Administration
http://www.osha-slc.gov/osha.html/

Note: Further readings appropriate to this chapter are listed on p. 597.

Environmental Engineer

Jim Bartlett

Jim Bartlett

When Jim Bartlett went off to college at the University of Iowa he was not at all settled on a career direction. He had a considerable interest in the outdoors coupled with an inclination to tinker, so he considered engineering a possibility. After his first year, he declared a major in industrial engineering. Soon after he found his heart wasn't really in his studies, so he switched to civil engineering, influenced, in part, by an emerging interest in the environment. He focused his efforts on building and design and got his bachelors degree in civil engineering with a balance of structural-related and environmental-related course work with the intention of seeking work with a consulting firm.

After graduation, Jim discovered that many consulting firmns were looking for people with advanced degrees. A friendly professor encouraged him to enroll in graduate school. After completing a master's degree in environmental engineering, Jim moved to San Francisco to pursue a career in this field. He accepted a job with Brown and Caldwell, a large firm providing environmental engineering, consulting, and analytical services. His firm is involved in a wide variety of projects, including environmental remediation efforts as well as design work. They worked with the Walnut Valley Water District to design a system for Pomona, CA, to use treated wastewater for irrigation of golf courses, school yards, and other large commercial and public landscapes. Reuse of the treated wastewater not only saved businesses money but avoided having to apply fertilizer to the sites being irrigated.

Jim's firm is also engaged in some exciting experimental work in the use of biofilter technology to remediate environmental contamination. In field tests, a common soil bacterium was able to degrade 98% of the commercial solvent, trichloroethylene, in some polluted groundwater. Efforts are now underway to commercialize that technology.

As a new employee, Jim's work consisted of performing lots of individual tasks on a variety of projects.

One day might be spent doing calculations, while the next week found him doing library research on a piece of equipment. Tasks were typically limited in scope and piecemeal, but did give Jim a sense of whole projects and the importance of the smaller pieces.

After the first few years, Jim has shifted to energy-related projects, with much of this work involved in developing energy-saving techniques. A recent project was to replace an inefficient heat pump system at a community college with a highly efficient system that runs at night to produce and store chilled water for use in cooling the buildings during the day. He has also worked on the new cogeneration system that uses heat from turbines to recover steam, producing significant energy savings for a large research hospital.

Jim admits that he could have been better prepared for his job, finding communication skills, writing in particular, to be of tremendous importance. He says he began his job with poor writing skills because he didn't work at developing them while in school. He feels very indebted to several mentors at Brown and Caldwell who taught him the basics of technical writing. He now works as a project manager, where he finds person-to-person communication with clients, contractors, vendors, and members of the project team a significant part of the job.

What are the prospects for employment in a field like Jim's? He says environmental consulting is very competitive. He feels there will be continued growth, but not nearly as rapidly as in the past. His firm has done lots of work that was regulation-driven, which has leveled off as well. The growth area is now found in retrofitting and remediation, particularly on federal sites. An example is the closure of McClellan Air Force Base. In order to make McClellan and other sites suitable for other uses, numerous site contamination problems must be remedied, providing increased work opportunities for environmental consultants even as employment in other sectors falls due to the facility closures.

Looking back on his environmental interests during his early college days, he feels he simply wasn't aware of the opportunities that existed for anyone with his environmental interests. He feels that in making career decisions, students need to focus on what things are really important to them, since that is where their real job satisfaction is going to come from.

$\mathcal{P}$art 3

Food, Land, and
Biological Resources

Field workers in Colombia harvest pineapples for export. Cash crops bring much-needed income to developing countries, but often reduce local production of basic foods.

$\mathcal{T}$here is no economic or social issue today that isn't also, by definition, an environmental one.

Mike Guerrero

Native Inupik fishermen haul in roe herring from Bristol Bay, AK. Worldwide, fish stocks have declined catastrophically due to overfishing.

CHAPTER 10: Food, Hunger, and Nutrition

Objectives

After studying this chapter, you should be able to:

- explain the major human nutritional requirements as well as sources for meeting those needs and the consequences of not doing so.

- describe the major crops on which humans depend and suggest some potential new crops for the future.

- sketch the sources of recent food supply increases and prospects for future growth of food production.

- differentiate between famine and chronic malnutrition and understand the sources and distribution of world hunger.

- analyze the role of cash crops and food trade policies in world hunger.

- suggest some solutions to the problems of malnutrition and world hunger.

Famine in China

Between 1959 and 1961, China suffered the worst famine in world history. Even today we don't know the full extent of the disaster, but most experts think that at least 30 million people—more than the current population of California—starved to death. Bad weather played a role, because droughts in some places and floods elsewhere ruined crops. But bureaucratic bungling made the calamity much worse. In 1958, Chairman Mao Zedong had ordered farm collectivization and planting of all land to grow food—regardless of its farming potential. Farmers were forced to plant the wrong crops at the wrong times. Steep hillsides stripped of forest cover eroded and smothered good bottomland in mud and silt. Food production fell disastrously.

*When this catastrophe became known in the West, many people saw it as only the first stage of permanent world food shortages. China will never be able to feed all its people, they said. Its huge population has outstripped the carrying capacity of the land. But, amazingly, only twenty years later, China was self-*sufficient in food production. After Mao's death, Chairman Deng Xiaopping broke up the communes and returned farms to family ownership. New rice strains were developed with double or triple the normal yield. A market system was reestablished and farmers were allowed to grow whatever they wanted. Between 1975 and 1985, the total rice harvest more than doubled—from 80 to 180 million metric tons per year. Almost half of that increase occurred in just four years, from 1980 to 1984. China now leads the world in both rice and wheat production.

Nearly all of China's 800 million farmers now cultivate small family plots, getting 50 percent more crops per hectare than did the communes (fig. 10.1). In addition to a diversified mix of crops, farmers are raising fish, chickens, pigs, ducks, and rabbits. Meat consumption has quadrupled. Now fewer than three percent of the Chinese people are underfed, a lower proportion than in the United States, where about one-tenth of the population goes hungry. China has even begun to export grain to other countries. The country that was thought to be beyond hope now seems to have enough food to feed everyone.

Figure 10.1

Recent decollectivization and a return to traditional farming practices have more than doubled China's farm yields and improved nutrition.

Figure 10.2

Even in the United States, a land of plenty, some people cannot afford food. Soup kitchens and relief shelters, such as this one, are often inadequate to meet the food needs of jobless and homeless people.

What does this example tell us about the future? For some people, the Chinese famine is a grim forecast of things to come for the whole world as population growth outpaces the world food supplies. In this worldview, the recent gains in Chinese crop production are only a temporary respite from inevitable doom. And as the world population surges toward 10 billion, pessimists warn us, the whole world soon will run out of water, fertilizer, and suitable land for farming. If current projections are correct, world food supplies must double or triple in the next thirty years to maintain current levels of consumption. Many experts regard this as impossible. Our attempts to feed an overpopulated world already are reducing biodiversity and threatening fragile ecosystems and ecological processes on which all life depends, they assert.

Optimists, on the other hand, view past successes in China and elsewhere as omens for a happy future. Even though the number of humans has doubled in the past fifty years, world agricultural output rose more than 2.5 times during that same period. We now have enough food to feed everyone in the world a healthy diet. And the potential exists, some agricultural experts contend, to raise much more. In theory, more than three times as much land as currently is farmed could be brought into production. Better crop varieties, more fertilizer, pesticides, and better management could double or triple yields in much of the developing world. We now consume only sixty percent of the food we grow worldwide. The rest is lost to spoilage, spillage, or vermin. Improved storage and distribution systems could almost double the amount of food available per person.

Still, in spite of current world food surpluses, some 15 to 20 million people (three-quarters of them children) starve to death or die of diseases associated with poor nutrition each year. Food supplies have risen faster than population in every continent except Africa; yet, at least 750 million people (half of them children) do not have enough calories to sustain normal growth and development or a healthy, productive working life. Even in rich countries like the United States, some 25 million people don't have enough to eat. If we can't feed everyone now, in this time of plenty, what will happen if population increases and resources become more limited?

Chapter 11 looks at soil and agriculture in more detail. In this chapter we will focus on human nutritional needs and world food supplies. We will examine where and how production has increased in recent years and what the prospects are for future crop growth. After surveying the world's current major food crops, we will investigate some potential developments with promise for future food production. We will examine the roles of economics and food trade in food security. Finally, we will look at some reasons why people are still hungry in a world with surplus food and what might be done about this tragic situation (fig. 10.2).

Human Nutrition

Our bodies require a constant supply of energy and raw materials to maintain vital functions and to rebuild tissues worn out in the day-to-day processes of living. In this section, we will study the major nutrients needed by humans to remain strong and healthy, and we will look at some of the more serious nutritional deficiencies from which people suffer.

Energy Needs

The amount of energy each of us needs to remain strong and healthy depends on body weight, climate, state of health, stress level, and basic metabolism. A small, sedentary adult living in a warm climate might require less than 2000 calories per day to remain healthy, whereas a large, muscular person living a vigorous outdoor life in a cold climate might need 6000 to 7000 calories per day to stay warm and active. The Food and Agriculture Organization (FAO) of the United Nations estimates that the average minimum daily caloric intake over the whole world is about 2500 calories per day.

People who receive less than 90 percent of their minimum dietary intake on a long-term basis are considered **undernourished.** While not starving to death, they tend not to have enough energy for an active, productive life. Lack of energy and nutrients also makes them more susceptible to infectious diseases. Poor diet and poverty create a vicious cycle. People who are weak or sick because of poor diet can't work; without an adequate income, they can't afford good food. This cycle extends from one generation to the next. Parents who can't work can't buy food for their children, who fail to grow properly and are likely to be impoverished when they become adults.

Those who receive less than 80 percent of their minimum daily caloric requirements are considered *seriously* undernourished. Children in this category are likely to suffer from permanently stunted growth, mental retardation, and other social and developmental disorders. Infectious diseases that are only an inconvenience for well-fed individuals become lethal threats to those who are poorly nourished. Diarrhea rarely kills a well-fed person, but children weakened by nutritional deficiencies are highly susceptible to this and a host of other diseases. One child in four in the developing world—around 13 million children per year—dies of diseases that could be prevented with a better diet, clean water, and simple medicines.

The *average* amount of food available per person worldwide has increased dramatically over the past 30 years. Grain supplies alone could provide everyone in the world with 2700 calories per day if that grain were equitably distributed, but it is not. At least 750 million people lack the energy to lead a healthy, productive life (table 10.1).

In the richer countries of the world, the most common dietary problem is too many calories. The average daily caloric intake in North America and Europe is above 3500 calories, nearly one-third more than is needed for adequate nutrition. At least 20 percent of Americans are seriously overweight. Overnutrition contributes to high blood pressure, heart attacks, strokes, and other cardiovascular diseases that have become the leading causes of death in most developed countries since infectious diseases have been reduced or controlled by better sanitation and health care.

Nutritional Needs

In addition to calories, we need specific nutrients in our diet, such as proteins, vitamins, and minerals. It is possible to have excess food and still suffer from **malnourishment,** a nutritional imbalance caused by a lack of specific dietary components or an inability to absorb or utilize essential nutrients. People in richer countries often eat too much meat, salt, and fat and too little fiber, vitamins, trace minerals, and other components lost from highly processed foods. In poorer countries, people often lack specific nutrients because they cannot afford more expensive foods such as meat, fruits, and vegetables that would provide a balanced diet. Let's look at some essential dietary requirements along with some important types of malnutrition.

Proteins

Proteins make up most of the metabolic machinery and cellular structure essential to life. The average adult human needs about 40 g (1.3 oz) of protein per day. Pregnant women, growing children, and adolescents need up to twice as much to build growing tissues. Not only is the total *amount* of protein crucial, but the *quality* of that protein also is of vital importance. We depend on our diet to supply us with the ten essential amino acids that we cannot make for ourselves: arginine, histidine, isoleucine, leucine, lysine, methionine, phenylalanine, threonine, tryptophane, and valine. These amino acids are not interchangeable; they must be present in a balanced ratio in order for us to synthesize functional proteins. Corn, for instance, has a reasonably high protein content but is generally low in lysine and tryptophane. Adding beans, peas, or milk to a corn-based diet can help balance the amino acid content.

The two most widespread human protein deficiency diseases are kwashiorkor and marasmus. **Marasmus** (from the Greek "to

	Effects of malnutrition		
Deficiency	**Effect**	**Prevalence[1]**	**Deaths per Year**
Protein and energy	Stunted growth, impaired lives	750 million	15–20 million
	Kwashiorkor and marasmus	1 million	
Iron	Anemia	350 million	
Iodine	Goiter	150 million	750,000 to 1 million[2]
	Cretinism	6 million	
Vitamin A	Blindness	6 million	

Source: T. A. Brun and M. C. Latham, (eds). *World Food Issues*, Cornell, 1990.
[1]At any given time.
[2]Deaths for all vitamin and mineral deficiencies together.

Food, Hunger, and Nutrition

Figure 10.3

Marasmus is caused by combined energy (calorie) and protein deficiencies. Children with marasmus have the wizened look and dry, flaky skin of an old person.

Figure 10.4

The swollen abdomen of this child is characteristic of kwashiorkor, a protein-deficiency disease. A diet containing a balanced mixture of amino acids can reverse these symptoms.

waste away") is caused by a diet low in both calories and protein. A child suffering from severe marasmus is generally thin and shriveled like a tiny, very old starving person (fig. 10.3). **Kwashiorkor** is a West African word meaning "displaced child." It occurs when people eat a starchy diet that is low in protein or has poor-quality protein, even though it may have plenty of calories. Children with kwashiorkor often have reddish-orange hair and puffy, discolored skin and a bloated belly (fig. 10.4). They become anemic and listless and have low resistance to even the mildest diseases and infections. Even if they survive childhood diseases, they are likely to suffer from stunted growth, mental retardation, and other developmental problems. Providing quality dietary protein for everyone is one of the world's greatest problems.

Carbohydrates

Most people in the world derive the bulk of daily calorie intake (up to 80 percent) from carbohydrates, mainly starches. A diet consisting primarily of starch such as white rice, cassava, or potatoes often leads to deficiencies in protein, vitamins, and minerals. Generally, as income goes up, the percentage of dietary protein, fat, and sugar increases and the percentage of calories from complex carbohydrates decreases. There are no indispensable dietary carbohydrates, because we synthesize all we need from other foods. Our digestive system seems to require complex carbohydrate and fiber, however, to absorb toxins and flush out wastes. Many of us in more-developed countries suffer from too much refined sugar and too little roughage in our diet.

Lipids and Oils

Lipids (fats, oils, and related compounds) are an important source of energy, containing about twice as many calories per gram as proteins or carbohydrates. Such foods as peanut butter, margarine, butter, and vegetable oils are rich in calories and important dietary components for those suffering from undernutrition. The main function of fat in our body—other than as a form of stored energy—is to make cellular membranes. Most of us get adequate levels of lipids in our diet, but very poor people in some areas of the world exist on a diet of starchy food, such as manioc, that does not provide enough oils and lipids. There are certain unsaturated lipids (lacking hydrogen atoms on some carbons in the hydrocarbon chain) that we can't synthesize for ourselves and must get from our diet.

Table 10.2
Some important minerals and their sources

Major Minerals	Food Sources	Roles in Body
Sodium (Na)	Table salt (NaCl), meat, processed foods	Transmission of electrochemical signals in muscles and nerves
Potassium (K)	Whole grains, fruits (esp. bananas), meat, legumes	Regulation of nerve and muscle action, protein synthesis
Calcium (Ca)	Milk, cheese, leafy green vegetables, egg yolk, nuts, whole grains, legumes	Matrix of bones and teeth, blood clotting, muscle relaxation, cell membrane function, intracellular signaling
Magnesium (Mg)	Whole grains, nuts, meat, legumes, milk	Constituent of bones and teeth, enzyme cofactor
Phosphorus (P)	Cheese, milk, meat, egg yolk, whole grains, nuts	Bone formation, energy metabolism, synthesis and use of carbohydrates, component of genetic material
Chlorine (Cl)	Table salt	Hydrochloric acid in stomach, water balance, membrane function
Sulfur (S)	Meat, eggs, cheese, milk, nuts, legumes	Component of some amino acids and proteins, regulates protein structure
Trace Minerals		
Iron (Fe)	Liver, egg yolk, whole grains, dark green vegetables, legumes	Component of hemoglobin in blood, energy metabolism in cells
Copper (Cu)	Meat, seafood, whole grains, legumes, nuts	Present in some enzymes, bone, maintenance of nervous tissue
Iodine (I)	Iodized salt, seafood	Component of thyroid hormone
Manganese (Mn)	Legumes, cereals, nuts, tea, coffee	Enzyme function, protein metabolism
Cobalt (Co)	Vitamin B_{12} in meat	Essential for red blood cell formation
Zinc (Zn)	Widely distributed in foods	Constituent of some enzymes
Molybdenum (Mo)	Organ meats, milk, leafy vegetables, grains, legumes	Constituent of some enzymes

Minerals

Humans also require inorganic nutrients, both for building cellular structures and for regulating many cellular reactions. Table 10.2 describes some important major and trace minerals along with their dietary sources. We generally need only small amounts of these minerals, but deficiencies can have serious health effects. The most common mineral deficiencies worldwide are for calcium, iodine, and iron. Calcium deficiency causes irritability, muscle cramps, and bone defects. Iron deficiency leads to **anemia** (low levels of hemoglobin in the blood), which more often is caused by an inability to absorb iron from food than a lack of iron in the diet. At least 350 million people—mostly women of childbearing age—suffer from anemia, making it one of the world's most common diseases.

The main symptoms of iodine deficiency are goiter (swollen thyroid glands) (fig. 10.5) and **hypothyroidism** (listlessness and other metabolic symptoms due to low thyroid hormone levels). Hypothyroidism in early childhood can cause developmental

Figure 10.5

Goiter, a swelling of the thyroid gland at the base of the neck, is often caused by an iodine deficiency. It is a common problem in many parts of the world, particularly where sea products are not frequently eaten.

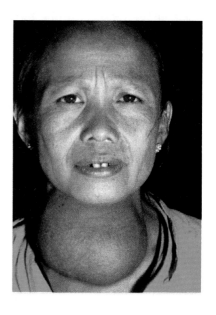

Food, Hunger, and Nutrition

abnormalities, such as mental retardation and deaf-mutism. Goiter was quite common in northern Europe and the United States before the introduction of iodized salt. It is estimated that 150 million people worldwide currently have symptoms of iodine deficiency and that 6 million suffer from cretinism (severe mental deficiency), mostly in South and Southeast Asia (see table 10.1). In some villages, 15 to 20 percent of the children are brain damaged due to iodine deficiency. Since the human body needs only a teaspoonful of iodine in a whole lifetime, this problem is both technically simple and inexpensive to solve. Adding potassium iodate to salt costs only a few cents per year per person and is highly successful.

The best cure for most mineral deficiencies is usually a well-balanced diet with a good variety of foods. Whole grains, legumes, milk, eggs, leafy vegetables, and fruits are all good sources of essential minerals and vitamins. The people most likely to have mineral deficiencies are those who eat a diet of highly processed foods or who subsist on a single starchy food, such as manioc or white rice.

Vitamins

Vitamins are organic molecules essential for life (*vita* = life). We cannot make vitamins for ourselves and so must get them from our diet. They generally act as cofactors of enzymes that metabolize energy or build essential cellular components. Table 10.3 shows the vitamins essential for humans. We usually require only minute amounts (milligrams per day) of vitamins, and get all we need from a varied diet of fruits, vegetables, whole grains, and dairy products. Highly processed foods (white bread, cane sugar, and snack foods) often have lost their nutrients and are supplemented with additional vitamins.

Vitamin deficiencies that once were common in the United States are now much rarer. In less-developed countries, however, vitamin deficiencies still are prevalent. Maize (called corn in North America) is not only deficient in tryptophan and lysine, but also low in usable niacin. A deficiency of tryptophan and niacin results in **pellagra,** the symptoms of which include lassitude, torpor, dermatitis, diarrhea, dementia, and death. Fifty years ago, pellagra was very common in the southern United States, where poor people subsisted mainly on corn. It still is tragically common in parts of India and Africa where jowah (a sorghum species) is the only food available to very poor people.

Vitamin B$_{12}$ is usually lacking in a strict vegetarian diet. Animal tissues and products are the ordinary sources of this essential dietary ingredient, but it is also present in tempeh, a cultured soybean food from Indonesia.

Vitamin A deficiency causes xerophthalmia (meaning dry eyes) and retinal degeneration, especially in children. About 1 million children in less-developed countries lose their sight every year because of these diseases. Three-quarters of those children die from

Table 10.3 Some important vitamins

Water-Soluble Vitamins	Food Sources	Deficiency Symptoms
Thiamine (B$_1$)	Fruits, cereal grains, milk, green vegetables	Beriberi, neuritis, fatigue, heart failure, edema
Riboflavin (B$_2$)	Milk, cheese, eggs	Sores on lips, bloodshot eyes
Niacin (B$_3$)	Whole grains, kidney, liver, fish, yeast	Pellagra, skin eruptions, fatigue, digestive disturbances
Pyridoxine (B$_6$)	Eggs, liver, yeast, milk, fish, grains	Anemia, convulsions, dermatitis, impaired immune response
Pantothenic acid	Most foods, meat	Retarded growth, mental instability
Folic acid	Egg yolk, liver, yeast	Anemia, impairment of immune system
Cobalamin (B$_{12}$)	Meat, fish, eggs, milk, cheese, tempeh	Pernicious anemia, defective DNA synthesis
Ascorbic acid	Citrus fruits, tomatoes, green leafy vegetables	Scurvy (loose teeth, hemorrhage, sterility)
Fat-Soluble Vitamins		
A (retinol)	Green and yellow vegetables, dairy products, fish oils	Night blindness, dry, flaky skin, and dry mucous membranes
D (calciferol)	Fish oils, liver, egg yolk, milk and dairy products, action of sunlight on lipids in skin	Rickets (deformed bones)
E (tocopherol)	Widely distributed: oils, grains, lettuce, eggs, beef	Red blood cell fragility, sterility in male rats, aging, susceptibility to environmental oxidants
K (menadione)	Green leafy vegetables, synthesis by intestinal bacteria	Slow blood clotting, hemorrhage

complications of malnutrition, but there are at least 6 million blind survivors at any given time. An additional 6 million to 7 million children show signs of moderate vitamin A deficiency and, therefore, are more vulnerable to infectious diseases and lost potential.

Other vitamin deficiency diseases, such as scurvy, beriberi, anemia, and rickets, are still a problem for poor people in many places. Enrichment of flour and milk with vitamins A and D has largely eliminated deficiency problems in developed countries. In fact, in richer countries there are now concerns about excess vitamins in our diets.

Eating a Balanced Diet

How can we avoid malnutrition and the ill effects of affluence such as obesity and cardiovascular diseases? Generally, it is as easy as consuming a balanced and varied diet with plenty of whole grains, fruits, and vegetables. For years Americans were advised to eat daily servings of four major food groups: meat, dairy products, grains, and fruits and vegetables. In 1992, in spite of intense protests from the meat and dairy industry, the U.S. Department of Agriculture revised these recommendations (fig. 10.6), suggesting that we eat more grains, fruits, and vegetables while decreasing our consumption of meat, milk, fats, oils, sugar, and salt.

World Food Resources

Of the thousands of edible plants and animals in the world, only about a dozen types of seeds and grains, three root crops, twenty or so common fruits and vegetables, six mammals, two domestic fowl, and a few fish and other forms of marine life make up almost all of the food humans eat. Table 10.4 shows annual production of some important foods in human diets. In this section, we will highlight sources and characteristics of those foods.

Major Crops

The three crops on which humanity depends for the majority of its nutrients and calories are wheat, rice, and maize. Together, about 1562 million metric tons of these three grains are grown each year, roughly half of all agricultural crops. Wheat and rice are especially important since they are the staple foods for most of the 4 billion people in the developing countries of the world. These two grass species supply around 60 percent of the calories consumed directly by humans. Both contain 8 to 15 percent protein and are good sources of vitamins and fiber if whole grains are consumed.

Potatoes, barley, oats, and rye are staples in mountainous regions and high latitudes (northern Europe, north Asia) because they grow well in cool, moist climates. Cassava, sweet potatoes, and other roots and tubers grow well in warm, wet areas and are staples in Amazonia, Africa, Melanesia, and the South Pacific. Sorghum and millet are drought resistant and are staples in the dry regions of Africa.

Fruits and vegetables—including vegetable oils—make a surprisingly large contribution to human diets. Altogether, they amount to nearly as large a quantity as maize. They are especially welcome because they typically contain high levels of vitamins, minerals, dietary fiber, and complex carbohydrates. There is considerable evidence that many fruits and vegetables contain anticancer agents that can help prevent this terrible disease.

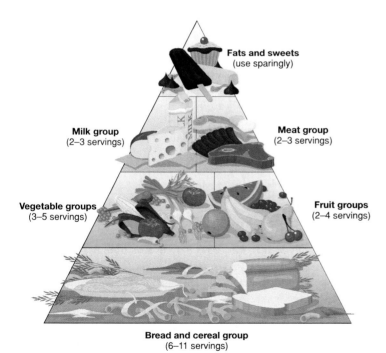

Figure 10.6

The food pyramid recommended by the U.S. Department of Agriculture. You should eat two to four times as much bread, cereal, rice, and pasta as milk, meat, eggs, or nuts. Fats, oils, and sweets should be eaten sparingly, if at all.

Table 10.4	Some important food resources
Crop	*1993 Yield (million metric tons)*
Wheat	564
Rice	527
Maize (corn)	471
Potatoes	288
Barley and oats	170
Cassava and sweet potato	277
Sugar (cane and beet)	111
Pulses (legumes)	58
Sorghum and millet	84
Vegetable oils	79
Vegetables and fruits	430
Meat and milk	150
Fish and seafood	102

Meat and Milk

Meat and milk are prized by people nearly everywhere, but their distribution is highly inequitable. Although the industrialized, more-developed countries of North America, Europe, and Japan make up only 20 percent of the world population, they consume 80 percent of all meat and milk in the world. The 80 percent of the world's people in less-developed countries raise 60 percent of the 3 billion domestic ruminants and 6 billion poultry in the world but consume only 20 percent of all animal products. The grazing lands that support many of these domestic animals are discussed further in chapter 14.

Fish and seafood contribute about 70 million metric tons of high-quality protein to the world's diet, about one-half as much as that from land animals. This is an important source of protein in many countries, contributing up to one-half of the animal protein and one-fourth of the total dietary protein in Japan, for instance.

There are indications that we have already surpassed the sustainable harvest of fish from most of the world's oceans. Annual catches of ocean fish rose by about 4 percent per year between 1950 and 1988. Since that time, however, catches have declined about 4 percent each year. Some species such as Atlantic cod, haddock, Pacific Ocean perch, and Peruvian anchoveta are showing signs of severe stress from overharvesting and destruction of young fish by wasteful, indiscriminate trawling (What Do *You* Think?, next page). Up to 10 tons of unwanted species are discarded for every ton of seafood brought to market. Much of what is discarded could be eaten, but it is the wrong size, too young, out of season, or a variety that we do not like to eat. Changing a commercial name from dogfish or hagfish to ocean perch or sea trout may make an unwanted species acceptable. Tragically, much of the catch that is tossed back into the sea is too badly injured by decompression or being crushed in the net to survive.

What are our alternatives? Obviously, we could use more selective harvest methods and expand our eating preferences. There also is great potential for raising fish and crustaceans in ponds or in estuaries. Under carefully controlled conditions, even environmentally sensitive fish like trout can be raised in high-density ponds. Genetic engineering techniques are being used to breed "superfish" that will grow faster and increase yields in much the same way the green revolution improved plants and plant yields.

Croplands

Where is most of the world's food grown? As might be expected, the biggest crops are grown in China, India, and the United States, which have the most land and the largest workforces. The United States has the world's greatest amount of arable land with 190 million hectares, or 16 percent of the total world cropland, to feed 255 million people. By contrast, China feeds 1.2 billion people from about 100 million hectares of land. India has only half as much land as the United States but has three times as many people and gets just one-third the yield from its land. If current population projections are correct, the world average of 0.28 hectare (0.7 acre) of

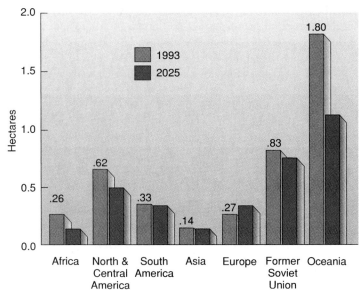

Figure 10.7

Cropland per capita by region, 1993 and 2025 (projected). Based on these projections, growing populations will mean less land per person everywhere, except Europe, in the next century.

Source: UNEP, *Environmental Data Report 1993–94*, and UN *World Population*, 1994.

cropland per person will decline to 0.17 hectare (0.4 acre) by the year 2025. Cropland in Asia will be even more scarce, amounting to only 0.09 hectare (0.2 acre or about the size of an average suburban lawn) per person (fig. 10.7).

Russia and Canada are the largest countries in the world but much of both lies too far north to be suitable for agriculture. In spite of its relative land shortage, China is the world's largest producer of both wheat and rice. Wheat (in the form of noodles) is the staple of the colder, drier regions of northern China; rice is the staple of the warmer, wetter, southern regions. The United States is the largest producer of maize, growing about 8 billion bushels (nearly 200 million metric tons) or a little less than half of the world total. Most of the grain we produce is not consumed directly by humans. About 90 percent is used to feed dairy and beef cattle, hogs, poultry, and other animals. It is used especially to fatten beef cattle during the last three months before they are slaughtered.

As chapter 3 shows, there is a great loss of energy with each step up the food chain. This means we could feed far more people if we ate more grain directly rather than feeding it to livestock. Every 16 kg of grain and soybeans fed to beef cattle in feedlots produce about 1 kg of edible meat. The other 15 kg are used by the animal for energy or body parts we do not eat or they are eliminated. If we were to eat the grain directly, we would get twenty-one times more calories and eight times more protein than we get by eating the meat it produces. Hogs and poultry are about two and four times as efficient, respectively, as cattle in converting feed to edible meat.

What Do You Think?
Collapse of the Canadian Cod Fishery

Inadequate dietary protein is responsible for some of our most common nutrient-deficiency diseases. Since global fisheries contribute about one-third of the high quality protein consumed worldwide, this resource is an important part of the effort to feed humanity.

Unfortunately, that resource is threatened. Seventy percent of the global fish stocks have been declared depleted or almost depleted by the United Nations. The collapse of the Canadian cod fishery in 1992 illustrates classic forces at work: a finite resource beset by escalating exploitation, where scientific understandings are incomplete, and decision-making is buffeted by conflicting political and sociological pressures. What went wrong in this fishery that was sustainably exploited for over 400 years, and what can we learn from it?

The northern cod stocks of eastern Canada provided the economic basis for settlement of Nova Scotia, Newfoundland and Labrador, ultimately providing work for thousands of people in coastal towns. Before 1870, cod fishing technology relied on stationary cod traps and handlines. A series of technological advances enabled fishermen to seek cod more aggressively with evermore efficient ships and gear. These changes enabled fishing to intensify over a huge section of the highly productive continental shelf off Newfoundland called the Grand Banks.

By the late 1960s and early 1970s the catch of older, larger cod had significantly diminished. The Canadian government came under considerable political pressure to save the fishery and the economy it supported. Quotas were first imposed in 1973, establishing the total allowable catch. Complicating Canadian management of the fisheries,

Figure 1

Fishing boats sit idle in Alert Bay, British Columbia. Since Alantic cod fishing was banned in 1991, this is a common sight in many Canadian ports.

Europeans were also engaged in intensive fishing on the Grand Banks.

Resolution of the problem seemed assured when, in 1977, Canada was able to establish a 200-mile territorial limit, giving it exclusive jurisdiction over fishing within 200 miles of its coast. The limit included most, but not all, of the productive Grand Banks.

By the late 1970s, the decline of cod stock had been stopped and prospects for a return to a healthy fishery seemed bright. But then, in 1991, the number of spawning age cod dropped precipitously. The decline was so severe that the government had no choice but to close the fishery, throwing 20,000 people out of work (fig. 1).

The full cause of the collapse is under dispute. Some argue that ecological factors in the ocean in 1991 that we don't yet understand, when added to the heavy fishing, caused the demise. Others argue that the large, highly efficient fleet that fished the offshore areas was responsible. Still others claim that inadequacy of scientific information and an inaccurate scientific model of the fishery set the total allowable catch too high. Inshore fishermen argue that the policy directed at reducing the number of fishermen through licensing was misguided, and that a

more appropriate policy would have been to reduce total fishing effort, which is more a function of technology than of the number of fishermen. Policy decisions did not rest on fishery science alone, however. Economic and political pressures were present as well.

Nature has great resiliency. Cod numbers have been low in the past, and have traditionally been able to rebound. There is no reason to expect they cannot do so again. But, past policies need to change to allow this to happen.

A basic tenet of effective thinking is that sound conclusions can only be built on sound premises. Some people claim that the basic problem besetting global fisheries is that it has developed along the industrial model. As stocks decline, each fish becomes more valuable. Stimulated by this higher price, the industrial approach is to catch fish even more efficiently, making the resource ever more scarce and ever more valuable.

How might Canadian fishery officials have foreseen this collapse? What policy changes might have prevented it?

Increasing Food Production

Contrary to what Thomas Malthus predicted in 1798 (chapter 7), crop yields have increased faster over the past two centuries than have human populations, even though we have experienced unprecedented population growth. World average daily caloric intake increased from an inadequate 2000 cal per person in 1965 to 2500 cal per person in 1995. Some people see this as a second agricultural revolution, a radical change that could affect the course of human history.

Over the past thirty years, world food supplies have doubled (fig. 10.8*a*). The centrally planned economies of Asia—China, North Korea, Mongolia, Cambodia, and Vietnam—have exhibited the greatest per capita gains during this period, increasing food supplies per person nearly 50 percent (fig. 10.8*b*). China and Indonesia experienced some of the most impressive food gains in history, tripling their output in little more than a decade. In only four years, Indonesia moved from the world's largest importer of rice to a country not only self-sufficient but boasting the world's biggest rice reserves. This success was due to a variety of factors including increased use of irrigation, fertilizers, and pesticides; expanded croplands; and new high-yielding crop varieties.

In some regions, notably Africa, food production has not kept pace with rapid population growth. Thirty-five of forty-seven African countries have had decreasing per capita food production over the past two decades (fig. 10.9). The worst declines have been in countries such as Angola, Ethiopia, Sudan, Somalia, and Mozambique where drought, war, grinding poverty, and governmental mismanagement have combined to produce heart-rending scenes of starvation and misery. If present trends continue, African populations will triple by 2025, and food supplies will fall even further below necessary levels.

The FAO projects that 64 countries (29 in Africa, 21 in Asia, and 14 in Central America) will be unable to feed their citizens by 2025 given current low levels of food production. Helping these countries stabilize populations, protect their environment, and feed the starving surely should be a world priority for security reasons if not for humanitarian ones.

Green Revolution

Table 10.5 shows that most recent growth in world food supply has come from higher yields per hectare of land. A few places such as sub-Saharan Africa and Latin America have increased yields by opening new lands to agriculture, but overall, developing countries have had to depend primarily on more intensive agriculture on existing lands to meet their needs. Notice that Europe and the former Soviet Union have actually lost croplands over the past 30 years. Much of this is due to degradation caused by inappropriate agricultural methods and industrial pollution.

Figure 10.10 shows increases in production of some major food groups between 1965 and 1992. Notice that cereal grains made the biggest total gains during this time, although fruits and vegetables increased by nearly as large a percentage. Root crops showed the lowest growth. To some extent this represents a lack of interest in these crops on the part of the richer countries where most agricultural research is done. Most of our research has been on crops that we raise ourselves such as maize, wheat, and tobacco. We haven't paid much attention to tropical crops except high-value

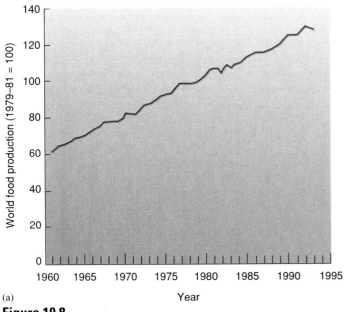

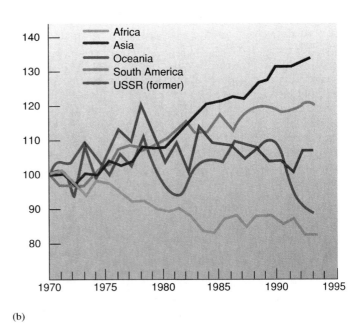

Figure 10.8

(*a*) Total world agricultural production has increased dramatically since 1960. (*b*) Relative per capita growth in food production by region. Index numbers are based on food production in each region in 1970 (1970=100).

Sources: Food and Agriculture Organization of the United Nations (FAO), March 1991, *Agrostat PC* on diskette (FAO, Rome, 1993) and Wood Mackenzie Consultants Limited, unpublished data, May 1993.

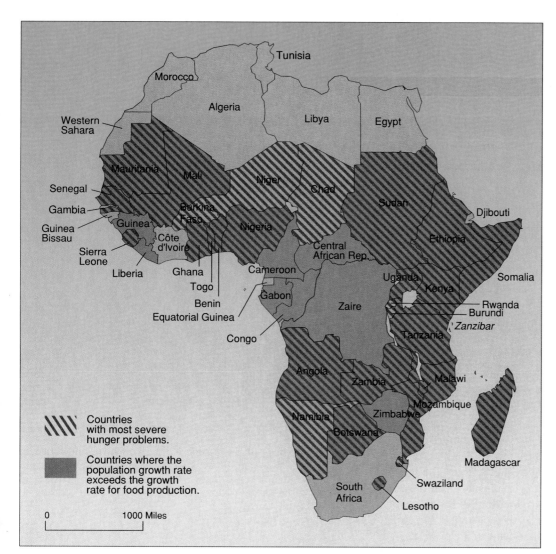

Figure 10.9

Thirty-one of forty-one countries in sub-Saharan Africa have population growth rates that exceed increases in food production. Note that some countries with high population growth rates have avoided severe food shortages while others, because of wars and other social factors, have had famines, even though food production has kept pace with population growth.

Source: World Bank.

Countries with most severe hunger problems.

Countries where the population growth rate exceeds the growth rate for food production.

0   1000 Miles

fruits and beverages that we import or staples such as rice that can be shipped internationally.

Maize yields in the United States provide an encouraging example of the potential for yield improvements through agricultural technology. Over the past century, maximum yields have risen from about 40 bushels per acre to more than 200 bushels per acre today. The two main contributors to this remarkable change have been better-yielding crop varieties and the use of fertilizer. Much of the progress in new hybrid maize varieties, farming procedures, and machines to carry out high-intensity agriculture came from research and training at U.S. land grant universities.

Starting shortly after the Second World War, the Ford and Rockefeller Foundations (along with a number of governments and other international agencies) set up agricultural research stations to breed tropical wheat and rice varieties that could provide yield gains similar to those obtained in the United States. Collectively, these stations are known as the Consultative Group on International Agricultural Research (CGIAR). The results have been spectacular for many crops. "Miracle" strains of rice and wheat have made it possible to triple or quadruple yields per hectare (fig. 10.11). This

dramatic yield increase from new crop varieties has been called the **green revolution.**

These new breeds are really "high responders" rather than high yielders. That is, they respond more efficiently to increases in fertilizer and water and have a higher yield under optimum conditions than do other varieties. Under poor conditions, on the other hand, high responders may not produce as well as traditional crops. Nevertheless, the average worldwide grain yields using these new strains have increased from 1.1 to 2.6 metric tons per hectare over the past twenty years. In East Asia, yields with bioengineered breeds approach that of more-developed countries (see table 10.5). These higher yields, however, are not available to everyone. Poor farmers usually cannot afford the seed, fertilizer, water, pesticides, fuel, and farm equipment necessary to cultivate the new strains. Often only the most prosperous farms are able to participate in the green revolution. The crop surpluses they produce are likely to drive prices down, so the marginal farmers are even worse off than before.

There is a worry about whether crop breeders can continue to produce new varieties that will maintain these high yields. Throughout the world, native crop varieties are being replaced by

Food, Hunger, and Nutrition

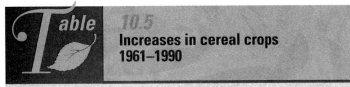

Table 10.5

Increases in cereal crops 1961–1990

Country Group or Region	Increase (percent)			Current Yields in Tons per Hectare
	Total	Due to More Cropland	Due to Higher Yields	
Sub-Saharan Africa	73	47	52	1.0
East Asia	189	6	94	3.7
South Asia	114	14	86	1.9
Latin America	111	30	70	2.1
Middle East and North Africa	68	23	77	1.4
Europe and the former U.S.S.R.	76	13	113	2.2
High-income countries	67	2	98	4.0
World	100	8	92	2.6

Source: *World Resources 1992–1993*, World Resources Institute.

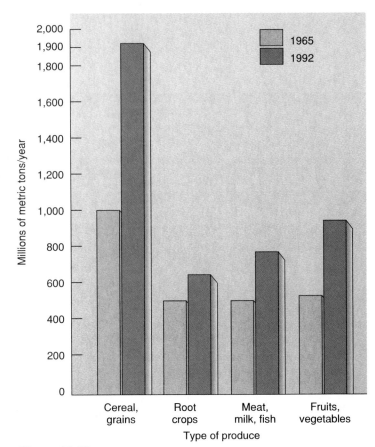

Figure 10.10

World crop production 1965–1992, in millions of metric tons per year. Production has increased by 70% over this 27-year period.

Source: Data from World Resources Institute, 1994.

these new types, and the genetic resource available to breeders is being seriously diminished. Fifty years ago, several hundred varieties of wheat were grown in the Middle East, each adapted through centuries of selection to a particular set of environmental conditions. Now, a few miracle varieties have displaced most of those indigenous species. There is a great danger that an epidemic might sweep through the fields when they are all planted with a single variety, resulting in a major famine. The United States experienced a hint of what this could mean in 1970, when nearly all hybrid corn in the country shared a set of Texas male sterile genes that made all the plants susceptible to a fungal disease called southern corn leaf blight. An epidemic swept across the country and threatened the entire crop.

Plant conservators are busily hunting for native crop varieties that might be disease resistant or able to grow in a unique set of environmental conditions. For instance, many irrigated soils in dry climates tend to accumulate surface salt and mineral deposits. Some wild strains are salt tolerant, and there is a possibility that they could

Figure 10.11

Short-stemmed "semidwarf" wheat (*center*) developed by Nobel prize winner, Norman Borlang, is a high responder; that is, it can utilize high levels of fertilizer and water to produce high yields without growing so tall that the stalks fall over and are impossible to harvest. Notice how much denser the seed heads are on the semidwarf than on its normal size relative (*left*).

Food, Land, and Biological Resources

Figure 10.12

Seed banks are part of an effort to preserve diversity of biological resources. Even the best seed bank is no match for a natural ecosystem in preserving species, however.

be foundation plants for salt-resistant domestic varieties. **Gene banks** (also called seed banks) have been set up to store many seed varieties for future breeding experiments (fig. 10.12).

Most of our commercial crops are thought to have originated in semiarid lands identified by Russian agronomist Nikolay Vavilov

as centers of high crop diversity (fig. 10.13). Note that while there is some overlap between these centers and the biodiversity priority areas in figure 13.2, regions where agriculture developed are generally semiarid and highly seasonal. These conditions favor plants with storable seeds, fruits, and tubers suitable for cropping. These areas are not always those with highest total biodiversity. Examples of Vavilov centers and crops, thought to have originated in each include:

- **China:** Naked oat, soybean, adzuki bean, leaf mustard, apricot, peach, orange, sesame, China tea.

- **India:** Rice, African millet, cucumber, tree cotton, pepper, jute, indigo.

- **Indo-Malaya:** Yam, pomelo, banana, coconut.

- **Central Asia:** Wheat, rye, pea, lentil, chickpea, sesame, flax, safflower, carrot, radish, pear, apple, walnut.

- **Near East:** Wheat, barley, rye, red oat, chickpea, pea, lentil, blue alfalfa, sesame, flax, melon, almond, fig, pomegranate, grape, apricot, pistachio.

- **Mediterranean:** Durum wheat, hulled oats, broad bean, cabbage, olive, lettuce.

- **Ethiopia:** Wheat, barley, chickpea, lentil, pea, teff, African millet, flax, sesame, castor bean, coffee.

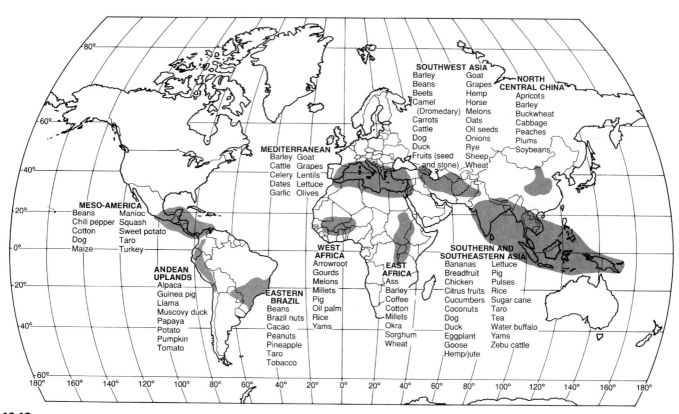

Figure 10.13

Vavilov centers are areas of high crop genetic diversity. They are thought to be ancestral homes of many of our current crop varieties.

Food, Hunger, and Nutrition

- **Southern Mexico and Central America:** Corn, common bean, pepper, cotton, sisal hemp, squash, pumpkin, gourd.

- **Peru, Ecuador, Bolivia:** Sweet potato, potato, lima bean, tomato, cotton, papaya, tobacco.

- **Chile:** Potato.

- **Brazil and Paraguay:** Cassava (manioc), peanut, cacao, rubber tree, pineapple, purple granadilla.

Some poorer countries with wild ancestors of cultivated crops object to having scientists from richer countries collecting samples of their flora and fauna. They feel entitled to a share of the profits that may come from the use of their native species for future crop development. There are threats of "gene wars" over control of this resource. What do you think? If Canadian scientists discover a gene in Pakistan to improve their wheat crop, do they owe a royalty payment to Pakistan for use of that gene?

New Food Sources

Although at least three thousand species of plants have been used for food at one time or another, most of the world's food now comes from only sixteen widely grown crops. Many new or unconventional crops might be valuable human food supplies, however, especially in areas where conventional crops are limited by climate, soil, pests, or other problems. Some plants now being studied as potential additions to our crop roster include the following:

Winged beans. A perennial plant that bears well in hot climates where other beans will not grow. It is totally edible (pods, mature seeds, shoots, flowers, leaves, and tuberous roots), resistant to diseases, and enriches the soil (fig. 10.14).

Amaranth. A staple seed crop of the Aztecs and the Incas with edible leaves and seeds; one of the most nutritious grains known that produces high yields in a variety of conditions.

Triticale. A hybrid between wheat (*Triticum*) and rye (*Secale*) that grows in light, sandy, infertile soil. It is drought resistant, has nutritious seeds, and is being tested for salt tolerance for growth in saline soil or irrigation with seawater.

Wax gourd vegetable. A creeping vine from Africa that looks like a pumpkin but is easier to grow, more nutritious, and can be eaten at any stage of growth as a cooked vegetable, soup base, or food extender. Its waxy coat resists microorganisms, and it can be stored up to one year without refrigeration.

Blue Revolution

Another change sweeping around the world may prove to be just as important as the green revolution. Fish farming, sometimes called the **blue revolution,** has the potential to contribute as much to human nutrition as did miracle cereal grains but may create similar social and environmental problems. Commercial fish farming already is a $5 billion per year industry, and domestic fish ponds supply as much as two-thirds of the protein consumed by subsistence farmers in many areas. Already more than one-half the salmon and trout sold in the United States is raised in captivity.

Worldwide, some ninety species of fin fish, thirteen species of shrimp and crawfish, and numerous species of shellfish are grown commercially. The total aquaculture production in 1990 was estimated to be 10,587 million metric tons or 15 percent of the world seafood market. About half this total was fin fish, while mollusks (oysters and clams) and aquatic plants (seaweeds) comprised nearly one quarter each. Shrimp and lobsters comprised only 2 percent of the total harvest but made up the most rapidly growing segment, having doubled in the past decade.

Fish farming involves growing organisms in enclosures such as ponds, tanks, or net pens anchored in an ocean bay or estuary. Food is supplied by humans, and fish can be grown to high densities under these artificial conditions. Fish ranching involves rearing young fish in a hatchery and then releasing them into a river, lake, or open ocean where they will feed on natural prey. The rancher hopes that after a few years the fish will return to their release point to be harvested. Some thorny issues are raised when commercial trawlers lurk just offshore to snare valuable salmon or trout returning to the hatchery where they were raised. How do you establish ownership of such a fugitive resource?

Social and ecological problems often accompany fish farming as well. Conversion of coastal wetlands into shrimp ponds and fish tanks is causing drastic losses of natural habitat in many areas. In Southeast Asia, for instance, which now supplies three-quarters of all farm-raised shrimp, millions of hectares of mangrove swamps have been cleared, endangering hundreds of species of wildlife that use these wetlands at some point in their life cycles. Birds and predators often are attracted to this rich food source. Farmers who see their profit margins eaten up by herons, cranes, terns, and other uninvited guests often respond with poisons or shotguns that take a deadly toll on wildlife populations.

The high densities of fish held in pens in some operations have severe impacts on coastal areas where tides are not strong enough to sweep away wastes and uneaten food. A typical 2-hectare salmon farm holds 75,000 fish and produces as much organic waste as a town of 20,000 people. In an enclosed bay or fjord, that waste

Figure 10.14

Winged beans bear fruit year-round in tropical climates and are resistant to many diseases that prohibit growing other bean species. The whole pod can be eaten when green or dried beans can be stored for later use. It is a good protein source in a vegetarian diet.

load can cause algal blooms and anoxic (oxygen-depleted) water that kills most other aquatic life and severely degrades aesthetic values. The high fish density and waste load encourage diseases that are then controlled with antibiotics and pesticides that escape into the surrounding environment. To prevent waste buildup in enclosed tanks, aquaculturists prefer flow-through water supplies that consume large amounts of water and often pollute the rivers or lakes into which they drain.

Agricultural Economics

Economic choices and relationships, whether monetary or not, strongly affect food availability on both the national and individual level. Even though the world has an abundance of food, some people do not have access to means of production to grow their own food nor enough money to buy what they need. In this section, we will look at international food trade, surpluses, subsidies, and agricultural aid.

Food Surpluses, Subsidies, and Agricultural Aid

U.S. food reserves reached historic highs of more than 200 million metric tons in the mid-1980s. This was nearly 60 percent of all the food stored in the world. Storing this excess is expensive, and finding places to put it is difficult in good harvest years (fig. 10.15). Price supports and subsidies prevent prices from falling too low,

Figure 10.15

After a good harvest in the Midwest there is often no place to store grain because so much storage space is already occupied by the previous year's surplus. When grain is piled on the ground in the open, it is subject to birds, insects, and spoilage.

yet rising costs of fertilizer, fuel, seed grain, and agricultural chemicals, combined with weak prices have driven an increasing number of family farms out of business. The conservation reserve program pays farmers to keep marginal lands out of production. This has had beneficial environmental effects by reducing erosion and providing wildlife habitat. Crop reductions have been modest, however, because farmers use rent from idle land to buy more fertilizer and farm more intensively on land still in production. The total cost of price supports, subsidies, and storage of surpluses in the United States was around $25 billion per year in the 1980s but are now scheduled to be eliminated. Many European countries spend even more to protect their national farming sectors. What to do about farm subsidies has been the sticking point in both the GATT (General Agreement on Tariff and Trade) talks and European unification negotiations.

Distribution of international **food aid** by the more-developed countries is often selective and highly political. We give aid to governments that support our policies rather than to the countries with the most hungry people. In 1990, for instance, Egypt received $5.6 billion in development aid—more than twice as much as any other country. Israel and Jordon each received by far the highest per capita aid, nearly $300 per person. In contrast, India and China received less than $2 per person, while the average for all poorer countries was less than $10 per capita. Much of the food aid we do give hurts rather than helps because food prices are driven down when markets are flooded with imported commodities. This makes food production by local farmers uneconomical and reduces indigenous food production, trapping some countries in permanent dependence on welfare. It also often helps support unpopular and repressive governments.

International Food Trade

One way to even out food distribution between areas with excess and those with shortages is through world food trade. International food shipments have risen dramatically in the past 50 years. In 1940, only ten countries imported more than 1 million tons of food. By 1990, this number had risen fivefold. More importantly, world food trade now accounts for more than one-quarter of all food consumed by humans. The world is becoming more and more interdependent.

World trade in agricultural commodities has brought both benefits and problems for farmers and planners in many countries besides the United States. On one hand, international trade makes up the food deficit in many countries, staving off famine and social disruption. On the other hand, nations that become dependent on imported food supplies may divert so much of their foreign earnings to buying expensive imports that they can never develop indigenous resources. They also may become politically indentured to the countries that sell them food. Rapidly fluctuating commodity prices (fig. 10.16) destabilize the economies of producing countries and make it difficult for them to plan for the future. Over the past forty years, the purchasing power of food and nonfood commodities has fallen sharply compared to manufactured goods.

Food-producing countries also can become captive to international trade. It may make sense for a country to specialize in a few

Food, Hunger, and Nutrition

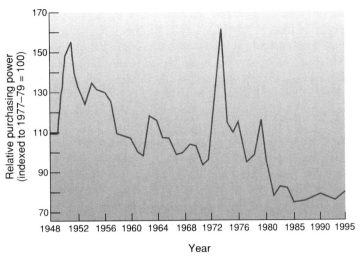

Figure 10.16

Purchasing power index of 33 primary commodities exported by developing countries, compared with manufactured goods imported by those countries. Although raw materials and agricultural product prices increased after the oil-price shocks of 1973, the general trend has been downward. A ton of rice or sugar would purchase twice as much in 1950 as in 1985. Commodities include beverages, cereals, fats, oils, other foods, nonfoods, metals, and minerals.

Source: The World Bank, 1995.

Figure 10.17

Cash crops, such as these coffee beans being picked in Colombia, often replace production of staples necessary to feed the local population of developing countries. Government policies encourage production of export commodities to raise foreign exchange.

crops that fit its climate and workforce. Economists call this "comparative advantage." Produce sold on world markets allows people to buy foods grown more efficiently elsewhere. The foreign exchange generated by export crops can provide funds necessary to buy tools, information, and technology needed for development. When a single crop dominates a national economy, however, it places the population at the mercy of fickle world markets. Honduras, for example, derives nearly 90 percent of its foreign exchange from bananas. Cuba is nearly as dependent on sugar cane. When prices of these commodities fall—sometimes by 50 percent in a few weeks—the effects on the local populace are devastating. With no basic food supplies of their own and no money to buy it abroad, they have few options.

Too often, the export crops grown by developing countries are not food supplies for needy neighbor nations, but rather luxury foods, drugs, and other nonessential items to be shipped to the rich countries of the world (fig. 10.17). Little of the money generated by these crops trickles down to the underclass of the producing countries. Most of it goes to landowners, money lenders, export brokers, or political leaders. These export-oriented policies contribute significantly to malnutrition problems in many countries. In Guatemala, for example, 97 percent of the citrus crop is exported, while a majority of the local population suffers from vitamin C deficiency. In Central America, beef production and export increased nearly sixfold in the 1960s and 1970s, while per capita meat consumption fell by 50 percent.

Whether growing **cash crops** (those sold rather than consumed or bartered) is positive or negative in terms of individual food security depends on local conditions and even on intrafamily

relationships. On one hand, growing cash crops provides income for farmers and farm workers that allows them to buy food and other essentials. This can be more important than you might suppose. Small children often can't eat enough of the staple crops grown in traditional agriculture to get the proteins and vitamins they need. Child health and welfare may be greatly improved if the family can afford to buy meat, milk, and other high-quality foods. Furthermore, diseases frequently prevent children and adults from benefiting from the food they eat. Having cash to pay for medicine and medical care—which may not be available by barter—may do more good for the family than a larger food supply.

On the other hand, conversion of cropland to cash crops can drive up local food prices so that even with higher incomes, poor people can't afford to buy food. Furthermore, cash crops like coffee, cocoa, bananas, and tea are perennials that don't bear for several years after planting. Small farmers can't afford to wait that long for income. Even annual crops bring a lump sum at harvest time, which may be wasted on alcohol or other nonfood items. By contrast, a home garden provides a steady supply of food that is not easily converted to luxury items.

One of the most important factors in a cash economy is intrahousehold distribution. Men often control the money and use it for their own personal consumption, leaving little for women and children. Because women do most of the gardening in many cultures, they control the food they produce and more is available for their children. Allowing women to manage the income from their crops is often the best way to improve family welfare. In Nigeria, for example, better roads and transportation have allowed rural women to take crops to market in Lagos. Because women are more independent in Nigeria than in many countries, they keep the income they make by marketing their produce. The benefits to family nutrition have been greater than any foreign aid project.

World Hunger

The nutritional deficiencies we discussed earlier can affect individuals anywhere in the world who don't get an adequate diet either because wholesome food isn't available, they can't afford it, or they don't know how to choose the right things to eat. What about larger-scale food shortages and their causes? Food and water are our barest necessities to remain alive. Can you imagine what it would be like to be without them?

Famines

The images that probably first come to your mind when you think about food shortages are the tragic scenes of mass starvation that come from places like Ethiopia, Somalia, Sudan, or Mozambique. These acute shortages, or **famines,** are characterized by large-scale loss of life, social disruption, and economic chaos. Starving people eat their seed grain and slaughter their breeding stock in a desperate attempt to keep themselves and their families alive (fig. 10.18). Even if better conditions return, they have sacrificed their productive capacity and will take a long time to recover. Famines are characterized by mass migrations as starving people travel to refugee camps in search of food and medical care. Many die on the way or fall prey to robbers.

What causes these terrible tragedies? Environmental conditions are usually the immediate trigger, but politics and economics are often equally important in preventing people from getting the food they need. Adverse weather, insect infestations, and other natural disasters cause crop failures and create food shortages. But these factors have generally been around for a long time, and local people usually have adaptations that get them through hard times if they are allowed to follow traditional patterns of migration and farming. Arbitrary political boundaries, however, along with wars and land seizures by the rich and powerful, block access to areas that once served as refuge during droughts, floods, and other natural disasters. Poor people can neither grow their own food nor find jobs to earn money to buy the food they need.

In 1974, for instance, a terrible famine struck Bangladesh and thousands died. The immediate cause was floods in June through August that interfered with rice planting, eliminating jobs on which many farm workers depended. Fears about impending rice shortages triggered panic buying, hoarding by speculators, and rapidly rising prices. The government began relief efforts, but it was too little, too late. By October it became apparent that rice harvests were actually higher than in previous years; prices dropped and so did mortality rates. The irony was that more food was available in 1974 than in any other year in the decade. Poor people simply couldn't afford to buy it. The United States played a role in this tragedy. We cut off aid to Bangladesh as punishment for selling jute fibers to Cuba. Withdrawal of food aid on which Bangladesh had become dependent only fueled the panic and price gouging that occurred later.

By contrast, the state of Maharashtra in central India suffered a similar drought in 1972–73 that reduced crop yields by 50 percent. As in Bangladesh, farm laborers were thrown out of work. The Indian government moved quickly, however, to employ workers building roads, wells, and other public projects. Although

Figure 10.18

Children wait for their daily ration of porridge at a feeding station in Somalia. When people are driven from their homes by hunger or war, social systems collapse, diseases spread rapidly, and the situation quickly becomes desperate.

wages were low, food prices didn't rise and people could afford an adequate if meager diet. Although the amount of food available per person was less than half that in Bangladesh, few Indians starved. Since they remained in their villages and rural infrastructure was improved during the drought by public works projects, farmers recovered quickly once the rains returned.

Similarly, the world was horrified by terrible photographs of starving children in Ethiopia and the Sudan in 1983 and 1984. More than 1 million people died in the Horn of Africa. Droughts triggered this famine, but neighboring countries that experienced comparable weather did not have problems as severe as these unfortunate countries did. In Sudan and Ethiopia, where total food supplies dropped about 10 percent, the famine was severe. Zimbabwe and Cape Verde, by contrast, which suffered worse droughts and lost about 40 percent of their normal harvest, had no famine. In fact, social welfare programs brought about mortality decreases in Zimbabwe and Cape Verde during this time. Author Amartya Sen points out that armed conflict and politics almost always are at the root of famine. No democratic country with a relatively free press, he says, has ever had a major famine.

The aid policies of rich countries often serve more to get rid of surplus commodities and make us feel good about our generosity than to get at the root causes of starvation (fig. 10.19). Herding people into feeding camps generally is the worst thing to do for them. The stress of getting there kills many of them, and the crowding and lack of sanitation in the camps exposes them to epidemic diseases. There are no jobs in the refugee camps, so people can't support themselves if they try. Social chaos and family breakdown expose those who are weakest to robbery and violence. Having left their land and tools behind, people can't replant crops when the weather returns to normal.

Figure 10.19
We rarely learn much from the media about the underlying causes of famine. Often our attention is focused on images of people from wealthy nations generously helping those who are suffering. This makes us feel good, but doesn't get to the root of the problems.

Table 10.6
Steps to eliminate world hunger

1. Remove distorting subsidies and tariffs.
2. Improve land-use planning.
3. Clarify resource ownership and land tenure.
4. Expand educational programs for girls and employment opportunities for women.
5. Increase investment in, and maintenance of, rural infrastructure.
6. Expand democracy and the free flow of information.
7. Eliminate poverty.

Source: *World Bank 1992 Development Report.*

Chronic Food Shortages

Although we most often think of hunger in terms of catastrophic famine and starvation, far more people in the world are affected by **chronic food shortages** in which they lack adequate nutrition on a long-term basis. About 1 million people are thought to have died in sub-Saharan Africa in the early 1980s as a result of the drought-related famine. By contrast, perhaps *twenty times* that many—mostly children—die every year from the effects of chronic undernutrition and malnutrition. How can this tragic situation persist in a world characterized more by abundance and surpluses than by scarcity?

In recent years, agricultural policymakers have shifted their attention from total food availability to **food security,** the ability to obtain sufficient food on a day-to-day basis. In the 1950s and 1960s, agricultural aid programs focused on expanding food production in developing countries. Aid workers found, however, that even when plenty of food is available in a country, some people have an excess while others don't have enough. In the 1970s, efforts were redirected to family food security by helping poor families improve their lot. But even when family income is adequate, women and children often don't get enough to eat because men spend the money on themselves. We now see that individual food security is what really matters. Programs for the 1990s work to protect the weakest members of society and call for new approaches to aid and governance. Table 10.6 presents some recent suggestions from the World Bank to alleviate world hunger.

How Many People Can the World Feed?

A number of studies have estimated the maximum number of people the world could support, given predicted supplies of fertilizers, water, arable cropland, and other factors of agricultural production. The results have ranged from pessimistic warnings that there are already more people than we can feed, to extreme claims that the world could support thirty to fifty times the present population. Ecologists estimate that humans already consume or usurp about one-third of all primary productivity on the earth. If we take away much more, they point out, there won't be room for many non-domestic species.

One set of estimates of maximum human populations is shown in figure 10.20. This study predicts that at a low level of agricultural inputs, developing countries could feed only about 2.5 billion people; the maximum world population would have to be less than the present 5.5 billion people if everyone is to be fed adequately.

As inputs are increased in this model, the maximum population also increases to around 24 billion people. This says nothing about the quality of life, however, or the ecological effects if all possible land were converted to crop production. Even with high levels of input, southwest Asia could not support many more people than it does now. Africa and South America, on the other hand, potentially could support far more people if agriculture were modernized and more land were converted to agricultural uses.

In chapter 11, we will continue this discussion with a survey of the agricultural systems that produce our food. As part of that discussion, we will look at the resource inputs of soil, water, fertilizer, and energy necessary to sustain crop production and support future human populations. In subsequent chapters, we will discuss other resource limitations that bear on this all-important question: How many people—and at what level of civilization and environmental quality—can the world support?

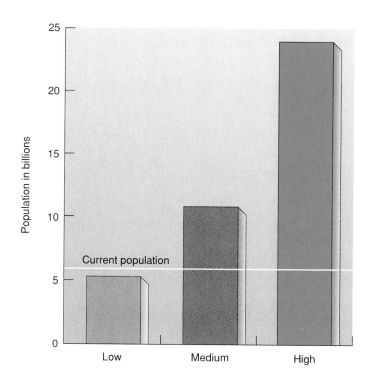

Figure 10.20

Estimates of potential human populations given maximum use of available cropland and varying agricultural inputs such as water, fertilizer, and energy. Notice that low inputs would not feed current populations, while high inputs might feed four times the present populations. High inputs are probably not sustainable, however.

Source: Data from the U.N. Food and Agriculture Organization.

Summary

In recent years, world food supplies have been rising at an unprecedented rate and have grown faster than populations in every continent except Africa. There now is enough food to supply everyone in the world with more than the minimum daily food requirements, but food is inequitably distributed. The FAO estimates that 750 million people are chronically undernourished or malnourished, and 15 to 20 million (mostly children) die each year from diseases related to malnutrition. Additional millions survive on a deficient diet, suffering from resulting stunted growth, mental retardation, and developmental disorders.

Among the essential dietary ingredients for good health are adequate calories, proteins, lipids (especially unsaturated ones), vitamins, and minerals. Marasmus and kwashiorkor are protein-deficiency diseases; anemia and goiter are caused by mineral deficiencies; and pellagra, scurvy, beriberi, and rickets are vitamin-deficiency diseases that affect millions of people worldwide.

The three major crops that are the main source of calories and nutrients for most of the world's people are rice, wheat, and maize. About a dozen other types of seeds and grains, a few root crops, twenty or so fruits and vegetables, six mammals (and their milk), a few domestic fowl, and a variety of seafoods comprise nearly all the food that humans eat. Some new crops or unrecognized traditional crops hold promise for increasing the nutritional status of the poorer people of the world. Scientific improvement of existing crops and modernization of agriculture (irrigation, fertilizer, and better management) are potential sources of greater agricultural production.

Over the past 30 years, the total amount of food in the world has increased faster than the average rate of population growth, so there is now more food per person than there was in 1960, even though the total number of people has doubled. The biggest gains have been in Asia, North America, and Latin America. The only major region in which food production has failed to keep pace with population growth has been sub-Saharan Africa, where adverse weather, insect infestations, wars, inept governments, social and religious factors, economics, and international politics have intervened.

World food trade and international food aid help transfer food from areas of abundance to areas of shortage, but they also undercut local food supplies by encouraging the conversion of land from production of food for local consumption to production of cash crops for export. They also widen economic and social disparities that increase food insecurity and make it even more difficult for the poorest people to feed themselves.

Hunger, poverty, population growth, environmental degradation, and social problems form a complex, interconnected web. Each is a cause, as well as a consequence, of the others. One of the most important questions in environmental science is: How many people, and at what level of civilization and environmental quality, can the world feed?

▼ Questions for Review

1. How many people in the world are chronically undernourished and how many die each year from starvation and nutritionally related diseases?

2. How many calories does the average person need per day to maintain a healthy, active life?

3. What are proteins, vitamins, lipids, and minerals, and why do we need them in our diet?

4. What are the three major crops of the world? List some alternative crops that might help feed the world.

5. What do we mean by green and blue revolutions?

6. List some potential benefits and dilemmas associated with biotechnology and genetic engineering.

7. Describe some problems associated with fish farming and fish ranching.

8. Which countries receive the greatest total amounts and the highest per capita food aid? What does this distribution reveal about food aid programs?

9. What are cash crops and how do they affect local and international food supplies?

10. How are interfamily relationships important in food security?

▼ Questions for Critical Thinking

1. Why has agricultural research paid so little attention to tropical crops like yams and cassava? How would you decide whether our priorities are fair or not?

2. What worldviews might make people believe that there are already too many people to be fed or that technological progress may allow us to feed double or triple current populations? Which side of this argument do you support?

3. Suppose that a seafood company wants to start a fish farming operation in a lake near your home. What regulations or safeguards would you want to see imposed on its operation? How would you weigh the possible costs and benefits of this operation?

4. Why do we have agricultural surpluses in the United States and Canada? Delve more deeply into this issue than the simple fact that we grow more food than we eat. Think about some underlying social, political, economic, and environmental implications of surpluses.

5. Some critics claim that international food aid creates more problems than it solves. List some arguments for and against this proposition.

6. Should poor countries like Guatemala grow cash crops? What are some advantages and disadvantages of doing so?

7. Debate the claim that famines are caused more by human actions (or inactions) than by environmental forces. What is the critical element or evidence in this debate?

8. How many people do you believe the world could feed or should feed? What changes might be necessary to reach an ideal food per person ratio?

▼ Key Terms

anemia 209	green revolution 215
blue revolution 220	hypothyroidism 209
cash crops 220	kwashiorkor 208
chronic food shortages 222	malnourishment 207
famines 221	marasmus 207
food aid 219	pellagra 210
food security 222	undernourishment 207
gene banks 217	vitamins 210

▼ Additional Information on the Internet

Eliminating Starvation/FeedingHumanity
http://www.pacificrim.net/~wginwrep/WorldGame/drfeed.html/

Food and Nutrition Web Sites
http://www.mother.com/agaccess/FoodNut.html/

Hunger Web
http://www.brown.edu/departments/world_hunger_program/html/

Pan-American Health Organization
http://www.paho.org/

Skip-A-Meal
http://www-relg-studies.scu.edu/facstaff/grassi/skipmeal.html/

United Nations Food and Agriculture Organization
http://www.fao.org/

World Crisis Network
http://www.ghn.org/wcn/

World Health Organization
http://www.who.ch/

World Hunger Program Gopher
gopher://gopher.brown.edu:70/11/brown/departs/worldhun/

World Hunger Relief Foundation
http://www.csi.nb.ca/geb/hunger1.htm/

World Hunger Year
http://www.iglou.com/why/

Note: Further readings appropriate to this chapter are listed on p. 598.

Environmental Consultant

Jeremy Sterk

*J*eremy Sterk

Despite growing up in the Midwest, Jeremy Sterk has been fascinated with oceans since childhood. His lifelong dream has been to become a marine biologist. But a hobby and a bit of serendipity intervened to have an important, and unplanned, influence on his career direction.

Jeremy attended college in the Midwest, where he pursued a major in aquatic biology. Following graduation, he planned to apply for graduate school at a coastal university to work on an advanced degree in marine biology. In addition to his interest in aquatics, however, Jeremy's earlier fascination with computers had become a serious hobby. Although he did not realize it at the time, the hours spent on the computer hobby would turn out to be an investment in his future employment.

One day, one of Jeremy's college biology professors described a project he had undertaken to create a computerized identification key to the freshwater algae. The project was laborious, however, because of the computer system and language the professor was working in. Because of his hobby, Jeremy quickly grasped what the professor was attempting to do and realized that Hypercard provided a simpler approach. Jeremy produced a demonstration model of the alternative that impressed the professor. Working in collaboration, the two of them completed the computerized key that has now been produced commercially by a major publisher.

After college, Jeremy decided to move to Florida. Although he had no immediate job prospects or contacts, Jeremy searched newspaper ads and followed other leads and found an opening with a small environmental consulting firm. The position involved field work, including plant identification, as well as other responsibilities. While he didn't have a strong botany background, his computer skills were an important asset and he got the job.

Much of the company's work involves helping individuals and land developers comply with various permitting requirements. As such, Jeremy typically serves as the go-between for company clients and the permitting agencies, such as the South Florida Water Management District and the Army Corps of Engineers.

Jeremy spends about half his time in the field doing a variety of tasks. He commonly does surveys of clients' property to determine if they contain wetlands and frequently has to identify wetlands boundaries. His responsibilities also include helping clients with mitigation plans to offset negative environmental impacts the project would otherwise have.

He performs transect surveys of plant and animal species, paying particular attention to those species listed as threatened or endangered. He has done quite a bit of work with gopher tortoises, eagles, and red-cockaded woodpeckers. He also does vegetation mapping for environmental assessment reports. He occasionally works with wildlife monitoring, such as installing infrared sensors in the wildlife crossings under Florida highways to obtain information on use of these crossings by wildlife, including the endangered Florida panther.

The other half of the time Jeremy is involved with paperwork: writing reports and corresponding with regulatory agencies and clients over permit conditions. The need for writing and communication skills was one of the abilities highlighted by the interviewer when Jeremy applied for his job. He has also updated his firm's computer operation.

Jeremy enjoys his job very much. He likes spending so much time in the woods and swamps, experiencing a side of Florida that few people ever see. For students interested in this line of work, Jeremy advises them to get a solid background in biology and as broad a base of other preparation as they can. He wishes he had taken more botany and formal training in soils science. He feels coursework in statistics is helpful in his and other related fields. A course in technical writing would have been useful as well. Not surprisingly, he also feels a person cannot have too much background in computers.

Jeremy enjoys his work so much that it has put his marine biology plans on hold. He is a bit uneasy about not pursuing a lifelong dream, and he will eventually go on to graduate school. But for now, he feels his present work could easily become a lifetime career.

A sunny counterpane landscape in Devon, England illustrates the bounty of nature. With careful stewardship, the land represents a sustainable resource for humans and other species.

CHAPTER 11: Soil Resources and Sustainable Agriculture

Objectives

After studying this chapter, you should be able to:

- describe soil composition and the role of soil particles, soil organisms, and soil chemistry in soil formation and productivity.

- understand soil types and soil profiles and what they mean for agriculture.

- differentiate between the causes and consequences of land degradation, including soil erosion, nutrient depletion, waterlogging, salinization, and other abuses that decrease soil fertility and crop production.

- analyze the agricultural inputs needed for sustained food production and tell how these resources may limit human activities.

- evaluate the principles of just and sustainable agriculture.

- explain what each of us can do to ensure a safe, secure food supply.

Dust Bowl Days

Sunday, April 14, 1935, dawned bright and clear over the city of Amarillo in the Texas panhandle. That afternoon, however, a huge black cloud of dust appeared on the northern horizon and quickly swept across the treeless plains. The dust swirled past, thick as falling snow, as cars stalled in the streets and pedestrians bumped into each other, unable to see things a few feet away. Terrified families huddled together with wet towels over their faces and rags stuffed in cracks around windows and doors, but still the dust seeped in. Tiny dunes formed on windowsills and doorjams and even the food in the refrigerator was covered with dust. Is this the end of the world they wondered? And where did all this dirt come from?

This storm became known as Black Sunday and coined the term "dust bowl" to describe both the decade of the 1930s and the high plains area from which the dirt had been blown. The heart of the dust bowl stretched from Texas to Manitoba but air-

borne dirt was often carried as far as the East Coast. Amarillo averaged nine serious dust storms per month from January to April—the main dust storm season—between 1933 and 1938. In April, 1934, it had "black blizzards" on 23 days. Homes, barns, tractors, and fields were buried under drifts up to 7 m (25 ft) high.

These dust storms were the worst human-caused environmental disaster the United States has ever experienced. The social, economic, and ecological costs were immense. The Soil Conservation Service, founded in 1935 to address this calamity, estimated that 10 billion tons of topsoil from the heart of the world's breadbasket had blown away on the wind. By 1938, farm losses had reached $25 million per day and more than half the rural families on the Southern Plains were on relief (fig. 11.1). Thousands of people died of "dust pneumonia," while millions joined the mass migration described by John Steinbeck in The Grapes of Wrath *(1939).*

Figure 11.1

Father and sons walking in a dust storm, Cimarron County, Oklahoma, 1936.

A prolonged drought beginning in 1931 was the immediate cause of the dust storms, but inappropriate agricultural practices allowed erosion to occur, exacerbating the situation. Early in the 20th century, American farmers were caught up in a specialized, market-driven system that encouraged all-out production and drove out diversified, subsistence farming. During World War I, rising wheat prices, unusually wet weather, and availability of tractors and combines encouraged speculators to expand cultivation into previously untouched land. Without prairie sod to protect the soil, the land blew away when drought came back in the 1930s.

To combat wind erosion, the Soil Conservation Service sponsored research and demonstration projects in alternative farming methods. It also helped finance shelterbelts (rows of trees planted as windbreaks), strip-cropping, reestablishment of grass on damaged cropland, and new tillage methods. Although it will take centuries to rebuild topsoil, most of the visible signs of this terrible erosion have been erased and huge dust storms rarely occur now (fig. 11.2). Still, this historic example raises questions

Figure 11.2

Environmental degradation isn't always irreversible. This farmer inspecting his field in Oklahoma in 1979 was the smaller boy in the 1936 photograph in figure 11.1. Irrigation and soil conservation have returned the land to fertility and stability in the 43 years between the pictures.

for current generations. Have we learned from our past mistakes? Are our agricultural policies and practices sustainable today?

In chapter 10, we examined the dramatic gains in agricultural productivity over the past century. Maintaining that productivity depends on healthy soils, water, fertilizer, energy, crop diversity, and farm workers. Worldwide, environmental degradation (soil erosion, salinization, pollution, biotic stress), urbanization, limits in essential inputs, and spread of weeds and pests are reducing crop yields. Establishing sustainable agricultural systems that provide an adequate diet and a wholesome environment for everyone remains one of the most important challenges we face. In this chapter, we will look at ways we use and abuse the soil, as well as some alternative agricultural approaches that help protect both our agricultural base and the rural communities that produce our food. ▶

What Is Soil?

Of all the earth's crustal resources, the one we take most for granted is soil. We are terrestrial animals and depend on soil for life, yet most of us think of it only in negative terms. English is unique in using "soil" as an interchangeable word for earth and excrement. "Dirty" has a moral connotation of corruption and impurity. Perhaps these uses of the word enhance our tendency to abuse soil without scruples; after all, it's only dirt.

The truth is that **soil** is a marvelous substance, a living resource of astonishing beauty, complexity, and frailty. It is a complex mixture of weathered mineral materials from rocks, partially decomposed organic molecules, and a host of living organisms. It can be considered an ecosystem by itself. Soil is an essential component of the biosphere, but it must be nurtured and cultivated to bring it to its highest potential.

There are at least 20,000 different soil types in the United States alone; perhaps hundreds of thousands worldwide. They vary

by origin, parent materials, age, and climate. There are young soils that, because they have not weathered much, are rich in soluble nutrients. There are old soils, like the red soils of the tropics, from which rainwater has washed away most of the soluble minerals and organic matter, leaving behind clay and rust-colored oxides. Some soils have exotic origins, such as the midwestern loess deposits that contain silts blown all the way from Asia.

To understand the potential for feeding the world on a sustainable basis we need to know how soil is formed, how it is being lost, and what can be done to protect and rebuild good agricultural soil.

A Renewable Resource

With careful husbandry, soil can be replenished and renewed indefinitely. Many farming techniques deplete soil nutrients, however, and expose the soil to the erosive forces of wind and moving water. As a result, in many places we are essentially mining this resource and using it much faster than it is being replaced.

Building good soil is a slow process. Under the best circumstances, good topsoil accumulates at a rate of about 10 tons per hectare (2.5 acres) per year—enough soil to make a layer about 1 mm deep when spread over a hectare. Under poor conditions, it can take thousands of years to build that much soil. Perhaps one-third to one-half of the world's current croplands are losing topsoil faster than it is being replaced. In some of the worst spots, erosion carries away about 2.5 cm (1 in) of topsoil per year. With losses like that, agricultural production has already begun to fall in many areas.

Soil Composition

Most soil is about half mineral. The rest is air and water together with a little organic matter from plant and animal residue. The mineral particles are derived either from the underlying bedrock or from materials transported and deposited by glaciers, rivers, ocean currents, windstorms, or landslides. The weathering processes that break rocks down into soil particles are described in chapter 16.

Particle sizes affect the characteristics of the soil (table 11.1). The spaces between sand particles give sandy soil good drainage and usually allow it to be well aerated (fig. 11.3), but also cause it to dry out quickly when rains are infrequent. Tight packing of small particles in silty or clay soils makes them less permeable to air and water than sandy soils. Tiny capillary spaces between the particles,

on the other hand, store water and mineral ions better than more porous soils. Because clay particles have a large surface area and a high ionic charge, they stick together tenaciously, giving clay its slippery plasticity, cohesiveness, and impermeability. Soils with a high clay content are called "heavy soils," in contrast to easily worked "light soils" that are composed mostly of sand or silt. Varying proportions of these mineral particles occur in each soil type (fig. 11.4). Farmers usually consider sandy loam the best soil type for cultivating crops.

The organic content of soil can range from nearly zero for pure sand, silt, or clay, to nearly 100 percent for peat or muck. Much of the organic material in soil is **humus,** a sticky, brown, insoluble residue from the partially decomposed bodies of dead plants and animals. Humus is much more important to soil quality than its proportion indicates. It gives soil its "structure," a description of how the soil particles clump together. Humus coats mineral particles and holds them together in loose crumbs, giving the soil a spongy texture that holds water and nutrients needed by plant roots, and maintains the spaces through which delicate root hairs grow.

Soil Organisms

Without soil organisms, the earth would be covered with sterile mineral particles far different from the rich, living soil ecosystems

Table 11.1

Soil particle sizes

Classification	Size
Gravel	2 to 64 mm
Sand	0.05 to 2 mm
Silt	0.002 to 0.05 mm
Clay	Less than 0.002 mm

 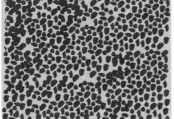

Figure 11.3

Soil characteristics depend, in part, on pore spaces and particle sizes. The soil on the left, composed of particles of various sizes, has spaces for both air and water. The soil on the right, composed of uniformly small particles, is more compacted and has less space for either air or water.

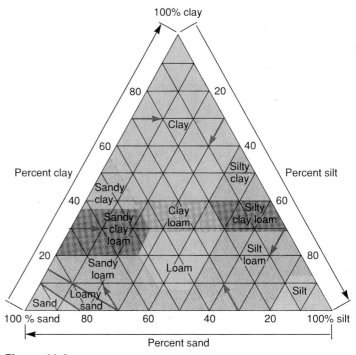

Figure 11.4

Soil texture is determined by the percentages of clay, silt, and sand particles in the soil. Soils with the best texture for most crops are loams, which have enough larger particles (sand) to be loose, yet enough smaller particles (silt and clay) to retain water and dissolved mineral nutrients.

Source: Soil Conservation Service.

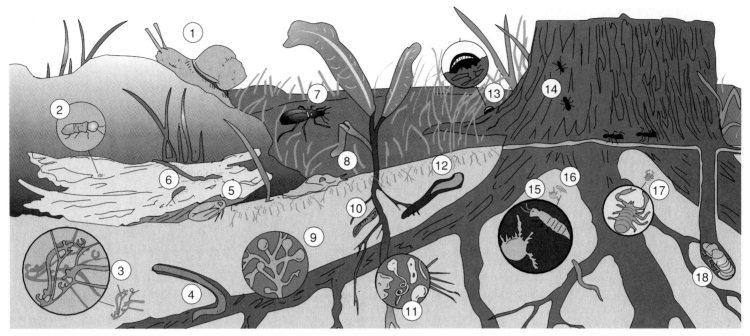

Figure 11.5

Soil ecosystems include numerous consumer organisms, as depicted here: (1) snail, (2) termite, (3) nematodes and nematode-killing constricting fungus, (4) earthworm, (5) wood roach, (6) centipede, (7) carabid (ground) beetle, (8) slug, (9) soil fungus, (10) wireworm (click beetle larva), (11) soil protozoans, (12) earthworm, (13) sow bug, (14) ants, (15) mite, (16) springtail, (17) pseudoscorpion, and (18) cicada nymph.

on which we depend for most of our food. The activity of the myriad organisms living in the soil help create structure, fertility, and tilth (structure suitable for tilling or cultivation) (fig. 11.5).

Soil organisms usually stay close to the surface, but that thin living layer can contain thousands of species and billions of individual organisms per hectare. Algae live on the surface, while bacteria and fungi flourish in the top few centimeters of soil. A single gram of soil (about one-half teaspoon) can contain hundreds of millions of these microscopic cells. Algae and blue-green bacteria capture sunlight and make new organic compounds. Bacteria and fungi decompose organic detritus and recycle nutrients that plants can use for additional growth. The sweet aroma of freshly turned soil is caused by actinomycetes, bacteria that grow in funguslike strands and give us the antibiotics streptomycin and tetracyclines.

Roundworms, segmented worms, mites, and tiny insects swarm by the thousands in that same gram of soil from the surface. Some of them are herbivorous, but many of them prey upon one another. Soil roundworms (nematodes) attack plant rootlets and can cause serious crop damage. A carnivorous fungus snares nematodes with tiny loops of living cells that constrict like a noose when a worm blunders into it. Burrowing animals, such as gophers, moles, insect larvae, and worms, tunnel deeper in the soil, mixing and aerating it. Plant roots also penetrate lower soil levels, drawing up soluble minerals and secreting acids that decompose mineral particles. Fallen plant litter adds new organic material to the soil, returning nutrients to be recycled.

Soil Profiles

Most soils are stratified into horizontal layers called **soil horizons** that reveal much about the history and usefulness of the soil. The thickness, color, texture, and composition of each horizon are used to classify the soil. A cross-sectional view of the horizons in a soil is called a soil profile. Figure 11.6 shows the series of horizons generally seen in a soil profile. Soil scientists give each horizon a letter name. Not all soils have all of these horizons. One or more may be missing, depending on the soil type and history of a specific area.

The soil surface is often covered with a layer of leaf litter, crop residues, or other fresh or partially decomposed organic material. Under this organic layer is the first true soil layer, called **topsoil** (or A horizon), where organic material is mixed with mineral particles. The topsoil horizon ranges from a thickness of 1 meter or more under virgin prairie to zero in some deserts. Topsoil contains most of the living organisms and organic material in the soil, and it is in this layer that most plants spread their roots to absorb water and nutrients. The topsoil horizon blends into a layer that is subject to leaching (removal of soluble nutrients) by water that percolates through it. This **zone of leaching** may have a very different appearance and composition from the layers above or below it.

Beneath the topsoil is the **subsoil,** which usually has a lower organic content and higher concentrations of fine mineral particles. Soluble compounds and clay particles carried by water percolating down from the layers above often accumulate in the subsoil. Subsoil particles can become cemented together to form hardpan, a dense, impermeable layer that blocks plant root growth and prevents water from draining properly.

Beneath the subsoil is the **parent material,** made of relatively undecomposed mineral particles and unweathered rock fragments with very little organic material. Weathering of this layer produces new soil particles for the layers above. About 97 percent of all the parent horizon material in the United States was transported to its

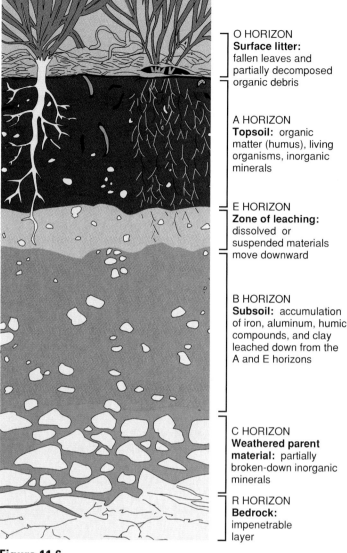

O HORIZON
Surface litter:
fallen leaves and
partially decomposed
organic debris

A HORIZON
Topsoil: organic
matter (humus), living
organisms, inorganic
minerals

E HORIZON
Zone of leaching:
dissolved or
suspended materials
move downward

B HORIZON
Subsoil: accumulation
of iron, aluminum, humic
compounds, and clay
leached down from the
A and E horizons

C HORIZON
**Weathered parent
material:** partially
broken-down inorganic
minerals

R HORIZON
Bedrock:
impenetrable
layer

Figure 11.6
Soil profile showing possible soil horizons. The actual number,
composition, and thickness of these layers varies in different soil types.

present site by geologic forces (glaciers, wind, and water) and is
not directly related to the **bedrock** below it.

Soil Types

Soils are classified according to their structure and composition into
orders, suborders, great groups, subgroups, families, and series. There
are hundreds of thousands of specific types within this taxonomic system! Figure 11.7 shows the major soil types for North and South
America. The richest farming soils are the mollisols (formed under
grasslands) and alfisols (formed under moist, deciduous forests). North
America is fortunate to have extensive areas of these fertile soils.

Ways We Use and Abuse Soil

Only about 10 percent of the earth's land area (14.78 million sq km
out of a total of 144.8 million sq km) is currently in agricultural

production. Perhaps four times as much land could potentially be
converted to cropland, but much of this land serves as a refuge for
cultural or biological diversity or suffers from constraints, such as
steep slopes, shallow soils, poor drainage, tillage problems, low
nutrient levels, metal toxicity, or excess soluble salts or acidity, that
limit the types of crops that can be grown there (fig. 11.8).

Land Resources

Table 11.2 shows the distribution of cropland by region. In parts of
Canada and the United States, temperate climates, abundant water,
and high soil fertility produce high crop yields that contribute to
high standards of living. Much of the former Soviet Union, although rich in land area, suffers from unreliable rainfall or short
growing seasons that often result in disastrous crop failures.

If current population projections are correct, the current world
average of 0.26 ha (0.64 ac) of cropland per person will decline to
0.17 ha (0.42 ac) by the year 2025. In Asia, cropland will be even
more scarce—0.09 ha (0.22 ac) per person—three decades from
now. If you live on a typical quarter-acre suburban lot, look at your
yard and imagine feeding yourself for a year on what you could
produce there.

In the developed countries, 95 percent of recent agricultural
growth in this century has come from improved crop varieties or
increased fertilization, irrigation, and pesticide use, rather than
from bringing new land into production. In fact, less land is being
cultivated now than one hundred years ago in North America, or
six hundred years ago in Europe. As more effective use of labor,
fertilizer, and water and improved seed varieties have increased in
the more-developed countries, productivity per unit of land has increased, and marginal land has been retired, mostly to forests and
grazing lands. In many developing countries, land continues to be
cheaper than other resources, and new land is still being brought
under cultivation, mostly at the expense of forests and grazing
lands. Still, at least two-thirds of recent production gains have
come from new crop varieties and more intense cropping rather
than expansion into new lands.

The largest increases in cropland over the last thirty years occurred in South America and Oceania where forests and grazing
lands are rapidly being converted to farms. Many developing countries are reaching the limit of lands that can be exploited for agriculture without unacceptable social and environmental costs, but
others still have considerable potential for opening new agricultural
lands. East Asia, for instance, already uses about three-quarters of
its potentially arable land. Most of its remaining land has severe restrictions for agricultural use. Further increases in crop production
will probably have to come from higher yields per hectare. Latin
America, by contrast, uses only about one-fifth of its potential land,
and Africa uses only about one-fourth of the land that theoretically
could grow crops. However, there would be serious ecological
trade-offs in putting much of this land into agricultural production.

While land surveys tell us that much more land in the world
could be cultivated, not all of that land necessarily *should* be farmed.
Much of it is more valuable in its natural state. The soils over much
of tropical Asia, Africa, and South America are old, weathered, and
generally infertile. Most of the nutrients are in the standing plants,

Figure 11.7

Soil types of North and South America. The alfisol and the mollisols of grasslands make the best farming.

Source: USDA.

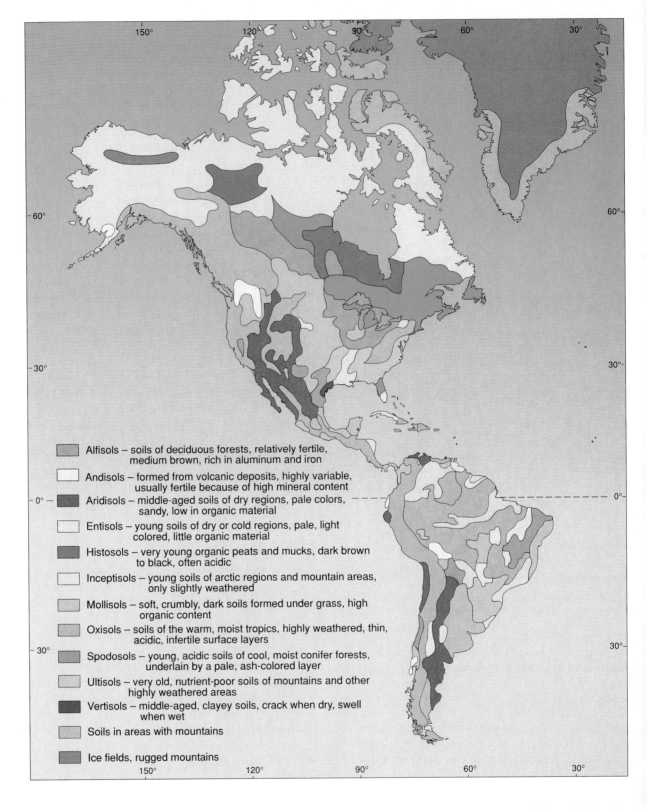

Alfisols – soils of deciduous forests, relatively fertile, medium brown, rich in aluminum and iron

Andisols – formed from volcanic deposits, highly variable, usually fertile because of high mineral content

Aridisols – middle-aged soils of dry regions, pale colors, sandy, low in organic material

Entisols – young soils of dry or cold regions, pale, light colored, little organic material

Histosols – very young organic peats and mucks, dark brown to black, often acidic

Inceptisols – young soils of arctic regions and mountain areas, only slightly weathered

Mollisols – soft, crumbly, dark soils formed under grass, high organic content

Oxisols – soils of the warm, moist tropics, highly weathered, thin, acidic, infertile surface layers

Spodosols – young, acidic soils of cool, moist conifer forests, underlain by a pale, ash-colored layer

Ultisols – very old, nutrient-poor soils of mountains and other highly weathered areas

Vertisols – middle-aged, clayey soils, crack when dry, swell when wet

Soils in areas with mountains

Ice fields, rugged mountains

not in the soil. In many cases, clearing land for agriculture in the tropics has resulted in tragic losses of biodiversity and the valuable ecological services that it provides. Ultimately, much of this land is turned into useless scrub or semidesert.

On the other hand, there are large areas of rich, subtropical grassland and forest that are well watered, have good soil, and could become productive farmland without unduly reducing the world's biological diversity. Argentina, for instance, has pampas grasslands about twice the size of Texas that closely resemble the American Midwest a century ago in climate and potential for agricultural growth. Some of this land could probably be farmed with relatively little ecological damage if it were done carefully.

Figure 11.8

In many areas, soil or climate constraints limit agricultural production. These farm workers in Chad weed a field that is too dry for good crop yields.

Table 11.3

Categories of soil degradation

Light: Part of topsoil removed; some rills or shallow gullies; 70 percent of natural vegetation remains. Soil loss greater than replacement rate.

Moderate: All topsoil removed, or moderately deep gullies; nutrients depleted, or toxic chemical buildup; soil no longer absorbs and retains water. Only 30 to 70 percent of natural vegetation remains.

Severe: Deeper and more frequent gullies; severe nutrient depletion and toxification; crops grow poorly, if at all. Less than 30 percent of natural vegetation remains. Restoration is difficult and expensive.

Extreme: No vegetation remains at all. Restoration is impossible.

Table 11.2

Cropland by region and per capita

Region	Population in Millions	Cropland in 10^6 Hectares	Hectares per Person
World	5629	1439	.26
Africa	708	178	.25
Central and North America	449	278	.62
South America	314	105	.33
Asia	3349	455	.14
Europe	513	137	.27
Former USSR	267	223	.83
Oceania	28	51	1.82

Source: Data from FAO 1994 and UN, 1994.

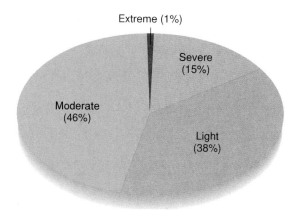

Figure 11.9

Most of the world's soils that have been surveyed show some level of degradation due to human activities. Estimated degrees of degradation worldwide are shown in this pie chart.

Source: *UNEP Environmental Data Report 1993–94.*

Land Degradation

Agriculture both causes and suffers from environmental degradation. The International Soil Reference and Information Centre in the Netherlands estimates that every year 3 million ha (7.4 million ac) of cropland are ruined by erosion, 4 million ha are turned into deserts, and 8 million ha are converted to nonagricultural uses such as homes, highways, shopping centers, factories, reservoirs, etc. Over the past 50 years, some 1.9 *billion* ha of agricultural land (an area greater than that now in production) have been degraded to some extent. About 300 million ha of this land is strongly degraded (table 11.3 and fig. 11.9), while 910 million ha—about the size of China—are moderately degraded. Nearly 9 million ha of former croplands are so degraded that they no longer support any crop growth at all. The causes of this extreme degradation vary: In Ethiopia it is water erosion, in Somalia it is wind, and in Uzbekistan salt and toxic chemicals are responsible.

Definitions of degradation are based on both biological productivity and our expectations about what the land should be like. Often this is a subjective judgment and it is difficult to distinguish between human-caused deterioration and natural processes like drought. We generally consider the land degraded when the soil is impoverished or eroded, water runs off or is contaminated more than is normal, vegetation is diminished, biomass production is decreased, or wildlife diversity diminishes. On farmlands this results in lower crop yields. On ranchlands it means fewer livestock can be supported per unit area. On nature reserves it means fewer species.

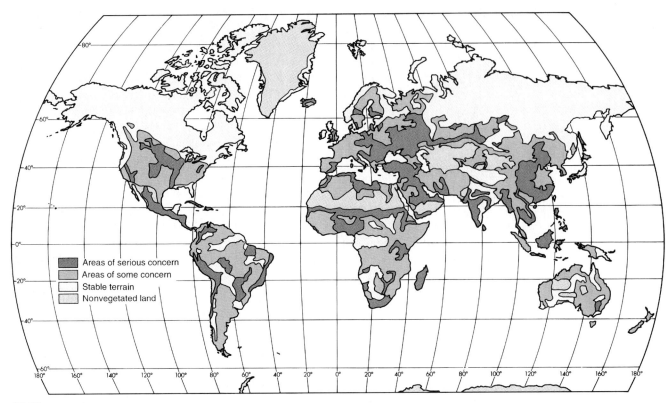

Figure 11.10

Areas of concern for soil degradation. In areas of serious concern, widespread moderate or localized severe degradation has already occurred, and resources are lacking or environmental conditions make rehabilitation difficult or impossible. In areas of some concern, current degradation is lighter and potential for rehabilitation is greater. How is your home area classified?

From Jerome Fellmann, et al., *Human Geography*, 4th ed., from data in *World Resources 1992–1993*, International Source Reference and Information Center. Copyright © 1995 Times Mirror Higher Education Group, Inc., Dubuque, Iowa. All Rights Reserved. Reprinted by permission.

The amount and degree of land degradation varies by region and country. About 20 percent of land in Africa and Asia is degraded, but most is in either the light or moderate category. In Central America and Mexico, by contrast, 25 percent of all vegetated land suffers moderate to extreme degradation. Figure 11.10 shows areas of concern for soil degradation. Figure 11.11 shows causes of soil degradation for all land categories and figure 11.12 shows mechanisms of degradation. Notice that agriculture is responsible for 28 percent of all land degradation, while grazing and forestry account for most of the rest. Water and wind erosion provide the motive force for the vast majority of all soil degradation, worldwide. Chemical deterioration includes nutrient depletion, salinization (salt accumulation), acidification, and pollution. Physical deterioration includes compaction by heavy machinery or trampling by cattle, waterlogging—water accumulation—from excess irrigation and poor drainage, and laterization—solidification of iron and aluminum-rich tropical soil when exposed to sun and rain.

Erosion: The Nature of the Problem

Erosion is an important natural process, resulting in the redistribution of the products of geologic weathering, and is part of both soil formation and soil loss. The world's landscapes have been sculpted by erosion. When the results are spectacular enough, we enshrine them in national parks as we did with the Grand Canyon. Where erosion has worn down mountains and spread soil over the plains, or deposited rich alluvial silt in river bottoms, we gladly farm it. Erosion is a disaster only when it occurs in the wrong place at the wrong time.

In some places, erosion occurs so rapidly that anyone can see it happen. Deep gullies are created where water scours away the soil, leaving fenceposts and trees sitting on tall pedestals as the land erodes away around them. In most places, however, erosion is more subtle. It is a creeping disaster that occurs in small increments. A thin layer of topsoil is washed off fields year after year until eventually nothing is left but poor-quality subsoil that requires more and more fertilizer and water to produce any crop at all.

The net effect, worldwide, of this general, widespread topsoil erosion is a reduction in crop production equivalent to removing about 1 percent of world cropland each year. Many farmers are able to compensate for this loss by applying more fertilizer and by bringing new land into cultivation. Continuation of current erosion rates, however, could reduce agricultural production by 25 percent in Central America and Africa and 20 percent in South America by the year 2000. The total annual soil loss from croplands is thought to be 25 billion metric tons. About twice that much soil is lost from rangelands, forests, and urban construction sites each year.

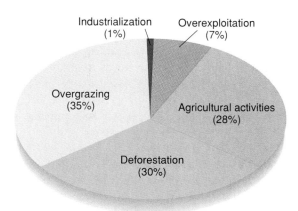

Figure 11.11

Causes of soil degradation worldwide for all land uses.

Source: UNEP 1993.

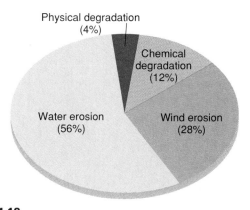

Figure 11.12

Mechanisms of soil degradation worldwide for all land uses.

Source: UNEP 1993.

In addition to reduced land fertility, this erosion results in sediment-loading of rivers and lakes, siltation of reservoirs, smothering of wetlands and coral reefs, and clogging of water intakes and waterpower turbines. It makes rivers unnavigable, increases the destructiveness and frequency of floods, and causes gullying that turns good lands into useless wastelands.

Mechanisms of Erosion

Wind and water are the main agents that move soil around. Thin, uniform layers of soil are peeled off the land surface in a process called **sheet erosion.** When little rivulets of running water gather together and cut small channels in the soil, the process is called **rill erosion** (fig . 11.13). When rills enlarge to form bigger channels or ravines that are too large to be removed by normal tillage operations, we call the process **gully erosion** (fig. 11.14). **Streambank erosion** refers to the washing away of soil from the banks of established streams, creeks, or rivers, often as a result of removing trees and brush along streambanks and by cattle damage to the banks.

Figure 11.13

Sheet and rill erosion caused by water flowing across an unprotected field. Cover crops, crop residue, and terracing are all effective means of reducing this problem.

Figure 11.14

Severe gullying is cutting deep trenches in this pastureland. Lost topsoil reduces productivity here while it causes siltation downstream.

Most soil erosion on agricultural land is sheet and rill erosion. Large amounts of soil can be transported by these mechanisms without being very noticeable. A farm field can lose 20 metric tons of soil per hectare during winter and spring runoff in rills so small that they are erased by the first spring cultivation. That represents a loss of only a few millimeters of soil over the whole surface of the field, hardly apparent to any but the most discerning eye. But it

Soil Resources and Sustainable Agriculture

doesn't take much mathematical skill to see that if you lose soil twice as fast as it is being replaced, eventually it will run out.

As the chapter introduction shows, wind can equal or exceed water in erosive force, especially in a dry climate and on relatively flat land. When plant cover and surface litter are removed from the land by agriculture or grazing, wind lifts loose soil particles and sweeps them away. Windborne dust is sometimes transported from one continent to another. Scientists in Hawaii can tell when spring plowing begins in China because dust from Chinese farmland is carried by winds all the way across the Pacific Ocean. Similarly, summer dust storms in the Sahara Desert of North Africa create a hazy atmosphere over islands in the Caribbean Sea, 5000 km (3000 mi) away. It has been estimated that winds blowing over the Mississippi River basin have one thousand times the soil-carrying capacity of the river itself.

Erosion in the United States and Canada

Figure 11.15 shows wind and water erosion rates by state and province. Note that these are averages over a whole state or province and mask local variations. Wind erosion per acre is highest in the western states and prairie provinces where soil is dry and winds are strong. Wind erosion rates in some areas of southern Saskatchewan are severe. Altogether, about 5 million hectares of cropland in Alberta and Saskatchewan are affected by wind erosion. Losses amount to some $500 to $700 million per year.

Water (sheet and rill) erosion rates in the United States are greatest in the hilly areas of Missouri, Tennessee, Kentucky, and the Palouse Hills of Washington state, where soils are soft, rainfall is abundant, and monoculture farming leaves the soil depleted and exposed to the elements. The highest water erosion rates in Canada

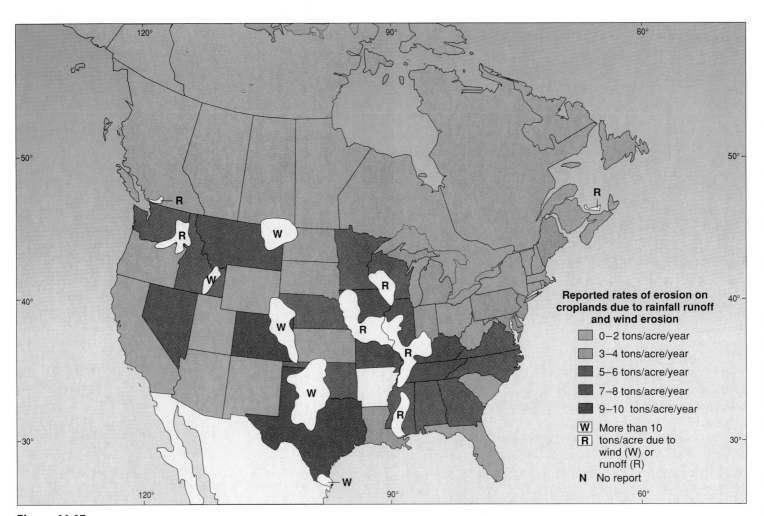

Figure 11.15

Wind and water easily erode soil exposed by agriculture. Although soil replacement rates vary with climate and vegetation, a general rule of thumb is that an erosion rate of 5 tons per acre per year represents soil depletion in humid regions.

Sources: R. Mason and M. Mattson, 1990, *Atlas of United States Environmental Issues,* Macmillan, NY; and *Environment Canada: The State of Canada's Environment, 1990.*

are in the Frasier River Valley, where losses up to 30 tons per hectare are reported under row crops. Potato fields on Prince Edward Island where rainfall is equally high also approach this rate.

The U.S. Department of Agriculture reports that 69 million ha (170 million ac) of U.S. farmland are eroding at rates that are reducing long-term productivity. About 2.5 billion metric tons of soil wash away or are blown from cultivated croplands in this country each year, and 3.3 billion metric tons are lost from forests, pastures, streambanks, and construction sites. Altogether, that's enough soil to fill about 50 million boxcars. Imagine 500,000 trains, each 100 cars long, carrying topsoil away from fields and forests and dumping it into rivers, lakes, and oceans. Soil erosion exceeds tolerable losses on nearly 40 percent of U.S. cropland. A frequently quoted estimate is that two bushels of soil are lost from most Iowa farms for every bushel of corn produced. We are essentially mining the soil to produce crops.

In a century of farming, we have lost, on average, half of the topsoil and perhaps two-thirds of the soil carbon that had been built up under forests and grasslands in the United States. In some places, there is no longer *any* topsoil left. Farmers plant their crops in rocky subsoil. Even with increasing amounts of expensive fertilizer and water, yields are getting smaller and smaller.

The rate of soil loss appears to be getting worse. When Hugh Bennett founded the Soil Conservation Service (now called the Natural Resources Conservation Service) in 1935 during the dust bowl era, the total soil loss each year was about 3 billion tons. It is now nearly twice that much. The U.S. General Accounting Office says that 84 percent of U.S. farms have soil losses greater than 5 tons per acre (11 metric tons per hectare). Five tons per acre is generally considered the soil tolerance level, or the rate at which the ecosystem can replace soil.

Intensive farming practices are largely responsible for this situation. Row crops, such as corn and soybeans, leave soil exposed for much of the growing season. Deep plowing and heavy herbicide applications create weed-free fields that look neat but are subject to erosion. Many farmers now plow in the fall to save a few days at spring planting time, leaving fields bare and exposed to high winds all winter. Because big machines cannot easily follow contours, they often go straight up and down the hills, creating ready-made gullies for water to follow. Farmers sometimes plow through grass-lined watercourses and have pulled out windbreaks and fencerows to accommodate the large machines and to get every last square meter into production. Consequently, wind and water carry away the topsoil.

Pressed by economic conditions, many farmers have abandoned traditional crop rotation patterns and the custom of resting land as pasture or fallow every few years. Continuous monoculture cropping can increase soil loss tenfold over other farming patterns. A soil study in Iowa showed that a three-year rotation of corn, wheat, and clover lost an average of only 2.7 short tons per acre (6 metric tons per hectare). By comparison, continuous wheat production on the same land caused nearly four times as much erosion,

and continuous corn cropping resulted in seven times as much soil loss as the rotation with wheat and clover.

Erosion in Other Countries

Data on soil condition and soil erosion in other countries are less complete and less easily accessible than U.S. data, but it is evident that many places have problems as severe as, or perhaps worse than, ours. China, for example, has a large area of loess (wind-blown silt) deposits on the North China Plain that once was covered by forest and grassland. The forests were cut down and the grasslands were converted to cropland. This plateau is now scarred by gullies 30 to 40 m (100 to 130 ft) deep, and the soil loss is thought to be at least 480 metric tons per hectare per year. This would be equivalent to 3 cm (1.2 in) of topsoil per year.

One way to estimate soil loss is to measure the sediment load carried by rivers draining an area. The highest concentration of sediment in any river is in the Huang (Yellow) River that originates in the loess plateau of China. Although its drainage basin is only one-fifth as big as that of the Mississippi River, the Huang carries more than four times as much soil each year. This suggests that the average soil loss *per hectare* in China may be twenty times that in the United States.

In its middle reaches, the Huang carries about 700 kg of silt per cubic meter of water, or 50 percent by weight—just under the level classified as liquid mud. As the river winds through northern China, much of this sediment settles out, raising the river bottom above the level of the surrounding countryside. Only by building dikes to contain the river have the Chinese been able to keep it in its course. In some places, the riverbed is now 10 m (30 ft) above the farmland through which it flows. If the dikes give way during the summer rice season, millions of people would starve to death as a result of lost crops.

Next after the Huang in annual sediment load is the Ganges River, which carries 1455 million metric tons of mud to the Bay of Bengal every year. Much of this sediment comes from the hill country of northern India, Nepal, and Bangladesh. Population pressures and preemption of good bottomlands for cash crop production have forced farmers to try to grow crops on steep, unstable slopes. Fuelwood shortages also cause local people to cut down the forests that stabilize mountain soils. When the monsoon rains come, they wash whole hillsides away, destroying villages and farmland below.

Perhaps the worst erosion problem in the world, per hectare of farmland, is in Ethiopia. Although Ethiopia has only 1/100 as much cropland in cultivation as the United States, it is thought to lose 2 billion metric tons of soil each year to erosion. This high rate of erosion is both a cause and consequence of famine, poverty, and continued social unrest in that country.

Haiti is another country with severely degraded soil. Once covered with lush tropical forest, the land has been denuded for firewood and cropland. Erosion has been so bad that some experts

Soil Resources and Sustainable Agriculture

now say the country has absolutely *no* topsoil left, and poor peasant farmers have difficulty raising any crops at all. Economist Lester Brown of Worldwatch Institute warns that the country may never recover from this ecodisaster.

Other Agricultural Resources

Soil is only part of the agricultural resource picture. Agriculture is also dependent upon water, nutrients, favorable climates to grow crops, productive crop varieties, and upon the mechanical energy to tend and harvest them.

Water

All plants need water to grow. Agriculture accounts for the largest single share of global water use (fig. 11.16). Some 73 percent of all fresh water withdrawn from rivers, lakes, and groundwater supplies is used for irrigation (chapter 19). This is about six times the total annual flow of the Mississippi River. Although estimates vary widely (as do definitions of irrigated land), about 15 percent of all cropland, worldwide, is irrigated.

Some countries are water rich and can readily afford to irrigate farmland, while other countries are water poor and must use water very carefully. The efficiency of irrigation water use is rather low in most countries. High evaporative and seepage losses from unlined and uncovered canals often mean that as much as 80 percent of water withdrawn for irrigation never reaches its intended destination. Farmers tend to over-irrigate because water prices are relatively low and because they lack the technology to meter water and distribute just the amount needed.

Most farmers assume it is better to overwater than to underwater. This may not be true. Profligate use not only wastes water; it often results in **waterlogging.** Waterlogged soil is saturated with

Figure 11.16

Irrigation allows us to grow crops on otherwise infertile land but causes problems of waterlogging and salinization. Worldwide irrigation accounts for about 70 percent of all human water use. This may become a limiting factor in food production as clean water becomes more scarce.

water, and plant roots die from lack of oxygen. **Salinization,** in which mineral salts accumulate in the soil and kill plants, occurs particularly when soils in dry climates are irrigated with saline water. As the water evaporates, it leaves behind a salty crust on the soil surface that is lethal to most plants. Flushing with excess water can wash away this salt accumulation but the result is even more saline water for downstream users.

Worldwide, irrigation problems are a major source of land degradation and crop losses. The Worldwatch Institute reports that 60 million ha (150 million ac) of cropland have been damaged by salinization and waterlogging. India has the world's largest total area of irrigated land (55 million ha). About one-third of that land is degraded and 7 million ha have been abandoned. China, the second largest irrigator, also has about 7 million ha (17 million ac) of saline and alkaline land. Pakistan and the former Soviet Union have about half as much land degraded by irrigation. The environmental and human health costs of irrigation abuse can be severe.

Water conservation techniques can greatly reduce problems arising from excess water use. Conservation also makes more water available for other uses or for expanded crop production where water is in short supply (see chapter 19).

Fertilizer

In addition to water, sunshine, and carbon dioxide, plants need small amounts of inorganic nutrients for growth. The major elements required by most plants are nitrogen, potassium, phosphorus, calcium, magnesium, and sulfur. Calcium and magnesium often are limited in areas of high rainfall and must be supplied in the form of lime. Lack of nitrogen, potassium, and phosphorus even more often limits plant growth. Adding these elements in fertilizer usually stimulates growth and greatly increases crop yields. A good deal of the doubling in worldwide crop production since 1950 has come from increased inorganic fertilizer use. In 1950, the average amount of fertilizer used was 20 kg per ha. In 1990, this had increased to an average of 91 kg per ha worldwide. This change represents an increase in total fertilizer use from 30 million metric tons in 1950 to 134 million metric tons in 1990.

Farmers may overfertilize because they are unaware of the specific nutrient content of their soils or the needs of their crops. Notice in figure 11.17 that while European farmers use more than twice as much fertilizer per hectare as do North American farmers, their yields are not proportionally higher. Overfertilization wastes money and degrades the environment. Phosphates and nitrates from farm fields and cattle feedlots are a major cause of aquatic ecosystem pollution. Nitrate levels in groundwater have risen to dangerous levels in many areas where intensive farming is practiced. England, France, Denmark, Germany, the Netherlands, and the United States have reported nitrate concentrations above the safe level of 11.3 mg per liter in the drinking water in farming areas. Young children are especially sensitive to the presence of nitrates. Using nitrate-contaminated water to mix infant formula can be fatal for newborns.

What are some alternative ways to fertilize crops? Manure, crop residues, ashes, composted refuse, and green manure (crops grown specifically to add nutrients to the soil) are important nat-

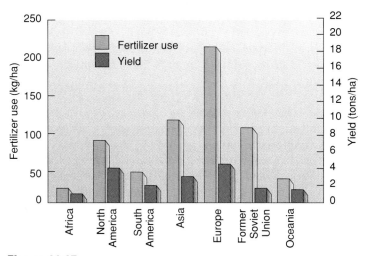

Figure 11.17

Fertilizer use and yield by region. Notice that high fertilizer applications in Europe increase yield, but not in a linear ratio like the yields in America. The excess fertilizer used in Europe causes serious water pollution problems as well as raises the cost of food.

Source: Data from *World Resources*, 1994–1995.

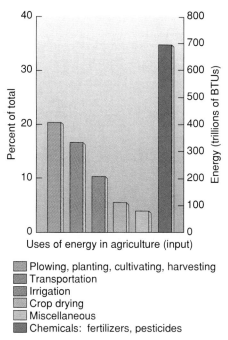

Plowing, planting, cultivating, harvesting
Transportation
Irrigation
Crop drying
Miscellaneous
Chemicals: fertilizers, pesticides

Figure 11.18

Energy use in agriculture. Energy is used to run farm machinery, irrigation pumps, and grain driers, but the largest energy input is in producing and applying synthetic fertilizers, pesticides, and other agricultural chemicals.

Source: Data from L. Brown, *State of the World*, 1987 Worldwatch Institute, Washington, DC.

ural sources of soil nutrients. Nitrogen-fixing bacteria living symbiotically in root nodules of legumes (a broad group of plants including peas, beans, vetch, alfalfa, and leucaena trees) are valuable for making nitrogen available as a plant nutrient (chapter 3). Interplanting or rotating beans or some other leguminous crop with such crops as corn and wheat are traditional ways of increasing nitrogen availability. In some cases, growing and then plowing under a leguminous green manure crop, such as alfalfa, clover, or vetch, every third or fourth year is necessary to maintain soil fertility.

There is considerable potential for increasing world food supply by increasing fertilizer use in low-production countries. Africa, for instance, uses an average of only 19 kg of fertilizer per ha (17 lb per ac), or about one-fourth of the world average. At the same time, African farmers obtain only about one-fourth the yield per hectare as do farmers in Europe and North America. India also has a relatively low average fertilizer level (30 kg per ha). It has been estimated that the developing world could at least triple its crop production by raising fertilizer use to the world average. Other increases could be achieved by using currently idle land, by introducing new high-yield crop varieties, and by investing in irrigation where water is available. All these steps, however, will provide only illusory relief unless careful thought is given to how agriculture is to be made stable and renewable, how food is to be distributed to those who need it, and how population growth can be brought into line with natural resource limitations.

Climate

Climate is a critically important ingredient in agriculture. Global climate change caused by the greenhouse effect (chapter 17) could have severe impacts on world food production. Crop yields could increase in some areas. For instance, Canada and the former Soviet

Union would benefit from warmer temperatures and longer growing seasons. Virtually all crops grow better with higher atmospheric carbon dioxide levels. Soybeans, for instance, may have up to 30 percent higher yields if carbon dioxide doubles. The most serious impact of climate warming would likely be decreasing rainfall at high latitudes and in midcontinent regions. More frequent and more severe droughts in the American Midwest, central Soviet Union, and China would surely have devastating effects on world food supplies. Some people fear terrible famines might ensue. The costs in social, ecological, and economic terms could be catastrophic.

Energy

Farming as it is generally practiced in the industrialized countries is highly energy-intensive. Fossil fuels supply almost all of this energy. The largest energy consumption is for liquid fuels for farm machinery used in planting, cultivating, harvesting, and transporting crops to market (fig. 11.18). The second largest energy cost is the energy contained in chemical stocks used to synthesize fertilizers, pesticides, and other agricultural chemicals. In the western United States, water pumping for irrigation is a major energy consumer, representing a *triple* mining of resources. We use diminishing fossil fuel supplies to pump scarce groundwater to grow crops that deplete the soil!

Total agricultural energy use in the United States, with a population of about 250 million, is now about 11.5 Quads (12 × 10^9GJ). By contrast, the 2.9 billion people in the less-developed

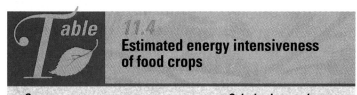

Table 11.4

Estimated energy intensiveness of food crops

Crop	Calories Invested per Calorie of Food Produced
Distant fishing	12
Feedlot beef	10
Fish protein concentrate	6
Grass-fed beef	4
Coastal fishing	2
Intensive poultry production	2
Milk from grass-fed cows	1
Range-fed beef	0.5
Intensive corn or wheat	0.5
Intensive rice	0.25
Hunting and gathering	0.1
Traditional wet-rice culture	0.05
Shifting agriculture	0.02

Source: U.N. Food and Agriculture Organization, 1987.
Note: Any crop with a ratio above 1 consumes more energy, in the form of tractor fuel, fertilizer, drying, processing, and shipping, than it yields. Crops with a ratio below 1 yield more energy than is put in from human-controlled sources.

countries of the world are supported by agricultural systems that use only about 16 Quads (17×10^9GJ). This means that U.S. agriculture uses about twelve times as much energy per person as do these poorer countries.

After crops leave the farm, additional energy is used in food processing, distribution, storage, and cooking. It has been estimated that the average food item in the American diet travels 2000 km (1250 mi) between the farm that grew it and the person who consumes it. The energy required for this complex processing and distribution system may be five times as much as is used directly in farming. Altogether the food system in the United States consumes about 16 percent of the total energy we use. Most of our foods require more energy to produce, process, and get to market than they yield when we eat them (table 11.4).

Clearly, unless we find some new sources of energy, our present system is unsustainable. As fossil fuels become more scarce, we may need to adopt farming methods that are self-supporting. Is it possible that we may need to go back to using draft animals that can eat crops grown on the farm? Could we reintroduce natural methods of pest control, fertilization, crop drying, and irrigation? Or can we develop alternative energy sources, like solar energy, to run our equipment and machinery?

Crop Diversity

As chapter 10 points out, much of the increase in world food production in recent years has come from new high-yield crop varieties. There is a worry, however, about whether crop breeders can continue to produce new cultivars. Throughout the world, traditional, locally-adapted crop types have been disappearing, pushed out by newer, more profitable, "miracle" strains. Furthermore, habitat destruction in centers of crop diversity (see fig. 10.13) is rapidly eliminating wild ancestral species that originally gave rise to our domestic crops. Preserving genetic diversity becomes especially important as diseases and weed species proliferate and become insensitive to chemical and cultural controls. In many cases, the genes that we need to confer desired traits to crops have been lost as traditional varieties and their wild precursors have become extinct.

One way to increase genetic diversity is through molecular engineering. Biotechnologists have made remarkable progress in discovering techniques for site-specific mutations and gene transfer from one species to another. It is now possible to move growth genes from a rat to a fish, for example, or to put light-producing genes from fireflies in a tobacco plant. We could create all kinds of synthetic chimeras, but there are grave ethical concerns about how far we should carry this new science (Ethical Considerations, p. 243).

Toward a Just and Sustainable Agriculture

How, then, shall we feed the world? Can we make agriculture compatible with sustainable ecological and social systems? Having discussed some of the problems that beset modern agriculture, we will now consider some suggested ways to overcome problems and make farming and food production just and lasting enterprises. This goal is usually termed **sustainable agriculture,** or **regenerative farming,** both of which aim to produce food and fiber on a sustainable basis and repair the damage caused by destructive practices. Some alternative methods are developed through scientific research; others are discovered in traditional cultures and practices nearly forgotten in our mechanization and industrialization of agriculture.

Soil Conservation

With careful husbandry, soil is a renewable resource that can be replenished and renewed indefinitely. Since agriculture is the area in which soil is most essential and also most often lost through erosion, agriculture offers the greatest potential for soil conservation and rebuilding. Some rice paddies in Southeast Asia, for instance, have been farmed continuously for a thousand years without any apparent loss of fertility. The rice-growing cultures that depend on these fields have developed management practices that return organic material to the paddy and carefully nurture the soil's ability to sustain life.

What can be done to put other agricultural systems on an equally permanent basis? Among the most important considerations in a soil conservation program are topography, ground cover, climate, soil type, and tillage system.

Managing Topography

Water runs downhill. The faster it runs, the more soil it carries off the fields. Comparisons of erosion rates in Africa have shown that a 5 percent slope in a plowed field has three times the water runoff volume and eight times the soil erosion rate of a comparable field with a 1 percent slope. Water runoff can be reduced by leaving grass strips in waterways and by **contour plowing,** that is, plowing across the hill rather than up and down. Contour plowing is often combined with **strip-farming,** the planting of different kinds of crops in alternating strips along the land contours (fig. 11.19). When one crop is harvested, the other is still present to protect the soil and keep water from running straight downhill. The ridges created by cultivation make little dams that trap water and allow it to seep into the soil rather than running off. In areas where rainfall is very heavy,

Figure 11.19
Contour plowing and strip cropping on these Wisconsin farms protect the soil from erosion and help maintain fertility, as well as providing a beautiful landscape. With care and good stewardship, we can increase the carrying capacity of the land and create a sustainable environment.

Figure 11.20
Rice terraces on Java, Indonesia. Some rice paddies have been cultivated for hundreds or even thousands of years without any apparent loss of productivity.

tied ridges are often useful. This method involves a series of ridges running at right angles to each other, so that water runoff is blocked in all directions and is encouraged to soak into the soil.

Terracing involves shaping the land to create level shelves of earth to hold water and soil (fig. 11.20). The edges of the terrace are planted with soil-anchoring plant species. This is an expensive procedure, requiring either much hand labor or expensive machinery, but makes it possible to farm very steep hillsides. The rice terraces in the Chico River Valley in the Philippines rise as much as 300 m (1000 ft) above the valley floor. They are considered one of the wonders of the world.

Planting **perennial species** (plants that grow for more than two years) is the only suitable use for some lands and some soil types. Establishing forest, grassland, or crops such as tea, coffee, or other crops that do not have to be cultivated every year may be necessary to protect certain unstable soils on sloping sites or watercourses (low areas where water runs off after a rain).

Providing Ground Cover

Annual row crops such as corn or beans generally cause the highest erosion rates because they leave soil bare for much of the year (table 11.5). Often, the easiest way to provide cover that protects soil from erosion is to leave crop residues on the land after harvest. They not only cover the surface to break the erosive effects of wind and water, but they also reduce evaporation and soil temperature in hot climates and protect ground organisms that help aerate and rebuild soil. In some experiments, 1 ton of crop residue per ac (0.4 ha) increased water infiltration 99 percent, reduced runoff 99 percent, and reduced erosion 98 percent. Leaving crop residues on the field also can increase disease and pest problems, however, and may require increased use of pesticides and herbicides.

Where crop residues are not adequate to protect the soil or are inappropriate for subsequent crops or farming methods, such **cover crops** as rye, alfalfa, or clover can be planted immediately after harvest to hold and protect the soil. These cover crops can be plowed under at planting time to provide green manure. Another method is to flatten cover crops with a roller and drill seeds through

Table 11.5 **Soil cover and soil erosion**

Cropping System	Average Annual Soil Loss (tons/hectare)	Percent Rainfall Runoff
Bare soil (no crop)	41.0	30
Continuous corn	19.7	29
Continuous wheat	10.1	23
Rotation: corn, wheat, clover	2.7	14
Continuous bluegrass	0.3	12

Source: Based on 14 years data from Missouri Experiment Station, Columbia, MO.

*I*n Depth: Bioengineering: Pandora's Box or Cornucopia?

In 1993, a new tomato variety appeared in some midwestern grocery stores. Although it didn't look much different than other tomatoes, its origins make it either an ominous sign of things to come or a fantastic breakthrough in agricultural technology, depending on your outlook. The Flavr Savr tomato, created and grown by Calgene, a genetic engineering firm in California, is the first food with laboratory-altered DNA to reach the commercial market. Genetic engineers and food producers, who spend more than $500 million every year working to develop plants and animals with designer genes, are confident that the Flavr Savr is just the beginning of a bioengineering food industry that will be worth billions of dollars in a few years. Already more than fifty genetically improved crops are being tested for commercial production, and more than three hundred field tests of foods, fibers, oil seeds, trees, and flowers have been carried out in the United States alone (fig.1). Soon you could be buying genetically designed apples, broccoli, lettuce, mel-

Figure 1

In 1987, during the first field trial of a genetically-engineered organism, a technician sprays ice-minus bacteria on strawberry plants to prevent frost damage.

ons, strawberries, and wheat flour. You could be wearing cotton colored by exotic genes and stirring sugar with no calories into your coffee.

The Flavr Savr tomato differs from other tomatoes because the gene that prompts it to soften and rot as it ripens has been taken out, turned around, and inserted backward into the plant cells. With this reversed gene, the fruit fails to produce an enzyme that normally breaks down the tissues of the ripening tomato. Fruit can stay on the vine a few days longer than most commercial tomatoes without becoming too soft to handle, and because it ripens on the vine, the new tomato can be as sweet and flavorful as a home-grown summer tomato.

Transgenic plants and animals—those with an altered gene or a gene borrowed from an entirely different species—have tremendous potential for improving agricultural production. We could have frost-resistant strawberries and cold-tolerant peaches that are hardier than current varieties. Healthier canola oil, high-yield beans, and lysine-rich corn and

the residue to provide a continuous protective cover during early stages of crop growth.

In some cases, interplanting of two different crops in the same field not only protects the soil but also is more efficient use of the land, providing double harvests. Native Americans and pioneer farmers, for instance, planted beans or pumpkins between the corn rows. The beans provided nitrogen needed by the corn, pumpkins crowded out weeds, and both crops provided foods that nutritionally balance corn. Traditional swidden (slash-and-burn) cultivators in Africa and South America often plant as many as twenty different crops together in small plots. The crops mature at different times so that there is always something to eat, and the soil is never exposed to erosion for very long.

Mulch is a general term for a protective ground cover that can include manure, wood chips, straw, seaweed, leaves, and other natural products. For some high-value crops, such as tomatoes, pineapples, and cucumbers, it is cost-effective to cover the ground with heavy paper or plastic sheets to protect the soil, save water, and prevent weed growth. Israel uses millions of square meters of plastic mulch to grow crops in the Negev Desert.

Using Reduced Tillage Systems

Farmers have traditionally used a moldboard plow to till the soil, digging a deep trench and turning the topsoil upside down. In the 1800s, it was shown that tilling a field fully—until it was "clean"—increased crop production. It helped control weeds and pests, reducing competition; it brought fresh nutrients to the surface, providing a good seedbed; and it improved surface drainage and aerated the soil. This is still true for many crops and many soil types, but it is not always the best way to grow crops. We are finding that less plowing and cultivation often makes for better water management, preserves soil, saves energy, and increases crop yields.

There are three major **reduced tillage systems.** *Minimum till* involves reducing the number of times a farmer disturbs the soil by

rice could improve human health substantially. Insect-resistant tobacco, blight-resistant potatoes, and herbicide-tolerant cotton would be easier to grow than conventional varieties. Salt-tolerant genes transferred from desert plants to wheat and soybeans could allow us to grow crops with less water on otherwise unsuitable land. Microbes have been developed that consume and clean up toxic wastes. Cows could produce more milk, pigs could have leaner meat, and fish could be bigger. Costly drugs, including insulin and anticlotting agents, can be cheaply produced by bacteria or in the milk of livestock.

But there are many concerns about bioengineering and transgenic species. A big concern for ecologists is whether we will introduce pest and disease resistance to weeds. Genes for robust growth or herbicide resistance may jump from domesticated varieties to their wild relatives or to unrelated weedy species with unwanted results. What such chimeras might do to ecological systems is hard to predict, but we have enough experience with weedy exotic species to know that effects can be disastrous.

A fear shared by many environmentalists is that while biotechnologists talk of developing pest-resistant crops, they are really putting their money into developing chemical-tolerant varieties. This could allow farmers to drench fields with heavy doses of herbicides that will kill everything but the tolerant crop species. Significantly, most genetic engineering firms operating today are owned or funded by corporations such as Monsanto, Upjohn, Dow, and others that produce agricultural chemicals.

Social planners worry that biotechnology could be a deeply destabilizing economic force. Farmers in both developed and developing countries could find themselves even more indebted to chemical and seed manufacturers. Genetically engineered bovine growth hormone (BGH) has been shown to boost milk production by about 10 percent. Only large farms will be able to afford this technology, however, and many marginal farmers could be driven out of business by rising costs and falling prices. The European Economic Community recently voted to ban BGH for six years specifically because of its potential economic effects. What do you think? Are higher prices for consumers justified to protect small farms or should we practice a kind of social Darwinism by letting the fittest survive?

Consumers are apprehensive about eating genetically altered food. Is it safe? Is it wholesome? In 1992, a well-publicized protest against genetic engineering resulted in a pledge by more than one thousand restaurant chefs and by Campbell Soup not to use genetically engineered tomatoes because of fears about unknown effects. Most studies of genetically engineered foods indicate, however, that they are as safe to eat as any other food. Geneticists argue that we have been moving genes around and modifying them by conventional breeding for thousands of years. Wheat is a combination of genes from three different species created by some neolithic experimenter about 10,000 years ago. New procedures simply make it easier and faster to move genes around and to modify them for new uses.

ETHICAL CONSIDERATIONS

How would you evaluate the ethics of this situation? Even if this technology is safe, are we justified in manipulating the genes of other organisms? Is there a significant difference between doing so by molecular biology or conventional breeding? Is engineering life different from other kinds of engineering? Would you eat transgenic food?

---

plowing, cultivating, etc. This often involves a disc or chisel plow rather than a traditional moldboard plow. A chisel plow is a curved chisel-like blade that gouges a trench in the soil in which seeds can be planted. It leaves up to 75 percent of plant debris on the surface between the rows, preventing erosion (fig. 11.21). *Conserv-till* farming uses a coulter, a sharp disc like a pizza cutter, which slices through the soil, opening up a furrow or slot just wide enough to insert seeds. This disturbs the soil very little and leaves almost all plant debris on the surface. *No-till* planting is accomplished by drilling seeds into the ground directly through mulch and ground cover. This allows a cover crop to be interseeded with a subsequent crop.

Farmers who use these conservation tillage techniques often must depend on pesticides (insecticides, fungicides, and herbicides) to control insects and weeds. Increased use of toxic agricultural chemicals is a matter of great concern. Massive use of pesticides is not, however, a necessary corollary of soil conservation. It is possible to combat pests and diseases with integrated pest management that combines crop rotation, trap crops, natural repellents, and biological controls (chapter 12).

Low-Input Sustainable Agriculture

Throughout the United States, Canada, and other crop-exporting countries such as Brazil and New Zealand, huge, highly mechanized farms are becoming increasingly common. Giant grain fields stretch over thousands of acres, while large numbers of cows, hogs, or chickens live their entire life in enormous, warehouse-size confinement barns. Often owned by multinational agricultural conglomerates, these corporate farms are operated by scientific standards that might make your local hospital seem backward. Milk cows, for instance, are tracked by computer, each one automatically given an individual diet, antibiotics, and hormone injections according to their pedigree, growth rate, or milk production. Farm managers travel by helicopter to survey vast spreads. One

Soil Resources and Sustainable Agriculture

Figure 11.21

In ridge-tilling, a chisel plow is used to create ridges on which crops are planted and shallow troughs filled with crop residue. Less energy is used in plowing and cultivation, weeds are suppressed and moisture is retained by the ground cover, or crop residue, left on the field.

Figure 11.22

On the Minar family's 230-acre dairy farm near New Prague, MN, cows and calves spend the winter outdoors in the snow, bedding down on hay. Dave Minar is part of a growing counterculture that is seeking to keep farmers on the land and bring prosperity to rural areas.

"farm" in California covers 2000 ha (almost 50,000 ac) and has an annual cash-flow of around $50 million.

In contrast to this trend towards industrialized agriculture, other farmers are going back to a size and style that their grandparents might have used a century ago. Finding that they can't—or don't want to—compete in capital-intensive production, these folks are making money and staying in farming by returning to small-scale, low-input farming. Rather than milking 1000 cows in a highly mechanized operation, some farmers now raise only a couple of dozen cows and let them graze freely in pastures. Animals live outdoors in the winter (fig. 11.22). Milk production follows natural yearly cycles rather than being controlled by synthetic hormones. The yield is lower, but so are costs.

Where 20,000 chickens living in a single immense shed—or 1000 hogs raised in an industrial warehouse—must constantly be dosed with antibiotics to keep them healthy, animals allowed free range tend to have fewer diseases and require less care. One hog farmer in Iowa whose 40 sows graze in the pasture along with his 80 dairy cows, spends 30 percent less per animal for feed, 70 percent less for veterinary bills, and half as much for buildings and equipment as his bigger neighbors.

Although there isn't complete agreement on what sustainable actually means (What Do *You* Think? next page), low-input farming generally involves more crop rotation and less pesticides (or none), as well as less inorganic fertilizer, water, heavy machinery, and fossil fuel energy than the high-intensity, monoculture operation practiced by conventional farms. Total revenue from this type of agriculture often is less than that of conventional operations, but profits may actually be higher. A study in North Dakota, for in-

stance, found that conventional farms averaged $20 more per acre in total yield but low-input farming had a net income that was $2 per acre *higher,* because total costs were less.

These alternative farming methods are good for our environment. By using fewer chemicals and less manipulation of the soil, erosion and pollution are reduced and the long-range health of the land is improved. Although this movement is too new to yet have good human health statistics, it seems likely that low-input farming will be good for the farmers too. And it is good for taxpayers, as well. A study in Minnesota found that conventional farmers received an average payment of $9214 in government subsidies in 1990, while sustainable farmers received an average of $3597 that same year.

Preserving small-scale, family farms also helps preserve rural culture. As Marty Strange of the Center for Rural Affairs in Nebraska asks, "Which is better for the enrollment in rural schools, the membership of rural churches, and the fellowship of rural communities—two farms milking 1000 cows each or 20 farms milking 100 cows each?" Family farms help keep rural towns alive by purchasing machinery at the local implement dealer, gasoline at the neighborhood filling station, and groceries at the mom-and-pop grocery store.

Public Policies to Encourage Fair and Ecologically Sound Farming Systems

There is a great need for agricultural research focused on sustainable systems, rather than exclusively on higher production. Some authors suggest that we need to be concerned more about ecology

WhatDo *You* Think?
Sustainability: What Does It Mean?

Agriculture is one of many enterprises that has become associated with the term sustainability. In addition to *sustainable agriculture,* we encounter *sustainable forestry, sustainable economy, sustainable yield, sustainable rangeland management,* and many others, all aspects of *sustainable development.*

Surprisingly, unlike other environmental issues, sustainability is widely embraced by people of otherwise starkly differing views. Why such harmony? Apparently because the word means different things to different people. Precise language is important for effective thinking. Fuzzy words produce imprecise thinking, leading to fuzzy communication. It is not surprising that clarification of terms is one step

in critical thinking. Contrast the quotes in the inset box. Do these writers share the same definition of *sustainable?*

Since each of these people has his own view of what the word means, it is easy to understand why both would readily say, "Why, of course, I'm for sustainable agriculture." But would either of them be so supportive under the other person's definition?

Sustainable agriculture is a central element of sustainable development. But disagreements over how sustainability is to be measured are traceable to differing definitions of "sustainable."

The language problem rests partly on the fact that agriculture operates in three overlapping environments. The ecological environment is the setting within which food production occurs. Sustainability here refers to the maintenance and regeneration of soil, water, and essential biotic resources, which in turn, depend on avoiding serious disruption of normal ecological processes. Soil erosion and nutrient loss, acidic precipitation, salinization, and desertification as well as changes in climate threaten agricultural sustainability in the ecological sense.

Secondly, the purpose of agriculture, of course, is to provide for human needs,

part of a social environment. Population increases, changing standards of living, and cultural attitudes create both demands and limits to which agriculture responds. Sustainability in this domain requires agriculture to provide a safe and secure food supply, in enough quantity to avoid hunger, and in enough variety to satisfy consumer demand.

Finally, agriculture also operates in an economic environment. Farmers must earn enough from the sale of their crops to pay the cost of production and to provide themselves with enough profit to farming feasible. Agriculture is not sustainable if it becomes unprofitable. Ecological and economic sustainability issues coincide if short-term practices cause degeneration of the land's productivity, jeopardizing future economic success. The issue is further complicated because sustainability can be viewed on different scales. Would sustainability issues at the individual farm level be the same as at the regional level? National level? Global level?

One's definition of sustainable agriculture is strongly influenced by the environment of closest association. The same practice that *enhances* sustainable agriculture in one person's mind because of its economic effects may actually *threaten* sustainable agriculture because of its ecological effects. Effective problem-solving requires precision in language. When people have different meanings for the same terms, they fail to speak the same language, producing communication analogous to the blind speaking to the deaf. Achieving sustainable agriculture will be difficult enough without being handicapped by language.

WHAT DOES SUSTAINABLE MEAN?

As of the early 1990s, there are no sustainable societies by any definition. All societies are still heavily under the influence of obsolete industrial-era thinking.

Mike Marien

Implicit in calls for 'sustainable development' is the contention that human society is currently unsustainable—if it were not, a new 'sustainable development' policy would hardly be necessary. Yet it is far from clear that human civilization is somehow unsustainable on its present course.

Jerry Taylor

and less about chemistry if we want to establish farming systems that will last for many generations. We need to explore new crops that may have useful nutritional attributes or pest resistance. We especially need to investigate tropical species that are suitable for low-input farming in Africa, southern Asia, and South America. Already, some progress has been reported in selecting and breeding such crops as yams, sorghum, millet, and rye that grow in particularly stressful environments or fit customary diets.

Some people believe the best way to stabilize both food production and population is to establish social, political, and eco-

nomic systems based on a just distribution of resources. They encourage land reform that would enable those who work the land to reap the benefits from their labor, providing incentives for farmers to increase productivity while protecting the land. Land reform and more personal freedom for rural people have boosted farm output in free-market and centrally-planned economies.

Paying a fair price for farm products and farm labor is another way to stabilize farming and stimulate food production. This not only is a more humane treatment of rural people, but it also gives them the capital necessary to improve their land and invest in better machinery, new crop varieties, conservation practices, and other ways to increase productivity and protect and sustain resources.

In much of the Third World, agricultural issues are inextricably associated with women's rights. Women grow most of the family food in many cultures and also do much of the field labor in cash crop production, yet they tend to lack economic and political equality with men. The health effects of pesticide use, fuelwood and water shortages, diversion of land and resources toward export crops, wages, and the costs of agricultural inputs are primarily women's issues in many places. Where women are denied access to credit, tools, education, respect, and power to participate in land-use decisions, families often go hungry.

Nutritional, economic, sociological, political, and population issues are interwoven. One way to increase nutritional levels for rural poor is to provide more rural jobs so people can have money to buy a variety of nutritional foods. The availability of jobs also seems to decrease birth rates. Women who have opportunities for paid employment outside the home are inclined to have fewer children. Parents with higher incomes and more financial security don't need to depend as much on having many children to take care of them in old age. Furthermore, a better educated population is more likely to be able to plan reproduction and avoid unwanted pregnancies.

According to the FAO, one of the most cost-effective ways to improve the nutritional status of the world is to provide clean water and better medical care for everyone. People who are sick from one of the many waterborne infectious diseases that inflict residents of the poorest countries cannot absorb the nutrients in the food they do have (see chapter 9). Furthermore, they don't have the energy to grow more food or to work to buy what they need.

For the foods that you need to buy, consider shopping at a farmer's market. The produce is fresh, and profits go directly to the person who grows the crop. A local food co-op or owner-operated grocery store also is likely to buy from local farmers and to feature pesticide-free foods. Many co-ops and buyers' associations sign contracts directly with producers to grow the types of food they want to eat. This benefits both parties. Producers are guaranteed a local market for organic food or specialty items. Consumers can be assured of quality and can even be involved in production.

If you belong to a co-op or grow your own food you may have to accept fruits and vegetables that are not absolutely perfect. You can trim away a few bad spots or wash off an insect or two. If you overlook a worm or a bug, it may be less harmful to you than the toxic chemicals that are used to make supermarket foods sterile and cosmetically perfect!

Summary

Fertile, tillable soil for growing crops is an indispensable resource for our continued existence on earth. Soil is a complex system of inorganic minerals, air, water, dead organic matter, and a myriad of different kinds of living organisms. There are hundreds of thousands of different kinds of soils, each produced by a unique history, climate, topography, bedrock, transported material, and community of living organisms.

We face a growing scarcity of good farmland to feed the world's rapidly growing population. Asia and Europe have little cropland per person and few opportunities to open new lands for crops. Africa and South America have extensive unused areas, but much of this land has environmental constraints or harbors cultural and biological diversity that deserves protection.

It is estimated that 25 billion tons of soil are lost from croplands each year because of wind and water erosion. Perhaps twice as much is lost from rangelands and permanent pastures. This erosion causes pollution and siltation of rivers, reservoirs, estuaries, wetlands, and offshore reefs and banks. The net effect of this loss is worldwide crop reduction equivalent to losing 15 million ha (37 million ac), or 1 percent of the world's cropland each year.

The United States has one of the highest total rates of soil erosion in the world. Soil erosion exceeds soil formation on at least 40 percent of U.S. cropland. About one-half of the topsoil that existed in North America before European settlement has been lost.

Other areas with high erosion rates are China, India, the former Soviet Union, and Ethiopia. Worldwide, about 25 percent of the land (about twice as much as we now use) has the potential for agricultural use. Putting much of that land into agricultural production would mean loss of valuable forests and grasslands and would cause loss of biodiversity and result in major ecological destruction. Large expanses of land, however, could be converted to agricultural use or could be used more intensively (but carefully) without causing great damage. It is possible that food production could be expanded considerably, even on existing farmland, given the proper inputs of fertilizer, water, high-yield crops, and technology. This will be essential if human populations continue to grow as they have during this century. Whether it will be possible to supply agricultural inputs and expand crop production remains to be seen. Global climate change could have devastating effects on world food supplies and might necessitate conversion of forests and grasslands to feed the world's population.

There are great differences in fertilizer use between the more-developed and less-developed countries. Many farmers in the developed countries use more fertilizer than is needed. Excess chemicals are a major source of water pollution and threaten health in some areas.

Many new and alternative methods could be used in farming to reduce soil erosion, avoid dangerous chemicals, improve yields, and make agriculture just and sustainable. Some alternative methods are developed through scientific research;

others are discovered in traditional cultures and practices nearly forgotten in our mechanization and industrialization of farming. Some authors advocate returning to low-input, regenerative, "organic" farming that may be more sustainable and more healthful than our current practices. Growing your own food or buying locally grown food at co-ops, farmer's markets, or through a producers' or buyers' association can provide healthy, wholesome food and also support sustainable agriculture.

▼ Questions for Review

1. What is the composition of soil? What is humus? Why are soil organisms so important?

2. Describe the differences between light, moderate, severe, and extreme soil degradation.

3. What are four kinds of erosion? Is erosion ever beneficial? Why is it a problem?

4. What are some possible effects of overirrigation?

5. What can farmers do to increase agricultural production without increasing land use?

6. What is the estimated potential for increasing world food supply by increasing fertilizer use in low-production countries?

7. What is sustainable agriculture?

8. What is a perennial species? Why is coffee a good crop for rugged and hilly country?

9. What is genetic engineering, or biotechnology, and how might they help or hurt agriculture?

▼ Questions for Critical Thinking

1. Should farmers be forced to use ecologically sound techniques that serve farmers' best interests in the long run, regardless of short-term consequences? How could we mitigate hardships brought about by such policies?

2. In a crisis, when small farms are in danger of being lost, should preserving the soil be the farmer's first priority?

3. Should we encourage (and subsidize) the family farm? What are the advantages and disadvantages (economic and ecological) of the small farm and the corporate farm?

4. How many people do you think the world could support? What do you think would be the ideal number?

5. Should we try to increase food production on existing farmland, or should we sacrifice other lands to increase farming areas?

6. Some rice paddies in Southeast Asia have been cultivated continuously for a thousand years or more without losing fertility. Could we, and should we, adapt these techniques to our own country?

7. Should we continue our current rate of pesticide use? What are the advantages and disadvantages of continuing such use?

▼ Key Terms

bedrock   230
contour plowing   241
cover crops   241
gully erosion   235
humus   229
mulch   242
parent material   230
perennial species   241
reduced tillage systems   242
regenerative farming   240
rill erosion   235
salinization   238

sheet erosion   235
soil   228
soil horizons   230
streambank erosion   235
strip-farming   241
subsoil   230
sustainable agriculture   240
terracing   241
tied ridges   241
topsoil   230
waterlogging   238
zone of leaching   230

▼ Additional Information on the Internet

American Farmland Trust
   http://farm.fic.niu.edu/aft/afthome.html/

FAIRS
   http://hammock.ifas.ufl.edu/

Institute for Agriculture and Trade Policy
   http://www.igc.apc.org/iatp/

National Agricultural Library
   http://www.nalusda.gov/

National Integrated Pest Management
   http://ipmwww.ncsu.edu/

National Soil Erosion Research Laboratory
   http://soils.ecn.purdue.edu:20002/

Sustainable Agriculture Research and Education Program
   http://www.sarep.ucdavis.edu/

United Nations Convention to Combat Desertification
   http://www.unep.ch/incd.html/

US Department of Agriculture
   http://www.usda.gov/

Note: Further readings appropriate to this chapter are listed on p. 598.

A helicopter sprays pesticides over a sugar beet field in California.

CHAPTER 12: Pest Control

Objectives

After studying this chapter, you should be able to:

- define the major types of pesticides and describe the pests they are meant to control.

- outline the history of pest control, including the changes in pesticides that have occurred in the last half of this century.

- appreciate the benefits of pest control.

- relate some of the problems of pesticide use.

- explain some alternative methods of pest control.

- discuss pesticide regulation and the acceptability of small amounts of weak carcinogens in our diet.

DDT and Fragile Eggshells

During the 1960s, peregrine falcons (fig. 12.1), bald eagles, osprey, brown pelicans, shrikes, and several other predatory bird species suddenly disappeared from former territories in eastern North America. What caused this sudden decline? Studies revealed that eggs laid by these birds had thin, fragile shells that broke before hatching. Eventually, these reproductive failures were traced to residues of DDT and its degradation product, DDE, which had concentrated through food webs until reaching toxic concentrations in top trophic levels such as these bird species.

DDT (dichloro-diphenyl-trichloroethane), an inexpensive and highly effective insecticide, had been used widely to control

Figure 12.1
Peregrine falcons disappeared from the eastern United States in the 1960s as a result of excess pesticide use.

mosquitoes, biting flies, codling moths, potato beetles, corn earworms, cotton bollworms, and a host of other costly and irritating pest species around the world. First produced commercially in 1943, more than 50 million pounds of DDT were sprayed on fields, forests, and cities in the United States by 1950.

But as we have since discovered, there are disadvantages to widespread release of these toxic compounds in the environment. In the case of falcons, eagles, and other top predators, DDT and DDE inhibit enzymes essential for deposition of calcium carbonate in eggshells, resulting in soft, easily broken eggs. As we will discuss later in this chapter, other compounds chemically related to DDT are now thought to be disrupting endocrine hormone functions and causing reproductive losses in many species other than birds (What Do You Think?, p. 260).

These effects, coupled with the discovery of a pervasive presence of various chlorinated hydrocarbons in human tissues worldwide, led to the banning of DDT in most industrialized countries in the early 1970s. Peregrine falcons, which had declined to only about 120 birds in the U.S. (outside of Alaska) in the mid-1970s, now number about 1400, most of them bred in captivity and then released into the wild. Bald eagle reproduction has increased from an average of 0.46 young per nest in 1974 to 1.21 young per nest in 1994 in eastern Canada. Peregrine falcons and bald eagles had recovered enough to be removed from the endangered species list in the eastern United States in 1994.

What have we learned from this experience? Chemical pesticides offer a quick, convenient, and relatively inexpensive way to eliminate annoying or destructive organisms. At the same time, however, excessive pesticide use can kill beneficial organisms and upset the natural balance between predator and prey species. Modern pesticides undoubtedly have saved millions of human lives by killing disease-causing insects and by increasing food supplies. But we must understand what these powerful chemicals are doing and use them judiciously. In this chapter, we will study the major types of pests and pesticides along with some of the benefits and problems involved in our battle against pests. ▶

What Are Pests and Pesticides?

A pest is something or someone that annoys us, detracts from some resource that we value, or interferes with a pursuit that we enjoy. In this chapter, we will concentrate on **biological pests,** organisms that reduce the availability, quality, or value of resources useful to humans. What's annoying or undesirable depends, of course, on your perspective. The mosquitoes that swarm in clouds over a marsh in the summer may be irritating to us, but they are an essential food source for swallows and bats that feed on them. You may regard dandelions in your yard as tenacious and obnoxious weeds, but in some countries dandelions are cultivated as beautiful, exotic flowers and as a food source. Of the millions of species of organisms only about one hundred plants, animals, fungi, and microbes cause 90 percent of all crop damage worldwide.

Insects tend to be the most frequent pests, in part, because they make up at least three-quarters of all species on the earth. Most pest organisms tend to be generalists, the opportunistic species that reproduce rapidly, migrate quickly into disturbed areas, and that are pioneers in ecological succession. They compete aggressively against more specialized endemic species and can often take over a biotic community, especially where humans have disrupted natural conditions and created an opening into which they can slip. Most Americans are familiar with dandelions, ragweed, English sparrows, starlings, European pigeons, and other "weedy" species that survive well in urban habitats. Chapter 13 describes how exotic aliens brought in by humans are crowding out native species in many places.

Some native animals—even large ones—can become pests (fig. 12.2). In Grand Teton National Park, for example, winter feeding programs and protection from hunters have allowed the elk herd to grow to 25,000 animals, at least ten times as many as the ecosystem can sustain. Elk block traffic, eat people's flowers and shrubs, and crowd out other native species such as prairie dogs, gophers, and the small predators that depend on them.

A **pesticide** is a chemical that kills pests. We generally think of toxic substances in this category, but chemicals that drive away

Figure 12.2

Even large native animals can become pests, as in the case of these elk at Mammoth Hot Springs in Yellowstone National Park, where protection from predators and hunters has allowed the population to grow far past the carrying capacity of the environment.

pests or prevent their development are sometimes included as well. Pest control can also include activities such as killing pests by burning crop residues or draining wetlands to eliminate breeding sites. A broad-spectrum pesticide that kills all living organisms is called a **biocide.** Fumigants, such as ethylene dibromide or dibromochloropropane, used to protect stored grain or sterilize soil fall into this category. Generally, we prefer narrower spectrum agents that attack a specific type of pest: **herbicides** kill plants; **insecticides** kill insects; **fungicides** kill fungi; acaricides kill mites, ticks, and spiders; nematicides kill nematodes (microscopic roundworms); rodenticides kill rodents; and avicides kill birds. Pesticides can also be defined by their method of dispersal (fumigation, for example) or their mode of action, such as an ovicide, which kills the eggs of pests. In a sense, the antibiotics used in medicine to fight infections are pesticides as well.

A Brief History of Pest Control

Humans have probably always used chemicals to protect themselves from pests, but in the past 50 years we have entered a new era of pesticide use. How did we develop these chemical agents and how do current uses differ from previous practices?

Early Pest Controls

Using chemicals to control pests may well have been among our earliest forms of technology. People in every culture have known that salt, smoke, and insect-repelling plants can keep away bothersome organisms and preserve food. The Sumerians controlled insects and mites with sulfur 5000 years ago. Chinese texts 2500 years old describe mercury and arsenic compounds used to control body lice and other pests. Greeks and Romans used oil sprays, ash and sulfur ointments, lime, and other natural materials to protect themselves, their livestock, and their crops from a variety of pests.

In addition to these metals and inorganic chemicals, people have used organic compounds, biological controls, and cultural practices for a long time. Alcohol from fermentation and acids in pickling solutions prevent growth of organisms that would otherwise ruin food. Spices were valued both for their flavors and because they deterred spoilage and pest infestations. Romans burned fields and rotated crops to reduce crop diseases. The Chinese developed plant-derived insecticides and introduced predatory ants in orchards to control caterpillars 1200 years ago. Many farmers still use ducks and geese to catch insects and control weeds (fig. 12.3).

Synthetic Chemical Pesticides

The modern era of chemical pest control began in 1934 with the discovery of the insecticidal properties of DDT (*D*ichloro-*D*iphenyl-*T*richloroethane) by Swiss chemist Paul Müller. DDT was used to control potato beetles in Switzerland in 1939 and commercial production began in 1943. It became extremely important during World War II in areas where tropical diseases and parasites posed greater threats to soldiers than enemy bullets.

Figure 12.3

Geese make good biological control agents. They eat weeds, grass, and insects but leave many crops alone. Their droppings enrich the soil, their down can be used to make garments and pillows, and a goose dinner makes a welcome protein source for many people.

DDT seemed like a wonderful discovery. It is cheap, stable, soluble in oil, and easily spread over a wide area. It is highly toxic to insects but relatively nontoxic to mammals. The oral LD50 for DDT for humans is 113 mg/kg, compared to 50 mg/kg for nicotine and 366 mg/kg for caffeine. Where other control processes act slowly and must be started before a crop is planted, DDT can save a crop even when pests already are well established. Its high toxicity for target organisms makes DDT very effective, often producing 90 percent control with a single application. DDT seemed like the magic bullet for which science had been searching. It was sprayed on crops and houses, dusted on people and livestock, and used to combat insects all over the world (fig. 12.4). Paul Müller received a Nobel prize in 1948 for his discovery.

As you will learn in this chapter, however, indiscriminate and excessive use of synthetic chemical pesticides has caused serious ecological damage and long-term harm to human health, and it has turned relatively innocuous species into serious pests.

Pesticide Uses and Types

Information on total world pesticide use is spotty and uncertain. The United Nations Environment Program reports sales totals rather than quantity because of difficulty in obtaining data from developing countries. In any case, it is clear that pesticide use has increased dramatically since the Second World War. Sales have grown from almost nothing in 1950 to $7.7 billion in 1970 to approximately $25 billion in 1990. About 90 percent of all pesticides worldwide are used in agriculture or food storage and shipping. The wealthier countries of the developed world consume some four-fifths of all pesticides, but rates of use in developing countries are rising 7 to 8 percent per year compared to 2 to 4 percent annual increases in the more-developed countries.

Figure 12.4
Before we realized the toxicity of DDT, it was sprayed freely on people to control insects as shown here at Jones Beach, NY, in 1948.

Pesticide Use in the United States and Canada

The U.S. Environmental Protection Agency reports that 527 million kilograms (1.16 billion pounds) of pesticides were used in the United States in 1990. Germany and Italy—which have the world's next highest consumption after the U.S.—each use about one-fifth as much. Canada uses one-tenth as much total pesticide as does the U.S. but has doubled its consumption over the past decade, whereas U.S. consumption has remained relatively stable. About 20 percent of all agricultural lands in Canada was treated with herbicides in 1970 and only 2 percent was sprayed with insecticides. Fifteen years later herbicide use had more than doubled to over half of all cropland, and insecticide use had grown tenfold, especially in the prairie provinces of Alberta and parts of Saskatchewan.

Nearly three-quarters of all pesticides applied in the United States are used in agriculture. Figure 12.5 shows the major categories of this use. Herbicides account for about 59 percent of the total quantity, insecticides 22 percent, fungicides 11 percent, and all other types together about 8 percent. Row crops such as corn, soybeans, and cotton account for about three-quarters of all herbicides used in the United States. When these crops are grown year after year in vast monoculture fields, they require heavy doses of insecticides as well. Fungicides are used mainly on fruits and vegetables. Household applications represent about 12 percent of all pesticide use in the United States but are almost 23 percent of insecticide use—higher than any other place in the world.

Pesticide Types

One way to classify pesticides is by their chemical structure. This is useful because environmental properties—such as stability, solubility, and mobility—and toxicological characteristics of members of a particular chemical group are often similar.

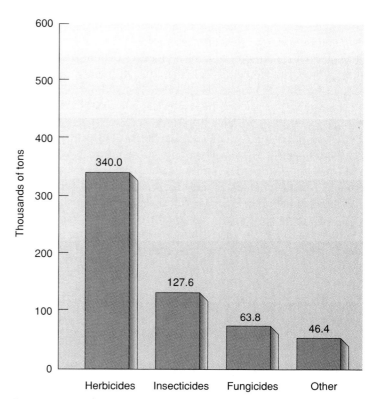

Figure 12.5
Agricultural pesticide use in the United States in 1992.

Source: Data from Environmental Protection Agency, 1994.

Inorganic pesticides include compounds of arsenic, copper, lead, and mercury. These broad-spectrum poisons are generally highly toxic and essentially indestructible, remaining in soil forever. Seeds are sometimes coated with a mercury or arsenic powder to deter insects and rodents during storage or after planting. Handling such seeds with bare hands can be very dangerous for farmers or gardeners. They are generally neurotoxins and even a single dose can cause permanent damage.

Natural organic pesticides, or "botanicals," generally are extracted from plants. Some important examples are nicotine and nicotinoid alkaloids from tobacco; rotenone from the roots of derris and cubé plants; pyrethrum, a complex of chemicals extracted from the daisylike *Chrysanthemum cinerariaefolium* (fig. 12.6); and turpentine, phenols, and other aromatic oils from conifers. All are toxic to insects, but nicotine is also toxic to a broad spectrum of organisms including humans. Rotenone is commonly used to kill fish. Turpentine, phenols, and other natural hydrocarbons are effective pesticides, but synthetic forms such as pentachlorophenol are more stable and more toxic than natural forms. They penetrate surfaces well and are used to prevent wood decay.

Fumigants are generally small molecules such as carbon tetrachloride, carbon disulfide, ethylene dichloride, ethylene dibromide, methylene bromide, and dibromochloropropane that gasify easily and penetrate rapidly into a variety of materials. They are used to sterilize soil and prevent decay or rodent and insect infes-

Figure 12.6

Harvesting pyrethrum-containing flowers in Kenya to extract natural pesticides.

tation of stored grain. Because these compounds are extremely dangerous for workers who apply them, use has been curtailed or banned altogether.

Chlorinated hydrocarbons, or organochlorines such as DDT, chlordane, aldrin, dieldrin, toxaphene, paradichlorobenzene (mothballs), and lindane, are synthetic organic insecticides that inhibit nerve membrane ion transport and block nerve signal transmission. They are fast acting and highly toxic in sensitive organisms. Toxaphene, for instance, kills goldfish at 5 parts per billion (5 μg/liter), one of the highest toxicities for any compound in any organism. Chlorinated hydrocarbons persist in soil up to 15 years, become concentrated through food chains, and are stored in fatty tissues of a variety of organisms. The chloriphenoxy herbicides 2,4 D and 2,4,5 T have hormone-like growth-regulating properties and are selective for broad-leaved flowering plants.

Organophosphates such as parathion, malathion, dichlorvos, dimethyldichlorovinylphosphate (DDVP), and tetraethylpyrophosphate (TEPP), are an outgrowth of nerve gas research during World War II. They inhibit cholinesterase, an enzyme essential for removing excess neurotransmitter from synapses in the peripheral nervous system. They are extremely toxic to mammals, birds, and fish (generally 10 to 100 times more poisonous than most chlorinated hydrocarbons). A single drop of TEPP on your skin can be lethal. Because they are quickly degraded, they are much less persistent in the environment than organochlorines, generally lasting only a few hours or a few days. These compounds are very dan-

gerous for workers such as grape pickers who often are sent into fields too soon after they have been sprayed.

Carbamates, or urethanes, such as carbaryl (Sevin), aldicarb (Temik), aminocarb (Zineb), carbofuran (Baygon), and Mirex share many organophosphate properties, including mode of action, toxicity, and lack of environmental persistence and low bioaccumulation. Carbamates generally are extremely toxic to bees and must be used carefully to prevent damage to these beneficial organisms.

Microbial agents and biological controls are living organisms rather than individual chemicals, but we include them here to complete our survey. Bacteria such as *Bacillus thuringiensis* or *Bacillus popilliae* kill caterpillars or beetles by rupturing the digestive tract lining when eaten. Parasitic wasps such as the tiny *Trichogramma* genus attack moths, while lacewings and ladybugs control aphids. Viral diseases also have been used against specific pests.

Pesticide Benefits

Like all organisms, humans compete with other species for food and shelter and struggle to protect ourselves from diseases and predators. Synthetic chemical pesticides are important weapons in this fight for survival.

Disease Control

Insects and ticks serve as vectors in the transmission of a number of disease-causing pathogens and parasites. Consider malaria, for example. About 500 million people suffer from this disease at any given time, and about 2 million die each year from *Plasmodium* protozoans spread to humans by *Anopheles* mosquitoes (fig. 12.7). It is estimated that insecticidal mosquito control has prevented at least 50 million deaths from malaria over the past 50 years. Sri Lanka is a classic example of pesticide benefits. In the early 1950s, more than 2 million cases of malaria were reported in Sri Lanka each year. After DDT spraying began in 1954, new malaria cases

Figure 12.7

Malaria, spread by the *Anopheles* mosquito, is one of the largest causes of human disease and premature death in the world. By controlling mosquitoes, pesticides save 1 million lives per year in tropical countries.

Figure 12.8

Elephantiasis is caused by parasitic worms (filaria) that block lymph vessels and cause fluid accumulation in various parts of the body.

almost completely disappeared. When DDT spraying was discontinued in 1964, however, malaria reappeared almost immediately. Within three years the annual incidence was more than 1 million cases per year. The Sri Lankan government resumed DDT spraying in 1968 and continues limited use of this insecticide despite environmental concerns. They concluded that the reductions in medical expenses, social disruption, pain and suffering, and lost work resulting from malaria outweighed the direct costs of insecticide spraying by a thousand to one.

Some other diseases spread by biting insects include yellow fever and related viral diseases such as encephalitis, also carried by mosquitoes, and trypanosomiasis or sleeping sickness caused by protozoans transmitted by the tsetse fly. Onchocerciasis (river blindness) and filariasis (one form of which is elephantiasis, fig. 12.8) are caused by tiny worms spread by biting flies that afflict hundreds of millions of people in tropical countries. Bubonic plague (or Black Death) is caused by bacteria transmitted by fleas shared by rats and humans. Typhoid fever, also caused by bacteria, is spread by body lice. All of these terrible diseases can be reduced by judicious use of pesticides. If you were faced with a choice of going blind before age 30 because of masses of worms accumulating in your eyeballs or a small chance of cancer due to pesticide exposures if you live to age 50 or 60, which would you choose?

Crop Protection

Although reliable data on crop losses are difficult to obtain, it is thought that plant diseases, insect and bird predation, and competition by weeds reduce crop yields worldwide by at least one-third. Postharvest losses to rodents, insects, and fungi may be as high as another 20 to 30 percent. Without modern chemical pesticides, these losses might be much higher. Although we said in chapter 10

that there is more than enough food in the world to adequately feed everyone now living—if food were equitably distributed—this most certainly would not be true, given current methods of production, if modern chemical pesticides were unavailable.

A commonly quoted estimate is that farmers save $3 to $5 for every $1 spent on pesticides. This means lower costs and generally better quality for consumers. In some cases, insects and fungal diseases cause only small losses in terms of the total crop quantity, but the cosmetic damage they cause greatly reduces the economic value of crops. For example, although codling moth larvae consume very little of the apples they infest, the brown trails they leave as they crawl through the fruit and the possibility of biting into a living worm often make an unsprayed crop unsalable. Some consumers have come to believe, however, that it may be better to trim a few visible worms or wormholes from their apples than to eat invisible but perhaps more dangerous pesticide residues.

Pesticide Problems

While synthetic chemical pesticides have brought us great economic and social benefits, they also cause a number of serious problems. In this section we will examine some of the worst of those problems.

Effects on Nontarget Species

It is estimated that up to 90 percent of the pesticides we use never reach their intended targets (fig. 12.9). Many beneficial organisms are poisoned unintentionally as a result. For instance, about 20 percent of all honeybee colonies in the United States are destroyed each year and another 15 percent are damaged by pesticide spray drift or residues on the flowers they visit. Direct losses to beekeepers amount to several million dollars per year. Losses to crops the bees would have pollinated may be ten times higher.

In some cases, the effects of poisoning nontarget species are immediate and unmistakable. In one episode in 1972, a single ap-

Figure 12.9

This machine sprays insecticide on orchard trees—and everything else in its path. An estimated 90 percent of all pesticides never reach target organisms and enter the environment instead.

plication of the insecticide Azodrin to combat potato aphids on a farm in Dade County, Florida, killed 10,000 migrating robins in three days. Similarly, a 1991 derailment of a Southern Pacific tanker car on a tricky canyon bridge just north of Dunsmuir, California, dumped 75,000 liters (20,000 gallons) of highly toxic metam sodium herbicide into the Sacramento River. The entire river ecosystem—including aquatic plants, insects, amphibians, and at least 100,000 trout—was completely wiped out for 45 kilometers (27 miles) downstream.

Pesticide Resistance and Pest Resurgence

Pesticides almost never kill 100 percent of a target species even under the most ideal conditions. As we discussed in chapter 4, every population contains some diversity in tolerance to adverse environmental factors. The most resistant members of a population survive pesticide treatment and produce more offspring like themselves with genes that enable them to withstand further chemical treatment. Because most pests propagate rapidly and produce many offspring, the population quickly rebuilds with pesticide-resistant individuals. We call this phenomenon **pest resurgence** or rebound (fig. 12.10). The United Nations Environment Program reports that at least 500 insect pest species and another 250 or so weeds and plant pathogens worldwide have developed chemical resistance (fig. 12.11). Of the twenty-five most serious insect pests in California, three-quarters are now resistant to one or more insecticides. This resistance means that it takes constantly increasing doses to get the same effect or that farmers who are caught on a **pesticide treadmill** must constantly try newer and more toxic chemicals in an attempt to stay ahead of the pests. Despite a 33-fold

increase in pesticide use in the United States since 1944, crop losses caused by pests have actually increased (table 12.1); some people argue that we are losing the battle to eradicate pests.

One of the most ominous developments in this race to find effective pesticides is that we increasingly find pests that are resistant to chemicals to which they have never been exposed. Apparently, genes for pesticide resistance are being transferred from one species to another by means of vectors such as viruses and plasmids (naked pieces of viral-like DNA). Often a whole cluster of genes jump between species so that multiple chemical-tolerance is inherited before a particular pest is ever exposed to any of the chemicals. This is probably a race we can't win. The more different chemicals we try, the more widespread the resistance is likely to be.

In a way, every pesticide has a very limited useful life span before target species become resistant to it or the pesticide builds up intolerable environmental concentrations. We have made a big mistake in broadcasting pesticides recklessly and extravagantly. DDT, for instance, is such a helpful insecticide that we should have used it sparingly and carefully, so that it would still be effective against the worst insect pests. It has been spread so widely that fifty of the sixty malaria-carrying mosquitoes are now resistant to it and the environmental side effects outweigh its benefits. You can think of the useful life of DDT as a nonrenewable resource that we squandered.

Creation of New Pests

Often the worst side effect of broadcast spraying a pesticide is that we kill beneficial predators that previously kept a number of pests

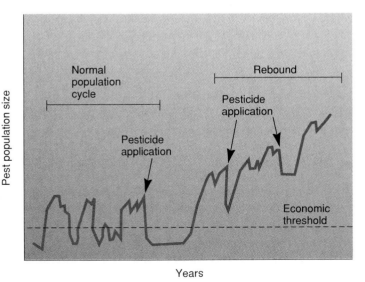

Figure 12.10

Hypothetical pest population response to pesticide application. Under natural conditions, pest numbers fluctuate, sometimes rising above economic damage thresholds and sometimes dropping below. The first use of a synthetic pesticide often is very successful, but pests quickly rebound because they develop pesticide resistance and because natural predators and competitors are removed. It becomes harder and harder to control pests.

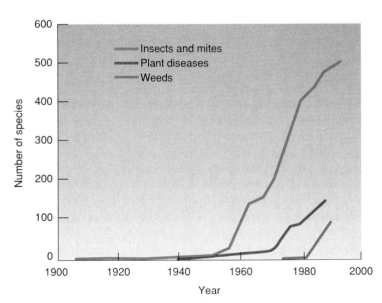

Figure 12.11

Many pests have developed resistance to pesticides. Because insecticides were the first class of pesticide to be used widely, selection pressures led insects to show resistance early. More recently, plant pathogens and weeds are also becoming insensitive to pesticides.

Source: Data from A. Z. Trappe, "The Impact of Agrochemicals on Human Health and Environment," *Industry and Environment*, vol. 8, p. 10, UNEP, Paris, 1985; and G. P. Georghiu, *Pest Resistance to Pesticides*, Plenum Press, New York, 1989.

Year	Insects	Diseases	Weeds
1944	7.1	10.5	13.8
1974	13.0	12.0	8.0
1989	13.0	12.0	12.0

Table 12.1 Percentage of crops lost annually to pests in the United States

Source: D. Pimentel, et al., "Environmental and Economic Effects of Reducing Pesticides," *Bioscience*, 41:6–12, June 1991.

Figure 12.12

The cotton boll weevil caused terrible devastation in southern U.S. cotton fields in the 1930s and 1940s. DDT and other insecticides controlled this pest briefly, but the beetle quickly became resistant.

under control. Many of the agricultural pests that we want to eliminate, for instance, are herbivores such as aphids, grasshoppers, or moth larvae that eat crop plants. Under natural conditions, their populations are kept under control by predators such as wasps, lady bugs, and praying mantises. As we discussed in chapter 3, there are generally far fewer predators in a food web than the species they prey upon. When we use a broad-spectrum pesticide, higher trophic levels are more likely to be knocked out than lower ones. This means that species that previously were insignificant can be released from natural controls and suddenly become major pests.

Consider the case of the Canete Valley in Peru. Before DDT was introduced in 1949, cotton yields were about 500 kg per ha. By 1952 yields had risen to nearly 750 kg per ha, but DDT-resistant boll weevils also had appeared (fig. 12.12). Toxaphene replaced DDT, but within two years it also became ineffective against boll weevils, which rebounded to higher levels than ever. Even worse, *Heliothis* worms—which had not previously been a problem—began increasing rapidly. The wasps that earlier had kept both organisms in check were poisoned by the increasing pesticide doses. By 1955, cotton yields were down to 330 kg per ha, one-third less than before any pesticides were used.

An even more complex chain of events was reported by ecologist Paul Ehrlich. In the early 1960s, DDT was sprayed on the thatched roofs and vegetation around villages in Borneo to kill mosquitoes and control malaria. The DDT killed flies and mosquitoes very successfully, but the poisoned insects were eaten by gecko lizards that inhabited the houses. The geckos accumulated so much DDT that they too died. Moths, which previously had been kept under control by the geckos, ate the palm thatch and caused the roofs to collapse. Village cats ate dead geckos and were themselves poisoned, allowing rats to descend on the villages, unleashing an epidemic of bubonic plague. To kill the rats, the government parachuted cats into some remote villages. Sometimes actions that are not well thought out have complex and unexpected consequences.

Persistence and Mobility in the Environment

The qualities that make DDT and other chlorinated hydrocarbons so effective—stability, high solubility, and high toxicity—make them environmental nightmares. Because they persist for years, even decades in some cases, and move freely through air, water,

and soil, they often show up far from the point of original application. A 1985 survey by the National Institute of Environmental Health found that 100 percent of Americans sampled have detectable amounts (in the 10 to 20 ppb range) of DDT or its daughter compound DDE. At least 90 percent of those sampled also had detectable amounts of other chlorinated compounds such as chlordane, heptachlor, aldrin, dieldrin, or hexachlorobenzene. In agricultural areas where these compounds have been heavily used within the past few decades, mother's milk often contains levels of these persistent chlorinated hydrocarbons that would make it illegal if sold on the market. As you will learn in the next section of this chapter, there are uncertainties about how dangerous these minute levels of pesticide residues are, but the evidence is sufficient to cause concern about the ubiquity of chemical contaminants.

In 1992, the U.S. Geological Survey sampled groundwater and rain in twenty-three agricultural states during the growing season. Detectable—although not necessarily dangerous—levels of pesticides were found in all states. Not surprisingly, the highest levels were found in the midwestern corn belt where pesticide use is generally highest. In Minnesota and Iowa, for example, 20 to 30 percent of all community wells and 30 to 60 percent of all private wells contained measurable pesticide residues. At least 60,000 rural wells were judged unsafe to drink. The most common residues were atrazine and alachlor, two herbicides used on corn and soybeans.

Even remote and presumably pristine locations like Isle Royale National Park in Lake Superior and Acadia National Park in Maine, however, were also found to be contaminated. Apparently, pesticides are carried long distances—hundreds or even thousands of kilometers—via air currents and deposited in rain or snow. This phenomenon of long-range transport is known as **pesticide rain,** a new concern to add to worries about acid rain and other air pollutants (see chapter 18). Pesticides can bioaccumulate through food chains. Beluga (white whales) in the St. Lawrence estuary have been shown to have 70,000 to 100,000 ppb DDT in their blubber even

CASE STUDY
California Medfly Wars

A small fruit-eating fly (*Ceratitis capitatia*) known as the Mediterranean fruit fly, or Medfly, was first detected in California in 1975. Since then the state has spent more than $170 million to rid itself of this pest. The stakes are high. Medflies attack more than 250 different fruits and vegetables. California's mild climate is virtually identical to the insect's Mediterranean home. More than 12.5 million hectares (31 million acres) of commercial orchards and farms in the state and millions of home gardens offer innumerable sites for flies to find food and shelter. A $17 billion per year industry—source of more than half of all fresh winter fruits and vegetables in the United States—is at risk.

Each female Medfly lays up to 1000 eggs in tiny, nearly invisible holes that she drills in fruit during her 40-day life span. When the eggs hatch, the little maggots munch their way through the fruit, turning the pulp into a disgusting jellylike brown goop that no one would eat. Perhaps worse than the total amount of produce that the flies ruin, however, is the threat that their presence poses for California's export business. Other states and foreign countries will refuse California produce if they believe it to be contaminated. China, for instance, is a potential $200 million annual market, but it may bar shipments from California for fear of introducing Medflies, which are not yet present in Asia.

California has responded vigorously—unnecessarily so, some say—every time a Medfly has been discovered. The entire area for a kilometer or so around the discovery site is cordoned off and sprayed both from the air and by troops of pest-fighters (fig. 1) who storm through homes and businesses, strip all fruit from trees, empty garbage cans, and spray everywhere they think a fly might be hiding. Hordes of sterilized male flies are released to interfere with breeding of any survivors. Since all the sightings so far have been

Figure 1
A Medfly commando prepares to engage the enemy in an all-out war.

in the densely occupied Los Angeles basin, much of this Medfly war has gone on in people's backyards. They, their houses, children, gardens, and pets are exposed to the organophosphate malathion, as are beneficial insects and other wildlife. Many people are understandably worried about the consequences of these tactics.

The official position of the California Department of Agriculture is that each sighting of Medflies represents imports from Mexico or Hawaii where the pests are common. The state insists that constant surveillance and rapid commando raids—each of which costs up to $40 million—have been successful in totally eradicating all Medflies that enter the state.

Not so, say critics, including entomologist James Carey. The frequency of sightings—nearly every year for 15 years—and the geographical clustering suggest to Carey that a small resident population of Medflies has existed in California for several decades. The warfare undertaken by the state never gets them all. The survivors rebound eventually and, Carey's maps suggest, are slowly moving out of the Los Angeles basin and into other areas.

The alternative approach suggested by Carey and others is to spend less money and energy trying to eradicate flies and concentrate instead on understanding where they are now, how they spread, and how populations might be controlled. His position might be summarized: We have flies; we have to live with them; how are we going to coexist? Is this fatalistic, or is the state's position that there are no resident flies simply denial?

ETHICAL CONSIDERATIONS

What ethical considerations apply in this case? Is it ethical to risk human health and ecosystem stability for the sake of economic gain? Given the uncertain effectiveness of mass spraying operations, how would you weigh the options for controlling this pest?

though the pesticide has been banned for two decades in both Canada and the United States. Dead whales must be treated as toxic waste.

Because many food crops now travel thousands of miles before they reach market, or they are stored for up to a year before being sold, treatment with fungicides, insecticides, and other potent chemicals is essential to preserve quality during shipping and storage. Ironically, for many years pesticides whose use was banned in the United States were still manufactured here and sold in Latin America or Asia where they were used on crops that are shipped back to be sold here. Export of banned pesticides is now prohibited but many of these compounds are still made and used elsewhere. In one survey by the Food and Drug Administration, forty-two of forty-five coffee samples and 100 percent of the bananas tested had pesticide residues, some at dangerous levels. We need to work with developing countries to ensure that they know about and use alternative and less dangerous forms of pest control.

Human Health Problems

Pesticide effects on human health can be divided into two categories: (1) short-term effects, including acute poisoning and illnesses caused by relatively high doses and accidental exposures, and (2) long-term effects suspected to include cancer, birth defects, immunological problems, Parkinson's disease, and other chronic degenerative diseases. The long-term health effects may be caused by very low doses of a variety of different chemicals and are difficult to tie to a specific source. Nevertheless, they probably affect far more people than tragic but localized accidents such as the deaths of some 2500 people and injury to tens of thousands more in the explosion of a Union Carbide pesticide-manufacturing plant in Bhopal, India, in 1984 (chapter 9).

The World Health Organization (WHO) estimates that some 1 million people suffer acute pesticide poisoning and at least 20,000 die each year. At least two-thirds of this illness and death results from occupational exposures in developing countries where people use pesticides without proper warnings or protective clothing (fig. 12.13).

In the United States, about 300,000 farmworkers suffer each year from pesticide-related illnesses. About 10 percent of those poisonings are acute and about twenty-five workers die every year. Many of the farmworkers exposed to the highest doses of the most toxic agricultural chemicals are migrant workers who handpick fruits and vegetables. Much of this harvest is treated with fungicides and insecticides to ensure blemish-free, long shelf-life produce for supermarkets across the country (fig. 12.14). Seven of the ten pesticides listed by the EPA as greatest health concerns in the United States are fungicides (table 12.2). Much, perhaps most, of the produce you buy in grocery stores has been treated with a fungicide either before or after harvest.

Every year about 20,000 Americans—mostly children—get sick from unsafe storage or home misuse of pesticides. Home owners tend to adopt a "scorched earth" policy toward all bugs, spiders, ants, and weeds, blasting lawns and home interiors with five to ten times as much pesticide per unit area as farmers use on their fields (fig. 12.15). Golf courses also are sites of intense pesticide use as owners attempt to create perfect lawns. A recent study of New York golf courses found that they used twenty-one herbicides, twenty

Figure 12.13

Farmworkers in developing countries often apply dangerous pesticides without any protective clothing or other safeguards. Warnings and instructions printed in English are often of little help.

Figure 12.14

The United Farm Workers of America estimates that 300,000 farmworkers in the United States suffer from pesticide-related illnesses each year.

fungicides, and eight insecticides at eight times the rate per unit area that farmers use. Some golfers have suffered massive allergic responses to these toxic chemicals and died within hours of walking on treated grass.

Long-term exposure to toxic pesticides, even at low doses, has long been suspected of having undesirable health effects. As we discussed in chapter 9, it is difficult to assess the effects of low-level, chronic exposures and to determine which of the many hazards to

Table 12.2 — Frequently detected pesticides in fruits and vegetables

Pesticide	Crops Affected	Potential Hazards
Captan	Grapes, peaches, strawberries, apples	Cancer
Carbaryl	Corn, bananas, grapes, peaches, oranges	Kidney damage
Dimethoate	Green beans, grapes, watermelon, cabbage, broccoli	Cancer, birth defects
Endosulfan	Spinach, lettuce, celery, strawberries, cauliflower	Liver, kidney damage
Methamidophos	Tomatoes, cauliflower, cabbage, melons, peppers	Nerve system effects

Source: Natural Resources Development Council.

Figure 12.15

Our approach to nature is to beat it into submission, but at what cost?

which we are exposed in a lifetime is responsible for a particular condition (see What Do *You* Think?, p. 189). After two decades of debate, authorities have finally conceded that Vietnam veterans who were exposed to the herbicide Agent Orange are likely to suffer some long-term adverse effects, including cancer. In a related development, farmers who use 2,4 D and 2,4,5 T (the main ingredients in Agent Orange) on their crops have been shown to be at least twice as likely to contract non-Hodgkin's lymphoma—a rare lymphoid cancer—as farmers who do not use these herbicides.

Alternatives to Current Pesticide Uses

Can we avoid using toxic pesticides or lessen their environmental and human-health impacts? In many cases, improved management programs can cut pesticide use between 50 and 90 percent without reducing crop production or creating new diseases. Some of these techniques are relatively simple and save money while maintaining disease control and yielding crops with just as high quality and quantity as we get with current methods. In this section, we will examine behavioral changes, biological controls, and integrated pest-management systems that could substitute for current pest-control methods.

Behavioral Changes

Crop rotation (growing a different crop in a field each year in a two-to six-year cycle) keeps pest populations from building up. For instance, a soybean/corn rotation is effective and economical against white-fringed weevils. Mechanical cultivation can substitute for herbicides. Flooding fields before planting or burning crop residues and replanting with a cover crop can suppress both weeds and insect pests. Habitat diversification, such as restoring windbreaks, hedgerows, and ground cover on watercourses, not only prevents soil erosion but also provides perch areas and nesting space for birds and other predators that eat insect pests. Growing crops in areas where pests are absent makes good sense. Adjusting planting times can avoid pest outbreaks, while switching from huge monoculture fields to mixed polyculture (many crops grown together) makes it more difficult for pests to find the crops they like. Tillage at the right time can greatly reduce pest populations. For instance, spring or fall plowing can help control overwintering corn earworms.

An important behavioral adjustment can occur in our attitudes and preferences. Farmers can be persuaded that the bare earth look in row crops made possible by powerful herbicides is not good for the soil or for themselves. Allowing a few weeds to creep in or planting a cover crop between rows may make more sense in the long run. Consumers may have to learn to accept fruits and vegetables that are less than perfect.

Biological Controls

Biological controls such as predators (wasps, ladybugs, praying mantises; figs. 12.16 and 12.17) or pathogens (viruses, bacteria, fungi) can control many pests more cheaply and safely than using massive amounts of pesticides. *Bacillus thuringiensis* or Bt, for example, is a naturally occurring bacterium that kills the larvae of lepidopteran (butterfly and moth) species but is harmless to mammals. A number of important insect pests such as tomato hornworm, corn rootworm, cabbage loopers, and others can be controlled by spraying bacteria on crops. Larger species are effective as well.

What Do *You* Think?
Environmental Estrogens

Figure 1

A shocking decrease in fertility and hatchling survival of alligators in Lake Apopka has been linked to pesticide pollution. Are humans being affected by similar toxins?

What might alligators in Florida, seals in the North Sea, salmon in the Great Lakes, and you have in common? All are at the top of their respective food chains and all appear to be accumulating threatening levels of toxic environmental chemicals in their body tissues. One of the most frightening possible effects of those chemicals is that they seem to be able to disrupt endocrine hormones that regulate many important bodily functions. Evidence for this seems quite convincing in some wildlife populations, but whether it also is true for humans is one of the most contentious and important questions in environmental toxicology today.

One of the first examples of hormone-disrupting chemicals in the environment was a dramatic decline in alligators a decade ago in Florida's Lake Apopka. Surveys showed that 90 percent of the alligator eggs laid each year were infertile and that of the few that hatched, only about half survived more than two weeks (fig. 1). Male hatchlings had shrunken penises and unusually low levels of the male hormone testosterone. Female alligators, meanwhile, had highly elevated estrogen levels and abnormal ovaries. The explanation seems to be that a DDT spill in the lake in the 1980s has led to high levels of DDE (a persistent breakdown product of DDT) in the reptiles' tissues and eggs. Because of a similarity in chemical structure, DDE appears to interfere with the action of androgens and estrogens, the normal sex hormones.

Researchers have begun to suspect that mysterious outbreaks of health and reproductive problems in other wildlife populations may have similar origins. Immune-system failures that killed thousands of seals along the coast of Europe and Scandinavia in 1992, for instance, are thought to have been caused by high levels of pesticides, PCBs, dioxins, and other toxins in their diet. Similarly, reproductive failures in fish and bird populations in the Great Lakes, fewer turtle hatchlings in farm ponds, abnormal thyroids and dramatic increases in tumors in fish, all are now thought to be related to hormone disturbances by exogenous chemicals.

But are humans affected as well? It is quite clear that people everywhere in the world have accumulated many of these same toxic chemicals in their bodies. Women who eat lots of fish from contaminated waters have been shown to have babies with elevated rates of mental, developmental, and behavioral disorders. Studies of women with estrogen-sensitive breast and vaginal cancers were found to have higher than normal levels of pesticides such as DDE in their tissues. Sperm counts in men appear to have decreased by about 50 percent over the past 50 years, while testicular and prostate cancers have increased dramatically during that same time.

Good evidence exists from controlled laboratory experiments that rats and mice exposed *in utero* or through mother's milk to very low levels of estrogen-like compounds develop physical, reproductive, and behavioral problems. We know that some of these chemicals act as synthetic hormones, others are antagonists that block normal hormone function. The question is whether these chemicals are linked to human health problems. Many of these compounds are hundreds or thousands of times less active than normal hormones, leading skeptics to doubt that they have any noticeable effects except in animals exposed to extremely high levels from a chemical spill. Furthermore, since some effects are positive while others are negative, they could cancel each other out. Furthermore, we may have protective mechanisms that are lacking in highly inbred laboratory rodents, and we can eat a highly varied diet that includes protective factors as well as toxins.

The bottom line is that we don't know (and we may never know for sure) whether falling sperm counts, increasing cancers, birth defects, immune diseases, and behavioral disorders in humans are caused by endocrine-disrupting environmental chemicals. Of course, we should do more research and testing of the physiological actions of these chemicals. In 1996, the EPA ordered pesticide manufacturers to begin testing for disrupting effects. Given the continuing uncertainty about the dangers we face, what more do you think we should do? Is this threat serious enough to warrant drastic steps to reduce our risk? If you were head of the Environmental Protection Agency or the Food and Drug Administration, how much certainty would you demand before acting to protect our environment and ourselves from this frightening potential threat?

Figure 12.16

The praying mantis looks ferocious and is an effective predator against garden pests, but it is harmless to humans. They can even make interesting and useful pets.

Figure 12.17

Ladybird beetles (ladybugs) prey on a variety of pests both as larvae and adults. For a few dollars you can buy several thousand of these hardy and colorful little garden protectors from organic gardening supply stores.

Ducks, chickens, and geese, among other species, are used to rid fields of both insect pests and weeds. These biological organisms are self-reproducing and often have wide prey tolerance. A few mantises or ladybugs released in your garden in the spring will keep producing offspring and protect your fruits and vegetables against a multitude of pests for the whole growing season. Plants, too, can be effective weapons against insect pests. Home gardeners often plant a border of insect-repelling plants, such as garlic or marigolds, around their crops. On a larger scale, farmers might also benefit from this approach.

Herbivorous insects also have been used to control weeds (fig. 12.18). For example, the prickly pear cactus was introduced to Australia about 150 years ago as an ornamental plant. This hardy cactus escaped from gardens and found an ideal home in the dry soils of the outback. It quickly established huge, dense stands that dominated 25 million ha (more than 60 million ac) of grazing land. A natural predator from South America, the cactoblastis moth, was introduced into Australia in 1935 to combat the prickly pear. Within a few years, cactoblastis larvae had eaten so much prickly pear that the cactus has become rare and is no longer economically important.

Genetics and bioengineering can help in our war against pests. Traditional farmers have long known to save seeds of disease-resistant crop plants or to breed livestock that tolerate pests well. Modern science can speed up this process through selection regimes or by using biotechnology to transfer genes between closely related or even totally unrelated species. There are concerns about the ecological dangers of moving too fast in this new area (see In Depth, pp. 242). Releasing sterile males to interfere with insect pest reproduction has been used successfully in some cases. Screwworms, for example, are the flesh-eating larvae of flies that lay their eggs in scratches or skin wounds of livestock. They were a terrible problem for ranchers in Texas and Florida in the 1950s, but release of massive numbers of radiation-sterilized males disrupted reproduction and eliminated this pest in Florida. In Texas, where flies continue to cross the border from Mexico, control has been more difficult, but continual vigilance keeps the problem manageable.

Figure 12.18

Biological pest control is illustrated by the Klamath weed (*foreground*), which had invaded millions of hectares of California rangeland in the 1940s. In the background, the weeds have been eliminated by the introduction of a natural predator, the *Chrysolina* beetle. Within a few years, the weeds were entirely eradicated.

    Pest Control

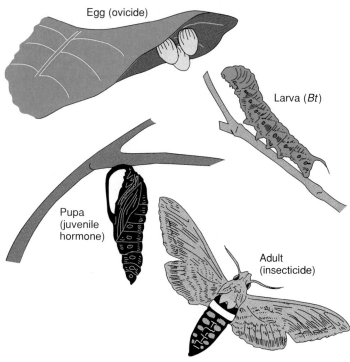

Egg (ovicide)

Larva (*Bt*)

Pupa
(juvenile
hormone)

Adult
(insecticide)

Figure 12.19

Stages in the life cycle of an insect and the control agents that kill or inhibit each stage. *Bacillus thuringiensis* (Bt) kills caterpillars when they eat leaves with these bacteria on the surface. The presence of excess juvenile hormone in the environment prevents maturation of pupae. Predators attack at all stages.

Other promising approaches are to use hormones that upset development or sex attractants to bait traps containing toxic pesticides. Many municipalities control mosquitoes with these techniques rather than aerial spraying of insecticides because of worries about effects on human health. Briquettes saturated with insect juvenile hormone are scattered in wetlands where mosquitoes breed. The presence of even minute amounts of this hormone prevent larvae from ever turning into biting adults (fig. 12.19). Unfortunately, many beneficial insects as well as noxious ones are affected by the hormone, and birds or amphibians that eat insects may be adversely affected when their food supply is reduced. Some communities that formerly controlled mosquitoes have abandoned these programs, believing that having naturally healthy wetlands is worth getting a few bites in the summer.

Integrated Pest Management

Integrated pest management (IPM) is a flexible, ecologically based pest-control strategy that uses a combination of techniques applied at specific times, aimed at specific crops and pests. It often uses mechanical cultivation and techniques such as vacuuming bugs off crops as an alternative to chemical application (fig. 12.20). IPM doesn't give up chemical pest controls entirely but rather tries to use the minimum amount necessary together with some of the steps outlined above to get the job done. Where there is no alterna-

Figure 12.20

This machine, nicknamed the "salad vac," vacuums bugs off crops as an alternative to treating them with toxic chemicals.

tive to using a chemical toxin for pest control, a single heavy dose of a nonpersistent pesticide might be applied just at the time insects or weeds are most vulnerable. Careful, scientific monitoring of pest populations to determine economic thresholds and the precise time, type, and method of pesticide application is critical in IPM.

Trap crops, small areas planted a week or two earlier than the main crop, are also useful. This plot matures before the rest of the field and attracts pests away from other plants. The trap crop then is sprayed heavily with enough pesticides so that no pests are likely to escape. The trap crop is destroyed so that workers will not be exposed to the pesticide and consumers will not be at risk. The rest of the field should be mostly free of both pests and pesticides.

IPM programs are already in use all over the United States on a variety of crops. Massachusetts apple growers who use IPM have cut pesticide use by 43 percent in the past ten years while maintaining per-acre yields of marketable fruit equal to that of farmers who use conventional techniques. Some of the most dramatic IPM success stories come from the Third World. In Brazil, pesticide use on soybeans has been reduced up to 90 percent with IPM. In Costa Rica, use of IPM on banana plantations has eliminated pesticides altogether in one region. In Africa, mealybugs were destroying up to 60 percent of the cassava crop (the staple food for 200 million people) before IPM was introduced in 1982. A tiny wasp that destroys mealybug eggs was discovered and now controls this pest in over 65 million ha (160 million ac) in thirteen countries.

One of the most important IPM programs now underway is in Indonesia, where an insect called the brown planthopper has developed resistance to virtually every insecticide and is threatening the country's hard-won self-sufficiency in rice. In 1986, President Suharto banned fifty-six of fifty-seven pesticides previously used in Indonesia and declared a crash program to educate farmers about IPM and the dangers of pesticide use. Researchers found that farmers were spraying their fields habitually—sometimes up to three times a week—regardless of whether fields were infested. By al-

Alternative Pest Control Strategies

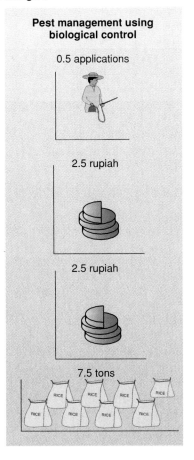

Pest management using chemicals

Number of times insecticide used in rice season — 4.5 applications

Cost to farmers per hectare — 7.5 rupiah

Cost to government per hectare — 27.5 rupiah

Rice yield per hectare — 6 tons

Pest management using biological control

0.5 applications

2.5 rupiah

2.5 rupiah

7.5 tons

Figure 12.21

Indonesia has one of the world's most successful integrated pest management (IPM) programs. Switching from toxic chemicals to natural pest predators has saved money while also increasing rice production.

Source: Data from Tolba, et al., *World Environment, 1972–1992*, p. 307, Chapman & Hall, © 1992 United Nations Environment Programme.

lowing natural predators to combat pests and spraying only when absolutely necessary with chemicals specific for planthoppers, Indonesian farmers using IPM have had higher yields than their neighbors using normal practices and they have cut pesticide costs by 75 percent. In 1988, only two years after its initiation, the program was declared a success and is being extended throughout the whole country. Since nearly half the people in the world depend on rice as their staple crop, this experiment could have important implications elsewhere (fig. 12.21).

Reducing Pesticide Exposure

The total numbers and amounts of different chemicals to which we are potentially exposed is overwhelming. The approximately one-half million metric tons of pesticides used in the United States every year contain about 600 active ingredients combined with more than 1200 presumably inactive carriers, solvents, preservatives, and other ingredients in about 25,000 commercial products. Less than 10 percent of the active pesticide ingredients have been subjected to a full battery of ten chronic health-effect tests, and studies of the inactive ingredients have started only recently (fig. 12.22). Since 1972, only forty pesticides have been banned. Critics charge that this slow pace puts all of us at risk and that assessment and regulation must proceed more quickly.

Regulating Pesticides

Three federal agencies share responsibility for regulating pesticides used in food production in the United States: the Environmental Protection Agency (EPA), the Food and Drug Administration (FDA), and the Department of Agriculture (USDA). The EPA regulates the sale and use of pesticides under the Federal Insecticide, Fungicide, and Rodenticide Act (FIFRA), which mandates the "registration" (licensing) of all pesticide products. Based on scientific studies, the EPA determines which pesticides will not pose significant risks to human health or the environment when used according to label directions. Under the Federal Food, Drug, and Cosmetic Act (FFDCA), the EPA also sets "tolerance levels" or limits for the amount of pesticide residues that lawfully may remain in or on foods marketed in the United States, whether grown domestically or imported from abroad. The FDA and USDA enforce pesticide use and tolerance levels set by the EPA. These agencies can seize and destroy food shipments found to contain pesticide residues in violation of limits set by the EPA.

In Canada, the Pest Control Products Act and Regulations, administered by Agriculture Canada, performs similar functions. Individual Canadian provinces also set local standards. The Ontario Ministry of Agriculture and Food, for instance, introduced a comprehensive program in 1988 that sets a goal of 50 percent reduction in pesticide use by the year 2002. Nonchemical pest control

Pest Control

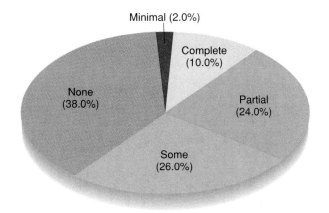

Figure 12.22

Level of testing for health effects of 3350 active and inert pesticide ingredients. Complete testing includes a battery of ten tests for both chronic and acute toxicity and genetic effects.

Source: Data from National Academy of Sciences.

methods, on-farm education, and biotechnology and the development of pest-resistant crop varieties are all planned to play a role in this program. Savings on chemical costs are expected to be about $100 million per year.

A controversial amendment to the U.S. FFDCA called the **Delaney Clause** was added in 1958, prohibiting the addition of any known cancer-causing agent to processed foods, drugs, or cosmetics. Processed foods include those that are canned, preserved, dehydrated, or frozen. Fresh fruits and vegetables are not covered by this regulation. Although this amendment is based on noble instincts, its absolute zero-risk approach produces some anomalous results. Because it is impossible to test every one of the millions of natural and synthetic chemicals in our food, the EPA has taken the position that additives in use before 1954 and "generally recognized as safe" at that time are permissible. Furthermore, pesticides registered after 1978 are subject to the Delaney Clause while those registered earlier must meet less stringent standards. One bizarre effect is that you can consume all the coffee—which has about twenty known carcinogens—you want, but, in theory, a food processor can't add a single drop of coffee to processed food.

One problem is that since the Delaney Clause went into effect nearly forty years ago, scientists have developed exquisitely sensitive techniques for measuring chemicals. Today some tests can detect one part per quintillion—roughly equal to a tablespoon of liquid in all the Great Lakes combined. At this level of sensitivity, probably everything we eat contains some carcinogens. The question becomes not one of zero risk but rather which risks are worth worrying about.

In 1987, a panel of experts from the National Academy of Sciences recommended a more realistic rule for pesticide regulation based on actual risk rather than zero tolerance. If researchers determine that a specific level of a pesticide is not expected to cause more than one extra case of cancer in 1 million people over a seventy-year life span, then that amount is considered a "negligible risk" and would be legal. Industry favors this change to the Delaney Clause while environmental and consumer groups generally oppose it. The Clinton administration supports this "health-based" approach to pesticide regulation, arguing that it makes more sense than an absolute ban. At the same time, EPA Administrator Carol Browner announced in June 1993 that the EPA, FDA, and USDA would push to reduce the use of all agricultural pesticides to protect children's health.

Some commonly used pesticides that should be banned under the Delaney Clause but are proposed for exceptions because they pose negligible risk are Propargite (a miticide used on most grapes), Dimethoate (an aphidicide used on apples), Atrazine (a herbicide used on sugarcane, corn, and soybeans), Linuron (a herbicide used on potatoes), and Acephate (an insecticide used on cotton). Industry officials claim that a 70 kg (154 lb) person would have to consume 11 tons of grape products daily to ingest the amount of Propargite found to cause cancer in laboratory animals.

Personal Safety

What can we do personally to reduce pesticide use and to protect ourselves from pesticide exposure? Table 12.3 offers some suggestions for consumer food safety. Many of the practices suggested earlier in this chapter for alternative methods of pest control can be used at home too.

Table 12.3

Food safety tips

To reduce the amount of pesticide residues and other toxic chemicals in your diet, follow these simple rules.

1. Wash and scrub all fresh fruits and vegetables thoroughly under running water.

2. Peel fruits and vegetable when possible. Throw away the outer leaves of leafy vegetables such as lettuce and cabbage.

3. Store food carefully so it doesn't get moldy or pick up contaminants from other foods. Use food as soon as possible to ensure freshness.

4. Cook or bake foods that you suspect have been treated with pesticides to break down chemical residues.

5. Trim the fat from meat, chicken, and fish. Eat lower on the food chain where possible to reduce bioaccumulated chemicals.

6. Don't pick and eat berries or other wild foods that grow on the edges of roadsides where pesticides may have been sprayed.

7. Grow your own fruits and vegetables without using pesticides or with minimal use of dangerous chemicals.

8. Ask for organically grown food at your local grocery store or shop at a farmer's market or co-op where you can get such food (fig. 12.23).

Figure 12.23
Concern about the possible effects of pesticide residues in our diet has led many consumers to request organically grown food that is produced without pesticides. Farmer's markets, such as the one shown here, can be a good place to find locally-produced oganic food.

- Plant ground cover in your yard that can compete successfully with weeds. Learn to appreciate a mixture of different plants rather than striving for a perfect, lush carpet of grass. What's wrong with a few dandelions, anyway?

- Install or repair screens on doors and windows to keep out insects. Caulk around windows, doors, and baseboards to seal routes of entry for ants, roaches, mice, and other pests. Put a few bay leaves in containers of flour or cereal to discourage flour weevils.

- Clean up spilled food and empty garbage regularly to eliminate food for ants or roaches. Keep food in tightly sealed containers. Sprinkle borax around drains and ducts where pests may be lurking. Use sticky flypaper to catch flies rather than spray dangerous chemicals inside your house or apartment.

- Get rid of aphids, scales, mites, and other houseplant pests by washing leaves and stems individually with rubbing alcohol, or spray plants with a dilute solution of dish soap in water. Simply washing vegetation with pure water will remove many pests and improve the health and appearance of your plants.

- A saucer of stale beer in your garden will attract and drown slugs. If you have only a few worms or bugs, pick them off by hand. Purchase some predators such as ladybugs or praying mantises to protect garden plants. If you have a lot of bugs in your house or apartment, a praying mantis can make an interesting house pet.

- Drain stagnant water in or near your yard that might serve as breeding sites for mosquitoes. Pick up dead branches, fallen leaves and other litter that provide a haven for termites and other pests. Put up houses for bug-eating purple martins or bats.

- Accept slightly blemished fruits and vegetables. If you garden, trim away bad parts or give up a portion of your crop rather than saturate the environment with dangerous chemicals. Interplant insect-repellent plants such as marigolds, basil, peppermint, or garlic together with your sensitive garden plants.

- If you must use toxic chemical pesticides, use them in the smallest possible amounts and apply them only where and when it is necessary. Rather than spray chemicals, use traps or applicators that limit dispersal into the environment.

- Read *Organic Gardening and Farming, Harrowsmith,* or some other magazine that reports regularly on environmental issues and alternative approaches to pest control.

Remember that being "natural" or organic does not necessarily make a chemical good, nor does being synthetic necessarily make one bad. Some natural products such as nicotine are highly toxic. Others such as mercury and lead are just as likely to bioconcentrate and persist in the environment as are synthetic compounds.

Pest Control

Summary

Biological pests are organisms that reduce the availability, quality, or value of resources useful to humans. Pesticides are chemicals intended to kill or drive away pests. Many nonchemical pest-control approaches perform these same functions more safely and cheaply than do toxic chemicals. Of the millions of species in the world, only about one hundred kinds of animals, plants, fungi, and microbes cause most crop damage. Many beneficial organisms are injured by indiscriminate pesticide use, including natural predators that serve a valuable function in keeping potential pests under control.

Humans have probably always known of ways to protect themselves from annoying creatures, but our war against pests entered a new phase with the invention of synthetic organic chemicals such as DDT. These chemicals have brought several important benefits, including increased crop production and control of disease-causing organisms. Indiscriminate and profligate pesticide use also has caused many problems, such as killing nontarget species, creating new pests of organisms that were previously not a problem, and causing widespread pesticide resistance among pest species. Often highly persistent and mobile in the environment, many pesticides move through air, water, and soil and bioaccumulate or bioconcentrate in food chains causing serious ecological and human health problems.

A number of good alternatives offer ways to reduce our dependence on dangerous chemical pesticides. Among these are behavioral changes such as crop rotation, cover crops, mechanical cultivation, and planting mixed polycultures rather than vast monoculture fields. Consumers may have to learn to accept less than perfect fruits and vegetables. Biological controls such as insect predators, pathogens, or natural poisons specific for a particular pest can help reduce chemical use. Genetic breeding and biotechnology can produce pest-resistant crop and livestock strains, as well. Integrated pest management (IPM) combines all of these alternative methods together with judicious use of synthetic pesticides under precisely controlled conditions.

Regulating pesticide use is a controversial subject. Many people fear that we are exposed to far too many dangerous chemicals. Industry claims that it could not do business without these materials. The Delaney Clause, an amendment to the Federal Food, Drug, and Cosmetic Act, prohibits willful addition of any known carcinogen to foods, drugs, or cosmetics. Although noble in intention, this amendment has become difficult to enforce as we find that many commonly used materials are carcinogens. Should we weaken the law and allow some carcinogens as long as the risk is "negligible"?

Many of the procedures and approaches suggested for agriculture and industry also work at home to protect us from pests and toxic chemicals alike. By using a little common sense, we can have a healthier diet, lifestyle, and environment.

▼ Questions for Review

1. What is a pest and what are pesticides? What is the difference between a biocide, a herbicide, an insecticide, and a fungicide?

2. How much pesticide is used worldwide and in your country? In your country, which of the general categories of use and which specific type accounts for the greatest use? Has use been increasing or decreasing in recent years?

3. What is DDT and why was it considered a "magic bullet"? What are its benefits and disadvantages?

4. Describe fumigants, botanicals, chlorinated hydrocarbons, organophosphates, carbamates, and microbial pesticides.

5. Explain why pests often resurge or rebound after treatment with pesticides and how they become pesticide resistant. What is a pesticide treadmill and pesticide rain?

6. Identify three major categories of alternatives to synthetic pesticides and describe, briefly, how each one works.

7. How did Australia fight prickly pear cactus? How did Florida eradicate screwworms?

8. What is the Delaney Clause and why is it controversial?

9. List nine things you could do to reduce pesticide use in your home.

10. List eight things you could do to reduce your dietary exposure to pesticides.

▼ Questions for Critical Thinking

1. In retrospect, do you think Paul Müller should have received a Nobel prize for discovering the insecticidal properties of DDT?

2. If you were a public health official in a country in which malaria, filariasis, or onchocerciasis were rampant, would you spray DDT to eradicate vector organisms? Would you spray it in your own house?

3. Pesticide rain, pesticide treadmill, and Medfly war are all highly emotional terms. Why would some people choose to use or not use these terms? Can you suggest alternative terms for the same phenomena that convey different values?

4. Many farmworkers who suffer from pesticide poisoning are migrants or minorities. Is this evidence of environmental racism? What evidence would you look for to determine whether environmental justice is being served?

5. Suppose that a developing country believes that it needs a pesticide banned in the United States or Canada to feed or protect the health of its people. Are we right to refuse to sell that pesticide?

6. How much extra would you pay for organically-grown food? How would you define organic in this context?

7. If alternative pest control methods are so effective and so much safer, why aren't farmers and consumers adopting them more rapidly?

8. If you were a member of Congress, would you vote to repeal the Delaney Clause to the FFDCA? Why or why not?

9. What would you personally consider a "negligible" risk? Would you eat grapes if you knew they had a measurable amount of Propargite? How small would the amount have to be?

10. Why is California reluctant to admit that it might have a resident population of Medflies? How might its control program change if it were to do so?

 Key Terms

▼ **Additional Information on the Internet**

BioControl Network
   http://www.usit.net/hp/bionet/Bionet.html/

Biological Control
   http://www.nysaes.cornell.edu/ent/biocontrol/

Biological Control of Pests Research Unit
   http://rsru2.tamu.edu/bcpru/bcpru.htm/

National IPM Network
   http://ipmwww.ncsu.edu/

National Agricultural Pest Information System
   http://www.ceris.purdue.edu/napis/

National Integrated Pest Management Network
   http://www.reeusda.gov/ipm/ipm-home.htm/

Pesticide Action Network North America
   http://www.panna.org/panna/

Pesticide Information Profiles
   http://sulaco.oes.orst.edu:70/1s/ext/extoxnet/pips/

USDA Integrated Pest Management Initiative
   http://raleigh.dis.anl.gov:83/

Note: Further readings appropriate to this chapter are listed on p. 599.

The American bald eagle, once threatened or endangered over much of its range in the lower forty-eight states by habitat destruction and pesticide poisoning, has made a remarkable comeback in many areas.

CHAPTER 13: Biodiversity

Objectives

After studying this chapter, you should be able to:

- define *biodiversity* and *species*.

- report on the total number and relative distribution of living species on the earth.

- summarize some of the benefits we derive from biodiversity.

- describe the ways humans cause biodiversity losses.

- evaluate the effectiveness of the Endangered Species Act and CITES in protecting endangered species.

- understand how gap analysis, ecosystem management, and captive breeding can contribute to preserving biological resources.

- propose ways we could protect endangered habitats and communities through large-scale, long-range, comprehensive planning.

Killing Lake Victoria

If you go into your local pet store, chances are, you'll see some cichlids (Haplochromis *sp.) for sale. These small, colorful, prolific fish come in a wide variety of colors and shapes from many parts of the world (fig. 13.1). The greatest cichlid diversity on earth—and probably the greatest vertebrate diversity any-where—are found in the three great African rift lakes: Victoria, Malawi, and Tanganyika. Together, these lakes once had about 1000 types of cichlids—more than all the fish species in Europe and North America combined. All these cichlids apparently evolved from a few ancestral varieties in the 15,000 years or so*

Figure 13.1

This harlequin cichlid guarding its young is a member of one of the most diverse groups of vertebrates in the world. Lakes Victoria, Malawi, and Tanganyika in Africa once had more cichlid species than all the fish species in Europe and North America combined. Exotic predators have destroyed many of these colorful little fish and now threaten the entire ecosystem in Lake Victoria.

since the lakes formed, one of the fastest and most extensive examples of vertebrate speciation known.

Unfortunately, a well-meaning but disastrous fish stocking experiment has wiped out at least half the cichlid species in these lakes in the last 20 years and set off a series of changes that is upsetting important ecological relationships. Lake Victoria, which lies between Kenya, Tanzania, and Uganda, has been particularly hard hit. Cichlids once made up 80 percent of the animal biomass in the lake and were the base for a thriving local fishery, supplying much-needed protein for native people. Colonial administrators, however, regarded the little, bony cichlids as "trash fish" and in the 1960s introduced the Nile perch (Lates niloticus), a voracious, exotic predator that can weigh 100 km (220 lbs) and grow up to 2 m long.

The perch gobbled up the cichlids so quickly that by 1980 two-thirds of the haplochromine species in the lake were extinct. Although there still are lots of fish in the lake, 80 percent of the animal biomass is now made up of perch, which are too large and powerful for the small boats, papyrus nets, and woven baskets traditionally used to harvest cichlids. International fishing companies now use large power boats and nylon nets to harvest great schools of perch, which are filleted, frozen, and shipped to markets in Europe and the Middle East. Because the perch are oily, local fishers can't sun dry them as they once did the cichlids.

Instead, the perch carcasses discarded by processing factories are cooked or smoked over wood fires for local consumption. Forests are being denuded for firewood, and protein malnutrition is common in a region that exports 200,000 tons of fish each year.

Perhaps worst of all, Lake Victoria, which covers an area the size of Switzerland, is dying. Algae blooms clot the surface, oxygen levels have fallen alarmingly, and thick layers of soft silt are filling in shallow bays. Untreated sewage, chemical pollution, and farm runoff are the immediate causes of this eutrophication, but destabilization of the natural community is ultimately responsible. The swarms of cichlids that once ate algae and rotting detritus were the lake's self-cleaning system. Eliminating them threatens the long-term ability of the lake to support any useful aquatic life.

As this example shows, biological diversity is important. Misguided management and development schemes that destroyed native species in Lake Victoria resulted in an ecosystem that no longer supports the natural community or the native people dependent on it. In this chapter, we will look at some important types of natural diversity, how human actions are threatening other species, what we are losing through our misuse of these biological resources, and some ways that we can protect and restore wild species and ecosystems on which we all rely. ▶

Biodiversity and the Species Concept

As far as we know, our planet is the only place in the universe that supports life, yet there are very few places on the earth that are not home to some kind of organism. From the driest desert to the dripping rainforests, from the highest mountain peaks to the deepest ocean trenches, life occurs in a marvelous spectrum of sizes, colors, shapes, life cycles, and interrelationships. Think for a moment how remarkable, varied, abundant, and important the other living creatures are with whom we share this planet. And consider how our lives would be impoverished if this biological diversity diminishes.

What Is Biodiversity?

Previous chapters of this book have described some of the fascinating varieties of organisms and complex ecological relationships that give the biosphere its unique, productive characteristics. Three kinds of **biodiversity** are essential to preserve these ecological systems: (1) *genetic diversity* is a measure of the variety of different versions of the same genes within individual species; (2) *species diversity* describes the number of different kinds of organisms within individual communities or ecosystems; and (3) *ecological diversity* assesses the richness and complexity of a biological community, including the number of niches, trophic levels, and ecological processes that capture energy, sustain food webs, and recycle materials within this system.

What Are Species?

As you can see in the discussion above, species occupy a crucial position in the concept of biodiversity, but what, exactly, do we mean by the term? In chapter 3, we defined species as all the organisms of the same kind able to breed in nature and produce live, fertile offspring. Underlying this commonly used definition is the idea that reproductive isolation caused by geography, physiology, or behavior prevents groups of otherwise similar organisms from exchanging genes, and therefore, gives them separate identities and evolutionary histories.

There are problems with species definitions based on reproductive isolation. Mating between species (hybridization) occurs in nature and may produce fertile offspring. Furthermore, investigators often cannot determine whether two groups that live in different places are capable of interbreeding. Sometimes, in fact, the only specimens available for study are dead. Species identification, therefore, is usually based on morphological characteristics such as size, shape, color, skeletal structure, or on either chemical or genetic traits.

Determining how much difference is necessary before two similar groups of organisms can be considered separate species is highly subjective, however, and taxonomists (scientists who study classification) engage in endless debates about which organisms are most closely related and which should be classified as separate species. Often these debates pit "lumpers" against "splitters." Given the same group of 100 closely related organisms, one scientist might split them into 50 or 100 different species, while another might lump them together into only 10 or 20 species.

Is all this simply a semantic debate or does it have practical significance? Consider the case of the red wolf (fig. 13.2). Two hundred years ago, these animals are thought to have ranged widely across the southeastern United States, but hunting and habitat destruction reduced their numbers so that by the 1970s only fourteen red wolves remained in the wild. The U.S. Fish and Wildlife Service (USFWS) established a captive breeding program to restore the population. Several hundred of these animals now exist and reintroduction to parks and nature preserves has begun. Total costs for this program are about $1 million per year. Recent DNA studies suggest, however, that red wolves are really hybrids between the more numerous gray (timber) wolves and coyotes. If so, they aren't protected under the Endangered Species Act and the millions we are spending on breeding and reintroduction might be better spent on other, real endangered species.

As this case illustrates, defining a species often is an important but difficult exercise. But if we want to study, discuss, and inventory the organisms with which we share the planet, we need some categories in which to sort them. Discussion and revision aside, we therefore group organisms that appear to be of the same type into species, as well as less restrictive subdivisions such as subspecies, varieties, ecotypes, races, or populations.

How Many Species Are There?

At the end of the great exploration era of the nineteenth century, some scientists confidently declared that every important kind of living thing on Earth would soon be found and named. Most of those explorations focused on charismatic species such as birds and mammals. Recent studies of less conspicuous organisms such as insects and fungi suggest that millions of new species and varieties remain to be studied scientifically.

We now believe that the 2.1 million species presently known (table 13.1) represent only a small fraction of the total number that exist. Based on the rate of new discoveries by research expeditions—especially in the tropics—taxonomists estimate that there may be somewhere between 3 million and 50 million different species alive today. In fact, there may be 30 million species of tropical insects alone. About 70 percent of all known species are invertebrates (animals without backbones such as insects, sponges, clams, worms, etc.). This group probably makes up the vast majority of organisms yet to be discovered and may constitute 95 percent of all species. What constitutes a species in bacteria and viruses is even less certain than for other organisms, but there are large numbers of physiologically or genetically distinct varieties of these organisms, and it seems likely that even more exist in nature that we have not yet discovered.

Of all the world's species, only 10 to 15 percent live in North America and Europe. It is a rare occurrence to find a new species of higher plant or animal in the developed countries of the world. By contrast, the centers of greatest biodiversity tend to be in the tropics (fig. 13.3). Many of the organisms in these megadiversity countries have never been studied by scientists. The Malaysian

Figure 13.2

Red wolves have been classified as a separate species from their canine relatives, but DNA analysis suggests it may be merely a hybrid between gray wolves and coyotes. Clarifying its taxonomic status will determine whether it deserves protection or not.

Table 13.1 Number of living species by taxonomic group

Group	Identified Species	Estimated Total
Bacteria and viruses	5800	10,000(?)
Protozoa and algae	100,000	250,000
Fungi	80,000	1,500,000
Invertebrates	1,500,000	7 to 50 million
Amphibians and reptiles	12,000	13,000
Fish	20,000	23,000
Birds	9100	9200
Mammals	4200	4300
Vascular plants	250,000	300,000
Nonvascular plants	150,000	200,000
Total	2,125,300	9 to 52 million

Source: UNEP 1993–94.

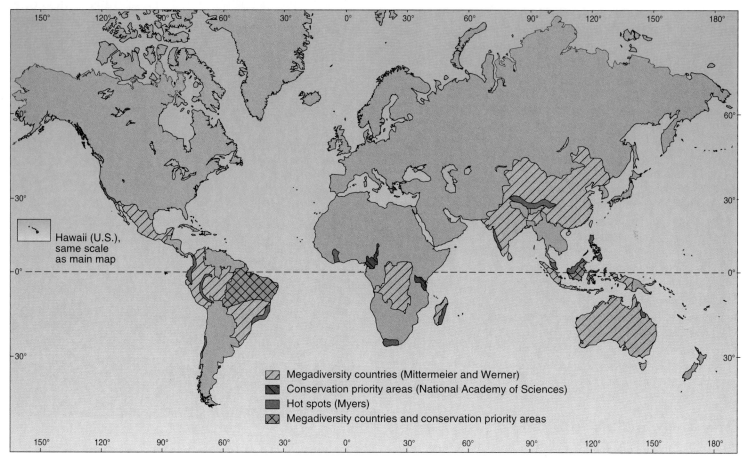

Figure 13.3

Priority areas for conservation of biodiversity. Twelve countries, mainly in the tropics, account for at least 60–70 percent of the world's biodiversity. A few islands, mountain chains, and coastal regions with Mediterranean-type climates are important because of their life endemism and rapid rates of habitat conversion. Together these areas cover less than 1 percent of the earth's land area but contain an estimated 20 percent of all species.

Sources: R. A. Mittermeier, "Primate Diversity and the Tropical Forest: Case Studies from Brazil and Madagascar and Importance of Megadiversity Countries," *Biodiversity,* 1988, National Academy Press; R. A. Mittermeier and T. B. Werner, "Wealth of Plants and Animals Unites 'Megadiversity' Countries" in *Tropicus* 4(2):1, 4–5, 1990; and National Academy of Sciences, *Research Priorities in Tropical Biology,* 1990 National Academy Press.

Peninsula, for instance, has at least 8000 species of flowering plants, while Britain, with an area twice as large, has only 1400 species. There may be more botanists in Britain than there are species of higher plants. South America, on the other hand, has fewer than one hundred botanists to study perhaps 200,000 species of plants.

How Do We Benefit from Biodiversity?

We benefit from other organisms in many ways, some of which we don't appreciate until a particular species or community disappears. Even seemingly obscure and insignificant organisms can play irreplaceable roles in ecological systems or be the source of genes or drugs that someday may be indispensable.

Food

All of our food comes from other organisms. Many wild plant species could make important contributions to human food supplies either as they are or as a source of genetic material to improve domestic crops. Noted tropical ecologist Norman Meyers estimates that as many as eighty thousand edible wild plant species could be utilized by humans. Villagers in Indonesia, for instance, are thought to use some four thousand native plant and animal species for food, medicine, and other valuable products. Few of these species have been explored for possible domestication or more widespread cultivation. A 1975 study by the National Academy of Science (U.S.) found that Indonesia has two hundred and fifty edible fruits, only forty-three of which have been cultivated widely (fig. 13.4).

Unfortunately, overgrazing, forest clearing, conversion of natural landscapes to agriculture, and other forms of human disturbance

Figure 13.4

Mangosteens from Indonesia have been called the world's best-tasting fruit, but they have never been cultivated on a large scale and are practically unknown beyond the islands where they grow naturally. There may be thousands of other traditional crops and wild food resources that could be equally valuable but are threatened by extinction.

able 13.2		Some natural medicinal products
Product	**Source**	**Use**
Penicillin	Fungus	Antibiotic
Bacitracin	Bacterium	Antibiotic
Tetracycline	Bacterium	Antibiotic
Erythromycin	Bacterium	Antibiotic
Digitalis	Foxglove	Heart stimulant
Quinine	Chincona bark	Malaria treatment
Diosgenin	Mexican yam	Birth-control drug
Cortisone	Mexican yam	Anti-inflammation treatment
Cytarabine	Sponge	Leukemia cure
Vinblastine, vincristine	Periwinkle plant	Anticancer drugs
Reserpine	Rauwolfia	Hypertension drug
Bee venom	Bee	Arthritis relief
Allantoin	Blowfly larva	Wound healer
Morphine	Poppy	Analgesic

are destroying potentially valuable food species and the wild ancestors of our domestic crops before they can be identified and their genes can be preserved. Later in this chapter, we will look at some of the programs underway to find useful wild species and preserve them in gene banks, botanical gardens, zoos, and nature preserves.

Drugs and Medicines

Living organisms provide us with many useful drugs and medicines (table 13.2). More than half of all prescriptions contain some natural products. The United Nations Development Programme estimates the value of pharmaceutical products derived from Third World plants, animals, and microbes to be more than $30 billion per year. Indigenous communities that have protected and nurtured the biodiversity on which these products are based are rarely acknowledged—much less compensated—for the resources extracted from them. Many consider this expropriation "biopiracy" and call for royalties to be paid for folk knowledge and natural assets.

Consider the success story of vinblastine and vincristine. These anticancer alkaloids are derived from the Madagascar periwinkle (*Catharanthus roseus*) (fig. 13.5). They inhibit the growth of cancer cells and are very effective in treating certain kinds of cancer. Twenty years ago, before these drugs were introduced, childhood leukemias were invariably fatal. Now the remission rate for some childhood leukemias is 99 percent. Hodgkin's disease was 98 percent fatal a few years ago, but is now only 40 percent fatal, thanks to these compounds. The total value of the periwinkle crop is roughly $15 million per year, although Madagascar gets little of those profits.

Figure 13.5

The rosy periwinkle from Madagascar provides anticancer drugs that now make childhood leukemias and Hodgkin's disease highly remissible. You may recognize it as a popular flowering ornamental plant.

Pharmaceutical companies are actively prospecting for useful products in many tropical countries. Merck, the world's largest biomedical company, is paying $1 million to the Instituto Nacional de Biodiversidad (INBIO) of Costa Rica for plant, insect, and microbe samples to be screened for medicinal applications. INBIO, a public/private collaboration, trains native people

Figure 13.6

Costa Rican taxonomists study insect collections as part of an ambitious project to identify and catalog all the species in this small, but highly diverse, tropical country. The knowledge gained may contribute toward valuable commercial products that will provide funds to help preserve biodiversity.

as practical "parataxonimists" to locate and catalog all the native flora and fauna—between 500,000 and 1 million species—in Costa Rica (fig. 13.6). Selling data and specimens will finance scientific work and nature protection. This may be a good model both for scientific information gathering and as a way for developing countries to share in the profits from their native resources.

Ecological Benefits

Although technological optimists may dream of someday living in completely artificial environments, so far, human life is inextricably linked to ecological services provided by other organisms. Soil formation, waste disposal, air and water purification, nutrient cycling, solar energy absorption, and management of biogeochemical and hydrological cycles all depend on the biodiversity of life (chapter 3). The Earth's ecosystems represent the culmination of historic evolutionary processes of immense antiquity and majesty. They have resulted from billions of years of evolution under conditions that may never have occurred anywhere else in the universe. Nature maintains ecological processes at no cost to us, and represents a genetic library of information we could never reproduce.

There has been a great deal of controversy about the role of biodiversity in ecosystem stability. Mathematical models suggest that simple ecosystems can be just as stable and resilient as more complex ones. Field studies by Minnesota ecologist David Tillman, have shown, however, that diverse biological plant communities withstand environmental stress better and recover more quickly than those with fewer species.

Because we don't fully understand the complex interrelationships between organisms, we often are surprised and dismayed at the effects of removing seemingly insignificant members of biological communities. For instance, wild species provide a valuable but often unrecognized service in suppressing pests and disease-carrying organisms. It is estimated that 95 percent of the potential pests and disease-carrying organisms in the world are controlled by other species that prey upon them or compete with them in some way. We find out how valuable natural predators are when we try to control systems with synthetic chemicals, because broad-spectrum biocides kill both pests and natural predators. As a result, pest populations often surge to higher levels than before (chapter 12). By preserving natural areas and conserving wild species, we utilize the stabilizing diversity of nature that keeps pest organisms in balance.

Aesthetic and Cultural Benefits

The diversity of life on this planet brings us many aesthetic and cultural benefits. Millions of people enjoy hunting, fishing, camping, hiking, wildlife watching, and other outdoor activities based on nature. These activities provide invigorating physical exercise and allow us to practice pioneer living skills. Contact with nature also can be psychologically and emotionally restorative. In some cultures, nature carries spiritual connotations, and a particular species or landscape may be inextricably linked to a sense of identity and meaning. Observing and protecting nature has religious or moral significance for many people.

Nature appreciation is economically important. The U.S. Fish and Wildlife Service estimates that Americans spend $18 billion every year watching wildlife. Some 8 million birdwatchers in the United States spend $5.2 billion per year, compared to $5.8 billion spent each year on movie tickets or the $5.9 billion spent on sporting events. Inexpensive intercontinental air flights now make it possible to visit remote and exotic places to enjoy nature and to experience other cultures. This can be a good form of sustainable economic development, but we have to be careful that we don't abuse the places and cultures we visit (see chapter 15).

For many people, the value of wildlife goes beyond the opportunity to shoot or photograph, or even see, a particular species. They argue that **existence value,** based on simply knowing that a species exists, is reason enough to protect and preserve it (fig. 13.7). We contribute to programs to save bald eagles, redwood trees, whooping cranes, whales, and a host of other rare and endangered organisms because we like to know they still exist somewhere, even if we may never have an opportunity to see them.

What Threatens Biodiversity?

Extinction, the elimination of a species, is a normal process of the natural world. Species die out and are replaced by others, often their own descendants, as part of evolutionary change. In undisturbed ecosystems, the rate of extinction appears to be about one species lost every decade. In this century, however, human impacts on populations and ecosystems have accelerated that rate, causing hundreds or perhaps even thousands of species, subspecies, and varieties to become extinct every year. If present trends continue, we may destroy *millions* of kinds of plants, animals, and microbes in the next few decades. In this section, we will look at some ways we threaten biodiversity.

Natural Causes of Extinction

Studies of the fossil record suggest that more than 99 percent of all species that ever existed are now extinct. Most of those species were gone long before humans came on the scene. Species arise through processes of mutation and natural selection and disappear the same way (chapter 4). Evolution can proceed gradually over millions of years or may occur in large jumps when new organisms migrate into an area or environmental conditions change rapidly. New forms replace their own parents or drive out less well-adapted competitors. In a sense, species that are replaced by their descendants are not completely lost. The tiny *Hypohippus,* for instance, has been replaced by the much larger modern horse, but most of its genes probably still survive in its distant offspring.

Periodically, mass extinctions have wiped out vast numbers of species and even whole families. The best studied of these events occurred at the end of the Cretaceous period when dinosaurs disappeared, along with at least 50 percent of existing genera and 15 percent of marine animal families. An even greater disaster occurred at the end of the Permian period about 250 million years ago

Figure 13.7
The mere existence of creatures like whales has value for many people. This may be expressed as a religious or cultural value or it may result in an emotional bonding to wildlife that transcends an intellectual appreciation of ecological roles.

when two-thirds of all marine species and nearly half of all plant and animal families died out over a period of about 10,000 years— a short time by geological standards. Current theories suggest that these catastrophes were caused by climate changes, perhaps triggered when large asteroids struck the earth. Many ecologists worry that global climate change caused by our release of "greenhouse" gases in the atmosphere could have similarly catastrophic effects (chapter 17).

Human-Caused Reductions in Biodiversity

The rate at which species are disappearing appears to have increased dramatically over the last 150 years. Between A.D. 1600 and 1850, human activities appear to have been responsible for the extermination of two or three species per decade. Harvard entomologist E. O. Wilson estimates that we now are pushing 20,000 species a year into extinction. We cannot be absolutely sure of these rates because many parts of the world haven't been thoroughly explored, and many species may have disappeared before they were studied and classified by biologists. Since 90 percent of all known species are tropical, it shouldn't be surprising that the greatest losses are in this group.

Habitat Destruction

The biggest reason for the current increase in extinctions is habitat loss (fig. 13.8). Habitat fragmentation divides populations into isolated groups that are vulnerable to catastrophic events. Very small populations may not have enough breeding adults to be viable even under normal circumstances (fig. 13.9). Destruction of forests, wetlands, and other biologically rich ecosystems around the world

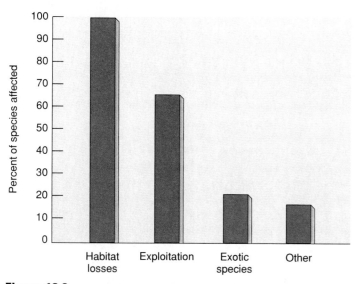

Figure 13.8

Threats to endangered mammals in Australasia and the Americas. For fish, exotic species represent a greater threat than shown here. For plants, trampling, grazing, collecting, flooding and other human actions constitute serious dangers. Notice that many species are endangered by multiple threats.

Source: UNEP 1994.

threatens to eliminate thousands or even millions of species in a human-caused mass extinction that could rival those of geologic history. By destroying habitat, we eliminate not only prominent species, but also many obscure ones of which we may not even be aware. For further discussion, see chapters 14 and 15.

Hunting and Fishing

Overharvesting is responsible for depletion or extinction of many species. A classic example is the extermination of the American passenger pigeon (*Ectopistes migratorius*). Even though it inhabited only eastern North America, 200 years ago this was the world's most abundant bird with a population of between 3 and 5 billion animals. It once accounted for about one-quarter of all birds in North America. In 1830, John James Audubon saw a single flock of birds estimated to be ten miles wide, hundreds of miles long, and thought to contain perhaps a billion birds. In spite of this vast abundance, market hunting and habitat destruction caused the entire population to crash in only about 20 years between 1870 and 1890. The last known wild bird was shot in 1900 and the last existing passenger pigeon, a female named Martha, died in 1914 in the Cincinnati Zoo (fig. 13.10).

Some other well-known overhunting cases include the near extermination of the great whales and the American bison (buffalo). In 1850, some 60 million bison roamed the western plains. Many were killed only for their hides or tongues, leaving millions of carcasses to rot. Much of the bisons' destruction was carried out by the U.S. Army to deprive native peoples who depended on bison for food, clothing,

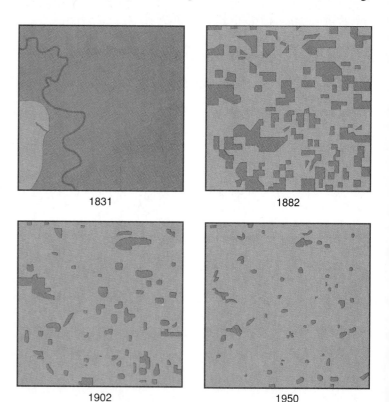

Figure 13.9

Decrease in wooded area of Cadiz Township in southern Wisconsin during European settlement. Shaded areas represent the amount of land in forest each year.

Figure 13.10

A stuffed passenger pigeon (*Ectopistes migratorius*) in a museum. The last member of this species died in the Cincinnati Zoo in 1914. Overhunting and habitat destruction caused their extinction.

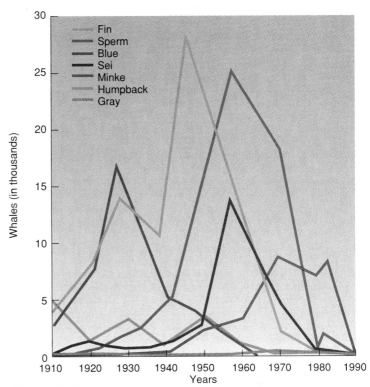

Figure 13.11

This world catch of whales shows a series of sharp peaks as each species in turn is hunted until commercially unprofitable. Some species such as minke, gray, and bowhead, have rebounded since commercial whaling stopped. Other species, such as blue and fin, remain rare and endangered.

Source: *UNEP Environmental Data Report 1993–1994.*

and shelter of these resources, thereby forcing them onto reservations. After forty years, there were only about one hundred and fifty wild bison left and another two hundred and fifty in captivity.

We think that about 2.5 million great whales once inhabited the world's oceans. Their blubber was highly prized as a source of oil. At first, only the slowest whales such as the right or bowhead could be caught by human-powered dories and hand-thrown harpoons. Introduction of steam ships and explosive harpoons in the 19th century, however, made it possible to catch and kill even the fastest whales and one species after another was driven into near extinction (fig. 13.11).

For the past decade, the International Whaling Commission has prohibited taking of great whales. This hunting ban seems to be having positive effects. With the exception of the North Atlantic right whales and southern blue whales, most populations appear to

be recovering. There are now about 24,000 California gray whales, more than twice as many as 20 years ago and close to the prehunting number. Around 3400 humpbacks now visit Hawaii each year compared to only one-third that many in the mid-1980s. Nearly one million minke whales—perhaps twice the pre-whaling number—now inhabit Arctic and Antarctic waters.

Iceland, Japan, and Norway argue that whale populations have rebounded enough to support limited hunting. They kill hundreds of whales each year mostly under the guise of "scientific research," although the meat and blubber of animals taken in these programs are still sold at a handsome profit.

It remains very difficult to count whales accurately. Whalers tend to come up with the highest estimates while environmental groups insist on more conservative numbers. For many people, the question is one of ethics. Is it acceptable to hunt and kill large, rare, intelligent, wild creatures like whales? What do you think? Where should we draw the line between what we eat and what we don't?

Fish stocks have been seriously depleted by overharvesting in many parts of the world. A huge increase in fishing fleet size and efficiency in recent years has led to a crash of many oceanic populations. Worldwide, 13 of 17 principal fishing zones are now reported to be commercially exhausted or in steep decline. Disputes over claims to offshore fisheries, allowable harvests, and acceptable

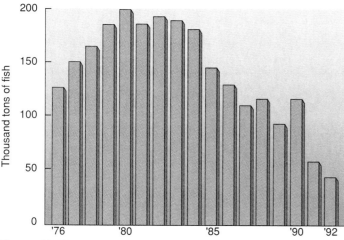

Figure 13.12

Overfishing of cod, haddock, and flounder on Canada's Grand Banks caused catches to decline 80 percent between 1980 and 1992, when all fishing was suspended.

Source: Data from Carl Safina, "The World's Imperiled Fish," *Scientific American*, vol. 273 (5): 46–53.

fishing techniques have led to international incidents. Iceland and Canada, for example, have cut nets or seized foreign boats violating their territorial waters. In 1990, the Canadian cod harvest was down 75 percent from its historic high (fig. 13.12). In response, Canada banned Atlantic cod fishing in 1992 putting 20,000 people out of work and closing down a $700 million-per-year industry (see What Do *You* Think?, p. 213). Similar population crashes and fishing bans have occurred in many of the world's oceans, and an important food source for many people is threatened.

Commercial Products and Live Specimens

In addition to harvesting wild species for food, we also obtain a variety of valuable commercial products from nature. Much of this represents sustainable harvest that may be a better option for some areas than agriculture or industry. Some forms of commercial exploitation are highly destructive, however, and represent a serious threat to certain rare species (fig. 13.13). Despite international bans on trade in products from endangered species, smuggling of furs, hides, horns, live specimens, and folk medicines amounts to millions of dollars each year (table 13.3).

Developing countries in Asia, Africa, and Latin America with the richest biodiversity in the world are the main sources of wild animals and animal products, while Europe, North America, and some of the wealthy Asian countries are the principal importers. Japan, Taiwan, and Hong Kong buy three-quarters of all cat and snake skins, for instance, while European countries buy a similar percentage of live birds. The United States imports 99 percent of all live cacti and 75 percent of all orchids sold each year.

The profits to be made in wildlife smuggling are enormous. Tiger or leopard fur coats can bring $100,000 in Japan or Europe. The population of African black rhinos dropped from approxi-

Figure 13.13

Endangered wildlife products were seized by United States Fish and Wildlife agents. Smugglers of products from endangered species are subject to heavy fines or jail terms or both. Unfortunately, however, as long as there are consumers for these products, some hunters and dealers will be encouraged to take the risk.

mately 100,000 in the 1960s to about 3000 today because of a demand for their horns (which are really hardened, compressed hairlike fibers). In Asia, where it is prized for its supposed medicinal properties, powdered rhino horn fetches $28,000 per kg. In Yemen, a rhino horn dagger handle can sell for up to $1000.

Similarly, bird collectors will pay $10,000 for a rare hyacinth macaw from Brazil or up to $12,000 for a pair of golden-shouldered parakeets from Australia. An endangered albino python might bring $20,000 in Germany. The mortality rate in this live animal trade is enormous. It is generally estimated that fifty animals are caught or killed for every live animal that gets to market. Buyers, many of whom say they love animals, keep this trade going. If you collect

Table 13.3 Yearly trade in wildlife and wildlife products	
Live primates	26,631
Cat skins	44,810
Live birds	933,672
Reptile skins	9,132,623
Live cacti	919,499
Live orchids	1,293,692

Source: *World Resources* 1994–95.

Food, Land, and Biological Resources

Figure 13.14

Soldiers load illegal elephant tusks seized from poachers. In 1989, Kenya burned 2400 confiscated tusks worth more than $3 million in an attempt to end the global ivory trade that threatens elephants with extinction.

Figure 13.15

A diver uses cyanide to stun tropical fish being caught for the aquarium trade. Many fish are killed by the method itself, while others die later during shipment. Even worse is the fact that cyanide kills the coral reef itself.

rare plants, fish, or birds, make sure that what you buy is commercially grown, not wild-caught. If you travel abroad, don't buy products made from rare or endangered species. You not only contribute to their decline by doing so; you also expose yourself to confiscation and serious fines when you get home.

Elephants are another important example of the pressures on wildlife. In 1980, there were an estimated 1.3 million elephants in Africa. Ten years later, only 625,000 remained. During that decade, at least 100,000 elephants were killed by poachers each year for the ivory trade (fig. 13.14). It is little wonder that people are tempted by this opportunity for wealth. When wholesale prices for ivory jumped from $10 to over $100 per kg (2.2 lbs) in the 1970s, a medium-sized set of tusks was equal to ten years' income for a subsistence farmer. The Convention on International Trade in Endangered Species (CITES) voted in 1989 to ban all ivory trade. The ban seems to be working and the world ivory market has collapsed.

Plants also are threatened by overharvesting. Wild ginseng has been nearly eliminated in many areas because of the Asian demand for roots that are used as an aphrodisiac and folk medicine. Cactus "rustlers" steal cacti by the ton from the American southwest and Mexico. TRAFFIC (a program of the World Wildlife Fund) reported that in 1990, nearly one million cactus plants were shipped from Mexico and the U.S. Southwest to both domestic and foreign markets. With prices ranging as high as $1000 for rare specimens, it's not surprising that many are now endangered.

The trade in wild species for pets is an enormous business. Worldwide, some 5 million live birds are sold each year for pets, mostly in Europe and North America. In the 1980s, pet traders imported (often illegally) into the United States some 2 million rep-

tiles, 1 million amphibians and mammals, 500,000 birds, and 128 million tropical fish each year. Keeping an aquarium is one of the most popular hobbies in America. About 75 percent of all saltwater tropical aquarium fish sold come from the marvelously rich coral reefs of the Philippines and Indonesia (fig. 13.15).

Many of these fish are caught by divers using plastic squeeze bottles of cyanide to stun their prey. Far more fish die with this technique than are caught. Worst of all, it kills the coral animals that create the reef. A single irresponsible diver can destroy all of the life on 200 square meters of reef in a day. Altogether, thousands of divers currently destroy about 50 km^2 of reefs each year. Net fishing would prevent this destruction, and it could be enforced if pet owners would insist on net-caught fish.

Predator and Pest Control

Some animal populations have been greatly reduced, or even deliberately exterminated, because they are regarded as dangerous to humans or livestock or because they compete with our use of resources. Every year, U.S. government animal control agents trap, poison, or shoot thousands of coyotes, bobcats, prairie dogs, and other species considered threats to people, domestic livestock, or crops. Many others are killed unintentionally by poisoned bait or misplaced traps or intentionally by private individuals for bounty or sport.

These animal control programs are controversial. In 1990, predator-related livestock losses (both confirmed and unconfirmed) in the U.S. were claimed to be $27.4 million. The cost of predator control that year by the Office of Animal Damage Control amounted to $38 million, mainly to kill 86,500 coyotes. Opponents of this program claim it would have been much cheaper to leave the wild animals alone and pay farmers and ranchers for their losses. Trappers argue that without animal control, the losses would have been much higher. Animal rights activists counter that the predators have as much right

CASE STUDY
Killing Baby Seals

Every spring, hundreds of thousands of female harp and hooded seals haul themselves out onto the floating sea ice along Canada's Arctic Coast to bear their young. Insulated by thick, pure white fur coats, the seal babies nap peacefully on the ice floes while their mothers dive for fish (fig. 1).

Their beautiful coats and relative helplessness make young seals an attractive quarry for hunters. Placid and trusting, unable to swim or protect themselves, the baby seals are easy to catch. Native Inuit people have always found the spring to be a good hunting season. Fragile skin kayaks couldn't penetrate far into the floating ice, however, and traditional hunters probably never had a serious impact on seal populations.

In this century, metal boats and powerful motors have made it possible for modern hunters to push farther into the ice pack and to bring out valuable furs. Dropped off on ice floes by large ocean-going vessels, squads of men simply walk up to the unprotected baby seals, club them on the head, and skin them out. By the mid-1980s, the annual harvest had reached some 200,000 animals.

Many people find this practice reprehensible. Their arguments are not so much ecological as ethical or emotional. This is hardly hunting in the sense of testing one's wits and skill against the superior speed and senses of a wily game animal. The baby seals are totally vulnerable. Furthermore, they are extremely cute and appealing. Many people who have no objections to killing rats, flies, or cockroaches, feel strongly that we should leave seals alone.

Fur hunters, on the other hand, argue that there is no evidence that managed harvesting of these seals is ecologically damaging. While seal populations have declined catastrophically in the eastern North Atlantic, Mediterranean, and northern Pacific oceans, the population off Canada's east coast remains robust and numbers in the hundreds of thousands. As is the case in many wild populations, more babies are born each year than can be supported by the available food supply. If the surplus isn't harvested, many in each generation will not survive their first year.

In the hunter's view, wild fur represents a manageable, renewable resource. Unlike synthetic fibers derived from non-renewable petrochemicals or cultivated crops such as cotton that require huge inputs of land, water, energy, fertilizer, and pesticides, harvesting seals doesn't harm the environment or deplete resources. Wool or leather from domesticated animals such as sheep or goats can be a sustainable way of using natural resources but overgrazing has caused enormous environmental damage in many places. By contrast, maintaining viable populations of wild furbearers requires preservation of large areas of relatively undisturbed land. And harvesting wild furs can employ more people over a wider area than does the production of cotton, wool, or synthetic fabrics. Perhaps most important to the seal hunters, it is an important source of income in this harsh northern climate where jobs are scarce.

Environmental groups such as Greenpeace, Sea Shepherd, and People for Ethical Treatment of Animals (PETA), which object to all fur hunting and trapping, find killing of baby seals particularly repugnant. Protests—some of which have turned into violent confrontations—and changing fashions reduced the take of baby seals by more than 90 percent in the mid-1980s. White-coat seal hunting increased again in the early 1990s, however, partly as a way to employ out-of-work fishers.

What do you think? Are the hunters "harvesting" a resource or "murdering" a beautiful, sentient animal? Is there a difference between killing a domestic animal such as a cow—which also is gentle, placid, and has big soulful eyes—and a wild animal? Is it more acceptable to kill an adult animal than a baby? How should we balance emotion, science, and human needs in making decisions such as this?

Figure 1

Every year hundreds of thousands of white-coated harp seal pups are born on the float ice off Canada's Arctic coast. For hunters, the luxuriant fur coats of the easily-captured seal pups are a renewable resource and a much-needed source of income. To animal rights activists, however, killing such appealing, helpless animals is barbaric. What do you think?

Figure 13.16

Sarah, a Hungarian Komondor guarding dog, protects a flock of goats from predators. Guarding dogs are not trained to herd sheep or goats; they merely stay with them to keep predators away. This not only saves livestock but avoids trapping, shooting, or poisoning the predators.

to inhabit the landscape as humans. From their perspective, killing more predators isn't the answer. Protecting flocks and herds with guard dogs or herders or keeping livestock out of areas that are home range of wild species would be a better solution (fig. 13.16).

Exotic Species Introductions

Exotics—organisms introduced into habitats where they are not native—are considered to be among the most damaging agents of habitat alteration and degradation in the world. They can be thought of as biological pollutants. Introducing species accidentally or intentionally from one habitat into another where they have never been before is risky business. Freed from the predators, parasites, pathogens, and competitors that normally keep their numbers in check, exotics often exhibit explosive population growth that crowds out native species. Their aggressive invasion might be considered a kind of ecological cancer. There now are more than 4500 alien species in the United States.

Species such as Nile perch, carp, and purple loosestrife are sometimes released intentionally because they are expected to be beneficial or attractive. Others, such as zebra mussels and spiny water fleas, are introduced accidentally, brought in on vehicles, animals, commercial goods, or even clothing. Once established, exotics are rarely eliminated, however. Alien species from around the world now threaten native species throughout the Great Lakes (fig. 13.17). Perhaps the most endangered group in the United States are freshwater mussels, of which more than half of the 297 known species are threatened, endangered, seriously declining, or already extinct. Pollution, siltation, and competition from exotics like the zebra mussel are largely responsible.

Many nuisance birds of our cities, such as starlings, English sparrows, and common pigeons, are aliens that were introduced by "acclimatizers" who spent considerable sums of money to import these species from Europe. These birds have prospered because they are opportunists that do well in urban areas, but they also have driven out more desirable native species in many places.

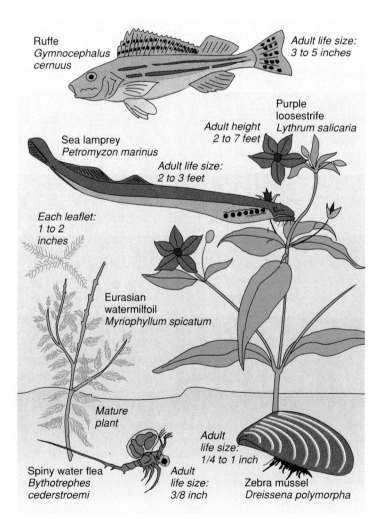

Figure 13.17

Some of the exotic organisms that have, with human help, invaded the Great Lakes. These aliens are driving out native species and irreparably damaging the ecosystem.

Biodiversity

Plants also can multiply explosively when introduced into a new environment. Kudzu, a cultivated legume in Asia, has run amok in the southern United States where it blankets trees, houses, and whole hillsides, smothering slower-growing native species. Water hyacinths are floating water plants with waxy green leaves and beautiful blue flower spikes. They were brought to the United States from Argentina in 1880 for an exhibition in New Orleans. People took plants home and either deliberately put them in streams and lakes or just tossed them out. Within ten years, this prolific weed had spread across the South from Florida to Texas, completely covering the surface of many waterways, blocking boat traffic, and smothering fish and native plants. Mechanical harvesters and poisons were tried, but the plant population doubled every two weeks. Ironically, there now appear to be some benefits to this pest that we have fought for nearly a century. The water hyacinth has been proposed as a source of biomass for energy production because it grows so fast. It is also useful in water purification, absorbing large amounts of contaminants, such as heavy metals and toxic organic compounds.

Diseases

Disease organisms, or pathogens, may also be considered predators. To be successful over the long term, a pathogen must establish a balance in which it is vigorous enough to reproduce, but not so lethal that it completely destroys its host. When a disease is introduced into a new environment, however, this balance may be lacking and an epidemic may sweep through the area.

The American chestnut was once the heart of many Eastern hardwood forests. In the Appalachian Mountains, at least one of every four trees was a chestnut. Often over 45 m (150 ft) tall, 3 m (10 ft) in diameter, fast growing, and able to sprout quickly from a cut stump, it was a forester's dream. Its nutritious nuts were important for birds (like the passenger pigeon), forest mammals, and humans. The wood was straight grained, light, rot-resistant and used for everything from fence posts to fine furniture and its bark was used to tan leather. In 1904, a shipment of nursery stock from China brought a fungal blight to the United States, and within forty years, the American chestnut had all but disappeared from its native range. Efforts are now underway to transfer blight-resistant genes into the few remaining American chestnuts that weren't reached by the fungus or to find biological controls for the fungus that causes the disease.

A similar disaster has devastated American elms. A shipment of European elm logs introduced a fungal disease fatal to the American elm. The loss is most noticeable in prairie towns of the Midwest where elms formed arching colonnades over the streets and provided an oasis of shade from the summer sun. As Dutch elm disease swept across the country, some towns lost all of their trees in just a few years. Des Moines, Iowa, had to remove some 25,000 trees in five years at a cost of about five million dollars.

Pollution

We have known for a long time that toxic pollutants can have disastrous effects on local populations of organisms. Pesticide-linked declines of fish-eating birds and falcons was well documented in the 1970s. Marine mammals, alligators, fish, and other declining populations suggest complex interrelations between pollution and health (chapter 12). Mysterious, widespread deaths of thousands of seals on both sides of the Atlantic in recent years are thought to be linked to an accumulation of persistent chlorinated hydrocarbons, such as DDT, PCBs, and dioxins, in fat, causing weakened immune systems that make animals vulnerable to infections. Similarly, mortality of Pacific sea lions, beluga whales in the St. Lawrence estuary, and striped dolphins in the Mediterranean are thought to be caused by accumulation of toxic pollutants.

Lead poisoning is another major cause of mortality for many species of wildlife. Bottom-feeding waterfowl, such as ducks, swans, and cranes, ingest spent shotgun pellets that fall into lakes and marshes. They store the pellets, instead of stones, in their gizzards and the lead slowly accumulates in their blood and other tissues. The U.S. Fish and Wildlife Service (USFWS) estimates that 3000 metric tons of lead shot are deposited annually in wetlands and that between 2 and 3 million waterfowl die each year from lead poisoning. Scavengers, such as condors and bald eagles, eat birds and mammals that have lead shot or bullet fragments in their bodies. In 1987, the last five wild California condors were trapped and put into zoos (fig. 13.18). Lead poisoning was thought to be a major cause of their high mortality in the wild. Captive breeding programs have rebuilt the condor population to about 100 birds but none have been successfully reintroduced to the wild as of 1995. Lead shot is now banned in hunting migratory waterfowl but large amounts of spent shot remain in bottom sediments where bottom feeders may consume it.

Figure 13.18

The California condor has the largest wingspan of any North American bird. Loss of habitat and food sources, shooting by hunters, and poisoning by lead shot and predator-control bait have made these giants extinct in the wild. Successful captive breeding in zoos raises hopes that a wild population can be reestablished someday.

Genetic Assimilation

Some rare and endangered species are threatened by **genetic assimilation** because they crossbreed with closely related species that are more numerous or more vigorous. Opportunistic plants or animals that are introduced into a habitat or displaced from their normal ranges by human actions may genetically overwhelm local populations. For example, hatchery-raised trout often are introduced into streams and lakes where they genetically dilute indigenous stocks. Similarly, black ducks have declined severely in the eastern United States and Canada in recent years. Hunting pressures and habitat loss are factors in this decline, but so is interbreeding with mallards forced into black duck habitat by destruction of prairie potholes in the west. As discussed earlier in this chapter, hybridization of endangered timber wolves with much more numerous coyotes or domestic dogs introduces foreign genes into an already dangerously small gene pool and reduces the likelihood that purebred animals will mate.

Endangered Species Management and Biodiversity Protection

Over the years, we have gradually become aware of the harm we have done—and continue to do—to wildlife and biological resources. Slowly, we are adopting national legislation and international treaties to protect these irreplaceable assets. Parks, wildlife refuges, nature preserves, zoos, and restoration programs have been established to protect nature and rebuild depleted populations. There has been encouraging progress in this area, but much remains to be done. While most people favor pollution control or protection of favored species such as whales or gorillas, surveys show that few understand what biological diversity is or why it is important.

Hunting and Fishing Laws

In 1874, a bill was introduced in the United States Congress to protect the American bison, whose numbers were already falling to dangerous levels. This initiative failed, however, because most legislators believed that all wildlife—and nature in general—was so abundant and prolific that it could never be depleted by human activity. As we discussed earlier in this chapter, however, by the end of the nineteenth century, bisons had plunged from some 60 million to only a few hundred animals.

By the 1890s most states had enacted some hunting and fishing restrictions. The general idea behind these laws was to conserve the resource for future human use rather than to preserve wildlife for its own sake. The wildlife regulations and refuges established since that time have been remarkably successful for many species. At the turn of the century, there were an estimated half a million white-tailed deer in the United States; now there are some 14 million—more in some places than the environment can support. Wild turkeys and wood ducks were nearly all gone fifty years ago. By restoring habitat, planting food crops, transplanting breeding stock, building shelters or houses, protecting these birds

during breeding season, and other conservation measures, populations of these beautiful and interesting birds have been restored to several million each. Snowy egrets, which were almost wiped out by plume hunters eighty years ago, are now common again.

The Endangered Species Act

Establishment of the U.S. Endangered Species Act of 1973 and the Committee on the Status of Endangered Wildlife in Canada (COSEWIC) in 1976 represented powerful new approaches to wildlife protection. Where earlier regulations had been focused almost exclusively on "game" animals, these programs seek to identify all endangered species and populations and to save as much biodiversity as possible, regardless of its usefulness to humans. **Endangered species** are those considered in imminent danger of extinction, while **threatened species** are those that have declined significantly in total numbers and may be on the verge of extinction in certain localities. **Vulnerable species** are naturally rare or have been depleted by human activities to a level that puts them at risk. Bald eagles, gray wolves, brown (or grizzly) bears, sea otters, and a number of native orchids and other rare plants are considered either vulnerable or threatened in many places even though they remain locally abundant over parts of their former range.

The United States currently has some 1350 species on its endangered and threatened species lists, and 4000 candidate species waiting to be considered. Canada, which generally has less diversity because of its boreal location, has designated a total of 46 endangered and 50 threatened species. The number of species listed in different taxonomic groups reflects much more on which organisms we consider interesting and desirable than the actual number in each category. Compare tables 13.1 and 13.4, for instance. Notice that the number listed as endangered bears little resemblance to percentage of total species each group represents.

Funding for the endangered species program in the United States has averaged about $39 million per year over the past two decades. This amounts to about 16 cents per person per year or enough to build about one mile of freeway every year. The way we spend this money reveals the political nature of wildlife conservation. In 1992, $55 million was spent on the top 12 species, more

Table 13.4 Worldwide endangered species	
Mammals	741
Birds	971
Reptiles	316
Amphibians	169
Fish	977
Invertebrates	2574
Plants	1100

Source: International Union for the Conservation of Nature, 1994.

than on all the other 1188 species together. Two of the top three species (spotted owl, $9.7 million; Bell's least vireo, $9.2 million; and grizzly bear, $5.9 million) are only threatened rather than endangered. The Florida panther—which may not be even a subspecies—got nearly $5 million (fig. 13.19). By contrast, the 81 endangered invertebrates and 239 endangered plant species got less than $5 million per year altogether. Strangely, we spent a total of about $6 million on jaguarundis and ocelots, both of which are native to Mexico and Central America but probably never occurred in the United States except immediately along the Mexican border. What explains this skewed distribution of funds? Support has much more to do with a species' charisma and the Congressional District in which it occurs than how unique it is.

Because of political opposition and limited funding, listing of species in the United States has been very slow. Hundreds of species were classified as "warranted (deserving of protection) but precluded" for lack of funds or local support. At least 18 species have gone extinct since being listed for protection and at least twice that number disappeared while waiting for consideration.

Some of the species protected by the Endangered Species Act are naturally rare and unknown, causing some people to grumble about spending money or disrupting commercial interests to save obscure organisms such as the Delhi Sands flower-loving fly, the Coachella Valley fringe-toed lizard, Mrs. Furbisher's lousewort, or the orange-footed pimple-back mussel. This raises some interesting ethical questions about the values and rights of seemingly minor species. Although uncelebrated, these species may be indicators of environmental health. Their loss may have implications that we don't yet understand and may portend changes that we may regret. Furthermore, endangered species have often been surrogates or champions for protecting whole communities threatened by our activities (fig. 13.20).

Challenges to the Endangered Species Act

The twenty-year-old U.S. Endangered Species Act (ESA) is up for reauthorization in what promises to be a major environmental battle. Real estate, business, agriculture, lumber and mineral interests are organizing to weaken or abolish it because they claim it impedes economic progress and favors the rights of plants and animals over those of business and private property. Ranchers, loggers, miners and others in the "Wise Use Movement" demand greater consideration of human needs and placing economic impacts and property rights above ecological considerations in listing endangered species (What Do *You* Think?, p. 286).

As part of the Republican "Contract with America," more than 70 bills intended to modify or abolish the ESA were introduced in Congress in 1995. Often written by industry lobbyists and given titles such as the "Farm, Ranch, and Homestead Protection Act" or the "Private Property Protection Act," these bills seek to repeal government power to restrict land use and to reduce the number of species listed as endangered. In spite of horror stories about poor farmers and ranchers unable to use their land because of excessive regulations, however, only 54 of the 98,000 projects reviewed by the U.S. Fish and Wildlife Service between 1987 and 1992 were terminated or withdrawn because of endangered species restrictions.

Figure 13.19

Only forty Florida panthers remain in the Everglades. We spend millions of dollars to protect them but the population may be too small to be viable. Furthermore, they may have hybridized with escaped pets and circus animals so they no longer represent a unique local race.

Judicial challenges to the ESA also have been numerous. A major victory occurred in 1995 when the U.S. Supreme Court ruled that "harming" species includes significant habitat modification or degradation, and that government intervention in private land use to protect endangered species is justified.

Environmentalists, too, want changes in the law. Criticizing the government for its slowness in listing endangered species, and emphasis on charismatic, "flagship" species such as tigers and elephants, many people call for a holistic, ecosystem approach to protecting wildlife and habitat. Calling themselves the "Endangered Species Coalition" a group of 40 conservation groups banded together to lobby for an even stronger ESA.

"DAMN SPOTTED OWL!"

Figure 13.20

Endangered species often serve as a barometer for the health of an entire ecosystem and as surrogate protector for a myriad of less well-known creatures. Abolishing the protection offered by the Endangered Species Act is the number one priority of the Wise Use Movement.

© 1990 by Herblock in the Washington Post.

In an effort to reach a compromise between these opposing camps, Interior Secretary Bruce Babbitt launched a National Biological Service in 1993 to provide scientifically accurate information about rare and endangered species in the U.S. as well as an inventory of the biological communities and ecosystems in which they live. By discovering in advance which species are threatened or endangered, the survey was intended to help avoid the expensive emergency recovery programs or what Secretary Babbitt called "national train wrecks" like the conflict over the northern spotted owl. Unfortunately, after only two years of work, funds for the Biological Service have been reduced and its future is in doubt.

Recent negotiations for preserving habitat of the threatened California gnatcatcher may serve as a model for dispute resolution under the ESA. This tiny blue-gray member of the thrush family, which formerly was plentiful but now is rare, lives exclusively in coastal sagebrush scrub between San Diego and Los Angeles. Only about 2800 pairs remain in the 100,000 ha (250,000 ac) of open land along the coast—some of the most valuable real estate in the world. In exchange for agreeing to work with environmentalists and local and state officials on a gnatcatcher conservation plan, builders are being allowed to destroy some of the bird's habitat if matched by replacement areas or other compensation. Neither side is completely happy with this compromise. Environmentalists wanted more protected land, developers wanted less. But by working together, the developers get to use most of their land and environmentalists get funding and land to save most of the gnatcatchers.

Recovery Programs

The Endangered Species Act not only prohibits projects or actions that jeopardize endangered species or critical habitats, it also requires the U.S. Fish and Wildlife Service to prepare a **species recovery plan** that spells out how the species can be restored to numbers that permit its delisting. A few of these plans have been gratifyingly successful. The American alligator was listed as endangered in 1967 because hunting (for meat, skins, and sport) and habitat destruction had reduced populations to precarious levels. Protection has been so effective that the species is now plentiful and has been delisted throughout its entire southern range. Florida alone estimates that it has at least 1 million alligators.

Twenty years ago, due to DDT poisoning, only 800 bald eagles remained in the contiguous United States. In 1994, the population had rebounded to more than 8000 and the eagle's status was reduced from endangered to threatened. Peregrine falcons, which had completely disappeared from the eastern U.S. because of DDT, have been bred in captivity and released back in their former range. Most American cities now have resident falcons swooping between office towers and feeding on plump urban pigeons.

Among the more controversial recovery plans now being considered is one for the northern spotted owl (chapter 14), whose protection may require a moratorium on harvesting of old-growth forest in the Pacific Northwest. An even more costly program may be needed to save native populations of Columbia River salmon and steelhead endangered by hydropower dams and water storage reservoirs that block their migration routes to the sea. Fish ladders were built at many dams to facilitate migration but have not been very effective.

Opening up floodgates to allow young fish to run downriver and adults to return to spawning grounds could have grave consequences for barge traffic, farmers, electric rate payers, and others who have come to depend on abundant water and cheap power. Electric rates in Washington and Oregon could increase 10 to 30 percent, costing somewhere around $200 million to $600 million per year. On the other hand, commercial and sport fishing for salmon is worth $1 billion per year and employed about 60,000 people in 1992.

Minimum Viable Populations

A critical question in all recovery programs is the minimum population size required for long-term viability of rare and endangered species. As chapter 6 shows, a species composed of a small

Biodiversity

What Do *You* Think?

Economic Impacts of the Endangered Species Act

Reauthorization of the Endangered Species Act, originally scheduled for 1992, has been delayed because of the heated debate that has developed over the Act's provisions. Some people want the Act repealed, others want it strengthened, while some think it should be modified to accommodate the concerns of both sides. Most people support the idea that decisions ought to flow from factual information. Not surprisingly, advocates on both sides of the Endangered Species Act debate attempt to present factual information to convince others of the soundness of their position.

Sound arguments and sound judgments can only be built on representative information, however. Unfortunately, anecdotal information is often used to project generalized conclusions. Because of the inherent variability of events, arguments based on anecdotes or small samples may not reflect generalized reality at all. That is why all sound scientific investigations include multiple replications—to avoid basing conclusions on a few outcomes that are not truly representative of the whole.

Use of the "generalization from anecdote" method of constructing an argument is common on both sides of many issues, including the Endangered Species Act debate.

Charles Oliver feels the Endangered Species Act ought to be repealed. His position is based, in part, on the argument that the Act significantly damages our economic well-being. He opens his argument with the following statements:

"In Washington state, logger Dean Hurn is worried. He could be among the estimated 40,000 to 100,000 timber workers who will lose their jobs to protect the Northern spotted owl. . . . In Florida's Polk and Highland counties, 600 landowners can't build the homes they planned because their property is home to the Florida scrub jay. In these and many other cases, the Endangered Species Act has halted the plans and dreams of thousands of ordinary Americans. The act is up for renewal this year, and one might expect a piece of legislation that has harmed so many people to be itself endangered. . . ."

The generalization implied is that the Endangered Species Act is responsible for massive economic dislocations. But other information calls this idea into question. An evaluation of 18,211 cases reviewed under ESA provisions between 1987 and 1991 revealed that 16,161 projects did not even require a biological opinion and proceeded without interference. In 1869 other cases, the biological opinion was that the proposal did not threaten an endangered species and the project could proceed as planned. Of the 181 remaining cases, most were completed, although with modifications in design. In all, over 99 percent of the projects examined under Endangered Species Act provisions proceeded unhindered or with marginal added cost or time delay.

By contrast, what generalization might be implied about the economic consequences of the Endangered Species Act based on the following anecdote as reported in the newsletter *Common Ground*?

"In Oregon, contrary to prediction, the sky isn't falling. Protecting northern spotted owls and old-growth forests, said many, would mean ruin to the state's economy. But Oregon, still the top timber-producing state despite less access to national forests, enjoys its lowest unemployment rate in years, well below the national average. Formerly timber-dependent counties report rising property values and a net increase in jobs. The 15,000 timber industry jobs lost by the state over 5 years have been offset by nearly 20,000 high tech jobs. Sony Corp., for instance, is coming to Springfield, Ore., lured in part by money from President Clinton's recovery program. "Owls versus jobs was just plain false," Mayor Bill Morrisette told The New York Times. "What we've got here is quality of life. As long as we don't screw that up, we'll always be able to attract people and business."

Can we generalize from this that the Endangered Species Act stimulates local economies? Hardly. Critical thinkers distrust anecdotes, regardless of the source, because valid generalizations cannot be built on small samples.

number of individuals can undergo catastrophic declines due to environmental change, genetic problems, or simple random events when isolated in a limited geographic range. This phenomenon was elegantly described as **island biogeography** in the work of R. H. MacArthur and E. O. Wilson in 1967. Noticing that small islands far from a mainland have fewer terrestrial species than larger, nearer islands, MacArthur and Wilson proposed that species diversity is a balance between colonization and extinction rates (fig. 13.21). An island far from a population source naturally has a lower rate of colonization than a nearer island because it is harder for terrestrial organisms to reach. At the same time, a large island can support more individuals of a given species and is, therefore, less likely to suffer extinction due to natural catastrophes, genetic problems, or demographic uncertainty—the chance that all the members of a single generation will be of the same sex.

Island biogeographical effects have been observed in many places. Cuba, for instance, is one hundred times as large and has about ten times as many amphibian species as its Caribbean neighbor, Monserrat. Similarly, in a study of bird species on the California Channel Islands, Jared Diamond observed that on islands with fewer than 10 breeding pairs, 39 percent of the populations went extinct over an 80-year period, while only 10 percent of populations numbering between 10 and 100 pairs went extinct in the same time (fig. 13.22). Only one species numbering between 100 and 1000 pairs went extinct and no species with over 1000 pairs disappeared over this time.

Small population sizes affect species in isolated landscapes other than islands. Grizzly bears (*Ursus arctos horribilis*) once roamed across most of Western North America. Hunting and habitat destruction reduced the number of grizzlies in the lower 48

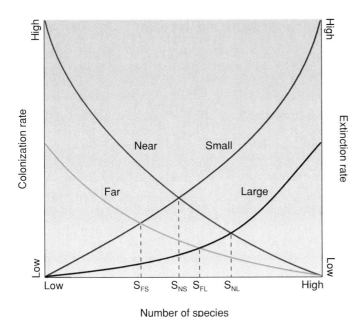

Figure 13.21

Predicted species richness on an island resulting from a balance between colonization (immigration) and extinction by natural causes. This island biogeography theory of MacArthur and Wilson (1967) is used to explain why large islands near a mainland (S_{NL}) tend to have more species than small, far islands (S_{FS}).

Source: Based on MacArthur and Wilson, *The Theory of Island Biogeography*, © 1967 Princeton University Press.

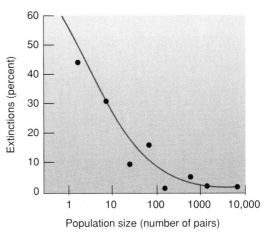

Figure 13.22

Extinction rates of bird species on the California Channel Islands as a function of population size over eighty years.

Source: Data from H.L. Jones and J. Diamond, "Short-term-base Studies of Turnover in Breeding Bird Populations on the California Coast Island," *Condor*, vol. 78: 526–549, 1976.

states from an estimated 100,000 in 1800 to less than 1000 animals in six separate subpopulations that now occupy less than 1 percent of the historic range. Recovery target sizes—based on estimated environmental carrying capacities—are less than 100 animals for some subpopulations. Conservation biologists predict that a completely isolated population of 100 bears cannot be maintained for more than a few generations. Even the 200 bears in Yellowstone National Park will be susceptible to genetic problems if completely isolated. Interestingly, computer models suggest that translocating only two unrelated bears into small populations every decade or so could greatly increase population viability.

Clearly loss of genetic diversity causes a variety of harmful effects that limit adaptability, reproduction, and species survival. How is diversity lost in small populations? (1) A *founder effect* occurs when a few individuals establish a new population. The limited genetic diversity from those original founders may not be enough to sustain the population. (2) A *demographic bottleneck* arises when only a few individuals survive some catastrophe. As the population replenishes itself, a limited genetic diversity similar to that in founder effect results. (3) *Genetic drift* is a reduction in gene frequency in a population due to unequal reproductive success, for example, some individuals breed more than others and their genes gradually come to dominate the population. (4) *Inbreeding* is mating of closely related individuals. Random, recessive, deleterious mutations—what we consider genetic diseases such as hemophilia in humans—that are usually hidden in a

widely outcrossed population can be expressed when inbreeding occurs. This is why we have laws and cultural taboos prohibiting mating between siblings or first cousins in humans.

Some examples of demographic bottlenecks and inbreeding problems in wildlife include the Hawaiian nene goose and the African cheetah. Flightless nene geese once were common in Hawaii, but by 1940 only 43 were left. A captive breeding program has rebuilt the population to 3000 birds (mostly in England), but male infertility has become a serious problem because of lack of genetic diversity. Similarly, DNA studies suggest that all existing cheetahs originated from a single female ancestor some time in the not-too-distant past. All male cheetahs, no matter where in the world they live, are essentially genetically identical. Deformed sperm and developmental problems reduce reproductive success to less than 30 percent. Despite our best efforts, these species may be genetically doomed.

Habitat Protection

Over the past decade, growing numbers of scientists, land managers, policymakers, and developers have been making the case that it is time to focus on a rational, continent-wide preservation of ecosystems that support maximum biological diversity rather than a species-by-species battle for the rarest or most popular organisms. By focusing on populations already reduced to only a few individuals, we spend most of our conservation funds on species that may be genetically doomed no matter what we do. Furthermore, by concentrating on individual species we spend millions of dollars to breed plants or animals in captivity that have no natural habitat where they can be released. While flagship species such as California condors or Indian tigers are reproducing well in zoos and wild animal parks, the ecosystems that they formerly inhabited have largely disappeared.

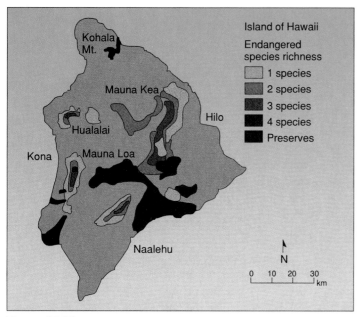

Figure 13.23

An example of the biodiversity maps produced by J. Michael Scott and the U.S. Fish and Wildlife Service. Notice that few of the areas of endangered species richness are protected in preserves, which were selected more for scenery or recreation than for biology.

A leader of this new form of conservation is J. Michael Scott, who was project leader of the California condor recovery program in the mid-1980s and had previously spent 10 years working on endangered species in Hawaii. In making maps of endangered species, Scott discovered that even Hawaii, where more than 50 percent of the land is federally owned, has many vegetation types completely outside of natural preserves (fig. 13.23). The gaps between protected areas may contain more endangered species than are preserved within them.

This observation has led to an approach called **gap analysis** in which conservationists and wildlife managers look for unprotected landscapes that are rich in species. Computers and geographical information systems (GIS) make it possible to store, manage, retrieve, and analyze vast amounts of data and create detailed, high-resolution maps relatively easily. This broad-scale, holistic approach seems likely to save more species than a piecemeal approach.

Conservation biologist, R. E. Grumbine suggests four remanagement principles for protecting biodiversity in a large-scale, long-range approach:

1 Protect enough habitat for viable populations of all native species in a given region.

2 Manage at regional scales large enough to accommodate natural disturbances (fire, wind, climate change, etc.).

3 Plan over a period of centuries so that species and ecosystems may continue to evolve.

4 Allow for human use and occupancy at levels that do not result in significant ecological degradation.

International Wildlife Treaties

The 1975 Convention on International Trade in Endangered Species (CITES) was a significant step toward worldwide protection of endangered flora and fauna. It regulated trade in living specimens and products derived from listed species, but has not been foolproof. Species are smuggled out of countries where they are threatened or endangered, and documents are falsified to make it appear they have come from areas where the species are still common. Investigations and enforcement are especially difficult in developing countries where wildlife is disappearing most rapidly. Still, eliminating markets for endangered wildlife is an effective way of stopping poaching. In the 1980s, as much as 500 metric tons of ivory were sold every year in world markets. The 1990 world ban on ivory trade dried up almost all of that traffic. Similar bans passed in 1992 on trade in endangered tropical birds and sea turtles appear to be having similarly beneficial effects.

Zoos, Botanical Gardens, and Captive Breeding Programs

Records of wildlife and exotic plant collections date back to the beginning of history. The ancient Chinese, Egyptians, Romans, and others kept menageries of exotic and bizarre wildlife. Many of these collections included humans as well as animals. War captives and strangers from distant lands were displayed in cages along with alien animals. Fights between animals or between animals and humans were common entertainment in earlier times.

In some places, low-life circuses and seedy roadside tourist attractions still exhibit sad collections of animals in cramped, unhealthy cages (fig. 13.24). In developed countries, however, these cruel and callous conditions have generally been eliminated by an-

Figure 13.24

Many zoos, such as this one in Central America, still house animals in cramped, unsanitary cages, where they are tormented by insensitive onlookers and deprived of any semblance of a normal, healthy life.

imal welfare laws. Most modern zoos and wildlife parks now keep animals in clean, humane conditions and have a strong commitment to species preservation and public education. Although they are held captive, many animals live longer in zoos and raise more young than they would in the wild. Often, animals that escape from their cages will voluntarily return because conditions are better there than on the outside.

Botanical gardens, such as the Kew Gardens in England, are repositories for rare and endangered plant species that sometimes have ceased to exist in the wild. Valuable genetic traits are preserved in these collections, and in some cases, plants with unique cultural or ecological significance may be reintroduced into native habitats after being cultivated for decades or even centuries in these gardens.

New Zoos and Game Parks

Around 1900, Karl Haugenbeck, a German wildlife dealer, introduced a new style of displaying animals. He created naturalistic enclosures that attempted to show creatures in something like their native habitat. By using moats rather than fences to contain animals, Haugenbeck gave patrons a better view and a more intimate experience of the wildlife. Gradually, these principles have spread to other zoos. Clever landscaping can create uncannily authentic terrain, and inconspicuous barriers confine humans rather than animals. Large enclosures and multi-species groupings allow more normal social behavior and encourage reproduction (fig. 13.25). Under these conditions, zoos can be important places for scientific research in wildlife biology.

Today in the United States some 150 municipal zoos provide the only experience that most people ever have with large wild animals. More than 110 million visitors per year learn about nature through visits to zoos. Many new zoos de-emphasize the crowd-pleasing exotic animals such as lions, tigers, and elephants, con-

Figure 13.25

Newer zoos and game parks are making special efforts to provide comfortable, naturalistic habitats. It can be difficult, however, to find a balance between accessibility to visitors and privacy for animals.

centrating instead on species that normally inhabit their climatic regime. Zoo visitors, however, often object that they want to see lions and tigers rather than indigenous species. They also complain that spacious, vegetated, modern enclosures hide animals from sight. If you were a zookeeper how would you balance entertainment and conservation?

Captive Breeding and Species Survival Plans

Until fairly recently, zoos depended on primarily wild-caught animals for most of their collections. This was a serious drain on wild populations, because up to 80 percent of the animals caught died from the trauma of capture and shipping. With better understanding of reproductive biology and better breeding facilities, most mammals in North American zoos now are produced by captive breeding programs. These programs have limitations, however. Bats, whales, and many reptiles rarely reproduce in captivity and still come mainly from the wild. Furthermore, we will never be able to protect the complete spectrum of biological variety in zoos. Worldwide, zoos house about 500,000 individual animals, but only about 900 species in total. Few of these animals are maintained in populations large enough to preserve the species indefinitely if wild populations were lost. According to one estimate, if all the space in U.S. zoos were used for captive breeding, only about one hundred species of large mammals could be maintained on a long-term basis.

These limitations lead to what is sometimes called the "Noah question": how many species can or should we save? How much are we willing to invest to protect the slimy, smelly, crawly things? Would you favor preserving disease organisms, parasites, and vermin or should we use our limited resources to protect only beautiful, interesting, or seemingly useful organisms?

Even given adequate area and habitat conditions to perpetuate a given species, continued inbreeding of a small population in captivity can lead to the same kinds of fertility and infant survival problems described earlier for wild populations. To reduce genetic problems, zoos often exchange animals or ship individuals long distances to be bred. It sometimes turns out, however, that zoos far distant from each other unknowingly obtained their animals from the same source. Computer data bases operated by the International Species Information System located at the Minnesota Zoo in Apple Valley, now keep track of the genealogy of many species. This system can tell the complete reproductive history of every animal in every zoo in the world for some species. Comprehensive species survival plans based on this genealogy help match breeding pairs and project resource needs.

Saving Rare Species in the Wild

Renowned zoologist George Schaller says that ultimately "zoos need to get out of their own walls and put more effort into saving the animals in the wild." An interesting application of this principle is a partnership between the Minnesota Zoo and the Ujung Kulon National Park in Indonesia, home to the world's few remaining Javanese rhinos. Rather than try to capture rhinos and move them to Minnesota, the zoo is helping to protect them in their

native habitat by providing patrol boats, radios, housing, training, and salaries for Indonesian guards (fig. 13.26). There are no plans to bring any rhinos to Minnesota and chances are very slight that any of us will ever see one, but we can gain satisfaction that, at least for now, a few Javanese rhinos still exist in the wild. As the great wildlife biologist, Aldo Leopold, said, "The first rule of intelligent tinkering is to save all the pieces."

Figure 13.26

The *KM Minnesota* anchored in Tamanjaya Bay in west Java. Funds raised by the Minnesota Zoo paid for local construction of this boat, which allows wardens to patrol Ujung Kulon National Park and protect rare Javanese rhinos from poachers.

Summary

In this chapter, we have briefly surveyed world biodiversity and the ways humans both benefit from and threaten it. Natural causes of wildlife destruction include evolutionary replacement and mass extinction. Among the threats from humans are overharvesting of animals and plants for food and commercial products. Millions of live wild plants and animals are collected for pets, houseplants, and medical research. Among the greatest damage we do to biodiversity are habitat destruction, the introduction of exotic species and diseases, pollution of the environment, and genetic assimilation.

The potential value of the species that may be lost if environmental destruction continues could be enormous. It is also possible that the changes we are causing could disrupt vital ecological services on which we all depend for life.

The first hunting and fishing laws in the United States were introduced more than a century ago to restrict overexploitation and to preserve species for future uses. The Endangered Species Act and CITES represent a new attitude towards wildlife in which we protect organisms just because they are rare and endangered. Now we are expanding our concern from individual species to protecting habitat, threatened landscapes, and entire biogeographical regions. Social, cultural, and economic factors must also be considered if we want to protect biological resources on a long-term, sustainable basis.

Zoos can be educational and entertaining while still serving important wildlife conservation and scientific functions. Modern zoos have greatly improved the living conditions for captive animals, resulting in improved breeding success; still, there are limits to the number and types of species that we could maintain under captive conditions. Zoos need to get out of their own walls and help save animals and plants in the wild.

▼ Questions for Review

1. What is the range of estimates of the total number of species on the earth? Why is the range so great?

2. What group of organisms has the largest number of species?

3. Define *extinction*. What is the natural rate of extinction in an undisturbed ecosystem?

4. What are rosy periwinkles and what products do we derive from them?

5. Describe some foods we obtain from wild plant species.

6. List three categories of damage to biological resources caused by humans.

7. What is the current rate of extinction and how does this compare to historic rates?

8. Compare the scope and effects of the Endangered Species Act and CITES.

9. Describe ten ways that humans directly or indirectly cause biological losses.

10. What is gap analysis and how is it related to ecosystem management and design of nature preserves?

Questions for Critical Thinking

1. One reviewer said that this chapter is the most biased in this book. Do you agree? How much moral outrage is appropriate in an issue such as this? Does emotion interfere with rational analysis or effective communication? What is the proper balance between emotion and objectivity in a subject such as this?

2. Many ecologists would like to move away from protecting individual endangered species to concentrate on protecting whole communities or ecosystems. Others fear that the public will only respond to and support glamorous "flagship" species such as gorillas, tigers, or otters. If you were designing conservation strategy, where would you put your emphasis?

3. Put yourself in the place of a fishing industry worker. If you continue to catch many species they will quickly become economically extinct if not completely exterminated. On the other hand, there are few jobs in your village and welfare will barely keep you alive. What would you do?

4. Only a few hundred grizzly bears remain in the contiguous United States, but populations are healthy in Canada and Alaska. Should we spend millions of dollars for grizzly recovery and management programs in Yellowstone National Park and adjacent wilderness areas?

5. How could people have believed a century ago that nature is so vast and fertile that human actions could never have a lasting impact on wildlife populations? Are there similar examples of denial or misjudgment occurring now?

6. In the past, mass extinction has allowed for new growth, including the evolution of our own species. Should we assume that another mass extinction would be a bad thing? Could it possibly be beneficial to us? To the world?

7. Some captive breeding programs in zoos are so successful that they often produce surplus animals that cannot be released into the wild because no native habitat remains. Plans to euthanize surplus animals raise storms of protests from animal lovers. What would you do if you were in charge of the zoo?

8. Debate with a friend or classmate the ethics of keeping animals captive in a zoo. After exploring the subject from one side, debate the issue from the opposite perspective. What do you learn from this exercise?

9. The United States and Russia have the last known remaining stocks of the smallpox virus. Should they be destroyed? Will this diminish biological diversity?

Key Terms

biodiversity	270	genetic assimilation	283
endangered species	283	island biogeography	286
existence value	274	species recovery plan	285
extinction	275	threatened species	283
gap analysis	288	vulnerable species	283

Additional Information on the Internet

Biodiversity: An Overview
   http://www.wcmc.org.uk/infoserv/biogen/biogen.html/

Biodiversity, Ecology, and the Environment
   http://golgi.harvard.edu/biopages/biodiversity.html/

Biodiversity from Around the World
   http://www.igc.apc.org/igc/ian.html/

Earth Island Institute
   http://www.earthisland.org/ei/

National Audubon Society
   http://www.audubon.org/audubon/

The NetVet and the Electronic Zoo
   http://netvet.wustl.edu/

United Nations Convention on Biological Diversity
   http://www.unep.ch/biodiv.html/

US Fish and Wildlife Service
   http://www.fws.gov/

World Conservation Monitoring Centre
   http://www.wcmc.org.uk/

World Wide Web Virtual Library: Environment-Biosphere
   http://ecosys.drdr.Virginia.EDU:80/bio.html/

   Note: Further readings appropriate to this chapter are listed on p. 600.

A summer flower display in a Colorado aspen forest shows that forests have values beyond production of paper pulp and lumber.

CHAPTER 14: Land Use: Forests and Rangelands

Objectives

After studying this chapter, you should be able to:

- discuss the major world land uses and how human activities impact these areas.

- summarize some forest types and the products we derive from them.

- report on how and why tropical forests are being disrupted as well as how they might be better used.

- understand the major issues concerning forests in more highly developed countries such as the United States and Canada.

- outline the extent, location, and state of grazing lands around the world.

- describe how overgrazing causes desertification of rangelands.

- evaluate landownership patterns and explain why land reform and recognition of indigenous rights are essential for social justice as well as environmental protection.

The Buffalo Commons

In 1990, Frank and Deborah Popper, two professors from New Jersey, outraged many Great Plains residents by proposing that 3.6 million sq km (139,000 sq mi) of farms and ranches in a swath stretching from Texas to North Dakota be turned into a nature preserve. Called the Buffalo Commons, this huge expanse would be managed for wildlife, recreation, and resource conservation. Much of it might even be given back to Native Americans who could earn money by harvesting wildlife and taking tourists on wild west camping trips.

This proposal shocked and offended many westerners whose families have lived on and worked the land for generations. But radical changes are necessary, the Poppers argued, because, except for a few oasis towns, the high plains were never suited for settlement. Depleted soil, dwindling water supplies, and degraded landscapes show how unsustainable our past practices have been. Abandoned homesteads dot the plains (fig. 14.1) as

Figure 14.1

This abandoned homestead on the American Great Plains symbolizes the human and environmental costs of inappropriate land use. Plowing prairie sod and overgrazing have led to Dust Bowl erosion and desertification in many semiarid regions of the world.

populations in hundreds of counties have fallen below levels necessary to support rural communities. And yet change is painful, and many people would find it very difficult to leave their present homes and occupations to start over in some new place.

This case illustrates some of the difficulties of establishing sustainable, socially just, ecologically sound land-use policies. Over the past 10,000 years, humans have transformed much of the earth's land area (table. 14.1). Careful stewardship can produce pleasant and productive landscapes—at least for humans—that are essential for civilized life. Increasingly, however, the ways we use—and abuse—the land are exhausting resources and disrupting ecosystems essential for future generations. How can we adapt our land-use practices in order to safeguard the interests and needs of both human and nonhuman species? More than that, how can we do so without trampling the rights and wishes of people whose lives and identities are tied to a specific place?

In this chapter, we will survey world land use, focusing especially on forests and rangelands (grasslands and open woodlands suitable for grazing livestock). We will examine some problems associated with land ownership and management as well as some ways land resources can be—and are being—protected. ▶

Table 14.1

Estimated changes in areas of major land cover types between preagricultural times and the present (in millions of km².

Land Cover Type	Preagricultural Area	Present Area	Percent Change
Tropical forest	12.8	12.3	-3.9
Other forest	34.0	27.0	-20.6
Woodland	9.7	7.9	-18.6
Shrubland	16.2	14.8	-8.6
Grassland	34.0	27.4	-19.4
Tundra	7.4	7.4	0.0
Desert	15.9	15.6	-1.9
Cultivation	0.0	17.6	+1760.0

Source: W. Meyer and B. L. Turner, (eds.) 1994. *Changes in Land Use and Land Cover: A Global Perspective.* Cambridge University Press, Cambridge, UK.

World Land Uses

The earth's total land area is about 144.8 million sq km (55.9 million sq mi), or about 29 percent of the surface of the globe. Figure 14.2 shows the area devoted to four major land-use categories. Much of the land that falls into the residual "other" category is naturally tundra, marsh, desert, scrub forest, bare rock, and ice or snow. About one-third of this land is so barren that it lacks plant cover altogether. While deserts and other unproductive lands are generally unsuitable for intensive human use, they play an important role in biogeochemical cycles and as a refuge for biological diversity. Presently,

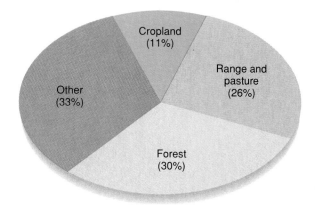

Figure 14.2

World land use. The "other" category includes tundra, desert, wetlands, and urban areas.

Source: FAO 1994.

only about 4 percent of the world's land surface is formally protected in parks, wildlife refuges, and nature preserves (chapter 15).

Notice that approximately 11 percent of the earth's landmass is now used for crops. Some agricultural experts claim that as much as half of the 7.2 billion ha (18 billion ac) of present forests and grazing lands—especially in Africa and South America—could be converted to crop production, given the proper inputs of water, fertilizer, erosion control, and mechanical preparation. Although this land could feed a vastly larger human population (perhaps ten times the present number), sustained intensive agriculture could result in serious environmental and social problems (chapter 11).

Rapidly increasing human populations and expanding forestry and agriculture have brought about extensive land-use changes throughout the world. Humans have affected every part of the globe, and we now dominate most areas with temperate climates and good soils. Table 14.1 shows land cover has changed over the past 10,000 years. Notice large reductions in forests, woodlands, and grasslands during this time. Most of these lands have been converted to cropland or permanent pasture, but overharvesting, erosion, pollution, and other forms of degradation have turned large areas to desert or useless scrub. Biodiversity losses resulting from disruption of natural ecosystems is of great concern (chapter 13). Air pollution-related forest declines are discussed in chapter 18.

Cutting down forests or plowing grasslands have immediate and obvious destructive impacts on landscapes and wildlife. Given enough time, however, nature can be surprisingly resilient. New England, for example, lost most of its native forests to agriculture by the mid-nineteenth century, but today the region is largely reforested. Vermont, which had only 35 percent of its woods standing in 1850, is now 80 percent forested. Upstate New York has

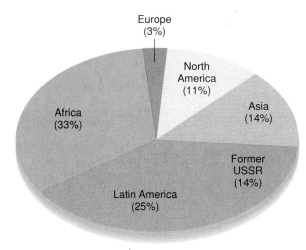

Figure 14.3
World forests by region.

Source: UNEP 1993–1994.

three times the population today that it had 150 years ago, yet it also has three times as much forest. Some rare species have been lost, but most of us would have difficulty distinguishing between second growth and primeval woods. Large mammals such as bears, moose, coyotes, and even wolves and mountain lions that had been absent for centuries now are reappearing. While this doesn't give us license to do anything we want to natural landscapes, there is hope that damage can be repaired, as we will discuss later in this chapter.

World Forests

Forests play a vital role in regulating climate, controlling water runoff, providing shelter and food for wildlife, and purifying the air. They produce valuable materials, such as wood and paper pulp, on which we all depend. Furthermore, forests have scenic, cultural, and historic values that deserve to be protected. In this section, we will look at forest distribution, use, and management.

Forest Distribution

Before large-scale human disturbances of the world began many thousands of years ago, forests and woodlands probably covered nearly 6 billion ha (15 billion ac). Since then, about 16 percent of that area has been converted to cropland, pasture, settlements, or unproductive wastelands. The 4.7 billion ha still forested covers around 30 percent of the earth's land surface, nearly three times as much as all croplands. About four-fifths of the forest is classified as **closed canopy** (where tree crowns spread over 20 percent or more of the ground) and has potential for commercial timber harvests. The rest is **open canopy** forest or **woodland,** in which tree crowns cover less than 20 percent of the ground.

Figure 14.3 shows the distribution of forest by region, while figure 14.4 shows the world's main vegetation zones. Russia, Canada, and the United States have vast areas of temperate deciduous or boreal coniferous forests (see chapter 4 for further description of these biomes). Together with Brazil, these countries account for 56 percent of all closed forests. South America and Central Africa have the largest remaining closed-canopy, broad-leaved, seasonal or deciduous tropical forests. Africa has the largest amount of open woodlands, mainly in the dry savannas and thorn brush of the sub-Saharan region.

Forest Products

Wood plays a part in more activities of the modern economy than does any other commodity. There is hardly any industry that does not use wood or wood products somewhere in its manufacturing and marketing processes. Think about the amount of junk mail, newspapers, photocopies, and other paper products that each of us in developed countries handles, stores, and disposes of in a single day. Total world wood consumption is about 3.7 billion metric tons or somewhat more than 3.7 billion cubic meters annually (table 14.2).

Industrial timber and roundwood (unprocessed logs) are used to make lumber, plywood, veneer, particleboard, and chipboard. Together, they account for slightly less than one-half of

worldwide wood consumption (about 1.66 billion tons per year). This exceeds the use of steel and plastics combined. International trade in wood and wood products amounts to more than $100 billion each year. Developed countries produce less than half of all industrial wood but account for about 80 percent of its consumption. Less-developed countries, mainly in the tropics, produce more than half of industrial wood but use only 20 percent.

The United States, the former Soviet Union, and Canada are the largest producers of both industrial wood (lumber and panels) and paper pulp. Although old-growth, virgin forest with trees large enough to make plywood or clear furniture lumber is diminishing everywhere, much of the industrial logging in North America and Europe occurs in managed forests where cut trees are replaced by new seedlings. In contrast, tropical hardwoods in Southeast Asia, Africa, and Latin America are being cut at an unsustainable rate, mostly from virgin forests.

Japan is by far the world's largest net importer of wood, purchasing about 43 million cubic meters per year or 87 percent of net Asian imports. Ironically, the United States is both a major exporter and importer of wood because we buy wood and paper pulp from Canada and finished wood products from Asia at the same time that we sell raw logs, rough lumber, and waste paper to Japan and other countries.

More than half of the people in the world depend on firewood or charcoal as their principal source of heating and cooking fuel. Consequently, **fuelwood** accounts for almost half of all wood harvested worldwide. Unfortunately, burgeoning populations and dwindling forests are causing wood shortages in many less-developed countries. About 1.5 billion people who depend on fuelwood as their primary energy source have less than they need. At present rates of population growth and wood consumption, the deficit is expected to increase from 400 million cubic meters (m^3) in 1990, to 2600 m^3 in 2025. At that point, the demand will be twice the available fuelwood supply. The average amount of wood used for cooking and heating in sixty-three less-developed countries is

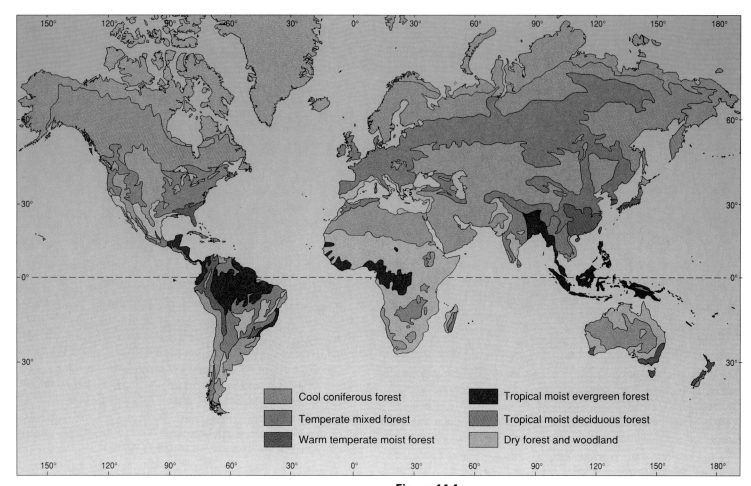

Figure 14.4

Main vegetation zones of the world's forests under natural conditions.

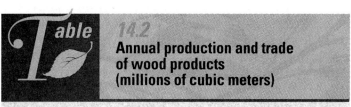

Annual production and trade of wood products (millions of cubic meters)

Region	Fuel & Charcoal	Roundwood	Paper	Net Trade
Africa	456	58	3	−4.1
Asia	817	254	56	+49.5
North & Central America	147	603	91	−25.0
South America	238	105	8	−5.8
Europe	53	314	67	+17.5
Former USSR	81	294	10	−15.9
Oceania	9	32	3	−10.8
World	1801	1661	238	—

Notes: A negative trade number indicates net exports, while a positive number indicates net imports. Roundwood includes lumber, panels, timbers, and other industrial wood products.
Source: *World Resources 1994–95.*

about 1 m^3 per person per year, roughly equal to the amount of wood that each American consumes as paper products alone.

Many people in poorer countries cook over open fires that deliver only about one-tenth of the available heat to cooking pots. Inexpensive metal stoves can double this efficiency, while locally-made ceramic stoves can be four times as efficient as open fires (fig. 14.5). These stoves can save up to 20 percent of household income for urban families.

Forest Management

Approximately 25 percent of the world's forests are managed scientifically for wood production. **Forest management** involves planning for sustainable harvests, with particular attention paid to forest regeneration. Aside from human use, what are some factors that contribute to forest loss? Fires, insects, and diseases damage up to one-quarter of the annual growth in temperate forests. Recently, reduced forest growth and sudden die-off of certain tree species in industrialized countries have caused great concern. It is thought that long-range transport of air pollutants (chapter 18) is contributing to this sudden forest death, but not all the causes and solutions are yet understood.

Figure 14.5

Locally made of inexpensive materials, this ceramic-lined metal Jiko stove and clay pot from Kenya are outstanding examples of appropriate technology. Using only one-fourth as much wood as an open fire, these stoves save forests, reduce air pollution, and save up to twenty percent of household income.

Most countries replant far less forest than is harvested or converted to other uses, but there are some outstanding examples of successful reforestation. China, for instance, cut down most of its forests one thousand years ago and has suffered centuries of erosion and terrible floods as a consequence. Recently, however, a massive reforestation campaign has been started. An average of 4.5 million ha per year were replanted during the last decade. South Korea also has had very successful forest restoration programs. After losing nearly all its trees during the civil war 30 years ago, the country is now about 70 percent forested again.

In spite of being the world's largest net importer of wood, Japan has increased forests to approximately 68 percent of its land area. Strict environmental laws and constraints on the harvesting of local forests encourage imports so that Japan's forests are being preserved while it uses those of its trading partners. It is estimated that two-thirds of all tropical hardwoods cut in Asia are shipped to Japan.

Many reforestation projects involve large plantations of single-species, single-use, intensive cropping called **monoculture forestry.** Although this produces high profits, a dense, single-species stand encourages pest and disease infestations. This type of management lends itself to mechanized clear-cut harvesting, which saves money and labor but tends to leave soil exposed to erosion. Monocultures eliminate habitat for many woodland species and often disrupt ecological processes that keep forests healthy and productive. When profits from these forest plantations go to absentee landlords or government agencies, local people have little incentive to prevent fires or keep grazing animals out of newly planted areas. In some countries, such as the Philippines, Israel, and El Salvador, government reforestation projects have been targets for destruction by antigovernment forces, with devastating environmental impacts.

Promising alternative agroforestry plans are being promoted by conservation and public service organizations such as The New Forest Fund and Oxfam. These groups encourage people to plant community woodlots of fast-growing, multipurpose trees such as *Leucaena.* Millions of seedlings have been planted in hundreds of self-help projects in Asia, Africa, and Latin America. *Leucaena* is a legume, so it fixes nitrogen and improves the soil. Its nutritious leaves are good livestock fodder. It can grow up to 3 meters per year and quickly provides shade, forage for livestock, firewood, and good lumber for building. A well-managed *Leucaena* woodlot can yield up to 50 tons per hectare on a sustained basis but pest problems have developed in some areas and *Leucaena* can be an aggressive, weedy exotic if it escapes from cultivation. Community woodlots can be planted on wasteland or along roads or slopes too steep to plow so they do not interfere with agriculture. They protect watersheds, create windbreaks, and, if composed of mixed species, also provide useful food and forest products such as fruits, nuts, mushrooms, or materials for handicrafts on a sustained-yield basis.

Tropical Forests

The richest and most diverse terrestrial ecosystems on the earth are the tropical forests. Although they now occupy less than 10 percent of the earth's land surface, these forests are thought to contain more than two-thirds of all plant biomass and at least one-half of all plant, animal, and microbial species in the world.

Diminishing Forests

While many temperate forests are expanding slightly due to reforestation and abandonment of marginal farmlands, tropical forests are shrinking rapidly. At the beginning of this century, an estimated 12.5 million square kilometers of tropical lands were covered with closed-canopy forest. This was an area larger than the entire United States. The Food and Agriculture Organization of the United Nations estimates that about 1 percent of the remaining tropical forest is cleared each year (fig. 14.6). The species extinction due to this loss of habitat is discussed in chapter 13.

There is considerable debate about current rates of deforestation in the tropics. In 1988, scientists at Brazil's National Space Research Institute (INPE) used weather satellite data to estimate the number of fires in the Amazon. They counted up to 7000 fires on a single day during the dry season (fig. 14.7). Assuming that 40 percent of these fires occurred on recently cleared forest, they calculated that 8 million ha (about 20 million ac) per year were being

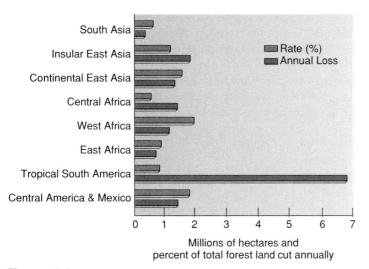

Figure 14.6

Rates of tropical forest losses in millions of hectares/year and rate of loss as a percent of total land area.

Source: FAO 1993.

Figure 14.7

Cutting and burning of tropical rainforest results in wildlife destruction, habitat loss, soil erosion, rapid water runoff, and waste of forest resources. It also contributes to global climate change.

cut. This number was widely circulated and led to widespread criticism of Brazil for allowing a priceless world heritage to be destroyed. More recent INPE studies of satellite images suggest that this original estimate is much too high. Scientists now calculate that the maximum rate of forest clearing in the Amazon in the 1980s was only 2.1 million ha per year and that the rate in the 1990s has fallen by about one-third.

Similar uncertainty about rates of forest destruction exist in many places, partly because of divergent definitions of deforestation. Some scientists and politicians insist that it means a complete change from forest to agriculture, urban areas, or desert. Others include any area that has been logged, even if the cut is selective and regrowth will be rapid. There also are difficulties in interpreting satellite images. Savannas, open woodlands, and early successional stages following natural disturbance are hard to distinguish from logged areas. Countries also have economic and political reasons to hide or exaggerate the extent of their activities. Consequently, estimates for total tropical forest losses range from about 5 million to more than 20 million ha per year. The FAO estimates of 14.5 million ha are generally the most widely accepted.

By most accounts, Brazil has the highest rate of deforestation in the world, but it also has by far the largest tropical forests. Other areas have lost a greater percentage of their original forests. The coastal forests of Ecuador, Sierra Leone, Ghana, Madagascar, Cameroon, and Liberia, for instance, already have been mostly destroyed. Haiti was once 80 percent forested; today, essentially all that forest has been destroyed and the land lies barren and eroded. India, Burma, Kampuchea, Thailand, and Vietnam all have little virgin lowland forest left. In Central America, nearly two-thirds of the original moist tropical forest has been destroyed, mostly within the last 30 years and primarily due to conversion of forest to cattle range (fig. 14.8).

A variety of causes lead to this deforestation. In Costa Rica, in addition to cattle ranching (fig. 14.9), banana plantations consume large areas of forest. In Brazil, a combination of land clearing for cattle ranching coupled with an invasion of immigrants from the south are responsible for forest losses. Elsewhere, gold mining in Ecuador, coca production in Colombia and Bolivia, and logging of hardwoods for export in Malaysia are some of the causes of forest destruction.

Swidden Agriculture

Indigenous forest people often are blamed for tropical forest destruction because they carry out shifting agriculture that requires forest clearing. Actually, this ancient farming technique can be an ecologically sound way of obtaining a sustained yield from fragile tropical soils if it is done carefully and in moderation. This practice is sometimes called "slash and burn" by people who don't realize how complex and carefully balanced this method of farming actually can be. The preferred terms of **milpa** or **swidden agriculture** are taken from local names for *field*.

In this system, farmers clear a new plot of about a hectare (2.5 acres) each year. Small trees are felled and large trees are killed by girdling (cutting away a ring of bark) so that sunlight can penetrate through to the ground. After a few weeks of drying, the branches, leaf litter, and fallen trunks are burned to prepare a rich seedbed of ashes. Fast-growing crops, such as bananas and papayas, are planted immediately to control erosion and to shade root crops, such as cassava and sweet potato, which anchor the soil. Maize, rice, and up to eighty other crops are planted in a riotous profusion. Although they would not recognize the terms, these indigenous people are practicing **mixed perennial polyculture.**

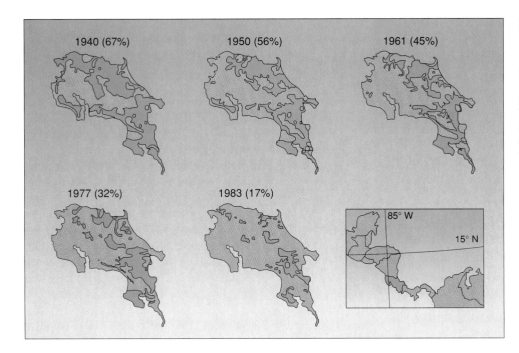

Figure 14.8
Loss of primary forest in Costa Rica 1940–1983. Percentages show forestland as a proportion of total land area.

Sources: Sader and Joyce, "Deforestation Rates and Trends in Costa Rica, 1940–1983," *Biotropica* 20:11–19; and T. C. Whitmore and G. T. Prance, *Biogeography and Quaternary History in Tropical America,* 1987 Clarendon Press, Oxford.

Figure 14.9

Cattle graze on recently cleared tropical rainforest land in Costa Rica. About two-thirds of the forest in Central America has been destroyed, mostly in the past few decades as land is converted to pasture or cropland. Unfortunately, the soil is poorly suited to grazing or farming, and these ventures usually fail in a few years.

The diversity of the milpa plot mimics that of the jungle itself, even though the species representation is more restricted. When managed well, the soil is covered with vegetation. This variety means that crops mature in a staggered sequence and there is almost always something to eat from the plot. It also helps prevent eruptive

insect infestations that would plague a monoculture crop. Annual yields from a single hectare can be as high as 6 tons of grain (maize or rice) and another 5 tons of roots, vegetable crops, nuts, and berries. This yield is comparable to the best results with intensive row cropping, and about one thousand times as much food is produced from the same land as when it is converted to cattle pasture.

After a year or two, the forest begins to take over the garden plot again. The farmer will continue to harvest perennial crops for a while and will hunt for small animals that are attracted to the lush vegetation. Ideally, the land then will be allowed to remain covered with rainforest vegetation for 10 to 15 years while nutrients accumulate before it is cleared and replanted again. In many places, population growth and displacement of farmers from other areas have forced shifting agriculturalists to reuse their traditional plots on shorter and shorter rotation. When plots are farmed every year or two, nutrients are lost faster than they can be replaced, the forest doesn't regrow as vigorously as it once did, and additional species are lost. Eventually, erosion and overuse reduce productivity so much that the land is practically useless.

Logging and Land Invasions

The other major source of forest destruction is usually the result of logging and subsequent invasion by land-hungry people from other areas (fig. 14.10). The loggers often are interested only in "creaming" the most valuable hardwoods, such as teak, mahogany, sandalwood, or ebony. Although only one or two trees per hectare might be taken, widespread devastation usually results. Because the canopy of tropical forests is usually strongly linked by vines and interlocking branches, felling one tree can easily bring down a dozen others. Tractors dragging out logs damage more trees, and construction of roads takes large land areas. Insects and infections

Figure 14.10

Tropical forests around the world are being cut at an alarming rate. Often hundreds of trees are smashed down to get to a single, valuable mahogany or teak tree. Logging roads open up access to landless settlers, miners, and hunters who drive away native species and indigenous people.

invade wounded trees. Tropical trees, which usually have shallow root systems, are easily toppled by wind and erosion when they are no longer supported by their neighbors. Up to three-fourths of the canopy may be destroyed for the sake of a few logs. Obviously, the complex biological community of the layered canopy (chapter 4) is severely disrupted by this practice.

What happens next? Bulldozed roads make it possible for large numbers of immigrants to move into the forest in search of land to farm. People with little experience or understanding of the complex rainforest ecosystem try to turn it into farms and ranches. Too often, the result is ecological disaster. Rains wash away the topsoil and the tropical sun bakes the exposed subsoil into an impervious hardpan that is nearly useless for farming.

Degradation of rivers is another disastrous result of forest clearing. Tropical rivers carry two-thirds of all freshwater runoff in the world. In an undisturbed forest, rivers are usually clear, clean, and flow year-round because of the "sponge effect" of the thick root mat created by the trees. When the forest is disrupted and the thin forest soil is exposed, erosion quickly carries away the soil, silting river bottoms, filling reservoirs, ruining hydroelectric and irrigation projects, filling estuaries, and smothering coral reefs offshore. In Malaysia, sediment yield from an undisturbed primary forest was about 100 cubic meters per square kilometer per year. After forest clearing, the same river carried 2500 cubic meters per square kilometer per year.

Forest Protection

What can be done to stop this destruction and encourage careful management? While much of the news is discouraging, there are some hopeful signs for tropical forest conservation. Many tropical countries have realized that forests represent a valuable resource and they are taking steps to protect them. Indonesia has announced

plans to preserve 100,000 square kilometers, one-tenth of its original forest. Zaire and Brazil each plan to protect 350,000 square kilometers (about the size of Norway) in parks and forest preserves. Costa Rica has one of the best plans for forest protection in the world. Attempts are being made there to not only rehabilitate the land (make an area useful to humans), but also restore the ecosystems to naturally occurring associations. One of the best known of these projects is Dan Janzen's work restoring the dry tropical forest of Guanacaste National Park (Case Study, next page).

People on the grassroots level also are working to protect and restore forests. Reforestation projects build community pride while also protecting the land (fig. 14.11). India, for instance, has a long history of nonviolent, passive resistance to protest unfair government policies. The *satyagrahas* go back to the beginning of Indian culture and often have been associated with forest preservation. Gandhi drew on this tradition in his protests of British colonial rule in the 1930s and 1940s. During the 1970s, commercial loggers began large-scale tree felling in the Garhwal region in the state of Uttar Pradesh in northern India. Landslides and floods resulted from stripping the forest cover from the hills. The firewood on which local people depended was destroyed, and the way of life of the traditional forest culture was threatened. In a remarkable display of courage and determination, the village women wrapped their arms around the trees to protect them, sparking the *Chipko Andolan* movement (literally, movement to hug trees). They prevented logging on 12,000 square kilometers of sensitive watersheds in the Alakanada basin. Today, the *Chipko Andolan* movement has grown to more than four thousand groups working to save India's forests.

Figure 14.11

School children plant trees in a community reforestation project in Haiti. Grassroots efforts, such as this, benefit both local people and their environment. Many top-down development projects benefit only the ruling elite and multinational corporations.

CASE STUDY
Restoring a Dry Tropical Forest

When the Spanish *conquistadores* arrived in Central America in the sixteenth century, about 5.5 million ha (21,000 sq mi) of dry tropical forest stretched along the Pacific coast from Colombia to Mexico. In contrast to the evergreen rainforests and cloud forests on the Atlantic side of the isthmus, dry forests have distinct seasons. During the wet summer months, the vegetation is dense and lush. In the winter, however, when rains are sparse, trees and bushes lose their leaves, and the whole forest becomes open and desertlike.

This dry forest was much easier to convert to farms and ranches than the moist forests. Its climate is healthier and its soil was more fertile and conducive to agriculture. Today, only about 1 percent of Central America's dry forest remains in anything like its original condition, making it one of the most threatened ecosystems in the world. As the forest has disappeared, many of its unique plant and animal species have become rare and endangered. If much more forest is lost, hundreds or even thousands of species will become extinct.

An exciting project is currently underway in Costa Rica, where scientists and local residents have joined together to restore about 700 sq km (28,000 ac) of dry tropical forest to its original condition (approximately). A new national park called Guanacaste (named after the Costa Rican national tree that once grew in this forest) is being created from private lands, an existing park, and other public land holdings. Under the leadership of entomologist Dan Janzen, attempts are being made to understand the ecosystem and to reintroduce native plants and animals in an effort to restore—rather than just rehabilitate—the forest.

How is this possible after the land has been abused and degraded for centuries? Isn't it long past the point at which it can be rescued? Fortunately, according to Janzen, most of the original flora and fauna have not been completely eliminated, only reduced. Small areas containing most of the indigenous species remain scattered across the countryside. The challenge is to find these species and create habitats where they can thrive and re-create the forest.

Fire is one of the greatest threats to the forest. Every year during the dry season local people accidentally or deliberately start fires that sweep across the land, destroying native species and converting forest to grassland full of non-native invaders. Creating breaks to control the spread of fire and persuading residents to fight fires rather than set them is the first step toward restoring the forest.

Contrary to what you might expect, grazing animals are not excluded from Guanacaste National Park. In fact, they are encouraged because they are efficient seed dispersers. Because the forest probably coevolved with a fauna that included large, hooved grazing animals before humans arrived, many plant species actually depend on animals for regeneration. Horses, monkeys, goats, birds, and even turtles eat fruits and pass their seeds through their digestive system days or weeks later. This not only distributes seeds to new locations, it also provides fertilizer for their initial growth. Furthermore, some seeds have tough outer coverings that are weakened by digestive acids and enzymes, facilitating germination. Being able to use the new national park for grazing during restoration makes the whole process much more attractive to its neighbors.

Involving local people in the project and making the park economically beneficial to them is another essential key to successful restoration (fig. 1). When they see how a park will help them, residents will be enthusiastic participants. Native people, with their knowledge of the forest and their skills as land stewards, can be an invaluable resource in the restoration process.

Once Guanacaste National Park is reconstituted, locals can work as guides and rangers or provide services to tourists who come to visit and view wildlife. Providing jobs in the area will help stem the tide of urbanization and also preserve local culture. Biodiversity and cultural heritage can be saved simultaneously. This exciting project may serve as an inspiration and guide to similar efforts in many areas of the world where bad land-use practices threaten both wildlife and indigenous people.

Figure 1

Former cowboys guard recovering pasturelands in the new Guanacaste National Park in Costa Rica where they once herded cattle. By involving local people in park management, the park service benefits both from their knowledge of the area and their support of ecosystem restoration.

Land Use: Forests and Rangelands

Debt-for-Nature Swaps

Those of us in developed countries can make a contribution toward saving tropical forests as well. Financing nature protection is often a problem in developing countries where the need is greatest. One promising approach is called **debt-for-nature swaps.** Banks, governments, and lending institutions now hold nearly $1 trillion in loans to developing countries. There is little prospect of ever collecting much of this debt, and banks are often willing to sell bonds at a steep discount—perhaps as little as 10 cents on the dollar. Conservation organizations buy debt obligations on the secondary market at a discount and then offer to cancel the debt if the debtor country will agree to protect or restore an area of biological importance.

The first such swap was made in 1987. Conservation International bought $650,000 of Bolivia's debt for $100,000—an 85 percent discount. In exchange for canceling this debt, Bolivia agreed to protect nearly 1 million ha (2.47 million ac) around the Beni Biosphere Reserve in the Andean foothills. Ecuador and Costa Rica have had a different kind of debt-for-nature swap. They exchanged debt for local currency bonds that are used to fund activities of local private conservation organizations in the country. This has the dual advantage of building and supporting indigenous environmental groups while protecting the land.

Agreements have been reached with Madagascar and Zambia to swap debts for nature, and negotiations are underway with Peru, Mexico, and Tanzania. Critics charge that these swaps compromise national sovereignty and that they will do little to reduce Third World debt or change the situations that led to environmental destruction in the first place.

Temperate Forests

Tropical countries are not unique in harvesting forests at unsustainable rates and in an ecologically damaging fashion. Northern countries have a long history of liquidating forest resources that continues today in many places. Perhaps the largest and most destructive harvest in the world today is taking place in Eastern Russia. Siberia is larger than Amazonia and contains one-fourth of the world's timber reserves. Four million ha (10 million ac) of Siberian taiga and deciduous forests are being felled annually—primarily by Korean loggers—to shore up Russia's faltering economy. China has announced plans for a giant hydroelectric dam on the Amur River, which forms its border with Siberia. This power source will allow China to move 100 million settlers into the region, which is home to the endangered Siberian tiger and several species of cranes. The results are likely to be similar to those we have just discussed in reference to tropical forests.

In the United States and Canada, the two main issues in timber management are (1) cutting of the last remnants of old-growth forest and (2) methods used in timber harvest.

Ancient Forests of the Pacific Northwest

Only a century ago, most of the coastal ranges of Washington, Oregon, northern California, British Columbia, and southeastern Alaska were clothed in a lush forest of huge, ancient trees (fig. 14.12). The moist, mild climate and rich soil of the lowland valleys nurtured magnificent stands of redwood in California and of western red cedar, Douglas fir, hemlock, and Sitka spruce along the rest of the coast. Everyone knows that redwoods can be huge and very old, but did you know that these other species can reach 3 to 4 m (9 to 12 ft) in diameter, 90 m in height (as high as a twenty-story building), and 1000 or more years in age? These temperate rainforests are probably second only to tropical rainforests in terms of terrestrial biodiversity, and they accumulate more total biomass in standing vegetation than any other ecosystem on earth.

These old-growth forests (where at least some trees are more than 200 years old) are extremely complex ecologically. Only in recent years have we begun to realize how many different species live there and how interrelated their life cycles are. Many endemic species such as the northern spotted owl (fig. 14.13), Vaux's swift, and the marbled murrelet are so highly adapted to the unique conditions of these ancient forests that they live nowhere else.

Before loggers and settlers arrived, there were probably 12.5 million ha (31 million ac) of virgin temperate rainforest in the Pacific Northwest. Less than 10 percent of that forest in the United States still remains, and 80 percent of what is left is scheduled to be cut down in the near future. British Columbia has felled at least 60 percent of its richest and most productive ancient forests and is now cutting some 240 million ha (600 million ac) annually, about ten times the rate of old-growth harvest in the United States. At

Figure 14.12

The Quinault Rainforest on the Olympic Peninsula in Washington is an excellent example of the old-growth forests of the Pacific northwest. Huge Sitka spruce, Douglas fir, hemlock, and western cedar form a dense, multilayered, closed canopy that shelters a number of rare and endangered plant and animal species. The forest floor is dark, moist, and quiet, cushioned with a layer of fallen needles, moss, and ferns. Downed tree trunks serve as a nursery bed for new growth. Unfortunately, these temperate rainforests are disappearing at an alarming rate and only scattered remnants remain.

Figure 14.13
Only about 2000 pairs of northern spotted owls remain in the temperate rainforests of the Pacific Northwest. Cutting old-growth forests threatens the endangered species but reduced logging threatens the jobs of many timber workers.

these rates, the only remaining ancient forests in North America in 50 years will be a fringe around the base of the mountains in a few national parks.

Wilderness and Wildlife Protection

Many environmentalists would like to save all remaining virgin forest in the United States as a refuge for endangered wildlife, a laboratory for scientific study, and a place for recreation and spiritual renewal. Economic pressures to harvest the valuable giant trees are considerable, however. The forest products industry employs about 150,000 people in the Pacific Northwest and adds nearly $7 billion annually to the economy. In Oregon, forest products account for one-fifth of the gross state product and many small towns depend almost entirely on logging for their economic life. On the other hand, an economic development study of Washington and Oregon suggests that recreation could provide sixteen jobs for every one lost by logging. The biggest problem is how people who are currently employed in the timber industry can make a living during a transition period.

In 1989, environmentalists sued the U.S. Forest Service over plans to clear-cut most of the remaining old-growth forest, arguing that spotted owls are endangered and must be protected under the Endangered Species Act. A federal judge agreed and ordered some 1 million ha (2.5 million ac) of ancient forest set aside to preserve the last 1000 pairs of owls. This would be about half the remaining virgin forest in Washington and Oregon. The timber industry claims that 40,000 jobs would be lost, although environmentalists dispute this number. Outrage in the logging communities was loud and clear. Convoys of logging trucks converged on protest sites while angry crowds burned environmentalists in effigy. Bumper stickers urged "Save a logger; eat an owl" and proclaimed "I love owls: poached, fried, or stewed."

Environmentalists agree that logging jobs are disappearing but claim it is due mostly to mechanization, a naturally dwindling resource base, and the shipping of raw logs to Japan. They argue that the big trees are disappearing anyway. The question is when to stop cutting: now while a few remain or in a few years when they are all gone (fig. 14.14)? The workers will have to be retrained anyway; why not sooner rather than later? In spite of complaints that logging restrictions on public land hurt business, companies with private land holdings have benefited from rising timber prices. Weyerhauser, for instance, reported a $90 million (81 percent) increase in the value of its timber holdings in 1992 as a result of the reduced cut in federal forests.

A compromise forest management plan is now in place that will allow some continued cutting in the Pacific Northwest but also will protect a high percentage of prime ancient forests. In this plan, 0.9 million ha (2.2 million ac) of riparian (streamside) preserves will be set aside to protect salmon and other aquatic wildlife, while 2 million ha (4.9 million ac) of "old-growth" reserves are established on national forest lands. The maximum annual allowable cut on public lands in the region is 1.2 billion board feet, more than environmentalists want, but far below the peak of 5 billion per year in the 1980s. Some of this logging is selective salvage and thinning in old-growth preserves.

Although current plans protect only a fraction of the original forest, local residents in many areas resent *any* restrictions on their access to resources on public or private lands. "Wise Use" groups have brought lawsuits and introduced federal, state, and local ordinances to prohibit programs that limit land use or to require repayment of losses that result from land-use regulations (What Do *You* Think?, p. 304).

Figure 14.14
What attitudes or assumptions underlie this cartoon? Do you agree with the implications it suggests?

Reprinted with special permission of North America Syndicate.

What Do *You* Think?
Regulations and Property Rights

Should private property owners be entitled to compensation when environmental regulations restrict how their land may be used? That is the subject of a growing debate in Congress and across the country. The issue is referred to as "takings." The matter is rooted in the Fifth Amendment to the United States Constitution, which states that ". . . nor shall private property be taken for public use, without just compensation."

This requirement imposed on the government was designed to make sure that when private land was needed for a public purpose, such as a highway, the government could not simply seize the land, but had to pay for it. A growing number of landowners are claiming that governmental restrictions, environmental or otherwise, that reduce the potential economic gain from their land have the effect of taking away their rights. They argue that this is comparable to "taking" their property and that the Fifth Amendment prohibits this "without just compensation."

Over the years, landowners have brought a variety of lawsuits seeking compensation for lost profits. Penn-

sylvania passed a mine safety act requiring coal companies to leave part of the coal seams in place underground beneath homes and other surface users to prevent highways, buildings, and gas pipelines from collapsing. A coal company claimed in court that this constituted a "taking" because of the profits lost on unmined coal, and that they should be compensated for these losses.

A rancher in Nevada sued the government to recover the value of grass and water eaten on public land by wild elk herds, claiming this use deprived his cattle of these resources. In California, agricultural interests are concerned about plans to use water to reestablish salmon runs. They are preparing to sue if this reduces subsidized irrigation water available to them. An Arkansas tavern owner even sued the state for income lost because its drunk driver checkpoints caused people to drink less!

To date, courts have rejected most of these suits. Property owners have never been viewed as having unrestricted rights of use. Court decisions have held that property owners do not have a right to use property in ways that harm the community interest as a whole, that the Fifth Amendment does not guarantee owners the right to use land in the most profitable way conceivable, and that use restrictions may be imposed for a wide range of environmental and land-use purposes.

More recently mining, logging, and land development interests, unhappy with court actions on takings, have sought legislation to restrict the scope of government regulations. Such bills

have passed in a number of state legislatures, and federal legislation is moving through Congress.

Some of these bills would require government agencies to do a thorough analysis of any action that would affect either the use or value of privately owned property. Other bills would require government to pay owners when regulations reduce property value in excess of some determined percentage.

These efforts are opposed by many environmental, consumer, civil rights, and labor groups. They contend that many of the proposals go far beyond constitutional provisions and would paralyze the government's ability to protect the public interest in a clean environment and healthful surroundings. They fear that the bills being proposed would prevent enforcement of many anti-pollution laws, zoning ordinances, and even obscenity laws (one dial-a-porn company sued the federal government for damages due to rules that restrict the access of children to its services).

In many respects, the controversy over takings represents a classic conflict over two kinds of freedoms: the freedom to act as you like and the right to be protected from the damaging actions of others.

Where should the balance point be between these two types of conflicting rights? To what extent should property rights include the right to negatively impact the rights of others? How would you decide?

Harvest Methods

Most lumber and pulpwood in the United States and Canada currently is harvested by **clear-cutting,** in which every tree in a given area is cut regardless of size (fig. 14.15). This method enables large machines to fell, trim, and skid logs rapidly, but wastes many small trees, often increases soil erosion, and eliminates habitat for many forest species. Early successional species, such as jack pine, lodgepole pine, or loblolly pine, on the other hand, flourish after clear-cutting. In a forest managed for these trees, or for raspberries, blueberries, deer, or grouse, this is a good method.

Size and shape of clear cuts vary depending on topography and management policies. Clearcuts can be in individual strips, alternating rows, small scattered patches, or areas as large as thousands

of acres. Natural regeneration is better in small patches or strips that resemble natural forest openings than in huge clearcuts. Small patches also are less disruptive for wildlife than are huge denuded spaces. It was once thought that good forest management required immediate removal of all dead trees and logging residue. Research has shown, however, that standing snags and coarse woody debris play important ecological roles, including soil protection, habitat for a variety of organisms, and nutrient recycling.

Other harvest practices offer variations on, or substitutes to, clear-cutting. *Coppicing* is used to encourage stump sprouts from species such as aspen, red oak, beech, or short-leaf pine and is usually accomplished by clear-cutting. In *seed tree harvesting,* some mature trees (generally 2 to 5 trees per hectare) are left standing to

Figure 14.15

This huge clear cut in Washington's Gilford Pinchot National Forest threatens species dependent on old-growth forest and exposes steep slopes to soil erosion. Restoring something like the original forest will take hundreds of years.

Figure 14.16

Selective logging with draft animals is much less damaging to forest soils and vegetation than using mechanical skidders.

serve as a seed source in an otherwise clear-cut patch. *Shelterwood harvesting* involves removing mature trees in a series of two or more cuts. This encourages regeneration of wind- and sun-sensitive species such as spruce and fir. **Strip cutting** entails harvesting all the trees in a narrow corridor (see Case Study, p. 113). For many forest types, the least disruptive harvest method is **selective cutting,** in which only a small percentage of the mature trees are taken in each ten- or twenty-year rotation. Ponderosa pine, for example, are usually selectively cut to thin stands that improve growth of the remaining trees. A forest managed by selective cutting can retain many of the characteristics of age distribution and ground cover of a mature, old-growth forest.

Logs are skidded out of the forest by a variety of means, depending on slope, economics, and value of remaining trees in the forest. In clearcuts with grades up to 30 percent, tractors or articulated skidders drag out logs. On steeper slopes, high-lead cables are used. In extremely rough or sensitive terrain, helicopters or balloons may lift out high-grade logs. Some loggers are finding it profitable and much less damaging to the forest to go back to old-fashioned horse or mule skidding (fig. 14.16). Animals make far narrower trails through the forest than big machines and cause less soil compaction, erosion, and scarring of trees that are left.

The lush Douglas fir and redwood forests of the rainy Pacific Coast Range provide a vivid illustration of why clear-cutting in old-growth forests is controversial. The rich soil and mild, moist climate of the region promote rapid regeneration after clear-cutting, but many of these forests are on steep slopes where erosion is a serious problem. Hillsides are stripped of soil, which fills streams and smothers aquatic life. British Columbia has felled at least 60 percent of its richest and most productive ancient forests. About 5 percent of the province—but only 2 percent of the temperate rainforest—is preserved in parks, ecological reserves, or recreational areas.

Vancouver Island, 384 km (240 mi) long, is being logged faster than any other part of British Columbia. Indigenous people have blocked roads and brought law suits to protest destruction of traditional lands and subsistence ways of life. People concerned about commercial and sport fishing have joined the battle, both in British Columbia and in the United States. They argue that salmon spawning depends on the clear cold streams of the native forests. Harvesting timber often destroys this valuable resource. The income from a single year's salmon run can outweigh all the profits from timber harvesting, and the salmon return year after year, while 1000-year-old trees will never be seen again.

Below-Cost Timber Sales

Through most of its history, the U.S. Forest Service has regarded its primary job as providing a steady supply of cheap logs to the nation's timber industry. Prices charged for timber sales have not been high enough to repay management costs, cleanup of logging debris, or replanting in many cases. The result is a hidden subsidy that has cost taxpayers hundreds of millions of dollars. In the 1992 fiscal year alone, 101 national forests failed to recoup costs on timber sales according to The Wilderness Society. Total losses for the year were more than $250 million. The Forest Service admits that costs were higher than income, but claims that the difference was less than one-fifth of what critics calculate. A part of the discrepancy is how roads are entered on the balance sheet. In the past 40 years, the Forest Service has expanded its system of logging roads more than tenfold to a current total of 547,000 km (340,000 mi) or more than ten times the length of the interstate highway system. Government economists regard building roads into *de-facto* wilderness as a benefit because it opens up the country to motorized recreation and industrial exploitation. Wilderness enthusiasts see this road-building program as destructive and costly.

In what may be a precedent-setting case, a scientific task force appointed by the Forest Service suggested in 1993 that two national forests in Wisconsin—the Chequamegon and the Nicolet—be managed for greater natural diversity rather than intensive timber and pulp production. The panel, which included ecologists from both inside and outside the Forest Service, recommended that the agency minimize forest fragmentation by logging roads and allow large blocks of forest to gradually recreate the natural, old-growth ecosystem that existed before commercial logging began. Environmentalists greeted this new management plan with enthusiasm, but industry opposed both reduction of timber availability and hands-off management practices.

Fire Management

For more than 70 years, firefighting has been a high priority for forest managers. Smokey the Bear has appeared on posters, brochures, and even postage stamps to tell us of the horrors of wildfires and to warn us that "only you can prevent forest fires." For most people, a blackened, smoking, burned forest appears ruined forever. We envision raging flames devouring helpless wildlife, threatening homes, and wasting valuable timber resources (fig. 14.17). Given such frightening images, it's no wonder that the public demands fire protection and government agencies make every effort to provide it. In 1994 alone, the U.S. spent over $1 billion and lost 33 lives in efforts to stop forest fires.

Recent studies of the ecological role of fire in forests, however, suggest that much of our horror of fire and our attempts to suppress it may be misguided. As discussed in chapter 3, many biological communities are fire-adapted and require periodic fires for regeneration. In the western United States, for instance, dry mon-

tane forests originally were dominated by big trees such as ponderosa pine, Douglas fir, and giant sequoias, whose thick, fire-resistant bark and lack of branches close to the ground protected them from frequent creeping ground fires. Historic accounts describe these forests as open and parklike, with little underbrush, luxuriant grass, and abundant wildlife.

Eliminating fire from these forests has allowed shrubs and small trees to fill the forest floor, crowding out grasses and forbes (herbs that are not grasses). As woody debris accumulates, the chances of a really big fire increase. Small trees act as "fire ladders" to carry flames up into the crowns of forest giants. By preventing low-intensity fires that once kept the forest open and free of fuel, we actually threaten the trees we intend to protect.

Our attempts to put fires out often cause more ecological damage than the fires themselves. Firefighters bulldoze fire breaks through sensitive landscapes such as tundra or wetlands, leaving scars that last far longer than the effects of the fire. Often the only thing that extinguishes a major fire is a change in the weather. Millions of dollars spent to dig fire lines and bomb outbreaks with chemicals and water have little effect as the fire goes where it will.

In 1970, the U.S. National Parks quietly began a program of allowing some natural fires to burn and of setting prescribed fires in ecosystems where scientists consider periodic burns beneficial. The U.S. Forest Service has considered a similar policy for wilderness areas but has been criticized by a public raised on Smokey's fearsome images of fire.

Another factor to consider is that firefighting is a lucrative business for many people. Arsonists sometimes set fires to create jobs or help businesses rent expensive equipment to government agencies. Deliberately starting a fire in a roadless area is a way to remove it from wilderness consideration and to make more timber available for salvage logging.

We are faced with a dilemma in many forests. After 70 years of fire suppression, fuel has now built up to a point where the next fire could be truly disastrous. The problem is how to remove excess fuel through controlled burning and thinning in a way that will return the forest to more natural conditions and yet protect property, human life, and important biological communities. We may need zoning regulations that restrict building or require fire-proof construction methods and land management in highly flammable areas. Public re-education programs to teach people the role of fire in natural systems is another important factor. Rather than seeing a burned forest as ruined, we can appreciate it as a natural stage in regeneration. This may require a very different environmental ethic than the one Smokey has been preaching and a modified view of our proper role in nature.

Rangelands

Pasture (generally enclosed domestic meadows or managed grasslands) and **open range** (unfenced, natural prairie and open woodlands) occupy about 26 percent of the world's land surface (fig. 14.18). The 3.8 billion ha (12 million sq mi) of permanent grazing lands in the world make up about twice the area of all agricultural crops. When you add to this about 4 billion ha of other lands (forest, desert, tun-

Figure 14.17

Is this fire in Yellowstone National Park destroying the forest or renewing it? By suppressing fires and allowing fuel to accumulate, we make major fires such as this more likely. The safest and most ecologically sound management policy for some forests may be to allow natural or prescribed fires that don't threaten property or human life to burn periodically.

Figure 14.18

Grasslands, such as this area in Badlands National Park, dominated much of the North American interior before they were converted to croplands and rangelands.

dra, marsh, and thorn scrub) used seasonally for raising livestock, more than half of all land is used at least occasionally as grazing lands. More than 3 billion cattle, sheep, goats, camels, buffalo, and other domestic animals turn plants into protein-rich meat and milk that make a valuable contribution to human nutrition.

Because grasslands and open woodlands are attractive for human occupation, they frequently are converted to cropland, urban areas, or other human-dominated landscapes. You probably have heard a great deal about tropical rainforest destruction, but worldwide only about 4 percent of all tropical forests have been completely modified by human use—compared to nearly 20 percent of all grasslands (see table 14.1). Although they may appear to be uniform and monotonous to the untrained eye, native prairies can be highly productive and species-rich. More threatened American plant species occur in rangelands than any other major biome (fig. 14.19).

Range Management

By carefully monitoring the numbers of animals and the condition of the range, ranchers and pastoralists (people who live by herding animals) can adjust to variations in rainfall, seasonal plant conditions, and nutritional quality of forage to keep livestock healthy and avoid overusing any particular area. Conscientious management can actually improve the quality of the range.

Some nomadic pastoralists who follow traditional migration routes and animal management practices produce admirable yields from harsh and inhospitable regions. They can be ten times more productive than dryland farmers in the same area and come very close to maintaining the ecological balance, diversity, and productivity of wild ecosystems on their native range. Nomadic herding requires large open areas, however, and wars, political problems, travel restrictions, incursions by agriculturalists, growing popula-

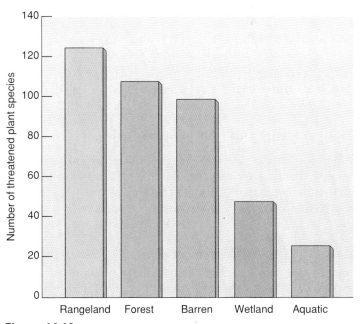

Figure 14.19

Threatened plant species in the United States, by ecosystem type, in 1990. Rangelands have more biodiversity and more threatened and endangered species than any other major biome.

Source: Curtis A. Flather, Linda A. Joyce, and Carol A. Bloomgarden, "Species Endangerment Patterns in the United States," USDA, Forest Service Technical Report 241, Fort Collins, CO, 1994, Table 3.

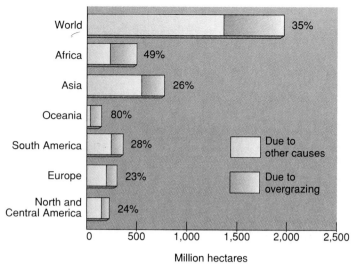

Figure 14.20

Rangeland soil degradation due to overgrazing and other causes. Notice that in Europe, Asia, and the Americas, farming, logging, mining, urbanization, etc., are responsible for about three-quarters of all soil degradation. In Africa and Oceania, where more grazing occurs and desert or semi-arid scrub make up much of the range, grazing damage is higher.

Source: World Resources 1994–1995.

tions, and changing climatic conditions on many traditional ranges have combined to disrupt an ancient and effective way of life. The social and environmental consequences often are tragic.

Overgrazing and Land Degradation

About one-third of the world's range is severely degraded by overgrazing, making this the largest cause of soil degradation (fig. 14.20). Among the countries with the most damage and the greatest area at risk are Pakistan, Sudan, Zambia, Somalia, Iraq, and Bolivia. Usually, the first symptom of improper range management is elimination of the most palatable herbs and grasses. Grazing animals tend to select species they prefer and leave the tougher, less tasty plants. When native plant species are removed from the range, weedy invaders move in. Gradually, the nutritional value of the available forage declines. As overgrazing progresses, hungry animals strip the ground bare and their hooves pulverize the soil, hastening erosion (fig. 14.21).

The process of denuding and degrading a once-fertile land initiates a desert-producing cycle that feeds on itself and is called **desertification.** With nothing to hold back surface runoff, rain drains off quickly before it can soak into the soil to nourish plants or replenish groundwater. Springs and wells dry up. Trees and bushes not killed by browsing animals or humans scavenging for firewood or fodder for their animals die from drought. When the earth is denuded, the microclimate near the ground becomes inhospitable to seed germination. The dry barren surface reflects more of the sun's heat, changing wind patterns, driving away moisture-laden clouds, and leading to further desiccation. Because of this process, deserts have been called the footprints of civilization.

Figure 14.21

Sheep grazing on land in Guatemala that was once tropical forest. Soils in the tropics are often thin and nutrient-poor. When forests are felled, heavy rains carry away the soil and fields quickly degrade to barren scrubland. In some areas, land clearing has resulted in climate changes that have turned once lush forests into desert.

This process is ancient, but in recent years it has been accelerated by expanding populations and political conditions that force people to overuse fragile lands. Those places that are most severely affected by drought are the desert margins, where rainfall is the single most important determinant in success or failure of both natural and human systems (fig. 14.22). In good years, herds and farms prosper and the human population grows. When drought comes, there is no reserve of food or water and starvation and suffering are widespread. Can we reverse this process? In some places, people are reclaiming deserts and repairing the effects of neglect and misuse.

As is the case in tropical forests, estimates of the extent of desertification around the world vary widely. Many arid lands are prone to highly variable rainfall with long periods of drought in which natural vegetation disappears and the land lies barren for months or years, making it difficult to distinguish between human-caused changes and normal climatic variations. A recent survey by the United Nations Environment Program found that about half of Africa, for instance, has become significantly drier over the past half century and that arid and hyperarid areas—which are most susceptible to desertification—have increased by about 50 million

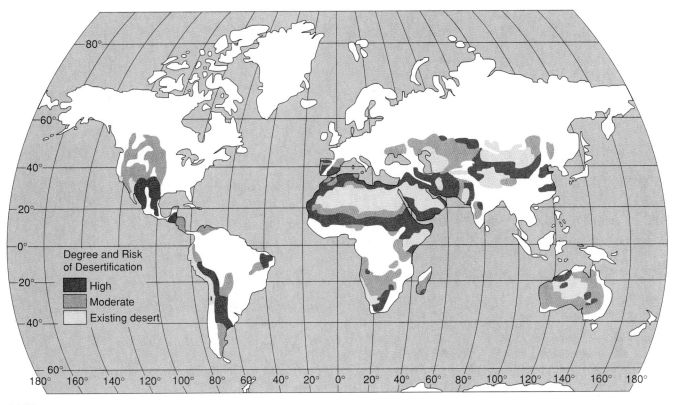

Figure 14.22

Degree and risk of desertification of arid lands due to human activities. Note that some existing deserts such as the Sahara in North Africa are so dry and lifeless that only a small amount of additional damage is possible.

hectares, while humid and semiarid lands have decreased by the same amount. According to the International Soil Reference and Information Centre in the Netherlands, nearly three-quarters of all rangelands in the world show signs of either degraded vegetation or soil erosion. The highest percentage of moderate, severe, and extreme land degradation is in Mexico and Central America.

Forage Conversion by Domestic Animals

Ruminant animals, such as cows, sheep, goats, buffaloes, camels, and llamas, are especially efficient at turning plant material into protein because bacterial digestion in their multiple stomachs allows them to utilize cellulose and other complex carbohydrates that many mammals (including humans) cannot digest. As a result, they can forage on plant material from which we could otherwise extract little food value. Many grazers have very different feeding preferences and habits. Often the most effective use of rangelands is to maintain small mixed-species herds so that all vegetation types are utilized equally and none is overgrazed. Cattle and sheep, for instance, prefer grass and herbaceous plants, goats will browse on low woody shrubs, and camels can thrive on tree leaves and larger woody plants.

Worldwide, 85 percent of the forage for ruminants comes from native rangelands and pasture. In the United States, however, only 15 percent of livestock feed comes from native grasslands. The rest is made up of crops grown specifically for feed, particularly alfalfa, corn, and oats. Grain surpluses and our taste for well-marbled (fatty) meat have shifted livestock growing in the developed countries to feedlot confinement and high-quality diets. In the United States, roughly 90 percent of our total grain crop is used for livestock feed.

Harvesting Wild Animals

A few people in the world still depend on wild animals for a substantial part of their food. About one-half of the meat eaten in Botswana, for instance, is harvested from the wild. There are good reasons to turn even more to native species for a meat source. On the African savanna, researchers are finding that springbok, eland, impala, kudu, gnu, oryx, and other native animals forage more efficiently, resist harsh climates, are more pest- and disease-resistant, and fend off predators better than domestic livestock. Native species also are members of natural biological communities and demonstrate niche diversification, spreading feeding pressure over numerous plant populations in an area. A study by the U.S. National Academy of Sciences concluded that the semiarid lands of the African Sahel can support only about 20 to 28 kg (44 to 62 lb) of cattle per hectare but can produce nearly three times as much meat from wild ungulates (hooved mammals) in the same area (fig. 14.23). In the United States, native bison and elk are raised as "novelty" meat sources; might there be a greater future role for ranching of wild animals on our own rangelands?

Land Use: Forests and Rangelands

Figure 14.23

Range grazers are a valuable part of grassland ecosystems; they also may be the best way to harvest biomass for human consumption. Native species often forage more efficiently, resist harsh climates, and are more pest- and disease-resistant than domestic livestock.

Figure 14.24

On the right side of this fence is a lightly grazed pasture consisting of a dense growth of grasses and small, broad-leaved plants. On the left side is overgrazed rangeland, where the vegetation has become sparse.

Rangelands in the United States

The United States has approximately 319 million ha (788 million ac) of rangeland. Most of this rangeland is in the West, and about 60 percent is privately owned. Of the 120 million cattle and 20 million sheep in the United States, only about 2 percent of the cattle and 10 percent of the sheep graze on public rangelands. Federal lands are not very important, overall, in livestock production, but they do have important local economic and environmental ramifications. The Bureau of Land Management (BLM) controls 84 million ha (200 million ac) of grazing lands and the U.S. Forest Service manages about one-fourth as much.

The BLM manages more land than any other agency in the United States, but it is little known outside of the western states where most of its lands are located. It was created in 1946 by a merger of the Grazing Service and the General Land Office. Its formation signaled an end to public land disposal and a commitment to permanent management by the government. The BLM has such a strong inclination toward resource utilization that critics claim the initials really stand for the "bureau of livestock and mining." While only 25 percent of BLM land is considered suitable for grazing, its policies have an important effect on the economy and environment of western states.

State of the Range

The health of most public grazing lands in the United States is not good. Political and economic pressures encourage managers to increase grazing allotments beyond the carrying capacity of the range. Lack of enforcement of existing regulations and limited funds for range improvement have resulted in overgrazing, loss of native forage species, and erosion. The National Resources Defense Council claims that only 30 percent of public rangelands are in fair condition, and 55 percent are poor or very poor (fig. 14.24).

Overgrazing has allowed populations of unpalatable or inedible species, such as sage, mesquite, cheatgrass, and cactus, to build up on both public and private rangelands. Furthermore, competing herbivores, such as grasshoppers, jackrabbits, prairie dogs, and feral burros and horses, further damage the range and reduce the available forage. (A **feral** animal is a domestic animal that has taken up a wild existence.) Control programs using poison baits, traps, and hunting to reduce the numbers of these competing native and feral herbivores have been highly controversial. They are seen as both inhumane and dangerous. The case of wild burros and mustangs is especially difficult. They reduce critical winter forage needed by both native wildlife and domestic stock. Most ecologists and land managers believe that the herds of feral animals must be controlled, but attempts to reduce their numbers or remove them from the range meet with vigorous opposition from those who view them as a symbol of the freedom and romance of the West.

Grazing Fees

Another controversial aspect of federal range management is that fees charged for grazing on public lands are far below market value and represent an enormous hidden subsidy to western ranchers. The 1995 charge for grazing permits on BLM or USFS land was $1.92 per animal unit month (AUM). (One AUM is enough to feed an average cow or five sheep for one month.) Comparable private land rented for an average of $7 that year. The 31,000 permits on federal range bring in only $15 million in grazing fees but cost $47 million per year for administration and maintenance. The $32 mil-

Figure 14.25

Landless rural peasants still work most of the land in Asia, Latin America, and Africa while they live in poverty.

lion difference amounts to a massive "cow welfare" system of which few people are aware.

Only 2 percent of the nation's ranchers use public lands, and the grazing program is dominated by large corporate ranches, not by the small, independent family operation revered in western myth. Only 10 percent of permit holders on BLM land account for almost two-thirds of the animals allowed to graze those lands; 6 percent of the permit holders on Forest Service lands are responsible for about half of the livestock. The Office of Management and Budget concluded that increasing grazing fees to a fair market value would reduce the livestock herd less than 10 percent and would result in removing animals only from the most marginal areas, which probably should not be grazed anyway.

Landownership and Land Reform

Many of the problems discussed in this chapter have their roots in landownership and public policies concerning resource use. Some land tenure patterns have supported or even created inequality, ignorance, and environmental degradation, while other systems of land use have promoted equality, liberty, and progress. In this final section, we will look at how fair access to the land and its benefits can help bring about better conditions for humans and their environment.

Who Owns How Much?

Whatever political and economic system prevails in a given area, the largest landowners usually wield the most power and reap the most benefits, while the landless and near landless suffer at the bottom of the scale. The World Bank estimates that nearly 800 million people live in "absolute poverty . . . at the very margin of existence." Three-fourths of these people are rural poor who have too little land to support themselves. Most of the other 200 million ab-

solute poor are urban slum dwellers, many of whom migrated to the city after being forced out of rural areas (chapter 24).

In many countries, inequitable landownership is a legacy of colonial estate systems in which landless peasants still work in virtual serfdom (fig. 14.25). For instance, the United Nations reports that only 7 percent of the landowners in Latin America own or control 93 percent of the productive agricultural land. By far the largest share of landless poor in the world are in South Asia. In Haiti, the poorest and most environmentally degraded country in the Western Hemisphere, 1 percent of the population owns 90 percent of the land.

The 86 million landless households in India comprise about 500 million people; they are two-thirds of both the total population of India and of all the landless rural people in the world. Not surprisingly, the countries with the widest disparity in wealth are often the most troubled by social unrest and political instability. As Adam Smith said in *The Wealth of Nations* (1776), "No society can surely be flourishing and happy, of which the far greater part of the members are poor and miserable."

Land tenure is not just a question of social justice and human dignity. Political, economic, and ecological side effects of inequitable land distribution affect those of us far from the immediate location of the problem. For example, people in developing countries forced from their homes by increasing farm mechanization and cash crop production often try to make a living on marginal land that should not be cultivated. The local ecological damage caused by their struggle for land can, collectively, have a global impact. The contribution of concentrated landownership to these environmental problems, however, receives less attention than do the threatening ecological trends themselves.

Land Reform

Throughout history, some **land reform** movements seeking redistribution of landownership have been successful and others have not. Significant land reforms have occurred in many countries only after violent revolutionary movements or civil wars. An important factor in revolutions in Cuba, Mexico, China, Peru, the former Soviet Union, and Nicaragua was the breaking of feudal or colonial landownership patterns. This struggle continues in many countries. Hundreds of large and small peasant movements demand land redistribution, economic security, and political autonomy in Latin America, Africa, and Asia. Entrenched powers respond with economic reprisals, intimidation, "disappearances," and even open warfare. Millions of people have been killed in such struggles in recent centuries alone.

Consider Brazil: Three hundred and forty rich landowners possess 46.8 million ha (117 million ac) of cropland, while 9.7 million families (70 percent of all rural people) are landless or nearly landless. Studies have shown that 13 percent of the land on the biggest estates is completely idle, while another 76 percent is unimproved pasture. These inequities and the tensions they create are important factors in rainforest invasions described earlier in this chapter.

In some countries, land reform has been relatively peaceful and also successful. In Taiwan, only 33 percent of the farm families owned the land they worked in 1949, whereas nearly 60 percent of rural families farm their own land today. South Korea also has carried out sweeping land redistribution. In the 1950s, more than half of all farmers were landless, but now 90 percent of all South Korean farmers own at least part of the land they till. In general, countries that have had successful land reform also have stabilized population growth and have begun industrialization and development.

What are the ecological implications of landownership? Absentee landlords have little personal contact with the land, may not know or care about what happens to it, and often won't let sharecroppers cultivate the same land from year to year for fear they may lay claim to it. This gives tenant farmers little incentive to protect or improve the land because they can't count on reaping any long-term benefits. In fact, if tenants do improve the land or increase their yields, their rents may be raised. Also, where land is aggregated into collective farms or large estates worked by landless peasants, productivity and health of the soil tend to suffer.

Far from being a costly concession to the idea of equality, land reform often can provide a key to agricultural modernization. In many countries, the economic case for land reform rivals the social case for redistributive policies. Many studies have shown that the productivity of owner-operated farms is significantly higher than that of corporate or absentee landowner farms. Independent farmers, especially those who have only a few hectares to grow crops, tend to lavish a great deal of effort and attention on their small plots, and their yields per hectare often are twice those of larger farms.

Indigenous Lands

No place on the earth is completely unoccupied. Every place, even Antarctica, is claimed by someone, although many places are very sparsely populated. Indigenous (native or tribal) people make up about 10 percent of the world's population but occupy about 25 percent of the land. Many of the approximately 5000 indigenous cultures that remain today possess ecological knowledge of their ancestral lands that is of vital importance to all of us. According to author Allen Durning, "encoded in indigenous languages, customs, and practices may be as much understanding of nature as is stored in the libraries of modern science." And indigenous people continue to play an important role as guardians of wildlife and forests. The Kuna Indians of Panama say, "Where there are forests there are indigenous people, and where there are indigenous people there are forests."

Nearly every country in the world has a sad history of decimation of indigenous cultures and exploitation or annexation of ancestral lands. Time after time, native people have been stripped of their rights through economic pressure, cunning and deceit, or outright theft (fig. 14.26). In the past, colonial governments appropriated tribal lands under the excuse that sparsely populated or uncultivated land was free for the taking. Many nations still fail to recognize indigenous rights. In the Philippines, Cameroon, and Tanzania, for example, the government claims ownership of all forest lands, and indigenous people are considered squatters, even on their ancestral territories.

Our appetite for natural resources and land puts both native people and the ecosystems in which they live at risk. Decimated by plagues, violence, and corruption—and overwhelmed by the onrush of materialistic Western culture—at least half of the world's indigenous cultures are at risk of disappearing within the next century. Consider the case of the 9000 Yanomami people who live along the border between Brazil and Venezuela (fig. 14.27). In 1991, Brazil recognized Yanomami claims to 17.6 million ha of land, but invasions by *garimpeiros* (miners) into this territory have introduced malaria, tuberculosis, flu, and respiratory diseases to which native people have no immunity. Placer mining and mercury efflu-

Figure 14.26

George Gillette, chairman of the Fort Berthold Indian Tribe Business Council, weeps as he watches the signing of a 1948 contract to sell the tribe's best land along the Missouri River for the Garrison Dam and Reservoir Project. After signing the contract, Gillette said, "Right now, the future does not look good to us."

Figure 14.27

Native tribal people, such as these Yanomami from Brazil, are threatened by tropical forest destruction. These people have lived in harmony with nature for thousands of years. If their culture is lost, valuable knowledge about the forest will be lost as well.

ents kill fish and poison the rivers. Wildlife is killed or driven off by miners and hundreds of the Yanomami themselves have been hunted down and shot.

From the Philippines to Labrador, indigenous people are fighting for their ancestral territories. Some countries with large native populations, such as Papua New Guinea, Ecuador, Canada, and Australia, have acknowledged indigenous titles or rights to extensive areas. Often, years of history and modern development complicate these indigenous claims. One Aboriginal band in Australia, for instance, claims ownership of Circular Quay, some of the most valuable land in the heart of Sidney. What would be a fair compensation after 150 years of occupation and building? In 1991, Canada created an Arctic territory called Nunavut spanning nearly 2 million square kilometers or about 20 percent of Canada's total landmass for Inuit (or Eskimo) people (fig. 14.28). In exchange for the land and about $1 billion, citizens of Nunavut agreed to relinquish all claims to oil, gas, and minerals elsewhere in the Canadian arctic. Perhaps the United States should consider a similar plan for the high plains.

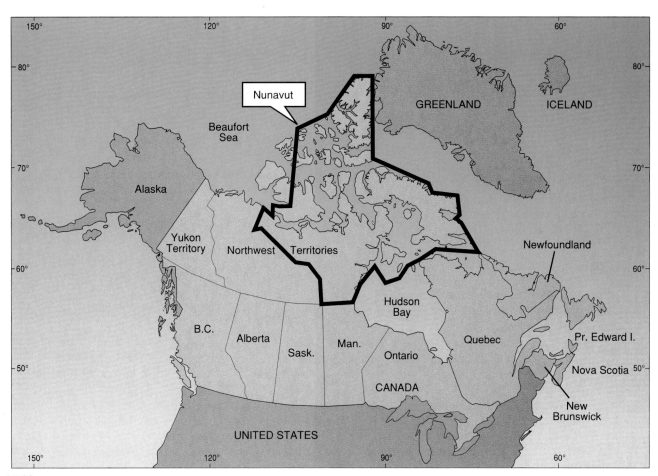

Figure 14.28

Nunavut, home to Canada's eastern Inuit people, was recognized in 1991 as one of the largest peaceful land settlements in history. In exchange for $1 billion and control of 2 million square kilometers of land, native people gave up their claim to oil, gas, and mineral deposits in the Arctic.

Summary

Land has traditionally been a source of wealth and power. The ways we use this limited resource shape our lives and futures. About one-third of the earth's surface is too inhospitable for agriculture, livestock, or forestry but is vital for such purposes as wilderness preservation and recreation. We grow crops on about 10 percent of the total land area. With proper preparation, we could expand cropland in some areas. Most of the earth's land, however, is inappropriate for agriculture and is ruined by attempts to cultivate it.

Forests cover about 30 percent of the earth's land area, providing a variety of useful products such as lumber, pulpwood, and firewood. Northern forests are growing faster in most areas of the world than they are being cut and seem in little danger of being exhausted. Tropical forests, on the other hand, are in critical danger. Irreplaceable ecosystems that are home to as many as half of all biological species are being destroyed. Quick profits encourage this exploitation, but hidden costs such as lost wildlife habitat, erosion and devaluation of exposed land, and other disastrous environmental damage will follow from small short-term gains.

A little more than one-quarter of the earth's land is used as range and pasture. Three billion grazing animals convert roughage to protein on poor land that could not otherwise be used to produce food for humans. When herds are managed properly, they actually can improve the quality of their pasture. Unfortunately, about one-third of the world's rangelands are degraded by overgrazing, with disastrous environmental consequences similar to forest destruction.

Land reform is an essential part of sound land-use management. Fair distribution of land and its benefits encourages good stewardship, increased food production, sustainable agriculture, and social justice. Inequitable land ownership, so common today in much of the world, forces the poor to use land unsuited to agriculture, while good land is monopolized by the rich. Dividing land more fairly could increase agricultural productivity and sustainability of the land because farmers who own their own land generally use it more efficiently and more carefully than do absentee landlords. Recognizing indigenous land rights is important both for preserving endangered cultures and for protecting ecological values.

▼ Questions for Review

1. Which type of land use occupies the greatest land area?

2. List some products that we derive from forests.

3. What are the advantages and disadvantages of monoculture forestry?

4. Describe milpa (or swidden) agriculture. Why are these techniques better for fragile rainforest ecosystems than other types of agriculture?

5. What are some results of deforestation?

6. What are clear-cutting and below-cost timber sales, and why are they controversial?

7. How does overgrazing encourage undesirable forage species to flourish?

8. Why can grazing animals generate food on land that would otherwise be unusable?

9. What is the relationship between fair land distribution and appropriate land use?

10. Give some examples of recognition of indigenous land titles.

▼ Questions for Critical Thinking

1. Some forestland and rangeland would be suitable as cropland. Do you think it should be converted? Why or why not?

2. What could we do to reduce or redirect the demand for wood products?

3. Brazil needs cash to pay increasing foreign debts and to fund needed economic growth. Why shouldn't it harvest its forests and mineral resources to gain the foreign currency it wants and needs? If we want Brazil to save its forests, what can or should we do to encourage conservation?

4. Thousands of landless peasants are mining gold or cutting trees to create farms in officially protected Brazilian rainforest. What might the government do to stop these practices?

5. What lessons do you think milpa (or swidden) agriculture has to offer large-scale commercial farming?

6. The U.S. government has kept timber sale prices and grazing leases low to maintain low prices of lumber and meat for consumers and to help support rural communities and traditional ways of life. How would you weigh those human interests against the ecological values of forests and rangelands?

7. Suppose that you live in an area of the Great Plains that is proposed for return to native prairie or that you are a logger in the Pacific Northwest whose job is threatened by a diminishing supply of old-growth timber. How would you feel about the changes discussed in this chapter?

8. Native Americans often were cheated out of their land or paid ridiculously low prices for it. Present owners, however, may not have been a part of earlier land deals and may have invested a great deal to improve the land. What would be a fair and reasonable settlement of these competing land claims?

9. There is considerable uncertainty about the extent of desertification of grazing lands or destruction of tropical rainforests. Put yourself in the place of a decision maker evaluating this data. What evidence would you want to see, or how would you appraise conflicting evidence?

▼ Key Terms

clear-cut 304
closed canopy 295
debt-for-nature swaps 302
desertification 308
feral 310
forest management 296
fuelwood 295
industrial timber 295
land reform 311
milpa agriculture 298

mixed perennial polyculture 298
monoculture forestry 297
open canopy 295
open range 306
pasture 306
selective cutting 305
strip cut 305
swidden agriculture 298
woodland 295

▼ Additional Information on the Internet

Canadian Forest Service
http://www.emr.ca/home/nrcan-cfs.html/

Forest and Rangeland Ecosystem Science Center
http://www.fsl.orst.edu/home/nbs/

METLA
http://www.metla.fi/

National Rangelands Program
http://www.dwe.csiro.au:80/local/comm/RangPrl.htm

Natural Resources Canada
http://www.emr.ca/

Rainforest Action Network
http://www.ran.org/ran/

Sustainable Forests Directory
http://www.together.net/~wow/Index.htm/

US Bureau of Land Management
http://www.blm.gov/

USDA Forest Service National Headquarters
http://www.fs.fed.us/

Note: Further readings appropriate to this chapter are listed on p. 601.

Reflection Lake, Mt. Rainier National Park, WA. Most national parks were chosen for scenic beauty rather than biological diversity.

CHAPTER 15: Preserving Nature

Objectives

After studying this chapter, you should be able to:

- understand the origins and current problems of national parks in America and other countries.

- recount some of the current problems in national parks in the United States and elsewhere in the world.

- evaluate the tension between conservation and economic development and how the Man and Biosphere program and ecotourism projects address this tension.

- explain the need for, and problems with, wildlife refuges and wilderness areas in the United States.

- demonstrate why wetlands are valuable ecologically and culturally and why they are currently threatened world-wide.

- report on current management policies and problems concerning floodplains, coastlines, and barrier islands.

Ecotourism on the Roof of the World

Rising dramatically from the steamy southern jungles of the Ganges River Valley to the icy peaks of the Himalayan mountains on the Tibetian border, Nepal is one of the most scenic countries in the world. Tourists savor the exotic culture of Katmandu or Namche Bazar or hike through lush mountain forests of rhododendron and pine. Offering spectacular scenery, friendly people, and low prices, this charismatic country has become a premiere destination for adventure travelers.

With an annual per capita income of only $170 per year, Nepal is among the poorest countries in the world. The phenomenal increase in visitors over the past twenty years has brought much-needed income but also has caused severe environmental degradation. Forests along popular trekking trails have been decimated to provide firewood for cooking and heating of water for the numerous wealthy outsiders, while tons of garbage and discarded gear litter popular campsites.

One of the most popular Nepalese trekking routes is a three-to four-week circuit of the Annapurna Range in the center of the Himalayan Range (fig. 15.1). Crossing rushing rivers on swaying suspension bridges, passing between the 8167-m Dhaulagiri and the 8091-m Annapurna I (the seventh and tenth highest mountains in the world, respectively), this ancient pilgrim trail follows the Kali Gandaki Valley to holy shrines at Muktinath. Surmounting the 5416-m (17,769-ft) Thorung La pass north of Annapurna, hikers follow the Marsyangdi valley back to the regional center at Pokhara. First opened to foreigners in 1977, this trail now attracts over 45,000 visitors each year.

Most Nepalese benefit very little from tourists who congest their villages, consume resources, and snap photographs incessantly, but the Annapurna region is different. An innovative project was launched in 1985 to alleviate the destructive impact of masses of trekkers and to maximize the income-generating potential of ecotourism. The Annapurna Conservation Area Project

Figure 15.1
The Annapurna Conservation Area Project directs money from visitors into development and environmental programs directed by, and of benefit to, local residents. This may be a model for preserving nature in other developing countries.

About $500,000 per year finances a variety of conservation, education, and development projects. More than 700 local entrepreneurs have been trained in lodge management, hygiene, and marketing. Forest guards have been hired, latrines built, trails repaired, and schools and clinics built for local people. Trekkers now are required to use kerosene rather than wood. Local tree nurseries provide stock for reforestation projects. Solar panels and water turbines provide renewable energy for both tourists and residents. The area is cleaner, healthier, and more enjoyable for everyone.

(ACAP) is a 2590-km² (1000-mi²) biosphere reserve that serves as an encouraging model for conservation and development in the Third World.

Far different from western ideals of parks composed of empty, virgin land, the ACAP is home to more than 100,000 people who continue to use resources in traditional ways. The area is divided into five different zones: intensive farming lands around the periphery, protected forest and seasonal grazing areas in the foothills, special management zones along tourist routes, protected regions with high biological or cultural richness, and wilderness areas in the high peaks.

Recognizing that there can be no meaningful conservation without the active involvement of local people, fees paid by visitors to ACAP go directly to residents to manage the preserve.

This unique and successful experiment gives us a different view of the meaning and purpose of parks and nature preserves than the ideal of pristine, unchanging nature conveyed by most American national parks. It raises some interesting questions about competing needs of human and nonhuman residents and how they might be balanced sustainably. It also provides a model of how protected areas might be designed and managed in other developing countries.

In this chapter, we will study the history of parks, preserves, and wildlife refuges around the world. We will examine other success stories as well as problems facing efforts to protect nature. Because of their great ecological importance, we will pay special attention to wetlands, floodplains, and coastal regions and how they need protection in many places. ▶

Parks and Nature Preserves

We can discern much about people's environmental ideals by looking at the gardens, parks, and pleasuring grounds that they create for themselves. Although parks and gardens generally had their earliest origins in human desires for comfort and pleasure or as demonstrations of wealth and power, we gradually have come to appreciate their value in protecting and preserving nature for its own sake.

Park Origins and History

Since ancient times, sacred groves have been set aside for religious purposes and hunting preserves or pleasuring grounds for royalty. As such, they have been reserved primarily for elite members of society. The imperial retreat of the Han emperors of China, for example, built in the second century B.C. near their capital Ch'ang-an, is the earliest landscaped park of which we have detailed description. Large enough to encompass mountains, forests, and marshes as well as palaces and formal gardens, the park reflected Taoist beliefs about the ideal landscape and our place in the cosmos. Great towers and mountaintop pavilions served as retreats from which the emperor could contemplate nature in tranquility. Not all was peace and serenity, however; the emperor and his entourage also enjoyed hunting herds of wildlife maintained for their enjoyment. Although much of the park appeared natural, it required just as much engineering and earth moving to achieve and maintain this wild appearance as in the formal geometric gardens of Renaissance Europe.

Natural landscaping became popular in England during the eighteenth century under the leadership of architects such as Lancelot Brown (known as Capability Brown because he saw a capability for improving nature everywhere). This new organic design rejected the straight lines and rigid symmetry of earlier gardens, opting for

Figure 15.2
Central Park in New York City was one of the first large urban parks designed to provide nature experience and healthful recreation for common people.

Figure 15.3
Yellowstone National Park, established in 1872, is regarded as the first national park in the world. Although focused initially on the spectacle of natural curiosities and wonders, it has come to be appreciated for its beauty and wilderness values.

sweeping vistas over rolling hills, meadows, forests, and natural-looking ponds and marshes. The illusion of wild nature was carefully contrived, however. Clumps of trees were sculpted to create vistas and miniature buildings were strategically placed to emphasize the receding perspective. Brown built moats and fences hidden in ditches to control access to private property without interrupting the vista, a concept rediscovered by modern zoos (chapter 13).

Perhaps the first public parks open to ordinary citizens were the grand esplanades and the tree-sheltered agora that served as a gathering place in the planned Greek city. Central Park in New York City is an important successor to both this democratic ideal and the naturalistic principles of romantic landscaping. Promoted in 1844 by newspaper editor William Cullen Bryant as a "pleasuring ground in the open air for the benefit of all," the park was to provide healthful open space and contact with nature for the crowded masses of the city (fig. 15.2). A worldwide competition for design of the park was won by Frederick Law Olmstead, who became the father of landscape architecture in the United States.

Olmstead left New York in 1864 to become the original commissioner of Yosemite Park in California, the first area set aside to protect truly wild nature in the United States. Yosemite was authorized by President Abraham Lincoln in the midst of the Civil War to protect its resources from the unbridled exploitation common in frontier areas. Because there was no mechanism for running a park at the national level at that time, Yosemite was deeded to the State of California. It was transferred back to the federal government as a national park in 1890.

In 1872, President Ulysses S. Grant signed an act designating about 800,000 ha (almost 2 million ac) of land in the Wyoming, Montana, and Idaho territories as Yellowstone National Park, the first *national* park in the world. Although the initial interest of both the founders and visitors to Yellowstone was the spectacular "curiosities" and natural "wonders" of the geysers, hot springs, and canyons (fig. 15.3), the park was large enough to encompass and

preserve real wilderness. Because the territories had no means to manage the area, Yellowstone was made a national park and guarded by the army until the National Park Service was founded in 1916.

As it became apparent that wild nature and places of scenic beauty and cultural importance were rapidly disappearing with the closing of the North American frontier, the drive to set aside more national parks accelerated. Canada's Banff National Park was established in 1885. In the United States, Mount Rainier was authorized in 1899, Crater Lake in 1902, Mesa Verde in 1906, Grand Canyon in 1908, Glacier Park in 1910, and Rocky Mountain National Park in 1915.

North American Parks

Parks serve a variety of purposes. They can teach us about our past and provide sanctuaries where nature is allowed to evolve in its own way. They are havens not only for wildlife but also for the human spirit. Canada and the United States have greater total amounts of land dedicated to protected areas than any country except Denmark (which protects vast areas of Greenland's ice and snow) and Australia (which has designated great expanses of outback as Aboriginal lands and parks). Although Mexico's parks are newer and less extensive than its wealthy neighbors', they contain far more biological and cultural diversity than do parks in either the United States or Canada.

Existing Systems

The U.S. national park system has grown to more than 280,000 sq km (108,000 sq mi) in 376 parks, monuments, historic sites, and recreation areas. Each year about 300 million visitors enjoy this system. The most heavily visited units are the urban recreation areas, parkways, and historic sites. The jewels of the park system, however, and what most people imagine when they think of a national park, are the great wilderness parks of the West. Passage of the

Alaska Lands Act of 1980 nearly doubled the national park system. State and local parks occupy only about one-sixteenth as much area as national parks yet have about twice as many visitors.

Canada has a total of 1,471 parks and protected areas occupying about 150,000 sq km. Among this group are national parks, provincial parks, outdoor recreation parks, and historic parks. They range in size from vast wilderness expanses such as Wood Buffalo National Park in northern Manitoba or Ellesmere Island National Park Reserve in the Northwest Territories, to tiny pockets of cultural or natural history occupying only a few hectares. Kluane National Park in the Yukon, the new Tatshenshini-Alsek Wilderness in British Columbia, the adjoining Wrangell-St. Elias National Park, and Glacier Bay National Park in Alaska together encompass an area of about 10 million ha, roughly the size of Belgium or ten times as big as Yellowstone National Park. While many ecological reserves in Canada enforce strictly controlled access, other protected areas encourage intensive recreation, allow hunting, logging, or mining, and permit environmental manipulation for management purposes.

Park Problems

Originally, the great wilderness parks of Canada and the United States were seen as fortresses protected from development or exploitation by legal boundaries and diligent park rangers. Most were buffered from human impacts by their remote location and the wild lands surrounding them. Today, the situation has changed. Many parks have become islands of nature surrounded and threatened by destructive land uses and burgeoning human populations that crowd park boundaries. Forests are clear-cut right up to the edges of some parks, while mine drainage contaminates streams and groundwater. Garish tourist traps clustered at park entrances detract from the beauty and serenity that most visitors seek.

Threats to park values come from within as well. Roads, trails, and buildings in U.S. parks have suffered from years of underfunding and neglect. While the number of park visitors increased by 25 percent over the past decade, park budgets remained flat in constant dollars. Hobbled by an unresponsive bureaucracy, inadequate budgets, and deteriorating infrastructure, many of the 12,000 employees are overworked, underappreciated, and thoroughly frustrated. The General Accounting Office estimates that the parks need $3 billion merely for repair and restoration. Unfortunately, when money for parks becomes available, members of Congress try to steer it to pet projects in their own districts rather than truly deserving areas elsewhere.

Yosemite National Park exemplifies these problems. On popular three-day weekends, as many as 25,000 visitors crowd into the 18-square-kilometer (7-square-mile) valley floor, which is less than 1 percent of the total park area. They fill the valley with noise and smoke, trample fragile meadows and riverbanks, and spend hours in traffic jams. You can buy a pizza, play video games, do your laundry, play golf or tennis, and shop for curios and souvenirs in Yosemite Valley today, but you are less and less likely to experience the solitude and peace of nature extolled by John Muir. Park rangers have become traffic cops and crowd-control specialists rather than naturalists. A general management plan for Yosemite directed that both Park Service and Curry Company—the conces-

Figure 15.4

Visitors crowd popular trails in Yosemite National Park on summer weekends, making quiet and solitude impossible to find.

sionaire for tourist services—headquarters be removed from the valley along with 370 buildings, employee housing, 17 percent of existing guest rooms, and all automobiles. Little progress has been made toward meeting these goals, however.

Other parks experience similar difficulties. The four contiguous mountain parks of the Canadian Rockies (Banff, Jasper, Yoho, and Kootenay) receive about 10 million visitors each year (fig. 15.4). Four towns are located within their boundaries and pressures are mounting to expand hotels, shops, downhill ski facilities, convention centers, and condominiums. On Cape Cod National Seashore and in the new California Desert Park, dune buggies, dirt bikes, and off-road vehicles (ORV) run over fragile sand dunes, disturbing vegetation and wildlife and destroying the aesthetic experience of those who come to enjoy nature (fig. 15.5). In Florida's Everglades National Park, water flow through the "river of grass" has been disrupted and polluted by encroaching farms and urban areas. Wading-bird populations have declined by 90 percent, down from 2.5 million in the 1930s to 250,000 now. In Yellowstone National Park, 70 years of fire suppression allowed a buildup of fuel that resulted in catastrophic fires in the unusually dry summer of 1988, raising debates about how actively we should manage wilderness parks and preserves (see chapter 14).

Air pollution is a serious threat to many parks. The haze over the Blue Ridge Parkway is no longer blue but gray-brown because of air pollution carried in by long-range transport. Sulfate concentrations in Shenandoah and Great Smoky Mountains National Parks are five times human health standards, and ozone levels in Acadia National Park in Maine exceed primary air quality standards by as much as 50 percent on some summer days. Visitors to the Grand Canyon once could see mountains 160 km (100 mi) away; now the air is so smoggy you can't see from one rim to the other during one-third of the year. The main culprits are power plants in Utah and Arizona that supply electricity to Los Angeles, Phoenix, and other urban areas. Acid rain threatens sensitive lakes in the high mountains of the West, as well as in New England and eastern Canada. Photochemical smog is damaging the giant red-

Figure 15.5
Off-road machines, such as motorcycles, dune buggies, and four-wheel drive vehicles can cause extensive and long-lasting damage to sensitive ecosystems. Tracks can persist for decades in deserts and wetlands where recovery is slow.

Figure 15.6
Wild animals have always been one of the main attractions in national parks. Many people lose all common sense when interacting with big, dangerous animals. This is not a petting zoo.

woods in California's Sequoia National Park and contributing to forest declines in the Adirondacks.

Mining and oil interests continue to push for permission to dig and drill in the parks, especially on the 3 million acres of private inholdings in the parks. These forces were successful in excluding mineral lands in the Misty Fjords and Cape Kruzenstern National Monuments in Alaska. Placer mines in Denali National Park wash sediments out of hillsides with high-pressure hoses, dumping thousands of tons of sediment each day into valuable salmon streams. Uranium mines at the edge of the Grand Canyon threaten to contaminate the Colorado River and the park's water supply with radioactive contaminants. In 1995, President Clinton temporarily blocked plans for an open pit gold mine on the Beartooth Plateau just outside Yellowstone National Park. The wreck of the *Exxon Valdez* in 1989 contaminated hundreds of miles of Alaskan coastline, including Katmai and Kenai National Parks.

Wildlife

Wildlife is at the center of many arguments regarding whether the purpose of the parks is to preserve nature or to provide entertainment for visitors (fig. 15.6). In the early days of the parks, "bad" animals (such as wolves and mountain lions) were killed so that populations of "good" animals (such as deer and elk) would be high. Rangers cut trees to improve views, put out salt blocks to lure animals to good viewing points close to roads, and otherwise manipulated nature to provide a more enjoyable experience for the guests.

Critics of this policy claim that favoring some species over others has unbalanced ecosystems and created a sad illusion of a natural system. They claim that excessively large elk populations in Yellowstone and Grand Teton National Parks, for instance, have degraded the range so badly that other species such as mice and ground squirrels are being crowded out. Park rangers tried hiring professional hunters to reduce the elk herd, but a storm of protest was raised. Sportsmen want to be able to hunt the elk themselves, animal lovers don't want them to be killed at all, and wilderness ad-

vocates don't like the precedent of hunting in national parks. The Park Service has retreated to a policy of "natural regulation," intended to let nature take its course. When elk starve to death, as thousands do in a hard winter, however, many people are appalled.

In 1995, after an absence of nearly 70 years, the haunting howls of gray wolves were heard once more in Yellowstone Park (fig. 15.7). As part of a larger restoration project in the Northern Rockies, biologists captured 29 wolves in Canada and translocated them to Wyoming, Montana, and Idaho—including 14 released in Yellowstone. The ultimate goal is to establish three populations of about 100 wolves each in wilderness areas in the Absorka, Bitterroot, Clearwater, and Salmon mountains. Ecologists see restoring the top predator to these ecosystems as essential in controlling prey species such as elk and deer, and a step toward recreating authentic biological communities.

This whole project was vehemently opposed by local ranchers, however, who see wolves as a deadly threat to livestock, children, and their whole way of life. Deliberately reintroducing a large, dangerous predator that had been systematically destroyed by government hunters in the early part of this century seems to many westerners as a step back into the dark ages. Although there has never been an authenticated case in which a healthy wolf has attacked a human in North America, centuries of stories about the big bad wolf have convinced many people that these animals are ruthless, evil killers and the embodiment of everything they fear and loathe in wild nature.

Most environmental groups, on the other hand, enthusiastically endorse the idea of restoring the wolves. To them, this animal is a symbol of wildness and freedom. Without its top predator, the wilderness is lamentably incomplete. To be able to hear the thrilling howls of a wild wolf pack is an ultimate wilderness experience for many people. By a 6-to-1 margin, Yellowstone visitors favor wolf reintroduction. For some reason, wolves have become extremely popular among the general public in recent years.

Figure 15.7

Are wolves beautiful, thrilling symbols of wild nature or ruthless killers of livestock and small children? Reintroduction of these top predators into Yellowstone National Park has enthusiastic support from environmental groups but passionate opposition from local ranchers and hunters. This raises fundamental questions about the purposes of parks.

How wolves will be managed is even more controversial than the reintroduction. Restored wolves are classified as "experimental, nonessential populations," which allows wildlife managers to kill all animals that leave designated areas or threaten livestock in any way. Wolf supporters believe that better livestock management and payment to ranchers for any authenticated losses can take care of problems without destroying wolves. Local residents insist that without this special classification, they will be forced to take matters into their own hands. "Shoot, shovel, and shut-up" is the preferred wolf management strategy for many who live nearby. As we saw in the introduction to this chapter, it is impossible to have effective conservation plans without the cooperation of local resi-

dents. How can we resolve conflicts such as this between the wishes and values of park neighbors and those of the broader public?

Management of feral species (escaped domestic animals) and rehabilitation of former human-dominated landscapes creates more complexities for park officials (Case Study, next page).

New Directions

What else is being done to enhance the visitor's experience of a national park while maintaining the natural ecosystem as much as possible? Several parks have removed facilities that conflict with natural values. In Yellowstone, a big laundry and a cabin ghetto next to Old Faithful have been torn down. Shabby hotels and filling stations that once stood at Norris Junction and Yancy's Hole are gone. The golf course, slaughterhouse, and tent camps at Mammoth Hot Springs also have been removed. In Yosemite, Grand Canyon, and Denali, tourists park their cars and take shuttle buses into the park to reduce congestion and pollution. A poorly performing concessionaire was denied renewal of a contract for Yosemite in 1979, something that had never before happened.

There are proposals that a number of parks be closed to cars, and some areas might be closed to tourists altogether to protect wildlife and fragile ecosystems. The International Union for the Conservation of Nature and Natural Resources (IUCN) divides protected areas into five categories with increasing levels of protection and decreasing human impacts (table 15.1). Many of our parks have tried to meet all these goals simultaneously but may have to select those of highest value. Most parks limit the number of overnight visitors. The time may come when park permits will need to be reserved years in advance and visits to certain parks will be limited to once in a lifetime! How would you feel about such a policy? Would you rather visit a pristine, uncrowded park once or a less perfect place whenever you wanted?

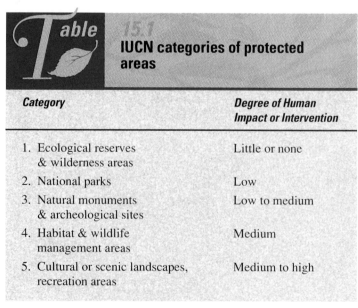

Table 15.1

IUCN categories of protected areas

Category	Degree of Human Impact or Intervention
1. Ecological reserves & wilderness areas	Little or none
2. National parks	Low
3. Natural monuments & archeological sites	Low to medium
4. Habitat & wildlife management areas	Medium
5. Cultural or scenic landscapes, recreation areas	Medium to high

Source: World Conservation Union, 1990.

CASE STUDY
Donkeys in the Virgin Islands

Deciding what belongs in a park and what doesn't can be difficult, especially where human occupation has had a heavy impact. In the Virgin Islands National Park, on the island of St. John in the U.S. Virgin Islands, for example, managers are struggling to rehabilitate a sample of the region's native biological communities in an overused landscape of old sugar plantations. Park ecologists work to identify and restore native species and to remove alien invaders. Not everyone agrees, however, on which exotics should be eliminated.

A small but popular population of donkeys is a case in point. Brought to the island as farm animals by European colonists in the early 1700s, the donkeys were abandoned when the sugar plantations shut down a century or more ago. They now roam the island roads, graze in the brush, and serve as a reminder of a former, agrarian landscape. Tourists love them because they're cute and furry (fig. 1). Residents generally are fond of them because they are amusing and harmless—except for a few that get in the way of cars—and an important part of local history and color. Park officials, meanwhile, would like to eradicate the donkeys, and their efforts have stirred up a heated debate.

For park rangers, the issue is ecological. Donkeys prefer to graze on the soft, moist leaves of native plants, threatening rare local species and encouraging thorny exotics. Furthermore, they spread weed seeds in their fur and feces, while their sharp hooves erode the soil. As an introduced species, they have no place in the island's native ecosystem, but it is impossible to keep them out of the park. Rounding up these agile, clever animals in the dense forest and steep topography of the park is nearly impossible. The most expedient way to remove donkeys would be to trap or shoot them.

For donkey lovers, this is a cruel and inhumane solution. How could we be so mean to these endearing creatures? Tourists would rather see donkeys and sugar mill ruins than some obscure, na-

Figure 1

Donkeys are not native to the Virgin Islands National Park and represent a threat to indigenous plants that lack defenses against hooved grazers. On the other hand, tourists think donkeys are cute and add charming local color to the park. If you were Park Superintendent would you eradicate the donkeys or let them be?

tive vegetation. Furthermore, supporters ask, what is "natural" on the island? The native community was so completely altered by farming that anything we recreate will be artificial to some extent. Who can say what the landscape might look like now if it hadn't been occupied for 300 years by sugar plantations? Species arrive and disappear naturally on islands. If humans facilitate the migration of some species or the extinction of others, is that unnatural?

The issue ultimately comes down to the purpose of national parks. Are they for the entertainment of tourists or refuges for rare and endangered species? Are we trying to preserve parks like museum specimens of some distant past? If so, at what point in the past should time be frozen? Or should we continue to allow the landscape to evolve and change? Is biological purity the highest priority in a park or is income for local residents a more important goal?

ETHICAL CONSIDERATIONS

1 Some people argue that the donkeys have a right to live on the island because they are sentient creatures and have been there much longer than the park has. Do we have a right to simply eliminate organisms we think don't belong?

2 What role should emotions and empathy play in decisions such as this? Would your answer to question 1 be different if we were talking about weedy plant invaders rather than mammals?

3 If all non-native species are to be removed from the park, then shouldn't humans, too, be excluded? After all, we do far more damage than the donkeys to the local environment.

4 Many tourists would be more interested in the park if it were a recreation of sugar plantation days than a native forest. The plantations were run, however, by slaves. How would you handle that part of a living history display?

Preserving Nature

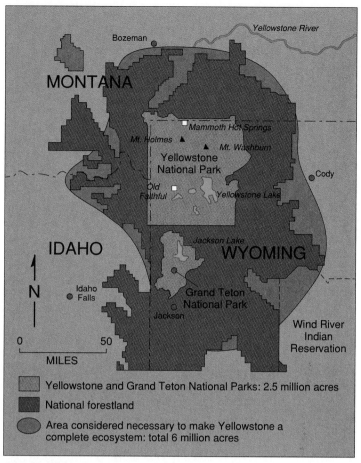

Figure 15.8

This map shows the Yellowstone ecosystem complex or biogeographical region, which extends far beyond the park boundaries. Park managers and ecologists believe that it is necessary to manage the entire region if the park itself is to remain biologically viable.

One of the biggest problems with managing parks and nature preserves is that boundaries usually are based on political rather than ecological considerations. Airsheds, watersheds, and animal territories or migration routes often extend far beyond official boundaries and yet profoundly affect communities that we are attempting to preserve. Yellowstone and Grand Teton Parks in northwestern Wyoming are examples of this concept. Although about 1 million ha (2.5 million ac) in total size, these parks probably cannot preserve viable populations of large predators such as grizzly bears. Management policies in the surrounding national forests and private lands seriously affect conditions in the park (fig. 15.8). The natural **biogeographical area** (an entire self-contained ecosystem and its associated land, water, air, and wildlife resources) must be managed as a unit if we are to preserve all its values.

New Parks

One solution to congestion and overuse of our parks is to create new ones to distribute the load. This would also allow us to protect and enjoy other magnificent areas that deserve preservation in their own right. The Great Basin Park in Nevada was established in 1988 to protect a portion of the intermountain desert and two groves of ancient bristlecone pines. The new Mojave National Preserve in California will protect sand dunes, Joshua tree "forests," Indian petroglyphs, fossils, and critical habitat for rare and endangered species such as the desert pupfish and desert tortoise, but has aroused fierce opposition from off-road vehicle clubs, hunters, ranchers, and others who resent restrictions on use of what they consider "their" land.

Some other areas that have been proposed for park status are the Tallgrass Prairie in Oklahoma and Kansas, the Columbia Gorge in Washington and Oregon, Big Sur and Lake Tahoe in California, some of the Florida Keys, and the Hudson Valley and Thousand Islands in New York.

As this is being written in 1995, however, the mood in Congress is strongly anti-environmental and opposed to any governmental ownership or management of public lands. In addition to sharp budget cuts in the park system, several bills are under consideration that would shrink existing parks and prohibit any new ones. Proposals have been made to completely eliminate Voyageurs National Park in Minnesota and the new Mojave National Preserve in California. Under the extremest scenarios, the entire park system might be privatized (sold to commercial interests) or given back to states or counties. Contact your local legislator to learn the current status of these measures.

Canada's Green Plan, released in 1990, called for a doubling of protected areas to a total of 12 percent of total land surface. Representative samples of every ecoregion should be included in this network, including cultural landscapes and marine ecosystems. The proposed Matamek Ecological Reserve in Quebec will be over 700 sq km (270 sq mi) in size and will protect an entire watershed.

World Parks and Preserves

The idea of setting aside nature preserves has spread rapidly over the past 50 years as people around the world have become aware of the growing scarcity of wildlife and wild places. So far, more than 530 million ha (nearly 4 percent of the earth's land) is designated as parks, wildlife refuges, and nature preserves worldwide (table 15.2 and fig. 15.9). The largest number of protected areas is in tropical dry forests—principally in Africa—and temperate deciduous forests—mainly in North America and Europe—but many of these preserves are too small to maintain significant biological populations over the long term. Vast stretches of arctic tundra in a few parks and preserves in Alaska, Canada, Greenland, and Scandinavia make up 26 percent of all protected areas. Deserts and tropical humid forests also are well represented, but grasslands, aquatic ecosystems, and islands are badly underrepresented.

According to the United Nations Environment Programme, North and Central America have the largest fraction—33 percent of all protected land or nearly 10 percent of their land area—designated for protection of any continent (fig. 15.10). The former Soviet Union, with 17 percent of the world's land, has only 3 percent of the officially protected area. The rapid destruction of Siberian forests (see chapter 14) and terrible pollution problems in Russia and its former allies raises serious concerns about these natural areas. The IUCN has identified an additional 3000 areas, to-

Table 15.2

World nature preserves by area and type

Biome	Number of Areas	Percent of Protected Area
Tropical dry forests	907	21
Tropical humid forests	355	12
Temperate deciduous forests	682	6
Temperate coniferous forests	114	7
Deserts	215	17
Tundra	31	26
Grasslands	116	3
Mountain regions	318	7
Islands	74	<1
Lakes and wetlands	10	<1

Source: Norman Meyers, 1993.

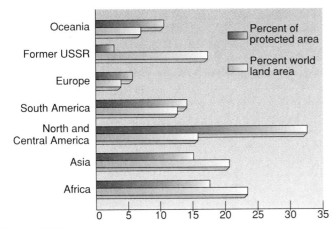

Figure 15.9

North and Central America have a far greater percentage of protected nature reserves than is represented by the continent's share of total world landmass. The former USSR, by contrast, has a far smaller percentage of its land area in protected status, and even those lands are subject to pollution and looting. We need a comprehensive, worldwide plan to save representative samples of our natural heritage everywhere.

Source: UNEP 1995

Figure 15.10

Plant collectors explore the mysteries of a tropical rainforest in search of new species. Who knows what useful properties might be found in the thousands of undiscovered species that live here?

taling about 3 billion hectares worthy for national park or wildlife refuge status. The most significant of these areas are designated world **biosphere reserves** or world heritage sites. Currently, about 300 of these special refuges have been designated in 75 countries.

Among the individual countries with the most admirable plans to protect natural resources are Costa Rica, Tanzania, Rwanda, Botswana, Benin, Senegal, Central Africa Republic, Zimbabwe, Butan, and Switzerland, each of which has designated 10 percent or more of its land as ecological protectorates. Brazil has even more ambitious plans, calling for some 231,600 sq km, or 18 percent of the country to be protected in nature preserves. So far, however, many of these areas are parks in name only. Lacking guards, visitor centers, administrative personnel, or even boundary fences, they are open to vandals and thieves to loot as they will.

Protecting Natural Heritage

Even parks with systems in place for protection and management are not always safe from exploitation or changes in political priorities. Many problems threaten natural resources and environmental quality in the parks. In Greece, the Pindos National Park is threatened by plans to build a hydroelectric dam in the center of the park. Furthermore, excessive stock grazing and forestry exploitation in the peripheral zone are causing erosion and loss of wildlife habitat. In Colombia, the Paramillo National Park also is threatened by dam building. Oil exploration along the border of the Yasuni National Park in Ecuador pollutes water supplies, while miners and loggers in Peru have invaded portions of Huascaran National Park. In Palau, coral reefs identified as a potential biosphere reserve are damaged by dynamiting, while on some beaches in Indonesia every egg laid by endangered sea turtles is taken by egg hunters. These are just a few of the many problems in parks around the world. Often countries with the most important biomes lack funds, trained personnel, and experience to manage some of the areas under their control.

The IUCN has developed a **world conservation strategy** for natural resources that includes the following three objectives: (1) to maintain essential ecological processes and life-support systems (such as soil regeneration and protection, recycling of nutrients, and cleansing of waters) on which human survival and development depend; (2) to preserve genetic diversity, which is the foundation of breeding programs necessary for protection and improvement of cultivated plants and domesticated animals; (3) to ensure that any utilization of species and ecosystems is sustainable.

These goals are further elaborated in the ecological plan of action adopted by the IUCN and shown in table 15.3. A promising approach for financing these objectives is debt-for-nature swaps (chapter 14).

Size and Design of Nature Preserves

What is the optimum size and shape of a wildlife preserve? For many years, conservation biologists have disputed whether it is better to have a *single large or several small* reserves (the SLOSS debate). Ideally, a reserve should be large enough to support viable populations of endangered species, keep ecosystems intact, and isolate critical core areas from damaging external forces (see chapter 13). But as gaps are opened in habitat by human disturbance (fig. 15.11), and eventually areas are fragmented into isolated islands, edge effects may eliminate core characteristics everywhere.

To satisfy the conflicting needs and desires of humans and nature, we may need a spectrum of preserves with decreasing levels of interference and management ranging from: (1) *recreation areas,* designed primarily for human entertainment, aesthetics, and enjoyment; (2) *historic areas,* intended to preserve a landscape as we imagine it looked in some previous time—such as pre-settlement or pioneer days; (3) *conservation reserves,* set aside to maintain essential ecological functions, preserve biodiversity, or protect a particular species or group of organisms; (4) *pristine research areas,* to serve as a baseline of undisturbed nature; and (5) *inviolable preserves,* for sensitive species from which all human entrance is strictly prohibited (What Do *You* Think?, p. 328).

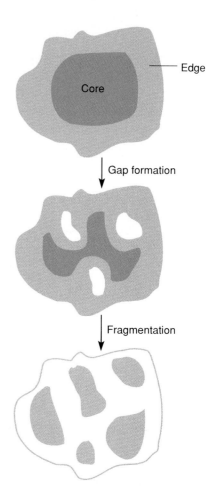

Figure 15.11

A patch or "island" of habitat becomes reduced and fragmented as vegetation gaps, or other human disturbances, expand until only small fragments remain. Note that core regions disappear early in this process. Finally, only edges are left in a matrix of disturbed habitat.

For some species with small territories, several small isolated refuges can support viable populations and provide insurance against a disease or other calamity that might wipe out a single population. But small preserves can't support species such as elephants or tigers that need large amounts of space. Given human needs and pressures, however, big preserves aren't always possible. Establishing **corridors** of natural habitat to allow movement of species from one area to another (fig. 15.12) can help maintain genetic exchange and prevent the high extinction rates often characteristic of isolated and fragmented areas.

An interesting experiment funded by the World Wildlife Fund and the Smithsonian Institution is being carried out in the Brazilian rainforest to determine the effects of shape and size on biological reserves. Some twenty-three test sites, ranging in size from one hectare (2.47 acres) to 10,000 hectares have been established. Some areas are surrounded by clear-cuts and newly created pastures (fig. 15.13), while others remain connected to the surrounding forest. Selected species are regularly inventoried to monitor their dynamics after disturbance. As was expected, some species disappear very

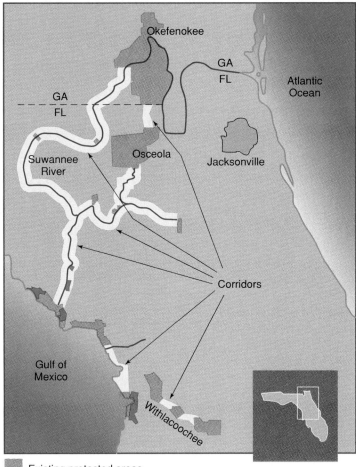

Existing protected areas

Already formally proposed acquisition

Presently proposed to establish functional system

Figure 15.12

Corridors serve as routes of migration, linking isolated populations of plants and animals in scattered nature preserves. Although individual preserves may be too small to sustain viable populations, connecting them through river valleys and coastal corridors can facilitate interbreeding and provide an escape route if local conditions become unfavorable.

Source: R. F. Noss and L. D. Harris, 1986, "Nodes, Networks and MUMs: Preserving Diversity at All Levels," *Environmental Management*, vol. 10:299–309.

quickly, especially from small areas. Sun-loving species flourish in the newly created forest edges, but deep-forest, shade-loving species move out, particularly when size or shape reduces the distance from the edge to the center below a certain minimum. This demonstrates the importance of surrounding some reserves with buffer zones that maintain the balance of edge and shade species.

Conservation and Economic Development

Many of the most seriously threatened species and ecosystems of the world are in the developing countries, especially in the tropics. This situation concerns us all because these countries are the guardians of biological resources that may be vital to all of us. Unfortunately, where political and economic systems fail to pro-

Figure 15.13

How small can a nature preserve be? In an ambitious research project, scientists in the Brazilian rainforest are carefully tracking wildlife in plots of various sizes, either connected to existing forests or surrounded by clear-cuts. As you might expect, the largest and most highly specialized species are the first to disappear.

vide people with land, jobs, and food, disenfranchised citizens turn to legally protected lands, plants, and animals for their needs. Immediate human survival always takes precedence over long-term environmental goals. Clearly the struggle to save species and unique ecosystems cannot be divorced from the broader struggle to achieve a new world order in which the basic needs of all are met.

The tropics are suffering the greatest destruction and species loss in the world, especially in humid forests and coastal ecosystems. People in some of the affected countries are beginning to realize that the biological richness of their environment may be their

What Do *You* Think?
Nature Preservation: How Much Is Enough?

Do we have enough parks and wilderness areas? This question is producing growing debate across the country. It is likely that conflicts over nature preservation will continue to increase. Our population continues to grow. The amount of undeveloped land is decreasing. Use of existing natural areas is increasing, while untapped sources of extractable, nonrenewable resources are being diminished. Interest in environmental ethics is growing. So, developing a calm, effective, method of resolving these conflicts is taking on growing importance.

From the critical thinking viewpoint, the issue of nature preservation is more complex than many other environmental questions and dilemmas.

Most environmental issues in dispute are contests between competing sets of factual claims. Global warming is, or is not, likely to produce serious consequences. Species extinction does, or does not, threaten ecological stability. The critical thinking strategies are clear in these cases. Be skeptical. Demand relevant, unbiased information. Evaluate critically. Base conclusions on sound judgments. Employ well-reasoned judgment where scientific uncertainty exists.

But these strategies presuppose that the premises underlying each side are fact-based, and that the central task is to steer clear of basing decisions on unsubstantiated opinion. But how can or ought the critical thinker respond when some of the premises are based on values? What should be the role of critical thinking when the issue is at least partly one of "should or should not" instead of a matter of "is or is not?"

Elsewhere in this chapter you have read the main arguments of those favoring and opposing additional nature preservation. When preservationists argue that natural areas serve as important baselines for ecological research, the premise is that understanding how nature works is important. The critical thinking process is well-suited to determining whether undisturbed nature is or is not important to such a goal. Evidence suggests that it is.

Preservationists also argue that economic development ought not automatically have pre-preemptory rights over use of an area to provide for psychological and recreational needs through experiences of solitude and primitive recreation. The critical thinking process can help evaluate the economic value of an area but it seems less well-suited to producing an objective decision about whether an area *should* be used for mining or primitive recreation. Such a question is not so amenable to a simple weighing of facts. It is more a question of values. The conflict of values over nature preservation represents, at its core, one of the skirmish lines between the expansionist and ecological worldviews introduced in chapter 2.

At issue is not whether we ought to incorporate values in making decisions. We cannot avoid doing so. The issue is the extent to which reason can weigh in as part of the decision-making process. Could reasoned thought contribute to decision making in value-laden issues? What do you think? Can reasoned thought help construct a framework for resolving issues such as the amount of minimally disturbed nature that ought to be available for primitive recreation?

Perhaps one constructive role of critical thinking is to lay out questions to be considered as we develop policy for addressing nature preservation matters.

1 *Should the goal be to seek a balance between competing interests? How could balance be defined?*

2 *Should the irreversibility of an action influence the decision? In other words, should a goal of keeping options open for future generations be a significant consideration?*

3 *Should decisions of this type be left purely to unrestricted political forces, or should the process include checks and balances that protect the interests of the side with the least political power?*

4 *Should ethical considerations play a role? What sort of role would that be? How could differing ethical viewpoints be fairly treated?*

Do you think these questions are appropriate and helpful? What additional questions would you propose be taken into account that would help society resolve conflicts of values in a thoughtful, reasoned way?

---

most valuable resource and that its preservation is vital for sustainable development. Tourism can be more beneficial to many of these countries over the long term than extractive industries such as logging and mining (In Depth, p. 330).

At the 1982 World Congress on National Parks in Bali, five hundred scientists, managers, and politicians discussed the design and location of biological reserves and the ecological, economic, and social factors that impinge on wildlife preservation. They concluded that conservation and rural development are not necessarily incompatible. In many cases, sustainable production of food, fiber, medicines, and water in rural areas depends on ecosystem services derived from adjacent conservation reserves. Tourism associated with wildlife watching and outdoor recreation can be a welcome source of income for underdeveloped countries. If local people share in the benefits of saving wildlife, they probably will cooperate and the programs will be successful. To reformulate Thoreau's famous dictum, "In broadly shared economic progress is preservation of the wild."

Table 15.3
IUCN ecological plan of action

1. Launch a consciousness-raising exercise to bring the issue of biological resources to the attention of policymakers and the public at large.

2. Design national conservation strategies that take explicit account of the values at stake.

3. Expand our network of parks and preserves to establish a comprehensive system of protected areas.

4. Undertake a program of training in the fields relevant to biological diversity to improve the scientific skills and technological grasp of those charged with its management.

5. Work through conventions and treaties to express the interest of the community of nations in the collective heritage of biological diversity.

6. Establish a set of economic incentives to make species conservation a competitive form of land use.

Source: International Union for Conservation and National Resources.

Indigenous Communities and Biosphere Reserves

Areas chosen for nature preservation are often traditional lands of indigenous people who cannot simply be ordered out. Finding ways to integrate human needs with those of wildlife is essential for local acceptance of conservation goals in many countries. In 1986, UNESCO initiated its **Man and Biosphere (MAB) program** that encourages division of protected areas into zones with different purposes. Critical ecosystem functions and endangered wildlife are protected in a central core region where limited scientific study is the only human access allowed. Ecotourism and research facilities are located in a relatively pristine buffer zone around the core, while sustainable resource harvesting and permanent habitation are allowed in multiple-use peripheral regions (fig. 15.14).

Mexico's 545,000-hectare (2100-square-mile) Sian Ka'an Reserve on the Caribbean coast is a good example of a MAB reserve. The core area includes 528,000 ha (1.3 million ac) of coral reef and adjacent bays, marshes, and lowland tropical forest. More than 335 bird species have been observed within the reserve, along with endangered manatees, five types of jungle cats, spider and howler monkeys, and four species of increasingly rare sea turtles. Approximately 25,000 people live in communities in peripheral regions around the reserve, and the resort developments of Cancun are located just to the north. In addition to tourism, the economic base of the area includes lobster fishing, small-scale farming, and coconut cultivation.

The Amigos de Sian Ka'an, a local community organization, played a central role in establishing the reserve and is working to protect the resource base while it improves living standards for local people. New intensive farming techniques and sustainable harvesting of forest products enable people to make a living without

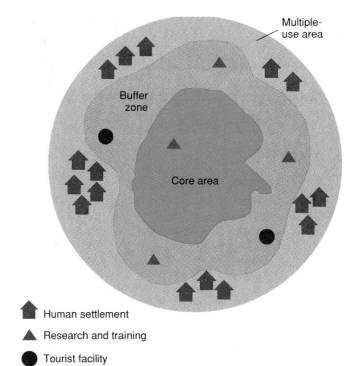

Human settlement

Research and training

Tourist facility

Figure 15.14

A model biosphere reserve. Traditional parks and wildlife refuges have well-defined boundaries to keep wildlife in and people out. Biosphere reserves, by contrast, recognize the need for people to have access to resources. Critical ecosystem is preserved in the core. Research and tourism are allowed in the buffer zone, while sustainable resource harvesting and permanent habitations are situated in the multiple use area around the perimeter.

destroying the resource base. Better lobster harvesting techniques developed at the reserve have improved the catch without depleting native stocks. Local people now see the preserve as a benefit rather than an imposition from outside. Unfortunately, the government has very limited funds to develop or patrol the reserve.

An even grander plan called *Passeo Pantera* (path of the panther) envisions a thousand-mile-long series of reserves and protected areas interconnected by natural corridors and managed buffer zones that would preserve both wildlife and native cultures along the entire Caribbean coast of Central America from the Yucatan to Panama. Whether the politically unstable and financially troubled countries involved could get together to establish such a plan is questionable. Competing goals and objectives of conservation groups, international development banks, and powerful neighbors such as the United States also complicate the picture, but the idea is a noble one whose time perhaps has come.

Wilderness Areas

Although indigenous people had lived in the Americas for thousands of years before the first Europeans arrived, introduced diseases killed up to 90 percent of the existing population so that the continent appeared a vast, empty wilderness to early explorers. As historian Frederick Jackson Turner pointed out in a series of articles and

*I*n Depth: Ecotourism

Travel is now the largest industry in the world, generating around $3 trillion per year in total revenues. A growing segment of this market is **ecotourism,** a combination of adventure travel, cultural exploration, and nature appreciation in wild settings. Trekking, hiking, bird watching, nature photography, wildlife safaris, camping, mountain climbing, fishing, snorkeling and scuba diving, river rafting and canoeing, and botanical study are some of the favorite forms of ecotourism (fig. 1). Experiencing other cultures, especially those of rural or native people who have traditional relationships to the land, is usually an important aspect of such travel.

Although only about 10 percent of the total vacation market, ecotourism currently generates some $20 billion per year in domestic and international receipts. This can provide both funding and an incentive for developing countries to preserve endangered wildlife and threatened habitats. Creating jobs for local people gives them alternatives to destructive harvesting practices that may previously have been their only source of income. Our interest in traditional customs and crafts serves to validate them in the eyes of young people who might be tempted to abandon their history. Dancers and artisans are paid to practice their art and keep it alive where it might otherwise be lost.

Hunting was the first form of tourism in many remote places. Kenya, for instance, developed an extensive infrastructure to serve big game safaris in the early part of this

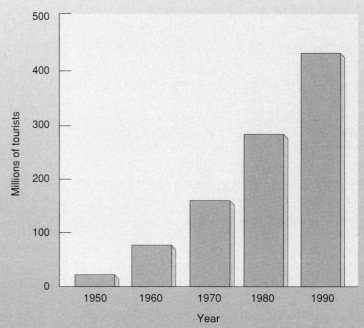

Figure 1

The number of international tourist arrivals worldwide has grown nearly twenty-fold in the past 40 years. At least 10 percent of these are cultural tourists or ecotourists.

Source: *Economic Review of World Tourism,* World Tourism Organization, Madrid, 1990.

century. By 1978, however, game was becoming scarce and a hunting ban eliminated the jobs of many safari guides and bearers. Happily, ecotourists—who shoot with cameras rather than rifles—have more than replaced big game hunters. Wildlife biologist Michael Soulé estimates that one maned lion in Kenya's Amboseli National Park is worth $515,000 for tourist viewing but is worth only $8500 for hunting. Another interesting comparison is that the economic yield from

tourists who come to see one lion in Amboseli is equal to the income from a herd of 30,000 cows, making ecotourism in this region more profitable than farming.

An outstanding example of ecotourism that benefits local people is the CAMPFIRE project (Communal Areas Management Programme For Indigenous Resources) in Zimbabwe. It is based on the idea that local residents should take responsibility for wildlife on their land and draw direct benefits from management of that

speeches around the turn of the century, a belief that wilderness was not only a source of wealth but also the origin of strength, self-reliance, wisdom, and character is deeply embedded in our culture. The frontier was seen as a place for continuous generation of democracy, social progress, economic growth, and national energy. A number of authors, including Henry Thoreau, Aldo Leopold, Sigurd

Olson, Edward Abbey, and Wallace Stegner, have written about the physical, mental, and social benefits for modern people of rediscovering solitude and challenge in wilderness (fig. 15.15).

Beginning with designated primitive and roadless areas in a few national forests in the 1920s, the United States has established a system of 264 wilderness areas encompassing nearly 36 million

wildlife. As its leaders say, "African elephants will have no chance until African people have a chance." Revenues generated from game viewing and hunting safaris go to villages to be used for game management and fencing to keep animals out of crop fields and other community projects. In its first year of operation, households around several game parks received a $200 cash payment. This was more than the normal yearly income for most families and served as a powerful incentive for residents to value wildlife and participate in the program.

Not all ecotourism is benign and beneficial, however. We who visit other countries must be careful to avoid alienating or humiliating local people by flaunting our wealth or treating them as subhuman curiosities. Too often, insensitive tourists consume resources at an exorbitant rate, drive up prices, defile holy places, and offend local sensibilities with unseemly behavior.

With care, however, these problems can be avoided and ecotourism can be helpful to you, your local hosts, and the environment (fig. 2). The following list of suggestions will help you plan a responsible, positive trip.

Figure 2
American students get a nature lesson in the Monteverde Cloud-forest Preserve in Costa Rica. Ecotourism can be fun as well as educational.

1 Pretrip preparation. *Learn about the history, geography, ecology, and culture of the area you will visit. Understand the do's and don't's that will keep you from violating local customs and sensibilities.*

2 Environmental impact. *Stay on designated trails and camp in established sites, if available. Take only photographs and memories and leave only good will wherever you go.*

3 Resource impact. *Minimize your use of scarce fuels, food, and water resources. Do you know where your wastes and garbage go?*

4 Cultural impact. *Respect the privacy and dignity of those you meet and try to understand how you would feel in their place. Don't take photos without asking first. Be considerate of religious and cultural sites and practices. Be as aware of cultural pollution as you are of environmental pollution.*

5 Wildlife impact. *Don't harass wildlife or disturb plant life. Modern cameras make it possible to get good photos from a respectful, safe distance. Don't buy products such as ivory, tortoise shell,*

animal skins, or feathers from endangered species.

6 Environmental benefit. *Is your trip strictly for pleasure or will it contribute to protecting the local environment? Can you combine ecotourism with work on clean up campaigns or delivery of educational materials or equipment to local schools or nature clubs?*

7 Advocacy and education. *Get involved in letter writing, lobbying, or educational campaigns to help protect the lands and cultures you have visited. Give talks at schools or to local clubs after you get home to inform your friends and neighbors about what you have learned.*

ha (88 million ac). The 1964 Wilderness Act, which is the basis for much of this system, defined **wilderness** as "an area of undeveloped land affected primarily by the forces of nature, where man is a visitor who does not remain; it contains ecological, geological, or other features of scientific or historic value; it possesses outstanding opportunities for solitude or a primitive and unconfined type of recreation; and it is an area large enough so that continued use will not change its unspoiled, natural conditions." Most of the areas meeting these standards are in the western states and Alaska.

Additional wilderness areas continue to be evaluated for protected status. The Forest Service was instructed by the 1964 act to carry out a roadless area review and evaluation (RARE) on all *de*

Figure 15.15

The Absaroka-Beartooth Wilderness, north of Yellowstone National Park, preserves a piece of relatively untouched natural ecosystem. Areas such as this serve as a refuge for endangered wildlife, a place for outdoor recreation, a laboratory for scientific research, and a source of wonder, awe, and inspiration.

facto wilderness areas under its jurisdiction. Using a deliberately "pure" interpretation that excluded all lands with any history of roads or development (even if all traces of human impact were long gone), it finally decided in 1979 that only about one-fourth of its 23 million ha (56 million ac) of roadless areas qualified for protection.

A prolonged battle has been waged over these *de facto* wilderness areas, pitting environmental groups who want more wilderness against loggers, miners, ranchers, and others who want less wilderness. The arguments for saving wilderness are that it provides (1) a refuge for endangered wildlife, (2) an opportunity for solitude and primitive recreation, (3) a baseline for ecological research, and (4) an area where we have chosen simply to leave things in their natural state. The arguments against more wilderness are that timber, energy resources, and critical minerals contained on these lands are essential for economic development.

To people who live in remote areas, jobs, personal freedom, and local control of resources seem more important than abstract values of wilderness. They often see themselves as an embattled minority trying to protect an endangered, traditional way of life against a wealthy elite who want to lock up huge areas for recreation or aesthetic purposes. Wilderness proponents point out that 96 percent of

the country already is open for resource exploitation; the remaining 4 percent is mostly land that developers didn't want anyway.

The last large areas still being studied for wilderness preservation in the United States are on Bureau of Land Management (BLM) land. In 1976 the Federal Land Policy and Management Act (FLPMA) ordered the BLM to review all its roadless areas for wilderness potential. Applying the same "pure" standards used by the Forest Service, the BLM found only one-fourth of its 112 million ha (276.6 million ac) suitable for wilderness. Preservationists argue that twice that much land should be considered.

For many people, especially those in developing countries, the idea of pristine wilderness untouched by humans is regarded as neither very important nor very interesting. In most places, all land is occupied fully—if sparsely—by indigenous people. To them the area is home no matter how empty it may look to outsiders. From this perspective, preserving biological diversity, scenic beauty, and other natural resources may be a good idea, but excluding humans and human features from the land does not necessarily make it more valuable. In fact, saving cultural heritage, working landscapes, and historical evidence of early human occupation can often be among the most important reasons to protect an area.

Figure 15.16
Wildlife refuges have contributed to the recovery of both game and non-game wildlife populations. In some cases, such as these elk on the National Elk Refuge near Jackson Hole, WY, protected species become so numerous that surplus population must be harvested or translocated.

Wildlife Refuges

In 1901 President Teddy Roosevelt established fifty-one national **wildlife refuges,** the first in an important but troubled system for wildlife preservation in the United States (fig. 15.16). There are now 504 wildlife refuges in this system, encompassing nearly 40 million hectares of land and water and representing every major biome in North America. They range in size from less than 1 ha (2.5 ac) for the tiny Mille Lacs Refuge in Minnesota to 7 million ha (18 million ac) in the Arctic Wildlife Refuge of Alaska. Altogether, about 1 percent of U.S. surface area is designated as wildlife refuge.

Major additions to the refuge system were made by President Franklin D. Roosevelt and Harold Ickes, his Secretary of the Interior, who took advantage of low prices during the Depression to add additional areas. The largest additions of land protected for wildlife came in 1980 when President Jimmy Carter signed the Alaska National Interest Land Act and added 22 million ha (54 million ac) of new refuges to the 12.5 million ha (31 million ac) already existing, about two-thirds of which was in Alaska.

Refuge Management

Although refuges were originally intended to be sanctuaries in which wildlife would be protected from hunting and other disturbance, a 1948 compromise allowed hunting in refuges in exchange for an agreement by hunters to purchase special duck stamps to raise money for wetland protection. Although a refuge that allows hunting seems like an oxymoron, and in spite of the fact that only 10 percent of refuge lands have been acquired with duck stamp funds, hunting has become firmly established in most units (fig. 15.17). Critics charge that the Fish and Wildlife Service, which adminis-

Figure 15.17
Hunters on their way to duck blinds in a wildlife refuge. Although it seems a contradiction, half of all refuges allow hunting.

ters the refuge system, is so strongly oriented toward hunting that many units have become little more than duck and goose farms.

Over the years, a number of improbable and incompatible uses have become accepted in wildlife refuges including oil drilling, cattle grazing, snowmobiling, motorboating, off-road vehicle use, timber harvesting, hay cutting, trapping, and camping. One Nevada refuge has a bombing and gunnery range, while another is the site of a brothel. A General Accounting Office report found that 60 percent of all refuges allow activities that are harmful to wildlife.

Refuges also face threats from external activities. More than three-quarters of all refuges in the United States have water pollution problems, two-thirds of which are serious enough to affect wildlife. A notorious example is the former Kesterson Wildlife Refuge in California, where selenium-contaminated irrigation water drained from farm fields turned the marsh into a death trap for wildlife rather than a sanctuary. Eventually, the marsh had to be drained and capped with clean soil to protect wildlife. Subsequent research has shown that at least twenty wildlife refuges in western states have toxic metal pollution caused by agriculture and industrial activities outside the refuge.

The biggest current battle over wildlife refuges concerns proposals for oil and gas drilling in the 7-million-hectare (18-million-acre) Arctic National Wildlife Refuge on the north slope of Alaska's Brooks Range (see Case Study, chapter 21).

International Wildlife Preserves

As we saw earlier in this chapter, most developing countries do not have separate systems of parks and wildlife refuges. Many nature preserves are set up primarily to protect wildlife, however. An outstanding example of both the promise and the problems in managing parks in the less-developed countries is seen in the Serengeti ecosystem in Kenya and Tanzania. This area of savanna, thorn woodland, and volcanic highland lying between Lake Victoria and the Great Rift Valley in East Africa is home to the highest density of ungulates (hooved grazing animals) in the world. Over 1.5 million wildebeests (or gnus) graze on the savanna in the wet season, when grass is available, and then migrate through the woodlands into the northern highlands during the dry season. The ecosystem also supports hundreds of thousands of zebras, gazelles, impalas, giraffes, and other beautiful and intriguing animals. The herbivores, in turn, support lions and a variety of predators and scavengers, such as leopards, hyenas, cheetahs, wild dogs, and vultures. This astounding diversity and abundance is surely one of the greatest wonders of the world.

Tanzania's Serengeti National Park was established in 1940 to protect 15,000 sq km (5700 sq mi), an area about the size of Connecticut or twice as big as Yellowstone Park. It is bordered on the east by the much smaller Ngorongoro Conservation Area and Lake Manyara National Park. Kenya's Masa Mara National Reserve borders the Serengeti on the north. Rapidly growing human populations push against the boundaries of the park on all sides. Herds of domestic cattle compete with wild animals for grass and water. Agriculturalists clamor for farmland, especially in the temperate highlands along the Kenya-Tanzania border. So many tourists flock to these parks that the vegetation is ground to dust by hundreds of sight-seeing vans, and wildlife find it impossible to carry out normal lives.

Perhaps the worst problem in Africa is **poachers,** illegal hunters who massacre wildlife for valuable meat, horns, and tusks. As recently as 1970, a healthy population of 65,000 black rhinoceroses roamed the continent, 1000 of which were in the Serengeti. In 1990, there were about 3000 black rhinos in Africa and only twenty in the Serengeti (fig. 15.18). The rest were killed for their horns.

Elephants are under a similar assault. Thirty years ago there were no elephants in the Serengeti, but perhaps 3 million in all of Africa. Since then, more than 80 percent of the African elephants have been killed—mainly for their ivory—at a rate of 100,000 each year. The 2000 elephants now in Serengeti National Park have been driven there by hunting pressures elsewhere. Fortunately, the elephants find refuge in the park and add to the pleasure of tourists who come to see the wildlife; however, they are changing the ecosystem. Crowded into inadequate space, the elephants smash down the acacia trees, turning the woodland and mixed savanna into continuous grassland. This is beneficial for some animals, but not for others.

The poachers continue to pursue the elephants and rhinos, even in the park. Armed with high-powered rifles and even machine guns and bazookas from the many African wars in the last decade, the poachers take a terrible toll on the wildlife. Park rangers try to stop the carnage, but they often are outgunned by the poachers. The parks themselves are beginning to resemble war zones, with fierce, lethal firefights rather than peace and tranquility.

Wetlands, Floodplains, and Coastal Regions

Although many wetlands are protected to some extent in wildlife refuges and nature preserves, this land category is so ecologically important and so underrepresented among world protected areas that it deserves special discussion here. What is a wetland? Although you might think it easy to do, defining wetlands is a subject of intense debate because not all lands that have important wetland characteristics are wet all the time. Roughly defined, however, a **wetland** is an area often covered by shallow water or one in which the ground is wet enough long enough to support plants specialized to grow under saturated soil conditions (fig. 15.19).

Figure 15.19

Wetlands provide irreplaceable biological functions. They are among the most diverse and productive of all ecosystems. Many species spend at least part of their life cycles in freshwater or saltwater wetlands.

Figure 15.18

A guard protects a black rhinoceros from poachers. His ancient rifle is no match, however, for the powerful automatic weapons with which the poachers are now armed.

Chapter 5 describes some important types of wetlands. Worldwide, Canada and Russia have the largest wetland areas.

Wetland Values

Wetlands provide a number of irreplaceable biological functions. They often are highly productive and provide food and habitat for a wide variety of species. Although wetlands currently occupy less than 5 percent of the 0.9 billion ha (2.2 billion ac) of land in the United States, the Fish and Wildlife Service estimates that one-third of all endangered species spend at least part of their lives in wetlands. Because open oceans are often biological deserts, coastal wetlands are even more important. Nearly two-thirds of all marine shellfish and finfish rely on salt marshes and estuaries for spawning and juvenile development. Wetland storage of flood waters is worth an estimated $3 to 4 billion per year. Wetlands also improve water quality by acting as natural water purification systems, removing silt and absorbing nutrients and toxins. The flow of groundwater through coastal marshes prevents saltwater intrusion that would otherwise contaminate wells. Coastal wetlands help stabilize shorelines and reduce storm damage. Finally, wetlands provide important recreational opportunities for hunters, fishers, birdwatchers, and many others who enjoy nature.

Wetland Destruction

For many people, wetlands are worthless, disagreeable, and dangerous places full of insects, spiders, snakes, and mud. This attitude was reflected in public policies such as the U.S. Swamp Lands Act of 1850 that allowed individuals to buy swamps and marshes for as little as 10 cents per acre. During the 1930s and 1940s, the U.S. government subsidized wetland drainage for conversion to farmland. Until 1977, dumping "landfill" into wetlands was considered a convenient way to dispose of waste as well as a good way to create building space for highways and housing developments. The first protective measures for wetlands in the United States were passed in 1899, to prevent dumping rubbish into navigable waterways. The 1972 Clean Water Act required discharge permits (called Section 404 permits) intended to protect surface water quality. The act was interpreted by the courts in 1977 to prohibit both pollution and filling (but not drainage) of wetlands unless an exception is granted. The 1985 Farm Bill went even further in its "swampbuster" provisions that withdraw agricultural subsidies from farmers who drain, fill, or damage wetlands.

As a result of our earlier land-use policies, only about 40 million ha (100 million ac) of wetlands remain out of the original 90 million ha in the lower forty-eight states. Some states have lost more wetlands than others (table 15.4). By some estimates, at least two-thirds of the millions of prairie potholes (small temporary ponds that furnish breeding habitat for waterfowl and shorebirds across the northern plains) have been drained or filled. In Louisiana, which has about 40 percent of the remaining coastal wetlands in the United States, canal dredging for oil and gas production and navigation—coupled with diversion of river sediment that once replenished marshes—are causing losses of about 13,000 ha (50 sq mi) per year (fig. 15.20). Between 1950 and 1980, wetland losses in the United

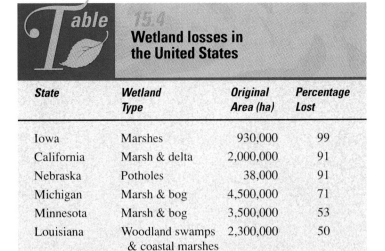

Table 15.4

Wetland losses in the United States

State	Wetland Type	Original Area (ha)	Percentage Lost
Iowa	Marshes	930,000	99
California	Marsh & delta	2,000,000	91
Nebraska	Potholes	38,000	91
Michigan	Marsh & bog	4,500,000	71
Minnesota	Marsh & bog	3,500,000	53
Louisiana	Woodland swamps & coastal marshes	2,300,000	50
Connecticut	Coastal marshes	6000	50
North Carolina	Pocosins	610,000	40
Wisconsin	Marshes	2,700,000	32

Source: United Nations Environment Program, 1992.

Figure 15.20

Louisiana has nearly 40 percent of the remaining coastal wetlands in the United States, but diversion of river sediments that once replenished these marshes and swamps, together with channels and boat wakes, are causing the loss of about 50 sq mi annually.

States were around 200,000 ha (500,000 ac) per year. A common estimate is that 87 percent of wetland losses are due to conversion to farmland, while 8 percent are due to urban development.

Although Section 404 permit applications are almost always approved and exemptions to the swampbuster provisions of the Farm Bill are almost always granted, these regulations are detested by farmers, real estate developers, miners, energy companies, and others who regard rules about what they can do with private lands as an infringement on owner's rights and an impediment to business. In 1991, in spite of campaign pledges of no further wetlands losses, the Bush administration attempted to weaken wetland definitions and ease restrictions on conversion and development. Environmental groups protested that as much as half of the remaining wetlands in the lower forty-eight states would lose protection under these new regulations. Ultimately, these attempts to reduce wetland protection failed, but there are sure to be new initiatives toward these ends in the future.

Floods and Flood Control

The summer of 1993 was the wettest on record over much of the Mississippi Valley (fig. 15.21). Torrential rains falling on already saturated ground caused enormous flooding of the Missouri and upper Mississippi Rivers. In mid-July, the Mississippi and its tributaries had inundated several million hectares of land in the upper basin; it was as if a new great lake had suddenly appeared. The Mississippi reached a historic high of 47 ft at St. Louis, 17 ft (5.23 m) above flood stage and five times the normal summer flow. Altogether, 8000 buildings were destroyed in the Mississippi Valley, 30,000 people were displaced, 26 lives were lost, and total losses exceeded $10 billion as a result of these floods.

The immediate cause of this disastrous flood was an unusual weather pattern—probably resulting from an El Niño event in the South Pacific (chapter 17)—that kept the jet stream low over the Midwest and pumped warm humid air up from the Gulf of Mexico and the Pacific to produce rains that were at least twice normal levels for the summer. Destruction of wetlands and natural vegetation cover over much of the watershed played a role in magnifying floods, as did a system of levees, flood walls, and channels that protect low-lying areas in normal years but failed catastrophically under these extreme conditions.

Before the natural vegetation of the Mississippi drainage basin was cut down, plowed up, and paved over by advancing civilization, the prairie grasses, forests, and natural wetlands acted like sponges to hold back rainfall and allow it to seep into the ground. Rivers tended to flow clear and uninterrupted, fed by a continuous supply released from this temporary storage. City streets, bare cropland, and drained wetlands lose their water-holding capacity, however, and rain that falls on them runs off immediately to create floods downstream.

Floodplains—low lands along riverbanks, lakes, and coastlines subjected to periodic inundation—are among the most valuable lands in many areas because of their rich soil, level topography, convenient water supply, access to river shipping, and recreational potential. For all these reasons, many cities are located at least in part in flood-prone areas. Public flood control efforts on the Mississippi began in 1879 when construction was started on levees (earthen berms along the river banks) to protect towns and low-lying farm fields. After a particularly disastrous flood in 1927, the Army Corps of Engineers began work on an extensive system of locks, dams, and channel dredging to improve navigation and provide further flood control (fig. 15.22).

The $25 billion river-control system built on the Mississippi and its tributaries has protected many communities over the past century. For others, however, this elaborate system has helped turn a natural phenomenon into a natural disaster. Deprived of the ability to spill out over floodplains, the river is pushed downstream to create faster currents and deeper floods until eventually a levee gives way somewhere. Hydrologists calculate that the floods of 1993 were about 3 m (10 ft) higher than they would have been given the same rainfall in 1900 before the flood-control structures were in place.

Under current rules, the government is obligated to restore any of the nearly 500 levees that meet federal standards and were dam-

Figure 15.21

The Mississippi River inundates downtown Davenport, IA, during the summer floods of 1993. Although sympathetic with the heartbreak and economic losses caused by this flooding, many people argue that floodplains such as this should never have been settled in the first place.

Figure 15.22

Locks, dams, levees, and other flood control and navigational structures such as this complex on the Mississippi River at Dubuque, IA, have been a godsend for commercial shippers but have also been disastrous for riparian ecosystems, backwater sloughs, marshes, and the wildlife that depends on them.

aged by flooding. Many people think that it would be much better to spend this money to restore wetlands, replace ground cover on water courses, build check dams on small streams, move buildings off the floodplain, and undertake other nonstructural ways of reducing flood danger. According to this view, floodplains should be used for wildlife habitat, parks, recreation areas, and other uses not susceptible to flood damage.

The National Flood Insurance Program administered by the Federal Emergency Management Agency (FEMA) was intended to aid people who cannot buy insurance at reasonable rates, but its effects have been to encourage building on the floodplains by making people feel that whatever happens, the government will take care of them. Many people would like to relocate homes and businesses out of harm's way after the recent floods or to improve them so they will be less susceptible to flooding, but owners of damaged property can collect only if they rebuild in the same place and in the same way as before. This serves to perpetuate problems rather than solve them.

Beaches, Barrier Islands, and Estuaries

Another group of lands of great biological importance threatened by our preference for living, working, and recreating near the water are beaches, barrier islands, coastal wetlands, and estuaries. An estuary is a bay or drowned valley where a river empties into the sea. Fresh water mingling with saltwater brings in sediment and nutrients and creates a gradient of salinity that makes estuaries among the most diverse and biologically productive ecosystems on the earth. As we pointed out earlier, the shoals and shallow reefs, salt marshes, tidal mudflats, mangrove swamps, and other coastal wetlands built up by these sediments are used during part of their life cycle by about two-thirds of all marine shellfish and finfish.

Where the continental shelf is relatively shallow, river sediments form beaches, sandbars, and barrier islands parallel to the shore. One of the world's longest and most spectacular sand beaches runs down the Atlantic coast of North America from New England to Florida and around the Gulf of Mexico. Much of this beach lies on some 350 long, thin barrier islands that stand between the mainland and the open sea. Behind these barrier islands lie shallow bays or brackish lagoons fringed by marshes or swamps. The sediments that make up these landforms are constantly moved and reshaped by winds, waves, and currents. A 50-meter-wide beach can be created or removed by a single violent storm. Similarly, islands and sandbars appear and disappear over time.

Early inhabitants recognized that the shore was a hazardous place to live and settled on the bay side of barrier islands or as far upstream on coastal rivers as was practical. Modern residents, however, place a high value on living where they have an ocean view and ready access to the beach. The most valuable and prestigious property is closest to the shore. Over the past 50 years more than 1 million acres of estuaries and coastal marshes have been filled to make way for housing or recreational developments.

Construction directly on beaches and barrier islands can cause irreparable damage to the whole ecosystem. Damaging the vegetation that holds the shifting sand in place often not only puts houses and human life at risk but also eliminates habitat for rare and en-

Figure 15.23

Winter storms have eroded the beach and undermined the foundations of homes on this barrier island. Breaking through protective dunes to build such houses damages sensitive plant communities and exposes the whole island to storm sand erosion. Coastal zone management attempts to limit development on fragile sites.

dangered species. Breaching dune systems to create roads or construction sites provides an avenue through which storms surge to wash away beaches or even whole islands. A single severe storm in 1962 caused $300 million in property damage along the East Coast and left hundreds of beach homes tottering into the sea (fig. 15.23).

As is the case for inland floodplains, government policies often encourage people to build where they probably shouldn't. Subsidies for road building and bridges, support for water and sewer projects, tax exemptions for second homes, flood insurance, and disaster relief are all good for the real estate and construction business but invite people to build in risky places. Flood insurance typically costs $300 per year for $80,000 of coverage. In 1992, FEMA paid out $30 billion in claims, 83 percent of which were flood related. Settlement usually requires that structures be rebuilt exactly where and as they were before. There is no restriction on how many claims can be made, and policies are never canceled, no matter what the risk. Some beach houses have been rebuilt—at public expense—three times in a decade. The General Accounting Office found that 2 percent of federal flood policies were responsible for 30 percent of the claims.

The Coastal Barrier Resources Act of 1982 prohibited federal support, including flood insurance, for development on sensitive islands and beaches. In 1992, however, the Supreme Court ruled that ordinances forbidding floodplain development amount to an unconstitutional "taking" or confiscation of private property (see What Do *You* Think?, p. 304).

As part of the general rejection of governmental control over resources by the "wise use" movement, dozens of "taking" bills have been introduced in state legislatures prohibiting regulations or restrictions on how private property can be used. This may be a tactic to repeal not only coastal and floodplain zoning but also park and greenbelt establishment, wildlife refuges, wilderness designation, and almost every other land use we have discussed in this chapter.

Preserving Nature

Summary

Parks, wildlife refuges, wilderness areas, and nature preserves occupy a small percentage of our total land area but protect valuable cultural resources and representative samples of the earth's species and ecosystems. We can discern much about people's environmental ideals by examining the gardens, parks, and recreation areas that they create. Yosemite in California was probably the first park in the world set aside to protect wild nature. Yellowstone, which was established in 1872, was the first official national park in the world.

Parks are havens for wildlife and places for healthful outdoor recreation. Many are overcrowded, misused, and neglected, however. Pollution and incompatible uses outside parks threaten the values that we seek to protect. Wildlife is at the center of many park controversies. Is the park's purpose to preserve wild nature or provide entertainment for visitors? How much management is acceptable? When should we intervene and when should we let nature take its course?

There are proposals to remove distracting or damaging uses from parks and to limit entrance permits. We may reach the point where each of us might be allowed to visit some of the more popular parks only once in a lifetime. One solution to congestion and overuse is to create additional parks in some of the many deserving but unprotected areas.

Worldwide, only about 4 percent of total land area has been protected in parks, wildlife refuges, and nature preserves. Some biomes such as dry tropical forests and tundra are well represented in this network, but others such as grassland and wetlands are underrepresented. The optimum size for nature preserves depends on the terrain and the values they are intended to protect, but—in general—the larger a reserve is, the more species it can protect. Establishing corridors to link separate areas can be a good way to increase effective space and to allow migration from one area to another. Economic development and nature protection can go hand in hand. Ecotourism may be the most lucrative and long-lasting way to use resources in many developing countries.

Areas chosen for preservation often are lands of indigenous people. Careful planning and zoning can protect nature and also allow sustainable use of resources. Man and Biosphere (MAB) Reserves provide for multiple use in some areas but strict conservation in others. Wildlife refuges were intended to be sanctuaries for wildlife, but over the years many improbable and damaging uses have become established in them.

Wetlands are areas periodically covered with water that support unique assemblages of plants that can survive in water-saturated soil. Much of our wildlife, including many endangered species, depend on wetlands for a part of their life cycles. Historically, we have regarded wetlands as worthless, disagreeable places and we have encouraged people to drain or fill them. About half our original wetlands in the United States have been destroyed or degraded. Flood control and flood insurance encourage people to build homes and businesses on flood plains, barrier islands, beaches, and other fragile, flood-prone areas. Human and ecological costs of these misguided policies have been high.

▼ Questions for Review

1. Why is Yosemite called the first park to preserve wilderness, while Yellowstone is the first national park?

2. List some problems and threats from inside and outside our national parks.

3. Why is the reintroduction of wolves into Yellowstone a controversial issue?

4. Describe the IUCN categories of protected areas and the amount of human impact or intervention allowed in each. Can you name some parks or preserves in each category?

5. Which biomes or landscape types are best represented among the world's protected areas and which are least represented?

6. List the three main points in the IUCN World Conservation Strategy and six steps in the action plan to meet these goals.

7. Draw a diagram of an ideal MAB reserve. What activities would be allowed in each zone?

8. What is the legal definition of wilderness in the United States?

9. What role did Teddy Roosevelt and his cousin Franklin play in establishing wildlife refuges?

10. Describe some ecological values of wetlands and how they are threatened by human activities.

▼ Questions for Critical Thinking

1. Is "contrived" naturalness a desirable feature in parks and nature preserves? How much human intervention do you think is acceptable in trying to make nature more beautiful, safe, comfortable, or attractive to human visitors? Think of some specific examples that you would or would not accept.

2. Suppose you were superintendent of Yellowstone National Park. How would you determine the carrying capacity of the park for elk? How would you weigh having more elk or more ground squirrels? If there are too many elk, how would you thin the herd?

3. Suppose that as park manager you know that building tourist facilities brings in needed funds to protect nature but at the same time more tourists destroy the natural values you want to protect. How do you balance these competing interests?

4. Why do you suppose that dry tropical forests and tundra are well represented in protected areas, whereas grasslands and wetlands are rarely protected? Consider social, cultural, and economic as well as biogeographical reasons in your answer.

5. Suppose that preserving healthy populations of grizzly bears and wolves requires that we set aside some large fraction of Yellowstone

Park as a zone into which no humans will ever again be allowed to enter for any reason. Would you support protecting bears even if no one ever sees them or could even be sure that they still existed?

6. Suppose that you had trespassed into the bear sanctuary and were attacked by a bear. Should the rangers shoot the bear or let it eat you?

7. Are there any conditions under which you would permit oil drilling in the Arctic National Wildlife Refuge?

8. People have built homes and businesses on floodplains, beaches, or barrier islands under the assumption that the government will protect them from floods and storms. Can we now tell them that we can't or won't protect them?

9. Is a zoning ordinance that says you can't use your land in ecologically damaging ways an unfair intrusion into private property rights?

 ## Key Terms

biogeographical area 324	natural landscaping 318
biosphere reserves 325	poachers 334
corridors 326	wetland 334
ecotourism 330	wilderness 331
floodplains 336	wildlife refuges 333
Man and Biosphere (MAB)	world conservation strategy 326
program 329	

 ## Additional Information on the Internet

Biodiversity
hHp://www.ftpt.br/structure/biodiversity.html/

CIESIN
http://infoserver.ciesin.org:80/

Earth Wise Journeys
http://www.teleport.com/~earthwyz/

Gaia Forest Conservation Archives
http://gaial.ies.wisc.edu/research/pngfores/

Instituto Nacional de Biodiversidat
http://www.inbio.ac.cr/english.html

Man and the Biosphere
http://ice.ucdavis.edu/MAB/

Missouri Botanical Gardens
http://www.mobot.org/

Natural Resources Defense Council
http://www.nrdc.org/nrdc/

Natural Resources Research Information Pages
http://sfbox.vt.edu:10021/Y/yfleung/nrrips.html/

National Wildlife Refuge System
http://bluegoose.arw.r9.fws.gov/

Protected Areas Virtual Library
http://www.wcmc.org.uk/~dynamic/pavl/

World Conservation Union
http://infoserver.ciesin.org:80/IC/iucn/IUCN.html

Note: Further readings appropriate to this chapter are listed on p. 601.

Aquatic Biologist

Lee Lebbon

*L*ee Lebbon

The Peace Corps offers a good way to travel, learn about different cultures, do humanitarian work in sustainable development, and you receive valuable experience that will help you enter the job market when you come back home. Lee Lebbon's experience illustrates some of these benefits. As an environmental biology major at the University of Colorado in Boulder, Lee received a broad background in science, but he wasn't prepared for any specific occupation. He developed an interest in aquatic ecology, however, and took courses in limnology, stream ecology, and aquatic ecosystems.

During his senior year, Lee learned about opportunities to work in aquaculture through the Peace Corps. He applied, was accepted, and was assigned to a rural development program in Thailand. After two months of training in rural aquaculture and another month of intensive language study in a village outside of Bangkok, Lee was on his way to his work station in a remote area in the northeastern part of the country.

Being immersed in foreign culture where no one speaks your language can be intimidating, but it can also be a great opportunity for growth and learning. For Lee, it was a matter of sink or swim at first. After a few months, however, he began to learn Thai and to understand how to function in his new surroundings. He found the Thai people to be extremely helpful, friendly, open, and curious about this foreigner in their midst.

Lee's job was to help rural families build and operate fish ponds as a way to add needed protein to their diets and as a cash crop to boost their income. Working with a variety of people ranging from poor farmers to government bureaucrats, Lee got invaluable experience in leadership, organization, building relationships, and getting along with others. Although there were many frustrations and setbacks in his project, he had the satisfaction of helping people learn how to help themselves. He also gained practical know-how and management skills that have given him a jump-start into his current career.

As a result of his two years in Thailand, Lee knew that he wanted to continue in commercial aquaculture when he returned to the United States. After sending his résumé to about 30 fish hatcheries, Lee was offered the job of manager of the Limestone Springs Trout Hatchery near Richland, PA. With a crew of about six people, Lee now manages the largest trout hatchery east of the Mississippi River. He is responsible for feeding, harvesting, and shipping about one million pounds of rainbow trout each year.

When asked what he would do differently, Lee says that he would have tried to learn more hands-on skills in college. Although the broad background of an environmental science major is beneficial, volunteering for a laboratory or field research project or doing an internship also provides useful training. He advises students to try a wide variety of things to get an idea of what you like to do and where you want to go. International travel and living in another country is a wonderful way to learn about your own values and to get a perspective on your life.

Finally, he says keep an open mind, learn all you can, and follow your interests. If you would like to learn more about the Peace Corps, ask the office nearest you for a list of former volunteers. They can give you a personal account of what you might experience and what you might gain from what the Peace Corps calls the "toughest job you'll ever love."

Part 4

Physical Resources

A South American forest burns, contributing to air pollution and global climate change as it destroys wildlife habitat.

The world was not left to us by our parents, it was lent to us by our children.

African proverb

The sharp peaks of the Canadian Rockies show their recent origins. What forces raise up and wear down landscape features such as these?

CHAPTER 16: The Earth and Its Crustal Resources

Objectives

After studying this chapter, you should be able to:

- understand some basic geologic principles, including how tectonic plate movements affect conditions for life on the earth.

- explain how the three major rock types are formed and how the rock cycle works.

- summarize economic mineralogy and strategic minerals.

- discuss the environmental effects of mining and mineral processing.

- recognize the geologic hazards of earthquakes, volcanoes, and tsunamis.

The Night the Earth Moved

At 5:46 A.M. on January 17, 1995, when most people in the Japanese city of Kobe were home asleep, the earth's crust under the nearby island of Awaji-shima shrugged and jerked as the Philippine Sea plate was shoved a little farther under the adjacent Eurasian plate. For 20 seconds, the land shook violently, reaching a magnitude of 7.2 on the Richter scale. The energy released was equivalent to the explosion of 240 kilotons of TNT, or about 24 times as much energy as was discharged by the atomic bomb dropped on Hiroshima in 1945.

No sizable earth tremors had struck Kobe in recent memory and most people didn't consider themselves in danger. Many buildings had not been upgraded to meet quake-proof construction standards enforced in other parts of Japan. As apartment buildings crumbled, tile house roofs caved in on people in their beds, and freeways toppled from elevated pylons (fig. 16.1), the city was plunged into chaos. Fires broke out as

gas lines ruptured. People were buried under tons of debris. At least 6300 people were killed and 300,000 were left homeless. Altogether the losses may amount to $100 billion and it may take 10 years to rebuild. Although this was not by any means the largest or most deadly earthquake in history, it reminds us of the powerful forces that shape the earth's crust as well as the need to understand something about geology and earth sciences.

Of course, the earth supplies us with many useful resources as well as geological hazards. In this chapter, we will look at some ways in which geological processes provide materials and creates landscapes. Like ecology, geology has a strong tradition of synthetic thinking and direct observations from nature as opposed to controlled laboratory experiments. Geologists refer to "ground truth" or evidence from the actual world as a reality check to abstract theory. Let's see what geology can tell us about the world in which we live. ▶

A Dynamic Planet

What causes dynamic circulation of molten material in the earth's core and the restless shifting of continents on the surface that result in earthquakes like the one in Kobe? In this section, we will look at the structure of our planet and the forces that shape it.

A Layered Sphere

The **core,** or interior of the earth (fig. 16.2), is composed of a dense, intensely hot mass of metal—mostly iron and nickel—thousands of kilometers in diameter. Solid in the center but more fluid in the outer core, this immense mass is stirred by convection currents that are thought to generate the magnetic field that shields us from cosmic radiation. These currents, and the heat that drives them, also provide much of the force that shapes and modifies the surface of our world. These dynamic processes create a significant amount of the geological resources on which we depend.

Surrounding the molten outer core is a hot, pliable layer of rock called the **mantle.** The mantle is much less dense than the core because it contains a high concentration of lighter elements, such as oxygen, silicon, and aluminum. Cooling as it nears the surface, the mantle becomes stiffer and more solid, eventually crystallizing into the lower margins of the lithosphere, or rock layer.

The outermost layer of the lithosphere is the cool, lightweight, brittle **crust** of rock that floats on the mantle something like the "skin" on a bowl of warm chocolate pudding. The oceanic crust, which forms the seafloor, has a composition somewhat like that of the mantle, but it is richer in silicon. Continents are thicker, lighter regions of crust rich in calcium, sodium, potassium, and aluminum. The continents rise above both the seafloor and the ocean surface. You might imagine them as marshmallows embedded in the surface of your chocolate pudding, surrounded by puddles of milk. Although the interior of the pudding is still soft, warm, and semi-solid, the surface has cooled to a crust that hardens, cracks, and traps

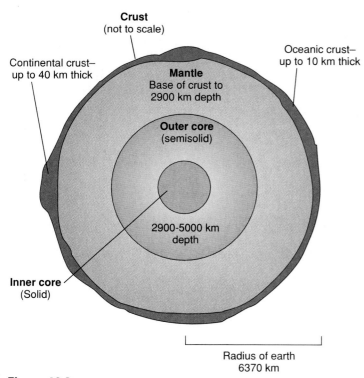

Figure 16.2

The layered Earth. The intensely hot, liquid or semisolid core is made mostly of molten metal. Around it is a mantle of lighter elements that cool and crystallize near the surface. Floating on top of the mantle is a thin crust of rock that breaks up into large, slowly moving tectonic plates. The crust appears ten times thicker in this drawing than it is in reality.

Whole Earth		Crust	
Iron	33.3	Oxygen	45.2
Oxygen	29.8	Silicon	27.2
Silicon	15.6	Aluminum	8.2
Magnesium	13.9	Iron	5.8
Nickel	2.0	Calcium	5.1
Calcium	1.8	Magnesium	2.8
Aluminum	1.5	Sodium	2.3
Sodium	0.2	Potassium	1.7

Table 16.1

Eight most common chemical elements (percent)

material floating on its surface. Table 16.1 compares the composition of the whole earth (dominated by the dense core) and the crust.

Tectonic Processes and Shifting Continents

Convection currents and uneven heat flows passing through the core and mantle break the overlying lithosphere into a mosaic of huge blocks called **tectonic plates** (fig. 16.3). These plates slide slowly across the earth's surface like immense icebergs, in some places breaking up into smaller pieces, in other places crashing ponderously into each other to create new, larger landmasses. Ocean basins form where continents crack and pull apart. Earthquakes are caused by the grinding and jerking as plates slide past each other. Mountain ranges are pushed up at the margins of colliding continental plates. The Atlantic Ocean is growing slowly as Europe and Africa drift away from the Americas. The Himalayas are still rising as the Indian subcontinent smashes into Asia. Southern California is slowly sailing north toward Alaska. In a few million years, Los Angeles will pass San Francisco, if either still exists by then.

When an oceanic plate collides with a continental landmass, the continental plate usually rides up over the seafloor, and the oceanic plate is subducted, or pushed down into the mantle, where it melts and rises back to the surface as **magma,** or molten rock

(fig. 16.4). Deep ocean trenches mark these subduction zones, and volcanoes form where the magma erupts through vents and fissures in the overlying crust. All around the Pacific Ocean rim from Indonesia to Japan to Alaska and down the West Coast of the Americas is the so-called "ring of fire" where the Pacific plate is being subducted under the continental plates. This ring is the source of more earthquakes and volcanic activity than any other place on the earth.

Over millions of years, the drifting plates can move long distances. Antarctica and Australia once were connected to Africa, for instance, somewhere near the equator, and supported luxuriant forests (fig. 16.5). Geologists suggest that several times in the earth's history most or all of the continents have gathered to form a single supercontinent surrounded by a single global ocean. Every few hundred million years, this supercontinent breaks up into many smaller pieces through a process called seafloor spreading. These massive rearrangements undoubtedly have profound effects on the earth's climate and may help explain the periodic mass extinctions of organisms marking the divisions between many major geologic periods.

The Rock Cycle

What could be harder and more permanent than rocks? Like the continents they create, rocks are also part of a relentless cycle of formation and destruction. They are made and then torn apart, cemented together by chemical and physical forces, crushed, folded, melted, and recrystallized by dynamic processes related to those that shape the large-scale features of the crust. We call this cycle of creation, destruction, and metamorphosis the **rock cycle** (fig. 16.6). Understanding something of how this cycle works helps explain the origin and characteristics of different types of rocks, as well as how they are shaped, worn away, transported, deposited, and altered by geologic forces.

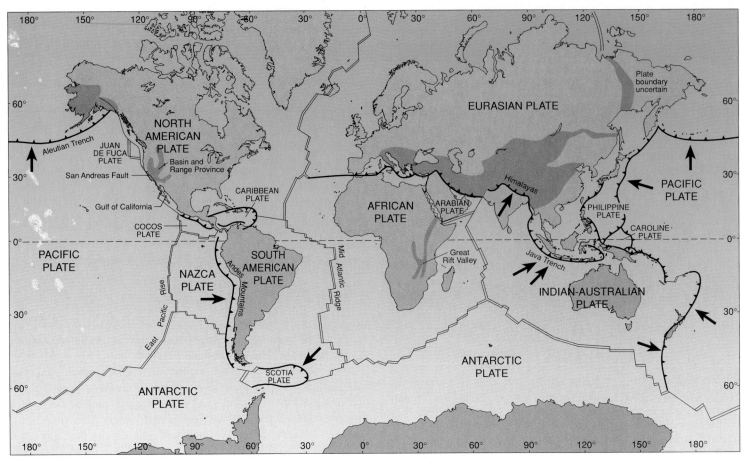

Figure 16.3

Map of tectonic plates. Plate boundaries are dynamic zones, characterized by earthquakes and volcanism and the formation of great rifts and mountain ranges. Arrows indicate direction of subduction where one plate is diving beneath another. These zones are sites of deep trenches in the ocean floor and high levels of seismic and volcanic activity.

Sources: U.S. Department of the Interior, U.S. Geological Survey.

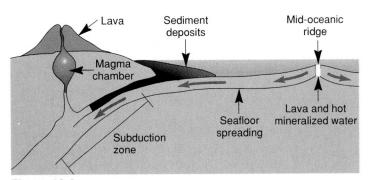

Figure 16.4

The rock cycle interpreted in plate-tectonic terms. Old seafloor and sediment deposits are melted in subduction zones. The magma rises to erupt through volcanoes or to recrystallize at depth into igneous rocks. Weathering breaks down surface rocks and erosion deposits residue in sedimentary formations. The pressure and heat caused by tectonic movements causes metamorphism of both sedimentary and igneous rocks.

Rocks and Minerals

The many different types of rocks are classified according to their internal structure, chemical composition, physical properties, and mode of formation. Minerals are the basic materials of all rocks. A **mineral** is a naturally occurring, inorganic, crystalline solid that has a definite chemical composition and characteristic physical properties. There are thousands of known and described minerals in the world, but most of the rocks we see every day are composed mainly of just a few dozen different minerals, such as quartz, feldspar, biotite, and calcite. These minerals, in turn, are composed mainly of just a few elements, such as silicon, oxygen, iron, magnesium, aluminum, and calcium.

Rock Types and How They Are Formed

There are three major rock classifications: igneous, sedimentary, and metamorphic. In this section, we will look at how they are made and some of their properties.

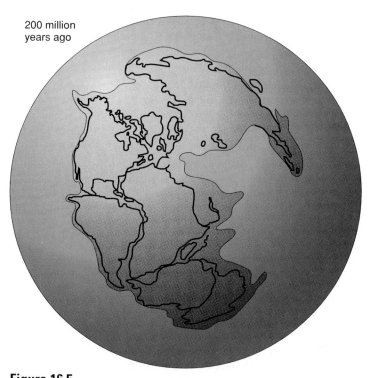

Figure 16.5

Pangaea, the ancient supercontinent of 200 million years ago, combined all the world's continents in a single landmass.

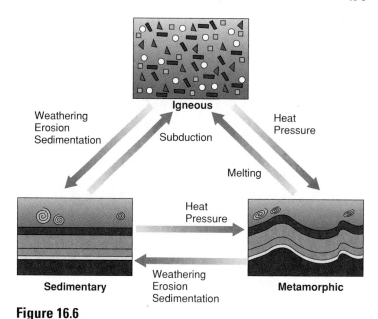

Figure 16.6

The rock cycle consists of processes of creation, destruction, and metamorphosis. Each of the three major rock types can be converted to either of the other types.

Igneous Rocks

Most rocks on the earth's surface are crystalline minerals solidified from magma, molten liquid from the earth's interior. These rocks are classed as **igneous rocks** (from *igni,* the Latin word for fire). Magma extruded to the surface from volcanic vents cools quickly to make basalt, rhyolite, andesite, and other fine-grained rocks. Magma that cools slowly in subsurface chambers or is intruded between overlying strata makes granite, gabro, or other coarse-grained crystalline rocks, depending on its specific chemical composition.

Weathering and Sedimentation

Most of these crystalline rocks are extremely hard and durable, but exposure to air, water, changing temperatures, and reactive chemical agents slowly breaks them down in a process called **weathering.** *Mechanical weathering* is the physical breakup of rocks into smaller particles without a change in chemical composition. You have probably seen mountain valleys scraped by glaciers or river and shoreline pebbles that are rounded from being rubbed against one another as they are tumbled by waves and currents. *Chemical weathering* is the selective removal or alteration of specific components that leads to weakening and disintegration of rock. Among the more important chemical weathering processes are oxidation (removal of electrons from atoms) and hydrolysis (addition of water to other molecules). The products of these reactions are more susceptible to both mechanical weathering and to dissolving in water. For instance, when carbonic acid (formed when CO_2 and H_2O

combine) percolates through porous limestone layers in the ground, it dissolves the calcium and creates caves and sinkholes.

Particles of rock loosened by wind, water, ice, and other weathering forces are carried downhill, downwind, or downstream until they come to rest again in a new location. The deposition of these materials is called **sedimentation.** Waterborne particles from sediments cover ocean continental shelves and fill valleys and plains. Most of the American Midwest, for instance, is covered with a layer of sedimentary material hundreds of meters thick in the form of glacierborne till (rock debris deposited by glacial ice), windborne loess (fine dust deposits), riverborne sand and gravel, and ocean deposits of sand, silt, and clay. Deposited material that remains in place long enough, or is covered with enough material to compact it, may once again become stone. Some examples of **sedimentary rock** are shale (compacted mud), sandstone (cemented sand), tuff (volcanic ash), and conglomerates (aggregates of sand and gravel).

Much of this settling is due to gravity, but chemical, evaporative, and biogenic sedimentation are also important mechanisms of rock formation. *Chemical sedimentation* occurs when soluble chemicals react to make insoluble products that precipitate from water, creating mineral deposits. Often, mineral-bearing groundwater seeps into cracks in existing strata, forming veins of valuable materials. *Evaporative sedimentation* occurs in warm, shallow bodies of water where evaporation is great and outflow is small. The evaporating water leaves behind any dissolved minerals it may have been carrying. Some economically important deposits include halite (rock salt) and gypsum ($CaSO_4$), a common building material. Evaporative sedimentation also can result in accumulation of harmful minerals, such as arsenic and selenium. Many soils

Figure 16.7

The biological origin of this limestone is easily visible in the seashells buried millions of years ago.

in the western United States have high levels of these toxic sediments. Irrigation run-off from these soils has contaminated rivers, lakes, and wetlands, including a number of wildlife refuges. The Kesterson marsh in California is one of the most dramatic and tragic examples, resulting in deaths and developmental deformities in waterfowl that feed and breed there. *Biogenic sedimentation* is caused by living organisms. The most important biogenic sedimentary rock is limestone, calcium carbonate ($CaCO_3$), formed from the skeletons and shells of marine organisms (fig. 16.7). Iron deposits also can be created by biogenic sedimentation.

Humans have become a major force in shaping landscapes. Geomorphologist Rodger Hooke, of the University of Minnesota, looking only at housing excavations, road building, and mineral production, estimates that we move somewhere around 30 to 35 gigatons (billion tons) per year worldwide. When combined with the 10 Gt each year that we add to river sediments through erosion, our earth-moving prowess is comparable to, or greater than, any other single geomorphic agent except plate tectonics.

Metamorphic Rocks

Pre-existing rocks can be modified by heat, pressure, and chemical reagents to create new forms called **metamorphic rock.** Deeply buried strata of igneous, sedimentary, and metamorphic rocks are subjected to great heat and pressure by deposition of overlying sediments or while they are being squeezed and folded by tectonic processes. Chemical reactions can alter both the composition and structure of the rocks as they metamorphose. Some common metamorphic rocks are marble (from limestone), quartzite (from sandstone), and slate (from mudstone and shale). Metamorphic rocks are often the source of metal ores, such as gold, silver, and copper.

Economic Mineralogy

Economic mineralogy is the study of minerals that are heavily used in manufacturing (mainly metal ores) and are, therefore, an impor-

tant part of domestic and international commerce. Nonmetallic economic minerals are mostly graphite, some feldspars, quartz crystals, diamonds, and many other crystals that are valued for their beauty and/or rarity. Minerals have been so important in human affairs that major epochs of human history are commonly known by the dominant materials and the technology to use them (Stone Age, Bronze Age, Iron Age, etc.). The mining, processing, and distribution of these minerals have broad and varied implications both for culture and our environment. Most economically valuable minerals exist everywhere in small amounts; the important thing is to find them concentrated in economically recoverable levels.

Public policy in the United States has encouraged mining on public lands as a way of boosting the economy and utilizing natural resources. Today these laws seem outmoded and in need of reform (What Do *You* Think, p. 350).

Metals

How has the quest for mineral supplies affected global development? We will focus first on world use of metals, earth resources that always have received a great deal of human attention. The availability of metals and the methods to extract and use them have determined technological developments, as well as economic and political power for individuals and nations. We still are strongly dependent on the unique lightness, strength, and malleability of metals.

The metals consumed in greatest quantity by world industry include iron and steel (740 million metric tons annually), aluminum (40 million metric tons), manganese (22.4 million metric tons), copper and chromium (8 million metric tons each), and nickel (0.7 million metric tons). Most of these metals are consumed in the United States, Japan, and Europe, in that order. They are produced primarily in South America, South Africa, and the former Soviet Union. It is easy to see how these facts contribute to a worldwide mineral trade network that has become crucially important to the economic and social stability of all nations involved (fig. 16.8). Table 16.2 shows the primary uses of these metals.

Nonmetal Mineral Resources

Nonmetal minerals are a broad class that covers resources from silicate minerals (gemstones, mica, talc, and asbestos) to sand, gravel, salts, limestone, and soils. Sand and gravel production comprise by far the greatest volume and dollar value of all nonmetal mineral resources. Sand and gravel are used mainly in brick and concrete construction, paving, as loose road filler, and for sandblasting. High-purity silica sand is our source of glass. These materials usually are retrieved from surface pit mines and quarries, where they have been deposited by glaciers, winds, or ancient oceans.

Limestone, like sand and gravel, is mined and quarried for concrete and crushed for road rock. It also is cut for building stone, pulverized for use as an agricultural soil additive that neutralizes acidic soil, and roasted in lime kilns and cement plants to make plaster (hydrated lime) and cement.

Evaporites are mined for halite, gypsum, and potash. These are often found at or above 97 percent purity. Halite, or rock salt, is used for water softening and melting ice on winter roads in some

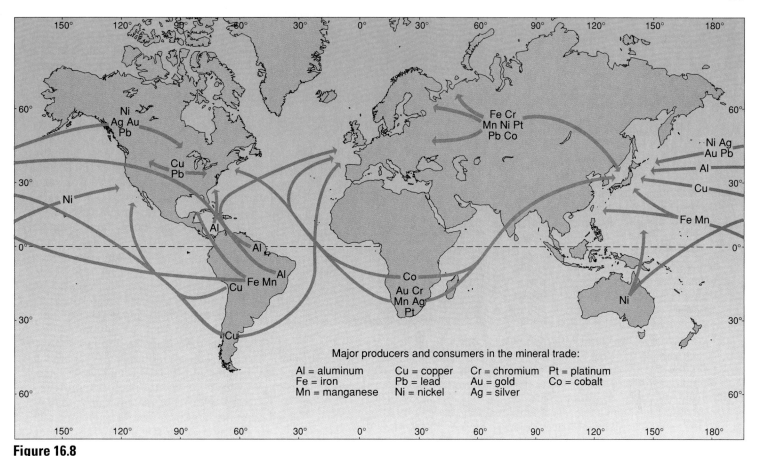

Figure 16.8

Global mineral trade. Metals produced in South Africa, South America, and the former Soviet Union are shipped to markets in the United States, Europe, and Japan, creating a global economic network on which both consumers and producers depend.

northern areas. Refined, it is a source of table salt. Gypsum (calcium sulfate) now makes our plaster wallboard, but it has been used for plaster ever since the Egyptians plastered the walls of their frescoed tombs along the Nile River some five thousand years ago. Potash is an evaporite composed of a variety of potassium chlorides and potassium sulfates. These highly soluble potassium salts have long been used as a soil fertilizer.

Sulfur, in the form of pyrite (FeS_2), is mined mainly for sulfuric acid production. In the United States, sulfuric acid use amounts to more than 200 lbs per person per year, mostly because of its use in industry, car batteries, and some medicinal products.

Strategic Minerals

World industry depends on about eighty minerals, some of which exist in plentiful supplies. Three-fourths of the eighty minerals are abundant enough to meet all of our anticipated needs or have readily available substitutes. At least eighteen minerals, including tin, platinum, gold, silver, and lead, are in short supply.

Of these eighty minerals, between one-half and one-third are considered "strategic" resources. **Strategic minerals** are those that a country uses but cannot produce itself. As the term strategic

Table 16.2

Primary uses of some major metals consumed in the United States

Metal	Use
Aluminum	Packaging foods and beverages (38%), transportation, electronics
Chromium	High-strength steel alloys
Copper	Building construction, electric and electronic industries
Iron	Heavy machinery, steel production
Lead	Leaded gasoline, car batteries, paints, ammunition
Manganese	High-strength, heat-resistant steel alloys
Nickel	Chemical industry, steel alloys
Platinum-group	Automobile catalytic converters, electronics, medical uses
Gold	Medical, aerospace, electronic uses; accumulation as monetary standard
Silver	Photography, electronics, jewelry

What Do *You* Think?
Reforming an Antiquated Mining Law

In 1872, the U.S. Congress passed the General Mining Law intended to encourage prospectors to open up the public domain and promote commerce. This law, which has been in effect more than a century, allows miners to stake an exclusive claim anywhere on public lands and to take—for free—any minerals they find. Claim holders can "patent" (buy) the land for $2.50 to $5 per acre (0.4 hectares) depending on the type of claim. Once the patent fee is paid, the owners can do anything they want with the land, just like any other private property. Although $2.50 per acre may have been a fair market value in 1872, many people regard it as ridiculously low today, amounting to a scandalous give-away of public property.

In Nevada, for example, a mining company is buying federal land for $9000 that contains an estimated $20 billion worth of precious metals. Quite a bargain! Similarly, Colorado investors bought about 7000 ha (17,000 ac) of rich oil-shale land in 1986 for $42,000 and sold it a month later for $37 million. You don't actually have to find any minerals to patent a claim. A Colorado company paid a total of $400 for 65 ha (160 ac) it claimed would be a gold mine. Ten years later, no mining has been done, but the property—which just happens to border the Keystone Ski Area—is being subdivided for condos and vacation homes.

Figure 1

Thousands of abandoned mines on public lands poison streams and groundwater with acid, metal-laced drainage. This old mine in Montana drains into the Blackfoot River, the setting of Norman Maclean's book *A River Runs Through It.*

According to the Bureau of Land Management (BLM), some $4 billion in minerals are mined each year on U.S. public lands. Under the 1872 law, mining companies don't pay a penny for the ores they take. Senator Dale Bumpers of Arkansas, who calls the antiquated mining law "a license to steal," estimates that the government could derive $320 million per year by charging an 8 percent royalty on all minerals and probably could save an equal amount by requiring a bond to be posted to clean up after mining is finished.

Mining companies claim they would be forced to close down if they had to pay royalties or post bonds.

Many people would lose jobs and the economies of western mining towns would collapse if mining becomes uneconomic. But other resource-based industries have been forced to pay royalties on materials they extract from public lands. Coal, oil, and gas companies pay 12.5 percent royalties on fossil fuels obtained from public lands. Timber companies—although they don't pay the full costs of the trees they take—have to bid on logging sales and clean up when they are finished. Even gravel companies pay for digging up the public domain. Ironically, we charge for digging up gravel, but give gold away free.

At the time of this writing, two radically different mining bills were being considered in Congress. The House of Representatives bill, which is enthusiastically supported by environmental groups, would require companies mining on federal lands to pay an 8-percent royalty on their production. It also would eliminate the patenting process, impose stricter reclamation requirements, and give federal managers authority to deny inappropriate permits. In contrast, the Senate bill leaves most provisions of the 1872 bill in place. It would charge a 2-percent royalty, but only after exploration, production, and other costs were deducted. Permitting processes would consider local economic needs before environmental issues in this version. What do you think we should do about this mining law? How could we separate legitimate public interest land use from private speculation and profiteering? Check with your Congressional representatives to find out what has happened.

suggests, these are minerals that a government considers capable of crippling its economy or military strength if unstable global economics or politics were to cut off supplies. For this reason, wealthy industrial nations stockpile strategic minerals in times when prices are low and a supply is available. Figure 16.9 shows some of the major stockpiles of strategic minerals in the United States.

For less wealthy mineral-producing nations, there is another side to strategic minerals. Many less-developed countries depend on steady mineral exports for most of their foreign exchange. Zambia, for instance, relies on cobalt production for 50 percent of its national income. If a steady international market is not maintained, such producer nations could be devastated. From the small-nation producer's point of view, mineral exports, like concentration on any single product or industry, are an unstable economic foundation. Often no option exists, however, if a producer is to participate in the world economy. Environmental consequences of mining may not be a high priority under these circumstances.

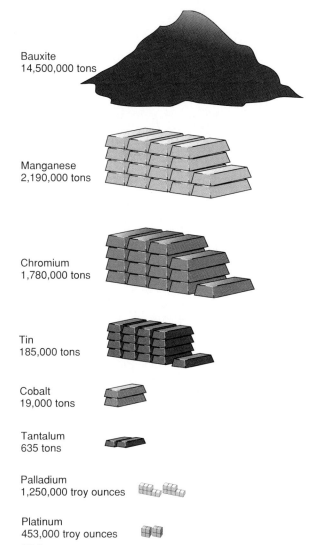

Bauxite
14,500,000 tons

Manganese
2,190,000 tons

Chromium
1,780,000 tons

Tin
185,000 tons

Cobalt
19,000 tons

Tantalum
635 tons

Palladium
1,250,000 troy ounces

Platinum
453,000 troy ounces

Figure 16.9
United States' stockpiles of strategic minerals.

Environmental Effects of Resource Extraction

Resource extraction involves physical processes of mining and physical and chemical processes of separating minerals from ores.

Mining

Mineral extraction is done by several different techniques depending on the accessibility of the ore. All of these methods have environmental hazards. Native metals deposited in the gravel of streambeds can be washed out hydraulically in a process called placer mining. This not only destroys stream beds but fills the water with suspended solids that smother aquatic life. Larger or deeper ore beds are extracted by strip mining or open pit mining where overlying material is removed by large earth-moving equipment (fig. 16.10). The resulting pits can be many kilometers across and

Figure 16.10
The world's largest open-pit mine is Bingham Canyon, near Salt Lake City, UT. More than 5 billion tons of copper ore and waste material have been removed since 1906 to create a hole 800 m (2640 ft) deep and nearly 4 km (2.5 mi) wide at the top.

hundreds of meters deep (Case Study, p. 352). Even deeper deposits are reached by underground tunneling, an extremely dangerous process for mine workers.

Old tunnels occasionally collapse, or subside. In coal mines, natural gas poses dangers of explosion. Uncontrollable fires, producing noxious smoke and gases, sometimes ignite coal-bearing scrap heaps stored inside or outside the mine. Surface waste deposits called tailings can cause acidic or otherwise toxic runoff when rainwater percolates through piles of stored material. Tailings from uranium mines give rise to wind scattering of radioactive dust.

Water leaking into mine shafts also dissolves metals and other toxic material. When this water is pumped out or allowed to seep into groundwater aquifers (chapter 20), pollution occurs. The

The Earth and Its Crustal Resources

CASE STUDY
The Great Canadian Diamond Rush

Diamonds in Canada? Who would have thought that the vast wilderness of tundra and muskeg north of Great Slave Lake in Canada's Northwest Territory might hold deposits of precious jewels? And what will be the environmental impacts of a stampede of miners trying to extract mineral wealth from this fragile ecosystem?

The story began in the 1980s when geologist Charles Fipke was hired to look for base metals, gold, and possibly diamonds in the Barren Lands of Canada's Far North. The search seemed hopeless. Although diamonds are found in many places, no commercial deposits had ever been discovered in the Western Hemisphere. Most diamonds are formed by the enormous pressures and temperatures deep in the earth's mantle. Ideal conditions for crystallization of carbon into diamonds seems to occur mainly in archons, the most ancient cores of continents. African and Siberian diamond fields—the only ones previously known—occur over these ancient formations. Seismologists, however, have recently mapped an archon containing rocks nearly 4 billion years old in the area where Fipke did his prospecting.

The best place to find diamonds is in kimberlite pipes, narrow channels through which magma rises from the mantle to the earth's surface. As the kimberlite rushes upward, it carries diamonds from deeper strata to the surface.

If the magma cools quickly enough, the pipe may remain filled with a diamond-rich deposit that can be commercially viable. But even the richest kimberlites contain very few diamonds—3 to 4 carats (0.6 to 0.8 grams) per ton.

Much to the experts' surprise, after many years of lonely, arduous prospecting, Fipke emerged from the bush with evidence of gem-quality diamonds from a remote region about 320 km (200 mi) north of the small frontier town of Yellowknife. Within a few months, more than 250 companies had staked claims on 22 million ha (53 million ac) of land. Suddenly the barren land is abuzz with helicopters and cargo planes as prefabricated buildings, massive drilling rigs, bulldozers, million-gallon fuel tanks, road-building equipment, and all the other equipment necessary for a huge industrial project are being air-lifted onto the tundra. Mining companies already have begun draining lakes and moving mountains of overburden to create open pits as deep as a 50-story skyscraper is tall. For 24 hours a day, 365 days a year, giant machines will grind up as much as 50 million metric tons of rock per year from each mine. The stakes are high: a single mine can produce billions of dollars in diamonds. But the environmental risks are equally large.

The diamond fields lie in one of the most remote and pristine places on earth. Wolves, grizzly bears, musk oxen,

snowy owls, and the rare gyrfalcon live there. Tundra lakes are the summer breeding grounds for millions of migratory waterfowl. Every year, 325,000 caribou of the Bathurst herd—one of the largest free-roaming groups of mammals on earth—migrate through the area. The scars left from mining and the other extractive industries bound to follow once roads and airstrips are built, will take centuries to heal. Some geologists say that it will take another ice age to remove the damage done to this landscape.

Native Cree and Athabaskan people who live a subsistence lifestyle are protesting the destruction of resources on which they depend. These first peoples are demanding control of their ancestral lands similar to Nunavut, the area granted to their Inuit neighbors to the East (see chapter 14). The mineral riches of this area are an impediment, however, to returning the land to native residents. What do you think? Is it ethical to spoil this great wild place to make a few speculators very rich? How would you weigh the rights of the wild animals and first nations against the economic gains that might result from exploiting these natural resources? Diamonds may be forever, but so too may be the damage caused by digging them up.

Mineral Policy Center in Washington, DC estimates that 25,000 abandoned mines remain in national forests alone. Some 16,000 km (10,000 mi) of rivers and streams in the United States are contaminated by mine drainage. The Environmental Protection Agency estimates that it will cost $2.4 billion to clean up the worst sites.

The process of strip-mining involves stripping off the vegetation, soil, and rock layers, removing the minerals, and replacing the fill. The fill is usually replaced in long ridges, called spoil banks, because this is the easiest way to dump it cheaply and quickly. Spoil banks are very susceptible to erosion and chemical weathering. Rainfall leaches numerous chemicals in toxic concentrations from the freshly exposed earth, and the water quickly picks up a heavy

sediment load. Chemical- and sediment-runoff pollution becomes a major problem in local watersheds. Acid runoff had contaminated 6700 miles of streams in the United States by 1980. Problems are made worse by the fact that the steep spoil banks are very slow to revegetate. Since the spoil banks do not have natural topsoil, succession, soil development, and establishment of a natural biological community occur very slowly.

The 1977 federal Strip-Mining Reclamation and Control Act (SMRCA) requires better restoration of strip-mined lands, especially of land classed as prime farmlands, but restoration is difficult and expensive. Even if soil is carefully replaced, it will take centuries, if not millenia, to regain its former fertility by natural processes. Topsoil is dispersed and often buried by the activity of heavy machinery working to resculpture the land. Compaction disrupts air and water flow through soil, restricts root growth, and causes poor drainage, resulting in wet, stagnant soil conditions. The difficulties of reestablishing vegetation in dry climates means that reclamation is essentially impossible where rainfall is less than 25 cm per year.

The monetary expense of reclamation is also high. Minimum reclamation costs about $1000 per acre, while "complete" restoration (where it is possible) costs $5000 an acre. Nevertheless, 50 percent of U.S. coal (225–270 million metric tons per year) is strip-mined. Nearly a million acres of land in the United States have been devastated by strip-mining, often on public lands for which mining companies pay little or nothing.

Processing

Minerals are extracted from ores by heating or treatment with chemical solvents. These processes often release large quantities of toxic materials that can be even more environmentally hazardous than mining. Smelting—roasting ore to release metals—is a major source of air pollution. One of the most notorious examples of ecological devastation from smelting is a wasteland near Ducktown, Tennessee (fig. 16.11). In the mid-1800s, mining companies began excavating the rich copper deposits in the area. To extract copper from the ore, they built huge open-air wood fires using timber from the surrounding forest. Dense clouds of sulfur dioxide released from sulfide ores poisoned the vegetation and acidified the soil over a 50-square-mile (13,000-hectare) area. Rains washed the soil off the denuded land, creating a barren moonscape where nothing could grow. Siltation of reservoirs on the Ocoee River impaired electric generation by the Tennessee Valley Authority (TVA).

Sulfur emissions from Ducktown smelters were reduced in 1907 after the U.S. Supreme Court ruled in Georgia's favor in a suit to stop interstate transport of air pollution. In the 1930s, the TVA began treating the soil and replanting trees to cut down on erosion. Recently, upwards of $250,000 per year has been spent on this effort. While the trees and other plants are still spindly and feeble, more than two-thirds of the area is considered "adequately" covered with vegetation; only 4 percent is still totally denuded. Similarly, smelting of copper-nickel ore in Sudbury, Ontario, a century ago caused widespread ecological destruction that is slowly being repaired following pollution control measures (see fig. 18.11).

A recent technique called **heap-leach extraction** (fig. 16.12) makes it possible to separate gold from extremely low-grade ores but has a high potential for water pollution. Typically, the process involves piling crushed ore in huge heaps and spraying it with a dilute alkaline-cyanide solution, which percolates through the pile to dissolve the gold. The gold-containing solution is pumped to a processing plant where gold is removed by electrolysis. A thick clay pad and plastic liner beneath the ore heap is supposed to keep the poisonous cyanide solution from contaminating surface or groundwater, but leaks are common.

Once all the gold is recovered, mine operators often simply walk away from the operation, leaving vast amounts of toxic effluent in open ponds behind earthen dams. A case in point is the

Figure 16.11

A luxuriant forest once grew on this now barren hillside near Ducktown, TN. Smelter fumes killed all the vegetation nearly a century ago, and erosion has washed away all the topsoil. Restoration projects are slowly bringing back ground cover and rebuilding soil.

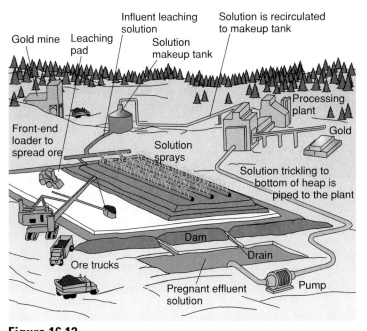

Figure 16.12

In a heap-leach operation, huge piles of low-grade ore are heaped on an impervious pad and sprayed continuously with a cyanide solution. As the leaching solution trickles through the crushed ore, it extracts gold and other precious metals. The "pregnant" effluent solution is then pumped to a processing plant where metals are extracted and purified. This technique is highly profitable but carries large environmental risks.

The Earth and Its Crustal Resources

Summitville mine near Alamosa, Colorado. After extracting $98 million in gold, the absentee owners declared bankruptcy in 1992, abandoning millions of tons of mine waste and huge leaking ponds of cyanide. The Environmental Protection Agency may spend more than $100 million trying to clean up the mess and keep the cyanide pool from spilling into the Alamosa River.

On a smaller scale, *garimperos* (independent placer miners) use mercury to extract gold from river gravels in the Brazilian Amazon. They are estimated to have lost or dumped over 100 metric tons of mercury into the Amazon and its tributaries in 1990.

Conserving Mineral Resources

There is great potential for extending our supplies of economic minerals and reducing the effects of mining and processing through recycling. The advantages of recycling are significant: less waste to dispose of, less land lost to mining, and less consumption of money, energy, and water resources.

Recycling

Some waste products already are being exploited, especially for scarce or valuable metals. Aluminum, for instance, must be extracted from bauxite by electrolysis, an expensive, energy-intensive process. Recycling waste aluminum, such as beverage cans, on the other hand, consumes one-twentieth of the energy of extracting new aluminum. Today, nearly two-thirds of all aluminum beverage cans in the United States are recycled, up from only 15 percent twenty years ago. The high value of aluminum scrap ($650 a ton versus $60 for steel, $200 for plastic, $50 for glass, and $30 for paperboard) give consumers plenty of incentive to deliver their cans for collection. Recycling is so rapid and effective that half of all the aluminum cans now on a grocer's shelf will be made into another can within two months. The energy cost of extracting other metals is shown in table 16.3

	Energy requirements in producing various materials from ore and raw source materials	
Product	**Energy Requirement (MJ/kg)**	
	New	**From Scrap**
Glass	25	25
Steel	50	26
Plastics	162	n.a.
Aluminum	250	8
Titanium	400	n.a.
Copper	60	7
Paper	24	15

Table 16.3

Source: E. T. Hayes, *Implications of Materials Processing.*

Platinum, the catalyst in automobile catalytic exhaust converters, is valuable enough to be regularly retrieved and recycled from used cars (fig. 16.13). Other metals commonly recycled are gold, silver, copper, lead, iron, and steel. The latter four are readily available in a pure and massive form, including copper pipes, lead batteries, and steel and iron auto parts. Gold and silver are valuable enough to warrant recovery, even through more difficult means. See chapter 23 for further discussion of this topic.

Steel and Iron Recycling: Minimills

While total U.S. steel production has fallen in recent decades—largely because of inexpensive supplies from new and efficient Japanese steel mills—a new type of mill subsisting entirely on a readily available supply of scrap/waste steel and iron is a growing

Figure 16.13

The richest ore we have—our mountains of scrapped cars—offers a rich, inexpensive, and ecologically beneficial resource that can be "mined" for a number of metals.

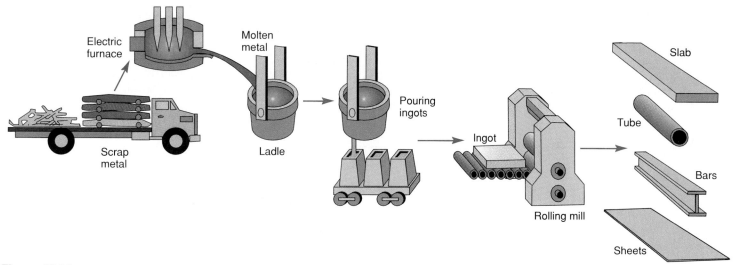

Figure 16.14

"Minimills" remelt and reshape scrap iron and steel. They not only extend our mineral resources by recycling discarded materials, they also conserve energy and are cheaper to operate than traditional integrated mills that depend on virgin ore.

industry. Minimills, which remelt and reshape scrap iron and steel, are smaller and cheaper to operate than traditional integrated mills that perform every process from preparing raw ore to finishing iron and steel products. Minimills produce steel at between $225 and $480 per metric ton, while steel from integrated mills costs $1425 to $2250 per metric ton on average (fig. 16.14). The energy cost is likewise lower in minimills: 5.3 million BTU/ton of steel compared to 16.08 million BTU/ton in integrated mill furnaces. Minimills are expected to produce 26 million metric tons in 1990, about 25 percent of U.S. steel production, and to peak at 30 to 40 percent sometime before 1995. Recycling is slowly increasing as raw materials become more scarce and wastes become more plentiful.

Substituting New Materials for Old

Mineral consumption can be reduced by new materials or new technologies developed to replace traditional minerals and mineral uses. This is a long-standing tradition, for example, bronze replaced stone technology and iron replaced bronze. More recently, the introduction of polyvinylchloride (PVC) plastic pipe has decreased our consumption of copper, lead, and steel pipes. In the same way, the development of fiber optic technology and satellite communication reduces the need for copper telephone wires.

Iron and steel have been the backbone of heavy industry, but we are now moving toward other materials. One of our primary uses for iron and steel has been machinery and vehicle parts. In automobile production, steel is being replaced by polymers (long-chain organic molecules similar to plastics), aluminum, ceramics, and new, high-technology alloys. All of these reduce vehicle weight and cost, while increasing fuel efficiency. Some of the newer alloys that combine steel with titanium, vanadium, or other metals wear much better than traditional steel. Ceramic engine parts provide heat insulation around pistons, bearings, and cylinders, keeping the rest of the engine cool and operating efficiently.

Plastics and glass fiber-reinforced polymers are used in body parts and some engine components.

Electronics and communications (telephone) technology, once major consumers of copper and aluminum, now use ultra-high-purity glass cables to transmit pulses of light, instead of metal wires carrying electron pulses. Once again, this technology has been developed for its greater efficiency and lower cost, but it also affects consumption of our most basic metals.

Geological Hazards

Earthquakes, volcanoes, floods, and landslides are normal earth processes, events that have made our earth what it is today. However, when they occur in proximity to human populations, their consequences can be among the worst and most feared disasters that befall us. For thousands of years people have been watching and recording these hazards, trying to understand them and learn how to avoid them.

Earthquakes

Earthquakes, such as the one in Kobe that opened this chapter, have always seemed mysterious, sudden, and violent, coming without warning and leaving in their wake ruined cities and dislocated landscapes (table 16.4). Earthquakes generally consist of a principal tremor followed by aftershocks from reverberations within the earth's crust. Sometimes the aftershocks are even more severe than the principal tremor because they can be amplified as they travel through the ground. Cities such as Kobe, Japan, Mexico City, or San Francisco, which are built on soft landfill or poorly consolidated soil usually suffer the greatest damage from earthquakes. Water-saturated soil can liquify when shaken. Buildings sometimes sink out of sight or fall down like a row of dominoes under these conditions.

Table 16.4
Worldwide frequency and effects of earthquakes of various magnitudes

Richter Scale Magnitude*	Description	Average Number per Year	Observable Effects
2–2.9	Unnoticeable	300,000	Detected by instruments, but not usually felt by people.
3–3.9	Smallest felt	49,000	Hanging objects swing, vibrations like passing of light truck felt.
4–4.9	Minor earthquake	6200	Dishes rattle, doors swing, pictures move, walls creak.
5–5.9	Damaging earthquake	800	Difficult to stand up. Windows, dishes break. Plaster cracks, loose bricks and tile fall, small slides along sand or gravel banks.
6–6.9	Destructive earthquake	120	Chimneys and towers fall, masonry walls damaged, frame houses move on foundations, small cracks in ground. Broken gas pipes start fires.
7–7.9	Major earthquake	18	General panic. Frame houses split and fall off foundations, some masonry buildings collapse, underground pipes break, large cracks in the ground.
8–8.9	Great earthquake	1 or 2	Catastrophic damage. Most masonry and frame structures destroyed. Roadways, dams, dikes collapse. Large landslides. Rails twist and bend. Underground pipes rupture.

Source: Data from B. Gutenberg in *Earth* by F. Press and R. Seiver. Copyright 1978 by W. H. Freeman & Company Publishers, New York.
*For every unit increase in the Richter Scale, ground displacement increases by a factor of 10, while energy release increases by a factor of 30. There is no upper limit to the scale, but the largest earthquakes recorded have been 8.9.

Frequent earthquakes occur along the edges of tectonic plates, especially where one plate is subducting, or diving down, beneath another. Earthquakes also occur in the centers of continents, however. In fact, the largest earthquake ever recorded in North America was one of magnitude 8.8 that struck the area around New Madrid, Missouri in 1812. Fortunately, few people lived there at the time and the damage was minimal. Among the most calamitous earthquakes in history are one that killed 10,000 people in Mexico City in 1985 (fig. 16.15) and another thought to have killed 242,000 people in China in 1976.

Modern contractors in earthquake zones are attempting to prevent damage and casualties by constructing buildings that can withstand tremors. The primary methods used are heavily reinforced structures, strategically placed weak spots in the building that can absorb vibration from the rest of the building, and pads or floats beneath the building on which it can shift harmlessly with ground motion.

One of the most notorious effects of earthquakes is the **tsunami.** These giant seismic sea swells (sometimes improperly called tidal waves) can move at 1000 km/hr (600 mph), or faster, away from the center of an earthquake. When these swells approach the shore, they can easily reach 15 m or more and some can be as high as 65 m (nearly 200 ft). A 1960 tsunami coming from a Chilean earthquake still caused 7-meter breakers when it reached Hawaii fifteen hours later. Tsunamis also can be caused by underwater volcanic explosions or massive seafloor slumping. The eruption of the Indonesian volcano Krakatoa in 1883 created a tsunami 40 m (130 ft) high that killed 30,000 people on nearby islands.

While most earthquakes occur in known earthquake-prone areas, sometimes they strike in unexpected places. In the United States, major quakes have occurred in South Carolina, Missouri, Massachusetts, Alaska, Nevada, Texas, Utah, Arizona, and Washington, as well as in California.

Volcanoes

Volcanoes and undersea magma vents are the sources of most of the earth's crust. Over hundreds of millions of years, gaseous emissions from these sources formed the earth's earliest oceans and atmosphere. Many of the world's fertile soils are weathered volcanic materials. Volcanoes have also been an ever-present threat to human populations. One of the most famous historic volcanic eruptions was that of Mount Vesuvius in southern Italy, which buried the cities of Herculaneum and Pompeii in A.D. 79. The mountain had been giving signs of activity before it erupted, but many citizens chose to stay and take a chance on survival. On August 24, the mountain buried the two towns in ash. Thousands were killed by the dense, hot, toxic gases that accompanied the ash flowing down from the volcano's mouth. It is still erupting.

Nuees ardentes (French for "glowing clouds") are deadly, denser-than-air mixtures of hot gases and ash like those that inundated Pompeii and Herculaneum. Temperatures in these clouds

It is not just a volcano's dust that blocks sunlight. Sulfur emissions from volcanic eruptions combine with rain and atmospheric moisture to produce sulfuric acid (H_2SO_4). The resulting droplets of H_2SO_4 interfere with solar radiation and can significantly cool the world climate. In 1991, Mt. Pinatubo in the Philippines emitted 20 million tons of sulfur dioxide that combined with water to form tiny droplets of sulfuric acid (fig. 16.16). This acid aerosol reached the stratosphere where it circled the globe for two years. This thin haze cooled the entire earth by 1° C and postponed global warming for several years. It also caused a 10 to 15 percent reduction in stratospheric ozone, allowing increased ultraviolet light to reach the earth's surface. One of several theories about the extinction of the dinosaurs 65 million years ago is that they died off in response to the effects of acid rain and climate changes caused by massive volcanic venting in the Deccan Plateau of India.

Figure 16.15

Collapsing like a house of cards, the Juarez hospital in Mexico City became a tomb for patients and hospital workers alike on September 19, 1985, when an earthquake of magnitude 8.1 hit the city.

may exceed 1000° C, and they move at more than 100 km/hour (60 mph). Nuees ardentes destroyed the town of St. Pierre on the Caribbean island of Martinique on May 8, 1902. Mount Pelee released a cloud of nuees ardentes that rolled down through the town, killing somewhere between 25,000 and 40,000 people within a few minutes. All the town's residents died except a single prisoner being held in the town dungeon.

Mudslides are also disasters sometimes associated with volcanoes. The 1985 eruption of Nevada del Ruiz, 130 km (85 mi) northwest of Bogata, Colombia, caused mudslides that buried most of the town of Amero and devastated the town of Chinchina. An estimated 25,000 people were killed. Heavy mudslides also accompanied the eruption of Mount St. Helens in Washington in 1980. Sediments mixed with melted snow and the waters of Spirit Lake at the mountain's base and flowed many kilometers from their source. Extensive damage was done to roads, bridges, and property, but because of sufficient advance warning, there were few casualties.

Volcanic eruptions often release large volumes of ash and dust into the air. Mount St. Helens expelled 3 km^3 of dust and ash, causing ash fall across much of North America. This was only a minor eruption. An eruption in a bigger class of volcanoes was that of Tambora, Indonesia, in 1815, which expelled 175 sq km of dust and ash, more than fifty-eight times that of Mount St. Helens. These dust clouds circled the globe and reduced sunlight and air temperatures enough so that 1815 was known as the year without a summer.

Figure 16.16

Mt. Pinatubo in the Philippines erupted in 1991, spewing hundreds of millions of tons of ash, rocks, and molten lava. Some 20 million tons of sulfur from the volcano created an acidic aerosol that circled the stratosphere for two years and cooled the global climate at least 1° C.

The Earth and Its Crustal Resources

Floods

In most moderately humid climates, stream channels adjust to accommodate average maximum stream flows. Much of the year, the water level may be well below the stream bank height, but heavy rains or sudden snow melt can deliver more water than the stream can carry. Excess water that overflows stream banks and covers adjacent land is considered a **flood.** The severity of floods can be described by the depth of water above the normal stream banks or by how frequently a similar event normally occurs—on average—for a given area. Note that these are *statistical averages* over long time periods. A "10-year flood" would be expected to occur once in every ten years; a "100-year flood" would be expected to occur once every century. But two 100-year floods can occur in successive years or even in the same year.

Many human activities increase both the severity of floods and their adverse consequences. Paving roads and parking lots reduces water infiltration into the soil and speeds the rate of runoff into streams and lakes. Clearing land for agriculture and filling cities with buildings similarly increase both the volume and rate of water discharge after a storm. Cities need water and usually are located near rivers or lakes for both practical and aesthetic reasons. Floodplains—the land near a stream inundated during normal floods—are generally the most fertile and easily farmed land in an area. These human preferences for riverfront or shoreline locations tend to increase the economic losses and human suffering when floods inevitably occur.

We often try to control floods and reduce their destructive impacts by building levees and flood walls to contain water within riverbanks or by channelization, that is, deepening and straightening stream channels to increase the velocity or volume of water it carries. These measures may protect one area but they often simply accentuate downstream problems (see chapter 15 for further discussion).

Landslides

Gravity constantly pulls downward on every material everywhere on earth, causing a variety of phenomena collectively termed "mass wasting," in which geological materials are moved downslope from one place to another. Erosion is one form of mass wasting; river transport of sediment is another. The resulting movement is often slow and subtle, but some slope processes such as rockslides, avalanches, and land slumping can be swift, dangerous, and very obvious. **Landslide** is a general term for rapid downslope movement of soil or rock. In the United States alone, over one billion dollars in property damage is done every year by landslides and related mass wasting.

In some areas, active steps are taken to control landslides or to limit the damage they cause. On the other hand, many human activities such as road construction, forest clearing, agricultural cultivation, and building houses on steep, unstable slopes increase both the frequency and the damage done by landslides. In some cases, people are unaware of the risks they face by locating on or

Figure 16.17

Parts of an expensive house slide down the hillside in Pacific Palisades, CA. Building at the edge of steep slopes made of unconsolidated sediment in an earthquake-prone region is a risky venture.

under unstable hillsides. In other cases, they simply deny clear and obvious danger. Southern California, where people build expensive houses on steep hills of relatively unconsolidated soil, is often the site of large economic losses from landslides. Chapparal fires expose the soil to heavy winter rains. Resulting mudslides carry away whole neighborhoods and bury downslope areas in debris flows (fig. 16.17). Generally these processes are slow enough that few lives are lost, but the property damage can be high.

Summary

The earth is a complex, dynamic system. Although it seems stable and permanent to us, the crust is in constant motion. Tectonic plates slide over the surface of the ductile mantle. They crash into each other in ponderous slow motion, crumpling their margins into mountain ranges and causing earthquakes. Sometimes one plate will slide under another, carrying rock layers down into the mantle where they melt and flow back toward the surface to be formed into new rocks.

Rocks are classified according to composition, structure, and origin. The three basic types of rock are igneous, metamorphic, and sedimentary. These rock types can be transformed from one to another by way of the rock cycle, a continuous process of weathering, transport, burying in sediments, metamorphosis, melting, and recrystallization.

During the cooling and crystallization process that forms rock from magma, minerals often distill into concentrated ores that become economically important reserves if they are close enough to the surface to be reached by mining. Hot, mineral-laden water flowing up through deep sea thermal vents also makes rich hydrothermal mineral deposits from dissolved minerals transported up from the mantle. Biogenic and chemical sedimentation, placer action, evaporation, and weathering of surface deposits also create valuable mineral deposits.

For reasons that are not entirely clear, a few places in the world are especially rich in mineral deposits. South Africa and the former Soviet Union contain most of the world's supply of several strategic minerals. Less-developed countries, most of which are in the tropics or the Southern Hemisphere, are often the largest producers of ores and raw mineral resources for the strategic materials on which the industrialized world depends. The major consumers of these resources are the industrialized countries.

Worldwide, only a small percentage of minerals are recycled, although it is not a difficult process technically. Recycling saves energy and reduces environmental damage caused by mining and smelting. It reduces waste production and makes our mineral supplies last much longer. Substitution of materials usually occurs when mineral supplies become so scarce that prices are driven up. Many of the strategic metals that we now stockpile may become obsolete when newer, more useful substitutes are found.

Both mining and extraction of minerals have environmental effects. Mine drainage has polluted thousands of kilometers of streams and rivers. Fumes from smelters kill forests and spread pollution over large areas. Surface mining results in removal of natural ecosystems, soil disruption, creation of trenches and open pits, and accumulation of tailings. It is now required that strip-mined areas be recontoured, but revegetation is often difficult and limited in species composition. Smelting and chemical extraction processes also create pollution problems.

Earthquakes and volcanic events are natural geological hazards that are a result of movements of the earth's restless core and mantle. Big earthquakes are among the most calamitous natural disasters that befall people, sometimes killing hundreds of thousands in a single cataclysm.

▼ Questions for Review

1. Describe the layered structure of the earth.

2. What heats the earth and keeps the core molten?

3. What are tectonic plates and why are they important to us?

4. Why are there so many volcanoes and earthquakes along the "ring of fire" that rims the Pacific Ocean?

5. Describe the rock cycle and name the three main rock types that it produces.

6. Distinguish between gravitational, chemical, and biogenic sedimentation. Give an example of each.

7. Give some examples of strategic minerals. Where are the largest supplies of these minerals located?

8. Give some examples of nonmetal mineral resources and describe how they are used.

9. What are some of the advantages of recycling minerals?

10. Describe some ways we recycle metals and other mineral resources.

11. What are some environmental hazards associated with mineral extraction?

12. Describe some of the leading geologic hazards and their effects.

▼ Questions for Critical Thinking

1. Look at the walls, floors, appliances, interior, and exterior of the building around you. How many earth materials were used in their construction?

2. What is the geologic history of your town or county?

3. Is your local bedrock igneous, metamorphic, or sedimentary? If you don't know, who might be able to tell you?

4. What would life be like without the global mineral trade network? Can you think of advantages as well as disadvantages?

5. Suppose a large mining company is developing ore reserves in the small, underdeveloped country where you live. How will revenues be divided fairly between the foreign company and local residents?

6. How could we minimize the destruction caused by geologic hazards? Should people be discouraged from building on floodplains or in volcanic or earthquake-prone areas?

7. What might be the climatic effects of having all the continents clustered together in one giant supercontinent?

8. How would your life be affected if your country were to run out of strategic minerals?

9. What effect do you think our need for strategic minerals has had on our foreign policy toward South Africa and the former Soviet Union?

10. How could our government encourage more recycling and more efficient use of minerals?

11. What is the potential for geologic hazards where you live?

12. What can you do to protect yourself from geologic hazards?

 ## Key Terms

core 344
crust 344
flood 358
heap-leach extraction 353
igneous rocks 347
landslide 354
magma 345
mantle 344
metamorphic rock 348

mineral 345
rock cycle 345
sedimentary rock 347
sedimentation 347
strategic minerals 349
tectonic plates 345
tsunami 356
weathering 347

 ## Additional Information on the Internet

Galaxy Einet Sites for Geosciences
http://galaxy.einet.net/galaxy/Science/Geosciences.html/
Tradewave Galaxy is copyright 1993, 1994, 1995, 1996 Enterprise Integration Network Corporation (Tradewave). All Rights Reserved. Tradewave is a trademark of Tradewave, Inc.

Geologic Information
http://geology.usgs.gov/

Lawrence Livermore National Laboratory Geological and Atmospheric Hazards Projects
http://www-ep.es.llnl.gov/www-ep/ghp.html/

Mine Net
http://www.microserve.net/~doug/

National Earthquake Information Center
http://wwwneic.cr.usgs.gov/

National Geophysical Data Center
http://www.ngdc.noaa.gov/

US Bureau of Mines
http://www.usbm.gov/

US Geological Survey: Earth and Environmental Science
http://www.usgs.gov/network/science/earth/earth.html/

World Wide Web Virtual Library: Environment-Lithosphere
http://ecosys.drdr.Virginia.EDU:80/lit.html/

Note: Further readings appropriate to this chapter are listed on p. 602.

Environmental Educator

*T*racy Williams (left)

Like many students with a general liberal arts education, Tracy Williams didn't see a clear path leading to an environmental career as she approached graduation. A degree in political science gave her good skills in critical thinking, writing, and interpersonal communications, but little technical training to help land a job.

During a spring vacation to south Texas, Tracy visited the Texas Natural Resources Conservation Commission office in Corpus Christi to see if she could get any ideas about jobs. Although she didn't know anyone there and wasn't even sure what she was looking for, she found that people were friendly and helpful. Several had good suggestions, including a tip about a new program in El Paso that might have a place for her.

Tracy sent a résumé, a transcript, and a writing sample to the El Paso office of the TNRCC. After a telephone interview, she was offered a paid summer internship. As is often the case, an internship was an excellent way to gain experience and get a head start toward permanent employment. When a full-time job opened up in her office, Tracy knew how to do the work and already had demonstrated her abilities. She was fortunate to have entered a rapidly growing field. With the passage of the North American Free Trade Agreement (NAFTA), transboundary pollution and development issues have become of great concern.

Tracy is now a full-time program coordinator. Her responsibilities include public outreach, environmental education, and industrial and governmental relations in El Paso and its sister city, Juarez, Mexico. She writes an English/Spanish newsletter to inform citizens about environmental issues, and she coordinates a volunteer water quality monitoring network. Local residents take water quality samples from the Rio Grande/Rio Bravo near their homes and form lakes and small streams that would be difficult for water quality professionals to assay on a regular basis. This program now includes people of all ages, from senior citizens to school children.

One of the things Tracy likes best about her job is the opportunity to continually learn new things and to work on different issues. One day she might speak to a group of local governmental officials or business people about state statutes and environmental quality regulations. The next day, she is out in the field, teaching children how to take water samples.

Tracy's work often takes her across the border. Both the state of Texas and the U.S. government are helping Mexico build sewage treatment plants and develop industrial pollution prevention programs. Coordination and cooperation between government agencies is essential. Although she is still working on her accent, being bilingual has really helped Tracy in her job. She strongly recommends language skills for anyone who wants to work on international—or sometimes even local—environmental issues.

For students interested in careers like hers, Tracy advises getting as much technical training and field experience as possible. A degree in science or engineering often opens more doors than does a general liberal education. At the same time, she recommends a strong background in communication, social sciences, and cultural discourse. Science might get you into a job, but you have to be able to write reports and make presentations once you are in a career. Tracy suggests getting an internship for experience, being flexible about where you are willing to go and what you are willing to do, and creatively seeking job possibilities. There are many opportunities; you just have to search them out.

Lightning flashes over the city as a summer thunderstorm approaches.

CHAPTER 17: Air, Climate, and Weather

Objectives

After studying this chapter, you should be able to:

- summarize the structure and composition of the atmosphere.

- understand how solar energy warms the atmosphere and creates circulation patterns.

- explain how the jet streams, prevailing winds, and frontal systems determine local weather.

- evaluate previous climatic catastrophes and the driving forces thought to bring about climatic change.

- debate the hypothesis that human actions may bring about global climate change.

Hurricane!

In the early morning hours of August 24, 1992, Hurricane Andrew roared ashore south of Miami. Born in Africa and nurtured by warm Caribbean seas, the swirling updrafts of this powerful low-pressure cell created steady winds of 240 km per hour (150 mph) and gusts as high as 320 kph (200 mph). As it passed over densely populated south Florida, the shrieking storm peeled off roofs, exploded houses, and piled boats and trailers in expensive scrap heaps (fig. 17.1). Cutting a swath 40 km (25 mi) wide, the storm flattened farm fields and decimated thousands of acres of the Everglades National Park. Two days later, after regaining strength over the Gulf of Mexico, the hurricane slammed ashore again in Louisiana. Weakening as it moved north, the storm finally died a week later in Ohio. Altogether, 43 people were killed, 80,000 dwellings were demolished, and 160,000 residents were left homeless. With total property damages amounting to more than $30 billion, this was the most destructive storm in United States history.

Weather and climate affect all of us. What will conditions be today where you are? If it's warm, is it the beginning of a global greenhouse effect? If it's cold, is a new ice age coming? We worry a great deal about what's happening atmospherically but our understanding of processes that control global climate is rather poor. In spite of this, however, we are making changes in our atmosphere that could have disastrous effects, not only for humans but also for the basic life-support systems that make our planet habitable. In this chapter, we will study the air around us. And we will look at the ways it affects us and is, in turn, affected by living organisms—including humans. ▶

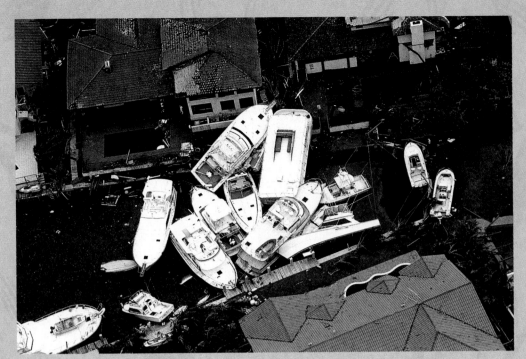

Figure 17.1
A million dollars worth of boats tossed in a pile by Hurricane Andrew. Weather and climate affect us in many ways, both positive and negative. What drives these atmospheric processes and how are we changing them? These are among the most important questions in environmental science today.

Weather is a description of the physical conditions of the atmosphere (moisture, temperature, pressure, and wind), all of which play a vital role in shaping ecosystems. The dynamism of the atmosphere is maintained by a ceaseless flow of solar energy. Winds generated by pressure gradients push large air masses of differing temperature and moisture content around the globe. Our daily weather is created by the movement of these air masses (fig. 17.2)

Climate is a description of the long-term pattern of weather in a particular area. Climates often undergo cyclic changes over decades, centuries, and millenia. Determining where we are in these cycles and predicting what may happen in the future is an important, but difficult, process. As human activities change the properties of the atmosphere, it becomes more difficult and more important to understand how the atmosphere works and what future weather and climate conditions may be.

Weather and climate are important, not only because they affect human activities, but because they are primary determinants of biomes and ecosystem distribution. Generalized, large-scale climate often is not as important in the life of an individual organism as is the microclimate of the atmosphere and soil in its specific habitat. Geographic boundaries that separate communities and ecosystems are established primarily by climatic boundaries created by temperature, moisture, and wind-distribution patterns (chapter 4). The movement and effects of pollutants also are strongly linked to weather and climatic conditions (chapter 18).

Composition and Structure of the Atmosphere

We live at the bottom of a virtual ocean of air. Extending upward about 1600 km (1000 mi), this vast, restless envelope of gases is far more turbulent and mobile than the oceans of water. Its currents and eddies are the winds.

Past and Present Composition

The composition of the atmosphere has changed drastically since it first formed as the earth cooled and condensed from interstellar gases. Most geochemists believe that the earth's earliest atmosphere was made up mainly of hydrogen and helium and was hundreds of times more massive than it is now. Over billions of years, most of that hydrogen and helium diffused into space. At the same time, volcanic emissions have added carbon, nitrogen, oxygen, sulfur, and other elements to the atmosphere.

The current composition of the earth's atmosphere is unique in our solar system. This is the only place we know of with free oxygen and water vapor. We believe that virtually all the molecular oxygen in the air was produced by photosynthesis in blue-green bacteria, algae, and green plants. If that oxygen were not present, heterotrophs (like us) who oxidize organic compounds as an energy source could not exist. As chapter 3 discusses, producers and

Table 17.1

Present composition of the lower atmosphere*

Gas	Symbol or Formula	Percent by Volume
Nitrogen	N_2	78.08
Oxygen	O_2	20.94
Argon	Ar	0.934
Carbon dioxide	CO_2	0.033
Neon	Ne	0.00182
Helium	He	0.00052
Methane	CH_4	0.00015
Krypton	Kr	0.00011
Hydrogen	H_2	0.00005
Nitrous oxide	N_2O	0.00005
Xenon	Xe	0.000009

*Average composition of dry, clean air

Figure 17.2

The atmospheric processes that purify and redistribute water, moderate temperatures, and balance the chemical composition of the air are essential in making life possible. The earth is the only planet we know with a habitable atmosphere. To a large extent, living organisms have created, and help to maintain, the atmosphere on which we all depend.

consumers create a balance between carbon dioxide and oxygen levels in the atmosphere. The Gaia hypothesis, first proposed by British chemist James Lovelock, suggests that the balance between atmospheric carbon dioxide and oxygen maintained by living organisms is responsible not only for creating a unique atmospheric chemical composition but also for other environmental characteristics that make life possible. We will discuss how carbon dioxide levels regulate temperatures later in this chapter. Organisms—especially humans—continue to modify the atmosphere, but at a

dramatically accelerated rate, and in ways that may be disastrous for other forms of life.

Table 17.1 presents the main components of clean, dry air. Water vapor concentrations vary from near zero to 4 percent, depending on air temperature and available moisture. Small but important concentrations of minute particles and droplets of material—collectively called **aerosols**—also are suspended in the air. The production of aerosols by human activities and their effects on human health and natural ecosystems are discussed in chapter 18.

A Layered Envelope

The atmosphere is layered into four distinct zones of contrasting temperature due to differential absorption of solar energy (fig. 17.3). Understanding how these layers differ and what creates them helps us understand atmospheric functions.

The layer of air immediately adjacent to earth's surface is called the **troposphere.** Ranging in depth from about 16 km (10 mi) over the equator to about 8 km (5 mi) over the poles, this zone is where most weather events occur. Due to the force of gravity and the compressibility of gases, the troposphere contains about 75 percent of the total mass of the atmosphere. The troposphere's composition is relatively uniform over the entire planet because this zone is strongly stirred by winds. Air temperature drops rapidly with increasing altitude in this layer, reaching about $-60°C$ ($-76°F$) at the top of the troposphere. A sudden reversal of this temperature gradient creates a sharp boundary, the tropopause, that limits mixing between the troposphere and upper zones.

The **stratosphere** extends from the tropopause up to about 50 km (31 mi). Air temperature in this zone is stable or even increases with

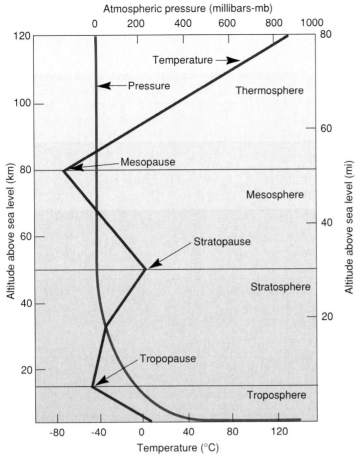

Figure 17.3

The atmosphere is layered into four distinct zones of contrasting temperature due to differential absorption of solar energy. Mixing between layers is inhibited by this temperature gradient. Note that pressure falls from 1000 mb at ground level to near that of outer space in the stratosphere. Although temperatures rise in the thermosphere, there are too few molecules present to measure much pressure rise.

higher altitude. Although more dilute than the troposphere, the stratosphere has a very similar composition except for two important components: water and ozone (O_3). The fractional volume of water vapor is about one thousand times lower, and ozone is nearly one thousand times higher than in the troposphere. Ozone is produced by lightning and solar irradiation of oxygen molecules and would not be present if photosynthetic organisms were not releasing oxygen. Ozone protects life on the earth's surface by absorbing most incoming solar ultraviolet radiation.

Recently discovered decreases in stratospheric ozone over Antarctica (and to a lesser extent over the whole planet) are of serious concern (see chapter 18). If these trends continue, we could be exposed to increasing amounts of dangerous ultraviolet rays, resulting in higher rates of skin cancer, genetic mutations, crop failures, and disruption of important biological communities. Unlike the troposphere, the stratosphere is relatively calm. There is so little mixing in the stratosphere that volcanic ash or human-caused contaminants can remain in suspension there for many years (chapter 16).

Above the stratosphere, the temperature diminishes again, creating the **mesosphere** or middle layer. The minimum temperature reached in this region is about $-80°C$ ($-120°F$). At an altitude of 80 km, another abrupt temperature change occurs. This is the beginning of the **thermosphere,** a region of highly ionized gases, extending out to about 1600 km (1000 mi). Temperatures are very high in the thermosphere because molecules there are constantly bombarded by high-energy solar and cosmic radiation. There are so few molecules per unit area, however, that if you were cruising through in a spaceship, you wouldn't notice the temperature increase.

The lower part of the thermosphere is called the **ionosphere.** This is where the aurora borealis (northern lights) appears when showers of solar or cosmic energy cause ionized gases to emit visible light. There is no sharp boundary that marks the end of the atmosphere. Pressure and density decrease gradually as one travels away from the earth until they become indistinguishable from the near vacuum of intrastellar space. The composition of the thermosphere also gradually merges with that of intrastellar space, being made up mostly of helium and hydrogen.

The Great Weather Engine

The atmosphere is a great weather engine in which a ceaseless flow of energy from the sun causes global cycling of air and water that creates our climate and distributes material through the environment.

Solar Radiation Heats the Atmosphere

The sun supplies the earth with an enormous amount of energy. Although it fluctuates from time to time, incoming solar energy at the top of the atmosphere averages about 1360 watts per sq meter. About half of this energy is reflected or absorbed by the atmosphere, and half the earth faces away from the sun at any given time. Still, the amount reaching the earth's surface is at least 10,000 times greater than all installed electric capacity in the world.

The absorption of solar energy by the atmosphere is selective. Visible light (see fig. 3.6) passes through almost undiminished, whereas ultraviolet light is absorbed mostly by ozone in the stratosphere. Infrared radiation is absorbed mostly by carbon dioxide (CO_2) and water (H_2O) in the troposphere. Scattering of light by water droplets, ice crystals, and dust in the air also is selective. Short wavelengths (blue) are scattered more strongly than long wavelengths (red). The blue of a clear sky or clean, deep water at midday, and the spectacular reds of sunrise and sunset are the result of this differential scattering. On a cloudy day, as much as 90 percent of insolation is absorbed or reflected by clouds. Figure 17.4 shows the average energy fluxes in the atmosphere.

Some solar energy is reflected from the earth's surfaces. **Albedo** is the term used to describe reflectivity. Fresh, clean snow can have an albedo of 90 percent, meaning that 90 percent of incident radiation falling on its surface is reflected. Dark surfaces, such as black topsoil or a dark forest canopy, absorb energy efficiently and might have an albedo of only 2 or 3 percent. The net average global albedo of the earth is about 30 percent. Clouds are responsible for most of that reflection. The earth's surface has a low av-

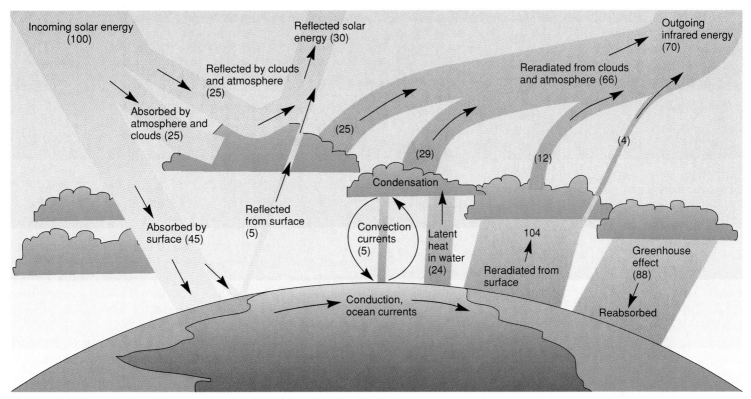

Figure 17.4

Energy balance between incoming and outgoing radiation. The atmosphere absorbs or reflects about half of the solar energy reaching the earth. Most of the energy reemitted from the earth's surface is long-wave, infrared energy. Most of this infrared energy is absorbed by aerosols and gases in the atmosphere and is reradiated toward the planet, keeping the surface much warmer than it would otherwise be. This is known as the greenhouse effect. The numbers shown are arbitrary units. Note that for 100 units of incoming solar energy, 100 units are reradiated to space, but more than 100 units are radiated from the earth's surface because of the greenhouse effect.

erage albedo (5 percent) due to the high energy absorbency of the oceans covering most of the globe.

Eventually, all the energy absorbed at the earth's surface is reradiated back into space. There is an important change in properties between incoming and outgoing radiation, however. Most of the solar energy reaching the earth is visible light, to which the atmosphere is relatively transparent; the energy reemitted by the earth is mainly infrared radiation (heat energy). These longer wavelengths are absorbed rather effectively in the lower levels of the atmosphere, trapping much of the heat close to the earth's surface. If the atmosphere were as transparent to infrared radiation as it is to visible light, the earth's surface temperature would be about 35°C (63°F) colder than it is now.

This phenomenon is called the "greenhouse effect" because the atmosphere, like the glass of a greenhouse, transmits sunlight while trapping heat inside. (The analogy is not totally correct, however, because glass is much more transparent to infrared radiation than is air; greenhouses stay warm mainly because the glass blocks air movement.) Increasing atmospheric carbon dioxide due to human activities appears to be causing a global warming that could cause major climatic changes, which we will discuss later in this chapter.

Because of cycling of infrared energy between the atmosphere and the planet, the amount of energy emitted from the earth's surface is about 30 percent greater than the total incoming solar radi-

ation. The amount of energy reflected or reradiated from the top of the atmosphere must balance with the total insolation if the earth is to remain at a constant temperature.

Convection Currents and Latent Heat

Air currents, especially those carrying large amounts of water vapor, also play an important role in shaping our weather and climate. As the sun heats the earth's surface, some of that heat is transferred to adjacent air layers, causing them to expand and become less dense. This lighter air rises and is replaced by cooler, heavier air, resulting in vertical **convection currents** that stir the atmosphere and transport heat from one area to another.

Much of the solar energy absorbed by the earth is used to evaporate water. Because of the unique properties of water (In Depth, p. 60), it takes a significant amount of energy to change water from liquid to vapor state, and this energy is stored in the water vapor as latent or potential energy. Latent energy is released as heat when the water condenses.

Water vapor carried into the atmosphere by rising convection currents transports large amounts of energy and plays an important role in the redistribution of heat from low to high altitudes, and from the oceans to the continental landmasses (fig. 17.5). As warm, moist air rises, it expands (due to lower air pressures at higher altitudes)

Air, Climate, and Weather

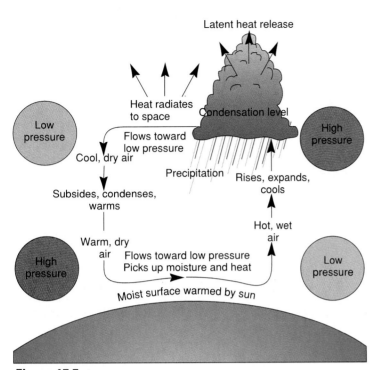

Figure 17.5
Convection currents and latent energy cause atmospheric circulation and redistribute heat and water around the globe.

and cools. If condensation nuclei are present or if temperatures are low enough, the water will condense to form water droplets or ice crystals and precipitation will occur. Releasing latent heat causes the air to rise higher, cool more, and lose more water vapor. Rising, expanding air creates an area of relatively high pressure at the top of the convection column.

Air flows out of this high pressure zone toward areas of low pressure where cool, dry air is sinking (subsiding). This subsiding air is compressed (and therefore warmed) as it approaches the earth's surface, where it piles up and creates a region of relatively high pressure at the surface. Air flows out of this region back toward the area of low surface pressure caused by rising air, thus closing the cycle. These are the driving forces of the hydrologic cycle (chapter 19).

The convection currents just described can be as small and as localized as a narrow column of hot air rising over a sun-heated rock, or as large as the desert low-pressure cell that covers the U.S. Southwest most of the summer. The circulation patterns they create can be as mild as a gentle onshore breeze moving from the warm ocean toward the cooling shoreline as the sun goes down in the evening, or they can create monster cyclonic storms that drive hurricanes hundreds of kilometers wide across oceans and over continents. The force generated by such a storm (powered by latent heat released from condensing water vapor) can be equivalent to hundreds of megaton-sized nuclear bombs. These circulation systems—both large and small—are the driving force behind our weather, the moment-by-moment changes in the atmosphere.

Weather

Even though we have been watching the skies for thousands of years, trying to forecast what the weather will be, the science of meteorology (weather studies) is still rather imprecise and uncertain. Weather forces still are so important in our lives that it makes sense to learn as much as we can about how the atmosphere makes weather.

Energy Balance in the Atmosphere

Solar energy doesn't strike the whole globe equally. At the equator, the sun is almost directly overhead all year long. Its rays are very intense because it shines through a relatively short column of air (straight down), and energy flux (flow) is high. At the poles, however, sunlight comes in at an oblique angle. The long column of air through which light must pass before it reaches the surface causes much greater energy losses from absorption and scattering. Moreover, when the light does reach the ground, it is spread over a larger area because of its angle of incidence, reducing surface heating even more.

Furthermore, seasonal tilting of the earth's axis means that there is no sunlight at the poles during much of the winter. The equator, by contrast, has days about the same length all year long; thus, compared to the poles, the equatorial regions have an energy surplus. This energy imbalance is evened out by movement of air and water vapor in the atmosphere, and by liquid water in rivers and ocean currents. Warm, tropical air moving toward the poles and cold, polar air moving toward the equator account for about half of this energy transfer. Latent heat in water vapor (mainly from the oceans) makes up about 30 percent of the global energy redistribution. The remaining 20 percent is carried mainly by ocean currents.

Hadley Cells and Prevailing Winds

As air warms at the equator, rises, and moves northward, it doesn't go straight to the pole in a single convection current. Instead, this air sinks and rises in several intermediate bands, forming circulation patterns called **Hadley cells** (fig. 17.6). Nor do the returning surface flows within these cells run straight north and south. Friction, drag, and momentum cause air layers close to the earth's surface to be pulled in the direction of rotation. This deflection is called the **Coriolis effect.** In the Northern Hemisphere, the Coriolis effect deflects winds about 30° to the right of their expected path, creating clockwise or anticyclonic spiraling patterns in winds flowing out of a high-pressure center, and cyclonic or counterclockwise winds spiraling into a low-pressure area. In the Southern Hemisphere, winds and water movements shift in the opposite direction. The earth's rotation affects only large-scale movements, however. Contrary to popular beliefs, water draining out of a bathtub in the Southern Hemisphere is controlled by the shape of the drain, and will not necessarily swirl in the opposite direction from what it would in the north.

A major zone of subsidence occurs at about 30° north latitude. Air flows into this region of low pressure both from the north and south. Air flowing back toward the equator is turned toward the west by the Coriolis effect, creating the steady northeast "trade winds" of

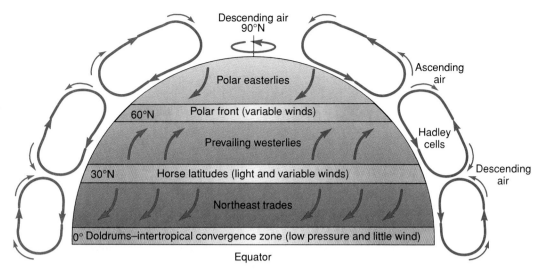

Figure 17.6

General circulation patterns over the Northern Hemisphere. The actual boundaries of the circulating Hadley cells vary from day to day and season to season, as do the local directions of surface winds. Surface topography also complicates circulation patterns, but within these broad regions, winds usually have a predominant and predictable direction.

subtropical oceans. Their name comes from the dependable routes they provided for merchant sailing ships in earlier days. Where this dry, subsiding air falls on continents, it creates broad, subtropical desert regions (chapter 19). Air flowing north from this region of subsidence turns eastward, giving rise to the prevailing westerlies of middle latitudes. (Notice that an eastward flowing wind is called a west wind or a westerly, due to the direction from which it originates.)

Winds directly under regions of subsiding air often are light and variable. They create the so-called horse latitudes because sailing ships bringing livestock to the New World were often becalmed here and had to throw the bodies of dead horses overboard. Rising air at the equator creates doldrums where the winds may fail for weeks at a time. Another band of variable winds at about 60° north, called the polar front, tends to block the southward flow of cold polar air. As we will see in the next section, however, all these boundaries between major air flows wander back and forth, causing great instability in our weather patterns, especially in midcontinent areas. The Southern Hemisphere has more stable wind patterns because it has more ocean and less landmass than the Northern Hemisphere.

Jet Streams

Superimposed on the major circulation patterns and prevailing surface winds are variations caused by large-scale upper air flows and shifting movements of the large air masses that they push and pull. The most massive of these rivers of air are the **jet streams,** powerful winds that circulate in shifting flows rivaling the oceanic currents in extent and effect. Generally following meandering paths from west to east, jet streams can be as much as 50 km wide and 5 km deep. The number, flowing speed, location, and size of jet streams all vary from day to day and place to place.

Wind speeds at the center of a jet stream are often 200 km/hr (124 mph) and may reach twice that speed at times. Located 6 to 12 km (3.7–7.5 mi) above the earth's surface, jet streams follow discontinuities in the tropopause (the boundary between the troposphere and the stratosphere), where they are broken into large, overlapping plates that fit together like shingles on a roof. The jet streams are probably generated by strong temperature contrasts where adjacent plates overlap.

There are usually two main jet streams over the Northern Hemisphere. The subtropical jet stream generally follows a sinuous path about 30° north latitude (the southern edge of the United States), while the northern jet stream follows a more irregular path along the edge of a huge cold air mass called the circumpolar vortex (fig. 17.7) that covers the earth's top like a cap with scalloped edges. This whole polar vortex rotates from west to east slightly

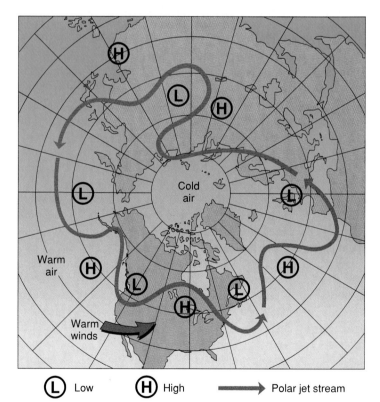

L ⃝ Low H ⃝ High ➡ Polar jet stream

Figure 17.7

A typical pattern of the arctic circumpolar vortex. This large, circulating mass of cold air sends "fingers," or lobes, across North America and Eurasia, spreading storms in their path. If the vortex becomes stalled, weather patterns stabilize, causing droughts in some areas and excess rain elsewhere.

Air, Climate, and Weather

faster than the planet's rotation. As it moves, the lobes, or fingers, of cold air that protrude south from the vortex sweep across Canada and the United States. The clash between cold, dry arctic air masses pushing south against warm, wet air masses moving north from the Gulf of Mexico or the Pacific Ocean brings winds, rains, and storms to the middle of the continent.

During the winter, as the Northern Hemisphere tilts away from the sun and the atmosphere cools, the polar air masses become stronger and push farther south, bringing snow and low temperatures across much of the United States. During the summer, as we tilt back toward the sun, warm air from the South pushes the polar jet stream back toward the pole.

Occasionally, the circumpolar vortex slows so that it rotates at nearly the same speed as the earth, stalling the motion of the lobes or air masses, and locking a huge ridge of hot, dry air over mid-America for months at a time. What causes air flow to be stalled like this—or to resume normal circulation patterns—is unknown; but the amount of heat in the atmosphere surely plays a role.

Frontal Weather

The boundary between two air masses of different temperature and density is called a front. When cooler air displaces warmer air, we call the moving boundary a **cold front.** Since cold air tends to be more dense than warm air, a cold front will hug the ground and push under warmer air as it advances. As warm air is forced upward, it cools adiabatically (without loss or gain of energy), and its cargo of water vapor condenses and precipitates. Upper layers of a moving cold air mass move faster than those in contact with the ground because of surface friction or drag, so the boundary profile assumes a curving, "bull-nose" appearance (fig. 17.8). Notice that the region of cloud formation and precipitation is relatively narrow. Cold fronts generate strong convective currents and often are accompanied by violent surface winds and destructive storms. An approaching cold front generates towering clouds called thunderheads that reach into the stratosphere where the jet stream pushes the cloud tops into a characteristic anvil shape. The weather after the cold front passes is usually clear, dry, and invigorating.

If the advancing air mass is warmer than local air, a **warm front** results. Since warm air is less dense than cool air, an advancing warm front will slide up over cool, neighboring air parcels, creating a long, wedge-shaped profile with a broad band of clouds and precipitation (fig. 17.8). Gradual uplifting and cooling of air in the warm front avoids the violent updrafts and strong convection currents that accompany a cold front. A warm front will have many layers of clouds at different levels. The highest layers are often wispy cirrus (mare's tail) clouds that are composed mainly of ice crystals. They may extend 1000 km (621 mi) ahead of the contact zone with the ground and appear as much as forty-eight hours before any precipitation. A moist warm front can bring days of drizzle and cloudy skies.

Cyclonic Storms

Few people experience a more powerful and dangerous natural force than cyclonic storms spawned by low-pressure cells over warm trop-

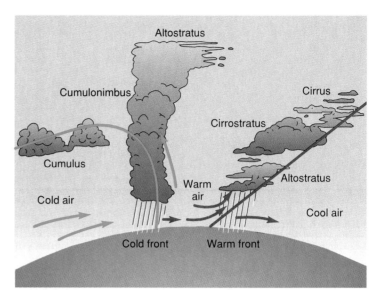

Figure 17.8

A cold front assumes a bulbous, "bull-nose" appearance because ground drag retards forward movement of surface air. As warm air is lifted up over the advancing cold front, it cools, producing precipitation. When warm air advances, it slides up over cooler air in front and produces a long, wedge-shaped zone of clouds and precipitation. The high cirrus clouds that mark the advancing edge of the warm air mass may be 1000 km and 48 hours ahead of the front at ground level.

ical oceans. As we discussed earlier in this chapter, low pressure is generated by rising warm air. Winds swirl into this low-pressure area, turning counterclockwise in the Northern Hemisphere due to the Coriolis effect. When rising air is laden with water vapor, the latent energy released by condensation intensifies convection currents and draws up more warm air and water vapor. As long as a temperature differential exists between air and ground and a supply of water vapor is available, the storm cell will continue to pump energy into the atmosphere. Called hurricanes in the Western Hemisphere or typhoons in the East, winds near the center of these swirling air masses can reach hundreds of kilometers per hour and cause tremendous suffering and destruction. In 1970, a killer typhoon brought torrential rains, raging winds, and a tidal surge that flooded thousands of square kilometers of the flat coastal area of Bangladesh, drowning more than a half million people.

Storms over the land never have as much water vapor to pump energy into the atmosphere as those over the ocean, but cyclonic storms over the land can be terribly destructive in localized areas. Sometimes when a strong cold front pushes under a warm, moist air mass over the land, the updrafts create small cyclones that we call **tornadoes** (fig. 17.9). These are most common in the spring and early summer, when temperature differentials are greatest. The southern plains states around Oklahoma are often referred to as "Tornado Alley" because of the frequency with which they experience tornadoes. Waterspouts over the ocean are also small cyclones caused by frontal weather patterns and temperature differentials.

Figure 17.9

Tornadoes are local cyclonic storms caused by rapid mixing of cold, dry air and warm, wet air. Wind speeds in the swirling funnel can be more than 160 km/hr (100 mph) and can be very destructive where the storm touches down on the ground.

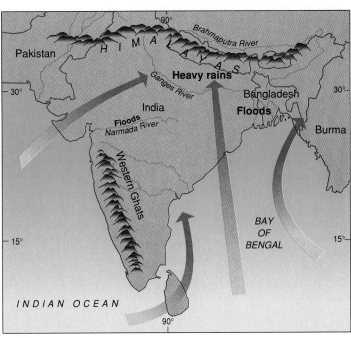

Figure 17.10

Summer monsoon air flows over the Indian subcontinent. Warming air rises over the plains of central India in the summer, creating a low-pressure cell that draws in warm, wet oceanic air. As this moist air rises over the Western Ghats or the Himalayas, it cools and heavy rains result. These monsoon rains flood the great rivers, bringing water for agriculture, but also causing much suffering.

Seasonal Winds

A **monsoon** is a seasonal reversal of wind patterns caused by different heating and cooling rates of the ocean and continents. The most dramatic examples of monsoon weather are found in tropical or subtropical countries where a large land area is cut off from continental air masses by mountain ranges and surrounded by a large volume of water. The Indian subcontinent is a good example (fig. 17.10). As the summer sun heats the Indian plains, a strong low-pressure system develops. Flow of cool air from the north is blocked by the Karakoram and Himalayan mountains. A continuous flow of moisture-laden air from the subtropical high-pressure area over the Arabian Sea sweeps around the tip of India and up into the Ganges Plain and the Bay of Bengal. As the onshore winds are driven against the mountains and rise, the air cools and drops its load of water. During the four months of the monsoon, which usually lasts from May until August, 10 m (390 in) of rain may fall on Nepal, Bangladesh, and western India. The highest rainfall ever recorded in a season was 25 m (82 ft), which fell in about five months on the southern foothills of the Himalayas in 1970. These heavy rains result in high rates of erosion and enormous floods in the flat delta of the Ganges River, but they also irrigate the farmlands that feed the second most populous country in the world. In winter, the Indian landmass is cooler than the surrounding ocean and the wind flow reverses. The northeast winds pick up moisture as they blow over the Bay of Bengal and bring winter monsoon rains to Indonesia, Australia, and the Philippines.

Africa also experiences strong seasonal monsoon winds. Hot air rising over North Africa in the summer pulls in moist air from the Atlantic Ocean along the Gulf of Guinea. The torrential rains released by this summer circulation nourish the tropical forests of Central Africa and sometimes extend as far east as the Arabian Ocean. How far north into the Sahara Desert these rain clouds penetrate determines whether crops, livestock, and people live or die in the desert margin (fig. 17.11). In the winter, the desert cools and air flows back out over the now warmer Atlantic. These shifting winds allowed Arabian sailors, like Sinbad, to sail from Africa to India in the summer and return in the winter.

There seems to be a connection between strong monsoon winds over West Africa and the frequency of killer hurricanes in the Caribbean and along the Atlantic coast of North and Central America. In years when the Atlantic surface temperatures are high, a stronger than average monsoon trough forms over Africa. This pulls in moist maritime air that brings rain (and life) to the Sahel (the margin of the Sahara). This trough gives rise to tropical depressions (low-pressure storms) that follow one another in regular waves across the Atlantic. The weak trade winds associated with "wet" years allow these storms to organize and gain strength. Evaporation from the warm surface water provides energy.

During the 1970s and 1980s when the Sahel had devastating droughts, the weather was relatively quiet in North America. Only one killer hurricane (winds over 70 km/hr or 112 mph) reached the United States in two decades. By contrast, between 1947 and 1969 when rains were plentiful in the Sahel, thirteen killer hurricanes hit the United States. A climatic shift appeared to have started between

Air, Climate, and Weather

Figure 17.11

Failure of monsoon rains brings drought, starvation, and death to both livestock and people in the Sahel desert margin of Africa. Although drought is a fact of life in Africa, many governments fail to plan for it, and human suffering is much worse than it needs to be.

1988 and 1990. When ocean temperatures were up, rains returned to North Africa, and storms were stronger and more frequent in the Caribbean. What drives these climatic changes is unknown, but it may be related to long-term shifts in ocean circulation and sea surface temperatures.

Weather Modification

As author Samuel Clemens (Mark Twain) said, "Everybody talks about the weather, but nobody does anything about it." People probably always have tried to influence local weather through religious ceremonies, dancing, or sacrifices. During the drought of the 1930s in the United States, "rainmakers" fleeced desperate farmers of thousands of dollars with claims that they could bring rain.

Some recent developments appear to be effective in local weather modification, at least in some circumstances, but they are not without drawbacks and controversy. Seeding clouds with dry ice or ionized particles, such as iodine crystals, can initiate precipitation if water vapor is present and air temperatures are near the condensation point (chapter 19). Dry ice also is very effective at dispersing cold fog (where supercooled water droplets are present). Warm fog (air temperatures above freezing) and ice fog (ice crystals in the air) are not usually amenable to weather modification. Hail suppression by cloud seeding also can be effective, but dissipation of the clouds that generate hail diverts rain from areas that need it, as well. There are concerns that materials used in cloud seeding could cause air, ground, and water pollution. North Dakota recently agreed to pay for an environmental impact assessment to convince Montana farmers that North Dakota is not stealing rain by seeding clouds. Similarly, Wyoming and Idaho have been arguing over cloud seeding to increase winter snowpack in the Tetons that would provide summer irrigation water downstream in Idaho.

Climate

If weather is a description of physical conditions in the atmosphere (humidity, temperature, pressure, wind, and precipitation), then climate is the *pattern* of weather in a region over long time periods. The interactions of atmospheric systems are so complex that climatic conditions are never exactly the same at any given location from one time to the next. While it is possible to discern patterns of average conditions over a season, year, decade, or century, complex fluctuations and cycles within cycles make generalizations difficult and forecasting hazardous (fig. 17.12). We always wonder whether anomalies in local weather patterns represent normal variations, a unique abnormality, or the beginnings of a shift to a new regime.

Climatic Catastrophes

Major climatic changes, such as those of the Ice Ages, have drastic effects on biotic assemblages. When climatic change is gradual, species may have time to adapt or migrate to more suitable locations. Where climatic change is relatively abrupt, many organisms are unable to respond before conditions exceed their tolerance limits. Whole communities may be destroyed, and if the climatic change is widespread, many species may become extinct.

Perhaps the most well-studied example of this phenomenon is the great die-off that occurred about 65 million years ago at the end of the Cretaceous period. Most dinosaurs—along with 75 percent of all previously existing plant and animal species—became extinct, apparently as a result of sudden cooling of the earth's climate. Geologic evidence suggests that this catastrophe was not an isolated event. There appear to have been several great climatic changes, perhaps as many as a dozen, in which large numbers of species were exterminated.

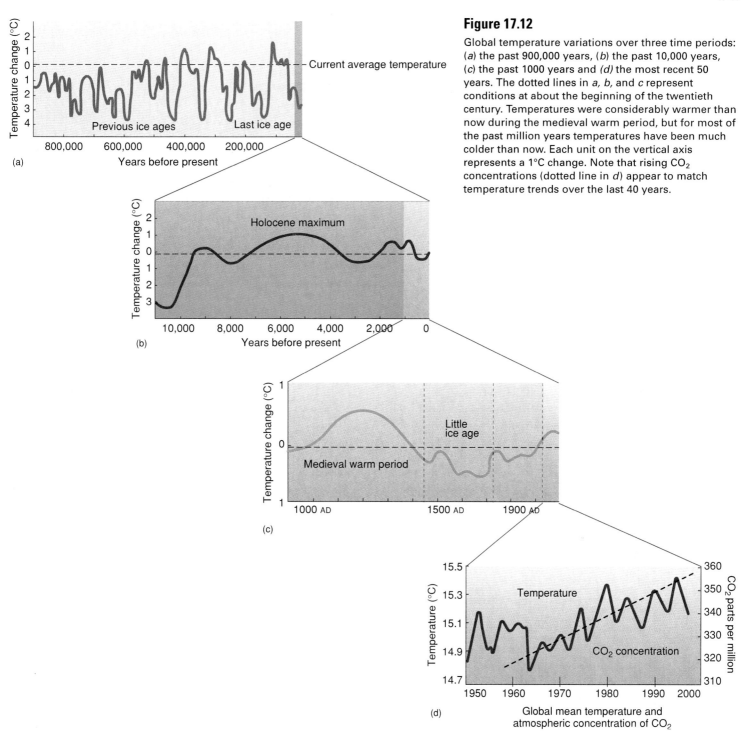

Figure 17.12

Global temperature variations over three time periods: (*a*) the past 900,000 years, (*b*) the past 10,000 years, (*c*) the past 1000 years and (*d*) the most recent 50 years. The dotted lines in *a, b,* and *c* represent conditions at about the beginning of the twentieth century. Temperatures were considerably warmer than now during the medieval warm period, but for most of the past million years temperatures have been much colder than now. Each unit on the vertical axis represents a 1°C change. Note that rising CO_2 concentrations (dotted line in *d*) appear to match temperature trends over the last 40 years.

Driving Forces and Patterns in Climatic Changes

What causes catastrophic climatic changes? As you can imagine, there are nearly as many theories as there are researchers studying this phenomenon. Some scientists believe that long-term climatic changes follow a purely random pattern brought about by chance interaction of unrelated events, such as asteroid impacts, cosmic radiation from exploding supernovas, massive volcanic eruptions, abrupt flooding of glacier meltwater into the ocean, and tectonic ocean spreading that changes patterns of ocean and wind circulation.

Other scientists discern periodic patterns in weather cycles. One explanation is that changes in solar energy associated with eleven-year sunspot cycles or twenty-two-year solar magnetic cycles might play a role. Another theory is that a regular 18.6-year cycle of shifts in the angle at which our moon orbits the earth alters

Air, Climate, and Weather

tides and atmospheric circulation in a way that affects climate. A theory that has received a great deal of attention in recent years is that orbital variations as the earth rotates around the sun might be responsible for cyclic weather changes.

Milankovitch cycles, named after Serbian scientist Milutin Milankovitch, who first described them in the 1920s, are periodic shifts in the earth's orbit and tilt (fig. 17.13). The earth's elliptical orbit stretches and shortens in a 100,000-year cycle, while the axis of rotation tilts more or less at a 40,000-year interval. Furthermore, over a 26,000-year cycle, the axis wobbles like an out-of-balance spinning top. These variations change the distribution and intensity of sunlight reaching the earth's surface and, consequently, global climate. Bands of sedimentary rock laid in the oceans (fig. 17.14) seem to match both these Milankovitch cycles and the periodic cold spells associated with worldwide expansion of glaciers every 100,000 years or so.

A historical precedent for cyclic temperature changes that had disastrous effects on humans was the "little ice age" that began in the 1300s. Temperatures dropped so that crops failed repeatedly in parts of northern Europe that once were good farmland. Scandinavian settlements in Greenland founded by Eric the Red during the warmer period around A.D. 1000 lost contact with Iceland and Europe as ice blocked shipping lanes. It became too cold to grow crops, and fish that once migrated along the coast stayed further south. The settlers slowly died out, perhaps having been attacked by Inuit people who were driven south from the high Arctic by colder weather.

Preliminary analysis from ice cores drilled in the Greenland ice cap suggest that world climate may be much less stable than previously thought. During the last major interglacial period 135,000

Figure 17.14

Milankovitch cycle periodicity? The light and dark bands in these seafloor sediments were laid down under different climatic regimes, each lasting 20,000 to 40,000 years. Now exposed along the French coast, these layers suggest the regularity of Milankovitch cycles.

to 115,000 years ago, it appears that temperatures flipped suddenly from warm to cold or vice versa over a period of years or decades rather than centuries. One possible explanation is that ocean currents acting as a conveyor belt to transport heat from the equator to the North Atlantic may have suddenly stopped or even reversed course when surges of fresh water diluted salty ocean waters. Volcanic eruptions may have played a role as well. When Mt. Pinatubo in the Philippines erupted in 1991, it ejected enough dust, aerosols, and gas into the atmosphere to cool the climate by at least 1°C. Larger volcanoes may have had greater effects.

El Niño/Southern Oscillations

What do sea surface temperatures near Indonesia, anchovy fishing off the coast of Peru, and the direction of tropical trade winds have to do with rainfall and temperatures over the midwestern United States? Strangely enough, they all may be interrelated. If the terms El Niño, La Niña, and Southern Oscillation are not yet a part of your vocabulary, perhaps they should be. They describe a connection between the ocean and atmosphere that appears to affect these factors as well as weather patterns throughout the world.

The core of this climatic system is a huge pool of warm surface water in the Pacific Ocean that sloshes slowly back and forth between Indonesia and South America like water in a giant bathtub. Most years, this mound of warm water is held in the western Pacific by steady equatorial trade winds pushing ocean surface currents westward (fig. 17.15). These surface winds are generated by a giant low-pressure cell formed by convection currents of moist air warmed by the ocean. Towering thunderheads created by rising air bring torrential summer rains to the tropical rainforests of Northern Australia and Southeast Asia. Winds high in the troposphere carry a return flow back to the eastern Pacific where dry subsiding currents create deserts from Chile to southern California. Surface waters driven westward by the trade winds are replaced by

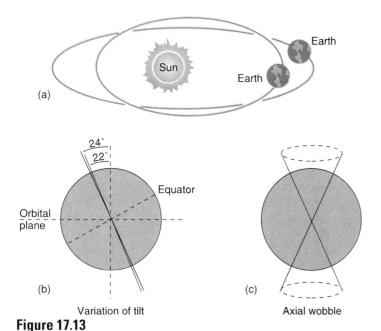

Figure 17.13

Milankovitch cycles, which may affect long-term climate conditions: (a) changes in the occupancy of the earth's orbit, (b) shifting tilt of the axis, and (c) wobble of the earth.

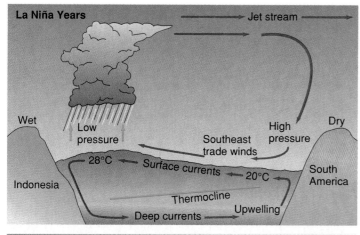

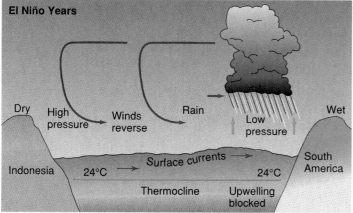

Figure 17.15

During La Niña years a large pool of warm surface water in the western Pacific near Indonesia heats the atmosphere and creates convection currents that carry water vapor aloft. Heavy rains fall in Southeast Asia, while dry subsiding air brings drought to western North and South America and drives prevailing southeasterly trade winds and ocean currents. In El Niño years, this pool of warm water and its associated tropical low shift toward the eastern Pacific, bringing wet weather to the Americas and drought to Australindonesia.

upwelling of cold, nutrient-rich, deep waters off the west coast of South America that support dense schools of anchovies and other finfish.

Every three to five years, for reasons that we don't fully understand, the Indonesian low collapses and the warm mass of surface waters surges back east across the Pacific. One theory is that the high cirrus clouds atop the cloud columns absorb enough solar radiation to cool the ocean surface and reverse trade winds and ocean surface currents so they flow eastward rather than westward. Another theory is that eastward-flowing deep currents called baroclinic waves periodically interfere with coastal upwelling, warming the sea surface off South America and eliminating the temperature gradient across the Pacific. At any rate, the shift in position of the tropical depression sets off a chain of events lasting a year or more with repercussions in weather systems across North and South America and perhaps around the world.

Peruvian fishermen were the first to notice irregular cycles of rising ocean temperatures that resulted in disappearance of the anchovy schools on which they depended. They named these events **El Niño** (Spanish for the Christ child) because they often occur around Christmas time. We have come to call the intervening years La Niña (or little girl). Together, this cycle is called the El Niño Southern Oscillation (ENSO).

How does the ENSO cycle affect us? During an El Niño year, the northern jet stream—which normally is over Canada—is drawn south over the United States. This pulls moist air from the Pacific and Gulf of Mexico inland, bringing intense storms and heavy rains from California across the midwestern states. The intervening La Niña years bring hot dry weather to these same areas. An unusually long El Niño event from 1991 to 1995 broke the seven-year drought over the western United States and resulted in floods of the century in the Mississippi Valley. Oregon, Washington, and British Columbia, on the other hand, tend to have warm, sunny weather in El Niño years rather than their usual rain. Severe droughts in Australia and Indonesia during an El Niño episode in 1982 and 1983 caused disastrous crop failures and forest fires, including one in Borneo that burned 3.3 million ha (8 million ac). There may even be a connection between monsoon patterns across Africa or South Asia and ENSO periodicity.

Are ENSO events becoming stronger or more irregular because of global climate change? It is difficult to say, but there are signs that warm ocean surface temperatures are expanding. In addition to the pool of warm water in the Western Pacific associated with La Niña years, investigators recently discovered a similar warm region in the Indian Ocean. A negative effect of warmer sea surface temperatures is larger and more violent storms. On the other hand, increased cloud cover would raise the albedo while upwelling convection currents generated by these storms could pump heat into the stratosphere. This might have an overall cooling effect and act as a safety valve for global warming.

Human-Caused Global Climate Change

Are we altering the atmosphere in ways that could lead to disastrous, worldwide climate change? Will increasing concentrations of infrared-absorbing gases released into the atmosphere by human activities trap heat and raise global temperature? Increasing temperatures in the second half of the twentieth century have closely matched greenhouse gas concentrations (fig. 17.12d). Some climatologists picture a grim future in which summer heat is unbearable, farms are turned to deserts, famines sweep the globe, melting polar ice caps raise sea levels and flood coastal regions, and thousands or even millions of species die that can't migrate or adapt to sudden climatic changes.

Greenhouse Gases

About half of this predicted warming would be due to carbon dioxide (CO_2) released by burning fossil fuels, making cement, and cutting and burning forests (fig. 17.16). Together, these activities release about 8.5 billion metric tons of CO_2 annually, causing atmospheric levels to rise about 0.4 percent each year. Two hundred years ago, at the be-

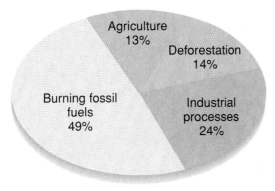

Figure 17.16

Contributions to global warming by different types of human activities in 1990.

Source: Data from World Resources Institute.

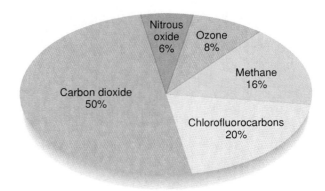

Figure 17.17

Relative contribution to global warming (percent of expected climate change) by anthropogenic (human-caused) releases of gases into the atmosphere. Notice that while far less methane and chlorofluorocarbons are released than carbon dioxide, they still are very powerful "greenhouse" gases.

Source: World Resources Institute and the United Nations Environment Programme.

ginning of the Industrial Revolution, the atmosphere contained about 280 parts per million (ppm) CO_2; now it contains about 355 ppm.

If current trends continue, preindustrial CO_2 concentrations will have doubled by the year 2075. Computer models predict that doubling atmospheric CO_2 could cause global temperatures to rise 1.5° to 4.5°C (3° to 9°F). This may not sound like much change, but the difference between the temperature now and the last ice age about 10,000 years ago when glaciers covered much of North America was only about 5°C.

Carbon dioxide is not the only gas that could cause climate warming. Methane, chlorofluorocarbons (CFCs), nitrous oxide, and other trace gases also absorb infrared radiation and warm the atmosphere (fig. 17.17). Although rarer than CO_2, some of these gases trap heat much more effectively. Methane, for instance, absorbs twenty to thirty times as much—molecule for molecule—as CO_2, and CFCs absorb approximately 20,000 times as much.

Methane is produced by intestinal bacteria in ruminant animals, anaerobic decomposition in wet-rice paddies, pipeline leaks, decaying wastes in landfills, and releases from coal mining. Atmospheric methane is increasing about 1 percent per year. Chlorofluorocarbons, used as spray propellants, degreasing agents, and refrigerants, have been accumulating at 5 percent each year. Nitrous oxide (N_2O), produced by burning organic material and soil denitrification, has been increasing at 0.2 percent annually.

Together, these minor greenhouse gases would have warming effects comparable to doubling CO_2 concentrations. That means that a 1.5° to 4.5°C temperature rise could occur in forty rather than eighty years if the models are accurate. The biggest share of greenhouse gases is produced by the developed countries of the world (fig. 17.18). Together, the United States, the former Soviet Union, Europe, and Japan are responsible for about two-thirds of all potential global warming.

Effects of Global Warming

If greenhouse warming occurs, it probably will not be distributed evenly around the globe. Additional ocean evaporation and cloud cover will likely keep tropical coastal areas about as they are now.

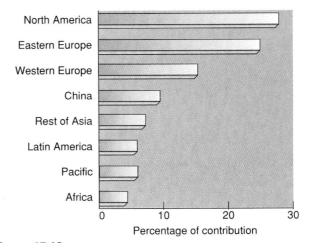

Figure 17.18

Contributions to world CO_2 production. North America produces nearly one-third of global CO_2 emissions.

Source: Data from *European Environmental Statistics Handbook 1993.*

The greatest temperature changes are predicted to be at high latitudes and in the middle of continents (fig. 17.19). Siberia and the Canadian arctic might experience increases of 10° to 12°C (18° to 22°F). Chicago might go from an average of 15 to 48 days each year above 32°C (90°F). Dallas may have 162 days per year that hot, rather than 100 as it does now. Calcutta, however, where the temperature is always hot, will not get much hotter.

Changing precipitation patterns might be one of the most serious consequences of the greenhouse effect. Some models predict that the midcontinents of North America, South America, and Asia will be significantly drier than they are now. This could have calamitous effects on world food supplies. The models don't agree

Temperature

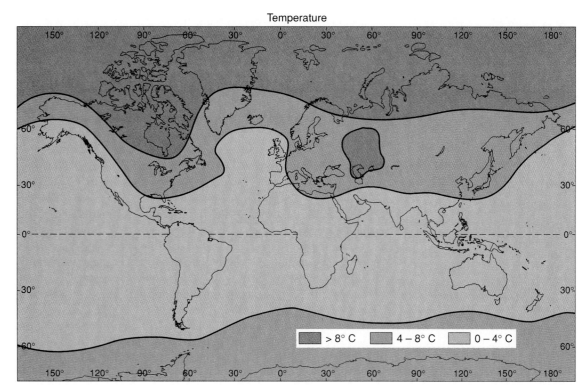

Figure 17.19

Predicted effects of global climate change: (*top*) change in temperatures in January, and (*bottom*) change in soil moisture in July. The most extreme cooling is expected to occur at high latitudes in the winter. Mid-continent regions will probably be drier—possibly causing severe effects on agriculture.

> 8° C   4 – 8° C   0 – 4° C

Moisture

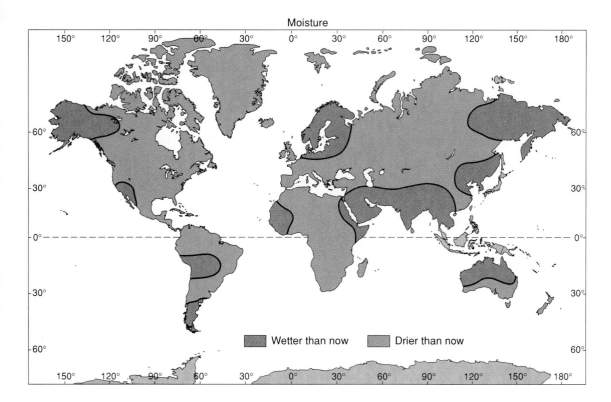

Wetter than now   Drier than now

on this point, however. While some show the American corn belt to be drier, others predict that it will be wetter than now. Model building is not yet an exact science.

Does anyone win in these scenarios? Residents of northern Canada, Siberia, and Scandinavia might benefit from warmer tem-peratures and longer growing seasons. More moisture in western Africa and South Asia also could bring higher crop production. But soils in some areas will not support farming, no matter how agree-able the climate. Ecological disruptions could be severe. Many wild plants and animals would be forced out of current ranges (fig. 17.20).

Air, Climate, and Weather

Figure 17.20

Change in suitable range for sugar maple, according to the climate-change model.

Source: Margaret Davis and Catherine Zabriskie, in *Global Warming and Biological Diversity*, ed. by Peters and Lovejoy, Yale University Press, 1991.

Given current barriers to migration, natural communities may never be reestablished.

Rising sea levels are among the most ominous potential results of global warming. Some oceanographers calculate that thermal expansion alone could raise the sea level by one meter or more in the next century. And if polar ice caps melt, oceans could rise by 30 m (90 ft). Most of the world's major cities are on coasts only a few meters above sea level. The homes and businesses of about half the world's population would be threatened if the ice caps melt. A large amount of valuable farmland would also be lost. The U.S. Gulf Coast, for instance, would lose about 5000 sq km, an area the size of Delaware. A few low-lying countries like the Maldive Islands in the Indian Ocean might disappear entirely.

A related concern is that global warming could cause an increase in violent storms such as Hurricane Andrew. The area of ocean with warm enough surface water (above 27°C or 81°F) to generate hurricanes has expanded about 20 percent in the past twenty years. If this trend continues, climatic anomalies such as the 500-year flood in the Mississippi River basin in 1993 or the severe winter storms that struck much of the United States in 1994 could become more frequent. Insurance companies worry that $2 trillion in insured property along U.S. coasts could be at risk from a combination of rising sea levels and increasingly catastrophic storms.

Many scientists are convinced that global climate change has already begun. They point to the fact that four of the five warmest years on record were in the late 1980s and early 1990s. As the aerosols released by Mt. Pinatubo are washed out of the atmosphere, they predict, high temperatures will return with a vengeance. Others remain skeptical, arguing that recent high temperatures could be just a random anomaly.

Unfortunately, the data are equivocal. Some studies show increasing temperatures; others show that temperatures are stable or even declining slightly. Often dissimilar data bases and analytical methods are used in climate studies. Other difficulties include inaccurate thermometers, urbanization (sampling stations that once were in the country now are in cities that are known to trap heat), and unrepresentative sampling locations (most official weather stations in the United States are at airports, which are significantly warmer than surrounding areas).

Furthermore, current climate models don't adequately represent effects of water vapor, clouds, air currents, ocean circulation, sulfate aerosols, decreasing solar radiation, biogeochemical ocean processes, or the possible growth stimulation of green plants and ocean plankton by higher CO_2. If the models' assumptions are changed slightly, the predictions change from positive (warming) to negative (cooling) trends. If some of these factors are built into the models, they would show another ice age rather than a greenhouse effect.

Some scientists claim that rising CO_2 levels could be beneficial because higher CO_2 stimulates growth of many plants and can result in more efficient water utilization. Areas now too dry for agriculture might come into production, and crop yields might increase rather than decrease. There seems to be some evidence of increased CO_2 uptake. In the early 1990s, atmospheric increases were about 2 billion tons per year less than emissions. Some scientists believe that the "missing" CO_2 is being taken up by increased photosynthesis in boreal forests. Others claim that growth of phytoplankton (single-celled photosynthetic organisms) in the ocean might be absorbing most of the additional CO_2. If true, this could help balance world temperatures and also make more seafood available for human consumption.

Cutting Emissions or Increasing Absorption

What should we do about greenhouse gases in the face of such uncertainty? Should we take steps now to reduce them or wait to see if their effects are as negative as some models suggest? Unfortunately, by the time we have evidence that a disaster is underway, it may be too late to make changes. On the other hand, it would be wasteful to spend hundreds of billions of dollars and disrupt people's lives to reduce these gases if the threats are only minor (What Do *You* Think?, next page).

There are, however, actions to reduce greenhouse effects that would be useful even if climate change is less damaging than we think. Climatologist Steven Schneider calls this a "no regrets" policy. For instance, the industrialized countries have agreed to phase out chlorofluorocarbons because they damage stratospheric ozone (chapter 18) as well as absorb heat. We could cut CO_2 emissions and save money by increasing energy efficiency in our homes, automobiles, factories, and offices (see chapter 22). Every gallon of gas you burn in your car produces 19 lbs of CO_2. Increasing your automobile mileage from 25 to 50 mpg would save about $3000 and reduce CO_2 production by 20 metric tons over the life of the vehicle.

Canada and some European countries have committed themselves to a 20- or 25-percent decrease and believe they can do so without undue sacrifice. In the United States, however, proposals to reduce speed levels, raise CAFE (corporate average fleet efficiency) for new cars and trucks to 45 mpg, or pass a carbon tax to discourage use of fossil fuels and encourage renewable energy sources have met with stiff opposition from corporations and conservatives. Even simple measures such as planting trees, investing

What Do *You* Think?
Reasoned Judgments and Scientific Uncertainty

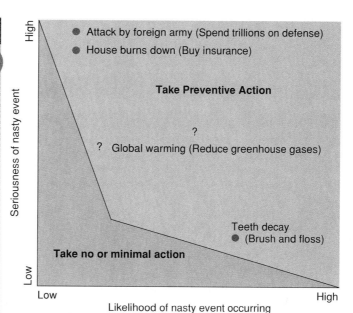

Figure 1

Society regularly takes preventive action to avoid or mitigate consequences of nasty events that may or may not happen.

Critical thinking is characterized by its skepticism and its reliance on factual information before drawing conclusions. You read in this chapter about the potential consequences if significant global warming were to occur. You have also read that data from computer models and other sources are equivocal, creating a degree of uncertainty about the timing and precise nature of the warming event.

Some people use this lack of absolute certainty to advocate that we not undertake preventive changes until the uncertainty has been removed. They claim that to do so without absolute proof would be unscientific. They say that reducing the production of greenhouse gases would impose significant economic costs, their argument being that such actions are unwise because they could turn out to have been unnecessary. This position fails to acknowledge that some degree of uncertainty surrounds the vast majority of decisions and judgments society must make, including issues in science. By casting decision making as purely a process of separating fact from opinion, this position omits an important element of critical thinking—reasoned judgment.

The threat of significant, rapid, global climate change is but one of many issues confronting us in which factual evidence does not establish absolute certainty. Will your home be destroyed by fire? Will you develop lung cancer if you smoke cigarettes? Like global warming, available facts are insufficient to establish that any of these events will be visited upon someone with certainty. However, that they are possible, or even probable, is not in doubt.

The generalized dilemma could perhaps be framed like this: In any of the events above, should we do nothing until certainty is established? The downside risk, of course, is that disaster may be unavoidable if we wait. An insurance company is unlikely to underwrite fire coverage on your house when the fire trucks are on their way. Any serious effort to curb release of greenhouse gases would require time to implement. There would also be a lag time before such efforts could influence global climate. Measures undertaken only when scientific certainty was finally established may be too late to be of much help.

Or, do we take preventive measures before an event is certain to occur, running the risk that our action may not have been needed? It certainly is true that your house may never burn down. And if it does not, the money spent for insurance premiums appears wasted.

In instances like these, critical thinking must come to a reasoned judgment. To do otherwise is to paralyze actions, preventing timely intervention. Society has long used reasoned judgment to address issues of uncertainty similar to global warming. Arriving at a reasoned judgment for most people involves a thought process that is very applicable to global warming. Two questions are evaluated:

1 *What is the likelihood that the nasty event actually will occur?*

2 *How serious would it be if the nasty event occurs?*

The more likely a nasty event is to happen, or the more serious the effect of the event, the greater the advisability of taking preventive action. This concept is modeled in figure 1 with some examples. Where on the graph would you plot global warming?

Thomas Lovejoy expresses the reasoned judgment of many ecologists when he argues against doing "total planet experiments that bet the biosphere, if there is even a small chance we may regret the result. After all, there is not even an experimental control planet to colonize if we lose at biosphere roulette."

The stakes seem high? Does reason require us to act or wait?

Air, Climate, and Weather

*I*n Depth: Climate Change and Justice

Imagine that you are the mediator at an international conference on controlling greenhouse gases. All the world's governments have finally agreed that CO_2 emissions must be cut. The problem is how to proceed. You are responsible for negotiating a plan acceptable to all parties, but many thorny issues of fairness and justice are raised when considering energy use and possible global climate change.

Western Europe and Canada have offered to reduce CO_2 emissions by the year 2005, but they want guarantees that others will do so, too. The United States and the oil-producing countries have, after much foot dragging, conceded that reductions are necessary. However, bowing to industry pressures, they refuse to endorse firm schedules or limits. These wealthy countries have the capital and technology to improve fuel efficiency without greatly reducing standards of living but they aren't willing to do so without some payback.

Developing nations, on the other hand, protest any planned limits. The current situation is not their fault, they argue; it was caused by the industrialized countries that have squandered fossil fuels for 200 years. Poorer nations now produce only small amounts of greenhouse gases. Having to limit energy use, agriculture, or industry would condemn them to low standards of liv-

ing and thwart their plans for a more comfortable future. Let the developed countries take responsibility for global warming, they say. They can take a turn at being poor, and let us have a chance to live an energy-consumptive, material-rich lifestyle.

What would you say to these disputing groups? Whose responsibility is it to solve this problem that affects us all? How do we represent the interests of future generations in this complex equation?

This exact debate took place in 1995 in Berlin at a United Nations conference on global climate change. Building on conventions agreed to at the 1992 Earth Summit in Rio

de Janeiro, the Berlin meeting produced the outlines of a workable solution. The wealthier countries agreed to stricter timetables and limits in exchange for emission credits. Now, when the United States helps China or India purchase pollution control equipment or upgrade a power plant, the U.S. will receive credits against its own emission limits. Developing countries are still responsible for limiting emissions, while wealthier countries—whose emissions are substantially greater—can continue to produce significantly more CO_2 than their poorer neighbors.

ETHICAL CONSIDERATIONS

Is this a fair resolution? Is it right for those in rich countries to continue to lead energy-consumptive lifestyles while profiting from the sale of pollution control technology to developing countries? Can we ask poorer people to lower their expectations and forego the luxuries that we already enjoy? And is this enough to ensure a habitable climate for our offspring?

Some people argue that it would be better to redistribute resources so that people in developing countries can enjoy a better life. Others maintain that the environment can't withstand billions of additional people trying to live at a high standard of living. Furthermore, they claim, if industrialized countries cut back on science and technology, there may not be innovations in conservation and renewable energy for the future.

What would you suggest for the next round of negotiations? Who should conserve and who should pay? And what is a reasonable set of options for us to leave to future generations?

in energy-efficient appliances, encouraging mass transit, and recovering methane from landfills as a means of reducing greenhouse gases have not been very successful.

Achieving a 50-percent emission reduction (enough to reverse current trends) will be much more difficult. It might require significant changes in our lifestyles to reach this level. Developing countries object to limits and regulations that could restrict their ability

to raise standards of living (In Depth, above). The nuclear power industry points out that their reactors are an alternative to fossil fuels and produce no CO_2. Some people suggest spreading fertilizer in the ocean to stimulate phytoplankton growth. We don't know what ecological complications might arise from such actions. Perhaps it is wise to wait until we know more before adopting some of these risky solutions.

ummary

The atmosphere and living organisms have evolved together so that the present chemical composition of the air is both suitable for, and largely the result of, biological processes. Compression concentrates most gas molecules in a thin layer (the troposphere) near the earth's surface. Upper layers of the atmosphere, while too dilute for life, play an important role in protecting the earth's surface by intercepting dangerous ultraviolet radiation from the sun. The atmosphere is relatively transparent to visible light that warms the earth's surface and is captured by photosynthetic organisms and stored as potential energy in organic chemicals.

Heat is lost from the earth's surface as infrared radiation, but fortunately for us, carbon dioxide and water vapor that are naturally present in the air capture the radiation and keep the atmosphere warmer than it would otherwise be. When air is warmed by conduction or radiation of heat from the earth's surface, it expands and rises, creating convection currents. These vertical updrafts carry water vapor aloft and initiate circulation patterns that redistribute energy and water from areas of surplus to areas of deficit. Pressure gradients created by this circulation drive great air masses around the globe and generate winds that determine both immediate weather and long-term climate.

Earth's rotation causes wind deflection called the Coriolis effect, which makes air masses circulate in spiraling patterns. Strong cyclonic convection currents fueled by temperature and pressure gradients and latent energy in water vapor can create devastating storms. Another source of storms are the seasonal winds, or monsoons, generated by temperature differences between the ocean and a landmass. Monsoons often bring torrential rains and disastrous floods, but they also bring needed moisture to farmlands that feed a majority of the world's population. When the rains fail, as they do in drought cycles, ecosystem disruption and human suffering can be severe.

Many procedures claiming to control the weather are ineffectual, but some human actions—both deliberate and inadvertent—may change local weather and long-term climate. Cloud seeding can induce rain or disperse fog under the right atmospheric conditions. Improving the local situation, however, often makes things worse somewhere else. It is not yet clear what the future of our weather and climate will be. Some scientists warn that the gaseous pollutants we release into the atmosphere may trap radiant energy and cause a global warming trend that could drastically disrupt human activities and natural ecosystems. Understanding and protecting this complex, vital aspect of our world is clearly essential.

▼ Questions for Review

1. What are the main constituents of air? What are their sources?
2. Name and describe the four layers of air in the atmosphere.
3. What is the greenhouse effect?
4. Describe the atmospheric heating and cooling cycle.
5. What are the jet streams? How do they influence weather patterns?
6. What are Hadley cells and how do they form?
7. Describe the Coriolis effect. What causes it?
8. What is the difference between weather and climate?
9. What are some theories explaining major climatic changes?
10. Would changes in the earth's orbit affect climate? How?

▼ Questions for Critical Thinking

1. What questions would you ask a climate modeler about the assumptions, compromises, and limitations of mathematical models?
2. Should humans try to control the weather? What would be the positive effects? What would be the dangers?
3. Can we avoid great climatic changes, such as another ice age or a greenhouse effect? Will we have the gumption to change our ways?
4. Has there been a major change recently in the weather where you live? Can you propose any reasons for such changes?
5. What forces determine the climate in your locality? Are they the same for neighboring states?
6. What was the weather like when your parents were young? Do you believe the stories they tell you? Why or why not?
7. Have you ever experienced a tornado or hurricane? What was it like? What omens or warnings told you it was coming?
8. From which direction does bad weather come in your region? Why?
9. What would you do to adapt to a permanent drought? What effects would it have on your life?
10. What should we do about coastal cities threatened by rising oceans—rebuild, enclose in dikes, or just move?
11. Would you favor building nuclear power plants to reduce CO_2 emissions?

▼ Key Terms

aerosols 365
albedo 366
climate 364
cold front 370
convection currents 367
Coriolis effect 368
El Niño 375
Hadley cells 368
ionosphere 366
jet streams 369

mesosphere 366
Milankovitch cycles 374
monsoon 371
stratosphere 365
thermosphere 366
tornadoes 370
troposphere 365
warm front 370
weather 364

▼ Additional Information on the Internet

The Daily Planet
http://www.atmos.uiuc.edu/

EcoNet's Climate Resources Directory
http://www.igc.apc.org/climate/

Environmental Technologies Program
http://www-ep.es.llnl.gov/www-ep/aet.html/

Global Warming
http://www.nbn.com/youcan/warm/warm.html/

Global Warming Update
http://www.ncdc.noaa.gov/gblwrmupd/global.html/

National Center for Atmospheric Research
http://http.ucar.edu/metapage.html/

National Climatic Data Center
http://www.ncdc.noaa.gov/ncdc.html/

National Oceanic and Atmospheric Administration
http://www.noaa.gov/

United Nations Convention on Climate Change
http://www.unep.ch/iucc.html/

US Global Change Research Program
http://www.usgcrp.gov/

World Wide Web Virtual Library: Environment-Atmosphere
http: //ecosys.drdr.Virginia.EDU:80/atm.html/

Note: Further readings appropriate to this chapter are listed on p. 603.

Land-Use Planning

𝓛isa **Vernegaard**

Lisa Vernegaard is a staff ecologist and planner at The Trustees of Reservations, a private, nonprofit land conversation organization in the Boston area. As with many people who find careers in environmental fields, Lisa didn't find an ideal situation immediately after college. Instead, she held a series of temporary, seasonal jobs, each of which gave her experience and training that ultimately have led to very satisfying work.

Lisa's first exposure to ecology and natural resources was a high school summer job with the Youth Conservation Corps that taught her to maintain trails, plant trees, and build cattle guards on a national forest in her native Utah. In college, Lisa majored in biology and worked in summer camps and parks as a naturalist. After college, she held several more short-term assignments while looking for a full-time appointment. Although the pay often was poor and she sometimes felt like a vagabond, each of these situations contributed in different ways toward building her career.

In retrospect, Lisa feels that working in environmental education was one of the best introductions she could have had to her present profession. Expanding her knowledge of plants and animals has helped greatly, but even more important was learning to work with people, including professional colleagues and the public. Every conservation problem results ultimately from something people have done or are now doing. Dealing with human behavior, Lisa says, is often the most important component of many environmental projects.

Be flexible in the kinds of jobs you consider, Lisa suggests. While you should not settle for just anything you are offered, make plans about the directions you want to go and the kinds of work you would be happy doing, and then be prepared to consider opportunities that take you along a general path even if they aren't exactly what you expected. Sometimes being prepared to move in unexpected directions can bring great benefits.

In her present assignment, Lisa works with a small group of professionals managing a system of seventy-six properties encompassing some 8,000 hectares (20,000 acres) in Massachusetts. As staff ecologist, she appraises ecological values, scenic qualities, environmental condition, and landscape potential of areas being considered for acquisition as well as property already owned or held in conservation restrictions by her organization.

When asked about the best aspects of her current job, Lisa says that being out in the field is what she enjoys most. A sense of satisfaction from having helped preserve some beautiful and unique places is also rewarding. Another important consideration is the opportunity to learn new things every day. During the summer, at least half of her time is spent in the field. The rest of her time (up to 75 percent in the winter) is devoted to analyzing field notes and integrating what she has learned into action plans for each property.

What are the prospects for jobs in land conservation and applied ecology? As the network of protected areas continues to grow, more people will be needed to manage them, Lisa predicts. She suggests that an excellent preparation for this profession is a combination of computer skills and ecology. There is a great need for people who can create ecosystem models or do geographical information system (GIS) landscape analysis. As is true in many environmental careers, technical training is important in conservation and land management, but communication and interpersonal skills are what help you advance in your profession. Most of all, Lisa suggests, you need perseverance, imagination, and perhaps willingness to accept a series of varied seasonal jobs to create a career for yourself as a field ecologist and land-use planner.

Steam billows from cooling towers of a nuclear plant on the Ohio River in Indiana. Although steam is not considered an air pollutant, other airborne materials from power plants represent significant health risks to nearby residents.

CHAPTER 18: Air Pollution

Objectives

After studying this chapter, you should be able to:

- describe the major categories and sources of air pollution.

- distinguish between conventional or "criteria" pollutants and unconventional types as well as explain why each is important.

- analyze the origins and dangers of some indoor air pollutants.

- relate why atmospheric temperature inversions occur and how they affect air quality.

- evaluate the dangers of stratospheric ozone depletion and radon in indoor air.

- understand how air pollution damages human health, vegetation, and building materials.

- compare different approaches to air pollution control and report on clean air legislation.

- judge how air quality around the world has improved or degraded in recent years and suggest what we might do about problem areas.

The Air Around Us

How does the air taste, feel, smell, and look in your home or your neighborhood? Chances are that wherever you live, the air is contaminated to some degree. Smoke, haze, dust, odors, corrosive gases, noise, and toxic compounds are present nearly every-where, even in the most remote, pristine wilderness. Air pollution is generally the most widespread and obvious kind of environmental damage. According to the Environmental Protection Agency (EPA), some 147 million metric tons of air pollution (not counting carbon dioxide or wind-blown soil) are released into the atmosphere each year in the United States by human activities (fig. 18.1). Total worldwide emissions of these pollutants are around 2 billion metric tons per year. The air in a typical industrial city can contain unhealthy concentrations of hundreds of different toxic substances; indoor air can be even worse.

Over the past twenty years, air quality has improved appreciably in most cities in Western Europe, North America, and Japan.

Many young people might be surprised to learn that 20 or 30 years ago most American cities were much dirtier than they are today. This is an encouraging example of improvement in environmental conditions. Our success in controlling some of the most serious air pollutants gives us hope for similar progress in other environmental problems. While developed countries have been making progress, however, air quality in the developing world has been getting much worse. Especially in the burgeoning megacities of rapidly industrializing coun-tries (chapter 24), air pollution often exceeds World Health Organization standards all the time. In Lahore, Pakistan, and Xian, China, for instance, airborne dust, smoke, and dirt often are ten times higher than levels considered safe for human health. For the four-fifths of the world who live in developing countries, air pollution often is not merely ugly or unpleasant but can be life-threatening, especially to sensitive members of the population.

Figure 18.1

Los Angeles on a smoggy day (*left*) is very different from days when the air is clear (*right*). Although much remains to be done, air quality in most American cities is significantly clearer now than when the Clean Air Act was passed in 1972.

In this chapter, we will examine the major types and sources of air pollution. We will study how they enter and move through the atmosphere and how they are changed into new forms, concentrated or dispersed, and removed from the air by physical and chemical processes. We also will look at some of the major *effects of air pollution on human health, ecosystems, and materials. Finally, we will survey some of the control methods available to reduce air pollution or mitigate its effects, and the results of air pollution control efforts on ambient air quality in the United States and elsewhere.* ▶

Natural Sources of Air Pollution

It is difficult to give a simple, comprehensive definition of pollution. The word comes from the Latin *pollutus,* which means made foul, unclean, or dirty. Some authors limit the use of the term to damaging materials that are released into the environment by human activities. There are, however, many natural sources of air quality degradation. Volcanoes spew out ash, acid mists, hydrogen sulfide, and other toxic gases. Sea spray and decaying vegetation are major sources of reactive sulfur compounds in the air. Forest fires create clouds of smoke that blanket whole continents. Trees and bushes emit millions of tons of volatile organic compounds (terpenes and isoprenes), creating, for example, the blue haze that gave the Blue Ridge Mountains their name. Pollen, spores, viruses, bacteria, and other small bits of organic material in the air cause widespread suffering from allergies and airborne infections. Storms in arid regions raise dust clouds that transport millions of tons of soil and can be detected half a world away. Bacterial metabolism of decaying vegetation in swamps and of cellulose in the guts of termites and ruminant animals is responsible for as much as two-thirds of the methane (natural gas) in the air.

In many cases, the chemical compositions of pollutants from natural and human-related sources are identical, and their effects are inseparable. Sometimes, however, materials in the atmosphere are considered innocuous at naturally occurring levels, but when humans add to these levels, overloading of natural cycles or disruption of essential processes can occur. While the natural sources of suspended particulate material in the air outweigh human sources at least tenfold worldwide, in many cities more than 90 percent of the airborne particulate matter is anthropogenic (human-caused).

Human-Caused Air Pollution

What are the major types of air pollutants and where do they come from? In this section, we will define some general categories and sources of air pollution and survey the characteristics and emission levels of the seven conventional pollutants regulated by the Clean Air Act.

Primary and Secondary Pollutants

Primary pollutants are those released directly into the air in a harmful form. **Secondary pollutants,** by contrast, are modified to a hazardous form after they enter the air or are formed by chemical reactions as components of the air mix and interact. Solar radiation often provides the energy for these reactions. Photochemical oxidants and atmospheric acids formed by these mechanisms are probably the most important secondary pollutants in terms of human health and ecosystem damage. We will discuss several important examples of such pollutants in this chapter.

Fugitive emissions are those that do not go through a smokestack. By far the most massive example of this category is dust from soil erosion, strip mining, rock crushing, and building construction (and destruction). In the United States, natural and anthropogenic sources of fugitive dust add up to some 100 million metric tons per year. The amount of CO_2 released by burning fossil fuels and biomass is nearly equal in mass to fugitive dust. Fugitive industrial emissions are also an important source of air pollution. Leaks around valves and pipe joints contribute as much as 90 percent of the hydrocarbons and volatile organic chemicals emitted from oil refineries and chemical plants.

Conventional or "Criteria" Pollutants

The Clean Air Act of 1970 designated seven major pollutants (sulfur dioxide, carbon monoxide, particulates, hydrocarbons, nitrogen oxides, photochemical oxidants, and lead) for which maximum **ambient air** (air all around us) levels are mandated. These seven **conventional** or **criteria pollutants** contribute the largest volume of air-quality degradation and also are considered the most serious threat of all air pollutants to human health and welfare. Figure 18.2 shows the major sources of the first five criteria pollutants. Table 18.1 shows an estimate of the total annual worldwide emissions of some important air pollutants. Now let's look more closely at the sources and characteristics of each of these major pollutants.

Sulfur Compounds

Natural sources of sulfur in the atmosphere include evaporation of sea spray, erosion of sulfate-containing dust from arid soils, fumes from volcanoes and fumaroles, and biogenic emissions of hydrogen sulfide (H_2S) and organic sulfur-containing compounds, such as dimethylsulfide, methyl mercaptan, carbon disulfide, and carbonyl sulfide. Total yearly emissions of sulfur from all sources amounts to some 182 million metric tons (fig. 18.3). Although anthropogenic sources represent only one-fourth of the total sulfur flux worldwide, in most urban areas they contribute as much as 90 percent of the sulfur in the air. The predominant form of anthropogenic sulfur is sulfur dioxide (SO_2) from combustion of sulfur-containing fuel (coal and oil), purification of sour (sulfur-containing) natural gas or oil, and industrial processes, such as smelting of sulfide ores. China and the United States are the largest sources of anthropogenic sulfur, primarily from coal burning.

Sulfur dioxide is a colorless corrosive gas that is directly damaging to both plants and animals. Once in the atmosphere, it can be further oxidized to sulfur trioxide (SO_3), which reacts with water vapor or dissolves in water droplets to form sulfuric acid (H_2SO_4). Very small solid particles or liquid droplets can transport

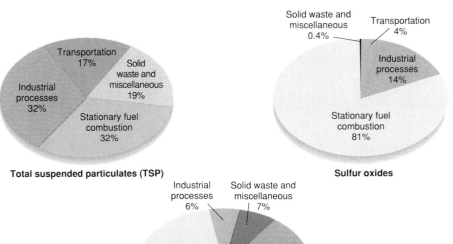

Figure 18.2

Sources of five major pollutants in the United States. Yearly emissions are 7.5 megatons TSP, 21.5 megatons $SO_2 + SO_4$, 79 megatons CO, 20.3 megatons NO_x, and 18.6 megatons VOC from anthropogenic sources. Total sum is not 100 percent due to rounding of figures.

Table 18.1

Estimated fluxes of pollutants and trace gases to the atmosphere

Species	Major Sources	Approximate Annual Flux (millions of tons)	Average Half-Life in Atmosphere (days)
CO_2 (Carbon dioxide)	Respiration	100,000	
CO_2 (Carbon dioxide)	Biomass, fossil fuel burning	10,000	2500
CO (carbon monoxide)	Biomass, fossil fuel burning	1000	2500
CH_4 (Methane)	Wetlands, rice paddies, termites, ruminant animals	400	75
VOC[a]	Human-made	100	3600
VOC	Isoprene, terpenes from plants	800	1–1000
NO_x (Nitrogen oxides)	Soils, burning biomass, fossil fuel	100	<1
N_2O (Nitrous oxide)	Fertilizer, tropical forests	10	4
NH_3 (Ammonia)	Industrial and biological nitrogen-fixation	100	60,000
SO_2 and SO_4^{-2} (Sulfur dioxide and sulfate)	Sea spray, fossil fuels, smelting	90	9
H_2S and organic sulfur[b] (Hydrogen sulfide)	Biogenic, anthropogenic	90	1–4
Metals[c]	Leaded gasoline, coal, industrial waste	3	1–900
SPM[d]	Wind erosion, fires, volcanoes, human sources	10,000	1–30
			1–1000

Reprinted with permission from H. A. Mooney, et al., *Science,* 238:926, 1987. Copyright 1987 American Association for the Advancement of Science.

[a]Volatile organic compounds (VOC) (other than methane), include benzene, formaldehyde, vinyl chloride, phenol, chloroform, trichloroethylene, gasoline ingredients, and chlorofluorocarbons.

[b]Organic sulfur includes methyl mercaptan, carbon disulfide, carbonyl sulfide, and dimethyl sulfide, among others.

[c]Metals include lead, cadmium, nickel, beryllium, mercury, and arsenic.

[d]Suspended particulate material (SPM) includes soot, dust, ash, pollen, bacterial and fungal spores, and algae.

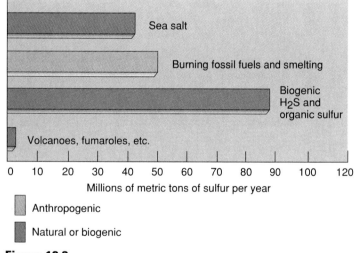

Figure 18.3

Sulfur fluxes into the atmosphere.

the acidic sulfate ion (SO_4^{-2}) long distances through the air or deep into the lungs where it is very damaging. Sulfur dioxide and sulfate ions are probably second only to smoking as causes of air pollution-related health damage. Sulfate particles and droplets reduce visibility in the United States as much as 80 percent.

Nitrogen Compounds

Nitrogen oxides are highly reactive gases formed when nitrogen in fuel or combustion air is heated to temperatures above 650°C (1200°F) in the presence of oxygen, or when bacteria in soil or water oxidize nitrogen-containing compounds. The initial product, nitric oxide (NO), oxidizes further in the atmosphere to nitrogen dioxide (NO_2), a reddish brown gas that gives photochemical smog its distinctive color. Because of their interconvertibility, the general term NO_x is used to describe these gases. Nitrogen oxides combine with water to make nitric acid (HNO_3), which is also a major component of atmospheric acidification.

The total annual emissions of reactive nitrogen compounds into the air are about 210 million metric tons worldwide (table 18.1). Anthropogenic sources account for 45 percent of these emissions

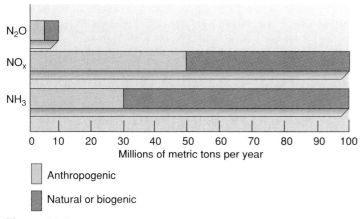

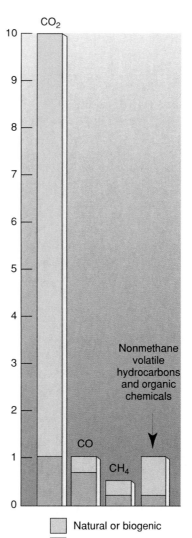

Figure 18.4

Fluxes of reactive nitrogen gases into the atmosphere.

Figure 18.5

Carbon fluxes into the atmosphere. Note that human activities contribute only 10% of CO_2, but this 10% exceeds the natural balance between photosynthesis and respiration and significantly increases atmospheric CO_2 concentrations.

(fig. 18.4). About 95 percent of all human-caused NO_x in the United States is produced by fuel combustion in transportation and electric power generation. Ammonia from fertilizer and decaying organic material is oxidized to NO_x and is an important source of nitrogen loading in rural areas. Nitrous oxide (N_2O) is an intermediate in soil denitrification that absorbs ultraviolet light and plays an important role in climate modification (chapter 17).

Carbon Oxides

The predominant form of carbon in the air is carbon dioxide (CO_2). It is usually considered nontoxic and innocuous, but increasing atmospheric levels (about 0.4 percent per year) due to human activities appear to be causing a global climate warming that may have disastrous effects on both human and natural communities. As table 18.1 and figure 18.5 show, 90 percent of the CO_2 emitted each year is from respiration (oxidation of organic compounds by plant and animal cells). These releases are usually balanced by an equal uptake by photosynthesis in green plants. Burning of fossil fuels and biomass each contribute about 5 billion metric tons per year to the air.

Carbon monoxide (CO) is a colorless, odorless, nonirritating but highly toxic gas produced by incomplete combustion of fuel (coal, oil, charcoal, or gas), incineration of biomass or solid waste, or partially anaerobic decomposition of organic material. CO inhibits respiration in animals by binding irreversibly to hemoglobin. About 1 billion metric tons of CO are released to the atmosphere each year, half of that from human activities. In the United States, two-thirds of the CO emissions are created by internal combustion engines in transportation. Land-clearing fires and cooking fires also are major sources. About 90 percent of the CO in the air is consumed in photochemical reactions that produce ozone.

Metals and Halogens

Many toxic metals are mined and used in manufacturing processes or occur as trace elements in fuels, especially coal. These metals are released to the air in the form of metal fumes or suspended particulates by fuel combustion, ore smelting, and disposal of wastes. Worldwide lead emissions amount to about 2 million metric tons

per year, or two-thirds of all metallic air pollution. Most of this lead is from leaded gasoline. Lead is a metabolic poison and a neurotoxin that binds to essential enzymes and cellular components and inactivates them. An estimated 20 percent of all inner-city children suffer some degree of mental retardation from high environmental lead levels.

Mercury is another dangerous neurotoxin that is widespread in the environment. The two largest sources of atmospheric mercury appear to be coal-burning power plants and waste incinerators. Mercuric fungicides in house paint were once a major source of this deadly pollutant but now are restricted. Long-range transport of lead and mercury through the air is causing bioaccumulation in aquatic ecosystems far from the emission sources. It is now dangerous to eat fish from some once-pristine lakes and rivers because of toxic metal contamination.

Other toxic metals of concern are nickel, beryllium, cadmium, thallium, uranium, cesium, and plutonium. Some 780,000 tons of arsenic, a highly toxic metalloid, are released from metal smelters, coal combustion, and pesticide use each year. Halogens (fluorine, chlorine, bromine, and iodine) are highly reactive and generally toxic in their elemental form. About 600 million tons of highly persistent chlorofluorocarbons (CFCs) are used annually worldwide in spray propellants, refrigeration compressors, and for foam blowing. They diffuse into the stratosphere where they release chlorine and fluorine atoms that destroy the ozone shield that protects the earth from ultraviolet radiation. We'll return to this topic later.

Particulate Material

An **aerosol** is any system of solid particles or liquid droplets suspended in a gaseous medium. For convenience, we generally describe all atmospheric aerosols, whether solid or liquid, as **particulate material.** This includes dust, ash, soot, lint, smoke, pollen, spores, algal cells, and many other suspended materials. Anthropogenic particulate emissions amount to about 100 million metric tons per year worldwide. Wind-blown dust, volcanic ash, and other natural materials may contribute one hundred times that much.

Particulates often are the most apparent form of air pollution since they reduce visibility and leave dirty deposits on windows, painted surfaces, and textiles. Respirable particles smaller than 2.5 micrometers are among the most dangerous of this group because they can be drawn into the lungs, where they damage respiratory tissues. Asbestos fibers and cigarette smoke are among the most dangerous respirable particles in urban and indoor air because they are carcinogenic.

Volatile Organic Compounds

Volatile organic compounds (VOCs) are organic chemicals that exist as gases in the air. Plants are the largest source of VOCs, re-

leasing an estimated 350 million tons of isoprene (C_5H_8) and 450 million tons of terpenes ($C_{10}H_{15}$) each year. About 400 million tons of methane (CH_4) are produced by natural wetlands and rice paddies and by bacteria in the guts of termites and ruminant animals. These volatile hydrocarbons are generally oxidized to CO and CO_2 in the atmosphere.

In addition to these natural VOCs, a large number of other synthetic organic chemicals, such as benzene, toluene, formaldehyde, vinyl chloride, phenols, chloroform, and trichloroethylene, are released into the air by human activities. About 28 million tons of these compounds are emitted each year in the United States, mainly unburned or partially burned hydrocarbons from transportation, power plants, chemical plants, and petroleum refineries. These chemicals play an important role in the formation of photochemical oxidants.

The EPA requires industries to report releases of some 332 toxic organic chemicals into the air. In the early 1990s, emissions totaled 2 *million* metric tons or 5 billion pounds (fig. 18.6). The largest carcinogen emission was 52,000 tons (115 million lbs) of dichloromethane, which is used as an industrial solvent and paint stripper.

Photochemical Oxidants

Photochemical oxidants are products of secondary atmospheric reactions driven by solar energy (fig. 18.7). One of the most important of these reactions involves formation of singlet (atomic) oxygen by splitting either molecular oxygen (O_2) or nitrogen dioxide (NO_2). This singlet oxygen then reacts with another molecule of O_2 to make **ozone** (O_3). Ozone formed in the stratosphere pro-

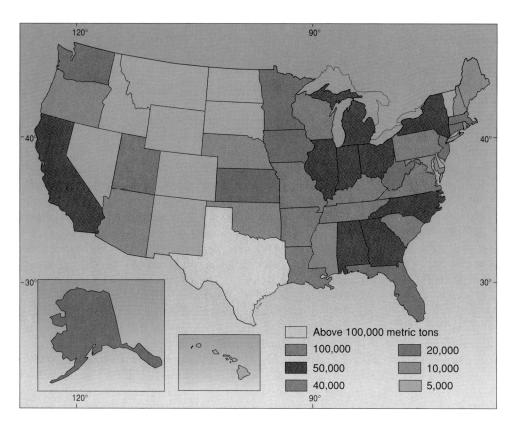

Figure 18.6

Annual toxic air pollution released in the United States. More than 300 toxic chemicals are released into the air. The 1990 Clean Air Amendments regulate these emissions for the first time.

Source: Data from Environmental Protection Agency.

Legend:
- Above 100,000 metric tons
- 100,000
- 50,000
- 40,000
- 20,000
- 10,000
- 5,000

Atmospheric Oxidant Production:

1. $O_2 + UV \longrightarrow O^* + O^*$ (oxygen free radical)
2. $NO_2 + UV \longrightarrow NO + O^*$
3. $O^* + O_2 \longrightarrow O_3$ (ozone)
4. $O_3 + NO \longrightarrow O_2 + NO_2$
5. $NO + VOC \longrightarrow NO_2 + PAN$ (peroxyacetylnitrate) + aldehydes[a]

Net results:

$$NO_2 + UV + VOC + O_2 \longrightarrow NO_2 + O_3 + PAN + aldehydes$$

[a]Examples of aldehydes are formaldehyde, acetylaldehyde, and benzaldehyde.

Figure 18.7

Some photochemical atmospheric reactions that contribute to smog formation.

vides a valuable shield for the biosphere by absorbing incoming ultraviolet radiation. In ambient air, however, O_3 is a strong oxidizing reagent and damages vegetation, building materials (such as paint, rubber, and plastics), and sensitive tissues (such as eyes and lungs). Ozone has an acrid, biting odor that is a distinctive characteristic of photochemical smog. Hydrocarbons in the air contribute to accumulation of ozone by removing NO in the formation of compounds, such as peroxyacetyl nitrate (PAN), which is another damaging photochemical oxidant. Notice that some of the chemical reactions shown in figure 18.7 are the same ones that produce stratospheric ozone discussed later in this chapter.

Unconventional Pollutants

The EPA has authority under the Clean Air Act to set **emission standards** (regulating the amount released) for certain **unconventional or noncriteria pollutants** that are considered especially toxic or hazardous. Among the materials regulated by emission standards are asbestos, benzene, beryllium, mercury, polychlorinated biphenyls (PCBs), and vinyl chloride. Most of these materials have no natural source in the environment (to any great extent) and are, therefore, only anthropogenic in origin.

In addition to these toxic air pollutants, some other unconventional forms of air pollution deserve mention. **Aesthetic degradation** includes any undesirable changes in the physical characteristics or chemistry of the atmosphere. Noise, odors, and light pollution are examples of atmospheric degradation that may not be life-threatening but reduce the quality of our lives. This is a very subjective category. Odors and noise (such as loud music) that are offensive to some may be attractive to others. Often the most sensitive device for odor detection is the human nose. We can smell styrene, for example, at 44 parts per billion (ppb). Trained panels of odor testers often are used to evaluate air samples. Factories that emit noxious chemicals sometimes spray "odor maskants" or perfumes into smokestacks to cover up objectionable odors.

In most urban areas, it is difficult or impossible to see stars in the sky at night because of dust in the air and stray light from build-

ings, outdoor advertising, and streetlights. This light pollution has become a serious problem for astronomers.

Indoor Air Pollution

We have spent a considerable amount of effort and money to control the major outdoor air pollutants, but we have only recently become aware of the dangers of indoor air pollutants. The EPA has found that indoor concentrations of toxic air pollutants are consistently higher than outdoors—up to twenty times higher for some toxins. Furthermore, people generally spend more time inside than out and therefore are exposed to higher doses of these pollutants.

Smoking is without doubt the most important air pollutant in the United States in terms of human health. The Surgeon General estimates that 400,000 people die each year in the United States from emphysema, heart attacks, strokes, lung cancer, or other diseases caused by smoking. Banning smoking probably would save more lives than any other pollution-control measure.

Other major indoor air pollution health hazards include asbestos, formaldehyde, vinyl chloride, radon, and combustion gases. Asbestos was widely used in floor and ceiling tiles, plaster, cement, insulation, and soundproofing. It is a serious concern in indoor air because of its carcinogenicity. Formaldehyde is used in more than three thousand products, including such building materials as particle board, waferboard, and urea-formaldehyde foam insulation. Vinyl chloride is used in plastic plumbing pipe, floor and wall coverings, and countertops. New carpets and drapes typically contain two dozen chemicals designed to kill bacteria and molds, resist stains, bind fibers, and retain colors.

In many cases, indoor air in homes has concentrations of chemicals that would be illegal outside or in the workplace. The EPA has found that concentrations of such compounds as chloroform, benzene, carbon tetrachloride, formaldehyde, and styrene can be seventy times higher in indoor air than in outdoor air. Many people are highly sensitive to these chemicals, and it is not uncommon to trace illness to a "sick house syndrome" caused by polluted indoor air. Next to smoking, radon gas leaking into homes from surrounding soil and rock (What Do *You* Think?, p. 392) is considered by the EPA to be the most serious indoor air pollutant in the United States.

In the less-developed countries of Africa, Asia, and Latin America where such organic fuels as firewood, charcoal, dried dung, and agricultural wastes make up the majority of household energy, smoky, poorly ventilated heating and cooking fires represent the greatest source of indoor air pollution (fig. 18.8). The World Health Organization (WHO) estimates that 2.5 billion people—half the world's population—are adversely affected by pollution from this source. Women especially spend long hours each day cooking over open fires or unventilated stoves in enclosed spaces. The levels of carbon monoxide, particulates, aldehydes, and other toxic chemicals can be one hundred times higher than would be legal for outdoor ambient concentrations in the United States. Designing and building cheap, efficient, nonpolluting energy sources for the developing countries would not only save shrinking forests but would make a major impact on health as well.

Air Pollution

What Do *You* Think?
Radon in Indoor Air •

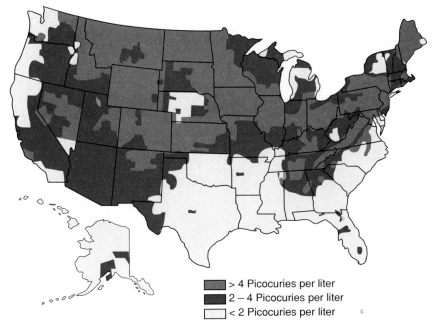

> ▨ > 4 Picocuries per liter
> ▩ 2 – 4 Picocuries per liter
> ▢ < 2 Picocuries per liter

Figure 1

Porous glacial till and granitic or basalt bedrock are among the geological features that make some regions more likely to have a radon gas hazard. This map shows an estimate of average radon levels in indoor air in the United States.

Source: EPA.

How safe is the air in your home? Should you be concerned about radioactive radon gas seeping through your foundation and contaminating your air? The Environmental Protection Agency (EPA) warns that one home in ten in the United States may exceed the recommended maximum radon level of 4 picocuries per liter (pCi/L). Continuously breathing this much radon, according to some experts, is equivalent to smoking about half a pack of cigarettes per day. By some estimates, as many as 10 percent of 136,000 yearly lung cancer deaths may be caused by radon. If true, this would make indoor radon second only to smoking as an air pollution hazard.

Radon is a colorless, odorless, tasteless gas produced naturally from uranium in rocks and soil. It is found almost everywhere, although some areas have geological formations that are especially high in radon (fig. 1). It is quite clear that hard rock miners who are exposed to high levels of radon have elevated lung cancer rates. During the 1950s and 60s, in poorly ventilated uranium mines in the American Southwest, as many as three-quarters of the miners—most of whom were Native Americans—died of lung cancer.

How does this evidence translate to risks for ordinary citizens? The EPA assumes that the dose/response relationship is strictly linear and that we can extrapolate directly from mine exposures to those of the general public. This assumption is controversial, however. Other factors, such as smoking by miners and high dust levels in mines may complicate the situation. It isn't clear how much these confounding factors affect the risk of radon in home exposures.

Epidemiological studies give contradictory results. A Swedish study suggested that radon in homes does cause lung cancer. A study in Iowa, on the other hand, concluded that radon may increase the risk of lung cancer in

smokers but does not pose a threat to nonsmokers. And a Manitoba study found that average radon exposure in a group of cancer victims was slightly *less* than that in a control group.

How can there be such a wide disagreement among experts about such an important issue such as this? Why can't they tell us with greater certainty how dangerous radon really is? By now you should recognize that environmental science often involves uncertainty. We can't deliberately expose a group of people to radon to see what will happen. What should you do about radon in your home then? It probably depends on your individual situation. If you have high levels of radon that you can easily eliminate, you probably should do so. On the other hand, other risks in your life may be more important than radon.

Nearly one-third of all counties in the United States are suspected of potentially having more than the recommended concentration of 4 pCi/L of radon. Remediating all the millions of homes with more than 4 pCi/L of radon could cost $50 billion. Some specialists argue that the "action threshold"

should be 20 pCi/L and that only one percent of the houses considered unsafe by the EPA really need to be treated.

How can you find out how much radon is in your home? Your state public health department probably has radon detectors or can tell you where to obtain them. What can you do if you find more radon in your house than you would like? Sealing foundation cracks and openings around pipes is a good first step. Lining walls with plastic may help. Gravel fill around foundations can allow radon to escape outside rather than seep into the house. Drilling holes in the basement floor and installing pipes to draw radon under the foundation can reduce infiltration. You also can improve ventilation of indoor spaces to remove radon.

Carrying out all of these steps can reduce indoor radon levels by 95 percent or more, but may cost more than the benefits are worth. On the other hand, a few hundred dollars invested wisely may be enough to give you a margin of safety and peace of mind.

Figure 18.8

Smoky cooking and heating fires may cause more ill health effects than any other source of indoor air pollution except tobacco smoking. Levels of carbon monoxide, particulates, and cancer-causing hydrocarbons can be one thousand times higher indoors than outdoors. Some 2.5 billion people, mainly women and children, spend hours each day in poorly ventilated kitchens and living spaces.

Climate, Topography, and Atmospheric Processes

Topography, climate, and physical processes in the atmosphere play an important role in transport, concentration, dispersal, and removal of many air pollutants. Wind speed, mixing between air layers, precipitation, and atmospheric chemistry all determine whether pollutants will remain in the locality where they are produced or go elsewhere. In this next section, we will survey some environmental factors that affect air pollution levels.

Inversions

Temperature inversions occur when a stable layer of warmer air overlays cooler air, reversing the normal temperature decline with increasing height and preventing convection currents from dispersing pollutants. Several mechanisms create inversions. When a cold front slides under an adjacent warmer air mass or when cool air subsides down a mountain slope to displace warmer air in the valley below, an inverted temperature gradient is established. These inversions are usually not stable, however, because winds accompanying these air exchanges tend to break up the temperature gradient fairly quickly and mix air layers.

The most stable inversion conditions are usually created by rapid nighttime cooling in a valley or basin where air movement is restricted. Los Angeles is a classic example of the conditions that create temperature inversions and photochemical smog (fig. 18.9). The city is surrounded by mountains on three sides and the climate is dry and sunny. Extensive automobile use creates high pollution levels. Skies are generally clear at night, allowing rapid radiant heat loss, and the ground cools quickly. Surface air layers are cooled by conduction, while upper layers remain relatively warm.

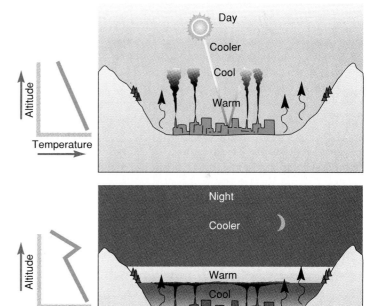

Figure 18.9

The Los Angeles Basin is a classic example of conditions that generate an atmospheric temperature inversion. During the day, the sun warms the ground and buildings. As air temperatures rise, dust and pollution are carried aloft, but mountains on three sides of the city trap the dirty air and keep it from spreading. At night, heat radiates quickly from bare surfaces and the air near ground level cools faster than that aloft, forming an inversion. Cool onshore ocean breezes bring clean air, which is trapped under the warm, polluted upper layers. The next morning, the sun reheats the surface and the air above it. The clean air is pushed back out over the coastline and the polluted layers descend back down to ground level.

Density differences retard vertical mixing. During the night, cool, humid, onshore breezes slide in under the contaminated air, squeezing it up against the cap of warmer air above and concentrating the pollutants accumulated during the day.

Morning sunlight is absorbed by the concentrated aerosols and gaseous chemicals of the inversion layer. This complex mixture quickly cooks up a toxic brew of hazardous compounds. As the ground warms later in the day, convection currents break up the temperature gradient and pollutants are carried back down to the surface where more contaminants are added. Nitric oxide (NO) from automobile exhaust is oxidized to nitrogen dioxide. As nitrogen oxides are used up in reactions with unburned hydrocarbons, the ozone levels begin to rise. By early afternoon, an acrid brown haze fills the air, making eyes water and throats burn. On a typical summer day, ozone concentrations in the Los Angeles basin reach 0.34 ppm or more by late afternoon and the pollution index can be 300, the stage considered a health hazard.

Dust Domes and Heat Islands

Even without mountains to block winds and stabilize air layers, many large cities create an atmospheric environment quite different from the surrounding conditions. Sparse vegetation and high levels of concrete and glass in urban areas allow rainfall to run off quickly and create high rates of heat absorption during the day and radiation at night. Tall buildings create convective updrafts that sweep pollutants into the air. Temperatures in the center of large cities are frequently 3° to 5°C (5° to 9°F) higher than the surrounding countryside. Stable air masses created by this "heat island" over the city concentrate pollutants in a "dust dome." Rural areas downwind from major industrial areas often have significantly decreased visibility and increased rainfall (due to increased condensation nuclei in the dust plume) compared to neighboring areas with cleaner air. In the late 1960s, for instance, areas downwind from Chicago and St. Louis reported up to 30 percent more rainfall than upwind regions.

Long-Range Transport

Air pollutants can be carried long distances by wind currents. In 1971, scientists at the Nagoya Water Research Institute observed dust passing over Japan from Asia. A few days later, the same dust was collected at Hawaii, some 10,000 km across the Pacific Ocean. Mineral content, timing of weather events, and calculated air paths suggested that this dust came as a single surge from a storm in the Gobi Desert. Similar dust storms in the Algerian Sahara have been traced to islands in the Caribbean.

Industrial pollutants are also transported great distances by wind currents. Some of the most toxic and corrosive materials delivered by long-range transport are secondary pollutants (such as sulfuric and nitric acids or ozone), produced by the mixing and interaction of atmospheric contaminants as they travel through the air. Tracing the sources of these chemically altered pollutants can be difficult. Lakes and forests in Sweden were showing evidence of sulfuric acid contamination years before the source of the acidity was traced to Germany, England, and other distant parts of Europe.

Controlling long-range pollutants is a highly political process. Germany and England were not very sympathetic about acid precipitation until their own forests began to die. In another case, ninety percent of the pollution falling into Lake Superior originates thousands of kilometers away in the United States, Canada, and even Mexico. Farms and industries in these distant regions resist spending money, however, to reduce emissions to protect someone else's environment.

Increasingly sensitive monitoring equipment has begun to reveal industrial contaminants in places usually considered among the cleanest in the world. Samoa, Greenland, and even Antarctica and the North Pole, all have concentrations of heavy metals, pesticides, and radioactive elements. Since the 1950s, pilots flying in the high Arctic have reported dense layers of reddish-brown haze clouding the arctic atmosphere. Aerosols of sulfates, soot, dust, and toxic heavy metals such as vanadium, manganese, and lead travel to the Pole from the industrialized parts of Europe and Russia (fig. 18.10). These contaminants, trapped by winds that circle the pole, concentrate at high latitudes and eventually, falling out in snow and ice, enter the food chain. The Inuit people of Broughton Island, well above the Arctic Circle, have higher levels of PCBs in their blood than any other known population, except victims of industrial accidents. Far from any source of this industrial by-product, these people accumulate PCBs from the flesh of fish, caribou, and other animals they eat.

Stratospheric Ozone

In 1985, the British Antarctic Atmospheric Survey announced a startling and disturbing discovery: ozone levels in the stratosphere over the South Pole were dropping precipitously during September and October every year as the sun reappears at the end of the long polar winter. This ozone depletion has been occurring at least since the 1960s but was not recognized because earlier researchers programmed their instruments to ignore changes in ozone levels that were presumed to be erroneous.

Each year the ozone "hole" has grown larger. In 1995, as much as 70 percent of the Antarctic stratospheric ozone was destroyed over an area about the size of North America (fig. 18.11). Ominously, this phenomenon is now spreading to other parts of the world as well. About 10 percent of all stratospheric ozone worldwide was destroyed during the winter of 1995 and losses over the western U.S. and Canada reached 20 percent, while Siberia exceeded 35 percent.

Why are we worried about stratospheric ozone? At ground level, ozone is a harmful pollutant, damaging plants, building materials, and human health; in the the upper atmosphere, however, where it screens out dangerous ultraviolet (UV) rays from the sun, ozone is an irreplaceable resource. Without this shield, organisms on the earth's surface would be subjected to life-threatening radiation burns and genetic damage. A 1 percent loss of ozone results in a 2 percent increase in UV reaching the earth's surface and could result in about a million extra human skin cancers per year worldwide if no protective measures are taken. Thus it is urgent that we learn what is attacking the ozone layer and find ways to reverse these trends if possible.

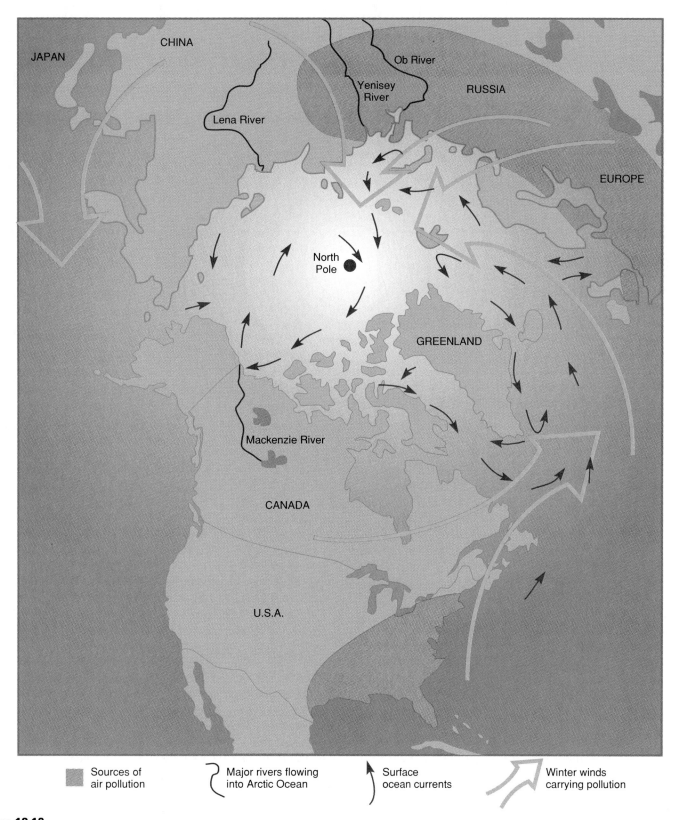

Figure 18.10

Air pollution from heavily industrialized regions in Europe and America are transported by circumpolar winds to the Arctic, where high levels of smog accumulate. The average transit time from Russia to Canada is only about three days.

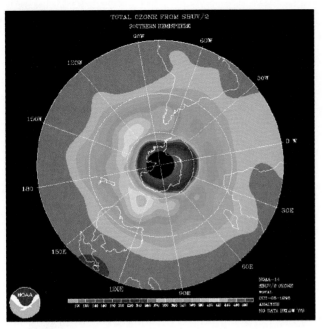

Figure 18.11

Ozone depletion over the South Pole is shown in this satellite image from October 5,1995. Outside the polar vortex, green and yellow areas show elevated ozone levels. Over Antarctica (outlined in white), however, stratospheric ozone levels are reduced by 80 percent or more from normal levels. The dark spot over the South Pole is an artifact of the measuring system.

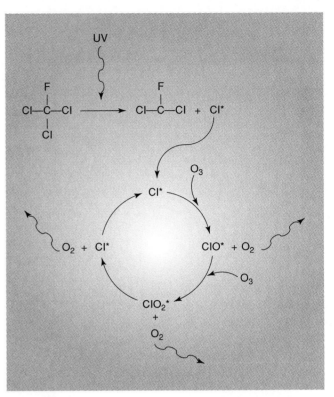

Figure 18.12

When exposed to UV radiation, CFCs release highly reactive chlorine radicals (Cl^*). These radicals then enter a repeating cycle of reacting with O_3, releasing O_2, and re-emerging as free chlorine radicals, ready to begin the cycle again.

The exceptionally cold temperatures (-85 to $-90°C$) in Antarctica play a role in ozone losses. During the long, dark winter months, the strong circumpolar vortex (chapter 17) isolates Antarctic air and allows stratospheric temperatures to drop low enough to create ice crystals at high altitudes—something that rarely happens elsewhere over the world. Ozone and chlorine-containing molecules are absorbed on the surfaces of these ice particles. When the sun returns in the spring and provides energy to liberate chlorine ions, destructive chemical reactions proceed quickly (fig. 18.12).

Humans release a variety of chlorine-containing molecules into the atmosphere. The ones suspected of being most important in ozone losses are chlorofluorocarbons (CFCs) and halon gases. CFCs were invented in 1928 by scientists at General Motors who were searching for a less toxic refrigerant than ammonia. Commonly known by the trade name Freon, CFCs were regarded as wonderful compounds. They are nontoxic, nonflammable, chemically inert, cheaply produced, and useful in a wide variety of applications.

Because these molecules are so stable, however, they persist for decades or even centuries once released. When they diffuse out into the stratosphere, the intense UV irradiation releases chlorine free radicals (Cl^*) that destroy ozone. Since the chlorine radicals are not themselves consumed in these reactions, they continue to destroy ozone for years until they finally drift into outer space.

Until 1978, aerosol spray cans used more CFCs than any other product. Although we didn't know about the special conditions in the Antarctic at that time, it was suspected that CFCs might threaten stratospheric ozone, so laws were passed in the United States, Canada, and some European countries to ban nonessential uses. Still, some 320,000 metric tons of CFCs were used worldwide every year until 1988 as refrigerants, solvents, spray propellants, and foam-blowing agents.

The discovery of stratospheric ozone losses has brought about a remarkably quick international response. At a 1989 conference in Helsinki, eighty-one nations agreed to phase out CFC production by the end of the century. As evidence accumulated showing that losses were larger and more widespread than previously thought, the deadline for the elimination of all CFCs (halons, carbon tetrachloride, and methyl chloroform) was moved up to 1996 and a $500 million fund was established to assist poorer countries to switch to non-CFC technologies. Fortunately, alternatives to CFCs for most uses already exist. The first substitutes will be hydrochlorofluorocarbons (HCFCs), which release much less chlorine per molecule. Eventually, we hope to develop halogen-free molecules that work just as well and are no more expensive than CFCs.

There is some evidence that the CFC ban is already having an effect. The buildup of CFCs in the atmosphere is declining more rapidly than expected, so that 1995 may be the first year in decades in which chlorine levels in the atmosphere declined rather than increased. In 70 years or so, stratospheric ozone levels are expected to be back to normal. Unfortunately, there may be a downside to stopping ozone destruction. Ozone is a potent greenhouse gas (chapter 17). Some models suggest that lower ozone levels have been offsetting the effects of increased CO_2. When ozone is restored, global warming may be accelerated. As so often is the case, when we disturb one environmental factor, we affect others as well.

Effects of Air Pollution

So far we have looked at the major types and sources of air pollutants. Now we will turn our attention to the effects of those pollutants on human health, physical materials, ecosystems, and global climate.

Human Health

The EPA estimates that people in the most polluted cities in the United States are 15 to 17 percent more likely to die prematurely than those in cities with the cleanest air. Heart attacks, respiratory diseases, and lung cancer all are significantly higher in people who breathe dirty air, compared to matching groups in cleaner environments. This can mean as much as a 5- to 10-year decrease in life expectancy if you live in the worst parts of Los Angeles or Baltimore, compared to a place with clean air. Of course your likelihood of suffering ill health from air pollutants depends on the intensity and duration of exposure as well as your age and prior health status. You are much more likely to be at risk if you are very young, very old, or already suffering from some respiratory or cardiovascular disease. Some people are super-sensitive because of genetics or prior exposure. And those doing vigorous physical work or exercise are more likely to succumb than more sedentary folks.

Conditions are often much worse in other countries than Canada or the United States. The United Nations estimates that at least 1.3 billion people around the world live in areas where the air is dangerously polluted. In the "black triangle" region of Poland, Hungary, the Czech Republic, and Slovakia, for example, respiratory ailments, cardiovascular diseases, lung cancer, infant mortality, and miscarriages are as much as 50 percent higher than in cleaner parts of those countries. And in China, city dwellers are 4 to 6 times more likely than country folk to die of lung cancer. As mentioned earlier, the greatest air quality problem is often in poorly ventilated homes in poorer countries where smoky fires are used for cooking and heating. Billions of women and children spend hours each day in these unhealthy conditions. The World Health Organization estimates that 4 million children under 5 die each year from acute respiratory diseases exacerbated by air pollution.

How does air pollution cause these health effects? The most common route of exposure to air pollutants is by inhalation, but direct absorption through the skin or contamination of food and water also are important pathways. Because they are strong oxidizing agents, sulfates, SO_2, NO_x, and O_3 act as irritants (see chapter 9) that damage delicate tissues in the eyes and respiratory passages. Fine suspended particulate materials (less than 10 μm) penetrate deep into the lungs and are both irritants and fibrotic agents. Inflammatory responses set in motion by these irritants impair lung function and trigger cardiovascular problems as the heart tries to compensate for lack of oxygen by pumping faster and harder. If the irritation is really se-

vere—see the story about Bhopal in chapter 9—so much fluid seeps into lungs through damaged tissues that the victim actually drowns.

Carbon monoxide binds to hemoglobin and decreases the ability of red blood cells to carry oxygen. Asphyxiants such as this cause headaches, dizziness, heart stress, and can even be lethal if concentrations are high enough. Lead also binds to hemoglobin and reduces oxygen-carrying capacity at high levels. At lower levels, lead causes long-term damage to critical neurons in the brain that results in mental and physical impairment and developmental retardation.

Some important chronic health effects of air pollutants include bronchitis and emphysema.

Bronchitis is a persistent inflammation of bronchi and bronchioles (large and small airways in the lung) that causes a painful cough and involuntary muscle spasms that constrict airways. Severe bronchitis can lead to **emphysema,** an irreversible obstructive lung disease in which airways become permanently constricted and alveoli are damaged or even destroyed. Stagnant air trapped in blocked airways swells the tiny air sacs in the lung (alveoli), blocking blood circulation. As cells die from lack of oxygen and nutrients, the walls of the alveoli break down, creating large empty spaces incapable of gas exchange (fig. 18.13). Thickened walls of the bronchioles lose elasticity and breathing becomes more difficult. Victims of emphysema make a characteristic whistling sound when they breathe. Often they need supplementary oxygen to make up for reduced respiratory capacity.

Irritants in the air are so widespread that about half of all lungs examined at autopsy in the United States have some degree of alveolar deterioration. The Office of Technology Assessment (OTA) estimates that 250,000 people suffer from pollution-related bronchitis and emphysema in the United States, and some 50,000 excess

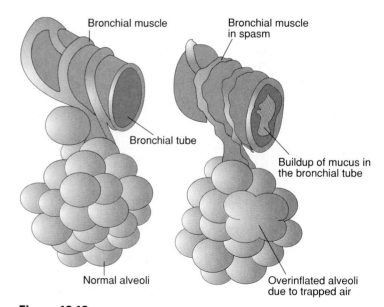

Figure 18.13

When an asthma victim encounters a "trigger," such as dust, cold air, or irritating chemicals, muscles around the bronchial tubes contract and secretory cells in the epithelia produce a thick mucus that blocks airways. Wheezing and difficulty in breathing result.

deaths each year are attributable to complications of these diseases, which are probably second only to heart attack as a cause of death.

Smoking is undoubtedly the largest cause of obstructive lung disease and preventable death in the world. The World Health Organization says that tobacco kills some 3 million people each year. This makes it rank with diarrhea and malaria as one of the world's leading killers. Because of cardiovascular stress caused by carbon monoxide in smoke and chronic bronchitis and emphysema, about twice as many people die of heart failure as die from lung cancer associated with smoking.

Plant Pathology

In the early days of industrialization, fumes from furnaces, smelters, refineries, and chemical plants often destroyed vegetation and created desolate, barren landscapes around mining and manufacturing centers. The copper-nickel smelter at Sudbury, Ontario, is a spectacular and notorious example of air pollution effects on vegetation and ecosystems. In 1886, the corporate ancestors of the International Nickel Company (INCO) began open-bed roasting of sulfide ores at Sudbury. Sulfur dioxide and sulfuric acid released by this process caused massive destruction of the plant community within about 30 km (18.6 mi) of the smelter. Rains washed away the exposed soil, leaving a barren moonscape of blackened bedrock. Super-tall, 400 m smokestacks were installed in the 1950s and sulfur scrubbers were added 20 years later. Emissions were reduced by 90 percent and the surrounding ecosystem is beginning to recover. The area near the factory is still a grim, empty wasteland, however (fig. 18.14). Similar destruction occurred at many other sites during the nineteenth century. Copperhill, Tennessee; Butte, Montana; and the Ruhr Valley in Germany are some well-known examples, but these areas also are showing signs of recovery since corrective measures were taken.

There are two probable ways that air pollutants damage plants. They can be directly toxic, damaging sensitive cell membranes much as irritants do in human lungs. Within a few days of exposure to toxic levels of oxidants, mottling (discoloration) occurs in leaves due to chlorosis (bleaching of chlorophyll), and then necrotic (dead) spots develop (fig. 18.15). If injury is severe, the whole plant may be killed. Sometimes these symptoms are so distinctive that positive identification of the source of damage is possible. Often, however, the symptoms are vague and difficult to separate from diseases or insect damage.

Another mechanism of action is exhibited by chemicals, such as ethylene, that act as metabolic regulators or plant hormones and disrupt normal patterns of growth and development. Ethylene is a component of automobile exhaust and is released from petroleum refineries and chemical plants. The concentration of ethylene around highways and industrial areas is often high enough to cause injury to sensitive plants. Some scientists believe that the devastating forest destruction in Europe and North America may be partly due to volatile organic compounds.

Certain combinations of environmental factors have **synergistic effects** in which the injury caused by exposure to two factors together is more than the sum of exposure to each factor individually. For instance, when white pine seedlings are exposed to subthreshold concentrations of ozone and sulfur dioxide individually, no visible injury occurs. If the same concentrations of pollutants are given together, however, visible damage occurs. In alfalfa, however, SO_2 and O_3 together cause less damage than either one alone. These complex interactions point out the unpredictability of future

Figure 18.14

Sulfur dioxide emissions and acid precipitation from the International Nickel Company copper smelter (*background*) killed all vegetation over a large area near Sudbury, Ontario. Even the pink granite bedrock has burned black. The installation of scrubbers has dramatically reduced sulfur emissions. The ecosystem farther away from the smelter is slowly beginning to recover.

Figure 18.15

Soybean leaves exposed to 0.8 parts per million sulfur dioxide for 24 hours show extensive chlorosis (chlorophyll destruction) in white areas between leaf veins.

effects of pollutants. Outcomes might be either more or less severe than previous experience indicates.

Pollutant levels too low to produce visible symptoms of damage may still have important effects. Field studies using open-top chambers and charcoal-filtered air show that yields in some sensitive crops, such as soybeans, may be reduced as much as 50 percent by currently existing levels of oxidants in ambient air. Some plant pathologists suggest that ozone and photochemical oxidants are responsible for as much as 90 percent of agricultural, ornamental, and forest losses from air pollution. The total costs of this damage may be as much as $10 billion per year.

Acid Deposition

Most people in the United States became aware of problems associated with **acid precipitation** (the deposition of wet acidic solutions or dry acidic particles from the air) within the last decade or so, but English scientist Robert Angus Smith coined the term "acid rain" in his studies of air chemistry in Manchester, England, in the 1850s. By the 1940s, it was known that pollutants, including atmospheric acids, could be transported long distances by wind currents. This was thought to be only an academic curiosity until it was shown that precipitation of these acids can have far-reaching ecological effects.

pH and Atmospheric Acidity

We describe acidity in terms of pH (the negative logarithm of the hydrogen ion concentration in a solution). The pH scale ranges from 0 to 14, with 7, the midpoint, being neutral (chapter 3). Values below 7 indicate progressively greater acidity, while those above 7 are progressively more alkaline. Since the scale is logarithmic, there is a tenfold difference in hydrogen ion concentration for each pH unit. For instance, pH 6 is ten times more acidic than pH 7; likewise, pH 5 is one hundred times more acidic, and pH 4 is one thousand times more acidic than pH 7.

Normal, unpolluted rain generally has a pH of about 5.6 due to carbonic acid created by CO_2 in air. Volcanic emissions, biological decomposition, and chlorine and sulfates from ocean spray can drop the pH of rain well below 5.6, while alkaline dust can raise it above 7. In industrialized areas, anthropogenic acids in the air usually far outweigh those from natural sources. Acid rain is only one form in which acid deposition occurs. Fog, snow, mist, and dew also trap and deposit atmospheric contaminants. Furthermore, fallout of dry sulfate, nitrate, and chloride particles can account for as much as half of the acidic deposition in some areas. These particles are converted to acids when they dissolve in surface water or contact moist tissues (like those in the lungs and eyes). Considerable evidence suggests that acid aerosols are a human health hazard.

Aquatic Effects

It has been known for about thirty years that acids—principally H_2SO_4 and HNO_3—generated by industrial and automobile emissions in northwestern Europe are carried by prevailing winds to Scandinavia where they are deposited in rain, snow, and dry precipitation. The thin, acidic soils and oligotrophic lakes and streams in the mountains of southern Norway and Sweden have been severely affected by this acid deposition. Some 18,000 lakes in Sweden are now so acidic that they will no longer support game fish or other sensitive aquatic organisms.

There has been a great deal of research on the mechanisms of damage by acidification. Generally, reproduction is the most sensitive stage in the life cycle. Eggs and fry of many fish species are killed when the pH drops to about 5.0. This level of acidification also can disrupt the food chain by killing aquatic plants, insects, and invertebrates on which fish depend for food. At pH levels below 5.0, adult fish die as well. Trout, salmon, and other game fish are usually the most sensitive. Carp, gar, suckers, and other less desirable fish are more resistant. There are several ways acids kill fish. Acidity alters body chemistry, destroys gills and prevents oxygen uptake, causes bone decalcification, and disrupts muscle contraction. Another dangerous effect (for us as well as fish) is that acid water leaches toxic metals, such as mercury and aluminum, out of soil and rocks. Which of these mechanisms is the most important is open to debate, but it is clear that acid deposition has had disastrous effects on sensitive aquatic ecosystems.

In the early 1970s, evidence began to accumulate suggesting that air pollutants are acidifying many lakes in North America. Studies in the Adirondack Mountains of New York revealed that about half of the high altitude lakes (above 1000 m or 3300 ft) are acidified and have no fish. Precipitation records show that the average pH of rain and snow has dropped significantly over a large area of the northeastern United States and Canada in the past two decades. Some 48,000 lakes in Ontario are endangered and nearly all of Quebec's surface waters, including about 1 million lakes, are believed to be highly sensitive to acid deposition. Figure 18.16 shows the location of acidic deposition in the United States and Canada. About 50 percent of the acid deposition in Canada comes from the United States, while only 10 percent of U.S. pollution comes from Canada. Canadians are understandably upset about this imbalance.

Much of the western United States has relatively alkaline bedrock and carbonate-rich soil, which counterbalance acids from the atmosphere. Recent surveys of the Rocky Mountains, the Sierra Nevadas in California, and the Cascades in Washington, however, have shown that many high mountain lakes and streams have very low buffering capacity (ability to resist pH change) and are susceptible to acidification.

Sulfates account for about two-thirds of the acid deposition in eastern North America and most of Europe, while nitrates contribute most of the remaining one-third. In urban areas, where transportation is the major source of pollution, nitric acid is equal to or slightly greater than sulfuric acids in the air. A vigorous program of pollution control has been undertaken by both Canada and the United States with promises of 50 percent reduction of SO_2 emissions and significant lowering of NO_x production. These programs already are producing significant reductions in acid depositions in many areas.

Forest Damage

In the early 1980s, disturbing reports appeared of rapid forest declines in both Europe and North America. One of the earliest was

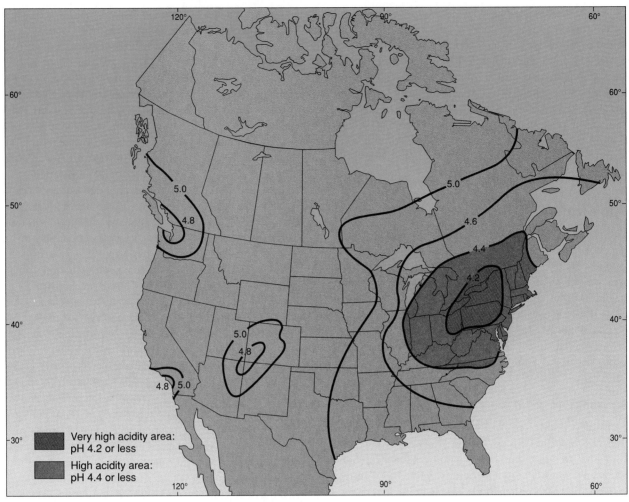

Figure 18.16

The acidity of precipitation over Canada and the United States. Numbers shown are average annual pH in 1982. Acid levels have changed little in the past decade.

Sources: Data from the National Atmospheric Deposition Program and the Canadian Network for Sampling of Precipitation.

a detailed ecosystem inventory on Camel's Hump Mountain in Vermont. A 1980 survey showed that seedling production, tree density, and viability of spruce-fir forests at high elevations had declined about 50 percent in 15 years. By 1990, almost all the red spruce, once the dominant species on the upper part of the mountain, were dead or dying. A similar situation was found on Mount Mitchell in North Carolina where almost all red spruce and Fraser fir above 2000 meters are in a severe decline. Nearly all the trees are losing needles and about half of them are dead.

European forests also are dying at an alarming rate. West German foresters estimated in 1982 only 8 percent of their forests showed air pollution damage. By 1983, some 34 percent of the forest was affected, and in 1985, more than 4 million hectares (about half the total forest) were reported to be in a state of decline. The loss to the forest industry is estimated to be about one billion DM (Deutsche marks) per year.

Similar damage is reported in Czechoslovakia, Poland (fig. 18.17), Austria, and Switzerland. Again, high elevation forests are

most severely affected. This is a disaster for mountain villages in the Alps that depend on forests to prevent avalanches in the winter. Sweden, Norway, the Netherlands, Romania, China, and the former Soviet Union also have evidence of growth reduction, defoliation, root necrosis, lack of seedling growth, and premature tree death. The species afflicted vary from place to place, but the overall picture is of widespread forest destruction.

This complex phenomenon probably has many contributing factors, but air pollution and deposition of atmospheric acids are thought to be leading causes of forest destruction in many areas. Considerable research has shown that acids are directly toxic to tender shoots and roots. High-altitude forests are subjected to especially intense doses of these acids because clouds saturated with pollutants tend to hang on mountaintops, bathing forests in a toxic soup for days or even weeks at a time.

Scientists have suggested that other mechanisms may play a role in forest decline. Overfertilization by nitrogen compounds may make trees sensitive to early frost. Essential minerals, such as

Figure 18.17

A forest killed by pollution in the Kakonoski National Park in southwest Poland. Combined effects of a toxic soup of many different air pollutants probably caused this damage.

Figure 18.18

Atmospheric acids, especially sulfuric and nitric acids, have almost completely eaten away the face of this medieval statue. Each year, the total losses from air pollution damage to buildings and materials amounts to billions of dollars.

magnesium, may be washed out of foliage or soil by acidic precipitation. Toxic metals, such as aluminum, may be solubilized by acidic groundwater. Plant pathogens and insect pests may damage trees or attack trees debilitated by air pollution. Fungi that form essential mutualistic associations (called mycorrhizae) with tree roots may be damaged by acid rain. Other air pollutants, such as sulfur dioxide, ozone, or toxic organic compounds may damage trees. Repeated harvesting cycles in commercial forests may remove nutrients and damage ecological relationships essential for healthy tree growth. Perhaps the most likely scenario is that all these environmental factors act cumulatively but in different combinations in the deteriorating health of individual trees and entire forests.

Buildings and Monuments

In cities throughout the world, some of the oldest and most glorious buildings and works of art are being destroyed by air pollution. Smoke and soot coat buildings, paintings, and textiles. Limestone and marble are destroyed by atmospheric acids at an alarming rate. The Parthenon in Athens, the Taj Mahal in Agra, the Colosseum in Rome, frescoes and statues in Florence, medieval cathedrals in Europe (fig. 18.18), and the Lincoln Memorial and Washington Monument in Washington, DC, are slowly dissolving and flaking away because of acidic fumes in the air. Medieval stained glass windows in Cologne's gothic cathedral are so porous from etching by atmospheric acids that pigments disappear and the glass literally crumbles away. Restoration costs for this one building alone are estimated at three to four billion German marks ($1.5 to $2 billion).

On a more mundane level, air pollution also damages ordinary buildings and structures. Corroding steel in reinforced concrete weakens buildings, roads, and bridges. Paint and rubber deteriorate due to oxidization. Limestone, marble, and some kinds of sand-

stone flake and crumble. The Council on Environmental Quality estimates that U.S. economic losses from architectural damage caused by air pollution amount to about $4.8 billion in direct costs and $5.2 billion in property value losses each year.

Visibility Reduction

Foul air obscuring the skies above industrialized cities has long been recognized as a problem, but we have realized only recently that pollution affects rural areas as well. Even supposedly pristine places like our national parks are suffering from air pollution. Grand Canyon National Park, where maximum visibility used to be 300 km (185 mi), is now so smoggy on some winter days that visitors can't see the opposite rim only 20 km (12.5 mi) across the canyon. Mining operations, smelters, and power plants (some of which were moved to the desert to improve air quality in cities like Los Angeles) are the main culprits. Similarly, the vistas from Shenandoah National Park just outside Washington, DC, are so hazy that summer visibility is often less than 1.6 km (1 mi) because of smog drifting in from nearby urban areas.

Historical records show that over the past four or five decades human-caused air pollution has spread over much of the United States. John Trijonis of the Santa Fe Corporation reports that a gigantic "haze blob" as much as 3000 km (about 2000 mi) across covers much of the eastern United States in the summer, cutting visibility as much as 80 percent. Smog and haze are so prevalent, Trijonis says, that it's hard for people to believe that the air once was clear. Studies indicate, however, that if all human-made sources of air pollution were shut down, the air would clear up in a few days and there would be about 150 km (90 mi) visibility nearly everywhere rather than the 15 km to which we have become accustomed.

Air Pollution Control

What can we do about air pollution? In this section we will look at some of the techniques that can be used to avoid creating pollutants or to clean up effluents before they are released. We also will look at some legislation that regulates pollutant emissions and ambient air quality.

Moving Pollution to Remote Areas

Among the earliest techniques for improving local air quality was moving pollution sources to remote locations and/or dispersing emissions with smokestacks. These approaches exemplify the attitude that "dilution is the solution to pollution." One electric utility, for example, ran newspaper and magazine ads in the early 1970s, claiming to be a "pioneer" in the use of tall smokestacks on its power plants to "disperse gaseous emissions widely in the atmosphere so that ground level concentrations would not be harmful to human health or property." The company claimed that their smoke would be "dissipated over a wide area and come down finally in harmless traces." Far from being harmless, however, those "traces" are the main source of many of our current problems. We are finding that there is no "away" to which we can throw our unwanted products. A far better solution to pollution is to prevent its release. We will now turn our attention to emission-control technology.

Particulate Removal

Filters remove particles physically by trapping them in a porous mesh of cotton cloth, spun glass fibers, or asbestos-cellulose, which allows air to pass through but holds back solids. Collection efficiency is relatively insensitive to fuel type, fly ash composition, particle size, or electrical properties. Filters are generally shaped into giant bags 10 to 15 meters long and 2 or 3 meters wide. Effluent gas is blown into the bottom of the bag and escapes through the sides much like the bag on a vacuum cleaner (fig. 18.19a). Every few days or weeks, the bags are opened to remove the dust cake. Thousands of these bags may be lined up in a "baghouse." These filters are usually much cheaper to install and operate than electrostatic filters.

Electrostatic precipitators (fig. 18.19b) are the most common particulate controls in power plants. Fly ash particles pick up an electrostatic surface charge as they pass between large electrodes in the effluent stream. This causes the particle to migrate to and accumulate on a collecting plate (the oppositely charged electrode). These precipitators consume a large amount of electricity, but maintenance is relatively simple and collection efficiency can be as high as 99 percent. Performance depends on particle size and chemistry, strength of the electric field, and flue gas velocity.

The ash collected by all of these techniques is a solid waste (often hazardous due to the heavy metals and other trace components of coal or other ash source) and must be buried in landfills or other solid waste disposal sites.

Sulfur Removal

As we have seen earlier in this chapter, sulfur oxides are among the most damaging of all air pollutants in terms of human health and

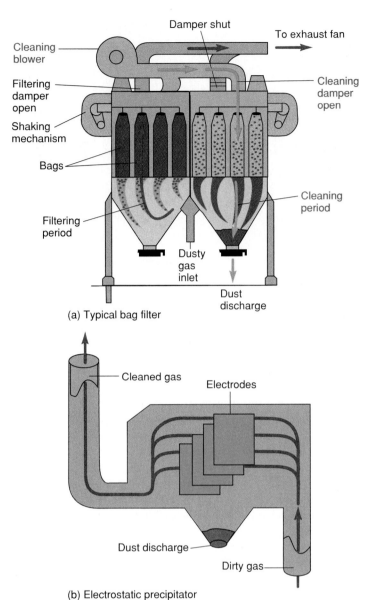

(a) Typical bag filter

(b) Electrostatic precipitator

Figure 18.19

Typical emission-control devices: (*a*) bag filter and (*b*) electrostatic precipitator. Note that two stages in the operational cycle are shown in (*a*), the filtering period on the left and the period of cleaning the filter bag on the right.

ecosystem damage. It is important to reduce sulfur loading. This can be done either by using low-sulfur fuel or by removing sulfur from effluents.

Fuel Switching and Fuel Cleaning

Switching from soft coal with a high sulfur content to low-sulfur coal can greatly reduce sulfur emissions. This may eliminate jobs, however, in such areas as Appalachia that are already economically depressed. Changing to another fuel, such as natural gas or nuclear energy, can eliminate all sulfur emissions as well as those of particulates and heavy metals. Natural gas is more expensive and more difficult to ship and store than coal, however, and many people prefer the sure dangers of coal pollution to the uncertain dangers of nuclear power (chapter 21). Alternative energy sources, such as wind and solar power, are preferable to either fossil fuel or nuclear power, and are becoming economically competitive (chapter 22) in many areas. In the interim, coal can be crushed, washed, and gassified to remove sulfur and metals before combustion. This improves heat content and firing properties but may replace air pollution with solid waste and water pollution problems.

Limestone Injection and Fluidized Bed Combustion

Sulfur emissions can be reduced as much as 90 percent by mixing crushed limestone with coal before it is fed into a boiler. Calcium in the limestone reacts with sulfur to make calcium sulfite ($CaSO_3$), calcium sulfate ($CaSO_4$), or gypsum ($CaSO_4 \cdot 2H_2O$). In ordinary furnaces, this procedure creates slag, which fouls burner grates and reduces combustion efficiency.

A relatively new technique for burning, called fluidized bed combustion, offers several advantages in pollution control. In this procedure, a mixture of crushed coal and limestone particles about a meter (3 ft) deep is spread on a perforated distribution grid in the combustion chamber (fig. 18.20). When high-pressure air is forced through the bed, the surface of the fuel rises as much as one meter and resembles a boiling fluid as particles hop up and down. Oil is sprayed into the suspended mass to start the fire. During operation, fresh coal and limestone are fed continuously into the top of the bed, while ash and slag are drawn off from below. The rich air supply and constant motion in the bed make burning efficient and prevent buildup of large slag clinkers. Steam generator pipes are submerged directly into the fluidized bed, and heat exchange is more efficient than in the water walls of a conventional boiler. More than 90 percent of SO_2 is captured by the limestone particles, and NO_x formation is reduced by holding temperatures around 800°C (1500°F) instead of twice that figure in other boilers. These low temperatures also preclude slag formation, which aids in maintenance. The efficient burning of this process makes it possible to use cheaper fuel, such as lignite or unwashed subbituminous coal, rather than higher priced hard coal.

Flue Gas Desulfurization

Crushed limestone, lime slurry, or alkali (sodium carbonate or bicarbonate) can be injected into a stack gas stream to remove sulfur after combustion. These processes are often called flue gas scrubbing. Spraying wet alkali solutions or limestone slurry is relatively inexpensive and effective, but maintenance can be difficult. Rock-hard plaster and ash layers coat the spray chamber and have to be chipped off regularly. Corrosive solutions of sulfates, chlorides, and fluorides erode metal surfaces. Electrostatic precipitators don't work well because of fouling and shorting of electrodes after wet scrubbing.

Dry alkali injection (spraying dry sodium bicarbonate into the flue gas) avoids many of the problems of wet scrubbing, but the expense of appropriate reagents is prohibitive in most areas. A hybrid procedure called spray drying has been tested successfully in pilot plant experiments. In this process, a slurry of pulverized limestone or slaked lime is atomized in the stack gas stream. The spray rate and droplet size are carefully controlled so that the water flash evaporates and a dry granular precipitate is produced. Passage through a baghouse filter removes both ash and sulfur very effectively.

As with coal washing, scrubbing often results in a trade-off of an air pollution problem for a solid waste disposal problem. Sulfur slag, gypsum, and other products of these processes can amount to three or four times as much volume as fly ash. A large power plant can produce millions of tons of waste per year.

Sulfur Recovery Processes

Instead of making a throwaway product that becomes a waste disposal problem, sulfur can be removed from effluent gases by processes that yield a usable product, such as elemental sulfur, sulfuric acid, or ammonium sulfate. Catalytic converters are used in these recovery processes to oxidize or reduce sulfur and to create chemical compounds that can be collected and sold. Markets have to be reasonably close for economic feasibility, and fly ash contamination must be reduced as much as possible.

Nitrogen Oxide Control

Undoubtedly the best way to prevent nitrogen oxide pollution is to avoid creating it. A substantial portion of the emissions associated with mining, manufacturing, and energy production could be eliminated through conservation (chapter 21).

Staged burners, in which the flow of air and fuel are carefully controlled, can reduce nitrogen oxide formation by as much as

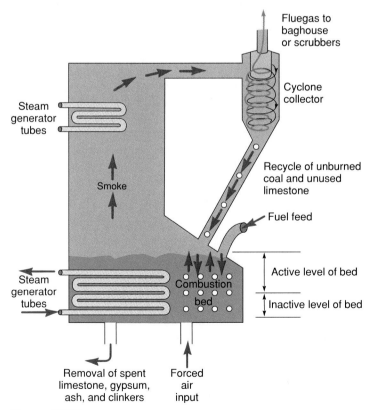

Figure 18.20

Fluidized bed combustion. Fuel is lifted by strong air jets from underneath the bed. Efficiency is good with a wide variety of fuels, and SO_2, NO_x, and CO emissions are much lower than with conventional burners.

50 percent. This is true for both internal combustion engines and industrial boilers. Fuel is first burned at high temperatures in an oxygen-poor environment where NO_x cannot form. The residual gases then pass into an afterburner where more air is added and final combustion takes place in an air-rich, fuel-poor, low-temperature environment that also reduces NO_x formation. Stratified-charge engines and new orbital automobile engines use this principle to meet emission standards without catalytic converters.

The approach adopted by U.S. automakers for NO_x reductions has been to use selective catalysts to change pollutants to harmless substances. Three-way catalytic converters use platinum-palladium and rhodium catalysts to remove up to 90 percent of NO_x, hydrocarbons, and carbon monoxide at the same time (fig. 18.21). Unfortunately, this approach doesn't work on diesel engines, power plants, smelters, and other pollution sources because of problems with back pressure, catalyst life, corrosion, and production of unwanted by-products, such as ammonium sulfate (NH_4SO_4), that foul the system.

Raprenox (*rap*id *r*emoval of *n*itrogen *ox*ides) is a new technique for removing nitrogen oxides that was developed by the U.S. Department of Energy Sandia Laboratory in Livermore, California. Exhaust gases are passed through a container of common, nonpoisonous cyanuric acid. When heated to 350° C (662°F), cyanuric acid releases isocyanic acid gas, which reacts with NO_x to produce CO_2, CO, H_2O, and N_2. In small-scale diesel engine tests, this system eliminated 99 percent of the NO_x. Whether it will work in full-scale applications, especially in flue gases contaminated with fly ash, remains to be seen.

Hydrocarbon Controls

Hydrocarbons and volatile organic compounds are produced by incomplete combustion of fuels or solvent evaporation from chemical factories, painting, dry cleaning, plastic manufacturing, printing, and other industrial processes that use a variety of volatile organic chemicals. Closed systems that prevent escape of fugitive gases can reduce many of these emissions. In automobiles, for instance, positive crankcase ventilation (PCV) systems collect oil

that escapes from around the pistons and unburned fuel and channels it back to the engine for combustion. Modification of carburetor and fuel systems prevents evaporation of gasoline. In the same way, controls on fugitive losses from valves, pipes, and storage tanks in industry can have a significant impact on air quality. Afterburners are often the best method for destroying volatile organic chemicals in industrial exhaust stacks. High air-fuel ratios in automobile engines and other burners minimize hydrocarbon and carbon monoxide emissions, but also cause excess nitrogen oxide production. Careful monitoring of air-fuel inputs and oxygen levels in exhaust gases can minimize all these pollutants.

Clean Air Legislation

Throughout history, there have been countless ordinances prohibiting emission of objectionable smoke, odors, and noise. Air pollution traditionally has been treated as a local problem, however, to be regulated by local authorities. The Clean Air Act of 1963 was the first national legislation in the United States aimed at air pollution control. This act called for research to be carried out by the U.S. Public Health Service on the sources and effects of air pollution. Federal grants were provided to states to combat pollution, but the act was careful to preserve states' rights to set and enforce air quality regulations. It soon became obvious that some pollution problems cannot be solved on a local basis. In 1965, amendments to the Clean Air Act called for national standards for automobile carbon monoxide and hydrocarbon exhausts.

On December 31, 1970, President Nixon signed an extensive set of amendments that essentially rewrote the Clean Air Act. These amendments identified the "criteria pollutants" discussed earlier in this chapter, and established national ambient air quality standards. These standards are divided into two categories. **Primary standards** (table 18.2) are intended to protect human health, while **secondary standards** are set to protect materials, crops, climate, visibility, and personal comfort. Primary and secondary standards are the same for all pollutants except total suspended particulates (TSP), which have a maximum annual geo-

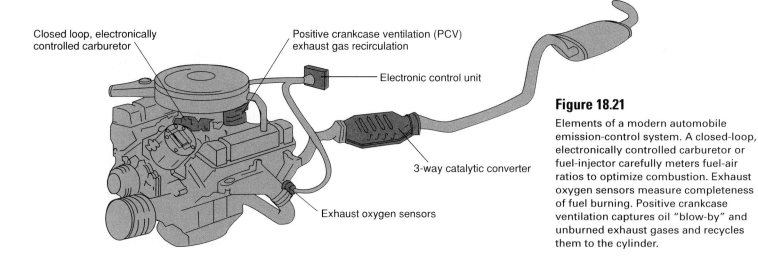

Figure 18.21

Elements of a modern automobile emission-control system. A closed-loop, electronically controlled carburetor or fuel-injector carefully meters fuel-air ratios to optimize combustion. Exhaust oxygen sensors measure completeness of fuel burning. Positive crankcase ventilation captures oil "blow-by" and unburned exhaust gases and recycles them to the cylinder.

Closed loop, electronically controlled carburetor

Positive crankcase ventilation (PCV) exhaust gas recirculation

Electronic control unit

3-way catalytic converter

Exhaust oxygen sensors

Table 18.2
National Ambient Air Quality Standards (NAAQS)

Pollutant	Primary (Health-Based) Averaging Time	Standard Concentration
TSP[a]	Annual geometric mean[b]	75 μg/m^3
	24 hours	260 μg/m^3
SO$_2$	Annual arithmetic mean[c]	80μg/m^3 (0.03 ppm)
	24 hours	365 μg/m^3 (0.14 ppm)
	3 hours	1300 μg/m^3 (0.5 ppm)
CO	8 hours	10 mg/m^3 (9 ppm)
	1 hour	40 mg/m^3 (35 ppm)
NO$_2$	Annual arithmetic mean	100 μg/m^3 (0.05 ppm)
O$_3$	Daily max 1 hour avg	235 μg/m^3 (0.12 ppm)
Lead	Maximum quarterly avg	1.5 μg/m^3
Hydrocarbons	3 hours	160 μg/m^3 (0.24 ppm)

[a]Total suspended particulates
[b]The geometric mean is obtained by taking the nth root of the product of n numbers. This tends to reduce the impact of a few very large numbers in a set.
[c]An arithmetic mean is the average determined by dividing the sum of a group of data points by the number of points.

metric mean of 60 μg/m^3. Ambient standards assume that pollutants have no adverse effects beyond certain thresholds. They also assume that pollutants arising from numerous diverse sources are more reasonably and effectively regulated by setting maximum total levels in the atmosphere than by regulating individual emissions. Some environmentalists disagree with both of these assumptions. These standards are the basis of a warning system called the Air Pollutant Standards Index (table 18.3).

In 1990, after many years of acrimonious debate and political maneuvering, Congress finally passed another set of amendments to the Clean Air Act to protect public health, property, and the environment. Among the most important provisions of this legislation are the following:

- *Acid rain.* Sulfur dioxide releases will be cut from 24 million tons in 1990 to 10 million tons in 2000 by requiring the 111 largest sulfur emitters to meet strict standards. Nitrogen oxide emissions were reduced from 6 million tons in 1990 to 4 million tons in 1992. Sulfur and nitrogen oxide controls already are having an effect. By 1993, sulfur dioxide emissions were down 30 percent and 26 of 33 monitoring stations in a nationwide U.S. system reported significant reductions in acid precipitation compared to previous years.

- *Urban smog.* Motor vehicle tailpipe emissions of hydrocarbons and nitrogen oxides were reduced 35 percent and 60 percent, respectively, in all new cars by 1996. Beginning in 1998, pollution control equipment on new cars must last ten years or 100,000 miles. Oil companies are required to offer alternative fuels, such as methanol or ethanol (sometimes called oxygenated fuels), hydrogen, or compressed natural gas (methane) in cities with the worst pollution problems. Automobile manufacturers will be required to produce 300,000 alternative-fuel cars per year by 1998. Cities not meeting air quality standards for ozone and smog are divided into five categories (marginal, moderate, serious, severe, and extreme); deadlines for attaining standards are set for three,

Table 18.3
Air Pollutant Standards Index

Rating	Description	Health Effects	Suggested Actions
500	Disaster	Very hazardous to all; serious injury and excess deaths, especially in sensitive persons	Stay inside with doors and windows closed; avoid all physical activity
400	Emergency	Hazardous to general population; grave injury possible	Avoid outdoor exercise; young, elderly, and ill should reduce all activity
300	Warning	Very unhealthy for all; serious threat to young, elderly, or ill	Elderly or those with heart or lung disease stay indoors
200	Alert	Irritation of eyes and lungs; aggravation of existing disease	Sensitive persons stay indoors
100	Moderate	NAAQS maximum permissible levels	Avoid traffic and congestion
50	Good	No known short-term effects	No restrictions

six, nine, fifteen, and twenty years, respectively. Pollution sources emitting 100 tons per year in marginal and moderate areas, 50 tons per year in serious areas, 25 tons per year in severe areas, and 10 tons per year in extreme areas are regulated.

- *Toxic air pollutants.* Although the EPA has had authority to set emission standards for air toxics since 1970, only seven (beryllium, mercury, asbestos, lead, vinyl chloride, benzene, and PCBs) were regulated. Now 189 chemicals are listed in about 250 source categories (chemical factories, dry cleaners, coke furnaces, printing plants, etc.). The largest polluters will be required to install the best available technology to reduce emissions 90 percent by 2003. The EPA estimates that toxic emissions will be reduced by about 500,000 tons per year if full compliance is achieved. The safety standard for acceptable cancer risk to nearby residents is set at 1 in 10,000, a rate much too high, according to most environmentalists.

- *Ozone protection.* Chlorofluorocarbons and carbon tetrachloride will be phased out by the year 2000. Recovery and recycling programs for existing CFCs will be instituted. Methyl chloroform will be outlawed by 2002. Hydrochlorofluorocarbons in aerosol cans and insulation will be phased out by 2030.

- *Marketing pollution rights.* Corporations are allowed to offset emissions by buying, selling, and "banking" pollution rights from other factories at an expected savings of $2 billion to $3 billion per year. This is a controversial free-market approach (Market Incentives, pp. 169) that may make economic sense for industry and environmental sense on average but may be disastrous for some localities.

- *Workers compensation.* A $250 million fund is set up to retrain and compensate workers displaced by provisions of this law. This is intended primarily for high-sulfur coal miners in Appalachia who will lose jobs due to fuel switching.

Industrial economists calculate that these regulations will double the current $30 billion per year cost of pollution control. They warn that consumers will pay more for electricity, space heating, transportation, and consumer goods. Some claim health and ecosystem benefits will amount to only about half the $30 billion price of controls. They also warn that these costs will reduce our competitiveness in foreign markets and damage the economy. Congress calculates that the cost of air pollution control will be between $4 to $10 billion per year. Proponents of these provisions argue that pollution control will be a lucrative new business that other countries will seek eagerly in coming years.

Although these revisions are a dramatic step toward cleaner skies, environmentalists didn't get everything they wanted. Electric utilities fought off regulations that would have reduced emissions of mercury and other toxic materials from coal-fired power plants. Steelmakers pleaded financial hardships and were given until the year 2020 to eliminate cancer-causing emissions from coke ovens, providing they take interim steps to reduce pollution.

California has gone further than the federal government in making specific plans for air pollution control. In 1990, the South Coast Air Quality Management District adopted 160 rules to clean the air in the Los Angeles Basin. If these measures are successful, smog-causing emissions could be reduced by 70 percent. By the year 2000, visibility would increase from a 10-mi current average to 60 mi. The number of days when the air is considered hazardous to breathe would decrease from 150 per year to 0 per year.

Reaching these goals will require substantial lifestyle changes for most Californians. Aerosol hair sprays, deodorants, charcoal lighter fluid, gasoline-powered lawnmowers, and drive-through burger stands could be banned. Paints and cleaning solutions would have to emit fewer volatile solvents. Radial tires and more stringent emission controls would be mandated for automobiles. Clean-burning oxygenated fuels or electric motors would be required for all vehicles. Car pooling would be encouraged, parking lots would be restricted, and limits would be placed on the number of cars a family could have.

California's land-use zones and housing codes might have to be changed to accommodate new commuting patterns. Substantial relocations could result. The cost is estimated to be about 60 cents per person per day. Opponents argue that the price tag could be as high as $15 billion a year and 30,000 lost jobs. Whether Californians, or any of us, care enough about health and the environment to make these changes and pay these costs remains to be seen.

Current Conditions and Future Prospects

Although we have not yet achieved the Clean Air Act goals in many parts of the United States, air quality has improved dramatically in the last decade in terms of the major large-volume pollutants. For twenty-three of the largest U.S. cities, the number of days in which air quality reached the hazardous level (PSI greater than 300) is down 93 percent from an average of 1.8 days/year a decade ago to 0.13 days/year now. Of 97 metropolitan areas that failed to meet clean air standards in the 1980s, 41 were in compliance in 1991–92. For many cities, this was the first time they met air quality goals in 20 years. Still, the EPA estimates that some 86 million Americans breathe unhealthy air at least part of the time.

The EPA estimates that emissions of particulate materials are down 60 percent, lead is down 80 percent, SO_2 and CO are down 30 percent, and O_3 is down 18 percent over the past two decades (fig. 18.22). Industrial cities, such as Chicago, Pittsburgh, and Philadelphia, that suffered "smokestack" pollution have had 90 percent reductions in number of days exceeding NAAQS maxima. Filters, scrubbers, and precipitators on power plants and other large stationary sources are responsible for most of the particulate and SO_2 reductions. Catalytic converters on automobiles are responsible for most of the CO and O_3 reductions. The only conventional "criteria" pollutant that has not dropped significantly is NO_x, which had risen 300 percent since 1940, and has dropped only slightly over the last ten years.

Because automobiles are the main source of NO_x, cities where pollution is largely from traffic have had increased PSI levels in recent years. Los Angeles, Anaheim, and Riverside, California, are

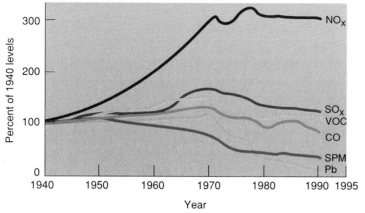

NOx = nitrogen oxides (1940 level: 6.8×10^6 metric tons/yr)
SOx = sulfur oxides (1940 level: 18.0×10^6 metric tons/yr)
VOC = volatile organic compounds (1940 level: 18.5×10^6 metric tons/yr)
SPM = suspended particulate matter (1940 level: 22.8×10^6 metric tons/yr)
Pb = lead (1940 level: 230×10^3 metric tons/yr)
CO = carbon monoxide (1940 level: 81.6×10^6 metric tons/yr)

Figure 18.22

Air pollution trends in the United States, 1940–1991. Emissions of six "criteria" pollutants are expressed as percent of 1940 levels.

Source: Data from the Environmental Protection Agency (EPA).

the only cities in the country in the extreme urban smog category, exceeding the ozone standards an average of 137.5 days per year between 1987 and 1989. Baltimore, New York City, Chicago, Gary, Houston, Milwaukee, Muskegon, Philadelphia, and San Diego are all in the severe category. Eighty-five other urban areas are still considered nonattainment regions. In spite of these local failures, however, 80 percent of the United States now meets the NAAQS goals. This improvement in air quality is perhaps the greatest environmental success story in our history.

The outlook is not so encouraging in other parts of the world, however. The major metropolitan areas of many developing countries are growing at explosive rates to incredible sizes (chapter 24), and environmental quality is abysmal in many of them. The composite average annual levels of SO_2 in Tehran, Iran, for instance, are more than 150 $\mu g/m^3$, and peak levels can be up to ten times higher. Mexico City remains notorious for bad air. Pollution levels exceed WHO health standards 350 days per year and more than half of all city children have lead levels in their blood sufficient to lower intelligence and retard development. Its 131,000 industries and 2.5 million vehicles spew out more than 5500 tons of air pollutants daily. Santiago, Chile, averages 299 days per year on which suspended particulates exceed WHO standards of 90 $\mu g/m^3$.

While there are few statistics on China's pollution situation, it is known that many of China's 400,000 factories have no air pollution controls. Experts estimate that home coal burners and factories emit 10 million tons of soot and 15 million tons of sulfur dioxide annually and that emissions have increased rapidly over the past 20 years. Sheyang, an industrial city in northern China, is thought to have the world's worst particulate problem with peak winter concentrations over 700 $\mu g/m^3$ (nine times U.S. maximum

standards). Airborne particulates in Sheyang exceed WHO standards on 347 days per year. Beijing, Xian, and Guangzhou are nearly as bad. The high incidence of cancer in Shanghai is thought to be linked to air pollution.

As political walls came down across Eastern Europe and the Soviet Union at the end of the 1980s, horrifying environmental conditions in these centrally-planned economies were revealed. Inept industrial managers, a rigid bureaucracy, and lack of democracy have created ecological disasters. Where governments own, operate, and regulate industry, there are few checks and balances or incentives to clean up pollution. Much of the Eastern bloc depends heavily on soft brown coal for its energy and pollution controls are absent or highly inadequate.

Southern Poland, northern Czech Republic, and Slovakia are covered most of the time by a permanent cloud of smog from factories and power plants. Acid rain is eating away historic buildings and damaging already inadequate infrastructures. The haze is so dark that drivers must turn on their headlights during the day. Residents complain that washed clothes turn dirty before they can dry. Zabrze, near Katowice in southern Poland, has particulate emissions of 3600 metric tons per square kilometer. This is more than seven times the emissions in Baltimore, Maryland, or Birmingham, Alabama, the dirtiest cities (for particulates) in the United States. Home gardening in Katowice has been banned because vegetables raised there have unsafe levels of lead and cadmium.

For miles around the infamous Romanian "black town" of Copsa Mica, the countryside is so stained by soot that it looks as if someone had poured black ink over everything. Birth defects afflict 10 percent of infants in northern Bohemia. Workers in factories there get extra hazard pay—burial money, they call it. Life expectancy in these industrial towns is as much as ten years less than the national average. Espenhain, in the industrial belt of the former East Germany, has one of the world's highest rates of sulfur dioxide pollution. One of every two children has lung problems, and one of every three has heart problems (fig. 18.23). Brass doorknobs and name plates have been eaten away by the acidic air in just a few months.

Not all is pessimistic, however. There have been some spectacular successes in air pollution control. Sweden and West Germany (countries affected by forest losses due to acid precipitation) cut their sulfur emissions by two-thirds between 1970 and 1985. Austria and Switzerland have gone even further. They even regulate motorcycle emissions. The Global Environmental Monitoring System (GEMS) reports declines in particulate levels in 26 of 37 cities worldwide. Sulfur dioxide and sulfate particles, which cause acid rain and respiratory disease, have declined in 20 of these cities.

Ten years ago, Cubatao, Brazil was described as the "Valley of Death," one of the most dangerously polluted places in the world. A steel plant, a huge oil refinery, and fertilizer and chemical factories churned out thousands of tons of air pollutants every year. Trees died on the surrounding hills. Birth defects and respiratory diseases were alarmingly high. Since then, however, the citizens of Cubatao have made remarkable progress in cleaning up their environment. The end of military rule and restoration of democracy allowed residents to publicize their complaints. The environment became an important political issue. The state of São Paulo invested

Figure 18.23
A Polish mother gives an oxygen treatment to her child who suffers from respiratory disease triggered by air pollution. Rapid industrialization and lack of pollution controls have left much of the former U.S.S.R. and its allies toxic wastelands. In some villages, three-fourths of all children suffer from pollution-related diseases.

about $100 million, and the private sector spent twice as much to clean up most pollution sources in the valley. Particulate pollution was reduced 75 percent. Ammonia emissions were reduced 97 percent, hydrocarbons that cause ozone and smog were cut 86 percent, and sulfur dioxide production fell 84 percent. Fish are returning to the rivers, and forests are regrowing on the mountains. Progress is possible! We hope that similar success stories will be obtainable elsewhere.

Summary

In this chapter, we have looked at major categories, types, and sources of air pollution. We have defined air pollution as chemical or physical changes brought about by either natural processes or human activities, resulting in air quality degradation. Air pollution has existed as long as there has been an atmosphere. Perhaps the first major human source of air pollution was fire. Burning fossil fuels, biomass, and wastes continues to be the largest source of anthropogenic (human-caused) air pollution. The six conventional large-volume pollutants are NO_x, SO_2, CO, lead, particulates, and volatile organic compounds. The major sources of air pollution are transportation, industrial processes, stationary fuel combustion, and solid waste disposal.

We also looked at some unconventional pollutants. Indoor air pollutants, including formaldehyde, asbestos, toxic organic chemicals, radon, and tobacco smoke may pose a greater hazard to human health than all of the conventional pollutants combined. Odors, visibility losses, and noise generally are not life-threatening but serve as indicators of our treatment of the environment. Some atmospheric processes play a role in distribution, concentration, chemical modification, and elimination of pollutants. Among the most important of these processes are long-range transport of pollutants and photochemical reactions in trapped inversion layers over urban areas.

Encouraging improvements have been made in ambient outdoor air quality over most of the United States in the last decade. We have made considerable progress in designing and installing pollution-control equipment to reduce the major conventional pollutants. There are many types of scrubbers, filters, catalysts, fuel modification processes, and new burning techniques for controlling pollution. The Clean Air Act regulates air quality in the United States through both ambient standards and emission limits, and its 1990 amendments promise a dramatic improvement in our atmosphere. There is much yet to be done, especially in developing countries and in Eastern Europe, but air pollution control is, perhaps, our greatest success in environmental protection and an encouraging example of what can be accomplished in this field.

▼ Questions for Review

1. What is the difference between bronchitis and emphysema? What causes these diseases?

2. What are the most important causes of human illness and death from air pollution?

3. What is acid deposition? What causes it?

4. What have been the effects of acid deposition on aquatic and terrestrial ecosystems?

5. How do electrostatic precipitators, baghouse filters, flue gas scrubbers, and catalytic converters work?

6. What is the difference between primary and secondary standards in air quality?

7. What is the difference between ambient standards and emission limits?

8. What are some of the major toxic air pollutants, and what are their sources?

9. Describe the health effects and suggested actions for each of the levels of the pollution standards index (PSI).

10. Which of the conventional pollutants has decreased most in the recent past and which has decreased least?

▼ Questions for Critical Thinking

1. How would you rank the risks of air pollution-related disease compared to other risks you may face?

2. What might be done to improve indoor air quality? Should the government mandate such changes?

3. Why do you suppose that air pollution is so much worse in Eastern Europe than in the West?

4. Suppose air pollution causes a billion dollars in crop losses each year but controlling the pollution would also cost a billion dollars. Should we insist on controls?

5. In 1984, David Stockman, director of the Office of Management and Budget for President Reagan, said that it would cost $1000 per fish to control acid precipitation in the Adirondack lakes and that it would be cheaper to buy fish for anglers than to put scrubbers on power plants in Ohio. Suppose that was true. Does it justify continuing pollution?

6. What will the ban on fluorocarbon production do to your life? Will it be worth it to save the ozone layer?

7. Is it possible to have zero emissions of pollutants? What does zero mean in this case?

8. If there are thresholds for pollution effects (at least as far as we know now), is it reasonable or wise to depend on environmental processes to disperse, assimilate, or inactivate waste products?

9. Catalytic converters on automobiles definitely improve air quality, but up to one-fourth of car owners disable the converters on their cars by using leaded gasoline. What should we do about this?

10. Do you think that we should continue to use ambient air quality standards or change to absolute emission standards for all pollutants?

▼ Key Terms

acid precipitation 399
aerosol 390
aesthetic degradation 391
ambient air 387
bronchitis 397
carbon monoxide 389
conventional or criteria
    pollutants 387
dry alkali injection 403
electrostatic precipitators 402
emission standards 391
emphysema 397
filters 402
fugitive emissions 386
nitrogen oxides 388

ozone 390
particulate material 390
photochemical oxidants 390
primary pollutants 386
primary standards 404
radon 392
secondary pollutants 386
secondary standards 404
sulfur dioxide 387
synergistic effects 398
temperature inversions 393
unconventional or noncriteria
    pollutants 391
volatile organic compounds 390

▼ Additional Information on the Internet

Acid Rain
    http://www.nbn.com/youcan/acid/acid.html/

Acid Rain FAQs
    http://www.ns.doe.ca/aeb/ssd/Acid/acidFAQ.html/

Air Pollution
    http://wwwwilson.ucsd.edu/education/airpollution/airpollution.html/

EcoNet's Acid Rain Resources
    http://www.econet.apc.org/acidrain/

EcoNet's Climate Resources Directory
    http://www.igc.apc.org/climate/

Measurement of Air Pollution from Satellites
    http://stormy.larc.nasa.gov/press.html/

Out of Breath
    http://www.nrdc.org/nrdc/publ/breath.html/

Ozone Action
    http://www.essential.org/orgs/Ozone_Action/Ozone_Action.html/

Ozone Layer
    http://www.nbn.com/youcan/ozone/ozone.html/

Radon in Earth, Air, and Water
    http://sedwww.cr.usgs.gov:8080/radon/radonhome.html/

World Wide Web Virtual Library: Environment-Atmosphere
    http://ecosys.drdr.Virginia.EDU:80/atm.html/

Note: Further readings appropriate to this chapter are listed on p. 604.

The Little River flows through the Great Smokey Mountain National Park. Without continuous replenishment by the hydrologic cycle, rivers and streams like this would soon dry up.

CHAPTER 19: Water Use and Management

Objectives

After studying this chapter, you should be able to:

- summarize how the hydrologic cycle delivers fresh water to terrestrial ecosystems and how the cycle balances over time.

- contrast the volume and residence time of water in the earth's major compartments.

- describe the important ways we use water and distinguish between withdrawal, consumption, and degradation.

- appreciate the causes and consequences of water shortages around the world and what they mean in people's lives in water-poor countries.

- debate the merits of proposals to increase water supplies and manage demand.

- apply some water conservation methods in your own life.

Would You Fight for Water?

In 1994, after nearly a half century of hostility, Jordan and Israel signed a historic peace agreement. A key issue was cooperative management of one of the region's most valuable resources: water. Israel agreed to supply 50 million cubic meters (13.2 billion gallons) of water annually to Jordan, and to help build dams on tributaries of the Jordan River to provide another 50 million cubic meters each year to Jordanian cities.

Since biblical times, competition for water has been a source of dissension and warfare in the arid Middle East (fig. 19.1). Syrian attempts to divert water from the Jordan River helped ignite the 1967 Six-Day War between Israel and its Arab neighbors. That war gave Israel control of two essential water resources: the Golan Heights, which form the watershed of the Jordan River, and the mountain aquifer under the occupied West Bank. Water from these sources—twice as much per person as is available in Jordan, and 16 times the average supply of Palestinians living directly above the mountain aquifer—has sustained the spectacular

economic growth of Israel's farms and factories, but also engenders envy and resentment among those who have less.

Although water has always been scarce in this region, the situation has become critical as populations have grown, water demand has risen, and cheap, uncontaminated supplies are being exhausted. Many countries already are pumping groundwater faster than it is being recharged, causing subsidence and salt water infiltration in depleted aquifers. Furthermore, farm runoff and urban waste have made surface waters in many areas too polluted to use. Together, Israel, the West Bank and Gaza, and Jordan are facing a combined water deficit of at least 300 million cubic meters per year. "If violence breaks out again in the Middle East," says British author Norman Myers, "it will not be over its most plentiful resource, oil, but over its scarcest, water."

Israel and its neighbors are not alone in struggles over this indispensable resource. Egypt has threatened war if anyone upstream interferes with its only water source, the Nile. Iraq

Figure 19.1

Water has always been the key to survival in the arid Middle East. Who has access to this precious resource and who doesn't has long been a source of tension and conflict. As water demand grows and sources become increasingly depleted, the chances for water wars become greater.

massed its armies at the border to rattle sabers in 1975, when Syria reduced the flow in the Euphrates to fill its new dam at Tabqu. Both Iraq and Syria watched the river apprehensively in 1990 when Turkey closed the gates of the giant Ataturk Dam upstream on the Euphrates. When filled, the reservoir behind this dam will hold ten times more water than the Sea of Galilee, the largest body of fresh water in the region.

Water from Ataturk soon will begin to rush through the world's two largest irrigation tunnels—each 8 m (25 ft) in diameter—to grow peaches, pecans, pomegranates, melons, and grapes on the semiarid Anatolian plains. But every drop that enriches Turkish farmers means less for drinking, growing crops, and producing power downstream. During the 1991 Gulf War, Turkey threatened to completely cut off water flowing to Iraq, a potential deathknell for this desert country. Iraq, in turn, threatened to blow up the Ataturk Dam if any moves were made to reduce its water supply.

As you can see, water—which is essential for life—can be a precious and contentious resource. It can drive nations to war or be used as a weapon against enemies.

Those of us in countries with abundant water supplies tend to think of water as an infinitely available, renewable resource because it is constantly purified and redistributed by the action of the sun, wind, and gravity. But in many parts of the world, water supply is increasingly limited. More people making demands on the resource, natural variations in rainfall, and wasteful or extravagant uses create shortages in many areas. To make matters worse, pollution makes whatever water is available unfit for many uses, further exacerbating supply problems. Eminent hydrologist Luna B. Leopold of the U.S. Geological Survey warns that water shortages might be the environmental crisis of the 1990s and that water conservation might be as much a priority in a few years as energy conservation was in the 1970s.

In this chapter, we will look at the processes that supply fresh water to the land and how humans access and use it. We will survey major water compartments of the environment and see how they are depleted by human uses and replenished by natural processes. Finally, we will examine some schemes for transferring water from one area to another, along with some techniques for conserving water. ▶

Water Resources

Water is a marvelous substance—flowing, rippling, swirling around obstacles in its path, seeping, dripping, trickling—constantly moving from sea to land and back again. Water can be clear, crystalline, icy green in a mountain stream (fig. 19.2) or black and opaque in a cypress swamp. Water bugs skitter across the surface of a quiet lake; a stream cascades down a stairstep ledge of rock; waves roll endlessly up a sand beach, crash in a welter of foam, and recede. Rain falls in a gentle mist, refreshing plants and animals. A violent thunderstorm floods a meadow, washing away stream banks.

Water is essential for life (chapter 3). It is the medium in which all living processes occur. Water dissolves nutrients and distributes them to cells, regulates body temperature, supports structures, and removes waste products. About 60 percent of your body is water. You could survive for weeks without food, but only a few days without water.

The earth is the only place in the universe, as far as we know, where liquid water exists in substantial quantities. Oceans, lakes, rivers, glaciers, and other bodies of liquid or solid water cover more than 70 percent of our world's surface. The total amount of water on our planet is immense—more than 1404 million cu km

Figure 19.2

Why do we find water so pleasing and so fascinating? Why are its many forms so central to natural and human systems? And what would life be like if we didn't have abundant supplies of clean, pure water?

Table 19.1

Some units of water measurement

One cubic kilometer (km^3) equals one billion cubic meters (m^3), one trillion liters, or 264 billion gallons.

One acre-foot is the amount of water required to cover an acre of ground one foot deep. This is equivalent to 325,851 gallons, or 1.2 million liters, or 1234 m^3, about the amount consumed annually by a family of four in the United States.

One cubic foot per second of river flow equals 28.3 liters per second or 449 gallons per minute.

(370 billion billion gal) (table 19.1). If the earth had a perfectly smooth surface, an ocean about 3 km (1.9 mi) deep would cover everything. Fortunately for us, continents rise above the general surface level, creating dry land over about 30 percent of the planet.

We generally assume that most of the earth's water has been formed from oxygen and hydrogen released from rocks by volcanic activity. Other planets have rocks and volcanic activity similar to the Earth's. Why don't they have oceans? Earth is unique in having an atmosphere to trap water vapor, and a temperature range that keeps most of it liquid.

The Hydrologic Cycle

"All rivers run into the sea, yet the sea is not full: Unto the place from which rivers come, thither they return again."

Ecclesiastes 1:7

The hydrologic cycle (water cycle) describes the circulation of water as it evaporates from land, water, and organisms; enters the atmosphere; condenses and is precipitated to the earth's surfaces; and moves underground by infiltration or overland by runoff into rivers, lakes, and seas (chapter 3). The total amount of water on earth remains about the same from year to year, and the hydrologic cycle simply moves it from one place to another (fig. 19.3). This process supplies fresh water to the land masses while also playing a vital role in creating a habitable climate and moderating world temperatures. Movement of water back to the sea in rivers and glaciers is a major geological force that shapes the land and redistributes material. Plants play an important role in the hydrologic cycle, absorbing groundwater and pumping it into the atmosphere by transpiration (transport plus evaporation). In tropical forests, as much as 75 percent of annual precipitation is returned to the atmosphere by plants.

Solar energy drives the hydrologic cycle by evaporating surface water. **Evaporation** is the process in which a liquid is changed to vapor (gas phase) at temperatures well below its boiling point. Water also can move between solid and gaseous states without ever becoming liquid in a process called **sublimation.** On bright, cold, windy winter days, when the air is very dry, snowbanks disappear by sublimation, even though the temperature never gets above freezing. This is the same process that causes "freezer burn" of frozen foods.

In both evaporation and sublimation, molecules of water vapor enter the atmosphere, leaving behind salts and other contaminants and thus creating purified fresh water. This is essentially distillation on a grand scale. We used to think of rainwater as a symbol of purity, a standard against which pollution could be measured. Unfortunately, increasing amounts of atmospheric pollutants are picked up by water vapor as it condenses into rain.

The amount of water vapor in the air is called humidity. Warm air can hold more water than cold air. When a volume of air contains as much water vapor as it can at a given temperature, we say that it has reached its **saturation point. Relative humidity** is the amount of water vapor in the air expressed as a percentage of the maximum amount (saturation point) that could be held at that particular temperature.

When the saturation concentration is exceeded, water molecules begin to aggregate in the process of **condensation.** If the temperature at which this occurs is above 0° C, tiny liquid droplets result. If the temperature is below freezing, ice forms. For a given amount of water vapor, the temperature at which condensation occurs is the **dew point.** Tiny particles, called **condensation nuclei,** float in the air and facilitate this process. Smoke, dust, sea salts, spores, and volcanic ash all provide such particles. Even apparently clear air can contain large numbers of these particles, which are generally too small to be seen by the naked eye. Sea salt is an excellent source of such nuclei, and heavy, low clouds frequently form in the humid air over the ocean. Some nucleating agents are so efficient at accumulating water that they can cause precipitation even when the air is far below its saturation point.

A cloud, then, is an accumulation of condensed water vapor in droplets or ice crystals. Normally, cloud particles are small enough to remain suspended in the air, but when cloud droplets and ice crystals become large enough, gravity overcomes uplifting air currents and precipitation occurs. Some precipitation never reaches the ground. Temperatures and humidities in the clouds where snow

Water Use and Management

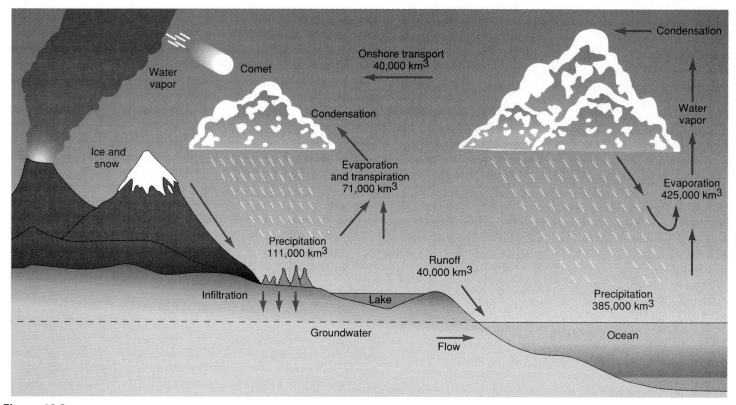

Figure 19.3

In the hydrologic cycle water moves constantly between aquatic, atmospheric, and terrestrial compartments driven by solar energy and gravity. The total annual runoff from land to the oceans is about 10.3×10^{15} gallons.

and ice form are ideal for their preservation, but as they fall through lower, warmer, and drier air layers, reevaporation occurs. Rising air currents lift this water vapor back into the clouds, where it condenses again; thus, liquid water and ice crystals may exist for only a few minutes in this short cycle between clouds and air (fig. 19.3).

Rainfall and Topography

Rain falls unevenly over the planet. In some places, it rains more or less constantly, while other areas get almost no precipitation of any kind. At Iquique, in the Chilean desert, for instance, no rain has fallen in recorded history. At the other end of the scale, 22 m (72 ft) of rain was recorded in a single year at Cherrapunji in India. Figure 19.4 shows broad patterns of precipitation around the world.

Very heavy rainfall is typical of tropical areas, especially where monsoon winds carry moisture-laden sea air onshore (chapter 17). Mountains act as both cloud formers and rain catchers. As air sweeps up the windward side of a mountain, pressure decreases and temperature falls, causing relative humidity to increase. Eventually, the air is supersaturated with moisture and condensation occurs. Further cooling of the air causes droplets of moisture to coalesce and become too heavy to remain suspended, so they fall as rain. The now cooler and drier air continues over the mountain to the leeward side. It descends and warms once again, reducing its relative humidity even further, not only preventing rainfall there, but absorbing moisture from other sources. As a result, a mountain range generally has two distinct climatic personalities. The wind-

ward side is usually cool, wet, and cloudy, while the leeward side is warm, dry, and sunny. We call the dry area on the downwind side of a mountain its **rain shadow.**

A striking example of this dichotomy is found in the Hawaiian Islands (fig. 19.5). The windward side of Mount Waialeale on the island of Kauai is one of the wettest places on earth, with an annual rainfall near 12 m (460 in). The leeward side, only a few kilometers away, is in the rain shadow of the mountain and has an average yearly rainfall of only 46 cm (18 in). On a broader scale, some mountain ranges cast rain shadows over vast areas. The Himalaya and Karakorum ranges of south Asia block moisture-laden monsoon winds from reaching central Asia. The Sierra Nevada of California and the coastal ranges of Oregon and Washington intercept moisture-laden Pacific winds, resulting in the arid intermountain Great Basin of the western United States.

Desert Belts

Rising and falling air masses that result from global circulation patterns also help create deserts in two broad belts on either side of the equator around the world. Evaporation is highest near the equator where direct rays of the sun produce the greatest heat budgets. Hot air over the equator rises, cools, and drops its moisture as rain; thus, equatorial regions are areas of high precipitation. As this cooler, drier air moves toward the poles, it condenses and sinks earthward again along the Tropics of Cancer and Capricorn (23° north and south latitude, respectively), warming as it descends.

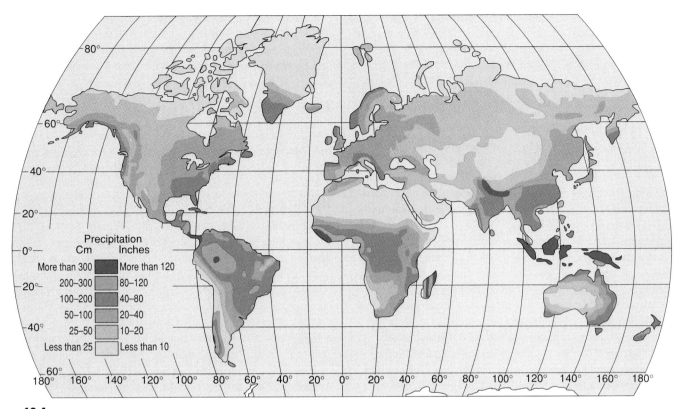

Figure 19.4

Mean annual precipitation. Note wet areas that support tropical rainforests occur along the equator while the major world deserts occur in subsidence zones between 20 and 40 degrees North and South.

From Jerome Fellmann, et al., *Human Geography*, 4th ed. Copyright © 1995 Times Mirror Higher Education Group, Inc., Dubuque, Iowa. All Rights Reserved. Reprinted by permission.

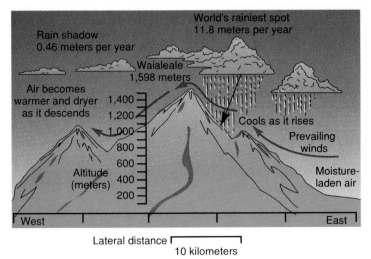

Figure 19.5

Rainfall on the east side of Mount Waialeale in Hawaii is more than twenty times as much as on the west side. Prevailing trade winds bring moisture-laden sea air onshore. The air cools as it rises up the flanks of the mountain and the water it carries precipitates as rain—11.8 m (38 ft) per year!

This hot, dry air causes high evaporative losses in these subtropical regions and creates great deserts on nearly every continent: the Sahara of North Africa, the Takla Makan and Gobi of China, the Sonoran and Chihuahuan of Mexico and the United States, the Kalahari and Namib of Southwest Africa, and Australia's Great Sandy Desert. Dry air cascading down the west side of the Andes creates some of the driest deserts in the world along the coasts of Chile and Peru.

Humans and domestic animals have expanded many of these deserts by destroying forests and stripping away protective vegetation from once fertile lands, exposing the bare soil to erosion. Local weather patterns and water supplies also are adversely affected by this process of desertification. Weather and climate were discussed in chapter 17.

Balancing the Water Budget

Everything about global hydrological processes is awesome in scale. Each year, the sun evaporates approximately 496,000 cu km of water from the earth's surface. More water evaporates in the tropics than at higher altitudes, and more water evaporates over the oceans than over land. Although the oceans cover about 70 percent of the earth's surface, they account for 86 percent of total evaporation. Ninety percent of the water evaporated from the ocean falls back on the ocean as rain. The remaining 10 percent is carried by prevailing winds over the continents where it combines with water evaporated from soil, plant surfaces, lakes, streams, and wetlands to provide a total continental precipitation of about 111,000 km^3.

What happens to the surplus water on land—the difference between what falls as precipitation and what evaporates? Some of it is incorporated by plants and animals into biological tissues. A large share of what falls on land seeps into the ground to be stored for a while (from a few days to many thousands of years) as soil moisture or groundwater. Eventually, all the water makes its way back downhill to the oceans. The 40,000 km^3 carried back to the ocean each year by surface runoff or underground flow represents the renewable supply available for human uses and sustaining freshwater-dependent ecosystems.

The global water budget thus is balanced by circulation systems on land, in the atmosphere, and in the oceans that move water from areas of excess to areas of deficit. Rivers that carry water from the land to the sea are balanced by wind currents flowing in great swirls above the earth, moving moisture-laden air from one region to another. Ocean currents are equivalent to vast rivers, carrying warm water from the equator to higher latitudes on the surface and returning cold, nutrient-rich waters in deep currents. The Gulf Stream, which flows along the east coast of North America at a steady rate of 10 to 12 km per hour (6 to 7.5 mph), carries more than one hundred times more water than all the rivers on land put together.

The redistribution of heat that results from the massive evaporation, precipitation, and transport of water is a major factor in keeping world temperatures relatively constant and making the world habitable. Without oceans to absorb and store heat, and wind currents to redistribute that heat in the latent energy of water vapor, the earth would probably undergo extreme temperature fluctua-

tions like those of the moon, where it is 100° C (212° F) during the day and −130° C (−200° F) at night. Water is able to perform this vital function because of its unique properties in heat absorption and energy of vaporization (chapter 3).

Major Water Compartments

The distribution of water often is described in terms of interacting compartments in which water resides for short or long times. Table 19.2 shows the major water compartments in the world.

Oceans

Together, the oceans contain roughly 97 percent of all the *liquid* water in the world. (The water of crystallization in rocks is far larger than the amount of liquid water.) While the ocean basins really form a continuous reservoir, shallows and narrows between them reduce water exchange, so they have different compositions, climatic effects, and even different surface elevations. Oceans play a crucial role in moderating the earth's temperature (fig. 19.6), but they are generally too salty for most human uses. Nevertheless, over 90 percent of the world's living biomass is contained in the oceans.

In tropical seas, surface waters are warmed by the sun, diluted by rainwater and runoff from the land, and aerated by wave action. In higher latitudes, surface waters are cold and much more dense. This dense water subsides or sinks to the bottom of deep ocean basins and flows toward the equator. Warm surface water of the tropics stratifies or floats on top of this cold, dense water like cream on an unstirred cup of coffee. Sharp boundaries form between different water densities, different salinities, and different temperatures, retarding mixing between these layers.

Table 19.2
Earth's water compartments—estimated volume of water in storage, percent of total, and average residence time

	Volume (thousands of km^3)	% Total Water	Average Residence Time
Total	1,403,377	100	2800 years
Ocean	1,370,000	97.6	3000 years to 30,000 years*
Ice and snow	29,000	2.07	1 to 16,000 years*
Groundwater down to 1 km	4000	0.28	From days to thousands of years*
Lakes and reservoirs	125	0.009	1 to 100 years*
Saline lakes	104	0.007	10 to 1000 years*
Soil moisture	65	0.005	2 weeks to a year
Biological moisture in plants and animals	65	0.005	1 week
Atmosphere	13	0.001	8 to 10 days
Swamps and marshes	3.6	0.003	From months to years
Rivers and streams	1.7	0.0001	10 to 30 days

Source: Data from U.S. Geological Survey.
*Depends on depth and other factors

Figure 19.6
The cold Japanese current keeps water temperatures low year round along the coast of Washington and Oregon but also delivers nutrients that make this a highly productive aquatic ecosystem.

While parts of the hydrologic cycle occur on a time scale of hours or days, other parts take centuries. The average **residence time** of water in the ocean (the length of time that an individual molecule spends circulating in the ocean before it evaporates and starts through the hydrologic cycle again) is about three thousand years. In the deepest ocean trenches, movement is almost nonexistent and water may remain undisturbed for tens of thousands of years.

Glaciers, Ice, and Snow

Of the 3 percent of all water that is fresh, about three-fourths is tied up in glaciers, ice caps, and snowfields. Glaciers are really rivers of ice flowing downhill very slowly (fig. 19.7). They now occur only at high altitudes or high latitudes, but as recently as 18,000 years ago about one-third of the continental landmass was covered by glacial ice sheets. Most of this ice has now melted and the largest remnant is in Antarctica. As much as 2 km (1.25 mi) thick, the Antarctic glaciers cover all but the highest mountain peaks and contain nearly 85 percent of all ice in the world.

An ice sheet that is similar in thickness but much smaller in volume covers most of Greenland. There is no landmass at the North Pole. A permanent ice pack made of floating sea ice covers much of the Arctic Ocean. Although sea ice comes from ocean water, salt is excluded in freezing so the ice is mostly fresh water. Together with the Greenland ice sheet, arctic ice makes up about 10 percent of the total ice volume. The remaining 5 percent of the world's permanent supply of ice and snow occurs mainly on high mountain peaks.

Groundwater

After glaciers, the next largest reservoir of fresh water is held in the ground as **groundwater.** Precipitation that does not evaporate back into the air or run off over the surface percolates through the soil

Figure 19.7
Glaciers are rivers of ice sliding very slowly downhill. Together polar ice sheets and alpine glaciers contain more than three times as much fresh water as all the lakes, ponds, streams, and rivers in the world. The dark streaks on the surface of this Alaskan glacier are dirt and rocks marking the edges of tributary glaciers that have combined to make this huge flow.

and into pores and hollows of permeable rocks in a process called **infiltration** (fig. 19.8). Upper soil layers that hold both air and water make up the **zone of aeration.** Moisture for plant growth comes primarily from these layers. Depending on rainfall amount, soil type, and surface topography, the zone of aeration may be very shallow or quite deep. Lower soil layers where all spaces are filled

Water Use and Management

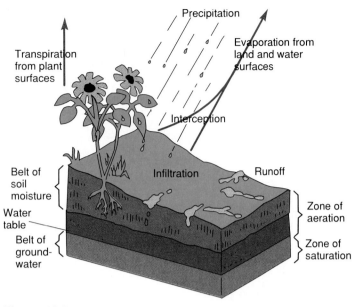

Figure 19.8

Precipitation that does not evaporate or run off over the surface percolates through the soil in a process called infiltration. The upper layers of soil hold droplets of moisture between air-filled spaces. Lower layers, where all spaces are filled with water, make up the zone of saturation or groundwater.

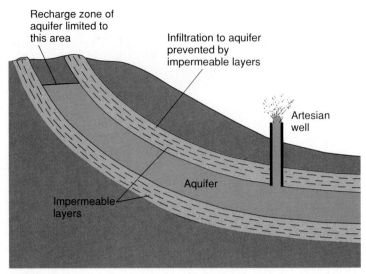

Figure 19.9

An aquifer is a porous, water-bearing layer of sand, gravel, or rock. This aquifer is confined between impermeable layers of rock or clay and bent by geological forces, creating hydrostatic pressure. A break in the overlying layer creates an artesian well or spring.

with water make up the **zone of saturation.** The top of this zone is the **water table.** The water table is not flat, but undulates according to the surface topography and subsurface structure. Nor is it stationary through the seasons, rising and falling according to precipitation and infiltration rates.

Porous, water-bearing layers of sand, gravel, and rock are called **aquifers.** Aquifers are always underlain by impermeable layers of rock or clay that keep water from seeping out at the bottom.

Folding and tilting of the earth's crust by geologic processes can create shapes that generate water pressure in confined aquifers (those trapped between two impervious rock layers). When a pressurized aquifer intersects the surface, or if it is penetrated by a pipe or conduit, an artesian well or spring results from which water gushes without being pumped.

Areas in which infiltration of water into an aquifer occurs are called **recharge zones** (fig. 19.9). The rate at which most aquifers are refilled is very slow, however, and groundwater presently is being removed faster than it can be replenished in many areas. Although water use and resource depletion are discussed later in this chapter, it is significant to refer to some present problems with aquifer recharging. Urbanization, road building, and other development often block recharge zones and prevent replenishment of important aquifers. Contamination of surface water in recharge zones and seepage of pollutants through wells has polluted aquifers in many places, making them unfit for most uses (chapter 20). Many cities protect aquifer recharge zones from pollution or development, both as a way to drain off rainwater and as a way to replenish the aquifer with pure water.

Some aquifers contain very large volumes of water. The groundwater within 1 km of the surface in the United States is more than thirty times the volume of all the lakes, rivers, and reservoirs on the surface. Water can flow through aquifers in massive underground rivers, and large springs sometimes produce billions of liters of water per day.

Rivers and Streams

Precipitation that does not evaporate or infiltrate into the ground runs off over the surface, drawn by the force of gravity back toward the sea. Rivulets accumulate to form streams, and streams join to form rivers. Although the total amount of water contained at any one time in rivers and streams is small compared to the other water reservoirs of the world (table 19.2), these surface waters are vitally important to humans and most other organisms. Most rivers, if they were not constantly replenished by precipitation, meltwater from snow and ice, or seepage from groundwater, would begin to diminish in a few weeks.

The speed at which a river flows is not a very good measure of how much water it carries. Headwater streams are usually small and fast, often tumbling downhill in a continuous cascade. As the stream reaches more level terrain, it slows and generally becomes deeper and more quiet. The best measure of the volume carried by a river is its **discharge,** the amount of water that passes a fixed point in a given amount of time. This is usually expressed as liters or cubic feet of water per second. The sixteen largest rivers in the world carry nearly half of all surface runoff on earth. The Mississippi River, which is the fourth longest river in the world and the largest in North America, carries an average of 14 million liters (450,000 cu ft) per second. Peak flow in the spring can be as high as 45 million liters per second.

Lakes and Ponds

Ponds are generally considered to be small temporary or permanent bodies of water shallow enough for rooted plants to grow over most of the bottom. Lakes are inland depressions that hold standing fresh water year-round. Maximum lake depths range from a few meters to over 1600 m (1 mi) in Lake Baikal in Siberia. Surface areas vary in size from less than one-half ha (one ac) to large inland seas, such as Lake Superior or the Caspian Sea, covering hundreds of thousands of square kilometers. Both ponds and lakes are relatively temporary features on the landscape because they eventually fill with silt or are emptied by cutting of the outlet stream through the barrier that creates them.

While lakes contain nearly one hundred times as much water as all rivers and streams combined, they are still a minor component of total world water supply. Their water is much more accessible than groundwater or glaciers, however, and they are important in many ways for humans and other organisms.

Wetlands

Bogs, swamps, wet meadows, and marshes play a vital and often unappreciated role in the hydrological cycle. Their lush plant growth stabilizes soil and holds back surface runoff, allowing time for infiltration into aquifers and producing even, year-long stream flow. In the United States, about 20 percent of the 1 billion ha of land area was once wetland. In the past 200 years, more than one-half of those wetlands have been drained, filled, or degraded. Agricultural drainage accounts for the bulk of the losses.

When wetlands are disturbed, their natural water-absorbing capacity is reduced and surface waters run off quickly, resulting in floods and erosion during the rainy season and dry, or nearly dry, stream beds the rest of the year. This has a disastrous effect on biological diversity and productivity, as well as on human affairs (chapter 14).

The Atmosphere

The atmosphere is among the smallest of the major water reservoirs of the earth in terms of water volume, containing less than 0.001 percent of the total water supply. It also has the most rapid turnover rate. An individual water molecule resides in the atmosphere for about ten days, on average. While water vapor makes up only a small amount (4 percent maximum at normal temperatures) of the total volume of the air, movement of water through the atmosphere provides the mechanism for distributing fresh water over the landmasses and replenishing terrestrial reservoirs.

Water Availability and Use

Clean, fresh water is essential for nearly every human endeavor. Perhaps more than any other environmental factor, the availability of water determines the location and activities of humans on earth. **Runoff** is the excess of precipitation over evaporation and infiltration. Of the 41,000 cubic km of annual runoff, however, about two-thirds is lost as seasonal floods. The remaining one-third is **stable** runoff (dependable year-round) and represents, in broad terms, the water available for human use. Of the stable runoff, 5000 km^3 is in sparsely inhabited regions, leaving only about 9000 km^3 readily accessible for human use. This is still a large amount, representing 1500 m^3 (about 400,000 gal) per person per year.

Water Supplies

The richest continents in terms of total water supply are South America and Asia. Each has about 12 percent of the total land area of the world but receives about one-fourth of the total global runoff. In terms of water available per person, South America has the most abundant supply. Its 27 percent of total runoff is shared by only 6 percent of the world population. However, most of the rainfall and runoff in South America occurs in the jungles of the Amazon basin, where infertile soil and inhospitable conditions limit human habitation. Much of the runoff in Asia does occur in areas suitable for agriculture, which is one of the reasons Asia has nearly 60 percent of all humans on earth.

The richest country in the world—in terms of per capita water supply—is Iceland, which has an annual renewable supply of 670,000 m^3 (177 million gal) per person. This is 6 times the per capita supply in Canada and 68 times the annual per capita supply in the United States. By contrast, Kuwait and Bahrain have *no* renewable water supply; they depend entirely on desalinized sea water or imports. Egypt, in spite of the fact that the Nile River flows through it, has only 30 cu m per capita per year on a renewable basis, 20,000 times less per person than Iceland.

Another important consideration is rainfall interannual variability. In some areas, such as the African Sahel region, abundant rainfall occurs some years but not others. Unless steps are taken to even out water flows, the lowest levels encountered usually limit both ecosystem functions and human activities. Some of the world's earliest civilizations, such as the Sumerians and Babylonians of Mesopotamia, the Harappans of the Indus Valley, and the early Chinese cultures, were based on communal efforts to control water, to divert floods and drain marshes during wet seasons or wet years, and to store water in reservoirs or divert it from streams so that it would be available during the dry seasons or dry years.

Drought Cycles

Rainfall is never uniform in either geographical distribution or yearly amount. Every continent has regions where rainfall is scarce because of topographic effects or wind currents. In addition, cycles of wet and dry years create temporary droughts. Water shortages have their most severe effect in semiarid zones where moisture availability is the critical factor in determining plant and animal distribution. Undisturbed ecosystems often survive extended droughts with little damage, but introduction of domestic animals and agriculture disrupts native vegetation and undermines natural adaptations to low moisture levels.

In the United States, the cycle of drought seems to be about 30 years. There were severe dry years in the 1870s, 1900s, 1930s, 1950s, and 1970s. The worst of these in economic and social terms were the 1930s. Wasteful farming practices and a series of dry years in the Great Plains combined to create the "dust bowl." Wind stripped topsoil from millions of hectares of land, and billowing dust clouds turned day into night (see chapter 11). Thousands of families were forced to leave farms and migrate to cities. There now is a great worry that the greenhouse effect (see chapter 17) will bring about major climatic changes and make droughts both more frequent and more severe than in the past.

Types of Water Use

In contrast to energy resources, which are consumed when used, water has the potential for being reused many times. In discussing water appropriations, we need to distinguish between different kinds of uses and how they will affect the water being appropriated.

Withdrawal is the total amount of water taken from a lake, river, or aquifer for any purpose. Much of this water is employed in nondestructive ways and is returned to circulation in a form that can be used again. **Consumption** is the fraction of withdrawn water that is lost in transmission, evaporation, absorption, chemical transformation, or otherwise made unavailable for other purposes as a result of human use. **Degradation** is a change in water quality due to contamination or pollution so that it is unsuitable for other desirable service. The total quantity available may remain constant after some uses, but the quality is degraded so the water is no longer as valuable as it was.

Worldwide, humans withdraw about 10 percent of the total annual runoff and about 25 percent of the stable runoff. The remaining three-quarters of the stable supply is generally either uneconomical to tap (it would cost too much to store, ship, purify, or distribute), or there are ecological constraints on its use. Consumption and degradation together account for about half the water withdrawn in most industrial societies. The other half of the water we withdraw would still be valuable for further uses if we could protect it from contamination and make it available to potential consumers.

We have always treated water as if there is an inexhaustible supply. It has been cheaper and more convenient for most people to dump all used water and get a new supply than to determine what is contaminated and what is not. The natural cleansing and renewing functions of the hydrologic cycle do replace the water we need if natural systems are not overloaded or damaged. Water is a renewable resource, but renewal takes time. The rate at which we are using water now may make it necessary to conscientiously protect, conserve, and replenish our water supply.

Quantities of Water Used

Human water use has been increasing about twice as fast as population growth over the past century (fig. 19.10). Water use is stabilizing in industrialized countries, but demand will increase in developing countries where supplies are available. The average amount of water withdrawn worldwide is about 646 cu m (170,616 gal) per person per year. This overall average hides great discrepancies in the proportion of annual runoff withdrawn in different areas. As you might expect, those countries with a plentiful water supply and a small population withdraw a very small percentage of the water available to them. Canada, Brazil, and the Congo, for instance, withdraw

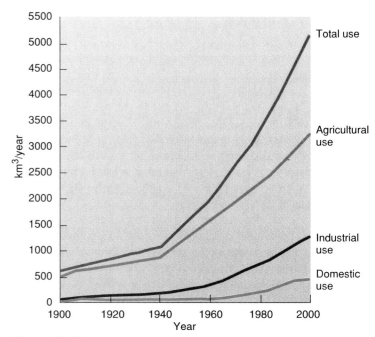

Figure 19.10

Growth of global water use 1900–2000.

Source: Data from L. A. Shiklomanov, "Global Water Resources" in *Nature and Resources,* vol. 26, p. 34–43, UNESCO, Paris.

less than 1 percent of their annual runoff. By contrast, in countries such as Libya and Israel, where water is one of the most crucial environmental resources, groundwater and surface water withdrawal together amount to more than 100 percent of their renewable supply.

The total runoff from precipitation in the United States amounts to an average of 10,430 cu m (2.7 million gal) per person per year. We now withdraw about one-fifth of that amount, or some 5400 l (1400 gal) per person per day. By comparison, the average water use in less-developed countries is only about 45 l per person per day.

Use by Sector

Water use can be analyzed by identifying three major kinds of use, or sectors: public, industry, and agriculture. Worldwide, agriculture claims about 69 percent of total water withdrawal, ranging from 93 percent of all water used in India to only 4 percent in Kuwait, which cannot afford to spend its limited water on crops. Canada, where the fields are well watered by natural precipitation, uses only 12 percent of its water for agriculture. As you can see in figure 19.11, water use by sector depends strongly on national wealth and degree of industrialization. Poorer countries with little industry and limited domestic supply systems use little of their water in these sectors.

In most places, agricultural water use is notoriously inefficient and highly consumptive. Typically, from 70 to 90 percent of the water withdrawn for agriculture never reaches the crops for which it is intended. The most common type of irrigation is to simply flood the whole field or run water in rows between crops. As much as half is lost through evaporation or seepage from unlined irrigation canals bringing water to fields. Most of the rest runs off, evap-

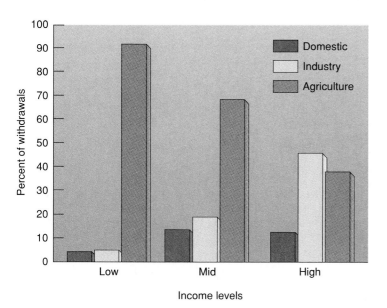

Figure 19.11

Water withdrawals by sector in low-, middle-, and high-income countries.

Source: Data from World Bank, 1992.

Figure 19.12

Center-pivot sprinklers allow farmers to irrigate crops on rolling terrain. Wheels driven by electric or gasoline engines circle around a central deep well and can water an area up to 2.6 km² (640 acres). Agriculture now uses 42 percent of all water withdrawn in high-income countries. In some areas, irrigation consumes 90 percent of all available water.

orates, or infiltrates into the field before it can be used. The water that evaporates or seeps into the ground is generally lost for other purposes, and the runoff from fields is often contaminated with soil, fertilizer, pesticides, and crop residues, making it low quality. Sprinklers (fig. 19.12) are more efficient in distributing water evenly over the field than flooding and can be used on uneven terrain, but they lose a great deal of water to evaporation. An interesting case of highly productive agricultural water management is seen on the Indonesian island of Bali (Case Study, p. 422).

Worldwide, industry accounts for about one-fourth of all water use, ranging from 70 percent of withdrawal in some European countries, such as Germany, to 5 percent in less industrialized countries, such as Egypt and India. Cooling water for power plants is by far the largest single industrial use of water, typically accounting for 50 to 100 percent of industrial withdrawal. Unlike agriculture, however, only a small fraction of this water is consumed or degraded. Most power plants have a "once-through" cooling system that returns water to its source after is passes through the plant. Typically, only 2 to 5 percent of the cooling water is lost through leaks or evaporation. If care is taken to avoid contamination, this water is not degraded and can be used for other purposes.

A few other industries account for the majority of the remaining industrial water use. In the United States, primary metal smelting and fabrication, petroleum refining, pulp and paper manufacturing, and food processing use about two-thirds of the industrial water not used by power plants. You would probably be surprised to learn how much water is used to manufacture some of the ordinary products that we consume. Table 19.3 shows a sample of products and the water used in their production. Much of the water used by these industries could be recycled and used over again in the factory. This would have benefits both in extending water supplies and in protecting water quality. Although Third World coun-

tries typically allocate only about 10 percent of their water withdrawal to industry, this could change rapidly as they industrialize. Water may be as important as energy in determining which countries develop and which remain underdeveloped.

Freshwater Shortages

Water is a major limiting factor of the environment, both for biological systems and human societies. Our growing world population is placing great demands upon natural freshwater sources. The world is faced with increasing pressure on water resources and widespread, long-lasting water shortages in many areas for three reasons: (1) rising demand, (2) unequal distribution of usable fresh water, and (3) increasing pollution of existing water supplies. The Russian hydrologist G. P. Kalinin predicts that by the year 2000 about half of all the earth's renewable water will be in use by humans.

A Scarce Resource

About 2 billion people, more than one-third of the world's population, lack safe drinking water or adequate sanitation. The World Health Organization considers 2000 cu m (53,000 gal) of good water per person per year to be the minimum for a healthful life. Some forty countries in the world fall below this level. The highest percentages of people in water-poor countries are in Africa and the Middle East (table 19.4). Some countries fare much worse than others. In Mali, for example, 88 percent of the population lacks clean water; in Ethiopia it is 94 percent. Rural people generally have less access to clean water than do city dwellers. In the thirty-three worst affected countries, 60 percent of urban people can get clean water as opposed to only 20 percent of those in the country.

Water Use and Management

CASE STUDY
The Engineered Landscape of Bali

Often regarded as one of the most beautiful places on earth, the Indonesian island of Bali fits most dreams of a South Sea paradise (see cover photo). Ringed by coral reefs in the shallow Java Sea just north of Australia, the island rises smoothly to the active volcanic cone of towering Mt. Agung. Patches of dense tropical forest alternate with carefully tended rice paddies that climb the mountain slopes like brilliant green stairsteps. Small villages sprinkled with coconut palms and banana trees dot the tranquil scene (fig. 1).

An overview doesn't reveal, however, that this small island is one of the most densely populated places in the world. Sustaining such a high density requires a great deal of social and environmental coordination. What appears to be a simple rural landscape is really a complex cultural, religous, and ecological system evolved over millennia to link people to each other and to their environment.

The heart of Balinese culture, religion, and economy is rice paddy agriculture. Rich volcanic soil, a perpetually warm climate, and abundant moisture generate some of the highest rice harvests in the world. Even more remarkable is that this productivity has been maintained for more than 1000 years without any apparent yield reductions. Careful water management is key in this highly productive landscape. Although Bali usually gets plenty of rain, it falls unevenly throughout the year. In the wet season, heavy rains threaten to wash away the rice terraces. During the dry season, there isn't enough water to fill the paddies if everyone plants and irrigates at once.

Farmers are organized into cooperatives called subaks. Each subak oversees a group of interconnected rice paddies sharing a common water source. Starting at a wier, a partial dam on one of the swift streams cascading down the mountainside, water is carried through tunnels or open channels to the uppermost terrace in the subak. From there,

Figure 1

The engineered landscape of Bali, based on religious beliefs and agroecological wisdom, inextricably links villagers to their community and to their environment.

water trickles from one paddy to the next until it is gathered into another channel to be carried to the next subak. Eventually, after passing through a dozen or more subaks and hundreds of individual rice paddies, the water returns to the river and continues on its way to the ocean below.

Each farmer is dependent on those upslope to release just enough water at the right time so that crops will grow and paddies won't be too dry or washed away by too much water. How is this managed? The answer is, through religious rituals. Water use is intimately connected with religious beliefs and practices. At the top of each subak is a small temple where farmers gather on important religious days. Higher up, a larger temple marks the wier or spring that supplies water to several subaks. The highest point in the landscape is dominated by a regional Masceti water temple that coordinates the activities of the entire watershed.

Before each planting season, representatives from each subak meet at the Masceti temple to consult with the priests, discuss planting and irrigation schedules, and arrange communal management of wiers, canals, and diversion systems. Planting times within the watershed are staggered so that an uphill farmer is emptying his or her paddies just as a downhill neighbor needs it.

Pest control also is regulated by the temple calendar. During the growing season, the paddies are fertile breeding grounds for pests and diseases. Fields must lie fallow for a time after harvest to reduce pest populations. But fallow periods must be coordinated within a subak to keep pests from simply migrating from one field to another.

Juggling the complex schedules so that every farmer gets enough water at the right time and no one is overwhelmed by pests is a mind-boggling problem. Scientists using computer models have found that they can't do any better scheduling than do the traditional religious system and the subtle negotiating skill of the villagers. When the Dutch conquered Bali in the later half of the 19th century, they assumed they knew all about water management and ignored the ancient Balinese system. The result was a disaster. Only recently, have they come to appreciate that the native system is both more complex and more successful than anything that we can offer in its place. This case raises questions about what else we might learn from traditional cultures elsewhere in the world.

Table 19.3
Examples of water use

	Liters	Gallons
Home use:		
Bath	100–150	30–40
Shower	20 per min	5 per min
Washing clothes	75–100	20–30
Cooking	30	8
Flushing toilet (once)	10–15	3–4
Watering lawn	40 per min	10 per min
Agriculture and food processing:		
1 egg	150	40
1 ear corn	300	80
1 loaf bread	600	160
1 pound beef	9500	2500
Industrial and commercial products:		
1 Sunday paper	1000	280
1 pound steel	110	32
1 pound synthetic rubber	1100	300
1 pound aluminum	3800	1000
1 automobile	380,000	100,000

Table 19.4
Water shortages by region (percentage of population living in countries with less than 2000 cubic meters per person per year)[1]

Sub-Saharan Africa	26
East Asia & Oceania	7
South Asia	0
East Europe & former USSR	22
Other Europe	20
Middle East & North Africa	71
Latin America & Caribbean	5
Canada & United States	0

Source: World Bank, 1992.
[1]This is the amount considered minimal for adequate hygiene and personal consumption.

More than two-thirds of the world's households have to fetch water from outside the home. For many people, this is time-consuming and heavy work; it may take up to 2 hours per day to gather a meager supply. Women and children do most of this work (fig. 19.13). Improved public systems bring many benefits to poor families. In Mozambique, for example, the World Bank reports that the average time women spent carrying water decreased from 2 hours per day to only 25 minutes when village wells were installed. The time saved could be used to garden, tend livestock, trade in the market, care for children, or even rest!

Figure 19.13

At least one billion people have inadequate or unsafe water supplies and women and children often spend several hours each day fetching what's available. With the help of a loan from international development agencies, residents of this Malawian village have a simple hand-pumped tube well that provides convenient access to safe, clean water.

The reasons for water shortages are many. In some cases, deficits are caused by natural forces: the rains fail; hot winds dry up reservoirs that normally would carry people through the dry season; rivers change their courses, leaving villages stranded. In other cases, shortages are human in origin: too many people compete for the resource; urbanization, overgrazing, and inappropriate agricultural practices allow water to run off before it can be captured; a lack of adequate sewage systems causes contamination of local supplies. Without money for wells, storage reservoirs, delivery pipes, and other infrastructure, people can't use the resources available to them.

Pure water is available in most countries—for those who can pay the price. Water from vendors often is the only source for many in the crowded slums and shanty towns around the major cities of the developing world. Although the quality is often questionable, this water generally costs about ten times more than a piped city supply. Naturally, sanitation levels decline when water is so expensive. A typical poor family in Lima, Peru, for instance, uses one-sixth as much water as a middle-class American family but pays three times as much for it. Following government recommendations that all water be boiled to prevent cholera would take up to one-third of the total income for such a poor family.

The United Nations declared the 1980s the decade to provide clean water and adequate sanitation to everyone. It estimated that a total of $300 billion—about the amount that the United States was spending each year on the military—would bring clean water to everyone in the world. Unfortunately, growing populations and stagnant economies meant that most countries only kept even or fell behind in the proportion of their people with acceptable water supplies.

Depleting Groundwater

Groundwater is the source of nearly 40 percent of the fresh water for agricultural and domestic use in the United States. Nearly half of all Americans and about 95 percent of the rural population depend on groundwater for drinking and other domestic purposes. Overuse of these supplies causes several kinds of problems, including drying of wells, natural springs, and disappearance of surface water features such as wetlands, rivers, and lakes.

In many areas of the United States, groundwater is being withdrawn from aquifers faster than natural recharge can replace it. On a local level, this causes a cone of depression in the water table, as is shown in figure 19.14. A heavily pumped well can lower the local water table so that shallower wells go dry. On a broader scale, heavy pumping can deplete a whole aquifer. The Ogallala Aquifer, for example, underlies eight states in the arid high plains between Texas and North Dakota. As deep as 400 m (1200 ft) in its center, this porous bed of sand and gravel once held more water than all the freshwater lakes, streams, and rivers on earth. Excessive pumping for irrigation and other uses has removed so much water that wells have dried up in many places and farms, ranches, even whole towns are being abandoned (see chapter 14).

Many aquifers have slow recharge rates, so it will take thousands of years to refill them once they are emptied. Much of the groundwater we now are using probably was left there by the glaciers thousands of years ago. It is fossil water, in a sense. When we pump water out of a reservoir such as the Ogallala that cannot be refilled in our lifetime, we essentially are mining a nonrenewable resource. Covering aquifer recharge zones with urban development or diverting runoff that once replenished reservoirs ensures that they will not refill.

Withdrawal of large amounts of groundwater causes porous formations to collapse, resulting in **subsidence** or settling of the sur-

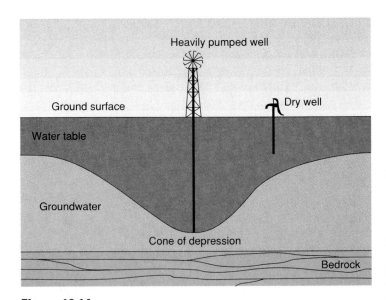

Figure 19.14

A cone of depression forms in the water table under a heavily pumped well. This may dry up nearby shallow wells or make pumping so expensive that it becomes impractical.

Figure 19.15
Sinkhole in Winter Park, Florida.
What damage was caused by the
subsidence?

face above. The U.S. Geological Survey estimates that the San Joaquin Valley in California has sunk more than 10 m in the last 50 years because of excessive groundwater pumping. Around the world, many cities are experiencing subsidence. Many are coastal cities, built on river deltas or other unconsolidated sediments. Flooding is frequently a problem as these coastal areas sink below sea level. Some inland areas also are affected by severe subsidence. Mexico City is one of the worst examples. Built on an old lake bed, it has probably been sinking since Aztec times. In recent years, rapid population growth and urbanization (chapter 24) have caused groundwater overdrafts. Some areas of the city have sunk as much as 8.5 m (25.5 ft). The Shrine of Guadalupe, the Cathedral, and many other historic monuments are sinking at odd and perilous angles.

Sinkholes form when the roof of an underground channel or cavern collapses, creating a large surface crater (fig. 19.15). Drawing water from caverns and aquifers accelerates the process of collapse. Sinkholes can form suddenly, dropping cars, houses, and trees without warning into a gaping crater hundreds of meters across. Subsidence and sinkhole formation generally represent permanent loss of an aquifer. When caverns collapse or the pores between rock particles are crushed as water is removed, it is usually impossible to reinflate these formations and refill them with water.

Another consequence of aquifer depletion is **saltwater intrusion.** Along coastlines and in areas where saltwater deposits are left from ancient oceans, overuse of freshwater reservoirs often allows saltwater to intrude into aquifers used for domestic and agricultural purposes (fig. 19.16).

Increasing Water Supplies

Where do present and impending freshwater shortages leave us now? On a human time scale, the amount of water on the earth is fixed, for all practical purposes, and there is little we can do to make more water. There are, however, several ways to increase local supplies.

Seeding Clouds and Towing Icebergs

In the dry prairie states of the 1800s and early 1900s, desperate farmers paid self-proclaimed "rainmakers" in efforts to save their withering crops. Centuries earlier, Native Americans danced and prayed to rain gods. We still pursue ways to make rain. Seeding clouds with dry ice or potassium iodide particles sometimes can initiate rain if water-laden clouds and conditions that favor precipitation are present. In a sense, cloud seeding, if it works, means "robbing Peter to pay Paul" because the rain that falls in one area decreases the

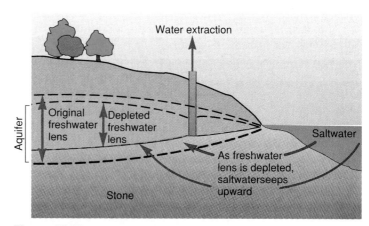

Figure 19.16
Saltwater intrusion into a coastal aquifer as the result of groundwater depletion. Many coastal regions of the United States are losing freshwater sources due to saltwater intrusion.

Water Use and Management

precipitation somewhere else. Furthermore, there are worries about possible contamination from the salts used to seed clouds.

Another scheme that has been proposed for supplying fresh water to arid countries is to tow icebergs from the Arctic or Antarctic. Icebergs contain tremendous quantities of fresh water, but whether they can be towed without excessive melting to places where water is needed is not clear. Nor is it clear whether the energy costs to move an iceberg thousands of miles (if that were, in fact, possible) would be worthwhile.

Desalination

A technology that might have great potential for increasing fresh-water supplies is **desalination** of ocean water. The most common methods of desalination are distillation (evaporation and recondensation) or reverse osmosis (forcing water under pressure through a semipermeable membrane whose tiny pores allow water to pass but exclude most salts and minerals). The number of desalination facilities producing more than 100 m^3 (26,400 gal) per day doubled in the 1980s to approximately 7500 plants worldwide. At the end of the decade, global capacity was estimated to be 13.3 million m^3 (3.4 billion gal) per day. Although desalination is still three to four times more expensive than most other sources of fresh water, it provides a welcome water supply in such places as Oman and Bahrain where there is no other access to fresh water. If a cheap, inexhaustible source of energy were available, however, the oceans could supply all the water we would ever need.

Dams, Reservoirs, Canals, and Aqueducts

People have been moving water around for thousands of years. Some of the great civilizations (Sumeria, Egypt, China, and the Inca culture of South America) were based on large-scale irrigation systems that brought river water to farm fields. In fact, some historians argue that organizing people to carry out large-scale water projects was the catalyst for the emergence of civilization. Roman aqueducts built two thousand years ago are still in use. Those early water engineers probably never even dreamed of moving water on a scale that is being proposed and, in some cases, being accomplished now.

It is possible to trap runoff with dams and storage reservoirs and transfer water from areas of excess to areas of deficit using canals, tunnels, and underground pipes. Some water transfer projects are truly titanic in scale. Los Angeles began importing water in 1913 through an aqueduct from the Owens Valley, 400 km (250 mi) to the north (fig. 19.17). This project led to a statewide program known as the California Water Plan in which a system of dams, reservoirs, aqueducts, and canals transfer water from the Colorado River on the eastern border and the Sacramento and Feather rivers in the north to Los Angeles and the San Joaquin Valley in the south.

There has been much controversy about this massive project. Some of the water now being delivered to southern California is claimed by other parts of the state, neighboring states, and even Mexico. Environmentalists claim that this water transfer upsets natural balances of streams, lakes, estuaries, and terrestrial ecosys-

Figure 19.17

Workers drive a team of 48 mules pulling a section of conduit for the Owens Valley aqueduct to Los Angeles in 1911. Water diversions such as this have allowed Los Angeles to grow to a metropolitan area of nearly 15 million people in what originally was a semiarid desert, but the ecological effects have been severe in the regions from which water has been diverted.

tems. Fishing enthusiasts, whitewater boaters, and others who enjoy the scenic beauty of free-running rivers mourn the loss of rivers drowned in reservoirs or dried up by diversion projects. These projects also have been criticized for using public funds to increase the value of privately held farmland and for encouraging agricultural development and urban growth in arid lands where other uses might be more appropriate.

Mono Lake, a salty desert lake in the Owens Valley east of Yosemite National Park, is an example of environmental consequences of the California Water Plan and is an important legal symbol as well (fig. 19.18). Diversion of tributary rivers has shrunk the surface area of this lake by one-third, threatening millions of resident and migratory wading birds, ducks, and gulls that seek shelter and food there. After years of legal wrangling, the California Water Resources Control Board ruled in 1994 that Los Angeles must restrict diversions and allow the lake's surface to rise 17 ft to 6391 ft above sea level. This is less than the 1941 level, but should improve conditions for birds and aquatic life.

In China, huge dams are being built on the Chang Jiang (Yangtze) River (What Do *You* Think?, p. 428). One stupendous scheme would transfer water from the Yangtze basin north to the dry plains around Beijing. The 1000 km (625 mi) main trunk canal in this system would move about 30 km^3 of water per year (three times the annual flow of the Colorado River in the United States). Among the engineering problems to be surmounted are the crossing of several mountain ranges and nearly 150 rivers in its path. The costs of this project include moving earth equivalent to that of about a dozen Panama Canals and an investment equal to 10 mil-

Figure 19.18

Mono Lake in the Owens Valley of eastern California. Diversion of tributary rivers to provide water for Los Angeles has shrunk the lake's surface area by one-third, threatening migratory bird flocks that feed here. These tuft formations were formed underwater where calcium-rich springs entered the brine-laden lake. They show how far the lake surface has been drawn down.

Figure 19.19

This dam is now useless because its reservoir has filled with silt and sediment.

lion working years. A big question about this project is whether the enormous distribution system required to deliver water to hundreds of thousands of farms and communes can be made to work. An even more disastrous situation has developed in the former Soviet Union where river diversions have dried up the Aral Sea to an alarming extent.

International water controversies are not limited to the few examples mentioned here. More than 200 rivers, draining over one-half of the planet, flow through two or more countries. The Niger crosses ten countries; the Nile flows through nine; the Zambezi is shared by eight countries; the Amazon drains seven countries. Pollution or diversion of these rivers by any one country affects the supplies available downstream.

Environmental Costs

Dams have been useful over the centuries for ensuring a year-round water supply, but they are far from perfect. In many cases, they reduce water availability and destroy both natural and human values. In this section, we will look at some of the disadvantages of dams.

Evaporation, Leakage, and Siltation

The main problem with dams is inefficiency. Some dams built in the western United States lose so much water through evaporation and seepage into porous rock beds that they waste more water than they make available. The evaporative loss from Lake Mead and Lake Powell on the Colorado River is about 1 km^3 per year, or about 10 percent of the annual river flow. This amounts to nearly

4500 l (1200 gal) for each person in the United States per year. The salts left behind by evaporation nearly double the salinity of the river and make its water unusable when it reaches Mexico. To compensate, the United States has built a $350 million desalination plant at Yuma, Arizona, to try to restore water quality.

As the turbulent Colorado River slows in the reservoirs created by Glen Canyon and Boulder Dams, it drops its load of suspended material. More than 10 million metric tons of silt per year collect behind these dams. Imagine a line of twenty thousand dump trucks backed up to Lake Mead and Lake Powell every day, dumping dirt into the water. Within as little as one hundred years, these reservoirs will be full of silt and useless for either water storage or hydroelectric generation (fig. 19.19).

The accumulating sediments that clog reservoirs and make dams useless also represent a loss of valuable nutrients. The Aswan High Dam in Egypt was built to supply irrigation water to make agriculture more productive. Although thousands of hectares are being irrigated, the water available is only about half that anticipated because of evaporation in Lake Nasser behind the dam and seepage losses in unlined canals which deliver the water. Controlling the annual floods of the Nile also has stopped the deposition of nutrient-rich silt on which farmers depended for fertility of their fields. This silt is being replaced with commercial fertilizer costing more than $100 million each year. Furthermore, the nutrients carried by the river once supported a rich fishery in the Mediterranean that was a valuable food source for Egypt. After the dam was installed, sardine fishing declined 97 percent. To make matters worse, growth of snail populations in the shallow permanent canals that distribute water to fields has led to an epidemic of schistosomiasis. This debilitating disease is caused by blood flukes (parasitic flatworms) spread by snails living in permanent ponds and irrigation canals. In some areas, 80 percent of the residents are infected (chapter 9).

Water Use and Management

What Do *You* Think?
Three Gorges Dam •

Midway on its long journey from the Tibetan plateau to the fertile Shanghai delta on the East China Sea, the Chang Jiang, or Yangtze, River plunges through three deep gorges in Hubei Province (fig. 1). Long renowned for their sheer cliffs and swirling currents, the gorges have been idealized by poets and painters as a mist-shrouded wilderness (fig. 2). Soon, however, that wild landscape will be profoundly changed. After 75 years of planning and debate, the Chinese government is now building what will be the largest dam in the world to tame the raging river and inundate the three gorges. Due for completion in 2009, the dam will be 185 m (607 ft) high and 2 km (1.25 mi) wide. It will create a lake 600 km (375 mi) long that will drown more than 150 villages and towns, along with hundreds of historic temples, rock carvings, and archeological sites. Altogether more than a million

Figure 2

These spectacular canyons will be flooded by the Three Gorges Dam.

people will be relocated. The total cost will be over $30 billion.

The two official reasons for the dam are hydroelectric generation and flood control. Housing the world's largest hydroelectric dynamo, the dam will generate 18,200 MW of electricity, the equivalent of 35–40 large coal-burning or nuclear power plants. This will significantly reduce China's dependence on coal, the major cause of greenhouse warming, acid rain, and the pall of smog that enshrouds Chinese cities and makes respiratory disease the leading cause of urban deaths. Yearly floods during the monsoon season have long afflicted the Yangtze Valley, which is home to 400 million people or about one-third of China's population. In this century alone, more than 500,000 people have drowned in Yangtze floods.

While agreeing that renewable energy and flood control are worthy goals, critics of this project argue that a huge dam is a poor answer to either problem. The main

cause of flooding is rapid forest clearing and crop production on unstable headwater slopes. Better land-use policies and soil conservation practices would hold back runoff more effectively than a single reservoir. Because of high soil erosion rates, the Yangtze often looks more like soup than water. Siltation as the water slows down behind the dam will quickly turn the reservoir into a giant mud puddle, useless for either water storage or power generation. A series of smaller dams upstream would be both cheaper and safer while generating more electricity than a single gargantuan dam.

Another worry about this dam is that it is being built directly above a geological fault line. The enormous weight of water behind the dam could trigger seismic activity that might crack the dam and unleash a flood of biblical proportions. Millions of lives would be lost if this happens. Chinese engineers claim that the dam is designed to withstand the maximum feasible earthquake, but skeptics are hardly convinced. Environmental groups have mounted a campaign to convince international lending banks to refuse $3 billion in loans needed to complete this monumental project.

What do you think? Do the benefits outweigh the risks of this dam? If you were a Chinese policymaker, what questions would you ask of the engineers to determine whether the dam will be safe and cost-effective? Recognizing that you will find some experts who claim that there is nothing to fear while others find serious problems in design and construction, how would you choose between opposing facts and figures? Finally, do we, as outsiders, have any right to tell China how to handle its own affairs? Does our concern about global warming outweigh the dangers to families who live downstream of this giant body of water?

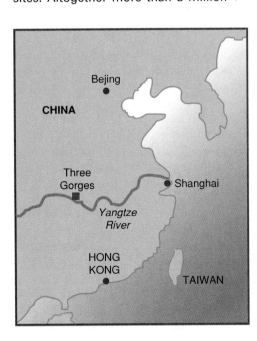

Figure 1

Site of the world's largest dam in China's Hubei Province.

Figure 19.20

The recreational and aesthetic values of free-flowing wild rivers and wilderness lakes may be their greatest assets. Competition between *in situ* values and extractive uses can lead to bitter fights and difficult decisions.

Loss of Free-Flowing Rivers

Among the environmental costs of many water projects are the loss of free-flowing rivers that are either drowned by reservoir impoundments or turned into linear, sterile irrigation canals. Conservation history records many battles between those who want to preserve wild rivers and those who would benefit from development (fig. 19.20).

One of the first and most divisive of these battles was over the flooding of the Hetch-Hetchy Valley in Yosemite National Park. In the early 1900s, San Francisco wanted to dam the Tuolumne River to produce hydroelectric power and provide water for the city water system. This project was supported by many prominent San Francisco citizens because it represented an opportunity for both clean water and municipal power. Leader of the opposition was John Muir, founder of the Sierra Club and protector of Yosemite Park. Muir said that Hetch-Hetchy Valley rivaled Yosemite itself in beauty and grandeur and should be protected. After a prolonged and bitter fight, the developers won and the dam was built. Hard feelings from this controversy persisted for many years. In 1987, Interior Secretary Donald Hodel suggested that the Hetch-Hetchy Valley might be drained and restored to its former pristine condition. Supporters and opponents are lining up to fight the same battle again. Meanwhile, impoundment schemes for the remaining free stretches of the Tuolumne River are being proposed.

Water Management and Conservation

Watershed management and conservation are often more economical and environmentally sound ways to prevent flood damage and store water for future use than building huge dams and reservoirs.

Watershed Management

After disastrous floods in the upper Mississippi Valley in 1993, it was suggested that, rather than allowing residential, commercial, or industrial development on flood plains, these areas should be reserved for water storage, aquifer recharge, wildlife habitat, and agriculture (chapter 15). Sound farming and forestry practices can reduce runoff. Retaining crop residue on fields reduces flooding, and minimizing plowing and forest cutting on steep slopes protects watersheds. Wetlands conservation preserves natural water storage capacity and aquifer recharge zones. A river fed by marshes and wet meadows tends to run consistently clear and steady rather than in violent floods.

A series of small dams on tributary streams can hold back water before it becomes a great flood. Ponds formed by these dams provide useful wildlife habitat and stock-watering facilities. They also catch soil where it could be returned to the fields. Small dams can be built with simple equipment and local labor, eliminating the need for massive construction projects and huge dams.

Domestic Conservation

Similarly, we could save as much as half of the water we now use for domestic purposes without great sacrifice or serious changes in our lifestyles. Simple steps, such as taking shorter showers, stopping leaks, and washing cars, dishes, and clothes as efficiently as possible, can go a long way toward forestalling the water shortages that many authorities predict. Isn't it better to adapt to more conservative uses now when we have a choice than to be forced to do it by scarcity in the future?

The use of conserving appliances, such as low-volume shower heads and efficient dishwashers and washing machines, can reduce water consumption greatly. If you live in an arid part of the country, you might consider whether you really need a lush green lawn that requires constant watering, feeding, and care. Planting native ground cover in a "natural lawn" or developing a rock garden or landscape in harmony with the surrounding ecosystem can be both ecologically sound and aesthetically pleasing. There are about 30 million ha (75 million ac) of cultivated lawns, golf courses, and parks in the United States. They receive more water, fertilizer, and pesticides per hectare than any other kind of land.

Our largest domestic water use is toilet flushing (fig. 19.21). We dispose of relatively small volumes of waste with very large volumes of water. In many cases it is much better to treat or dispose of waste at its origin before it is diluted or mixed with other materials. For instance, each person in the United States uses about 50,000 l (13,000 gal) of drinking-quality water annually to flush toilets. This is more than one-third of the amount supplied to homes each year. There are now several types of waterless or low-volume toilets. The Swedish-made Clivus Multrum (fig. 19.22) digests both human and kitchen wastes by aerobic bacterial action, producing a rich, nonoffensive compost that can be used as garden fertilizer. There are also low-volume toilets that use recirculating

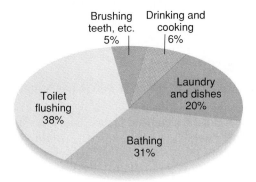

Figure 19.21

Typical household water use in the United States.

oil or aqueous chemicals to carry wastes to a holding tank, from which they are periodically taken to a treatment plant. Anaerobic digesters use bacterial or chemical processes to produce usable methane gas from domestic wastes. These systems provide valuable energy and save water but are more difficult to operate than conventional toilets. Few cities are ready to mandate waterless toilets, but in 1988 a number of cities (including Los Angeles, California; Orlando, Florida; Austin, Texas; and Phoenix, Arizona) ordered that water-saving toilets, showers, and faucets be installed in all new buildings. The motivation was twofold: to relieve overburdened sewer systems and to conserve water.

An example of the savings that can be obtained through conservation was demonstrated in the 1977 drought in California. Through a combination of rationing, laws against nonessential uses, and public appeals for conservation, Marin County, north of San Francisco, was able to temporarily reduce water use by 65 percent. In response to droughts in the 1980s, many communities ordered bans on all outdoor lawn sprinkling, car washing, and other nonessential uses. Whether we can make the attitudinal and behavioral changes necessary for conservation a permanent part of our lives remains to be seen.

Industrial and Agricultural Conservation

Perhaps half of all the agricultural water used is lost to leaks in irrigation canals, application to areas where plants don't grow, runoff, and evaporation. Better farming techniques, such as minimum tillage, leaving crop residue on fields and ground cover on

Figure 19.22

The Clivus Multrum waterless toilet. Wastes decompose and compost into an odorless, safe, rich fertilizer as they slowly slide down to the bottom compartment.

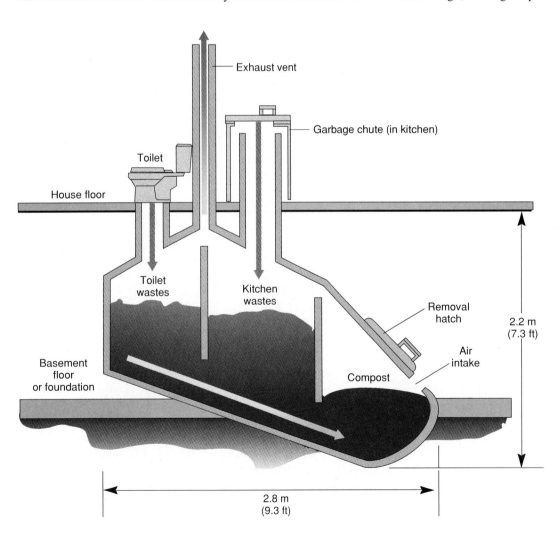

Figure 19.23

Drip irrigation delivers measured amounts of water exactly where the plants need and can use it. This technique can save up to 90 percent of irrigation water usage and reduces salt buildup. It is more expensive to install and operate, however, than simple flood irrigation.

drainage ways, intercropping, use of mulches, and trickle irrigation could reduce these water losses dramatically (fig. 19.23).

Nearly half of all water use is for cooling of electric power plants and other industrial facilities. Some of this water use could be avoided by installing dry cooling systems similar to the radiator of your car. In many cases, cooling water could be reused for irrigation or other purposes in which water does not have to be drinking quality. The waste heat carried by this water could be a valuable resource if techniques were developed for using it.

Price Mechanisms

We have traditionally treated water as if there were an endless supply and as if the water itself had no intrinsic value. Federal water supply projects charge customers only for the immediate costs of delivery. The cost of building projects is usually subsidized, and the discount value of future supplies and foregone opportunities is ignored. Farmers in California's Central Valley, for example, pay only about one-tenth of what it costs the government to supply water to them. The subsidy represented by this underpriced water averaged almost $500,000 per farm per year in some areas.

Over the whole nation, the cost of federal water projects amounts to about $3.5 billion each year. Much of the water supplied by these projects is used to grow crops, such as corn, wheat, and cotton, that we have in embarrassing surplus. How ironic that we spend great sums of money to deliver water to arid lands in the Southwest to grow crops that we have paid farmers not to grow in the well-watered East and Midwest. Since water is so cheap in most western states, there is little incentive for farmers to practice

conservation or efficient use. Agriculture accounts for 90 percent of the water consumed west of the 100th meridian. If farmers could save only 10 percent of the water they use, the amount available for all other uses would double. In 1992, Congress passed a landmark water bill that requires water from federal projects heretofore reserved for agriculture to be set aside for environmental uses such as restoring fish and waterfowl habitat or improving water quality.

Numerous municipalities also have unreasonably low prices for water. In New York City, for example, water was supplied to homes and businesses for many years at a flat rate. There were no meters because it was considered more expensive to install meters and read them than the water was worth. With no incentive to restrict water use or repair leaks, 750,000 cu m (200 million gal) of water were wasted each year from leaky faucets, toilets, and water pipes. The drought of 1988 convinced the city to begin a ten-year, $290 million program to install meters and reduce waste.

If water users were charged the real cost for environmental damage, future use, and public subsidies, conservation would be more attractive. One way to establish true water cost is to allow it to be marketed in interstate commerce. Laws intended to protect agriculture often prevent municipalities and industries from bidding on water supplies. This policy also protects inefficient and wasteful uses. Allowing the market to determine a price for water can encourage efficiency that makes more water available as if a new supply were being created. In 1982, the U.S. Supreme Court ruled that water is subject to the Interstate Commerce Clause of the Constitution so that state and local laws cannot interfere with its marketing.

Water Use and Management

It will be important, as water markets develop, to be sure that environmental, recreational, and wildlife values are not sacrificed to the lure of high-bidding industrial and domestic uses. Given prices based on real costs of using water and reasonable investments in public water supplies, pollution control, and sanitation, the World Bank estimates that everyone in the world could have an adequate supply of clean water by the year 2030 (fig. 19.24). We will discuss the causes, effects, and solutions for water pollution in chapter 20.

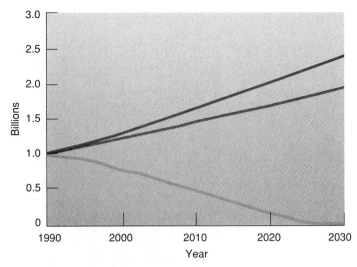

Figure 19.24

Three scenarios for government policies on clean water and sanitation services, 1990–2030.

Source: World Bank estimates based on research paper by Dennis Anderson and William Cavendish, "Efficiency and Substitution in Pollution Abatement: Simulation Studies in Three Sectors."

—— "Business as usual" scenario
—— Scenario with accelerated investment in water supply and sanitation services
—— Scenario with accelerated investment and efficiency reforms

Summary

The hydrologic cycle constantly purifies and redistributes fresh water, providing an endlessly renewable resource. The physical processes that make this possible—evaporation, condensation, and precipitation—depend upon the unusual properties of water, especially its ability to absorb and store solar energy. Roughly 98 percent of all water in the world is salty ocean water. Of the 33,400 km^3 that is fresh, 99 percent is locked up in ice or snow or buried in groundwater aquifers. Lakes, rivers, and other surface freshwater bodies make up only about 0.01 percent of all the water in the world, but they provide habitat and nourishment for aquatic ecosystems that play a vital role in the chain of life.

Water is essential for nearly every human endeavor. In the United States, direct personal use constitutes about one-tenth of the water we withdraw from our resources. Our two largest water uses are agricultural irrigation and industrial cooling. Altogether, the amount of water we withdraw from underground aquifers and surface reservoirs represents about 19 percent of the surplus runoff in the United States. Only about half the water we withdraw is consumed or degraded so that it is unsuitable for other purposes; much could be reused or recycled. Water conservation and recycling would have both economic and environmental benefits.

Water shortages in many parts of the world result from rising demand, unequal distribution, and increased contamination. Arid zones are especially vulnerable to the effects of natural droughts and land abuse by humans and domestic animals. Lakes, rivers, and groundwater reservoirs are being depleted at an alarming rate, leading not only to water shortages, but also to subsidence, sinkhole formation, saltwater intrusion, and permanent loss of aquifers.

Water storage and transfer projects are a response to flooding and water shortages. Giant dams and diversion projects can have environmental and social costs that are not justified by the benefits they provide. Among the problems they pose are evaporation and infiltration losses, siltation of reservoirs, and loss of recreation and wildlife habitat. Watershed management and small dams are preferred by many conservationists as means of flood control and water storage.

There is much we can do to save water. Charging users the true cost of water is a good start toward conservation. We can each use less water in our personal lives, and society can encourage development of water-saving appliances, natural yards, recycling, and efficient water use. Perhaps the most important change we can make is to treat wastes at their sources rather than use precious water resources for waste disposal. Not everyone can live upstream.

▼ Questions for Review

1. What is the difference between withdrawal, consumption, and degradation of water?

2. How does water use by sector differ between rich and poor countries?

3. What is subsidence? What are its results?

4. Describe some problems associated with dam building and water diversion projects.

5. Describe the path a molecule of water might follow through the hydrologic cycle from the ocean to land and back again.

6. Define *evaporation, sublimation, condensation, precipitation,* and *infiltration.* How do they work?

7. How do mountains affect rainfall distribution? Does this affect your part of the country?

8. What are the major water reservoirs of the world?

9. How much water is fresh (as opposed to saline) and where is it?

10. Define *aquifer.* How does water get into an aquifer?

▼ Questions for Critical Thinking

1. What changes might occur in the hydrologic cycle if our climate were to warm or cool significantly?

2. Why does it take so long for the deep ocean waters to circulate through the hydrologic cycle? What happens to substances that contaminate deep ocean water or deep aquifers in the ground?

3. Where would you most like to spend your vacations? Does availability of water play a role in your choice? Why?

4. Why do we use so much water? Do we need all that we use?

5. Are there ways you could use less water in your own personal life? Would that make any difference in the long run?

6. Should we use up underground water supplies now or save them for some future time?

7. How much should the United States invest to provide clean water to people in less-developed countries?

8. How should we compare the values of free-flowing rivers and natural ecosystems with the benefits of flood control, water diversion projects, hydroelectric power, and dammed reservoirs?

9. Would it be feasible to change from flush toilets and using water as a medium for waste disposal to some other system? What might be the best way to accomplish this?

10. How does water differ from other natural liquids? How do the properties of water make the hydrologic cycle, and life, possible? (You may need to review chapter 2 to answer this question.)

▼ Key Terms

aquifer 418
condensation 413
condensation nuclei 413
consumption 420
degradation 420
desalination 426
dew point 413
discharge 418
evaporation 413
groundwater 417
infiltration 417
rain shadow 414

recharge zones 418
relative humidity 413
residence time 417
runoff 419
saltwater intrusion 429
saturation point 413
sinkholes 425
stable runoff 419
sublimation 413
subsidence 424
water table 418
withdrawal 420
zone of aeration 417
zone of saturation 418

▼ Additional Information on the Internet

Ceres
http://agency.resource.ca.gov/

Hydrology Web
http://terrassa.pnl.gov:2080/HydroWeb.html/

SeaWifs Project
http://SeaWifs.gcfc.nasa.gov/

Texas Environmental Center Encyclopedia of Water Terms
http://www.tec.org/tec/terms2.html/

Universities Water Information Network
http://www.c-wr.siu.edu/

US Water News
http://www.mother.com/uswaternews/

The Water FAQ
http://www.siouxlan.com/water/faq.html/#imp/5/

Water On Line
http://agency.resource.ca.gov/ceres/WOL/home.html/

Water Resources of the United States
http://h2o.usgs.gov/

Water Web Home Page
http://www.waterweb.com/

Water Wiser
http://www.waterwiser.org/

World Wide Web Virtual Library: Environment-Hydrosphere
http://ecosys.drdr.Virginia.EDU:80/hyd.html/

World Wide Web Virtual Library: Oceanography
http://www.mth.uea.ac.uk/ocean/oceanography.html/

Note: Further readings appropriate to this chapter are listed on p. 604.

Water pollution is probably the single greatest cause of human
disease world-wide.

CHAPTER 20: Water Pollution

Objectives

After studying this chapter, you should be able to:

- define *water pollution* and describe the sources and effects of some major types.

- appreciate why access to sewage treatment and clean water are important to people in developing countries.

- discuss the status of water quality in developed and developing countries.

- delve into groundwater problems and suggest ways to protect this precious resource.

- fathom the causes and consequences of ocean pollution.

- weigh the advantages and disadvantages of different human waste disposal techniques.

- judge the impact of water pollution legislation and differentiate between best available and best practical technology.

Black Sea in Crisis

Until about 25 years ago, the Black Sea supported a diverse and productive ecosystem with five times as many fish per square kilometer as the adjacent Mediterranean. Black Sea commercial fishing provided an important food source for neighboring countries, while popular beaches and seaside resorts made Crimea the Russian equivalent of south Florida.

In recent years, however, the Black Sea has experienced severe pollution problems that illustrate the potential for catastrophic collapse of some ecosystems. Eutrophication and toxification have caused fisheries to fail abruptly. Untreated sewage washing up on beaches has forced closure of many resort areas. Massive fish kills and algal blooms have turned many sheltered bays into stinking cesspools that no one wants to go near.

Reckless energy development, unrestrained industrial expansion, and rapidly growing human populations in the watersheds surrounding the sea lie at the roots of these environmental problems. Every year millions of tons of sediment

and pollutants—including untreated sewage, industrial wastes, oil, heavy metals, and radioactive substances—flow into the Black Sea (fig. 20.1). The Danube River, for instance, carries chrome, copper, mercury, lead, zinc, and oil to the Black Sea at 20 times the levels that the Rhine River transports those contaminants to the North Sea.

One city—Bratislava, the capital of Slovakia—dumps 73 million cubic meters of industrial and municipal wastes every year into the Danube. Since 1970, the Danube's nitrate and phosphate load have increased sixfold and fourfold, respectively. Levels of these same chemicals in the Dniester River, which originates in Ukraine and Moldavia, are up 700 percent over this same period. Tanker dumping and production spills cause higher oil pollution levels in the Black Sea than in the busy Persian Gulf. Less than one percent of Turkey's population is served by any kind of sewage treatment.

Fed by the Danube and more than 30 other rivers from Eastern Europe and Western Asia, but landlocked except for a

Figure 20.1

Two Gypsy children from Copşa Micã, Romania, play in a polluted river across from the carbon factory that turns everything in town black. Eventually this filthy water will be flushed down the Danube and into the Black Sea, which, in only 25 years, has deteriorated from a highly productive ecosystem to one nearly devoid of all life.

narrow outlet through the Bosporus and Dardanelles to the Mediterranean, the Black Sea's unique hydrology adds to pollution woes. The top 100 meters of the sea are less salty than most oceans, but are nutrient rich and well oxygenated. Beneath this upper layer is a deep, essentially lifeless zone of cold, highly saline water with little oxygen and high levels of poisonous hydrogen sulfide. These density differences, which have existed for centuries, prevent mixing between layers. Pollutants entering the sea are trapped in the shallow surface layer and quickly reach toxic concentrations.

In 1986, fish catches in the Black Sea amounted to 900,000 metric tons, but by 1992, less than one-tenth of this amount was caught. Overfishing may be partly to blame for this catastrophic decline, but toxic pollutants and oxygen depletion have killed many species that once flourished in the sea. Another problem is invasion by an exotic jellyfish-like ctenophore Mnemiopis leidya *from the East Coast of North America. With no natural enemies, this predator—which feeds on zooplankton, fish eggs, and larvae—has undergone explosive population growth. During some times of the year, these comb jellies make up more than 95 percent of all biomass in the Black Sea.*

Is there any hope in this dismal situation? Perhaps. In 1992, in spite of fierce religious, racial, political, and economic divisions, Bulgarians, Georgians, Rumanians, Russians, Turks, Armenians, and Ukrainians met in Bucharest to hammer out a draft convention to protect the Black Sea. Following precedents set by international conventions on other regional seas, this agreement will focus on preventing land-based pollution, vessel dumping, and deposition of atmospheric contaminants. Watershed protection is especially important, and the eight Central European countries in the Danube drainage basin have developed a separate understanding to clean up this historic river.

Few of the states around the Black Sea have the institutional capacity, manpower, or funds to do much at present. It may take decades before any tangible improvements will be seen. Still, if agreements for environmental protection can be reached in this deeply divided region, perhaps there is hope for other places as well. ▶

What Is Water Pollution?

Any physical, biological, or chemical change in water quality that adversely affects living organisms or makes water unsuitable for desired uses can be considered pollution. Often, however, a change that adversely affects one organism may be advantageous to another. Nutrients that stimulate oxygen consumption by bacteria and other decomposers in a river or lake, for instance, may be lethal to fish, but will stimulate a flourishing community of decomposers. Whether the quality of the water has suffered depends on your perspective. There are natural sources of water contamination, such as poison springs, oil seeps, and sedimentation from erosion, but in this chapter we will focus primarily on human-caused changes that affect water quality or usability.

Pollution control standards and regulations usually distinguish between point and nonpoint pollution sources. Factories, power plants, sewage treatment plants, underground coal mines, and oil wells are classified as **point sources** because they discharge pollution from specific locations, such as drain pipes, ditches, or sewer outfalls (fig. 20.2). These sources are discrete and identifiable, so they are relatively easy to monitor and regulate. It is generally pos-

Figure 20.2

Sewer outfalls, industrial effluent pipes, acid draining out of abandoned mines, leaking underground storage tanks, and other point sources of pollution are generally easy to recognize.

Figure 20.3

This bucolic scene looks peaceful and idyllic, but allowing cows to trample stream banks is a major cause of bank erosion and water pollution. Nonpoint sources such as this have become the leading unresolved cause of stream and lake pollution in the United States.

sible to divert effluent from the waste streams of these sources and treat it before it enters the environment.

In contrast, **nonpoint sources** of water pollution are scattered or diffuse, having no specific location where they discharge into a particular body of water. Nonpoint sources include runoff from farm fields and feedlots (fig. 20.3), golf courses, lawns and gardens, construction sites, logging areas, roads, streets, and parking lots. Whereas point sources may be fairly uniform and predictable throughout the year, nonpoint sources are often highly episodic. The first heavy rainfall after a dry period may flush high concentrations of gasoline, lead, oil, and rubber residues off city streets, for instance, while subsequent runoff may have much less of these pollutants. Spring snowmelt carries high levels of atmospheric acid

deposition into streams and lakes in some areas. The irregular timing of these events, as well as their multiple sources and scattered location, makes them much more difficult to monitor, regulate, and treat than point sources.

Perhaps the ultimate in diffuse, nonpoint pollution is **atmospheric deposition** of contaminants carried by air currents and precipitated into watersheds or directly onto surface waters as rain, snow, or dry particles. The Great Lakes, for example, have been found to be accumulating industrial chemicals such as PCBs and dioxins, as well as agricultural toxins such as the insecticide toxaphene that cannot be accounted for by local sources alone. The nearest sources for many of these chemicals are sometimes thousands of kilometers away.

Amounts of these pollutants can be quite large. Altogether, it is estimated that there are 600,000 kg of the herbicide Atrazine in the Great Lakes, most of which is thought to have been deposited from the atmosphere. Concentration of these chemicals through the food chain can produce high levels in top predators. Several studies have indicated health problems among people who regularly eat fish from the Great Lakes (see Case Study, p. 456).

Ironically, lakes also can be pollution sources as well. In the past 12 years, about 26,000 metric tons of PCBs have "disappeared" from Lake Superior. Apparently, these compounds have evaporated from the lake surface and moved to other areas where they are re-deposited.

Types and Effects of Water Pollution

Although the types, sources, and effects of water pollutants are often interrelated, it is convenient to divide them into major categories for discussion (table 20.1). Let's look more closely at some of the important sources and effects of each type of pollutant.

Infectious Agents

The most serious water pollutants in terms of human health worldwide are pathogenic organisms (chapter 9). Among the most important waterborne diseases are typhoid, cholera, bacterial and amoebic dysentery, enteritis, polio, infectious hepatitis, and schistosomiasis.

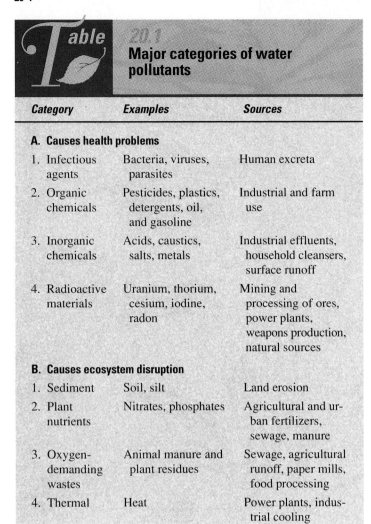

Table 20.1

Major categories of water pollutants

Category	Examples	Sources
A. Causes health problems		
1. Infectious agents	Bacteria, viruses, parasites	Human excreta
2. Organic chemicals	Pesticides, plastics, detergents, oil, and gasoline	Industrial and farm use
3. Inorganic chemicals	Acids, caustics, salts, metals	Industrial effluents, household cleansers, surface runoff
4. Radioactive materials	Uranium, thorium, cesium, iodine, radon	Mining and processing of ores, power plants, weapons production, natural sources
B. Causes ecosystem disruption		
1. Sediment	Soil, silt	Land erosion
2. Plant nutrients	Nitrates, phosphates	Agricultural and urban fertilizers, sewage, manure
3. Oxygen-demanding wastes	Animal manure and plant residues	Sewage, agricultural runoff, paper mills, food processing
4. Thermal	Heat	Power plants, industrial cooling

Malaria, yellow fever, and filariasis are transmitted by insects that have aquatic larvae. Altogether, at least 25 million deaths each year are blamed on these water-related diseases. Nearly two-thirds of the mortalities of children under 5 years old are associated with waterborne diseases (see In Depth, p. 184).

The main source of these pathogens is from untreated or improperly treated human wastes. Animal wastes from feedlots or fields near waterways and food processing factories with inadequate waste treatment facilities also are sources of disease-causing organisms.

In the more-developed countries of the world, sewage treatment plants and other pollution-control techniques have reduced or eliminated most of the worst sources of pathogens in inland surface waters. Furthermore, drinking water is generally disinfected by chlorination so epidemics of waterborne diseases are rare in these countries. The United Nations estimates that 90 percent of the people in the more-developed countries have adequate (safe) sewage disposal, and 95 percent have clean drinking water.

The situation is quite different in the less-developed countries of the world. The United Nations estimates that at least 2.5 billion

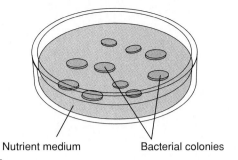

Nutrient medium   Bacterial colonies

Figure 20.4

Coliform bacteria growing in a petri dish. More than one colony per sample indicates water is unsafe for drinking.

people in these countries lack adequate sanitation, and that less than half of these people also lack access to clean drinking water. Conditions are especially bad in remote, rural areas where sewage treatment is usually primitive or nonexistent, and purified water is either unavailable or too expensive to obtain. The World Health Organization estimates that 80 percent of all sickness and disease in less-developed countries can be attributed to waterborne infectious agents.

If everyone had pure water and satisfactory sanitation, the World Bank estimates that 200 million fewer episodes of diarrheal illness would occur each year, and 2 million childhood deaths would be avoided. Furthermore, 450 million people would be spared debilitating roundworm or fluke infections. Surely these are goals worth pursuing.

Detecting specific pathogens in water is difficult, time-consuming, and costly; thus, water quality control personnel usually analyze water for the presence of **coliform bacteria,** any of the many types that live in the colon or intestines of humans and other animals. If large numbers of these organisms are found in a water sample, recent contamination by untreated feces is indicated. Exposure to an alien strain of coliform bacteria is usually the cause of upset stomach and diarrhea that strike tourists. It is usually assumed that if coliform bacteria are present in a water sample, infectious pathogens are present also.

To test for coliform bacteria, a 100 ml (4 oz) sample of water is passed through a filter that removes bacterial cells. The filter is placed in a dish containing a liquid nutrient medium that supports bacterial growth. After twenty-four hours at the appropriate temperature, each living cell will have produced a small colony of cells on the filter (fig. 20.4). If more than one colony per sample is found in a drinking water sample, the U.S. Environmental Protection Agency considers the water unsafe and requiring chlorination. The EPA-recommended maximum coliform count for swimming water is 200 colonies per 100 ml, but some cities and states allow higher levels. If the limit is exceeded, the contaminated pool, river, or lake usually is closed to swimming (fig. 20.5).

Oxygen-Demanding Wastes

The amount of oxygen dissolved in water is a good indicator of water quality and of the kinds of life it will support. Water with an

Figure 20.5

Our national goal of making all surface waters in the United States "fishable and swimmable" has not been fully met, but scenes like this have been reduced by pollution control efforts.

oxygen content above 8 parts per million (ppm) will support game fish and other desirable forms of aquatic life. Water with less than 2 ppm oxygen will support only worms, bacteria, fungi, and other detritus feeders and decomposers. Oxygen is added to water by diffusion from the air, especially when turbulence and mixing rates are high, and by photosynthesis of green plants, algae, and cyanobacteria. Oxygen is removed from water by respiration and chemical processes that consume oxygen.

The addition of certain organic materials to water, such as sewage, paper pulp, or food-processing wastes, stimulates oxygen consumption by decomposers. The impact of these materials on water quality can be expressed in terms of **biological oxygen demand (BOD):** a standard test of the amount of dissolved oxygen consumed by aquatic microorganisms over a five-day period. A new method, called the chemical oxygen demand (COD), uses a strong oxidizing agent (dichromate ion in 50 percent sulfuric acid) to completely break down all organic matter in a water sample. This method is much faster than the BOD test, but normally gives much higher results because it oxidizes compounds not ordinarily metabolized by bacteria. A third method of assaying pollution levels is to measure **dissolved oxygen (DO) content** directly, using an oxygen electrode. The DO content of water depends on factors other than pollution (for example, temperature and aeration), but it is usually more directly related to whether aquatic organisms survive than is BOD.

The effects of oxygen-demanding wastes on rivers depends to a great extent on the volume, flow, and temperature of the river water. Aeration occurs readily in a turbulent, rapidly flowing river, which is, therefore, often able to recover quickly from oxygen-depleting processes. Downstream from a point, such as a municipal sewage plant discharge, a characteristic decline and restoration of water quality can be detected either by measuring dissolved oxygen content or by observing the flora and fauna that live in successive sections of the river.

The oxygen decline and rise downstream are called the **oxygen sag** (fig. 20.6). Above the pollution source, oxygen levels support normal populations of clean-water organisms. Immediately below the source of pollution, oxygen levels begin to fall as decomposers metabolize waste materials. Rough fish, such as carp, bullheads, and gar, are able to survive in this oxygen-poor environment where they eat both decomposer organisms and the waste itself. Further downstream, the water may become anaerobic (without oxygen) so that only the most resistant microorganisms and invertebrates can survive. Eventually, most of the nutrients are used up, decomposer populations are smaller, and the water becomes oxygenated once again. Depending on the volumes and flow rates

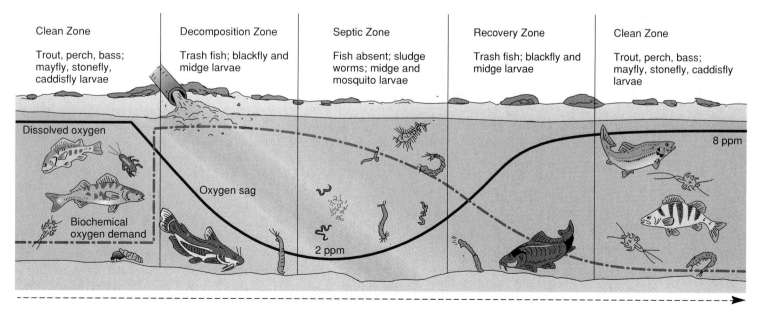

Clean Zone	Decomposition Zone	Septic Zone	Recovery Zone	Clean Zone
Trout, perch, bass; mayfly, stonefly, caddisfly larvae	Trash fish; blackfly and midge larvae	Fish absent; sludge worms; midge and mosquito larvae	Trash fish; blackfly and midge larvae	Trout, perch, bass; mayfly, stonefly, caddisfly larvae

Figure 20.6

Oxygen sag downstream of an organic source. A great deal of time and distance may be required for the stream and its inhabitants to recover.

Water Pollution

of the effluent plume and the river receiving it, normal communities may not appear for several miles downstream.

Plant Nutrients and Cultural Eutrophication

Water clarity (transparency) is affected by the abundance of plankton organisms and is a useful measure of water quality and water pollution. Rivers and lakes that have clear water and low biological productivity are said to be **oligotrophic** (oligo = little + trophic = nutrition). By contrast, **eutrophic** (eu + trophic = truly nourished) waters are rich in organisms and organic materials. Eutrophication, an increase in nutrient levels and biological productivity, is a normal part of successional changes (chapter 5) in most lakes. Tributary streams bring in sediments and nutrients that stimulate plant growth. Over time, the pond or lake tends to fill in, eventually becoming a marsh and then a terrestrial biome (fig. 20.7). The rate of eutrophication and succession depends on water chemistry and depth, volume of inflow, mineral content of the surrounding watershed, and the biota of the lake itself.

Human activities can greatly accelerate eutrophication. An increase in biological productivity and ecosystem succession caused by human activities is called **cultural eutrophication.** Cultural eutrophication can be brought about by increased nutrient flows, higher temperatures, more sunlight reaching the water surface, or a number of other changes. Increased productivity in an aquatic system sometimes can be beneficial. Fish and other desirable species may grow faster, providing a welcome food source. Often, however, eutrophication has undesirable results. An oligotrophic lake or river usually has aesthetic qualities and species of organisms that we value.

The high biological productivity of eutrophic systems is often expressed as "blooms" of algae or thick growths of aquatic plants and high levels of sediment accumulation. Bacterial populations also increase, fed by larger amounts of organic matter. The water often becomes opaque and has unpleasant tastes and odors. The deposition of silt and organic sediment caused by cultural eutrophication can accelerate the "aging" of a water body enormously over natural rates. Lakes and reservoirs that normally might exist for hundreds or thousands of years can be filled in a matter of decades.

Eutrophication also occurs in marine ecosystems, especially in near-shore waters and partially enclosed bays or estuaries. Blooms of minute organisms called dinoflagellates produce toxic **red tides** that kill fish. Partially enclosed seas such as the Black Sea, the Baltic, and the Mediterranean tend to be in especially critical condition. During the tourist season, the coastal population of the Mediterranean, for example, swells to 200 million people. Eighty-five percent of the effluents from large cities go untreated into the sea. Extensive beach pollution, fish kills, and contaminated shellfish result. Massive algal blooms suffocate the Venice lagoons. The stench is unbearable. Swarms of flies feeding on dead plants infest the city. The lagoons and the sea along much of the adjacent Adriatic coast are essentially devoid of oxygen during most of the summer.

Toxic Inorganic Materials

Some toxic inorganic chemicals are released from rocks by weathering, are carried by runoff into lakes or rivers, or percolate into groundwater aquifers. This pattern is part of natural mineral cycles (chapter 3). Humans often accelerate the transfer rates in these cycles thousands of times above natural background levels through the mining, processing, using, and discarding of minerals.

In many areas, toxic, inorganic chemicals introduced into water as a result of human activities have become the most serious form of water pollution. Among the chemicals of greatest concern are heavy metals, such as mercury, lead, tin, and cadmium. Supertoxic elements, such as selenium and arsenic, also have reached hazardous levels in some waters. Other inorganic materials, such as acids, salts, nitrates, and chlorine, that normally are not toxic in low concentrations may become concentrated enough to lower water quality or adversely affect biological communities.

Heavy Metals

Many metals such as mercury, lead, cadmium, and nickel are highly toxic. Levels in the microgram range—so little that you cannot see or taste them—can be fatal. Because metals are highly persistent, they accumulate in food chains and have a cumulative effect in humans. A famous case of mercury poisoning in Japan in the 1950s was one of our first warnings of this danger (fig. 20.8).

Another mercury-poisoning disaster appears to be in process in South America. Since the mid-1980s, a gold rush has been under way in Brazil, Ecuador, and Bolivia. Forty thousand *garimperios* or prospectors have invaded the jungles on the Amazon River and its tributaries to pan for gold. They use mercury to trap the gold and separate it from sediments. Then, the mercury is boiled off with a blow torch. Miners and their families suffer nerve damage from breathing the toxic fumes. Estimates are that 130 tons of mercury per year are deposited in the Amazon. It will probably be impossible to clean up this huge river system.

Figure 20.7
Eutrophic lake. Nutrients from agriculture and domestic sources have stimulated growth of algae and aquatic plants in this lake. This reduces water quality, alters species composition, and lowers recreational and aesthetic values of the lake.

Figure 20.8

A mother from Minamata, Japan, bathes her daughter, who suffered permanent brain damage and birth defects from mercury-contaminated seafood the mother ate while pregnant. This kind of poisoning is now known as Minamata Disease.

We have come to realize that other heavy metals released as a result of human activities also are concentrated by hydrological and biological processes so that they become hazardous to both natural ecosystems and human health. A condition known as Itai-Itai (literally ouch-ouch) disease that developed in Japanese living near the Jintsu River was traced to cadmium poisoning. Bacteria forming methylated tin have been found in sediments in Chesapeake Bay, leading to worries that this toxic metal also may be causing unsuspected health effects. The use of tin compounds as antifouling agents on ship bottoms has been criticized as a potential source of dangerous pollution.

Lead poisoning has been known since Roman times to be dangerous to human health. Lead pipes are a serious source of drinking water pollution, especially in older homes or in areas where water is acidic and, therefore, leaches more lead from pipes. Even lead solder in pipe joints and metal containers can be hazardous. In 1990, the EPA lowered the maximum limit for lead in public drinking water from 50 parts per billion (ppb) to 20 ppb. Some public health officials argue that lead is neurotoxic at any level, and the limits should be less than 10 ppb.

Mine drainage and leaching of mining wastes are serious sources of metal pollution in water. A recent survey of water quality in eastern Tennessee found that 43 percent of all surface streams and lakes and more than half of all groundwater used for drinking supplies was contaminated by acids and metals from mine drainage. In some cases, metal levels were two hundred times higher than what is considered safe for drinking water.

Nonmetallic Salts

Desert soils often contain high concentrations of soluble salts, including toxic selenium and arsenic. You have probably heard of poison springs and seeps in the desert where these compounds are brought to the surface by percolating groundwater. When the water evaporates, they are left in increasing concentrations. Irrigation and drainage of desert soils mobilize these materials on a larger scale and can result in serious pollution problems, as in Kesterson Marsh in California where selenium poisoning killed thousands of migratory birds in the 1980s.

Such salts as sodium chloride (table salt) that are nontoxic at low concentrations also can be mobilized by irrigation and concentrated by evaporation, reaching levels that are toxic for plants and animals. Salt levels in the San Joaquin River in central California rose from 0.28 gm/l in 1930 to 0.45 gm/l in 1970 as a result of agricultural runoff. Salinity levels in the Colorado River and surrounding farm fields have become so high in recent years that millions of hectares of valuable croplands have had to be abandoned. The United States has built a huge desalinization plant at Yuma, Arizona, to reduce salinity in the river. In northern states, millions of tons of sodium chloride and calcium chloride are used to melt road ice in the winter. The corrosive damage to highways and automobiles and the toxic effects on vegetation are enormous. Leaching of road salts into surface waters has a similarly devastating effect on aquatic ecosystems.

Acids and Bases

Acids are released as by-products of industrial processes, such as leather tanning, metal smelting and plating, petroleum distillation, and organic chemical synthesis. Coal mining is an especially important source of acid water pollution. Sulfides in coal are solubilized to make sulfuric acid. Thousands of kilometers of streams in the United States have been acidified by acid mine drainage, some so severely that they are essentially lifeless.

Coal and oil combustion also leads to formation of atmospheric sulfuric and nitric acid (chapter 18), which are disseminated by long-range transport processes and deposited via precipitation (acid rain, acidic snow, acid fog, or dry deposition) in surface waters. Where soils are rich in such alkaline material as limestone, these atmospheric acids have little effect because they are neutralized. In high mountain areas or recently glaciated regions where crystalline bedrock is close to the surface and lakes are oligotrophic, however, there is little buffering capacity (ability to neutralize acids) and aquatic ecosystems can be severely disrupted. These effects were first recognized in the mountains of northern England and Scandinavia about thirty years ago. In recent years, aquatic damage due to acid precipitation has been reported in about two hundred lakes in the Adirondack Mountains of New York state and in several thousand lakes in Eastern Quebec, Canada. Game fish, amphibians, and sensitive aquatic insects are generally the first to be killed by increased acid levels in the water. If acidification is severe enough, aquatic life is limited to a few resistant species of mosses and fungi. Increased acidity may result in leaching of toxic metals, especially aluminum, from soil and rocks, making water unfit for drinking or irrigation, as well.

Organic Chemicals

Thousands of different natural and synthetic organic chemicals are used in the chemical industry to make pesticides, plastics, pharmaceuticals, pigments, and other products that we use in everyday life. Many of these chemicals are highly toxic (chapter 9). Exposure to

Figure 20.9

The deformed beak of this young cormorant is thought to be due to dioxins, DDT, and other toxins in its mother's diet.

Figure 20.10

Sediment and industrial waste flow from this steel mill into Lake Erie.

very low concentrations (perhaps even parts per quadrillion in the case of dioxins) can cause birth defects, genetic disorders, and cancer. They also can persist in the environment because they are resistant to degradation and toxic to organisms that ingest them. Contamination of surface waters and groundwater by these chemicals is a serious threat to human health.

The two most important sources of toxic organic chemicals in water are improper disposal of industrial and household wastes and runoff of pesticides from farm fields, forests, roadsides, golf courses, and other places where they are used in large quantities. The EPA estimates that about 410,000 metric tons of pesticides are used in the United States each year. Much of this material washes into the nearest waterway, where it passes through ecosystems and may accumulate in high levels in certain nontarget organisms. The bioaccumulation of DDT in aquatic ecosystems was one of the first of these pathways to be understood. Dioxins, and other chlorinated hydrocarbons (hydrocarbon molecules that contain chlorine atoms) have been shown to accumulate to dangerous levels in the fat of salmon, fish-eating birds, and humans and to cause health problems similar to those resulting from toxic metal compounds (fig. 20.9).

Hundreds of millions of tons of hazardous organic wastes are thought to be stored in dumps, landfills, lagoons, and underground tanks in the United States (chapter 23). Many, perhaps most, of these sites are leaking toxic chemicals into surface waters or groundwater or both. The EPA estimates that about 26,000 hazardous waste sites will require cleanup because they pose an imminent threat to public health, mostly through water pollution.

Sediment

Sediment and suspended solids make up the largest volume of water pollution in the United States and most other parts of the world. Rivers have always carried sediment to the oceans, but erosion rates in many areas have been greatly accelerated by human activities. As chapter 11 shows, some rivers carry astounding loads of

sediment. Erosion and runoff from croplands contribute about 25 billion metric tons of soil, sediment, and suspended solids to world surface waters each year. Forests, grazing lands, urban construction sites, and other sources of erosion and runoff add at least 50 billion additional tons. This sediment fills lakes and reservoirs, obstructs shipping channels, clogs hydroelectric turbines, and makes purification of drinking water more costly. Water with high levels of suspended solids is less suitable for fish and other forms of aquatic life. Sunlight is blocked so that plants cannot carry out photosynthesis and oxygen levels decline. Murky, cloudy water also is less attractive for swimming, boating, fishing, and other recreational uses (fig. 20.10).

Sediment also can be beneficial. Mud carried by rivers nourishes floodplain farm fields. Sediment deposited in the ocean at river mouths creates valuable deltas and islands. The Ganges River, for instance, builds up islands in the Bay of Bengal that are eagerly colonized by land-hungry people of Bangladesh. In Louisiana, lack of sediment in the Mississippi River (it is being trapped by locks and dams upstream) is causing biologically rich coastal wetlands to waste away. Sediment also can be harmful. Excess sediment deposits can fill estuaries and smother aquatic life on coral reefs and shoals near shore. As with many natural environmental processes, acceleration as a result of human intervention generally diminishes the benefits and accentuates the disadvantages of the process.

Thermal Pollution and Thermal Shocks

Raising or lowering water temperatures from normal levels can adversely affect water quality and aquatic life. Water temperatures are usually much more stable than air temperatures, so aquatic organisms tend to be poorly adapted to rapid temperature changes. Lowering the temperature of tropical oceans by even one degree can be lethal to some corals and other reef species. Raising water temperatures can have similar devastating effects on sensitive organisms. Oxygen solubility in water decreases as temperatures increase, so species requiring high oxygen levels are adversely affected by warming water.

Humans cause thermal pollution by altering vegetation cover and runoff patterns, as well as by discharging heated water directly into rivers and lakes. As chapter 19 shows, nearly half the water we withdraw is used for industrial cooling. Metal smelters and processing mills use and release large amounts of cooling water, as do petroleum refineries, paper mills, food-processing factories, and chemical manufacturing plants. The electric power industry uses about three-quarters of all cooling water in the United States. Steam-driven turbogenerators of both nuclear and fossil fuel power plants are only about 40 percent efficient. That means that the other 60 percent of the energy released from the fuel—almost entirely heat energy—must be gotten rid of in some way.

The cheapest way to remove heat from an industrial facility is to draw cool water from an ocean, river, lake, or aquifer, run it through a heat-exchanger to extract excess heat, and then dump the heated water back into the original source. A **thermal plume** of heated water is often discharged into rivers and lakes, where raised temperatures can disrupt many processes in natural ecosystems and drive out sensitive organisms. To minimize these effects, power companies frequently are required to construct artificial cooling ponds or wet- or dry-cooling towers in which heat is released into the atmosphere and water is cooled before being released into natural water bodies. Wet cooling towers are cheaper to build and operate than dry systems, but lose large quantities of water to evaporation.

In some circumstances, introducing heated water into a water body is beneficial. Warming catfish-rearing ponds, for instance, can increase yields significantly. Warm water plumes from power plants often attract fish, birds, and marine mammals that find food and refuge there, especially in cold weather. This artificial environment can be a fatal trap, however. Organisms dependent on the warmth may die if they leave the plume or if the flow of warm water is interrupted by a plant shutdown. The manatee, for example, is an endangered marine mammal species that lives in Florida. Manatees are attracted to the abundant food supply and warm water in power plant thermal plumes and are enticed into spending the winter much farther north than they normally would. On several occasions, a midwinter power plant breakdown has exposed a dozen or more of these rare animals to a sudden thermal shock that they could not survive.

Water Quality Today

In 1990, the EPA announced that 17,365 segments of surface water in the United States and its territories were contaminated by toxic chemicals, sewage, or other pollutants. This contamination affects about 10 percent of the river, stream, coastal water, lake, and estuary mileage in the country. In addition, between 1 to 2 percent of the groundwater near the surface is also polluted. How does this situation compare to past pollution levels? How do the United States and Canada compare to other countries? In the next section, we will look at areas of progress and remaining problems in water pollution control.

Surface Waters in the United States and Canada

Water pollution problems in surface waters are often both highly visible and a direct threat to environmental quality. Consequently, more has been done to eliminate surface water pollution than any other type. This is probably the greatest success story in our antipollution efforts. Much remains to be done, however.

Areas of Progress

Like most developed countries, the United States and Canada have made encouraging progress in protecting and restoring water quality in rivers and lakes over the past forty years. In 1948, only about one-third of Americans were served by municipal sewage systems, and most of those systems discharged sewage without any treatment or with only primary treatment (the bigger lumps of waste are removed). Most people depended on cesspools and septic systems to dispose of domestic wastes.

The 1972 Clean Water Act established a National Pollution Discharge Elimination System (NPDES), which requires an easily revoked permit for any industry, municipality or other entity dumping wastes in surface waters. The permit requires disclosure of what is being dumped and gives regulators valuable data and evidence for litigation. As a consequence, only about 10 percent of our water pollution now comes from industrial or municipal point sources. One of the biggest improvements has been in sewage treatment.

Since the Clean Water Act was passed in 1972, the United States has spent more than $100 billion in public funds and perhaps ten times as much in private investments on water pollution control. Most of that effort has been aimed at point sources, especially to build or upgrade thousands of municipal sewage treatment plants. As a result, nearly everyone in urban areas is now served by municipal sewage systems and no major city discharges raw sewage into a river or lake except as overflow during heavy rainstorms.

This campaign has had significant results on surface water quality in many places. The EPA reports that gross pollution of rivers, lakes, and coastal waters by sewage and industrial wastes is largely a thing of the past. Fish and aquatic insects have returned to waters that formerly were depleted of life-giving oxygen. Swimming and other water-contact sports are again permitted in rivers, lakes, and at ocean beaches that once were closed by health officials.

The national goal of making all U.S. surface waters "fishable and swimmable" has not been fully met, but nearly 90 percent of the river miles and lake acres that are assessed for water quality fully or partly support their designated uses (fig. 20.11). There is some ambiguity in these statistics, however. Note that 82 percent of rivers and 54 percent of lakes are not now monitored. This could mean they have no pollution problems and don't need monitoring, or it could mean that we know they are polluted and don't consider monitoring helpful. Similarly, the "designated uses" could mean clean enough to drink in some cases or merely liquid enough to get a boat through in others. Evaluation of data such as these can be tricky (What Do *You* Think?, p. 445).

Percentage of lake acres supporting designated uses

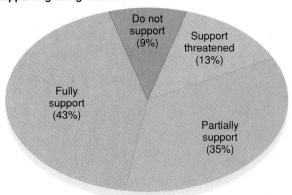

Do not support (9%)

Support threatened (13%)

Fully support (43%)

Partially support (35%)

Percentage of river miles supporting designated uses

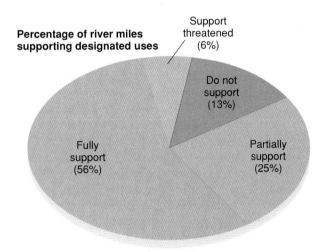

Support threatened (6%)

Do not support (13%)

Fully support (56%)

Partially support (25%)

Figure 20.11

Percentages of assessed river miles and lake acres in the United States supporting designated uses. Water quality has improved over the past 20 years and more areas now meet designated standards but only 18 percent of the 3.5 million river miles and 46 percent of the 40 million lake acres are currently assessed for water quality. Note that 13 percent of lake acres and 6 percent of river miles now support designated uses but are threatened and may become impaired by current pollution rates.

Source: EPA 1994.

Passage of the 1970 Water Act in Canada has produced comparable results. Seventy percent of all Canadians in towns over 1000 population are now served by some form of municipal sewage treatment. In Ontario, the vast majority of those systems include tertiary treatment. After 10 years of controls, phosphorus levels in the Bay of Quinte in the northeast corner of Lake Ontario have dropped nearly by half, and algal blooms that once turned waters green are less frequent and less intense than they once were. Elimination of mercury discharges from a pulp and paper mill on the Wabigoon-English River system in western Ontario has resulted in a dramatic decrease in mercury contamination that produced Minamata-like symptoms in local native people 20 years ago. Extensive flooding associated with hydropower projects have raised mercury levels in fish to dangerous levels elsewhere, however.

Remaining Problems

The greatest impediments to achieving national goals in water quality in both the United States and Canada are nonpoint discharges of pollutants. These sources are harder to identify and to reduce or treat than are specific point sources. About three-fourths of the water pollution in the United States comes from soil erosion, fallout of air pollutants, and surface runoff from urban areas, farm fields, and feedlots. In the United States, as much as 25 percent of the 46,800,000 metric tons (52 million tons) of fertilizer spread on farmland each year is carried away by runoff (fig. 20.12).

Cattle in feedlots produce some 129,600,000 metric tons (144 million tons) of manure each year, and the runoff from these sites is rich in viruses, bacteria, nitrates, phosphates, and other contaminants. A single cow produces about 30 kg (14 lb) of manure per day, or about as much as that produced by ten people. Some feedlots have 100,000 animals with no provision for capturing or treating runoff water. Imagine drawing your drinking water downstream from such a facility. Pets also can be a problem. It is estimated that the wastes from about a half million dogs in New

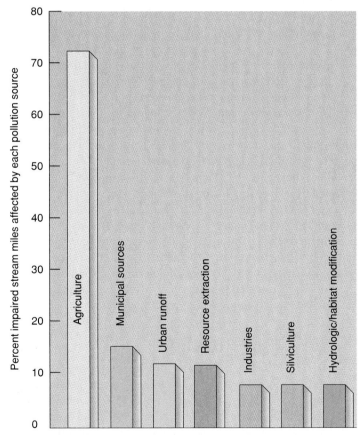

Figure 20.12

Percentage of impaired river miles in the United States by cause of damage. Totals add up to more than 100 percent because a single river can be affected by many pollution sources. Hydrologic/habitat modification refers to dams, channels, fillings, diversions, and other modifications that affect water quality.

Source: *EPA National Water Quality Inventory 1992.*

What Do *You* Think?
Assessing Progress in Water Quality

Figure 1

Excess lawn fertilizer and herbicide runs off into nearby lakes and streams after a rain, adding to pollution loads and eutrophication.

Efforts to clean up the nation's waters have been among the most popular environmental efforts of the last quarter century. Because water is present in different forms, is used in diverse ways, and can be degraded by so many different human activities, there is a huge diversity of data about its condition. Interpreting this volume of data poses special challenges.

The Clean Water Act has been effective at reducing water pollution from large individual sources such as industries and city sewage outfalls, but there has been only minimal progress in resolving nonpoint source water pollution. Nonpoint pollution is due principally to runoff, carrying myriad pollutants off the land that reflect the cumulative daily actions of hundreds of millions of individuals. In addition to agriculture, urban runoff is also a significant source.

Urbanization greatly increases the volume and rate of runoff. Also, greater quantities of pollutants become available as the intensity of land use increases. Heavy metals, salt, organic chemicals dripped from vehicles on streets and parking lots, lawn fertilizers and pesticides, sediment from construction sites, pet wastes, leaves and grass clippings are all part of the pollutant load carried off urban lands into surrounding streams and lakes (fig. 1).

Many sets of data can be used to size up the condition of the nation's waters. But different data sets addressing the same question can leave conflicting impressions. The following information is all taken from a single source, the most recently published National Water Quality Inventory, The 1992 Report to Congress (published in 1994). This report, produced every two years by the Environmental Protection Agency, is based on information reported to it by the states. Notice in the examples below the care needed in interpreting numerical information.

EPA reports that 456,317 lake acres are classified as having *improving* water quality. That sounds encouraging. But, how do you feel upon learning that that figure accounted for only 10 percent of the lakes reported?

The report also states that the water quality in 82 percent of our lakes is *stable or improving.* Again, that sounds like pollution is under control. But how should the information be interpreted? A significant percentage of our lakes could have low water quality but are not getting worse. A significantly higher percentage of our lakes could be deteriorating rather than improving in quality. In fact, 31 percent of the lake *acreage* assessed is reported to be degrading.

Numbers can also be less informative than meets the eye. The report states that 50 percent of the nation's wetlands, over 5 million out of 10.5 million acres assessed, fully support their designated use. But the proper interpretation of those numbers depends on understanding their context. Many states have wetlands but, for various reasons, only five states reported information to EPA in this category. North Dakota alone accounted for 10.3 million out of the 10.5 million acres reported. Although a statement that 50 percent of the nation's wetlands that were assessed fully support their designated use would be true, it would be totally unjustified to conclude that this was a valid reflection of conditions in the nation's wetlands as a whole. Nationally, wetlands could be much better (or much worse) than those in North Dakota.

The proper interpretation of numerical data that in themselves are valid is made even more difficult because of our tendency to attach more significance to numbers that support our point of view than those in conflict with it. Yet, accurate assessments are necessary if proper public policies are to be established. What are some things you could incorporate into your own thinking to help you effectively evaluate numerical information about the quality of the nation's waters?

York City are disposed of primarily through storm sewers, and therefore do not go through sewage treatment.

Loading of both nitrates and phosphates in surface water have decreased from point sources but have increased about fourfold since 1972 from nonpoint sources. Fossil fuel combustion has become a major source of nitrates, sulfates, arsenic, cadmium, mercury, and other toxic pollutants that find their way into water. Carried to remote areas by atmospheric transport, these combustion products now are found nearly everywhere in the world. Toxic organic compounds, such as DDT, PCBs, and dioxins, also are transported long distances by wind currents. The highest concentrations of PCBs in human tissues and mother's milk in the world are reported to be found in Inuit people in Canada's high arctic, thousands of kilometers from any known industrial source. Carried through the atmosphere and concentrated in marine food chains, these toxins reach dangerous levels in top predators such as humans, seals, and bears.

Surface Waters in Other Countries

Japan, Australia, and most of Western Europe also have improved surface water quality in recent years. Sewage treatment in the wealthier countries of Europe generally equals or surpasses that in the United States. Sweden, for instance, serves 98 percent of its population with at least secondary sewage treatment (compared with 70 percent in the United States), and the other 2 percent have primary treatment. Denmark and Germany have municipal sewage treatment for 90 percent and 84 percent of their populations, respectively. The poorer countries have much less to spend on sanitation. Spain serves only 18 percent of its population with even primary sewage treatment. In Ireland, it is only 11 percent, and in Greece, less than 1 percent of the people have even primary treatment. Most of the sewage, both domestic and industrial, is dumped directly into the ocean.

This lack of pollution control is reflected in inland water quality as well. In Poland, 95 percent of all surface water is unfit to drink. The Vistula River, which winds through the country's most heavily industrialized region, was so badly polluted in 1978 that only 432 of its 1068 km were suitable even for industrial use. It was reported to be "utterly devoid of life." In 1980, however, the Polish government instituted an ambitious program to build domestic and industrial waste treatment plants and to clean up the river. In the former Soviet Union, the lower Volga River is reported to be on the brink of disaster due to the 300 million tons of solid waste and 20 trillion l (5 trillion gal) of liquid effluent dumped into it annually.

There are also some encouraging pollution control stories. One of the most outstanding is the Thames River in London. Since the beginning of the Industrial Revolution, the Thames had been little more than an open sewer, full of vile and toxic waste products from domestic and industrial sewers. In the 1950s, however, England undertook a massive cleanup of the Thames. More than $250 million in public funds plus millions more from industry were spent to curb pollution. By the early 1980s, the river was showing remarkable signs of rejuvenation. Some ninety-five species of fish had returned, including pollution-sensitive salmon, which had not

been seen in London for three hundred years. In the Rhine River, levels of toxic metals such as cadmium, chromium, copper, mercury, lead, nickel, and zinc have fallen between two- and twentyfold in the past decade as a result of international cooperation in pollution control.

The less-developed countries of South America, Africa, and Asia have even worse water quality than do the poorer countries of Europe. Sewage treatment is usually either totally lacking or woefully inadequate. In urban areas, 95 percent of all sewage is discharged untreated into rivers, lakes, or the ocean. Low technological capabilities and little money for pollution control are made even worse by burgeoning populations, rapid urbanization, and the shift of much heavy industry (especially the dirtier ones) from developed countries where pollution laws are strict to less-developed countries where regulations are more lenient.

Appalling environmental conditions often result from these combined factors (fig. 20.13). Two-thirds of India's surface waters are contaminated sufficiently to be considered dangerous to human health. The Yamuna River in New Delhi has 7500 coliform bacteria per 100 ml (thirty-seven times the level considered safe for swimming in the United States) *before* entering the city. The coliform count increases to an incredible 24 *million* cells per 100 ml as the river leaves the city! At the same time, the river picks up some 20 million l of industrial effluents every day from New Delhi. It's no wonder that disease rates are high and life expectancy is low in this area. Only 1 percent of India's towns and cities have any sewage treatment, and only eight cities have anything beyond primary treatment.

In Malaysia, forty-two of fifty major rivers are reported to be "ecological disasters." Residues from palm oil and rubber manu-

Figure 20.13

Ditches in this Haitian slum serve as open sewers into which all manner of refuse and waste are dumped. The health risks of living under these conditions are severe.

facturing, along with heavy erosion from logging of tropical rainforests, have destroyed all higher forms of life in most of these rivers. In the Philippines, domestic sewage makes up 60 to 70 percent of the total volume of Manila's Pasig River. Thousands of people use the river not only for bathing and washing clothes but also as their source of drinking and cooking water. China treats only 2 percent of its sewage. Of seventy-eight monitored rivers in China, fifty-four are reported to be seriously polluted. Of forty-four major cities in China, forty-one use "contaminated" water supplies, and few do more than rudimentary treatment before it is delivered to the public.

Groundwater and Drinking Water Supplies

About half the people in the United States, including 95 percent of those in rural areas, depend on underground aquifers for their drinking water. This vital resource is threatened in many areas by overuse and pollution. In even more places, the precious remaining reserves are being made unfit for use by a wide variety of industrial, agricultural, and domestic contaminants. For decades it was widely assumed that groundwater was impervious to pollution because soil would bind chemicals and cleanse water as it percolated through. Springwater or artesian well water was considered to be the definitive standard of water purity, but that is no longer true in many areas.

The Office of Technology Assessment estimates that every day some 4.5 trillion l (1.2 trillion gal) of contaminated water seep into the ground in the United States from septic tanks, cesspools, municipal and industrial landfills and waste disposal sites, surface impoundments, agricultural fields, forests, and wells (fig. 20.14). The most important of these in terms of toxicity are probably waste disposal sites. The largest total volume of pollutants and area affected are agricultural chemicals. It usually takes hundreds to thousands of years for most deep aquifers to turn over their water content, and many contaminants are extremely stable once underground. It is possible, but expensive, to pump water out of aquifers, clean it, and then pump it back. For very large aquifers, pollution may be essentially irreversible.

We don't know exactly how contaminated our aquifers already are because we have access to few aquifers, and testing is expensive. Results of recent groundwater studies have been alarming. In Iowa, pesticides and other synthetic chemicals were detected in half of all wells tested. One-fifth of these wells had nitrate levels from fertilizer infiltration that exceeded federal standards.

In farm country, especially in the Midwest's Corn Belt, fertilizers and pesticides commonly contaminate aquifers and wells (fig. 20.15). Herbicides such as Atrazine and Alachlor are widely used on corn and soybeans and show up in about half of all wells tested in Iowa, for example. Nitrates from fertilizers often exceed safety standards in rural drinking water. These high nitrate levels are dangerous to infants (nitrate combines with hemoglobin in the blood and results in "blue-baby" syndrome). They also are transformed into cancer-causing nitrosamines in the human gut. In Florida, one thousand drinking water wells were shut down by

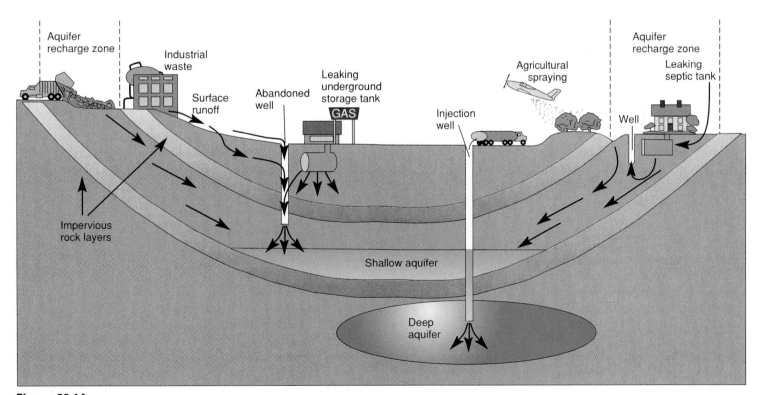

Figure 20.14

Sources of groundwater pollution. Septic systems, landfills, and industrial activities on aquifer recharge zones leach contaminants into aquifers. Wells provide a direct route for injection of pollutants into aquifers.

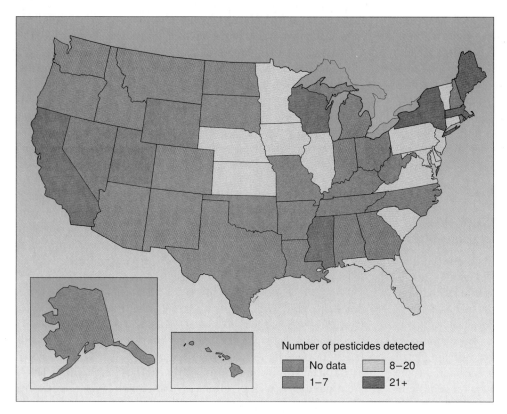

Figure 20.15

Pesticide residues are now found in groundwater in most states, according to the EPA.

Source: Environmental Protection Agency, *EPA Journal*, September/October 1990, p. 21.

Number of pesticides detected

- No data
- 1–7
- 8–20
- 21+

state authorities because of excessive levels of toxic chemicals, mostly ethylene dibromide (EDB), a pesticide used to kill nematodes (roundworms) that damage plant roots.

The United States has at least 2.5 million underground chemical storage tanks. Many of these tanks were left behind at abandoned gasoline stations or industrial sites, their exact whereabouts and contents unknown. The EPA estimates that about 42 million l (11 million gal) of gasoline are lost each year from leaking underground storage tanks (LUST). Considering that a single gallon of gasoline can make an aquifer unsuitable for drinking, these old, rusting, forgotten tanks represent a problem of tremendous proportions. The EPA now requires that all new tanks have double walls or be placed in concrete vaults to help prevent leaks into groundwater.

Aquifers in the United States also are threatened by direct injection of wastes. Every year, 38 billion l (10 billion gal) of liquid wastes, such as oil field brine, effluents from chemical plants, and treated sewage, are pumped down deep wells as an alternative to incineration or other treatment. The EPA estimates that 58 percent of all hazardous wastes generated are injected into deep wells. No permits are required, nor are there any limits on where or how wastes are pumped. Opponents of this disposal method argue that we don't know exactly how underground aquifers are connected or how toxic substances might flow through them, but it seems likely that some of those wastes have made, or will make, their way into aquifers used for domestic and municipal water supplies.

Abandoned wells represent another major source of groundwater contamination. Most domestic wells have no casings to prevent surface contaminants from leaking directly into aquifers that they penetrate. When these wells are no longer in use, they often are not capped adequately, and people forget where they are. They become direct routes for drainage of surface contaminants into aquifers. Oil wells and municipal water wells have casings to prevent leakage into aquifers, but these casings corrode and crack as they age.

In addition to groundwater pollution problems, contaminated surface waters and inadequate treatment make drinking water unsafe in many areas. A 1995 survey concluded that at least half of all public drinking water systems in the United States expose consumers to contaminants such as lead, pesticides, and pathogens at levels that violate EPA rules. A vast majority of these systems are small, serving fewer than 3000 people, but altogether some 50 million people are sometimes at risk. Problems often occur because small systems can't afford modern purification and distribution equipment, regular testing, and trained operators to bring water quality up to acceptable standards. Some aging central cities find themselves in a similar situation. Boston, for instance, still uses some wooden pipelines for water distribution originally installed more than a century ago.

An outbreak of a pathogen called cryptosporidum in April, 1993, that made 400,000 people in Milwaukee sick and killed more than 100 underscores the importance of clean water. Epidemiologists estimate that around 1.5 million Americans fall ill each year from infections caused by fecal contamination. The total costs of these diseases amount to billions of dollars per year. Preventive measures such as protecting water sources and aquifer recharge zones, providing basic treatment for all systems, installing modern technology and distribution networks, consolidating small systems, and strengthening the Clean Water Act and the Safe Drinking Water Act would cost far less. Unfortunately, in the present climate of budget-cutting and anti-regulation, these steps seem unlikely.

Relatively little information is available about groundwater quality in most countries, especially in the less-developed countries, because (1) it is expensive to drill test wells and to monitor pollutants and (2) this is not yet a major priority. In Europe, where fertilizer use is even more intensive than in the United States (chapter 11), nitrate levels in groundwater are reported to be alarmingly high in many areas. Britain, for instance, calculates that half of its underground reservoirs have been contaminated by fertilizer nitrates. China reports that forty-one of forty-four large cities suffer from polluted groundwater. As agricultural modernization increases the use of fertilizer and pesticides in less-developed countries, we may see more groundwater pollution there as well.

Ocean Pollution

During the summer of 1988, bathers from New Jersey to Massachusetts experienced unwelcome firsthand evidence of increasing levels of ocean pollution. They found floating garbage, ranging from untreated sewage to used drug paraphernalia and medical wastes, washing up on their favorite beaches. One author reported that the experience was as safe and appealing as bathing in an unflushed toilet.

This distressing situation is only one aspect of a global problem. Near-shore zones around the world, especially bays, estuaries, shoals, and reefs near large cities or the mouths of major rivers, are being overwhelmed by human-caused contamination. Suffocating and sometimes poisonous blooms of algae regularly deplete ocean waters of oxygen and kill enormous numbers of fish and other marine life. High levels of toxic chemicals, heavy metals, disease-causing organisms, oil, sediment, and plastic refuse are adversely affecting some of the most attractive and productive ocean regions. The potential losses caused by this pollution amount to billions of dollars each year. In terms of quality of life, the costs are incalculable. Oceanographer Jacques Cousteau warns that the oceans are dying and that our own survival is threatened.

One of the most massive and least understood sources of this pollution is agricultural and urban runoff. Fertilizers, manure, pesticides, and crop residues from farm fields combine with oil, rubber, metals, salts, and other urban contaminants and are carried by rivers to the ocean. Industrial wastes and municipal sewage effluents are also chronic pollution sources of near-shore ocean zones. In the United States, 1300 major industrial and 600 municipal facilities dump untreated wastewater directly into estuaries and coastal regions. Thousands of other facilities discharge a variety of toxic wastes into rivers that run into the oceans.

Discarded plastic flotsam and jetsam are becoming an ubiquitous mark of human impact on the oceans (fig. 20.16). Since plastic is lightweight and nonbiodegradable, it is carried thousands of miles on ocean currents and lasts for years. Even the most remote beaches of distant islands are likely to have bits of styrofoam containers or polyethylene packing material that were discarded half a world away by some careless person. It has been estimated that some 6 million metric tons of plastic bottles, packaging material, and other litter are tossed from ships every year into the ocean where they ensnare and choke seabirds, mammals (fig. 20.17), and even fish. Sixteen states now require that six-pack yokes be made

of biodegradable or photodegradable plastic, limiting their longevity as potential killers. The amount of municipal and industrial plastic that finds its way to the ocean is unknown but immense. In one day, volunteers in Texas gathered more than three hundred tons of plastic refuse from gulf coast beaches.

Few coastlines in the world remain uncontaminated by oil or oil products. Tar granules and sticky crude oil droplets stick to feet on beaches everywhere. Because oil floats on water, it is easily detected on the open ocean. Figure 20.18 shows locations where

Figure 20.16
Plastic litter covers this beach on the island of Niihau in the Hawaiian Islands.

Figure 20.17
A deadly necklace. Marine biologists estimate that castoff nets, plastic beverage yokes, and other packing residue kill hundreds of thousands of birds, mammals, and fish each year.

Water Pollution

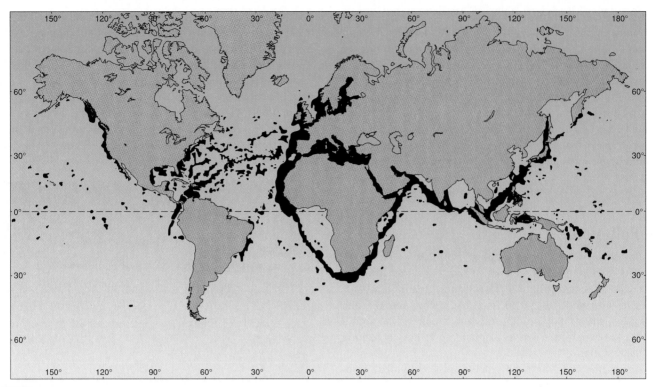

Figure 20.18

Oil slicks are visible in every ocean and along the coasts of most continents. Note the prevalence of oil spills along major shipping routes.

Source: Organization for Economic Cooperation and Development (OECD), *The State of the Environment 1985*, OECD, Paris, 1985, p. 84.

sailors reported visible oil slicks in the early 1980s. Oceanographers estimate that somewhere between 3 million and 6 million metric tons of oil are discharged into the world's oceans each year from both land- and sea-based operations. About half of this amount is due to maritime transport. Most of the 40 million liters of discharge (nearly 11 million gal) is not from dramatic, headline-making accidents, such as the 1989 oil spill from the *Exxon Valdez* in Prince William Sound, Alaska, but from routine open-sea bilge pumping and tank cleaning, which are illegal but, nonetheless, are carried out once ships are beyond sight of land. Much of the rest comes from land-based municipal and industrial runoff or from atmospheric deposition of residues from refining and combustion of fuels.

The transport of huge quantities of oil creates opportunities for major pollution episodes through a combination of human and natural hazards. Military conflict in the Middle East and increasing amounts of oil being pumped and shipped from off-shore drilling areas in inhospitable places, such as the notoriously rough North Sea and the Arctic Ocean, make it likely that more oil spills will occur. Plans to drill for oil along the seismically active California and Alaska coasts have been controversial because of the damage that oil spills could cause to these biologically rich coastal ecosystems.

The toxic chemicals of all sorts that we dump into the ocean are having deadly effects on marine life. In 1987, more than 750 dolphins died mysteriously along the Atlantic coast. Some had snouts, flippers, and tails pocked with blisters and craters. Others had huge patches of skin sloughed off. The most likely cause of

these distressing conditions is viral infections. It is thought that exposure to pesticides and other water pollutants may have weakened the mammals' immune systems and made them susceptible to infections. Harbor seals in the Gulf of Maine have the highest pesticide levels of any U.S. mammals on land or sea. Fishers bring up lobsters and crabs with gaping holes in their shells, and fish have rotted fins and ulcerous lesions. In Louisiana, 35 percent of the oyster beds have been closed because of sewage contamination. Japan's heavily used inland sea has more than two hundred toxic red tides each year. One of these episodes killed more than 1 million yellowtail that would have been worth $15 million in the market.

We often don't know exactly which pollutant is causing these distressing biological effects. In many cases, the cause may not be a single pollutant but a complex series of interactions in marine ecosystems that have many manifestations. Sometimes symptoms of shifting balances may not be noticeable until a disastrous population crash occurs. We have always assumed that the oceans are so vast that their capacity to absorb and neutralize contaminants would be inexhaustible. Marine ecosystems do have an enormous ability to recover from pollution episodes and to regenerate biological communities. Some enormous oil spills, such as the wreck of the *Amoco Cadiz* on the coast of France in 1978 or the blowout of a Mexican oil well in the Gulf of Mexico in 1979, had far less disastrous consequences than had been feared. Many scientists feel that localized, short-term pollution episodes in warm tropical waters may not cause serious long-term damage. Spills in cold arctic waters, like that of the *Exxon Valdez*, may take much longer to dissipate and may cause much more damage. Attention is now turning more to the chronic, land-based pollution from industrial, municipal, and agricultural wastes that slowly build up until they overwhelm natural systems and destroy the ocean's regenerative capacity.

Water Pollution Control

Appropriate land-use practices and careful disposal of industrial, domestic, and agricultural wastes are essential for control of water pollution.

Source Reduction

In many cases, the cheapest and most effective way to reduce pollution is to avoid producing it or releasing it to the environment in the first place. Elimination of lead from gasoline has resulted in a widespread and significant decrease in the amount of lead in surface waters in the United States. Studies have shown that as much as 90 percent less road deicing salt can be used in many areas without significantly affecting the safety of winter roads. Careful handling of oil and petroleum products can greatly reduce the amount of water pollution caused by these materials. Although we still have problems with persistent chlorinated hydrocarbons spread widely in the environment, the banning of DDT and PCBs in the 1970s has resulted in significant reductions in levels in wildlife (fig. 20.19).

Industry can modify manufacturing processes so fewer wastes are created. Recycling or reclaiming materials that otherwise might be discarded in the waste stream also reduces pollution. Both of these approaches usually have economic as well as environmental benefits. In some states, "sewering" of heavy metals by industry

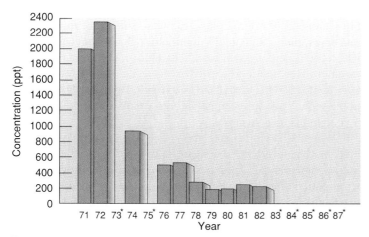

Figure 20.19

Dioxin concentrations in herring gull eggs on Scotch Bonnet Island in Lake Ontario have fallen sharply since 1971. Discontinued production and use of the herbicide 2,4,5-T are major reasons for this decline.

*Data not available.

Source: Data from *State of Canada's Environment*, Minister of the Environment.

has been outlawed. Producers are required instead to separate their wastes. It turns out that a variety of valuable metals can be recovered from wastes and reused or sold for other purposes. The company benefits by having a product to sell, and the municipal sewage treatment plant benefits by not having to deal with highly toxic materials mixed in with millions of gallons of other types of wastes.

Nonpoint Sources and Land Management

Among the greatest remaining challenges in water pollution control are diffuse, nonpoint pollution sources. Unlike point sources, such as sewer outfalls or industrial discharge pipes, which represent both specific locations and relatively continuous emissions, nonpoint sources have many origins and numerous routes by which contaminants enter ground and surface waters. It is difficult to identify—let alone monitor and control—all these sources and routes. Some main causes of nonpoint pollution are:

- *Agriculture:* The EPA estimates that between 50 to 70 percent of all impaired or threatened surface waters are affected by sediment from eroded fields and overgrazed pastures; fertilizers, pesticides, and nutrients from croplands; and animal wastes from feedlots.

- *Urban runoff:* Pollutants carried by runoff from streets, parking lots, and industrial sites contain salts, oily residues, rubber, metals, and many industrial toxins. Yards, golf courses, parklands, and urban gardens often are treated with far more fertilizers and pesticides per unit area than farmlands. Excess chemicals are carried by storm runoff into waterways.

- *Construction sites:* New buildings and land development projects such as highway construction affect relatively small areas but produce vast amounts of sediment, typically 10 to 20 times as much per unit area as farming.

- *Land disposal:* When done carefully, land disposal of certain kinds of industrial waste, sewage sludge, and biodegradable garbage can be a good way to dispose of unwanted materials. Some poorly run land disposal sites, abandoned dumps, and leaking septic systems, however, contaminate local waters.

Often the best way to control nonpoint pollution is through improved land-use practices. Measures can be taken to decrease runoff and erosion by maximizing cover on croplands, terracing slopes, increasing soil water retention, prohibiting logging or cultivation on steep slopes, protecting vegetation in and along watercourses, and banning clear cutting of forests. Generally these methods are the same ones that conserve soil (see chapter 11). Applying precisely determined amounts of fertilizer, irrigation water, and pesticides saves money and reduces contaminants entering the water. Preserving wetlands that act as natural processing facilities for removing sediment and contaminants helps protect surface and groundwaters.

In urban areas, reducing materials carried away by storm runoff is helpful. Citizens can be encouraged to recycle waste oil and to minimize use of fertilizers and pesticides. Regular street sweeping greatly reduces contaminants. Runoff can be diverted

Figure 20.20

Erosion on construction sites produces a great deal of sediment and is a major cause of nonpoint water pollution. Builders generally are required to install barriers to contain sediments, but these measures are often ineffectual.

away from streams and lakes. Many cities are separating storm sewers and municipal sewage lines to avoid overflow during storms. Contractors generally are required to place erosion barriers around construction sites to contain sediments (fig. 20.20).

One of the best examples of watershed management now in place in the United States is that for the Chesapeake Bay, America's largest estuary. Once fabled for its abundant oysters, crabs, shad, striped bass, and other valuable fisheries, the Bay had deteriorated seriously by the early 1970s. Citizens' groups, local communities, state legislatures, and the federal government together established an innovative pollution control program in 1975 that made the Bay the first estuary in America targeted for protection and restoration.

Among the principal objectives of this plan is reducing nutrient loading through land-use regulations in the six watershed states to control agricultural and urban runoff. Pollution prevention measures such as banning phosphate detergents also are important, as are upgrading wastewater treatment plants and improving compliance with discharge and filling permits. Efforts are underway to replant thousands of hectares of seagrasses and to restore wetlands that filter out pollutants. Between 1985 and 1991, annual phosphorous discharges into the Bay dropped 40 percent. Nitrogen levels, however, have remained constant or have even risen in some tributaries. Although progress has been made, the goals of reducing both nitrogen and phosphate levels by 40 percent and restoring viable fish and shellfish populations are still decades away. Still, as EPA Administrator Carol Browner says, it demonstrates the "power of cooperation" in environmental protection.

Human Waste Disposal

As we have already seen, human and animal wastes usually create the most serious health-related water pollution problems. More than 500 types of disease-causing (pathogenic) bacteria, viruses, and parasites can travel from human or animal excrement through water. In this section, we will look at how to prevent the spread of these diseases.

Natural Processes

In the poorer countries of the world, most rural people simply go out into the fields and forests to relieve themselves as they have always done. Where population densities are low, natural processes eliminate wastes quickly, making this an effective method of sanitation. The high population densities of cities make this practice unworkable, however. Even major cities of many less-developed countries are often littered with human waste which has been left for rains to wash away or for pigs, dogs, flies, beetles, or other scavengers to consume. This is a major cause of disease, as well as being extremely unpleasant. Studies have shown that a significant portion of the airborne dust in Mexico City is actually dried, pulverized human feces.

Where intensive agriculture is practiced—especially in wet rice paddy farming in Asia—it has long been customary to collect "night soil" (human and animal waste) to be spread on the fields as fertilizer. This waste is a valuable source of plant nutrients, but it is also a source of disease-causing pathogens in the food supply. It is the main reason that travelers in less-developed countries must be careful to surface sterilize or cook any fruits and vegetables they eat. Collecting night soil for use on farm fields was common in Europe and America until about one hundred years ago when the association between pathogens and disease was recognized.

Until about fifty years ago, most rural American families and quite a few residents of towns and small cities depended on a pit toilet or "outhouse" for waste disposal. Untreated wastes tended to seep into the ground, however, and pathogens sometimes contaminated drinking water supplies. The development of septic tanks and properly constructed drain fields represented a considerable improvement in public health (fig. 20.21). In a typical septic system, wastewater is first drained into a septic tank. Grease and oils rise to the top and solids settle to the bottom, where they are subject to bacterial decomposition. The clarified effluent from the septic tank is channeled out through a drainfield of small perforated pipes embedded in gravel just below the surface of the soil. The rate of aeration is high in this drainfield so that pathogens (most of which are anaerobic) will be killed, and soil microorganisms can metabolize any nutrients carried by the water. Excess water percolates up through the gravel and evaporates. Periodically, the solids in the septic tank are pumped out into a tank truck and taken to a treatment plant for disposal.

Where land is available and population densities are not too high, this can be an effective method of waste disposal. It is widely used in suburban areas, but as suburban densities grow, groundwater pollution often becomes a problem, indicating the need to shift to a municipal sewer system. It doesn't work well in cold, rainy climates where the drainfield may be too wet for proper evaporation, or where the water table is close to the surface.

Municipal Sewage Treatment

Over the past one hundred years, sanitary engineers have developed ingenious and effective municipal wastewater treatment systems to protect human health, ecosystem stability, and water quality. This topic is an important part of pollution control, and is a central focus of every municipal government; therefore, let's

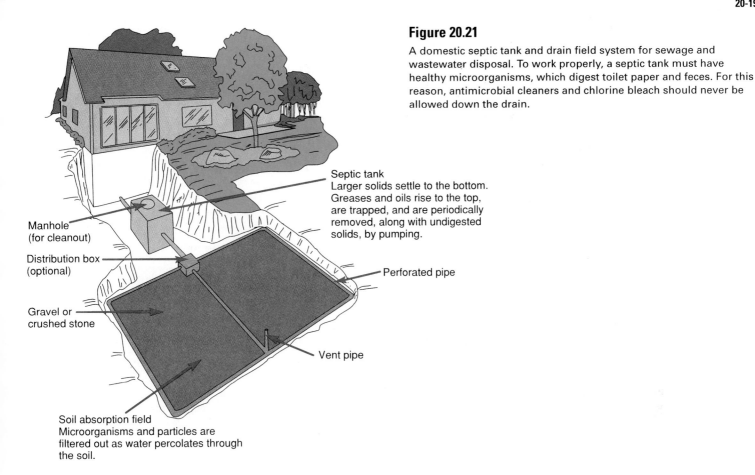

Figure 20.21

A domestic septic tank and drain field system for sewage and wastewater disposal. To work properly, a septic tank must have healthy microorganisms, which digest toilet paper and feces. For this reason, antimicrobial cleaners and chlorine bleach should never be allowed down the drain.

Septic tank
Larger solids settle to the bottom. Greases and oils rise to the top, are trapped, and are periodically removed, along with undigested solids, by pumping.

Manhole (for cleanout)

Distribution box (optional)

Perforated pipe

Gravel or crushed stone

Vent pipe

Soil absorption field
Microorganisms and particles are filtered out as water percolates through the soil.

look more closely at how a typical municipal sewage treatment facility works.

Primary treatment is the first step in municipal waste treatment. It physically separates large solids from the waste stream. As raw sewage enters the treatment plant, it passes through a metal grating that removes large debris (fig. 20.22a). A moving screen then filters out smaller items. Brief residence in a grit tank allows sand and gravel to settle. The waste stream then is pumped into the primary sedimentation tank where about half the suspended, organic solids settle to the bottom as sludge. Many pathogens remain in the effluent and it is not yet safe to discharge into waterways or onto the ground.

Secondary treatment consists of biological degradation of the remaining suspended solids. The effluent from primary treatment is pumped into a trickling filter bed, an aeration tank, or a sewage lagoon. The trickling filter is simply a bed of stones or corrugated plastic sheets through which water drips from a system of perforated pipes or a sweeping overhead sprayer (fig. 20.22b). Bacteria and other microorganisms in the bed catch organic material as it trickles past and aerobically decompose it.

Aeration tank digestion is also called the activated sludge process. Effluent from primary treatment is pumped into the tank and mixed with a bacteria-rich slurry. Air pumped through the mixture encourages bacterial growth and decomposition of the organic material. Water is siphoned off the top of the tank and sludge is removed from the bottom. Some of the sludge is used as an inoculum

for incoming primary effluent. The remainder would be valuable fertilizer if it were not contaminated by metals, toxic chemicals, and pathogenic organisms. The toxic content of most sewer sludge necessitates disposal by burial in a landfill or incineration. Sludge disposal is a major cost in most municipal sewer budgets (fig. 20.23).

Where space is available for sewage lagoons, the exposure to sunlight, algae, aquatic organisms, and air does the same job more slowly but with less energy costs. Effluent from secondary treatment processes is usually treated with chlorine to kill harmful bacteria before it is released.

Tertiary treatment removes plant nutrients, especially nitrates and phosphates, from the secondary effluent. Although wastewater is usually free of pathogens and organic material after secondary treatment, it still contains high levels of inorganic nutrients, such as nitrates, phosphates, iron, potassium, and calcium. When discharged into surface waters, these nutrients stimulate algal blooms and eutrophication. To preserve water quality, these nutrients also must be removed. Passage through a wetland or lagoon can accomplish this or chemicals often are used to bind and precipitate nutrients. (see fig. 20.22c)

In most American cities, sanitary sewers are connected to storm sewers, which carry runoff from streets and parking lots. A large line called an interceptor delivers the combined stream of storm runoff and domestic and industrial waste to the municipal treatment plant. Storm sewers are routed to the treatment plant rather than discharged into surface waters because runoff from

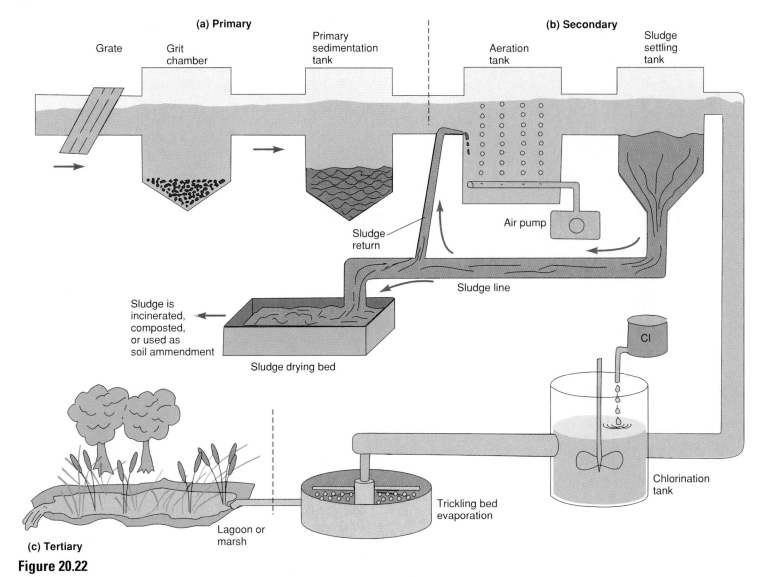

(a) Primary
Grate
Grit chamber
Primary sedimentation tank

(b) Secondary
Aeration tank
Sludge settling tank

Sludge return

Air pump

Sludge line

Sludge is incinerated, composted, or used as soil ammendment

Sludge drying bed

Cl

Chlorination tank

Trickling bed evaporation

Lagoon or marsh

(c) Tertiary

Figure 20.22

(*a*) Primary sewage treatment removes only solids and suspended sediment. (*b*) Secondary treatment, through aeration of activated sludge followed by sludge removal and chlorination of effluent, kills pathogens and removes most organic material. (*c*) During tertiary treatment passage through a trickling bed evaporator and/or a lagoon or marsh further removes inorganic nutrients, oxidizes any remaining organics, and reduces effluent volume.

streets, yards, and industrial sites generally contains a variety of refuse, fertilizers, pesticides, oils, rubber, tars, lead (from gasoline), and other undesirable chemicals. During dry weather, this plan works well. Heavy storms often overload the system, however, causing bypass dumping of large volumes of raw sewage and toxic surface runoff directly into receiving waters. To prevent this overflow, cities are spending hundreds of millions of dollars to separate storm and sanitary sewers. These are huge, disruptive projects. When they are finished, surface runoff will go directly into the river and cause another pollution problem.

Low-Cost Waste Treatment

The municipal sewage systems used in developed countries are often too expensive to build and operate in the developing world where low-cost, low-tech alternatives for treating wastes are

needed. One option is **effluent sewerage,** a hybrid between a traditional septic tank and a full sewer system. A tank near each dwelling collects and digests solid waste just like a septic system. Rather than using a drainfield, however, to dispose of liquids—an impossibility in crowded urban areas—effluents are pumped to a central treatment plant. The tank must be emptied once a year or so, but because only liquids are treated by the central facility, pipes, pumps, and treatment beds can be downsized and the whole system is much cheaper to build and run than a conventional operation.

Another alternative is to use natural or artificial wetlands to dispose of wastes. Arcata, California, for instance, needed an expensive sewer plant upgrade. By transforming a 65-hectare (160-acre) garbage dump into a series of ponds and marshes that serve as a simple, low-cost waste treatment facility, the city saved millions of dollars and improved the environment simultaneously. Sewage is piped

Figure 20.23

"Well, if *you* can't use it, do you know anyone who *can* use 3000 tons of sludge every day?"

© 1977 by Sidney Harris *What's So Funny About Science 7*, Wm. Kaufmann, Inc.

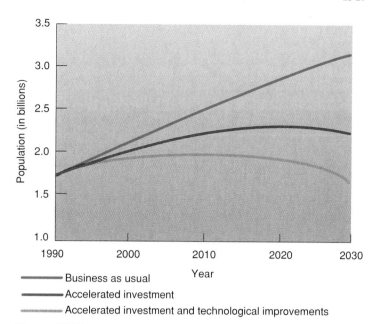

Figure 20.24

World population without adequate sanitation—three scenarios in the year 2030. If business as usual continues, more than 3 billion people will lack safe sanitation. Accelerated investment in sanitation services could lower this number. Higher investment, coupled with technological development, could keep the number of people without adequate sanitation from growing even though the total population increases.

Source: World Bank estimates based on research paper by Dennis Anderson and William Cavendish, "Efficiency and Substitution in Pollution Abatement: Simulation Studies in Three Sectors."

to holding ponds where solids settle out and are digested by bacteria and fungi. Effluent flows through marshes where it is filtered and cleansed by aquatic plants. The marsh is a haven for wildlife and has become a prized recreation area for the city (see chapter 5). Eventually, the purified water flows into the bay where marine life flourishes.

Similar wetland waste treatment systems are now operating in many developing countries. Effluent from these operations can be used to irrigate crops or raise fish for human consumption if care is taken to first destroy pathogens. Usually 20 to 30 days of exposure to sun, air, and aquatic plants is enough to make the water safe. These systems make an important contribution to human food supplies. A 2500-hectare (6000-acre) waste-fed aquaculture facility in Calcutta, for example, supplies about 7000 metric tons of fish annually to local markets. The World Bank estimates that some 3 billion people will be without sanitation services by the middle of the next century under a business-as-usual scenario (fig. 20.24). With investments in innovative programs, however, sanitation could be provided to about half those people and a great deal of misery and suffering could be avoided.

Water Legislation

Water pollution legislation has been among the most popular and most effective of all environmental legislation. In this section, we will look at some major water quality laws and their provisions (table 20.2).

The Clean Water Act of 1972 is the farthest reaching of all federal water legislation in the United States to date. It regulates release of "conventional pollutants" (dirt, organic wastes, and sewage), toxic chemicals (from the Toxic Substances Control Act), and "nonconventional pollutants" (those whose toxicity is suspected but not yet determined). For such point sources as industries, municipal facilities, and other discrete sources of conventional pollutants, the act requires discharge permits and **best practical control technology (BPT)** to reduce pollution. It sets national goals of **best available, economically achievable technology (BAT)** for toxic substances and zero discharge for 129 priority toxic pollutants. An important provision of the act provided $52 billion in federal grants to help states and communities construct sewage treatment plants. By 1988, all municipalities were required to have at least secondary sewage treatment.

Reauthorization of the Clean Water Act is shaping up to be one of the biggest environmental battles of the mid-1990s for the U.S. Congress. On one side is a coalition of more than 400 environmental groups known as the Clean Water Network that wants to strengthen and expand water protection. Opposing them is an association of industries, developers, farmers, local governmental officials, and others calling themselves the Water/Environment Federation who regard the current regulations as oppressive and

Water Pollution

CASE STUDY
Cleaning Up the Great Lakes

The Great Lakes—Lakes Superior, Huron, Michigan, Erie, and Ontario—constitute the largest freshwater lake system in the world, containing 20 percent of the world's supply of fresh surface water. Historically, the Great Lakes and their tributary rivers have provided transportation, fish, drinking water, energy, and recreation that have been critical in regional economies. The current population of the Great Lakes basin is about 35 million, including 8 million Canadians—almost one-third of Canada's total population. Although the north shores of Lakes Superior and Huron are densely forested and sparsely populated, most of the remaining region is devoted to agriculture and industry. In Canada, 84 percent of Great Lakes residents live in urban areas; a similar rate of urbanization is found on the U.S. side of the basin.

The watershed or drainage basin of the Great Lakes is small in comparison to the area and volume of the lakes themselves. No place in the basin is more than 320 km (200 mi) from one of the lakes. This means that the rivers flowing into the lakes generally are small. Much of the water input comes from rain or snow falling directly on the lakes themselves. Similarly, much of the pollution—90 percent of the PCBs in Lake Superior, for instance—come from direct atmospheric deposition.

Because of its huge volume—more than half the water in all five lakes—and small inflow, Lake Superior has a residence time (the average time between when a molecule enters and leaves the lake) of 191 years. By contrast, the residence time of Lake Erie, which is relatively small and shallow, is only 2.6 years. This means that Lake Erie can be cleaned up rather quickly, but once Superior is contaminated, it will take a long time to flush out pollutants.

Although it has long been apparent that local bays and harbors have been polluted by human activities, many people once thought the whole system too vast to ever be seriously polluted. In the 1960s, however, algal blooms turned large areas of Lakes Erie and Ontario pea-soup green, and windrows of dead fish, urban garbage, and rotting vegetation on beaches showed that entire lakes could be in danger. Petrochemical fires on the Cuyahoga River prompted ominous warnings that Lake Erie was already dead and that the other lakes were dying.

In 1972, Canada and the United States signed the first Great Lakes Water Quality Agreement to control and clean up pollution. The first agenda item was to reduce phosphorus inputs that stimulate algal blooms. Phosphate levels in detergents were reduced and more than $9 billion was spent to upgrade sewage treatment. The results are encouraging. A decade of control efforts have resulted in a 60 percent decline in phosphate levels in open water areas of Lake Erie. Eutrophication is decreasing and fish populations are rebuilding. Similarly, phosphate contamination in the Bay of Quinte in northeastern Lake Ontario has fallen by nearly half and algal blooms are down correspondingly.

More recently, attention has turned to toxic industrial and agricultural chemicals in the lakes. The International Joint Commission has identified 362 toxic chemicals of concern in the Great Lakes. Eleven critical chemicals—dioxins, furans, benzo(a)pyrene, DDT, dieldrin, HCB (from combustion of chlorinated fuels and wastes), lead, mirex, mercury, PCBs, and toxaphene—are of special concern. Seventy others are listed for regular monitoring. Most of these chemicals are unintentionally leaked from industrial sites, landfills, urban runoff, and farms.

Fish consumption warnings have been issued for all the Great Lakes. Nursing mothers, pregnant women, women who anticipate bearing children, female children of any age, and male children under 15 should not eat fish, especially large game fish such as lake trout, salmon, and walleye; or bottom feeders, such as catfish and carp. The further downstream you are, the fewer fish you should eat. Don't eat anything from the St. Lawrence or Niagara Rivers if you happen to be in that area.

Issues of environmental justice are also involved in these exposures, since many of the people who eat contaminated fish are poor or minorities. In Wisconsin, for instance, about 1 in 6 Ojibwa who live near Lake Superior or Michigan have blood mercury levels greater than 5 mg/l, an amount that can cause birth defects and developmental problems in children. In a survey of river bank anglers in Detroit, nearly 60 percent of those who ate the fish caught were children or women of childbearing age. Ninety-five percent of those anglers were minorities, mostly African Americans.

Controlling these chemicals is difficult and complex, but progress has been made. As a result of outright bans on use and disposal of PCBs, for instance, concentrations of these chemicals in lake trout in Lake Michigan in 1988 were one-third those found in 1978. Similarly, discontinued use of herbicide 2,4,5-T has resulted in a 90 percent decline in dioxin concentrations in herring gull eggs on Scotch Bonnet Island in Lake Ontario (see fig. 20.17). Closing landfills, controlling farm runoff, and requiring better handling and disposal of industrial wastes has reduced inputs of other toxic chemicals as well.

Lakes Superior and Huron, at the top of the drainage system, have the cleanest water of all the lakes. While local bays and harbors are contaminated, open water quality is still very good. Parts of Lake Michigan still have fairly good water quality, but levels of toxic metals are high in the south around Chicago and Milwaukee. The Saint Clair and Detroit Rivers run from Lake Huron to Lake Erie. As they pass through highly industrialized, urbanized areas of Ontario on one side and Michigan on the other, these rivers pick up high concentrations of metals, organic solvents, and chlorinated hydrocarbons, making

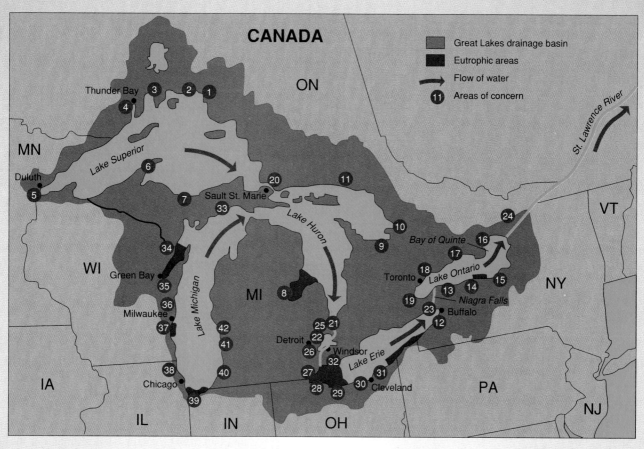

Figure 1

Pollution "hot spots" of greatest concern as identified by the International Joint Commission, a boundary-dispute- resolution body formed by the United States and Canada. Most occur in bays or river mouths.

Redrawn from Michael Morgan, et al., *Environmental Science*, Copyright © 1993 Times Mirror Higher Education Group, Inc., Dubuque, IA. All Rights Reserved. Reprinted by permission.

them one of the most polluted segments of the whole system.

The western basin of Lake Erie is badly degraded and eutrophic. Levels of persistent contaminants in fish are less than might be expected, however, perhaps due to high fish die-offs during eutrophication episodes in the 1970s. Because of its short flushing time, Lake Erie has shown the greatest improve-

ments of any of the Great Lakes. Water quality there is now much better than it was 20 years ago.

Lake Ontario, which receives contaminated inflow from all the upstream lakes as well as urban areas—such as Toronto—along its shores, remains the most polluted of all the Great Lakes in terms of both the amount and variety of pollutants it contains.

The International Joint Commission has identified 42 areas of special concern around the Great Lakes (fig. 1). Cleaning up all these areas will cost an estimated $12 billion to $25 billion. This is a large amount of money, but is dwarfed by the $1 trillion annual gross domestic product of the Great Lakes region.

ETHICAL CONSIDERATIONS

1 *Who should pay for cleaning up the Great Lakes? Is it a local problem or do we all bear some responsibility?*

2 *If poor people persist in eating contaminated fish even after being warned of health risks, should they be prevented from doing so by regulation? Do we have a right to tell people that they can't take risks? Do parents have a right to expose their children to risks?*

Table 20.2
Some important U.S. water quality legislation

1. *Federal Water Pollution Control Act* (1972). Established uniform nationwide controls for each category of major polluting industries.

2. *Marine Protection Research and Sanctuaries Act* (1972). Regulates ocean dumping and established sanctuaries for protection of endangered marine species.

3. *Ports and Waterways Safety Act* (1972). Regulates oil transport and the operation of oil handling facilities.

4. *Safe Drinking Water Act* (1974). Requires minimum safety standards for every community water supply. Among the contaminants regulated are bacteria, nitrates, arsenic, barium, cadmium, chromium, fluoride, lead, mercury, silver, pesticides; radioactivity and turbidity also regulated. This act also contains provisions to protect groundwater aquifers.

5. *Resource Conservation and Recovery Act* (RCRA) (1976). Regulates the storage, shipping, processing, and disposal of hazardous substances and sets limits on the sewering of toxic chemicals.

6. *Toxic Substances Control Act* (TOSCA) (1976). Categorizes toxic and hazardous substances, establishes a research program, and regulates the use and disposal of poisonous chemicals.

7. *Clean Water Act* (1985) (amending the 1972 Water Pollution Control Act). Sets as a national goal the attainment of "fishable and swimmable" quality for all surface waters in the United States.

8. *London Dumping Convention 1990.* Calls for an end to all ocean dumping of industrial wastes, tank washing effluents, and plastic trash.

unnecessary restrictions on business as well as an infringement on private property rights and local self-regulation.

The EPA proposes a major change in water pollution control from an "end of the pipe" focus on removing specific pollutants from effluents to more attention to changing industrial processes to prevent toxic substances from being produced in the first place. Another important proposed change is increased attention to non-point pollution (runoff from city streets, construction sites, farmland, mines, and waste dumps), which has become the major source of pollution for many waterways.

Environmentalists generally support both of these proposals but opponents find them highly objectionable. Industrial managers may agree that better water quality is desirable, but they recoil at the prospect of having the government tell them how to run their businesses. Farmers bristle at being told how to manage their farms, while developers are outraged at the prospect that they may not be able to use their land as they please.

Some other contentious proposals include stricter enforcement of regulations, mandatory minimum penalties for violations, strengthening of community right-to-know provisions, and increased powers for citizen lawsuits against polluters. Under the current law, these provisions have been powerful tools for environmentalists. Using data that polluters themselves are required to submit, groups such as the Natural Resources Defense Council and the Sierra Club have won million-dollar settlements in civil law suits (the proceeds are generally applied to clean-up projects) and some transgressors even have been sent to jail. Stronger wetland protection proposals limiting draining, filling, or converting wetlands to other purposes also are supported by environmentalists but opposed strongly by the Water/Environment Federation.

Draft legislation is now making its way through Congress that would slash federal authority to impose uniform water quality standards or to protect wetlands. By some estimates, 60 to 80 percent of the nation's remaining wetlands could be threatened by what environmentalists call the "Dirty Water Act." Contact your local congressional representative to find out what is happening in this important battle and to express your views on what should be done.

Among the most important international agreements on water quality is the 1972 Great Lakes Water Quality Agreement between Canada and the United States. The agreement has produced encouraging progress in cleaning up the world's largest freshwater system (What Do *You* Think?, p. 456). Another important international agreement is the 1990 London Dumping Convention, which calls for phasing out all ocean dumping of industrial waste, tank-washing effluent, and plastic trash by 1995. The sixty-four nations that have signed the Law of the Sea Treaty are bound by this agreement. The United States has already passed legislation to support its provisions. Whether this can be enforced remains to be seen, however.

Is it safe to assume that we are well on our way to solving our water pollution problems? We are better off in terms of legislation, policy, and practice than we were in 1960. Laws, however, are only as good as (1) the degree to which they are not weakened by subsequent amendments and exceptions and (2) the degree to which they are funded for research and enforcement. Economic interests cause continued pressure on both of these points, so that the importance of an overriding national attitude to maintain the intent of protective legislation must continually be stressed and retaught.

Summary

Worldwide, the most serious water pollutants, in terms of human health, are pathogenic organisms from human and animal wastes. We have traditionally taken advantage of the capacity of ecosystems to destroy these organisms, but as population density has grown, these systems have become overloaded and ineffective. Effective sewage treatment systems are needed that purify wastewater before it is released to the environment.

In industrialized nations, toxic chemical wastes have become an increasing problem. Agricultural and industrial chemicals have been released or spilled into surface waters and are seeping into groundwater supplies. The extent of this problem is probably not yet fully appreciated.

Ultimately, all water ends up in the ocean. The ocean is so large that it would seem impossible for human activities to have a significant impact on it, but pollution levels in the ocean are increasing. Major causes of ocean pollution are oil spills from tanker bilge pumping or accidents and oil well blowouts. Surface runoff and sewage outfalls discharge fertilizers, pesticides, organic nutrients, and toxic chemicals that have a variety of deleterious effects on marine ecosystems. We usually think of eutrophication (increased productivity due to nutrient addition) as a process of inland waterways, but this can occur in oceans as well.

The major water pollutants in terms of quantity are silt and sediments. Biomass production by aquatic organisms, land erosion, and refuse discharge all contribute to this problem. Addition of salts and metals from highway and farm runoff and industrial activities also damage water quality. In some areas, drainage from mines and tailings piles deliver sediment and toxic materials to rivers and lakes. Water pollution is a major source of human health problems. As much as 80 percent of all disease and some 25 million deaths each year may be attributable to water contamination.

Appropriate land-use practices and careful disposal of industrial, domestic, and agricultural wastes are essential for control of water pollution. Natural processes and living organisms have a high capacity to remove or destroy water pollutants, but these systems become overloaded and ineffective when pollution levels are too high. Municipal sewage treatment is effective in removing organic material from wastewater, but the sewage sludge is often contaminated with metals and other toxic industrial materials. Reducing the sources of these materials is often the best solution to our pollution problems.

▼ Questions for Review

1. Define *water pollution.*

2. List eight major categories of water pollutants and give an example for each category.

3. Describe ten major sources of water pollution in the United States. What pollution problems are associated with each source?

4. Name some waterborne diseases. How do they spread?

5. What is eutrophication? What causes it?

6. What are the origins and effects of siltation?

7. Describe primary, secondary, and tertiary processes for sewage treatment. What is the quality of the effluent from each of these processes?

8. Why do combined storm and sanitary sewers cause water quality problems? Why does separating them also cause problems?

9. What pollutants are regulated by the 1985 Clean Water Act? What goals does this act set for abatement technology?

10. What is the difference between best practical control technology and best available technology? Which do you think should be required? Why does the Clean Water Act start with BPT?

▼ Questions for Critical Thinking

1. How precise is the estimate that 2 billion people lack access to clean water? What difference would it make if the estimate is off by 10 percent or 50 percent?

2. How would you define *adequate* sanitation? Think of some situations in which people might have different definitions for this term.

3. Do you think that water pollution is worse now than it was in the past? What considerations go into a judgment like this? How do your personal experiences influence your opinion?

4. What additional information would you need to make a judgment about whether conditions are getting better or worse? How would you weigh different sources, types, and effects of water pollution?

5. Imagine yourself in a developing country with a severe shortage of clean water. What would you miss most if *your* water supply were suddenly cut by 90 percent?

6. What is the practical difference between best available, economically achievable technology (BAT) and best practical control technology (BPT)? Why do environmentalists and industry fight over these terms?

7. Proponents of deep well injection of hazardous wastes argue that it will probably never be economically feasible to pump water out of aquifers more than 1 kilometer below the surface. Therefore, they say, we might as well use those aquifers for hazardous waste storage. Do you agree? Why or why not?

8. Under what conditions might sediment in water or cultural eutrophication be beneficial? How should we balance positive and negative effects?

9. Suppose that part of the silt in a river is natural and part is human-caused. Is one pollution but the other not?

10. Suppose that you own a lake but it is very polluted. An engineer offers options for various levels of cleanup. As you increase water quality, you also increase costs greatly. How clean would you want the water to be—fishable, swimmable, drinkable—and how much would you be willing to pay to achieve your goal? Make up your own numbers. The point is to examine your priorities and values.

▼ Key Terms

atmospheric deposition 437
best available, economically achievable technology (BAT) 455
best practical control technology (BPT) 455
biological oxygen demand (BOD) 439
coliform bacteria 438
cultural eutrophication 440
dissolved oxygen (DO) content 439

effluent sewerage 454
eutrophic 440
nonpoint sources 437
oligotrophic 440
oxygen sag 439
point sources 436
primary treatment 453
red tides 440
secondary treatment 453
tertiary treatment 453
thermal plume 443

▼ Additional Information on the Internet

American Water Works Association
http://www.awwa.org/

National Oceanic and Atmospheric Administration Office of Public and Constituent Affairs
http://www.noaa.gov/public-affairs/

National Pollution Prevention Center for Higher Education
http://www.umich.edu/~nppcpub/

Ocean Planet
http://seawifs.gsfc.nasa.gov/ocean-planet.html/

Soil and Water Science Handbook
http://hammock.ifas.ufl.edu/txt/fairs/19835

Texas Environment Center
http://www.tec.org/

Water Pollution
http://www.tribenet.com/environ/env_watp.htm/

Water Quality Association
http://www.wqa.org/

Water Quality Information Center
http://www.inform.umd.edu:8080/Educational_Resources/Topic/AgrEnv/Water/

World Wide Web Virtual Library: Wastewater Engineering
http://www.cleanh2o.com/cleanh2O/ww/welcome.html/

Note: Further readings appropriate to this chapter are listed on p. 605.

Environmental Researcher

*S*arah Lloyd

Following personal interests not only is a way to do something you can feel enthusiastic about but also can help you develop unique skills and experiences that will open career doors. Sarah Lloyd, for example, credits international travel, an unpaid summer internship, and an unusual blend of interests with leading to a job right out of college that is just what she hoped to find.

While in high school, Sarah had two strong interests: Russia and environmental issues. Spending her junior year in Sweden on an American Field Study program, Sarah learned both Swedish and Russian and had the opportunity to visit Leningrad (now St. Petersburg) with her high school class. In college, Sarah majored in environmental studies but continued studying Russian. Her major focused on policy-based, social science aspects of environmental science but included a broad science background of biology, geology, ecology, and economics. In her junior year in college, Sarah returned to St. Petersburg for a semester of intensive Russian language study.

During the next summer, Sarah found an unpaid internship with Friends of the Earth (FOE) in Washington, D.C. As an intern, Sarah had valuable opportunities to learn abut environmental issues and how government works. She did many of the routine office maintenance tasks, such as stuffing envelopes and answering telephones, that interns often do, but her knowledge of Russian language and society enabled her to do more professional work as well, such as writing a lengthy position paper on the Russian environment.

Although her unpaid internship was difficult financially, Sarah now believes that it was a very worthwhile investment in her career. She made contacts in Washington that have helped immensely both in her subsequent studies and in finding the job that she wanted. Her internship also gave her an excellent insight into the workings of NGOs (non-governmental organizations).

After returning to college in the fall, Sarah had to choose a topic for a senior thesis. She decided to study Russian energy issues, especially how development of the natural gas industry might benefit or harm the environment. She was advised that the topic was too difficult, and that the break-up of the former Soviet Union would make it impossible to get good information. Sarah persisted, however, because she really cared about this subject.

The lack of information and expertise on Russian environmental issues at her college forced Sarah to search out data from a variety of international environmental organizations and governmental agencies. At the end of informational interviews she often was able to ask for advice about how best to pursue her own career. As a result of this wide network of contacts, Sarah learned of an opening at the Pacific Environment and Resources Center (PERC) in Sausalito, California. Immediately after graduation, Sarah joined this small research-oriented nonprofit organization to gather information on deforestation and sustainable forestry initiatives in Siberia and the Russian Far East and to support Russian environmentalists working to defend the forests.

Although PERC has some of the drawbacks of any NGO, it operates at a less frenzied pace than some activist groups. Its small size also offers more responsibilities and opportunities for leadership than would be possible in a larger setting. Sarah says that she sometimes can hardly believe how lucky she is to work for a great organization in a wonderful location on exactly the kind of topics that most fascinate her. Her advice to other students is to find something that you feel passionate about and follow it. If you really care about something and persevere, you too will find a way to accomplish your dreams.

Retreating Iraqi troops set more than 600 oil wells on fire as they left Kuwait in the 1990 Gulf War. At least 10 percent of Kuwait's oil was burned or spilled. Our insatiable appetite (some would say addiction) for oil plays a powerful role in international politics and our balance of payments deficit.

CHAPTER 21: Conventional Energy

Objectives

After studying this chapter, you should be able to:

- summarize our current energy sources and explain briefly how energy use has changed through history.

- compare our energy consumption with that of other people in the world.

- itemize the ways we use energy.

- analyze the resources and reserves of fossil fuels in the world.

- evaluate the costs and benefits of using coal, oil, and natural gas.

- understand how nuclear reactors work, why they are dangerous, and how they might be made safer.

- discuss our options for radioactive waste storage and disposal.

- summarize public opinion about nuclear power and how you feel about this controversial energy source.

The Night the Lights Went Out in New York City

On the evening of September 18, 1981, New York City was plunged into darkness. The crisis started when a transformer overheated and exploded somewhere upstate. Within minutes, an escalating cascade of tripped switches and blown equipment caused the regional power system to crash. Elevators stalled, computers died, traffic lights failed, and motors groaned to a standstill as the biggest city in the U.S. became strangely dark and quiet. People poured into the streets asking each other what happened. There was some vandalism and crime, but for the most part, people were more neighborly and cooperative than ever before. Many people saw stars in the night sky for the first time in their lives. Others built bonfires in the streets where they sat cooking food and telling stories like nomads living in the ruins of some lost civilization.

It took hours to restore the power and days to repair all the damage that resulted from this fiasco. Jeremiahs warned that this was only foretaste of what will happen when the world runs out

of fossil fuels in a few years. So far, this hasn't happened. In fact, known reserves of oil, gas, and coal are larger now than they were in 1980, but so is the rate at which we are depleting them (fig. 21.1). This incident serves as a reminder of how dependent we have become on energy supplies. Think for a moment how your life would change if a constant supply of energy were no longer available.

What is energy and how is it measured? **Energy** *is the capacity to do work, changing the physical state or motion of an object.* **Work** *is the application of force through a distance. Energy is measured in joules, calories, ergs, or British thermal units (table 21.1).* **Power** *is the rate of energy delivery and is measured in horsepower or watts.*

In this chapter, we will examine the current primary sources of energy in the world and how the energy from those sources is extracted and used. The relationship between economic development and energy use is an important aspect

of energy policy. We will look briefly, therefore, at the disparities in energy use between the more-developed and less-
developed countries of the world and ways in which economics
and energy access are linked to environmental conditions,

quality of life, and future development in the world. Chapter 22
examines some energy conservation options and alternative,
renewable energy sources. ▶

Figure 21.1

Fossil fuels provided the energy source for the Industrial Revolution
and built the world in which we now live. We have used up much of
the easily accessible supply of these traditional fuels, however, often
through wasteful practices such as the flaring shown here.

Figure 21.2

Sudden price shocks in the 1970s caused by anticipated oil shortages
showed Americans that our energy-intensive lifestyles may not
continue forever.

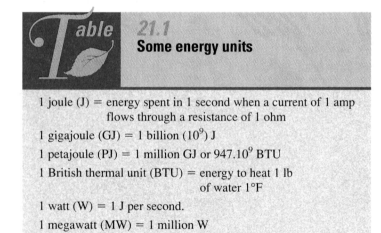

Table 21.1
Some energy units

1 joule (J) = energy spent in 1 second when a current of 1 amp
    flows through a resistance of 1 ohm

1 gigajoule (GJ) = 1 billion (10^9) J

1 petajoule (PJ) = 1 million GJ or 947.10^9 BTU

1 British thermal unit (BTU) = energy to heat 1 lb
    of water 1°F

1 watt (W) = 1 J per second.

1 megawatt (MW) = 1 million W

1 metric ton of oil = 6.66 barrels of 42 gal each

A Brief Energy History

Fire was probably the first human energy technology. Charcoal
from fires has been found at sites occupied by our early ancestors
1 million years ago. Muscle power provided by domestic animals
has been important at least since the dawn of agriculture some
10,000 years ago. Wind and water power have been used nearly as
long. The invention of the steam engine, together with diminishing
supplies of wood in industrializing countries, caused a switch to
coal as our major energy source in the nineteenth century. Coal, in
turn, has been replaced by oil in this century due to the ease of shipping, storing, and burning liquid fuels.

World oil use peaked in 1979, when daily production passed
66 million barrels per day. An Arab oil embargo in 1973 and the
Iranian revolution in 1977 caused crude oil prices to rise nearly tenfold during the 1970s. These sudden price shocks were a major
source of the crushing debt burdens that still hold back many developing countries. The results were less disruptive in richer countries, but Americans waiting in long lines at gas stations became
aware for the first time that much of our luxurious lifestyle is dependent on a limited unstable oil supply (fig. 21.2).

The early 1980s saw an increased concern about conservation
and development of renewable energy resources. This concern
didn't last long, unfortunately, as reduced demand and increased
production in the mid-1980s caused an oil glut that made prices fall
almost as rapidly as they had risen a decade earlier. By 1994, the
price of a barrel of crude oil had fallen to $13.75. This was less,
when adjusted for inflation, than before prices started to rise in
1973. Now, for the first time ever, the United States imports more
than half of its oil. These imports now cost some $50 billion each
year, more than one half our total foreign trade deficit.

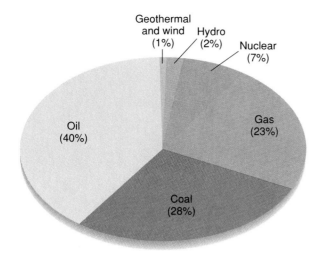

Figure 21.3

Worldwide commercial energy production.

Source: *World Resources* 1994–1995.

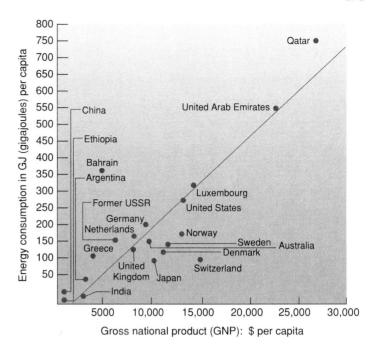

Figure 21.4

Per capita energy use and GNP. In general, higher energy use correlates with a higher standard of living. Scandinavia and Switzerland, however, use about half as much energy as we do and have a higher standard of living by most measures. Oil-rich states, such as Qatar, Bahrain, and the United Arab Emirates, use two to three times as much energy as we do but are still relatively undeveloped in most areas.

Source: Data from World Resources Institute, 1990.

Current Energy Sources

Fossil fuels (petroleum, natural gas, and coal) now provide about 95 percent of all commercial energy in the world (fig. 21.3). Sustainable or renewable energy resources—such as solar, biomass, hydroelectric, and other less-developed types of power production—currently provide less than 3 percent of world power needs. Worldwide, nuclear power provides about twice as much energy as renewable sources. We have enough uranium to fuel many more nuclear plants. Nuclear energy has the benefit of not contributing to global warming (chapter 17), but as we discuss later in this chapter, safety concerns make this option unacceptable to most people.

You may have noticed a conspicuous hole in the energy consumption data given for the United States and the world: use of biomass, especially fuelwood. The preceding figures concentrated on *commercial* energy sources, whereas most biomass burning occurs for residential use, mostly by individuals in the Third World. The importance of biomass to billions of people can't be neglected. Many countries use fuelwood (including charcoal) for their main nonmuscle energy source. In the poorest countries, such as Haiti, Mali, Malawi, and Burkina Faso, biomass supplies more than 90 percent of total energy for heating and cooking. This is a serious cause of forest destruction and soil loss (chapters 11 and 14).

Per Capita Consumption

Perhaps the most important facts about fossil fuel consumption are that we in the 20 richest countries consume nearly 80 percent of the natural gas, 65 percent of the oil, and 50 percent of the coal produced each year. Although we make up less than *one-fifth* of the world's population, we use more than *two-thirds* of the commercial energy supply. The United States and Canada, for instance, constitute only 5 percent of the world's population but consume about one-quarter of the available energy. To be fair, however, some of

that energy goes to grow crops or manufacture goods that are later shipped to developing countries.

How much energy do you use every year? Most of us don't think about it much, but maintaining the lifestyle we enjoy requires an enormous energy input. On average, each person in the United States and Canada uses about 300 GJ (equivalent to about 60 barrels of oil) per year. By contrast, in the poorest countries of the world, such as Ethiopia, Kampuchea, Nepal, and Bhutan, each person generally consumes less than one GJ per year. This means that each of us consumes, on average, almost as much energy in a single day as a person in one of these countries consumes in a year.

Clearly, there is a link between energy consumption and the comfort and convenience of our lives (fig. 21.4). Those of us in the richer countries enjoy many amenities not available to most people in the world. The linkage is not absolute, however. Several European countries, including Sweden, Denmark, and Switzerland, have higher standards of living than does the United States by almost any measure but use only half as much energy as we do. These countries have had effective energy conservation programs for many years. Japan, also, has a far lower energy consumption rate than might be expected for its industrial base and income level. Because Japan has few energy resources of its own, it has developed very efficient energy conservation measures.

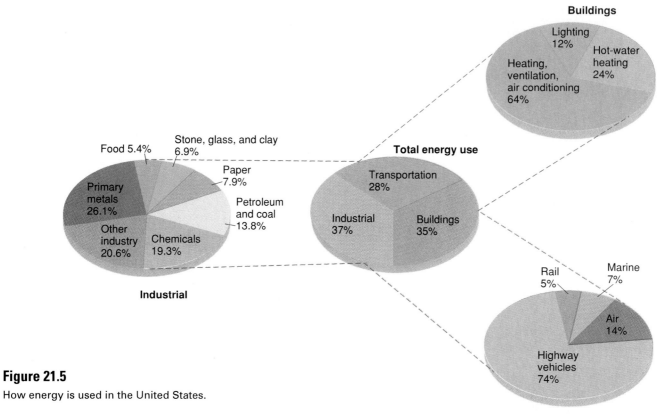

Figure 21.5

How energy is used in the United States.

Source: Data from U.S. Energy Agency.

How Energy Is Used

The largest share (37 percent) of the energy used in the United States is consumed by industry (fig. 21.5). Mining, milling, smelting, and forging of primary metals consume about one-quarter of the industrial energy share. The chemical industry is the second largest industrial user of fossil fuels, but only half of its use is for energy generation. The remainder is raw material for plastics, fertilizers, solvents, lubricants, and hundreds of thousands of organic chemicals in commercial use. The manufacture of cement, glass, bricks, tile, paper, and processed foods also consumes large amounts of energy. Residential and commercial buildings use some 35 percent of the primary energy consumed in the United States, mostly for space heating, air conditioning, lighting, and water heating. Small motors and electronic equipment take an increasing share of residential and commercial energy. In many office buildings, computers, copy machines, lights, and human bodies liberate enough waste heat so that a supplementary source is not required, even in the winter.

Transportation consumes 28 percent of all energy used in the United States each year. About 98 percent of that energy comes from petroleum products refined into liquid fuels, and the remaining 2 percent is provided by natural gas and electricity. Almost three-quarters of all transport energy is used by motor vehicles. Nearly 3 trillion passenger miles and 600 billion ton miles of freight are carried annually by motor vehicles in the United States. About 75 percent of all freight traffic in the United States is carried

by trains, barges, ships, and pipelines, but because they are very efficient, they use only 12 percent of all transportation fuel.

Finally, analysis of how energy is used has to take into account waste and loss of potential energy. About *half* of all the energy in primary fuels is lost during conversion to more useful forms, while it is being shipped to the site of end use, or during its use. Electricity, for instance, is generally promoted as a clean, efficient source of energy because when it is used to run a resistance heater or an electrical appliance almost 100 percent of its energy is converted to useful work and no pollution is given off. What happens before then, however? We often forget that huge amounts of pollution are released during mining and burning of the coal that fires power plants. Furthermore, nearly two-thirds of the energy in the coal that generated that electricity was lost because of inefficient conversion in the power plant. About 10 percent more is lost during transmission and stepping down to household voltages. Similarly, about 75 percent of the original energy in crude oil is lost during distillation into liquid fuels, transportation of that fuel to market, storage, marketing, and combustion in vehicles.

Natural gas is our most efficient fuel. Only 10 percent of its energy content is lost in shipping and processing since it moves by pipelines and usually needs very little refining. Ordinary gas-burning furnaces are about 75 percent efficient, and high-economy furnaces can be as much as 95 percent efficient. Because natural gas is more efficient to produce and burn than oil or coal, it produces about half as much carbon dioxide—and therefore half as much contribution to global warming—per unit of energy.

Coal

Coal is fossilized plant material preserved by burial in sediments and altered by geological forces that compact and condense it into a carbon-rich fuel. Coal is found in every geologic system since the Silurian Age 400 million years ago, but graphite deposits in very old rocks suggest that coal formation may date back to Precambrian times. Most coal was laid down during the Carboniferous period (286 million to 360 million years ago) when the earth's climate was warmer and wetter than it is now. Because coal takes so long to form, it is essentially a nonrenewable resource.

Coal Resources and Reserves

World coal deposits are vast, ten times greater than conventional oil and gas resources combined. Coal seams can be 100 m thick and can extend across tens of thousands of square kilometers that were vast swampy forests in prehistoric times. The total resource is estimated to be 10 trillion metric tons. If all this coal could be extracted, and if coal consumption continued at present levels, this would amount to several thousand years' supply. At present rates of consumption, these proven-in-place reserves—those explored and mapped but not necessarily economically recoverable—will last about 200 years.

Where are these coal deposits located? They are not evenly distributed throughout the world (fig. 21.6). Notice that the countries with the largest land area have the most coal. North America has one-fourth of all proven reserves located in large deposits across the central and mountain regions (fig. 21.7). China and the former Soviet Union have nearly as much. Eastern and Western Europe have fairly large coal supplies in spite of their relatively small size. Africa and Latin America have very little coal despite their large size. Antarctica is thought to have large coal deposits, but they would be difficult, expensive, and ecologically damaging to mine.

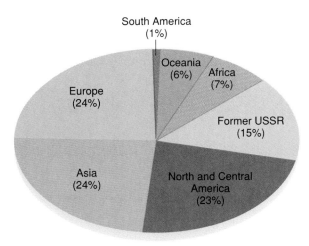

Figure 21.6

Proven-in-place coal reserves, by region as percent of a total 1.96 trillion metric tons. *Note:* total is more than 100 percent due to rounding.

Source: World Resources Institute 1994–95.

It would seem that the abundance of coal deposits is a favorable situation. But do we really want to use all of the coal? What are the environmental and personal costs of coal extraction and use? In the next section, we will look at some of the disadvantages and dangers of mining and burning coal.

Mining

Coal mining is a hot, dirty, and dangerous business. Underground mines are subject to cave-ins, fires, accidents, and accumulation of poisonous or explosive gases (carbon monoxide, carbon dioxide, methane, hydrogen sulfide). Between 1870 and 1950, more than 30,000 coal miners died of accidents and injuries in Pennsylvania alone, equivalent to one man per day for 80 years. Untold thousands have died of respiratory diseases. In some mines, nearly every miner who did not die early from some other cause was eventually disabled by **black lung disease,** inflammation and fibrosis caused by accumulation of coal dust in the lungs or airways (chapter 9). Few of these miners or their families were compensated for their illnesses by the companies for which they worked. The U.S. Department of Labor now compensates miners and their dependents under the Black Lung Benefits Program, but workers often have difficulty proving that health problems are occupationally related. Even when compensation is provided, black lung remains a wretched occupational hazard of coal mining.

Strip mining or surface mining is cheaper than underground mining (fig. 21.8) but often makes the land unfit for any other use. Mine reclamation is now mandated in the United States, but efforts often are shallow and ineffective. Coal mining also contributes to water pollution. Sulfur and other water soluble minerals make mine drainage and runoff from coal piles and mine tailings acidic and highly toxic. Thousands of miles of streams in the United States have been poisoned by coal-mining operations.

Air Pollution

Many people aren't aware that coal burning releases radioactivity and many toxic metals. Uranium, lead, cadmium, mercury, rubidium, thallium, and zinc—along with a number of other elements—are absorbed by plants and concentrated in the process of coal formation. These elements are not destroyed when the coal is burned, instead they are released as gases or concentrated in fly ash and bottom slag. You are likely to get a higher dose of radiation living next door to a coal-burning power plant than a nuclear plant under normal (nonaccident) conditions. Coal combustion is responsible for about 25 percent of all atmospheric mercury pollution in the United States.

Coal contains up to 10 percent sulfur (by weight). Unless this sulfur is removed by washing or flue-gas scrubbing, it is released during burning and oxidizes to sulfur dioxide (SO_2) or sulfate (SO_4) in the atmosphere. The high temperatures and rich air mixtures ordinarily used in coal-fired burners also oxidize nitrogen compounds (mostly from the air) to nitrogen monoxide, dioxide, and trioxide. Every year the 900 million tons of coal burned in the United States (83 percent for electric power generation) releases 18 million metric tons of SO_2, 5 million metric tons of nitrogen oxides (NO_x),

Conventional Energy

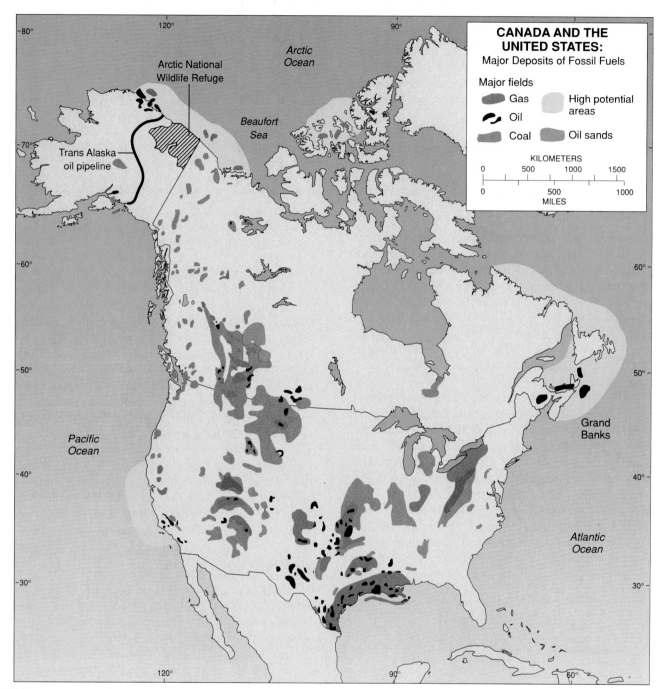

Figure 21.7

Canadian and U.S. fossil fuel deposits. Colored areas show off-shore sedimentary deposits with high potential for oil or gas. The ecological risks of drilling in these areas are high; many environmentalists argue that drilling should be prohibited in these places.

From *Geography: Regions and Concepts*, 6th ed., by H. J. DeBlij and P. O. Muller. Copyright © 1992 John Wiley & Sons, Inc. Redrawn by permission of John Wiley & Sons, Inc.

4 million metric tons of airborne particulates, 600,000 metric tons of hydrocarbons and carbon monoxide, and close to a trillion metric tons of CO_2. This is about three-quarters of the SO_2, one-third of the NO_x, and about half of the industrial CO_2 released in the United States each year.

These air pollutants have many deleterious effects, including human health costs, injury to domestic and wild plants and animals, and damage to buildings and property (chapter 18). Total losses from air pollution are estimated to be between $5 billion and $10 billion per year in the United States alone. By some accounts,

Figure 21.8

A giant power shovel scoops up coal in an open pit mine in Montana. Careful recontouring and restoration can make this land productive once again, but costs are high and many old mines remain desolate and bare for decades; if not centuries.

at least 5000 excess human deaths per year can be attributed to coal production and burning.

Sulfur can be removed from coal before it is burned, or sulfur compounds can be removed from the flue gas after combustion. Formation of nitrogen oxides during combustion also can be minimized. Perhaps the ultimate limit to our use of coal as a fuel will be the release of carbon dioxide into the atmosphere. As we discussed in chapter 17, carbon traps heat in the atmosphere and is a major contributor to global warming. It is difficult to imagine how we could trap and dispose of the 1 trillion tons of CO_2 generated each year in the United States by burning coal, not to mention the amount formed when we burn petroleum products. Alternative energy sources, such as solar and wind power (chapter 22), may be our only options.

Oil

Like coal, petroleum is derived from organic molecules created by living organisms millions of years ago and buried in sediments where high pressures and temperatures concentrated and transformed them into energy-rich compounds. Depending on its age and history, a petroleum deposit will have varying mixtures of oil,

gas, and solid tarlike materials. Some very large deposits of heavy oils and tars are trapped in porous shales, sandstone, and sand deposits in the western areas of Canada and the United States.

Liquid and gaseous hydrocarbons can migrate out of the sediments in which they formed through cracks and pores in surrounding rock layers. Oil and gas deposits often accumulate under layers of shale or other impermeable sediments, especially where folding and deformation of systems create pockets that will trap upward-moving hydrocarbons (fig. 21.9). Contrary to the image implied by its name, an oil pool is not usually a reservoir of liquid in an open cavern but rather individual droplets or a thin film of liquid permeating spaces in a porous sandstone or limestone, much like water saturating a sponge.

Pumping oil out of a reservoir is much like sucking liquid out of a sponge. The first fraction comes out easily, but removing subsequent fractions requires increasing effort. We never recover all the oil in a formation; in fact, a 30 to 40 percent yield is about average. There are ways of forcing water or steam into the oil-bearing formations to "strip" out more of the oil, but at least half the total deposit usually remains in the ground at the point at which it is uneconomical to continue pumping. Methods for squeezing more oil from a reservoir are called **secondary recovery techniques** (fig. 21.9).

Conventional Energy

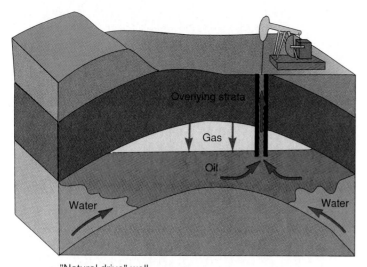

"Natural drive" well

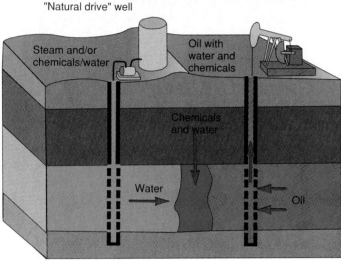

Enhanced recovery or "stripping" well

Figure 21.9

Recovery process for petroleum. In a "natural drive" well, water or gas pressure forces liquid petroleum into the well. In an enhanced recovery or "stripping" well, steam and/or water mixed with chemicals are pumped down a well behind the oil pool, forcing oil up the extraction well.

Oil Resources and Reserves

The total amount of oil in the world is estimated to be about 4 trillion barrels (600 billion metric tons), half of which is thought to be ultimately recoverable. Some 465 billion barrels of oil already have been consumed. In 1990, the proven reserves were roughly 1 trillion bbls, enough to last only 50 years at the current consumption rate of 20 billion barrels per year. It is estimated that another 800 billion barrels either remain to be discovered or are not recoverable at current prices with present technology. As oil resources become depleted and prices rise, it probably will become economical to find and bring this oil to the market unless alternative energy sources are developed. This estimate of the resource does not take into account the very large potential from unconventional liquid hydrocarbon resources, such as shale oil and tar sands,

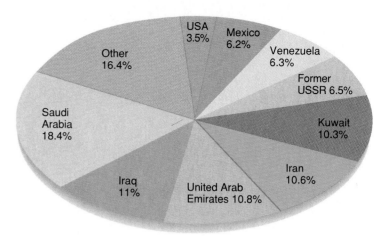

Figure 21.10

World proven recoverable oil reserves by country (1991). Annual world consumption is about 20 billion barrels, or 2.4% of the 823.6 billion barrel proven recoverable reserve.

Source: Data from World Resources Institute, 1991.

which might double the total reserve if they can be mined with acceptable social, economic, and environmental impacts.

By far the largest supply of proven-in-place oil is in Saudi Arabia, which has 168.8 billion barrels, nearly one-fifth of the total world reserve (fig. 21.10). Kuwait had about 10 percent of the proven world oil reserves before Iraq invaded in 1990. Some 600 wells were blown up and set on fire by the retreating Iraqis. Although the consequences were not as catastrophic as first feared, at least 10 percent of Kuwait's oil was burned, spilled, or otherwise lost (see p. 462). Together, the Persian Gulf countries in the Middle East contain nearly two-thirds of the world's proven petroleum supplies. With our insatiable appetite for oil (some would say addiction), it is not difficult to see why this volatile region plays such an important role in world affairs.

Although oil consumption in Europe and North America has been relatively constant in recent years, Asian demand has been growing rapidly as economies there expand and people become more affluent. China, for instance, is expanding oil use 10 percent per year. By the year 2000, if current trends continue, about half of all oil produced by the Organization of Petroleum Exporting Countries (OPEC) will be going to Asia.

Oil Imports and Domestic Supplies

Originally, the United States is estimated to have had about 200 billion barrels of recoverable oil, 10 percent of the total world oil resource. This large supply of relatively accessible oil played an important part in development of the country's economic and industrial power during the twentieth century. Until 1947, the United States was the leading oil export country in the world. By 1995, the U.S. was importing nearly 9 million barrels per day, more than half of total consumption. The major suppliers of this oil were Mexico and Venezuela in the 1970s but shifted to Saudi Arabia and other Middle Eastern countries in the 1990s, making the U.S. vulnerable to the unstable political situations in that volatile region.

Altogether, the United States has already used about 40 percent of its original recoverable resource of 200 billion barrels. Of the 120 billion barrels thought to remain, about 58 billion barrels are proven-in-place. If we stopped importing oil and depended exclusively on indigenous supplies, our proven reserve would last only *ten years* at current rates of consumption.

The regions of North America with the greatest potential for substantial new oil discoveries are portions of the continental shelf along the California coast, the Beaufort Sea, and the Grand Banks (see fig. 21.7), all of which are prime wildlife areas. Proposals to do exploratory drilling in these areas have been strongly opposed by many people concerned about the potential for long-term environmental damage from drilling activities and oil spills like the *Exxon Valdez* in Prince William Sound. Other people argue that having an inexpensive, plentiful energy supply, even if only for a few decades, is worth the social and environmental costs. (Case Study, p. 472).

Oil Shales and Tar Sands

Estimates of our total oil supply usually do not reflect the very large potential from **unconventional oil** resources, such as shale oil and tar sands, which might double the total reserve if they can be extracted with reasonable social, economic, and environmental costs. Canada, for example, has an estimated 270 billion cubic meters of bituminous tar sands, mostly in northern Alberta (fig. 21.11). Liquid petroleum can be extracted from these sands with hot water, chemicals, or other stripping processes. If even part of this resource is recoverable, it would be the world's largest known oil deposit. There are severe environmental constraints, however. A typical plant producing 125,000 barrels of oil per day creates about 15 million cubic meters of toxic sludge and consumes or contaminates billions of liters of water each year.

Similarly, vast deposits of oil shale occur in the western United States. Actually, oil shale is neither oil nor shale but a fine-grained sedimentary rock rich in solid organic material called kerogen. When heated to about 480°C (900°F), the kerogen liquefies and can be extracted from the stone. Oil shale beds up to 600 m (1800 ft) thick occur in the Green River Formation in Colorado, Utah, and Wyoming, and lower-grade deposits are found over large areas of the eastern United States. If these deposits could be extracted at a reasonable price and with acceptable environmental impacts, they might yield the equivalent of several trillion barrels of oil.

Mining and extracting shale oil creates many problems, however. It is expensive; it uses vast quantities of water, a scarce resource in the arid west; it has a high potential for air and water pollution; and it produces enormous quantities of waste. In the early 1980s, when the search for domestic oil supplies was at fever pitch, serious discussions occurred about filling whole canyons, rim to rim, with oil shale waste. One experimental mine used a nuclear explosion to break up the oil shale. All the oil shale projects dried up when oil prices fell in the mid-1980s, however. What do you think? Should we declare some western states sacrifice areas so we can have cheap gasoline?

Figure 21.11

Canada has huge reserves of tar sands such as the one being mined here in Alberta. Mining and extraction of the petroleum products causes great environmental damage, however, and may never be safe or practical.

Natural Gas

Natural gas is the world's third largest commercial fuel (after oil, and coal), making up 24 percent of global energy consumption. It is the most rapidly growing energy source because it is convenient, cheap, and clean burning. Because natural gas produces only half as much CO_2 as an equivalent amount of coal, substitution could help reduce global warming (chapter 17). Natural gas is difficult to ship across oceans or to store in large quantities. North America is fortunate to have an abundant, easily available

CASE STUDY
Oil Versus Wildlife

Beyond the jagged Brooks Range in Alaska's far northeastern corner lies one of the world's largest nature preserves, the 7-million ha (18-million ac) Arctic National Wildlife Refuge (ANWR). Geologists suspect that the narrow coastal plain bordering the Beaufort Sea contains oil and gas-bearing strata similar to the nearby fields of Prudhoe Bay (see fig. 21.7). Debates have raged for years over whether to do exploratory drilling and seismic mapping of the area. A harsh and inhospitable wilderness in winter, this arctic tundra biome is a summer wetland of innumerable potholes and marshes that teem with wildlife. It is the central calving ground of the 180,000 caribou of the Porcupine herd, thought to be the world's second largest caribou herd. It also is an important habitat for thousands of snow geese and other migratory waterfowl, musk oxen, Alaskan brown bears, polar bears, and arctic wolves, along with unimaginable clouds of insects and many other species of rare plants and animals.

Energy companies are extremely interested in this area because any oil found there could be pumped out through the existing Trans-Alaska oil pipeline, thus extending the life of their multibillion dollar investment. The state of Alaska hopes that revenues from ANWR will replenish dwindling coffers as the oil supply from Prudhoe Bay dries up. Automobile companies, tire manufacturers, filling stations, and others who depend on our continued use of petroleum point out that domestic oil supplies are rapidly being depleted and that imported oil accounts for more than half of our balance of payments deficit.

Figure 1

Alaska's Arctic National Wildlife Refuge, home to one of the world's largest caribou herds, has been called the crown jewel of the entire refuge system. Oil companies have proposed exploratory drilling here, but environmentalists fear irreparable damage to the fragile ecosystem.

Opponents to this project point out that scars left by mechanized equipment, blasting for seismic mapping, and drilling will persist in the tundra for centuries. A Department of the Interior environmental impact statement (EIS) predicted the loss of 40 percent of the caribou, 25 to 50 percent of the musk oxen, and 50 percent of the snow geese if drilling is allowed. Native Gwich'in people who depend on wildlife for subsistence have protested disruption of their traditional lifestyle.

Environmentalists argue that wildlife and wilderness values that the refuge was set up to protect should be paramount over short-term economic interests. Others feel that the potential for a valuable energy source overrides biological importance, and that, furthermore, wildlife will not be adversely affected if drilling is done carefully. They claim that caribou have actually increased at Prudhoe Bay since oil extraction began.

Few people ever visit this remote wilderness. Most of us, in fact, probably wouldn't do so even if we had a chance. Frigid winds howling off the Arctic Ocean make this bleak, barren landscape uninviting much of the year and swarming clouds of insects make it unbearable in the summer. The question becomes one of values: the rights of wildlife and ecosystems to exist undisturbed versus our desires to exploit resources. Ironically, we could easily do without this oil. The Department of the Interior estimates

only a 19-percent chance of finding any oil, which at best would amount to only a 200-day supply if it were our only source. If we were to raise the average fuel efficiency of American automobiles to the levels now common in Europe, or if everyone in the United States who drives alone to school or work were to share the ride with one other person, we would save more oil in 6 months than we will ever pump out of ANWR even at the most optimistic estimates.

ETHICAL CONSIDERATIONS

What do you think about the ethics of this situation? How would you weigh the potential for finding oil in the wildlife refuge against wilderness values and wildlife? Do human interests take precedence over nature? Would you feel differently about this issue if the oil there were the last available supply in the world? If we're going to have to wean ourselves from dependence on oil, why not do it sooner rather than later?

---

supply of gas and a pipeline network to deliver it to market. Many developing countries cannot afford such a pipeline network, and much of the natural gas produced in conjunction with oil pumping is simply burned (flared off), a terrible waste of a valuable resource (see fig. 21.1).

Natural Gas Resources and Reserves

The republics of the former Soviet Union have 44 percent of known natural gas reserves (mostly in Siberia and the Central Asian republics) and account for 36.5 percent of all production. Both

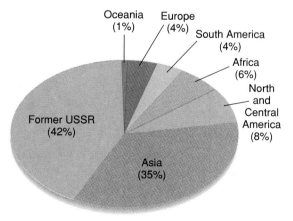

Figure 21.12

Percent of proven-in-place natural gas reserves, by region (1991).

Source: Data from World Resources Institute, 1994–95.

Eastern and Western Europe buy substantial quantities of gas from these wells. Figure 21.12 shows the distribution of proven natural gas reserves in the major producing countries of the world.

The total ultimately recoverable natural gas resources in the world are estimated to be 10,000 trillion cubic feet, corresponding to about 80 percent as much energy as the recoverable reserves of crude oil. The proven world reserves of natural gas are 3200 trillion cubic feet (73,000 million metric tons). Because gas consumption rates are only about half of those for oil, current gas reserves represent roughly a 60-year supply at present usage rates. Proven reserves in the United States are about 200 trillion cubic feet, or 6 percent of the world total. This is a 10-year supply at current rates of consumption. Known reserves are more than twice as large.

Unconventional Gas Sources

Natural gas resources have not been as extensively investigated in most places as have oil resources. Until recently, gas has been a by-product of crude oil drilling and production. There may be many sources of "unconventional" gas, such as Devonian shales, "tight-sand" gas, and coal-seam methane, that have not been extensively explored but could provide large and valuable fuel sources. Some geologists theorize that natural methane may come from abiogenic origins (inorganic carbon in rocks) deep within the earth and not, as generally supposed, primarily from degeneration of biological materials. If this is true, resources of methane within the earth's crust may be vastly larger than current estimates predict.

Unlike coal and oil, natural gas is presently being formed as a by-product of human activities. Some municipalities heat their waste

Conventional Energy

treatment plants with methane produced during sewage digestion. Methane is also collected from landfills. Some farmers in both China and the United States generate methane for heating, lighting, and cooking with animal waste digesters. Such domestic methane sources have potential for small-scale, low-cost, local applications (chapter 22).

Nuclear Power

In 1953, President Dwight Eisenhower presented his "Atoms for Peace" speech to the United Nations. He announced that the United States would build nuclear-powered electrical generators to provide clean, abundant energy. He predicted that nuclear energy would fill the deficit caused by predicted shortages of oil and natural gas. It would provide power "too cheap to meter" for continued industrial expansion of both the developed and the developing world. It would be a supreme example of "beating swords into plowshares." Technology and engineering would tame the evil genie of atomic energy and use its enormous power to do useful work.

Glowing predictions about the future of nuclear energy continued into the early 1970s. Between 1970 and 1974, American utilities ordered 140 new reactors for power plants (fig. 21.13). Some advocates predicted that by the end of the century there would be 1500 reactors in the United States alone. The International Atomic Energy Agency (IAEA) projected worldwide nuclear power generation of at least 4.5 million megawatts (MW) by the year 2000, eighteen times more than our current nuclear capacity and twice as much as present world electrical capacity from all sources.

Rapidly increasing construction costs, declining demand for electric power, and safety fears have made nuclear energy much

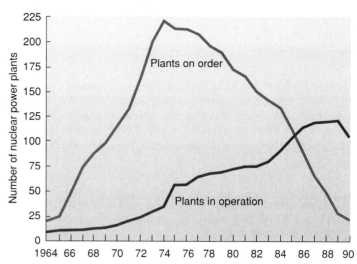

Figure 21.14

The changing fortunes of nuclear power in the United States are evident in this graph showing the number of nuclear plants on order and plants in operation. Since the early 1970s, few utilities have ordered new plants and many previous orders have been canceled.

Source: Data from Richard K. Lester, "Rethinking Nuclear Power," *Scientific American*, March 1986.

less attractive than promoters expected. Electricity from nuclear plants was about half the price of coal in 1970 but twice as much by 1990. Wind energy is already cheaper than nuclear power in many areas and solar power or hydropower are becoming cheaper as well (chapter 22).

The United States led the world into the nuclear age, and it appears that we are leading the world out of it. After 1975, only thirteen orders were placed for new nuclear reactors in the United States, and all of those orders subsequently were canceled (fig. 21.14). In fact, 100 of the 140 reactors ordered before 1975 were canceled. It is beginning to look as if the much-acclaimed nuclear power industry may have been a very expensive wild goose chase that may never produce enough energy to compensate for the amount invested in research, development, mining, fuel preparation, and waste storage.

After the 1991 Persian Gulf war, the nuclear industry began advertising again, promoting reactors as environmentally friendly because they do not contribute carbon dioxide to the greenhouse effect (chapter 17). Whether solutions to the problems of reactor safety and waste storage will be found so that nuclear power can serve as an interim energy source until we develop renewable sources remains to be seen.

How Do Nuclear Reactors Work?

The most commonly used fuel in nuclear power plants is U^{235}, a naturally occurring radioactive isotope of uranium. Ordinarily, U^{235} makes up only about 0.7 percent of uranium ore, generally too little to sustain a chain reaction. It must be purified and concentrated by mechanical or chemical procedures (fig. 21.15). When the U^{235} concentration reaches about 3 percent, the uranium is formed into cylindrical pellets slightly thicker than a pencil and

Figure 21.13

The nuclear power plant at Three Mile Island near Harrisburg, PA, was the site of a partial meltdown of the reactor core in 1979. Most of the radioactive material was kept inside the containment buildings (short, cylindrical buildings in center). The four large open cylinders are cooling towers.

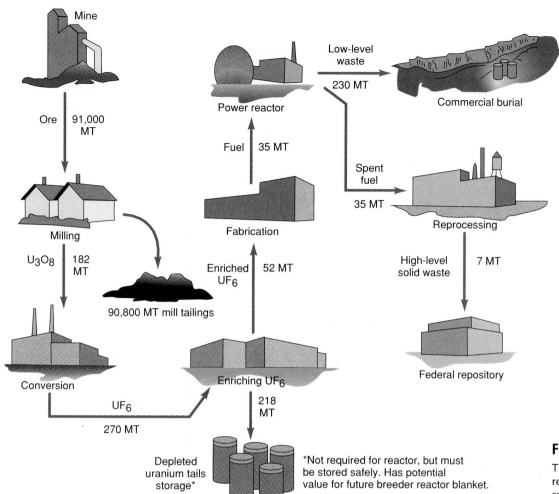

Figure 21.15

The nuclear fuel cycle. Quantities represent the average annual fuel requirements for a typical 1000 MW light water reactor (MT=metric tons).

about 1.5 cm long. Although small, these pellets pack an amazing amount of energy. Each 8.5-gram pellet is equivalent to a ton of coal or four barrels of crude oil.

The pellets are stacked in hollow steel rods approximately 4 m long. About one hundred of these rods are bundled together to make a **fuel assembly.** Thousands of fuel assemblies containing 100 tons of uranium are bundled together in a heavy steel vessel called the reactor core. Radioactive uranium atoms are unstable—that is, they spontaneously release high-energy subatomic particles called neutrons. When uranium is packed tightly in the reactor core, the neutrons released by one atom will trigger the **nuclear fission** (splitting) of another uranium atom and the release of heat and still more neutrons (fig. 21.16). Thus a self-sustaining **chain reaction** is set in motion and vast amounts of heat are released.

The chain reaction is moderated (slowed) in a power plant by a neutron-absorbing cooling solution that circulates between the fuel rods. In addition, **control rods** of neutron-absorbing material, such as cadmium or boron, are inserted into spaces between fuel assemblies to shut down the fission reaction or are withdrawn to allow it to proceed. Water or some other coolant is circulated between the fuel rods to remove excess heat.

The greatest danger in one of these complex machines is a cooling system failure. If the pumps fail or pipes break during operation, the nuclear fuel quickly overheats and a "meltdown" can result that releases deadly radioactive material. (What Do *You* Think? p. 478). Although nuclear power plants cannot explode like a nuclear bomb, the radioactive releases from a worst-case disaster like Chernobyl are just as devastating as a bomb.

Kinds of Reactors in Use

Seventy percent of the nuclear plants in the United States and in the world are Pressurized Water Reactors (PWR) (fig. 21.17). Water is circulated through the core, absorbing heat as it cools the fuel rods. This primary cooling water is heated to 317°C and reaches a pressure of 2235 psi. It then is pumped to a steam generator where it heats a secondary water-cooling loop. Steam from the secondary loop drives a high-speed turbine generator that produces electricity. Both the reactor vessel and the steam generator are contained in a thick-walled concrete and steel containment building that prevents radiation from escaping and is designed to withstand high pressures and temperatures in case of accidents. Engineers operate

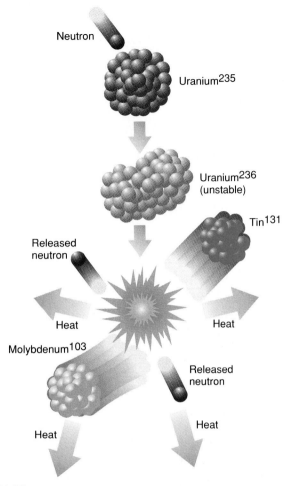

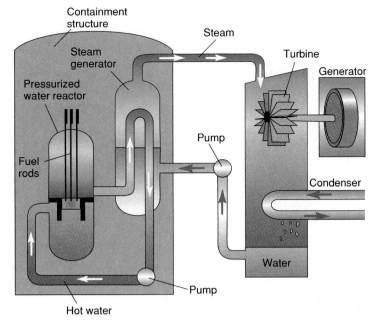

Figure 21.17

Pressurized water nuclear reactor. Water is superheated and pressurized as it flows through the reactor core. Heat is transferred to nonpressurized water in the steam generator. The steam drives the turbogenerator to produce electricity.

Figure 21.16

The process of nuclear fission is carried out in the core of a nuclear reactor. In the sequence shown here, the unstable isotope, uranium-235, absorbs a neutron and splits to form tin-131 and molybdenum-103. Two or three neutrons are released per fission event and continue the chain reaction. The total mass of the reaction product is slightly less than the starting material. The residual mass is converted to energy (mostly heat).

the plant from a complex, sophisticated control room containing many gauges and meters to tell them how the plant is running.

Overlapping layers of safety mechanisms are designed to prevent accidents, but these fail-safe controls make reactors very expensive and very complex. A typical nuclear power plant has 40,000 valves compared to only 4000 in a fossil fuel-fired plant of similar size. In some cases, the controls are so complex that they confuse operators and cause accidents rather than prevent them. Under normal operating conditions, a PWR releases very little radioactivity and is probably less dangerous for nearby residents than a coal-fired power plant.

A simpler, but dirtier and more dangerous reactor design is the Boiling Water Reactor (BWR). In this model, water from the reactor core boils to make steam, which directly drives the turbine-generators. This means that highly radioactive water and steam leave the containment structure. Controlling leaks is difficult, and the chances of releasing radiation in an accident are very high.

Canadian nuclear reactors use deuterium (H^2 or ^2H, the heavy, stable isotope of hydrogen) as both a cooling agent and a moderator. These *Can*adian *deu*terium (CANDU) reactors operate with natural, unconcentrated uranium (0.7 percent U^{235}) for fuel, eliminating expensive enrichment processes. "Heavy water," deuterium-containing water (^2H_2O), is expensive, however, and these reactors are susceptible to overheating and meltdown if cooling pumps fail.

In Britain, France, and the former Soviet Union, a common reactor design uses graphite, both as a moderator and as the structural material for the reactor core. In the British MAGNOX design (named after the magnesium alloy used for its fuel rods), gaseous carbon dioxide is blown through the core to cool the fuel assemblies and carry heat to the steam generators. In the Soviet design, called RBMK (the Russian initials for a graphite-moderated, water-cooled reactor), low-pressure cooling water circulates through the core in thousands of small metal tubes.

These designs were originally thought to be very safe because graphite has high capacity for both capturing neutrons and dissipating heat. Designers claimed that these reactors could not possibly run out of control; unfortunately, they were proven wrong. The small cooling tubes are quickly blocked by steam if the cooling system fails and the graphite core burns when exposed to air. The two most disastrous reactor accidents in the world, so far, involved fires in graphite cores that allowed the nuclear fuel to melt and escape into the environment. In 1956, a fire at the Windscale Plutonium Reactor in England released roughly 100 million curies of radionuclides and contaminated hundreds of square kilometers of countryside. Similarly, burning graphite in the Chernobyl nuclear plant in Ukraine made the fire much more difficult to control than it might have been in another reactor design. The worst nuclear accident in U.S. history, for instance, in 1979 at Three Mile Island near Harrisburg, Pennsylvania, released less than 1/100 as much radiation as Chernobyl or Windscale, (see fig. 21.13).

Alternative Reactor Designs

Several other reactor designs are inherently safer than the ones we now use. Among these are the modular High-Temperature, Gas-Cooled Reactor (HTGCR) and the Process-Inherent Ultimate-Safety (PIUS) reactor.

In HTGCR, uranium is encased in tiny ceramic-coated pellets; gaseous helium blown around these pellets is the coolant. If the reactor core is kept small, it cannot generate enough heat to melt the ceramic coating, even if all coolant is lost; thus, a meltdown is impossible and operators could walk away during an accident without risk of a fire or radioactive release. Fuel pellets are loaded into the reactor from the top, shuffle through the core as the uranium is consumed, and emerge from the bottom as spent fuel. This type of reactor can be reloaded during operation. Since the reactors are small, they can be added to a system a few at a time, avoiding the costs, construction time, and long-range commitment of large reactors. Only two of these reactors have been tried in the United States: the Brown's Ferry reactor in Alabama and the Fort St. Vrain reactor near Loveland, Colorado. Both were continually plagued with problems (including fires in control buildings and turbine-generators), and both were closed without producing much power.

A much more successful design has been built in Europe by General Atomic. In West German tests, a HTGCR was subjected to total coolant loss while running at full power. Temperatures remained well below the melting point of fuel pellets and no damage or radiation releases occurred. These reactors might be built without expensive containment buildings, emergency cooling systems, or complex controls. They would be both cheaper and safer than current designs.

The PIUS design features a massive, 60-meters-high pressure vessel of concrete and steel, within which the reactor core is submerged in a very large pool of boron-containing water (fig. 21.18). As long as the primary cooling water is flowing, it keeps the borated water away from the core. If the primary coolant pressure is lost, however, the surrounding water floods the core, and the boron poisons the fission reaction. There is enough secondary water in the pool to keep the core cool for at least a week without any external power or cooling. This should be enough time to resolve the problem. If not, operators can add more water and evaluate conditions further.

The Canadian "slow-poke" is a small-scale version of the PIUS design that doesn't produce electricity but might be a useful substitute for coal, oil, or gas burners in district heating plants. The core of this "mini-nuke" is only about half the size of a kitchen stove and generates only 1/10 to 1/100 as much power as a conventional reactor. The fuel sits in a large pool of ordinary water, which it heats to just below boiling and sends to a heat exchanger. A secondary flow of hot water from the exchanger is pumped to nearby buildings for space heating. Promoters claim that a runaway reaction is impossible in this design and that it makes an attractive and

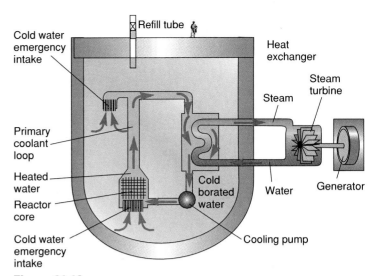

Figure 21.18

A PIUS reactor consists of a core and primary cooling system immersed in a very large pool of cold, borated water. As long as the reactor is operating, the cool water is excluded from the cooling circuit by the temperature and pressure differential. Any failure of the primary cooling system would allow cold water to flood the core and shut down the nuclear reaction. The large volume of the pool would cool the reactor for several days without any replenishment.

cost-efficient alternative to fossil fuels. Despite a widespread aversion to anything nuclear, Switzerland and Germany are developing similar small nuclear heating plants.

If these reactors had been developed initially, the history of nuclear power might have been very different. Neither reactor type is suitable, however, for mobile power plants, such as nuclear submarines, which tells you something about the history of nuclear power and the motivation for its development. Aside from being inherently safer in case of coolant pressure loss, neither of these alternative reactor designs eliminates other problems that we will discuss later in this chapter. Perhaps nuclear power in one or more of these alternate forms will once again be an important energy source in the developed world.

Breeder Reactors

For more than thirty years, nuclear engineers have been proposing high-density, high-pressure, **breeder reactors** that produce fuel rather than consume it. These reactors create fissionable plutonium and thorium isotopes from the abundant, but stable, forms of uranium (fig. 21.19). The starting material for this reaction is plutonium reclaimed from spent fuel from conventional fission reactors. After about ten years of operation, a breeder reactor would produce enough plutonium to start another reactor. Sufficient plutonium currently is stockpiled in the United States to produce electricity for one hundred years at present rates of consumption, if breeder reactors can be made to work safely and dependably.

Several problems have held back the breeder reactor program in the United States. One problem is the concern about safety. The reactor core of the breeder must be at a very high density for the breeding reaction to occur. Water does not have enough heat

What Do You Think?
Chernobyl: Could It Happen Here?

In the early morning hours of April 26, 1986, residents of the Ukrainian village of Pripyat saw a spectacular and terrifying sight. A glowing fountain of molten nuclear fuel and burning graphite was spewing into the dark sky through a gaping hole in the roof of the Chernobyl Nuclear Power Plant only a few kilometers away. Although officials assured them that there was nothing to worry about in this "rapid fuel relocation," the villagers knew that something was terribly wrong. They were witnessing the worst possible nuclear power accident, a "meltdown" of the nuclear fuel and rupture of the containment facilities, releasing enormous amounts of radioactivity into the environment (fig. 1).

The accident was a result of a risky experiment undertaken by the plant engineers in violation of a number of safety rules and operational procedures. They were testing whether the residual energy of a spinning turbine could provide enough power to run the plant in an emergency shutdown if off-site power were lost. Reactor number four had been slowed down to only 6 percent of its normal operating level. To conserve the small amount of electricity being generated, they then disconnected the emergency core-cooling pumps and other safety devices, unaware that the reactor was dangerously unstable under these conditions.

The heat level in the core began to rise, slowly at first, and then faster and faster. The operators tried to push the control rods into the core to slow the reaction, but the graphite pile had been deformed by the heat so that the rods wouldn't go in. In 4.5 seconds, the power level rose 2000-fold, far above the rated capacity of the cooling system. Chemical explosions (probably hydrogen gas released from the expanding core) ripped open the fuel rods and cooling tubes. Cooling water flashed into steam and blew off the 1000-ton-concrete cap on top of the reactor. Molten uranium fuel puddled in the bottom of the reactor, creating a critical mass that accelerated the nuclear fission reactions. The metal superstructure of the containment building was ripped apart and a column of burning graphite, molten uranium, and radioactive ashes billowed 1000 m (3000 ft) into the air.

Panic and confusion ensued. Officials first denied that anything was wrong. The village of Pripyat was not evacuated for 36 hours. There was no public announcement for three days. The first international warning came, not from Soviet authorities, but from Swedish scientists 2000 km away who detected unusually high levels of radioactive fallout and traced airflows back to the southern Soviet Union.

There were many acts of heroism during this emergency. Firefighters climbed to the roof of the burning reactor building to pour water into the blazing inferno. Engineers dived into the suppression pool beneath the burning core to open a drain to prevent another steam explosion. Helicopter pilots hovered over the gaping maw of the ruined building to drop lead shot, sand, clay, limestone, and boron carbide onto the burning nuclear core to smother the fire and suppress the nuclear fission reactions. More than 600,000 workers participated in putting out the fire and cleaning up contamination. Thousands who were exposed to high radiation doses already have died and many more have dim prospects for the future.

The amount of radioactive fallout varied from area to area, depending on wind patterns and rainfall. Some places had heavy doses while neighboring regions had very little. One band of fallout spread across Yugoslavia, France, and Italy. Another crossed Germany and Scandinavia. Small amounts of radiation even reached North America. Altogether, about 7 tons of fuel containing 50 to 100 million curies were released, roughly 5 percent of the reactor fuel.

For several years after the accident, the Soviet government tried to suppress information and deny the conse-

Figure 1

The explosion and fire in Unit 4 (*circled*) at the Chernobyl nuclear power plant in the Soviet Ukraine in 1986 spread radioactive fallout across the entire Northern Hemisphere. Millions of people in Europe and Asia were exposed to dangerously high radiation levels.

quences. The dissolution of the USSR has allowed many of these secrets to come to light. As of 1995, at least 125,000 people have died because of Chernobyl and another 576,000 are on the official registry of potential victims. Most tragic are effects in children of the Ukraine and neighboring Belarus. Thyroid cancers have increased 200-fold among children in the most contaminated areas and leukemia rates are now soaring. It is possible that millions of people worldwide may suffer ill health or prematurely shortened lives because of Chernobyl.

For the present, the damaged reactor has been entombed in a giant, steel-reinforced concrete "sarcophagus." Unfortunately, this containment structure was hastily built of inferior materials, and already has begun to deteriorate. It will have to be replaced soon at great cost and danger.

So far, more than 250,000 people have been relocated from the contaminated area. More than 70 villages have been destroyed and millions of hectares of the richest farmland in the Commonwealth of Independent States has been abandoned. The immediate direct costs were roughly $3 billion; total costs might be one hundred times that much.

The Ukraine would like to shut down the remaining reactors at Chernobyl but fuel shortages and a crippled economy make building replacement power plants prohibitively expensive. The United States, Canada, and the European Community are currently negotiating with Ukraine and Russia to close down all remaining Chernobyl-type reactors. What do you think the U.S. should contribute to help bring this about? How important is it to prevent further catastrophes?

Proponents of nuclear power point out that the Soviet-style reactors were much more dangerous than most of those now in the U.S. or Canada. Furthermore, they argue, the accident was caused by operator error and bureaucratic bungling. Something similar could never happen here, they claim. Opponents respond that we should learn from this tragedy and abandon this dangerous technology. What do you think? How likely is a similar accident in your neighborhood? Even if the chances are very low, does the potential harm make this an unacceptable risk? How would you weigh a low probability against a terrible potential outcome?

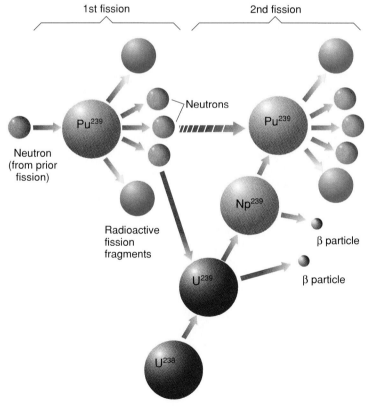

Figure 21.19

Reactions in a "breeder" fission process. Neutrons from a plutonium fission change U^{238} to U^{239} and then to Pu^{239} so that the reactor creates more fuel than it uses.

capacity to carry away the high heat flux in the core, so liquid sodium is used as a coolant. Liquid sodium is very corrosive and difficult to handle. It burns with an intense flame if exposed to oxygen, and it explodes if it comes into contact with water. Because of its intense heat, a breeder reactor will melt down and self-destruct within a few seconds if the primary coolant is lost, as opposed to a few minutes for a normal fission reactor.

Another very serious concern about breeder reactors is that they produce excess plutonium that can be used for bombs. It is essential to have a spent-fuel reprocessing industry if breeders are used, but the existence of large amounts of weapons-grade plutonium in the world would surely be a dangerous and destabilizing development. The chances of some of that material falling into the hands of terrorists or other troublemakers are very high. Japan plans to purchase 30 tons of this dangerous material from France and ship it half way around the world through some of the most dangerous and congested shipping lanes on the planet. Worldwide protests have been launched in opposition to this plan.

A proposed $1.7 billion breeder-demonstration project in Clinch River, Tennessee, has been on and off for fifteen years. At last estimate, it would cost up to five times the original price if it is ever completed. The United Kingdom built a breeder reactor at Dounray, Scotland, but it didn't work well and was shut down in 1985 with great financial loss. In 1986, France put into operation a full-sized commercial breeder reactor, the SuperPhenix, near Lyons. It cost three times the original estimate to build and produces electricity at twice the cost per kilowatt of conventional nuclear power. In 1987, a large crack was discovered in the inner containment vessel of the SuperPhenix, and it had to be shut down for repairs.

Conventional Energy

Radioactive Waste Management

One of the most difficult problems associated with nuclear power is the disposal of wastes produced during mining, fuel production, and reactor operation. How these wastes are managed may ultimately be the overriding obstacle to nuclear power.

Ocean Dumping of Radioactive Wastes

Until 1970, the United States, Britain, France, and Japan disposed of radioactive wastes in the ocean. Dwarfing all these dumps, however, are those of the former Soviet Union, which has seriously—and some fear permanently—contaminated the Arctic ocean. Rumors of Soviet nuclear waste dumping had circulated for years, but it was not until after the collapse of the Soviet Union that the world learned the true extent of what happened. Starting in 1965, the Soviets disposed of eighteen nuclear reactors—seven loaded with nuclear fuel—in shallow waters of the Kara Sea off the eastern coast of Novaya Zemlya island, and millions of liters of liquid waste in the nearby Barents Sea (fig. 21.20). Two other reactors were sunk in the Sea of Japan.

Altogether, the former Soviet Union dumped 2.5 million curies of radioactive waste into the oceans, more than twice as much as the combined amounts that twelve other nuclear nations have reported dumping over the past 45 years. In 1993, despite protests from Japan, Russia dumped 900 tons of additional radioactive waste into the Sea of Japan. With aging storage tanks overflowing with millions of liters of highly radioactive wastes and no funds to maintain or monitor them, Russia is desperate for someplace to dispose of its nuclear wastes. This is an international crisis that threatens all of us.

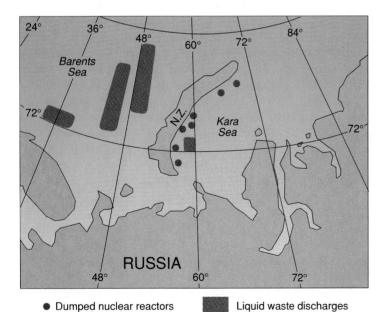

Dumped nuclear reactors   **Liquid waste discharges**

Figure 21.20

Sites of dumped nuclear reactors east of Novaya Zemlya. Dark blue areas indicate liquid waste discharges.

Land Disposal of Nuclear Waste

Enormous piles of mine wastes and abandoned mill tailings in uranium-producing countries represent another serious waste disposal problem. Production of 1000 tons of uranium fuel typically generates 100,000 tons of tailings and 3.5 million liters of liquid waste. There now are approximately 200 million tons of radioactive waste in piles around mines and processing plants in the United States. This material is carried by the wind and washes into streams, contaminating areas far from its original source. Canada has even more radioactive mine waste on the surface than does the United States.

In addition to the leftovers from fuel production, there are about 100,000 tons of low-level waste (contaminated tools, clothing, building materials, etc.) and about 15,000 tons of high-level (very radioactive) wastes in the United States. The high-level wastes consist mainly of spent fuel rods from commercial nuclear power plants and assorted wastes from nuclear weapons production. For the past 20 years, spent fuel assemblies from commercial reactors have been stored in deep water-filled pools at the power plants. These pools were originally intended only as temporary storage until the wastes were shipped to reprocessing centers or permanent disposal sites.

The pools are now full but neither reprocessing nor permanent storage is yet available in the United States. In 1993, Consumers Power Company began storing nuclear wastes in large metal dry casks outside its Palisades plant in southwestern Michigan (fig. 21.21). A number of other utilities have requested permission for similar storage. These plans are meeting with fierce opposition from local residents who fear that the casks will leak. Most nuclear power plants are built near rivers, lakes, or seacoasts. Extremely toxic radioactive materials could quickly spread over a very large area if these wastes escape.

In 1987, the U.S. Department of Energy announced plans to build the first **high-level waste repository** at a cost of $6 billion to $10 billion on a barren desert ridge near Yucca Mountain, Nevada. Intensely radioactive wastes are to be buried deep in the ground where it is hoped that they will remain unexposed to groundwater and earthquakes for the tens of thousands of years required for the radioactive materials to decay to a safe level (fig. 21.22). Although the area is very dry now, we can't be sure that it will always remain that way. We also don't know how long the metal canisters will contain the wastes, or where they might migrate over thousands of years.

Some nuclear experts believe that **monitored, retrievable storage** would be a much better way to handle wastes. This method involves holding wastes in underground mines or secure surface facilities where they can be watched. The storage sites would be kept open and inspected regularly. If canisters begin to leak, they could be removed for repacking. Safeguarding the wastes would be expensive and the sites might be susceptible to wars or terrorist attacks. We might need a perpetual priesthood of nuclear guardians to ensure that the wastes are never released into the environment.

have to be stored just like other wastes. This includes not only the reactor and pipes but also the meter-thick, steel-reinforced concrete containment building. The pieces have to be cut apart by remote-control robots because they are too dangerous to be worked on directly.

Only a few plants have been decommissioned so far, but it has generally cost two to ten times as much to tear them down as to build them in the first place. That means that the one hundred reactors now in operation might cost somewhere between $200 billion and $1 trillion to decommission. No one knows how much it will cost to store the debris for thousands of years or how it will be done. However, we would face this problem, to some degree, even without nuclear electric power plants. Plutonium production plants and nuclear submarines also have to be decommissioned. Originally, the Navy proposed to just tow old submarines out to sea and sink them. The risk that the nuclear reactors would corrode away and release their radioactivity into the ocean makes this method of disposal unacceptable, however.

Public Opinion about Nuclear Power

Opposing nuclear power has been a high priority for many conservation organizations in recent years. A federation of antinuclear alliances (such as Clamshell Alliance, Northern Sun Alliance) has rallied thousands of people to oppose nuclear power. Protests and civil disobedience at some sites have gone on for a decade or more, and several plants have been abandoned because there was so much opposition to them. Proponents of nuclear power argue that abandoning this energy source—especially plants that are already built—is foolish since the risks may be lower than is commonly held (fig. 21.23). Where do you stand on this issue?

In recent years, public opinion in many countries has swung strongly against nuclear power (fig. 21.24). In France, more than 20,000 protesters battled police and one person was killed in 1985 at the site of the SuperPhenix near Lyons. In Hong Kong, more than 1 million people (20 percent of the population) signed a petition opposing the 1200-MW reactor built by China at Daya Bay, only 50 km from the center of this densely populated city. Hong Kong is at the end of a narrow, rocky peninsula, so evacuation would be impossible in case of a nuclear emergency.

A few countries remain strongly committed to nuclear power. France generates 60 percent of its electricity with nuclear plants and plans to build more. The United Kingdom and Japan both depend on nuclear power for about 20 percent of their electricity. After the Chernobyl disaster, officials of the former Soviet Union declared a construction moratorium on nuclear power plants. In 1993, however, the new government of Russia announced plans to build at least thirty new nuclear stations—several of the RBMK design. Russian environmentalists denounced this plan as "unacceptable legally, ecologically, economically, and politically."

Many other countries, however, have decided to postpone building reactors, to phase out already built reactors, or simply to

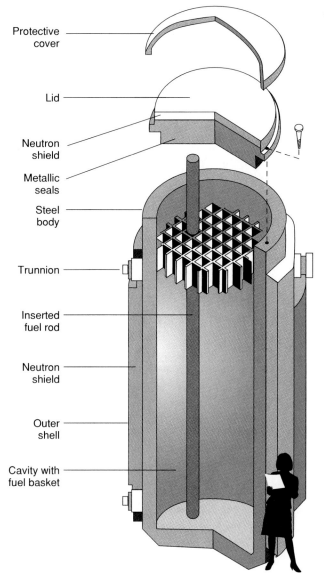

Protective cover
Lid
Neutron shield
Metallic seals
Steel body
Trunnion
Inserted fuel rod
Neutron shield
Outer shell
Cavity with fuel basket

Figure 21.21
Dry cask storage for nuclear waste. Each cask is 17 ft tall, 8.5 feet in diameter, has steel walls 10 in thick, and weighs 122 tons when fully loaded with 40 fuel assemblies. The casks are expected to last 40 years and cost $700,000 each. Critics are worried that the casks will leak, but proponents point out that they can be monitored easily and replaced if necessary.

Decommissioning Old Nuclear Plants

Old power plants themselves eventually become waste when they have outlived their useful lives. Most plants are designed for a life of only 30 years. After that, pipes become brittle and untrustworthy because of the corrosive materials and high radioactivity to which they are subjected. Plants built in the 1950s and early 1960s already are reaching the ends of their lives. You don't just lock the door and walk away from a nuclear power plant; it is much too dangerous. It must be taken apart, and the most radioactive pieces

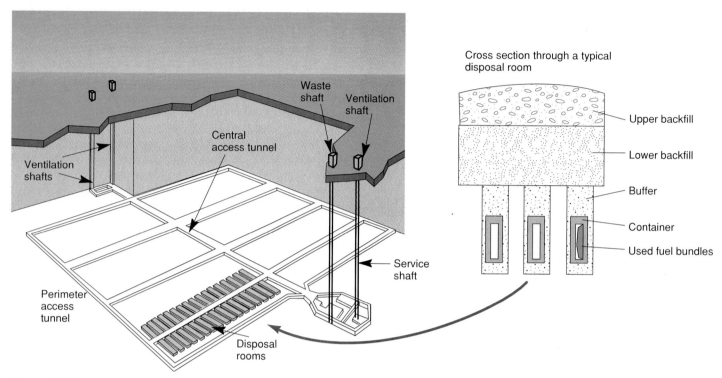

Figure 21.22

One proposal for the permanent disposal of used fuel is to bury the used fuel containers 500 to 1000 m deep in the stable rock of the Canadian Shield. Spent fuel, in the form of ceramic fuel pellets, are sealed inside the corrosion-resistant containers. Glass beads are compacted around the used fuel bundles and into the spaces between the fuel and the container shell. The containers are then buried in specially built vaults that are packed and sealed with special materials to further retard the migration of radioactive material.

Figure 21.23

The Seabrook nuclear reactor in New Hampshire has been the site of protests and counter-protests for twenty years. Here pipefitters march in support of the plant and their jobs.

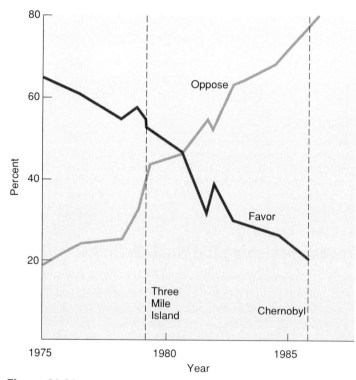

Figure 21.24

United States public opinion on building more nuclear power plants, 1975 to 1986.

abandon reactors that have been built but not yet been put into service. Altogether, about half the nations with nuclear power have chosen to reduce or dismantle existing plants, and few new plants have been ordered in recent years. Opponents of nuclear power claim that the nuclear fission experiment may have been a very brief episode in our history, but one whose legacy will remain with us for a long time. Proponents of nuclear power argue that it may be our only near-term alternative to fossil fuels. What do you think?

Nuclear Fusion

Fusion energy is an alternative to fission energy that many scientists believe has virtually limitless potential. **Nuclear fusion** energy is the energy released when two smaller atomic nuclei fuse into one larger nucleus. Nuclear fusion reactions, the energy source for the sun and for hydrogen bombs, have not yet been harnessed by humans to produce useful net energy. The fuels for these reactions are deuterium and tritium, two heavy isotopes of hydrogen that are so plentiful in seawater that fusion often is cited as a possible ultimate energy source for the future.

It has been known for forty years that if temperatures in an appropriate fuel mixture are raised to 100 million degrees Celsius and pressures of several billion atmospheres are obtained, fusion of deuterium and tritium will occur. Under these conditions, the electrons are stripped away from atoms and the forces that normally keep nuclei apart are overcome. As nuclei fuse, some of their mass is converted into energy, some of which is in the form of heat. There are two main schemes for creating these conditions: magnetic confinement and inertial confinement.

Magnetic confinement involves the containment and condensation of plasma, a hot, electrically neutral gas of ions and free electrons in a powerful magnetic field inside a vacuum chamber.

Compression of the plasma by the magnetic field should raise temperatures and pressures enough for fusion to occur. The most promising example of this approach, so far, has been a Russian design called *tokomak* (after the Russian initials for "tordial magnetic chamber"), in which the vacuum chamber is shaped like a large donut (fig. 21.25).

Inertial confinement involves a small pellet (or a series of small pellets) bombarded from all sides at once with extremely high-intensity laser light. The sudden absorption of energy causes an implosion (an inward collapse of the material) that will increase densities by one thousand to two thousand times and raise temperatures above the critical minimum. These conditions will be maintained for a fraction of a second before the pellet flies apart in a miniature hydrogen explosion, but there should be enough reaction to release some harnessable energy. So far, no lasers powerful enough to create fusion conditions have been built, but researchers are hopeful that this may be achieved in a few years.

In both of these cases, high-energy neutrons escape from the reaction and are absorbed by molten lithium circulating in the walls of the reactor vessel. The lithium absorbs the neutrons and transfers heat to water via a heat exchanger, making steam that drives a turbine generator, as in any steam power plant.

The advantages of fusion reactions, if they are ever feasible, include production of fewer radioactive wastes, the elimination of fissionable products that could be made into bombs, and a fuel supply that is much larger and less hazardous than uranium. Even with breeder reactors that create fissionable isotopes (such as plutonium^{239}) from stable atoms (such as uranium^{238}), uranium and thorium supplies would last for approximately one hundred years, while hydrogen isotopes are plentiful enough to last for thousands of years. There would be some radioactivity and radioactive waste from fusion because the high neutron flux would irradiate

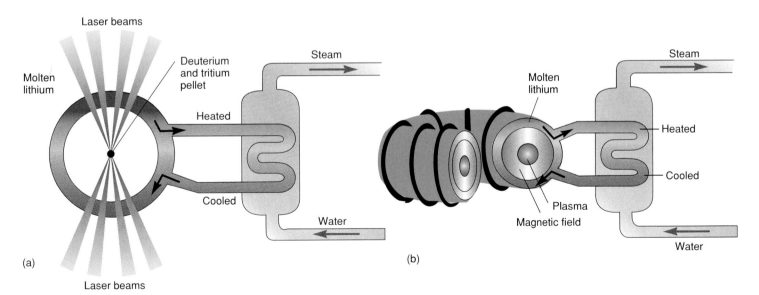

(a)

(b)

Figure 21.25

Nuclear fusion devices. (*a*) Inertial confinement is created by laser beams that bombard and ignite fuel pellets. Molten lithium absorbs energy from the fusion and transfers it as heat to a steam generator. (*b*) A powerful magnetic field confines the plasma and compresses it so that critical temperatures and pressures are reached. Again, molten lithium absorbs energy and transfers it to a steam generator.

the reactor vessel and create radionucleides, but nothing on the order of fission by-products.

After forty years and $20 billion of research, plasmas have been created with temperatures above 230 million degrees Celsius for a fraction of a second. Laser beams have achieved power densities of 10^{13} watts per sq cm, about one-tenth of what is needed.

Proponents of fusion power predict that within ten years we will pass the break-even point at which a fusion reaction can be sustained long enough to produce more energy than it takes to start it. The discovery of new, high-temperature superconducting materials may enhance progress in this field, but opponents fear that this is just another expensive wild-goose chase.

Summary

Energy is the capacity to do work. Power is the rate of energy delivery. Today, nearly 95 percent of all commercial energy is generated by fossil fuels, about 39 percent coming from petroleum. Next are coal, with 32 percent, and natural gas (methane), with 24 percent. Petroleum and natural gas were not used in large quantities until the beginning of this century, but supplies are already running low. Coal supplies will last several more centuries at present rates of usage, but it appears that the fossil fuel age will have been a rather short episode in the total history of humans. Nuclear power provides only about 2.5 percent of commercial energy worldwide.

Energy is essential for most activities of modern society. Its use generally correlates with standard of living, but there are striking differences between energy use per capita in countries with relatively equal standards of living. The United States, for instance, consumes about twice as much energy per person as does Switzerland, which is higher in many categories that measure quality of life. This difference is based partly on level of industrialization and partly on policies, attitudes, and traditions in the United States that encourage extravagant or wasteful energy use.

The environmental damage caused by mining, shipping, processing, and using fossil fuels may necessitate cutting back on our use of these energy sources. Coal is an especially dirty and dangerous fuel, at least as we currently obtain and use it. Some new coal treatment methods remove contaminants, reduce emissions, and make its use more efficient (so less will be used). Coal combustion is a major source of acid precipitation that is suspected of being a significant cause of environmental damage in many areas. We now recognize that CO_2 buildup in the atmosphere has the potential to trap heat and raise the earth's temperature to catastrophic levels.

Nuclear energy offers an alternative to many of the environmental and social costs of fossil fuels, but it introduces serious problems of its own. In the 1950s, there was great hope that these problems would be overcome and that nuclear power plants would provide energy "too cheap to meter." Recently, however, much of that optimism has been waning. No new reactors have been started in the United States since 1975. Many countries are closing down existing nuclear power plants, and a growing number have pledged to remain or become "nuclear-free."

The greatest worry about nuclear power is the danger of accidents that release hazardous radioactive materials into the environment. Several accidents, most notably the "meltdown" at the Chernobyl plant in the Soviet Ukraine in 1986, have convinced many people that this technology is too risky to pursue.

Other major worries about nuclear power include where to put the waste products of the nuclear fuel cycle and how to ensure that it will remain safely contained for the thousands of years required for "decay" of the radioisotopes to nonhazardous levels. Yucca Mountain, Nevada, was chosen for a high-level waste repository, but many experts believe that burying these toxic residues in nonretrievable storage is a mistake.

Nuclear fusion occurs when atoms are forced together in order to create new, more massive atoms. If fusion actions could be created under controlled conditions in a power plant, they might provide an essentially unlimited source of energy that would avoid many of the worst problems of both fossil fuels and fission-based nuclear power. So far, however, no one has been able to sustain controlled fusion reactions that produce more energy than they consume. It remains to be seen whether this will ever be possible, and whether other unforeseen problems will arise if it does become a reality.

None of our current major energy sources appear to offer security in terms of stable supply or environmental considerations. Neither coal nor nuclear power is a good long-term energy source with our present level of technology. We urgently need to develop alternative sources of sustainable energy.

▼ Questions for Review

1. What is energy? What is power?

2. What are the major sources of commercial energy worldwide and in the United States? Why is data usually presented in terms of commercial energy?

3. How is energy used in the United States?

4. How does our energy use compare with that of people in other countries?

5. How much coal, oil, and natural gas are in proven reserves worldwide? Where are those reserves located?

6. What are the most important health and environmental consequences of our use of fossil fuels?

7. Describe how a nuclear reactor works and why they are dangerous.

8. What are the four most common reactor designs? How do they differ from each other?

9. What are the advantages and disadvantages of the breeder reactor?

10. Describe methods proposed for storing and disposing of nuclear wastes.

▼ Questions for Critical Thinking

1. We have discussed a number of different energy sources and energy technologies in this chapter. Each has advantages and disadvantages. If you were an energy policy analyst, how would you compare such different problems as the risk of a nuclear accident versus air pollution effects from burning coal?

2. If your local utility company were going to build a new power plant in your community, what kind would you prefer?

3. The nuclear industry is placing ads in popular magazines and newspapers claiming that nuclear power is environmentally friendly since it doesn't contribute to the greenhouse effect. How do you respond to that claim?

4. Our energy policy effectively treats some strip-mine lands as national sacrifice areas since we know that they will never be restored to their original state when mining is finished. How do we decide who wins and who loses in this transaction?

5. Storing nuclear wastes in dry casks outside nuclear power plants is highly controversial. Opponents claim the casks will inevitably leak. Proponents claim they can be designed to be safe. What evidence would you consider adequate or necessary to choose between these two positions?

6. Since the environmental impact of a nuclear power plant accident cannot be limited to national boundaries (as in the case of Chernobyl), should there be international regulations for power plants? Who would define these regulations? How could they be enforced?

7. Although we have wasted vast amounts of energy resources in the process of industrialization and development, some would say that it was a necessary investment to get to a point at which we can use energy more efficiently and sustainably. Do you agree? Might we have followed a different path?

8. The policy of the United States has always been to make energy as cheap and freely available as possible. Most European countries charge three to four times as much for gasoline as we do. Think about the ramifications of our energy policy. Why is it so different from that

in Europe? Who benefits and who or what loses in these different approaches? How have our policies shaped our lives? What does existing policy tell you about how our government works?

▼ Key Terms

black lung disease 46	magnetic confinement 483
breeder reactor 477	monitored, retrievable storage 480
chain reaction 475	nuclear fission 475
control rods 475	nuclear fusion 483
energy 463	power 463
fossil fuels 464	secondary recovery techniques 469
fuel assembly 475	unconventional oil 471
high-level waste repository 480	work 463
inertial confinement 483	

▼ Additional Information on the Internet

Energy Information Administration
http://www.eia.do.gov/

Fossil Fuel Issues
http://zebu.uoregon.edu/fossil.html/

Nuclear Information World Wide Web Server
http://nuke.WestLab.com/

RiSo National Laboratory
http://www.risoe.dk/

University of California Energy Institute
http://www-ucenergy.eecs.berkeley.edu/UCENERGY/

US Department of Energy
http://www.doe.gov/

US Department of Energy—Office of Fossil Energy
http://www.fe.doe.gov/

US Department of Transportation
http://www.dot.gov/

US Nuclear Regulatory Commission
http://www.nrc.gov/

World Wide Web Virtual Library: Energy
http://solstice.crest.org/online/virtual-library/VLib-energy.html/

Note: Further readings appropriate to this chapter are listed on p. 605.

Parabolic mirrors focus sunlight on steam-generating tubes at this power plant in the California desert. Renewable energy sources could replace much of our dependence on fossil fuels while helping to reduce air and water pollution.

CHAPTER 22: Sustainable Energy

Objectives

After studying this chapter, you should be able to:

- appreciate the opportunities for energy conservation available to us.

- understand how active and passive systems capture solar energy and how photovoltaic collectors generate electricity.

- comprehend why diminishing fuelwood supplies are a crisis in less-developed countries and how that crisis affects us.

- evaluate the use of dung, crop residues, energy crops, and peat as potential energy sources.

- explain how hydropower, wind, and geothermal energy contribute to our power supply.

- describe how tidal and wave energy and ocean thermal gradients can be used to generate electrical energy.

- compare and contrast different options for storing electrical energy from intermittent or remote sources.

A Photovoltaic Village in Java

Until a few years ago, life in the remote village of Sukatani in western Java was not much different than it had been for centuries. As the sun went down in the evening, families gathered around flickering candles or smoky oil lamps to eat supper. The rusty pulley on the village well squeaked as someone drew a pail of water for the night. A few people walked home through the growing dusk as fireflies blinked above the rice paddies. Here on the equator, nights are 12 hours long year round—a long time to be in the dark. There wasn't much to do and not much money to spend on luxuries like batteries for a portable radio or a flashlight. Isolated by mountainous terrain and bumpy, unpaved roads, Sukatani wasn't likely to be hooked into a national power grid or telecommunications system anytime soon.

Everything changed in 1989 when Sukatani was chosen as the first of 50 rural Indonesian villages to participate in an innovative renewable power program. Jointly designed and financed by a Dutch firm and the Indonesian government, this project demonstrates the promise of sustainable energy systems. Today, photovoltaic panels on tall utility poles convert sunshine into electricity (fig. 22.1). Homeowners pay only about $2.50 per month for power, less than they previously paid for kerosene, batteries, and battery charging.

Compact fluorescent bulbs now provide light so that children can do homework after supper, and a new motorized pump on the village well provides a steady supply of water for better sanitation. Adults use evening hours sewing, weaving, or carving items to sell at the market. A few households have a "warung" (shop) that now can be open after dark. The village even has a few television sets that provide evening entertainment and a window on the wider world. The village has joined the global telecommunications network decades earlier than was expected.

Figure 22.1
Solar panels on tall poles rise above the rooftops of Sukatani, a remote village in western Java. This demonstration project shows how renewable energy can promote economic development and make life more comfortable without destroying the environment.

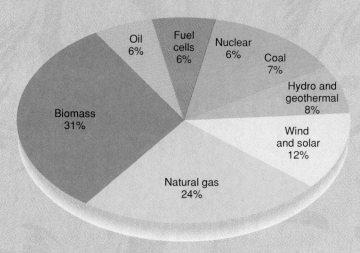

Figure 22.2
An alternative energy future for the United States in 2050. This scenario projects a 50 percent drop in total energy consumption and a 75 percent reduction in CO_2 production. Compare this to figure 21.3.

Source: Environmental Protection Agency, 1992.

This project offers hope for how both developing and developed countries can benefit from sustainable energy. As chapter 21 shows, nonrenewable sources (fossil fuels and nuclear power) currently provide nearly 90 percent of all the commercial energy used worldwide. If the rest of the world were to use energy at the rate that some of us do, known fuel supplies wouldn't last

long. Furthermore, our environment couldn't withstand the assault of extracting and burning all that fuel.

Fortunately, as Sukatani demonstrates, fossil fuels and nuclear power are not our only energy options. A combination of energy conservation and renewables could reduce our dependence on coal and oil to about 15 percent of our energy supply rather than the 65 percent they now provide (fig. 22.2). In this chapter, we will look at some options for conservation and renewable energy sources that might allow a real increase in the standard of living for the poorest people of the world and also help preserve our global environment. ▶

Conservation

One of the best ways to avoid energy shortages and to relieve environmental and health effects of our current energy technologies is simply to use less. Conservation offers many benefits both to society and to the environment.

Utilization Efficiencies

Much of the energy we consume is wasted. This statement is not a simple admonishment to turn off lights and turn down furnace thermostats in winter; it is a technological challenge. Our energy conversion technologies are so inefficient that most potential energy in fuel is lost as waste heat, becoming a form of environmental pollution. Of the energy we do extract from primary resources, however, much is used for frankly trivial or extravagant purposes. As chapter 21 shows, several European countries have higher stan-

dards of living than the United States, and yet use 30 to 50 percent less energy.

Many conservation techniques are relatively simple and highly cost effective. More efficient and less energy-intensive industry, transportation, and domestic practices could save large amounts of energy. Improved automobile efficiency, better mass transit, and increased railroad use for passenger and freight traffic are simple and readily available means of conserving transportation energy. In response to the 1970s oil price shocks, automobile gas-mileage averages in the United States more than doubled from 13 mpg in 1975 to 28.8 mpg in 1988. Unfortunately, the oil glut and falling fuel prices of the late 1980s discouraged further conservation. Between 1990 and 1995, the average slipped to only 27.6 mpg.

Much more could be done. Prototype high-efficiency automobiles that get about 100 mpg at highway speeds and over 65 mpg at normal city traffic speeds already have been road tested (fig. 22.3). Amory B. Lovins of the Rocky Mountain Institute in Colorado

Figure 22.3

This experimental automobile with a superefficient turbo-charged diesel engine gets over 100 mpg at a constant 65 mph and 40 mpg in city driving conditions. In addition to standard diesel fuel, it can burn methanol, corn oil, kerosene, and a variety of other nontraditional fuels. Its acceleration, safety, and driving characteristics are comparable to standard autos.

estimated that raising the average fuel efficiency of the United States car and light truck fleet by one mile per gallon would cut oil consumption about 295,000 barrels per *day*. In one year, this would equal the total amount the Interior Department hopes to extract from the Arctic National Wildlife Refuge in Alaska (see chapter 21).

Many improvements in domestic energy efficiency have occurred in the past decade. Today's average new home uses one-half the fuel required in a house built in 1974, but much more can be done. Household energy losses can be reduced by one-half to three-fourths through better insulation, double or triple glazing of windows, thermally efficient curtains or window coverings, and by sealing cracks and loose joints. Reducing air infiltration is usually the cheapest, quickest, and most effective way of saving energy be-

cause it is the largest source of losses in a typical house. It doesn't take much skill or investment to caulk around doors, windows, foundation joints, electrical outlets, and other sources of air leakage.

For even greater savings, new houses can be built with extra thick superinsulated walls, air-to-air heat exchangers to warm incoming air, and even double-walled sections that create a "house within a house." The R-2000 program in Canada details how energy conservation can be built into homes. Special double-glazed windows that have internal reflective coatings and that are filled with an inert gas (argon or xenon) have an insulation factor of R11, the same as a standard 4-inch thick insulated wall or ten times as efficient as a single-pane window. Superinsulated houses now being built in Sweden require 90 percent less energy for heating and cooling than the average American home.

Orienting homes so that living spaces have passive solar gain in the winter and are shaded by trees or roof overhang in the summer also helps conserve energy. Earth-sheltered homes built into the south-facing side of a slope or protected on three sides by an earth berm are exceptionally efficient energy savers because they maintain relatively constant subsurface temperatures (fig. 22.4). In addition to building more energy-efficient homes, there are many personal actions we can take to conserve energy (In Depth, p. 494).

Industrial energy savings are another important part of our national energy budget. More efficient electric motors and pumps, new sensors and control devices, advanced heat-recovery systems, and material recycling have reduced industrial energy requirements significantly. In the early 1980s, U.S. businesses saved $160 billion per year through conservation. When oil prices collapsed, however, we returned to wasteful ways.

Energy Conversion Efficiencies

Energy efficiency is a measure of energy produced compared to energy consumed. Table 22.1 shows the typical energy efficiencies of a variety of energy-conversion devices. Thermal-conversion machines, such as steam turbines in coal-fired or nuclear power

Figure 22.4

Earth-sheltered structures retain much of their heat during cold winters because the earth surrounding them remains considerably warmer than the outside air temperature. They also remain cool in the hot summer.

Table 22.1
Typical Net Efficiencies of Energy-Conversion Devices

	Yield (percent)
Electric power plants	
Hydroelectric (best case)	90
Combined-cycle steam	90
Fuel cell (hydrogen)	80
Coal-fired generator	38
Oil-burning generator	38
Nuclear generator	30
Photovoltaic generation	10
Transportation	
Pipeline (gas)	90
Pipeline (liquid)	70
Waterway (no current)	65
Diesel-electric train	40
Diesel-engine automobile	35
Gas-engine automobile	30
Jet-engine airplane	10
Space heating	
Electric resistance	99*
High-efficiency gas furnace	90
Typical gas furnace	70
Efficient wood stove	65
Typical wood stove	40
Open fireplace	−10
Lighting	
Sodium vapor light	60*
Fluorescent bulb	25*
Incandescent bulb	5*
Gas flame	1

Source: U.S. Department of Energy.
*Note that 60–70% of the energy in the original fuel is lost in electric power generation.

Table 22.2
Typical Net Useful Energy Yields

Energy Source	Yield/Cost Ratio
Nonrenewable sources	
Coal (space or process heat)	20/1
Natural gas (as heat source)	10/1
Gasoline and fuel oil	7/1
Coal gasification (combined cycle)	5/1
Oil shale (as liquid fuel)	1/1
Nuclear (excluding waste disposal)	2/1*
Renewable sources	
Hydroelectric (best case)	20/1**
Wind (electric generation)	2/1
Biomass methane	2/1
Solar electric (10% efficient)	1/1
Solar electric (20% efficient)	2/1

*Decommissioning of old nuclear plants and perpetual storage of wastes may consume more energy than nuclear power has produced.
**Hydropower yield depends on availability of water and life expectancy of dam and reservoir. Most dams produce only about 40% of rated capacity due to lack of water. Some dams have failed or silted up without producing any net energy yield.

elimination, its efficiencies can approach 80 percent with such fuel as hydrogen gas or methane.

Another way to look at energy efficiency is to consider the total **net energy yield** from energy-conversion devices (table 22.2). Net energy yield is based on the total useful energy produced during the lifetime of an entire energy system minus the energy required to make useful energy available. To make comparisons between different energy systems easier, the net energy yield is often expressed as a *ratio* between the output of useful energy and the energy costs for construction, fuel extraction, energy conversion, transmission, waste disposal, etc.

Nuclear power is a good case for net energy yield studies. We get a large amount of electricity from a small amount of fuel in a nuclear plant, but it also takes a great deal of energy to extract and process nuclear fuel and to build, operate, and eventually dismantle power plants. As chapter 21 points out, we may never reach a break-even point where we get back more energy from nuclear plants than we put into them, especially considering the energy that may be required to decommission nuclear plants and guard their waste products in secure storage for thousands of years.

Net yields and overall conversion efficiencies are not the only considerations when we compare different energy sources. The yield/cost ratio and conversion-cycle efficiency is much higher for coal burning, for instance, than for photovoltaic electrical production, making coal appear to be a better source of energy than solar

plants, can turn no more than 40 percent of the energy in their primary fuel into electricity or mechanical power because of the need to reject waste heat. Does this mean that we can never increase the efficiency of fossil fuel use? No. Some waste heat can be recaptured and used for space heating, raising the net yield to 80 or 90 percent. In another kind of process, fuel cells convert the chemical energy of a fuel directly into electricity without an intermediate combustion cycle. Since this process is not limited by waste heat

radiation. Solar energy, however, is free, renewable, and nonpolluting. If we can use solar energy to get electrical energy, it doesn't matter how *efficient* the process is, as long as we get more out of it than we put in.

Negawatt Programs

Utility companies are finding it much less expensive to finance conservation projects than to build new power plants. Rather than buy megawatts of new generating capacity, power companies are investing in "negawatts" of demand avoidance. Each of us can help save energy in everyday life. Pacific Gas and Electric in California and Potomac Power and Light in Washington, D.C., both have instituted large conservation programs. They have found that conservation costs about $350 per kilowatt (kW) saved. By contrast, a new nuclear power plant costs between $3000 and $8000 per kW of installed capacity. New coal-burning plants with the latest air pollution-control equipment cost at least $1000 per kW. By investing $200 million to $350 million in public education, home improvement loans, and other efficiency measures, a utility can avoid building a new power plant that would cost a billion dollars or more. Furthermore, conservation measures don't consume expensive fuel or produce pollutants.

Can application of this approach help alleviate energy shortages in other countries as well? Yes. For example, Brazil could cut its electricity consumption an estimated 30 percent with an investment of $10 billion (U.S.). It would cost $44 billion to build new power plants to produce that much electricity. South Korea has instituted a comprehensive conservation program with energy-saving building standards, efficiency labels on new household appliances, depreciation allowances, reduced tariffs on energy-conserving equipment, and loans and tax breaks for upgrading homes and businesses. It also forbids some unnecessary uses, such as air conditioning and elevators between the first and third floors.

Cogeneration

One of the fastest growing sources of new energy is **cogeneration,** the simultaneous production of both electricity and steam or hot water in the same plant. By producing two kinds of useful energy in the same facility, the net energy yield from the primary fuel is increased from 30–35 percent to 80–90 percent. In 1900, half the electricity generated in the United States came from plants that also provided industrial steam or district heating. As power plants became larger, dirtier, and less acceptable as neighbors, they were forced to move away from their customers. Waste heat from the turbine generators became an unwanted pollutant to be disposed of in the environment. Furthermore, long transmission lines, which are unsightly and lose up to 20 percent of the electricity they carry, became necessary.

By the 1970s, cogeneration had fallen to less than 5 percent of our power supplies, but interest in this technology is being renewed. The capacity for cogeneration more than doubled in the 1980s to about 30,000 megawatts (MW). District heating systems are being rejuvenated, and plants that burn municipal wastes are being studied. New combined-cycle coal-gasification plants or "mini-nukes"

(chapter 21) offer high efficiency and clean operation that may be compatible with urban locations. Small neighborhood- or apartment building-sized power-generating units are being built that burn methane (from biomass digestion), natural gas, diesel fuel, or coal. The Fiat Motor Company makes a small generator for about $10,000 that produces enough electricity and heat for four or five houses.

Tapping Solar Energy

The sun serves as a giant nuclear furnace in space, constantly bathing our planet with a free energy supply. Solar heat drives winds and the hydrologic cycle. All biomass, as well as fossil fuels and our food (both of which are derived from biomass), results from conversion of light energy (photons) into chemical bond energy by photosynthetic bacteria, algae, and plants.

A Vast Resource

The average amount of solar energy arriving at the top of the atmosphere is 340 watts per square meter. About half of this energy is absorbed or reflected by the atmosphere (more at high latitudes than at the equator), but the amount reaching the earth's surface is some 8000 times all the commercial energy used each year. However, this tremendous infusion of energy comes in a form that, until this century, has been too diffuse and low in intensity to be used except for environmental heating and photosynthesis. Figure 22.5 shows solar energy levels over the United States for a typical summer and winter day.

Passive Solar Heat

Our simplest and oldest use of solar energy is **passive heat absorption,** using natural materials or absorptive structures with no moving parts to simply gather and hold heat. For thousands of years, people have built thick-walled stone and adobe dwellings that slowly collect heat during the day and gradually release that heat at night. After cooling at night, these massive building materials maintain a comfortable daytime temperature within the house, even as they absorb external warmth.

A modern adaptation of this principle is a glass-walled "sunspace" or greenhouse on the south side of a building. Incorporating massive energy-storing materials, such as brick walls, stone floors, or barrels of heat-absorbing water into buildings also collects heat to be released slowly at night (fig. 22.6). An interior, heat-absorbing wall called a trombe wall is an effective passive heat collector. Some trombe walls are built of glass blocks enclosing a water-filled space or water-filled circulation tubes so heat from solar rays can be absorbed and stored, while light passes through to inside rooms.

Active Solar Heat

Active solar systems generally pump a heat-absorbing, fluid medium (air, water, or an antifreeze solution) through a relatively small collector, rather than passively collecting heat in a stationary medium like masonry. Active collectors can be located adjacent to

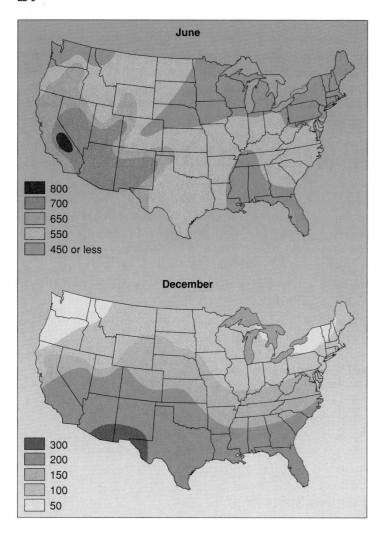

June

800
700
650
550
450 or less

December

300
200
150
100
50

Figure 22.5

Average daily solar radiation in the United States in June and December. One langley, the unit for solar radiation, equals 1 calorie per square centimeter of earth surface (3.69 Btu/square foot).

Source: National Weather Bureau, U.S. Department of Commerce.

or on top of buildings rather than being built into the structure. Because they are relatively small and structurally independent, active systems can be retrofitted to existing buildings.

A flat black surface sealed with a double layer of glass makes a good solar collector. A fan circulates air over the hot surface and into the house through ductwork of the type used in standard forced-air heating. Alternatively, water can be pumped through the collector to pick up heat for space heating or to provide hot water (fig. 22.7). Water heating consumes 15 percent of the United States' domestic energy budget, so savings in this area alone can be significant. A simple flat panel with about 5 sq m of surface can reach 95° C (200° F) and can provide enough hot water for an average family of four almost anywhere in the United States. In California, 650,000 homes now heat water with solar collectors. In Greece, Italy, Israel, Asia, and Africa where fuels are more expensive, up to 70 percent of domestic hot water comes from solar collectors.

Sunshine doesn't reach us all the time, of course. How can solar energy be stored for times when it is needed? There are a number of options. In a climate where sunless days are rare and seasonal variations are small, a small, insulated water tank is a good solar energy storage system. For areas where clouds block the sun for days at a time or where energy must be stored for winter use, a large, insulated bin containing a heat-storing mass, such as stone, water, or clay, provides solar energy storage (fig. 22.8). During the

South

Figure 22.6

Stone floors or barrels of water are simple, heat-absorbing devices. Insulated shutters keep heat inside at night.

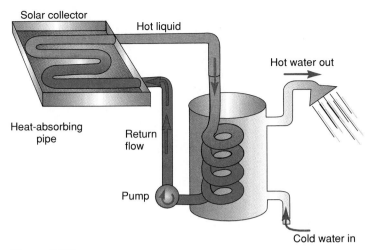

Figure 22.7

An active, closed-cycle solar water heating system. A cycling liquid, such as sodium or antifreeze, can reach hundreds of degrees Celsius in a parabolic reflector. Alternatively, a flat black panel enclosed with three layers of glass can absorb heat for the collecting pipes.

summer months, a fan blows the heated air from the collector into the storage medium. In the winter, a similar fan at the opposite end of the bin blows the warm air into the house. During the summer, the storage mass is cooler than the outside air, and it helps cool the house by absorbing heat. During the winter, it is warmer and acts

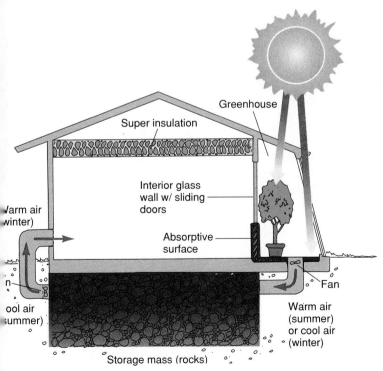

Figure 22.8

Underground massive heat storage unit. Heated air collected behind double- or triple-glazed windows is pumped down into a storage medium of rock, water, clay, or similar material, where it can be stored for a number of months.

as a heat source by radiating stored heat. In many areas, six or seven months' worth of thermal energy can be stored in 10,000 gallons of water or 40 tons of gravel, about the amount of water in a very small swimming pool or the gravel in two average-sized dump trucks.

Another heat storage system uses **eutectic** (phase-changing) **chemicals** to store a large amount of energy in a small volume. Heating melts these chemicals and cooling returns them to a solid state. This is an important way the hydrologic cycle stores solar energy in the natural environment, converting water from its solid phase to its liquid and gaseous phases. Most eutectic chemical systems (such as salts) do not swell when they solidify (as does water) and undergo phase changes at higher temperatures than water and ice, making them more convenient for storing heat. Since the latent heat of crystallization is usually much greater than the specific heat required to raise temperatures, a eutectic system can store more energy in a smaller volume than can a simple, one-phase system.

Heat also can be captured in chemical reactions in the storage material. Some clays, for instance, undergo reactions that release or store heat when they absorb or release water. An insulated bin filled with a clay, such as bentonite, can be baked dry (absorbing heat) by blowing warm air into it in the summer. Sprinkling water on the clay to rewet it will release heat in winter.

High-Temperature Solar Energy

Parabolic mirrors are curved mirrors that focus light from a large area onto a single, central point (fig. 22.9). Use of parabolic reflectors to focus intense heat on a central tube containing air, water, sodium, or antifreeze produces a much higher quality (high temperature) heat than does the basic flat panel collector. Temperatures in the collection medium can reach 500°C (1000°F). Ideally, such collectors could reach 100 MW per 0.5 km^2 of reflectors. An array

Figure 22.9

Parabolic mirrors focus solar energy on pipes carrying super-heated fluid to a steam-powered electric generator in Jerusalem, Israel.

In Depth: Personal Energy Efficiency

For some people, home energy efficiency means designing and building a new house that incorporates methods and materials that have been proven to conserve energy. Most of us, however, have to live in houses or apartments that already exist. How can we practice home energy conservation?

Think about the major ways we use energy. The biggest factor is space heating. Heat conservation is among the simplest, cheapest, and most effective ways to save energy in the home (table 1). Easy, surprisingly effective, and relatively inexpensive measures include weatherstripping, caulking, and adding layers of plastic to windows. Insulating walls, floors, and ceilings increases the energy conservation potential. Simply lowering your thermostat, especially at night, is a proven energy and money saver.

Water heating is the second major user of home energy, followed by large electric or gas appliances, such as stoves, refrigerators, washing machines, and dryers. Careful use of these appliances can save significant amounts of energy. Consider using less hot water, lowering the thermostat of your water heater, and buying an insulating blanket designed to wrap around it. Be sure that the refrigerator door

Energy Saving Potential

	Model Average	Best on Market	Best Prototype	Saving (percent)
Automobile (mi/gal)	18	55	107	83
Home (1000 J/m^2)	190	68	11	94
Refrigerator (kW hr/day)	4	2	1	75
Gas furnace (million J/day)	210	140	110	48
Air conditioner (kW hr/day)	10	5	3	70

Source: U.S. Department of Energy.

doesn't stand open, and defrost it regularly (if it is not frost-free) to reduce the ice buildup that prevents the heat exchange system from working well. Make sure your oven door has a tight seal, and plan ahead when you use it; bake several things at once rather than reheating it repeatedly. Wash your clothes in cold or cool water rather than hot. Air dry your laundry, especially in the summer when sun-dried clothes are so pleasant to wear.

Some energy-efficient appliances can save substantial amounts of energy. Air conditioners and refrigerators with better condensers and heat exchangers use one-half to one-fourth as much energy as older models. New, improved furnaces can offer 95 percent efficiency, compared with conventional 70 percent effi-

of solar collectors in Australia produces electricity for 4 cents per kilowatt hour.

The reliability and durability of large-scale active solar projects are issues of concern. Solar-powered electric motors must keep the mirrors properly aligned as the sun moves across the sky. We don't have enough experience with solar collectors to know what their life expectancy will be, but breakdowns and malfunctions have been significant problems at some prototype operations. Another concern is that these projects occupy large land areas. If the entire present U.S. electrical output came from central tower solar steam generators, 60,000/km^2 of collectors would be needed. This is an area about one-half the size of South Dakota, or one-quarter of the land now used by military bases. It is less land, however, than would be strip-mined in a thirty-year period if all our en-

ergy came from coal or uranium. And we can put solar collectors wherever we choose (such as lands unsuited for agriculture, grazing, or habitation), whereas strip-mining occurs wherever coal or uranium exist, regardless of other values associated with the land.

Solar Cookers

Parabolic mirrors have been tested for home cooking in tropical countries where sunshine is plentiful and other fuels are scarce. They produce such high temperatures and intense light that they are dangerous, however. A much cheaper, simpler, and safer alternative is the solar box cooker (fig. 22.10). An insulated box costing only a few dollars with a black interior and a glass lid serves as a passive solar collector. Several pots can be placed inside at the same time. Temperatures only reach about 120°C (250°F) so cooking

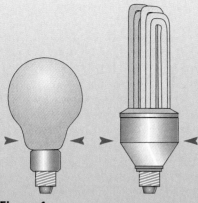

Figure 1

Comparison of an ordinary incandescent bulb with a new, compact fluorescent lamp. Both bulbs have standard screw-in bases, but the fluorescent ballast (arrows) is wider than the base and may require a socket "extender" to fit in some fixtures. A 15-watt compact fluorescent bulb puts out as much light as a 60-watt incandescent bulb and lasts about ten times longer.

ciency in older models. The payback period may be as little as two to three years if you trade an old wasteful appliance for a newer, more efficient one. Some lightbulbs are made to save energy (fig. 1). High-efficiency fluorescent lights emit the same amount of light but use only one-fourth as much energy as conventional incandescent bulbs. They cost ten times more than an ordinary bulb, but last ten times as long. Total life-time savings can be $30 to $50 per lamp. Almost all types of appliances, large and small, are available in efficient models and brands; we simply need to shop for them.

The easiest way to save energy is to keep an eye on energy-use habits. Turning off lights, televisions, and other devices when they are not in use is elementary. Line drying clothes and hand washing dishes are not difficult. In some cases, the most beneficial result of living a frugal, conservative life is not so much the total amount of energy you save, as the effect that this lifestyle has on you and the people around you. When you live conscientiously, you set a good example that could have a multiplying effect as it spreads through society. Making conscious ethical decisions about this one area of your life may stimulate you to make other positive decisions. You will feel better about yourself and more optimistic about the future when you make even small, symbolic gestures toward living as a good environmental citizen.

Some Things You Can Do To Save Energy

1   Drive less: make fewer trips, use telecommunications and mail instead of going places in person.

2   Use public transportation, walk, or ride a bicycle.

3   Use stairs instead of elevators.

4   Join a car pool or drive a smaller, more efficient car; reduce speeds.

5   Insulate your house or add more insulation to the existing amount.

6   Turn thermostats down in the winter and up in the summer.

7   Weatherstrip and caulk around windows and doors.

8   Add storm windows or plastic sheets over windows.

9   Create a windbreak on the north side of your house; plant deciduous trees or vines on the south side.

10 During the winter, close windows and drapes at night; during summer days, close windows and drapes if using air conditioning.

11 Turn off lights, television sets, and computers when not in use.

12 Stop faucet leaks, especially hot water.

13 Take shorter, cooler showers; install water-saving faucets and showerheads.

14 Recycle glass, metals, and paper; compost organic wastes.

15 Eat locally grown food in season.

16 Buy locally made, long-lasting materials.

takes longer than an ordinary oven. Fuel is free, however, and the family saves hours each day usually spent hunting for firewood or dung. These solar ovens help reduce tropical forest destruction and reduce the adverse health effects of smoky cooking fires.

Photovoltaic Solar Energy

The photovoltaic cell offers an exciting potential for capturing solar energy in a way that will provide clean, versatile, renewable energy. This simple device has no moving parts, negligible maintenance costs, produces no pollution, and has a lifetime equal to that of a conventional fossil fuel or nuclear power plant.

Photovoltaic cells capture solar energy and convert it directly to electrical current by separating electrons from their parent atoms and accelerating them across a one-way electrostatic barrier formed by the junction between two different types of semiconductor material (fig. 22.11). The photovoltaic effect, which is the basis of these devices, was first observed in 1839 by French physicist Alexandre Edmond Becquerel, who also discovered radioactivity. His discovery didn't lead to any practical applications until 1954, when researchers at Bell Laboratories in New Jersey learned how to carefully introduce impurities into single crystals of silicon.

These handcrafted single-crystal cells were much too expensive for any practical use until the advent of the U.S. space program. In 1958, when Vanguard I went into orbit, its radio was powered by six palm-sized photovoltaic cells that cost $2000 per peak watt of output, more than two thousand times as much as conventional energy at the time. Since then, prices have fallen by about a

Figure 22.10

A simple box of wood, plastic, and foil can help reduce tropical deforestation, improve women's lives, and avoid health risks from smoky fires in Third World countries. These inexpensive solar cookers could revolutionize energy use in developing tropical countries.

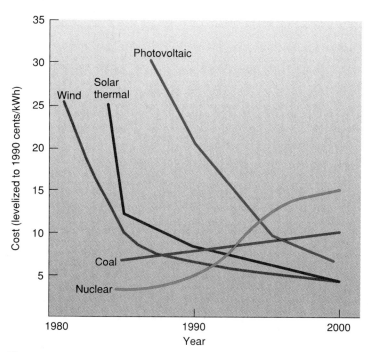

Figure 22.12

Costs for alternative and renewable energy sources have fallen dramatically in recent years. Wind power and solar-thermal electric power are already cheaper than coal in many places. Photovoltaic electricity should be less expensive than coal-fired electrical generation by the end of this decade.

Source: Union of Concerned Scientists, *Nucleus,* vol. 3(1), Spring 1991.

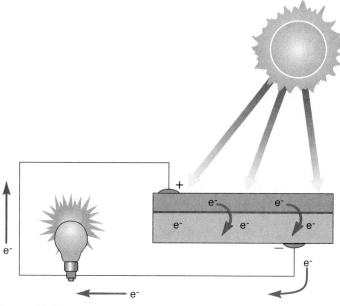

Figure 22.11

The operation of a photovoltaic cell. Boron impurities incorporated into the upper silicon crystal layers cause electrons (e⁻) to be released when solar radiation hits the cell. The released electrons move into the lower layer of the cell, thus creating a shortage of electrons, or a positive charge, in the upper layer and an oversupply of electrons, or negative charge, in the lower layer. The difference in charge creates an electric current in a wire connecting the two layers.

factor of ten each decade. In 1970, they cost $100 per watt; in 1990 they were less than $5 per watt. This makes solar energy cost-competitive with other sources in some areas (fig. 22.12).

By the year 2000, photovoltaic cells could be less than $1 per watt of generating capacity. With 15 percent efficiency and thirty-year life, they should be able to produce electricity for around 6¢ per kilowatt hour. At that time, coal-fired steam power will probably cost about half again as much and nuclear power will likely cost twice as much as photovoltaic energy.

During the last twenty-five years, the efficiency of energy capture by photovoltaic cells has increased from less than 1 percent of incident light to more than 10 percent under field conditions and over 30 percent in the laboratory. Promising experiments are under way using exotic metal alloys, such as gallium arsenide, cadmium telluride, and silicon germanium, which are more efficient in energy conversion than silicon crystals. Every year college students from all over North America race solar-powered cars across the country to test solar technology and publicize its potential (fig. 22.13). Perhaps you and your classmates would like to build a solar car and enter the race.

One of the most promising developments in photovoltaic cell technology in recent years is the invention of amorphous silicon collectors. First described in 1968 by Stanford Ovshinky, a self-taught inventor from Detroit, these noncrystalline silicon semiconductors can be made into lightweight, paper-thin sheets that require much less material than conventional photovoltaic cells. They also are vastly cheaper to manufacture and can be made in a variety of shapes and sizes, permitting ingenious applications. Roof tiles with

Figure 22.13

A solar car built by Mankato University students is on display after the 1993 transcontinental Sunrayce USA. This exciting event is both a good learning experience and a lot of fun.

Figure 22.14

Leading the way toward a renewable-energy, zero-emissions future, the Sacramento Municipal Utility District has built the first solar-powered electric charging station in the western United States. Photovoltaic panels in the back charge a variety of electric vehicles.

photovoltaic collectors layered on their surface already are available. Even flexible films can be coated with amorphous silicon collectors. Silicon collectors already are providing power to places where conventional power is unavailable, such as lighthouses, mountaintop microwave repeater stations, villages on remote islands, and ranches in the Australian outback.

You probably already use amorphous silicon photovoltaic cells. They are being built into light-powered calculators, watches, toys, photosensitive switches, and a variety of other consumer products. Japanese electronic companies presently lead in this field, having foreseen the opportunity for developing a market for photovoltaic cells. This market is already more than $100 million per year. Within a few years, Japanese companies expect to have home-roof arrays capable of providing all the electricity needed for a typical home at prices competitive with power purchased from a utility. By the end of this century, Japan plans to meet a considerable part of its energy requirements with solar power, an important goal for a country lacking fossil and nuclear fuel resources.

The world market for solar energy was about $500 million in 1995, and is expected to grow rapidly in the near future. The Dutch company that installed the pilot project in Sukatani described earlier in this chapter sees a huge potential market in 30,000 unelectrified Indonesian villages. Altogether, some 1.5 billion people around the world now have no access to electricity. Most would like to have a modern power source if it were affordable.

Think about how solar power could affect your future energy independence. Imagine the benefits of being able to build a house anywhere and having a cheap, reliable, clean, quiet source of energy with no moving parts to wear out, no fuel to purchase, and little equipment to maintain. You could have all the energy you need without commercial utility wires or monthly energy bills. Coupled with modern telecommunications and information technology, an independent energy source would make it possible to live in the countryside and yet have many of the employment and entertain-

ment opportunities and modern conveniences available in a metropolitan area (What Do *You* Think?, p. 498).

Storing Electrical Energy

Electrical energy is difficult and expensive to store. This is a problem for photovoltaic generation as well as other sources of electric power. Traditional lead-acid batteries are heavy and have low energy densities; that is, they can store only moderate amounts of energy per unit mass or volume. Acid from batteries is hazardous and lead from smelters or battery manufacturing is a serious health hazard for workers who handle these materials. A typical lead-acid battery array sufficient to store several days of electricity for an average home would cost about $5000 and weigh 3 or 4 tons. All the components for an electric car are readily available, but battery technology limits how far they can go between charges. Still, some communities are encouraging use of electric vehicles because they produce zero emissions (fig. 22.14).

Other types of batteries also have drawbacks. Metal-gas batteries, such as the zinc-chloride cell, use inexpensive materials and have relatively high-energy densities, but have shorter lives than other types. Sodium-sulfur batteries have considerable potential for large-scale storage. They store twice as much energy in half as much weight as lead-acid batteries. They require an operating temperature of about 300°C (572°F) and are expensive to manufacture. Alkali-metal batteries have a high storage capacity but are even more expensive. Lithium batteries have very long lives and store more energy than other types, but are the most expensive. Recent advances in thin-film technology may hold promise for future energy storage but often require rare, toxic elements that are both expensive and dangerous.

Another strategy is to store energy in a form that can be turned back into electricity when needed. Pumped-hydro storage involves pumping water to an elevated reservoir at times when excess

What Do *You* Think?
Living Off the Grid

Many of us dream of living at the end of the road in some beautiful, remote place. We fantasize about being free of the utility grid that ties us to urban living, yet few of us are willing or able to give up entirely the comforts of modern life. Which of the many gadgets that you now own could you do without?

Lots of modern pioneers and "end of the roaders" escape to the wilderness intending to live off the land but end up filling their lives with more machines and internal combustion engines than they ever had in the city. It starts with some power tools to build your dream house and a gasoline-powered generator to run them. Then you need a chainsaw, a woodsplitter, a truck or tractor to haul firewood out of the woods and to make trips into town for essentials, a three-wheeler to get around on rough trails, a rototiller for the garden, a lawnmower, a weed-eater, an outboard motor for your boat—the list goes on and on. You need propane for heating, cooking, and lights. After a while you need a bigger generator to run the TV, the VCR, and the new computer you're going to use to write your masterpiece.

Pretty soon living the simple, rustic life takes a lot more energy and causes a lot more pollution and disruption of nature than living in an apartment in the city where you shared walls—and heat—with your neighbors, and where you could walk to work or take mass transit. If all of us decided to live off the land there would be little wild nature anywhere. On the other hand, some of the new technologies described in this chapter now make it possible to live at the end of the road with much less impact than we might have had before.

In a cool climate, the biggest energy need is for space heating in the winter. Cooking and water heating are next in energy consumption for most people. In forested areas, wood is generally the cheapest and most abundant source of heat for all these applications. Although wood is a renewable resource, burning it is not environmentally cost-free. Heating a single house with a typical wood-burning stove creates more soot and smoke than heating 1000 houses with natural gas. Modern low-emission stoves can cut pollution considerably, however, and using passive solar energy, earth-sheltering, super insulation, and other renewable energy sources and conservation techniques can greatly reduce the amount of wood you have to burn.

What about appliances and entertainment? You need electricity for many modern conveniences. Here, again, technology might come to the rescue. With prices dropping for photovoltaics, wind generators, and hydropower, you can have power without exorbitant expense. If you live in a reasonably sunny area, two 50-watt photovoltaic panels should provide you with about 500 watt-hours per day. That's enough to run a radio or tape player, a couple of 18-watt compact fluorescent lights in the evening, a radio-telephone, some water pumping, and a laptop computer. A complete set of solar panels, charge controller, deep-cycle batteries, and wiring cost about $1000 at 1996 prices.

What if you want more power for more gadgets? For about $2500 you can buy four 50-watt solar panels and associated control equipment to produce around 1000 watt-hours per day. This would be enough to run a small TV, a VCR, a desktop computer, a cellular phone, half a dozen lights, a water pump, a small microwave and other kitchen appliances. For about $5000 you can get eight 50-watt solar panels that will produce about 2000 watt-hours per day. With this much energy you could run a large color TV, a high-efficiency refrigerator, a vacuum cleaner, a pressurized water system, and most power tools along with the other appliances listed above. You probably won't eliminate all internal combustion engines from your life but you can greatly diminish the amount of noise, odor, and pollution you create.

If you live in an area with dependable winds or falling water, you can produce the same amount of energy much more cheaply right now than with solar photovoltaic. For instance, a wind generator that produces 2000 watt-hours per day—if you have 25 km/hr (15 mph) average wind—would cost about half as much as eight solar panels. A microhydro turbine system would cost only about $1500—roughly one-quarter as much as solar—for a similar amount of energy. Since few of us can create these devices ourselves, we still wouldn't be living totally independently, but they would allow us to live in a relatively pristine, remote place and still have a rather modern lifestyle without having as much negative impact as we might once have had.

What do you think? Is this vision of the future an answer to our fondest wishes or merely escapism and an excuse to consume more resources? How would you weigh the relative costs and benefits of living a conventional life in the city versus being off the grid?

electricity is available. The water is released to flow back down through turbine generators when extra energy is needed. Using a similar principle, pressurized air can be pumped into such reservoirs as natural caves, depleted oil and gas fields, abandoned mines, or special tanks. An Alabama power company currently uses off-peak electricity to pump air at night into a deep salt mine. By day, the air flows back to the surface through turbines, driving a generator that produces electricity. Cool night air is heated to 1600°F by compression plus geothermal energy, increasing pressure and energy yield.

An even better way to use surplus electricity is in electrolytic decomposition of water to H_2 and O_2. These gases can be liquefied (like natural gas) at $-252°C$ ($-423°F$), making them easier to store and ship than most forms of energy. They are highly explosive, however, and must be handled with great care. They can be burned in internal combustion engines, producing mechanical en-

ergy, or they can be used as fuel in fuel cells to produce more electrical energy. Liquid hydrogen cars are already being produced. They could be very attractive in smoggy cities because they produce no carbon monoxide, smog-forming hydrocarbons, carcinogenic chemicals, or soot.

Flywheels are the subject of current experimentation for energy storage. Massive, high-speed flywheels, spinning in a nearly friction-free environment, store large amounts of mechanical energy in a small area. This energy is convertible to electrical energy. It is difficult, however, to find materials strong enough to hold together when spinning at high speed. Flywheels have a disconcerting tendency to fail explosively and unexpectedly, sending shrapnel flying in all directions.

New high-temperature superconducting ceramics may offer yet another way to store electricity. Giant superconducting underground rings might be able to store massive amounts of energy as electricity circulates without resistance through this marvelous material.

Energy from Biomass

Photosynthetic organisms have been collecting and storing the sun's energy for more than 2 billion years. Plants capture about 0.1 percent of all solar energy that reaches the earth's surface. That kinetic energy is transformed, via photosynthesis, into chemical bonds in organic molecules (chapter 3). A little more than half of the energy that plants collect is spent in such metabolic activities as pumping water and ions, mechanical movement, maintenance of cells and tissues, and reproduction; the rest is stored in biomass.

The magnitude of this resource is difficult to measure. Most experts estimate useful biomass production at fifteen to twenty times the amount we currently get from all commercial energy sources. It would be ridiculous to consider consuming all green plants as fuel, but biomass has the potential to become a prime source of energy. It has many advantages over nuclear and fossil fuels because of its renewability and easy accessibility. Renewable energy resources account for about 18 percent of total world energy use, and biomass makes up three-quarters of that renewable energy supply. Biomass resources used as fuel include wood, wood chips, bark, branches, leaves, starchy roots, and other plant and animal materials.

Burning Biomass

Wood fires have been a primary source of heating and cooking for thousands of years. As recently as 1850, wood supplied 90 percent of the fuel used in the United States. Wood now provides less than 1 percent of the energy in the United States, but in many of the poorer countries of the world, wood and other biomass fuels provide up to 95 percent of all energy used. The 1500 million cubic meters of fuelwood collected in the world each year is about half of all wood harvested.

In northern industrialized countries, wood burning has increased since 1975 in an effort to avoid rising oil, coal, and gas prices. Most of these northern areas have adequate wood supplies to meet demands at current levels, but problems associated with wood burning may limit further expansion of this use. Inefficient and incomplete burning of wood in fireplaces and stoves produces smoke laden with fine ash and soot and hazardous amounts of carbon monoxide (CO) and hydrocarbons. In valleys where inversion layers trap air pollutants, the effluent from wood fires can present a major source of air quality degradation and health risk. Polycyclic aromatic compounds produced by burning are especially worrisome because they are carcinogenic (cancer-causing).

In Oregon's Willamette Valley or in the Colorado Rockies, where woodstoves are popular and topography concentrates contaminants, as much as 80 percent of air pollution on winter days is attributable to wood fires. Several resort towns, such as Vail, Aspen, and Telluride, Colorado, have banned installation of new woodstoves and fireplaces because of high pollution levels. Oregon, Colorado, and Vermont now have emission standards for new woodstoves. The Environmental Protection Agency ranks wood burners high on a list of health risks to the general population, and standards are being considered to regulate the use of woodstoves nationwide.

Highly efficient and clean-burning woodstoves are available but expensive. Running exhaust gases through heat exchangers recaptures heat that would escape through the chimney, but if flue temperatures drop too low, flammable, tarlike creosote deposits can build up, increasing fire risk. A better approach is to design combustion chambers that use fuel efficiently, capturing heat without producing waste products. Brick-lined fireboxes with afterburner chambers to burn gaseous hydrocarbons do an excellent job. Catalytic combusters, similar to the catalytic converters on cars, also are placed inside stovepipes. They burn carbon monoxide and hydrocarbons to clean emissions and to recapture heat that otherwise would have escaped out the chimney in the chemical bonds of these molecules.

Wood chips, sawdust, wood residue, and other plant materials (fig. 22.15) are being used in some places in the United States and Europe as a substitute for coal and oil in industrial boilers. In Vermont, for instance, where fossil fuels are expensive and 76 percent of the land is covered by forest, 250,000 cords of unmarketable cull wood are burned annually to fuel a 50-MW power plant in Burlington. Michigan's Public Service Commission estimated that the state's surplus forest growth could provide 1300 MW of electricity each year. Pollution-control equipment is easier to install and maintain in a central power plant than in individual home units. Wood burning also contributes less to acid precipitation than does coal. Because wood has little sulfur, it produces few sulfur gases and burns at lower temperatures than coal; thus, it produces fewer nitrogen oxides. Burning wood as a renewable crop doesn't produce any net increase in atmospheric carbon dioxide (and, therefore, doesn't add to global warming) because all the carbon released by burning biomass was taken up from the atmosphere when the biomass was grown.

Fuelwood Crisis in Less-Developed Countries

Two billion people—about 40 percent of the total world population—depend on firewood and charcoal as their primary energy

Figure 22.15

Harvesting marsh reeds (*Phragmites sp.*) in Sweden as a source of biomass fuel. In some places, biomass from wood chips, animal manure, food-processing wastes, peat, marsh plants, shrubs, and other kinds of organic material make a valuable contribution to energy supplies. Care must be taken, however, to avoid environmental damage in sensitive areas.

Figure 22.16

The firewood shortage in less-developed countries means that women and children must spend hours each day searching for fuel. Destruction of forests and removal of ground cover result in erosion and desertification.

source. Of these people, three-quarters (1.5 billion) do not have an adequate, affordable supply. Most of these people are in the less-developed countries where they face a daily struggle to find enough fuel to warm their homes and cook their food. The problem is intensifying because rapidly growing populations in many Third World countries create increasing demands for firewood and charcoal from a diminishing supply.

As firewood becomes increasingly scarce, women and children, who do most of the domestic labor in many cultures, spend more and more hours searching for fuel (fig. 22.16). In some places, it now takes eight hours, or more, just to walk to the nearest fuelwood supply and even longer to walk back with a load of sticks and branches that will only last a few days.

For people who live in cities, the opportunity to scavenge firewood is generally nonexistent and fuel must be bought from merchants. This can be ruinously expensive. In Addis Ababa, Ethiopia, 25 percent of household income is spent on wood for cooking fires. A circle of deforestation has spread more than 160 km (100 mi) around some major cities in India, and firewood costs up to ten times the price paid in smaller towns.

As you might expect, increasing demand for diminishing fuelwood resources has many adverse environmental impacts. As fallen wood becomes scarce, people lop off branches, fell trees, and even uproot seedlings, bushes, and herbaceous plants. Destroying forests decreases wildlife habitat, depletes groundwater supplies, exposes soil to erosion, and contributes to desertification and climate change.

These problems are expected to worsen unless steps are taken immediately to provide alternative energy sources. It is predicted that by the end of this century, some 2.9 billion people—about half the world's population—will find their minimum fuel needs unmet. The worst areas will continue to be the arid and semiarid regions of Africa, the populous regions of Southeast Asia, and the mountainous areas of Latin America, where demand is already as much as six times the available supply.

The 1500 million cubic meters of fuelwood now harvested annually is about 500 million cubic meters short of the needed amount. By 2025, it is estimated that the worldwide demand for fuelwood will be 4400 million cubic meters, while supplies will not expand much beyond present levels. This means that demand will be more than double the available supply. In many countries, the situation will be much worse than the world average. In some African countries, such as Mauritania, Rwanda, and the Sudan, firewood demand is already ten times the sustainable yield of remaining forests (fig. 22.17).

Dung and Methane as Fuels

Where wood and other fuels are in short supply, people often dry and burn animal manure. This may seem like a logical use of waste biomass, but it can intensify food shortages in poorer countries. Not putting this manure back on the land as fertilizer reduces crop production and food supplies. In India, for example, where fuelwood supplies have been chronically short for many years, a limited manure supply must fertilize crops and provide household fuel. Cows in India produce more than 800 million tons of dung per year, more than half of which is dried and burned in cooking fires (fig. 22.18). If that dung were applied to fields as fertilizer, it could boost crop production of edible grains by 20 million tons per year, enough to feed about 40 million people.

When cow dung is burned in open fires, more than 90 percent of the potential heat and most of the nutrients are lost. Compare that to the efficiency of using dung to produce methane gas, an excellent fuel. In the 1950s, simple, cheap methane digesters were designed for villages and homes, but they were not widely used. In China, 6 million households use biogas for cooking and lighting. Two large municipal facilities in Nanyang will soon provide fuel for more than 20,000 families. Perhaps other countries will follow China's lead.

Methane gas is a "natural" gas. It is produced by anaerobic decomposition (digestion by anaerobic bacteria) of any moist organic material. Many people are familiar with the fact that swamp gas is

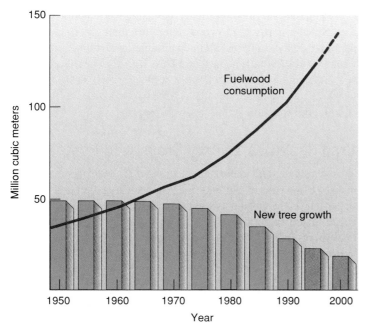

Figure 22.17

Fuelwood consumption and new tree growth in the Sudan, with projections to the year 2000. Firewood demand in the Sudan is already two to three times the sustainable yield of the forests. In Mauritania and Rwanda, demand is more than ten times the sustainable yield.

Source: Data from *State of the World, 1986*, ed. by Lester Brown. Copyright 1986 by Worldwatch Institute.

Figure 22.18

In many countries, animal dung is gathered by hand, dried, and used for fuel. This girl in Mali is carrying home fuel for cooking. Using dung as fuel deprives fields of nutrients and reduces crop production.

explosive. Swamps are simply large methane digesters, basins of wet plant and animal wastes sealed from the air by a layer of water. Under these conditions, organic materials are decomposed by anaerobic (oxygen-free) rather than aerobic (oxygen-using) bacteria, producing flammable gases instead of carbon dioxide. This same process may be reproduced artificially by placing organic wastes in a container and providing warmth and water (fig. 22.19). Bacteria are ubiquitous enough to start the culture spontaneously.

Burning methane produced from manure provides more heat than burning the dung itself, and the sludge left over from bacterial digestion is a rich fertilizer, containing healthy bacteria as well as most of the nutrients originally in the dung. Whether the manure is of livestock or human origin, airtight digestion also eliminates some health hazards associated with direct use of dung, such as exposure to fecal pathogens and parasites.

All sizes and forms of containment have been used for methane production, from coffee cans to oil drums to specially designed tanks. Like other bacteria and yeast cultures, such as those used for bread, yogurt, and beer, methane-producing cultures need heat to thrive. The highest gas production is generally at 35°C (95°F), but once a culture is established, it produces enough of its own heat to continue. This warmth can be felt in all backyard compost heaps. It also accounts for early melting of snow over septic tanks.

How feasible is methane—from manure or from municipal sewage—as a fuel resource in developed countries? Methane is a

clean fuel that burns efficiently. It is produced in a low-technology, low-capital process from any kind of organic waste material: livestock manure, kitchen and garden scraps, and even municipal garbage and sewage. In fact, municipal landfills are active sites of methane production, contributing as much as 20 percent of the annual output of methane to the atmosphere. This is a waste of a valuable resource and a threat to the environment because methane absorbs infrared radiation and contributes to the greenhouse effect (chapter 17). Some municipalities are drilling gas wells into landfills and garbage dumps. Cattle feedlots and chicken farms in the United States are a tremendous potential fuel source. Collectible crop residues and feedlot wastes each year contain 4.8 billion gigajoules (4.6 quadrillion BTUs) of energy, more than all the nation's farmers use. Municipal sewage treatment plants routinely use anaerobic digestion as a part of their treatment process, and many facilities collect the methane they produce and use it to generate heat or electricity for their operations. Although this technology is well-developed, its utilization could be much more widespread.

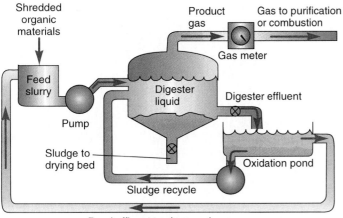

Figure 22.19

Continuous unit for converting organic material to methane by anaerobic fermentation. One kilogram of dry organic matter will produce 1–1.5 m^3 of methane, or 2500–3600 million calories per metric ton.

Source: *Solar Energy as a National Energy Resource,* NSF/NASA Solar Energy Panel, National Science Foundation, December 1972.

Alcohol from Biomass

Ethanol (grain alcohol) and methanol (wood alcohol) are produced by anaerobic digestion of plant materials with high sugar content, mainly grain and sugarcane. Ethanol can be burned directly in automobile engines adapted to use this fuel, or it can be mixed with gasoline (up to about 10 percent) to be used in any normal automobile engine. A mixture of gasoline and ethanol is often called **gasohol.** Ethanol in gasohol raises octane ratings and is a good substitute for lead antiknock agents, the major cause of lead pollution. It also helps reduce carbon monoxide emissions in automobile exhaust. Gas stations in U.S. cities that exceed EPA air quality standards have been ordered to sell so-called oxygenated fuels containing ethanol or methanol.

Ethanol production could be a solution to grain surpluses and bring a higher price for grain crops than the food market offers. It also offers promise for reduced dependence on gasoline, which is refined from petroleum. Brazil has instituted an ambitious national program to substitute crop-based ethanol for imported petroleum. In 1985, the Brazilian sugar harvest produced 2.5 billion gallons of ethanol, but falling oil prices have undercut plans to expand ethanol production to 4 billion gallons per year.

Researchers at the Solar Energy Research Institute in Golden, Colorado, have developed a new biomass-to-gasoline process that yields a high-octane, high-energy product from wood scraps, crop residues, and paper or municipal wastes. In this one-step process, organic materials are vaporized and passed over a zeolite (silica alumina) catalyst, which converts hydrocarbons to long-chain, gasolinelike fluids, carbon dioxide, and water. A ton of biomass will produce about 225 l (60 gal) of liquid fuel suitable for direct use in engines, or to be blended with other fuels to raise octane levels.

Methanol might have many advantages as a motor vehicle fuel. Since it burns at a lower temperature than gasoline or diesel, the bulky, heavy radiator might be eliminated, making automobiles sleeker and more efficient. Combined with flywheel-energy storage, intermittent burning engines, turbochargers, and other high-tech devices, we might be able to have personal transportation that is both energy efficient and less polluting.

Crop Residues, Energy Crops, and Peat

Crop residues, such as cornstalks, corncobs, or wheat straw, can be used as a fuel source, but they are expensive to gather and often are better left on the ground as soil protection (chapter 11). The residue accumulated in food processing, however, can be a useful fuel. Hawaiian sugar growers burn bagasse, a fibrous sugarcane residue, to produce the state's second-largest energy source, surpassed only by petroleum.

Some crops are being raised specifically as an energy source. Fast-growing trees, such as *Leucaena,* and shrubs, such as alder and willow, are grown to provide wood for energy. Milkweeds, sedges, marsh grasses, cattails, and other biomass crops also might become useful energy sources, grown on land that is otherwise unsuitable for crops (see fig. 22.15). Unfortunately, there may be unacceptable environmental costs from disruption of wetlands and forests that are converted to energy crop plantations. Some plant species produce hydrocarbons that can be used directly as fuel in internal combustion engines without having to be fermented or transformed to hydrogen or methane gas. Sunflower oil, for instance, can be burned directly in diesel engines. In most cases, these oils and other high-molecular-weight hydrocarbons are more valuable for other purposes, such as food, plastics, and industrial chemicals; however, they might be an attractive energy source in some circumstances.

Several other forms of biomass are actual or proposed sources of energy. People in northern climates have been digging up peat (partially decomposed plant residues) from bogs and marshes for centuries. There has been a resurgence of interest in exploiting the vast Minnesota and Canadian peat bogs, which some people regard as wastelands. After mining some or all of the peat, the land could be used for energy crops, such as cattails. Critics of these schemes point out the damaging effects of bog draining on ecosystems, watersheds, and wildlife. Burning of municipal garbage ("waste to watts") is a popular proposal; however, many people are worried about air pollution from these facilities (chapter 23). Garbage may contain hazardous, volatile, or explosive materials that are difficult or dangerous to sort out. Effects on air quality from all forms of direct combustion of biomass are a serious concern.

Energy from the Earth's Forces

The winds, waves, tides, ocean thermal gradients, and geothermal areas are renewable energy sources. Although available only in selected locations, these sources could make valuable contributions to our total energy supply.

Hydropower

Falling water has been used as an energy source since the Middle Ages. The invention of water turbines in the nineteenth century greatly increased the efficiency of hydropower dams. By 1925, falling water generated 40 percent of the world's electric power. Since then, hydroelectric production capacity has grown fifteen-fold, but fossil fuel use has risen so rapidly that water power is now only one-quarter of total electrical generation. Still, many countries produce most of their electricity from falling water (fig. 22.20). Norway, for instance, depends on hydropower for 99 percent of its electricity; Brazil, New Zealand, and Switzerland all produce at least three-quarters of their electricity with water power. Canada is the world's leading producer of hydroelectricity, running 400 power stations with a combined capacity exceeding 60,000 MW.

The total world potential for hydropower is estimated to be about 3 million MW. If all of this capacity were put to use, the available water supply could provide between 8 and 10 terrawatt hours (10^{12} Whr) of electrical energy. Currently, we use only about 10 percent of the potential hydropower supply. The energy derived from this source in 1994 was equivalent to about 500 million tons of oil, or 8 percent of the total world commercial energy consumption.

Much of the hydropower development in recent years has been in enormous dams. There is a certain efficiency of scale in giant dams, and they bring pride and prestige to the countries that build them, but they can have unwanted social and environmental effects (Case Study, p. 504 and What Do *You* Think?, p. 428). The largest hydroelectric dam in the world at present is the Itaipu Dam on the Parana River between Brazil and Paraguay. Designed to generate 12,600 MW of power, this dam should produce as much energy as thirteen large nuclear power plants when completed. The lake that it is creating already has flooded about 1300 sq km (500 sq mi) of tropical rainforest and displaced many thousands of native people and millions of other creatures.

There are still other problems with big dams, besides human displacement, ecosystem destruction, and wildlife losses. Dam failure can cause catastrophic floods and thousands of deaths. Sedimentation often fills reservoirs rapidly and reduces the usefulness of the dam for either irrigation or hydropower. In China, the Sanmenxia Reservoir silted up in only four years, and the Laoying Reservoir filled with sediment before the dam was even finished. Rotting vegetation in artificial impoundments can have disastrous effects on water quality. When Lake Brokopondo in Suriname flooded a large region of uncut rainforest, underwater decomposition of the submerged vegetation produced hydrogen sulfide that killed fish and drove out villagers over a wide area. Acidified water from this reservoir ruined the turbine blades, making the dam useless for power generation.

Floating water hyacinths (rare on free-flowing rivers) have already spread over reservoir surfaces behind the Tucurui Dam on the Amazon River in Brazil, impeding navigation and fouling machinery. Herbicides sprayed to remove aquatic vegetation have contaminated water supplies. Herbicides used to remove forests before dam gates closed caused similar pollution problems.

Figure 22.20

Hydropower dams like this one in New South Wales, Australia, can prevent floods and provide useful energy, but often they destroy valuable farmland, forests, villages, and historic areas when valleys are flooded for water storage. In addition, dams can also lose large amounts of water to seepage and evaporation, or they could fail and thus be a threat to downstream areas.

Schistosomiasis, caused by parasitic flatworms called blood flukes (chapter 9), is transmitted to humans by snails that thrive in slow-moving, weedy tropical waters behind these dams. It is thought that 14 million Brazilians suffer from this debilitating disease.

As mentioned before, dams displace indigenous people. The Narmada Valley project in India will drown 150,000 ha of tropical forest and displace 1.5 million people, mostly tribal minorities and low-caste hill people. China's Three Gorges project, now under construction, will displace one million people. The Akosombo Dam built on the Volta River in Ghana nearly twenty years ago displaced 78,000 people from seven hundred towns. Few of these people ever

CASE STUDY
James Bay Hydropower

In the 1970s, Hydro-Quebec, a government-owned electric utility, began to divert rivers flowing into James Bay, flooding more than 10,000 square kilometers (4000 square miles) of tundra and coastal wetlands along the eastern shore of Hudson Bay. If the whole project is ever finished, more than 600 dams and dikes will block 19 large rivers and damage a pristine watershed area the size of Germany. Rivers as large as the Grand Canyon's Colorado will be dried up and diverted into new paths. The total cost may well be $100 billion.

The area affected is the traditional home of about 12,000 Crees and 6000 Inuit Eskimos who live in hunting camps and small villages scattered across the rocky land. Under Phase I of this project, a series of reservoirs, dams and dikes diverted several rivers into the La Grande River. One of these was the Caniapiscau River, which used to flow east into the Atlantic but now flows west into James Bay. The 10,300 megawatts generated by this massive diversion is sent over high voltage transmission lines to homes and businesses in Quebec, New York, and the New England states. Phase I was expected to cost $1 billion, but

overruns pushed the final total to more than 15 times that much.

Phase II, now under construction, is diverting the Little Whale and Nastapoca Rivers to feed hydroelectric plants on the Great Whale River. Phase III, if built, will divert the Nottaway and the Rupert Rivers into the Broadback River, storing the water in seven new reservoirs and transforming hundreds of kilometers of the diverted rivers into dry bedrock. The whole scheme would generate

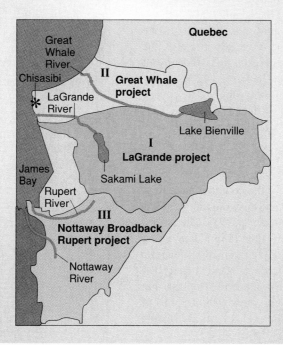

Figure 1

The James Bay project may eventually encompass an area the size of Germany and produce three times as much electricity as all of Africa. Native cultures, wildlife, and the natural ecosystem already are suffering from this grandiose scheme.

---

found another place to settle, and those still living remain in refugee camps and temporary shelters.

In tropical climates, large reservoirs often suffer enormous water losses. Lake Nasser, formed by the Aswan High Dam in Egypt, loses 15 billion cubic meters each year to evaporation and seepage. Unlined canals lose another 1.5 billion cubic meters. Together, these losses represent one-half of the Nile River flow, or enough water to irrigate 2 million ha of land. The silt trapped by the Aswan Dam formerly fertilized farmland during seasonal flooding and provided nutrients that supported a rich fishery in the Delta region. Farmers now must buy expensive chemical fertilizers, and the fish catch has dropped almost to zero. As in South America, schistosomiasis is an increasingly serious problem.

If big dams—our traditional approach to hydropower—have so many problems, how can we continue to exploit the great potential of hydropower? Fortunately, there is an alternative to gigantic dams and destructive impoundment reservoirs. Small-scale, **low-head hydropower** technology can extract energy from small

headwater dams that cause much less damage than larger projects. Some modern, high-efficiency turbines can even operate on **run-of-the-river flow.** Submerged directly in the stream and small enough to not impede navigation in most cases, these turbines don't require a dam or diversion structure and can generate useful power with a current of only a few kilometers per hour. They also cause minimal environmental damage and don't interfere with fish movements, including spawning migration. **Micro-hydro generators** operate on similar principles but are small enough to provide economical power for a single home. If you live close to a small stream or river that runs year-round and you have sufficient water pressure and flow, hydropower is probably a cheaper source of electricity for you than solar or wind power (fig. 22.21).

Small-scale hydropower systems also can cause abuses of water resources. The Public Utility Regulatory Policies Act of 1978 included economic incentives to encourage small-scale energy projects. As a result, thousands of applications were made to dam or divert small streams in the United States. Many of these projects

26,400 megawatts of electrical power, three times more than all the electrical capacity now existing in Africa.

This project already has had adverse cultural and environmental impacts. In September 1984, shortly after Phase I was completed, 10,000 caribou drowned while trying to follow their traditional migration route across the swollen Caniapiscau River, which was flooded by releases from an upstream reservoir. Even more threatening to native people is poisoning caused by mercury leached from newly flooded land. Bacteria change inorganic mercury to a soluble organic form that concentrates as it passes from plankton to fish to humans. In the Cree village of Chisasibi, scientists found that two-thirds of the residents had mercury levels in their bodies higher than is considered safe by the World Health Organization. Changing their diet from locally caught fish to canned or frozen food has reduced people's mercury levels, but has disrupted their economy and culture.

The extensive coastal marshes and estuaries of James Bay provide breeding grounds and migration routes for millions of shorebirds and waterfowl. Few other areas provide the biological richness capable of feeding such a large population during the brief summer months. Changes in river hydrology already have begun to degrade this vital ecosystem. Eel grass, one of the keystone species in this food web has disappeared from large areas of once-fertile tideflats. Furthermore, dams and river diversions have blocked salmon and other anadromous species from reaching their spawning grounds.

A critical source of funding for the whole James Bay venture is long-range contracts with utilities and cities in the United States. Native groups and environmental organizations had little success trying to persuade the Quebec government to reconsider this massive project. Attempts to influence public opinion south of the border, however, have been much more effective. New York, Vermont, and Maine recently canceled $19 million in long-term contracts for electric power from Hydro-Quebec because of both cost and ethics.

When these contracts were negotiated 20 years ago, American utilities were projecting continued growth at a rate of 2 to 3 percent per year. Lower economic growth rates, however, coupled with greater than expected savings from conservation and an upsurge in non-utility generation have lowered demand forecasts. New England is unlikely to need new energy sources before the end of this century, and energy consumption in New York is expected to decline until sometime in the next century.

ETHICAL CONSIDERATIONS

Is this a case of developing a "clean" renewable energy source that otherwise would be wasted? Or is it tantamount to exporting environmental and cultural destruction to northern forests, the ultimate not-in-my-backyard solution to New York's energy habits and pollution problems? As so often happens, remote areas and indigenous people bear the brunt of consuming desires of the majority. Hydropower doesn't release CO_2 or create nuclear waste. But does reducing air pollution for millions of people in cities and averting global warming justify disrupting this remote but important wilderness area? What are the rights of the native people and the fish and wildlife? How much effort and sacrifice should we be expected to shoulder to avoid disrupting ecosystems and traditional cultures?

have little merit. All too often, fish populations, aquatic habitat, recreational opportunities, and the scenic beauty of free-flowing streams and rivers are destroyed primarily to provide tax benefits for wealthy investors.

Wind Energy

The air surrounding the earth has been called a 20-billion-cubic-kilometer storage battery for solar energy. The World Meteorological Organization has estimated that 20 million MW of wind power could be commercially tapped worldwide, not including contributions from windmill clusters at sea. This is about fifty times the total present world nuclear generating capacity.

Wind power has advantages and disadvantages, as do other nontraditional technologies. Like solar power and hydropower, wind power taps a natural physical force. Like solar power (its ultimate source), wind power is a virtually limitless resource, is nonpolluting, and causes minimal environmental disruption. Like solar power, however, it requires expensive storage during peak production times to offset nonwindy periods.

Windmills played a crucial role in the settling of the American West. The Great Plains had abundant water in underground aquifers, but little surface water. The strong, steady winds that blew across the prairies provided the energy to pump water that allowed agriculture to move west across the prairies. By the end of the nineteenth century, nearly every farm or ranch west of the Mississippi River had at least one windmill. The manufacture, installation, and repair of windmills was a major industry in America. Even today, some 150,000 windmills still spin productively in the Great Plains (fig. 22.22).

Windmill technology was flourishing early in this century. Efficiencies increased ten times or more with the invention of aerodynamic propellers and affordable electric-wind generators in the 1920s. This promising technology was nipped in the bud, however, by passage of the Rural Electrification Act of 1935. This act brought many benefits to rural America, but it effectively killed wind-power

Sustainable Energy

Figure 22.21

Solar collectors capture power only when the sun shines, but hydropower is available 24 hours a day. Small turbines such as this one can generate enough power for a single-family house with only 15 m (50 ft) of head and 200 l (50 g) per minute flow. The turbine can have up to four nozzles to handle greater water flow and generate more power.

Figure 22.22

Millions of traditional windmills once pumped water, generated electricity, and provided power for farms and ranches across America. Most have been displaced by internal combustion engines and rural electric power systems, but there is still potential for capturing large amounts of renewable energy from the sun, wind, and other sustainable sources.

development and encouraged the building of huge hydropower dams and coal-burning power plants. It is interesting to speculate what the course of history might have been if we had not spent trillions of dollars on fossil fuel and nuclear energy, but instead had invested that money on small-scale, renewable energy systems.

As the world's conventional fuel prices rise, interest in wind energy is resurging. The United States and Denmark are presently the world's largest producers, but planners and manufacturers in other European countries also are studying wind power. Enthusiasm also is evident in Asia, India, and Australia, as well as the island countries of the Pacific and Caribbean, where energy costs are high.

Pilot projects already have demonstrated the economy of wind power. Theoretically 60 percent efficient, windmills typically produce at 35-percent efficiency under field conditions. General Electric estimates the wind could provide 14 percent of U.S. electrical demand by the year 2000. One thousand MW of installed wind power could meet the energy needs of 400,000 households at present use levels and save 3.7 million barrels of oil per year ($55.5 million each year at current world oil prices). A General Electric study also found more than 556,400 sq km of land in the United States alone that are suitable for wind generators (average wind speeds in excess of 25 km/hr) and do not involve competing and conflicting uses, such as cities or national parks.

The standard modern wind turbine uses only two or three propeller blades. More blades on a windmill provide more torque in low-speed winds, so that the traditional Midwestern windmill, with twenty or thirty blades, was most appropriate for small-scale use in less reliable wind fields. Fewer blades operate better in high-speed winds, providing more energy for less material cost at wind speeds of 25–40 km/hr (15–25 mph). A two-blade propeller can extract most of the available energy from a large vertical area and has less material to weaken and break in a storm. Three-bladed propellers often are preferred because they are easier to balance and spin more smoothly.

In 1929, the French engineer G. J. M. Darrieus proposed an alternative to the traditional horizontal-axis windmill. Developed in Canada in the 1960s, the Darrieus rotor, a vertical-axis "eggbeater" machine, is durable and inexpensive (fig. 22.23). It needs no adjustment to wind direction and operates in any wind stronger than 19 km/hr (12 mph), compared to conventional turbines that generally require at least 24 km/hr (15 mph). Quantity manufacturing of Darrieus generators has reduced the price rapidly, bringing the current cost to around 5.4 cents per kWhr for a conservatively estimated fifteen-year life span.

Wind farms are large-scale public utility efforts to take advantage of wind power. The first major wind farm started in New Hampshire in 1981, but California quickly took the lead in wind farm development. By 1982, the first one hundred-turbine farm

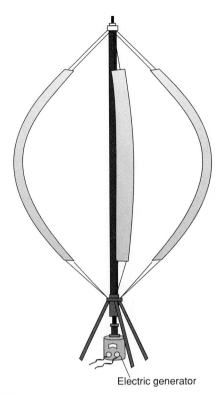

Figure 22.23

The Darrieus rotor. This "eggbeater" wind machine is a low-cost device capable of producing electricity at relatively low wind speeds. It has been in use since 1927 when it was patented by G. J. M. Darrieus in France.

Electric generator

Figure 22.24

Together, these wind turbines in Altamont Pass, CA, produce as much electricity as a large coal-fired or nuclear power plant.

stretched across Altamont Pass in the rolling ranch country east of San Francisco. The windmills, mainly medium-sized, American- and Danish-built two- and three-bladed machines, stand along the ridge following a contour line of highest wind availability. By 1995, more than 7000 windmills stood at Altamont (fig. 22.24). Nine thousand more are installed at California's Tehachapi and San Gorgino Passes. The existing project already produces well over 1500 MW, the equivalent of three large nuclear reactors. Altogether, California's 16,000 commercial turbines produce 95 percent of U.S. wind energy and 75 percent of the world's production.

Construction of wind farms is not limited to mountain ridges. The Great Plains are legendary for their strong, incessant winds. Promoters claim that the prairie states and provinces could be the OPEC of wind power. Seacoasts also provide excellent wind farm sites. Offshore wind farms have been proposed, with large (up to 8 MW each) generators to be anchored a short distance out at sea where competition for real estate development is low and winds are high year-round. According to such plans, collected energy would probably be stored in electrolyzed water (discussed earlier as storage for photovoltaic power) for transportation to shore.

Are there negative impacts of wind farms? They generally occupy places with wind and weather too severe for residential or other development. The wind speed at Altamont, for instance, averages 35 km/hr (22 mph) from March to September. Livestock graze undisturbed beneath the windmills. Wind farm opponents complain, however, of visual and noise pollution (the low, rhythmic whoosh of the blades); proponents claim these effects are negligible. Most wind farms are too far from residential areas to be heard or seen. But they do interrupt the view in remote isolated places and destroy the sense of isolation and natural beauty. They also can be a wildlife hazard. Hawks, eagles, and other birds fly into the whirling blades. If California condors are ever released again into the wild, they may be threatened by windmills on their favorite mountain passes.

Wouldn't wind power take up a huge land area if we were to depend on it for a major part of our energy supply? As table 22.3 shows, the actual space taken up by towers, roads, and other structures on a wind farm is only about one-third as much as would be consumed by a coal-fired power plant or solar thermal energy system to generate the same amount of energy over a 30-year period. Furthermore, the land under windmills is more easily used for grazing or farming than is a strip-mined coal field or land under solar panels. Wind power also generates many more jobs per unit energy produced than do most other technologies even though its total cost is generally lower.

When a home owner or community invests independently in wind generation, the same question arises as with solar energy: What should be done about energy storage when electricity production exceeds use? Besides the storage methods mentioned earlier, many private electricity producers believe the best use for excess electricity is in cooperation with the public utility grid. When private generation is low, the public utility runs electricity through the meter and into the house or community. When the wind generator or photovoltaic systems overproduce, the electricity runs back into the grid and the meter runs backward. Ideally, the utility reimburses individuals for this electricity, for which other consumers pay the company. The 1978 Public Utilities Regulatory Policies

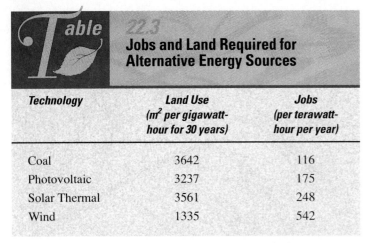

Technology	Land Use (m² per gigawatt-hour for 30 years)	Jobs (per terawatt-hour per year)
Coal	3642	116
Photovoltaic	3237	175
Solar Thermal	3561	248
Wind	1335	542

Source: Lester R. Brown, et al. *Saving the Planet.* Copyright 1991 W. W. Norton & Co., Inc.

Act required utilities to buy power generated by small hydro, wind, cogeneration, and other privately owned technologies at a fair price. Not all utilities yet comply, but some—notably in California, Oregon, Maine, and Vermont—are purchasing significant amounts of private energy.

Geothermal Energy

The earth's internal temperature can provide a useful source of energy in some places. High-pressure, high-temperature steam fields exist below the earth's surface. Around the edges of continental plates or where the earth's crust overlays magma (molten rock) pools close to the surface, this energy is expressed in the form of hot springs, geysers, and fumaroles (chapter 16). Yellowstone National Park is undoubtedly the most famous geothermal region in the United States. Iceland, Japan, and New Zealand also have high concentrations of geothermal springs and vents. Depending on the shape, heat content, and access to groundwater, these sources produce wet steam, dry steam, or hot water.

Until recently, the main use of this energy source was in baths built at hot springs. More recently, **geothermal energy** has been used in electric power production, industrial processing, space heating, agriculture, and aquaculture. In Iceland, most buildings are heated by geothermal steam. A few communities in the United States, including Boise, Idaho, use geothermal home heating. Recently, geothermal steam also has been developed for electrical generation.

California's Geysers project is the world's largest geothermal electric-generating complex, with 200 steam wells that provide some 1300 MW of power. This system functions much like other heat-driven generators. A shaft sunk into the sub-surface steam reservoir brings pressurized steam up to a turbine at the surface. The steam spins the turbine to produce electricity and then is piped off to a condensing unit and then to a cooling pond (fig. 22.25). The remaining brine (water heavily laden with dissolved minerals) is pumped back down to the reservoir. When the steam source has high mineral concentrations that would corrode delicate turbine

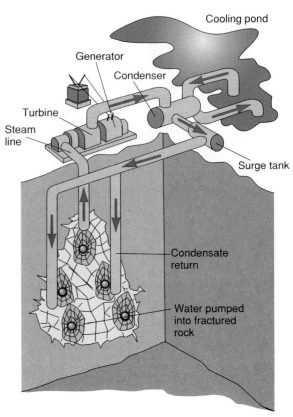

Figure 22.25

Geothermal steam extraction system for electricity production. Once the steam has spun the turbine, it is condensed, cooled, and pumped back into the underground steam field. The steam source for this relatively low-capital electric plant should last for several decades.

Sources: American Oil Shale Corporation and U.S. Atomic Energy Commission, *A Feasibility Study of a Plowshare of Geothermal Power Plant,* April 1971.

blades, a heat exchanger is used to generate clean steam as is done in a nuclear power plant.

Among the advantages of geothermal generators are a reasonably long life span (at least several decades), no mining or transporting of fuels, and little waste disposal. The main disadvantages are the potential danger of noxious gases in the steam and noise problems from steam-pressure relief valves.

While few places have geothermal steam, the entire earth is underlain by warm rocks that can be reached by drilling down far enough into the crust. Some heating systems pump groundwater from underground aquifers or inject surface water into deep strata and then withdraw it after it has been warmed, as a way of extracting geothermal energy. There have been experiments with pumping surplus heated water into underground reservoirs during the summer when the heat is not needed, and then withdrawing it in the winter as make-up water for steam-heating systems. A major concern with such systems, however, is to ensure that contaminants aren't injected into aquifers, which are important sources of drinking water.

Tidal and Wave Energy

Ocean tides and waves contain enormous amounts of energy that can be harnessed to do useful work. Tidal power exploitation is not new. The earliest recorded tide-powered mills were built in about 1100 in England. One built in Woodbridge, England, in 1170 functioned productively for eight hundred years. Through the seventeenth century, similar mills were built by the Dutch in the Netherlands and in New York.

The Rance River Power Station in France, in operation since 1966, was the first large tidal electric generation plant, producing 160 MW. A **tidal station** works like a hydropower dam, with its turbines spinning as the tide flows through them (fig. 22.26). It requires a high-tide low-tide differential of several meters to spin the turbines. Unfortunately, the tidal period of thirteen and one-half hours causes problems in integrating the plant into the electric utility grid, as it seldom coincides with peak-use hours. Nevertheless, demand has kept the plant running for more than two decades.

The first North American tidal generator, producing 20 MW, was completed in 1984 at Annapolis Royal, Nova Scotia. A much larger project has been proposed to dam the Bay of Fundy and produce 5000 MW of power on the Bay's 17-meter tides. The total flow at each tide through the Bay of Fundy theoretically could generate energy equivalent to the output of 250 large nuclear power plants. The environmental consequences of such a gargantuan project, however, may prevent its ever being built. The main worries are saltwater flooding of freshwater aquifers when seawater levels rise behind the dam and the flooding and destruction of rich shoals and salt flats, breeding grounds for aquatic species and a vital food source for millions of migrating shorebirds. There also would be heavy siltation, as well as scouring of the seafloor as water shoots through the dam. However, it appears that modest-sized plants like Annapolis Royal's will avoid most of these dangers. Three other medium-sized projects are under consideration for sites within the Bay of Fundy.

Ocean wave energy can easily be seen and felt on any seashore. The energy that waves expend as millions of tons of water are picked up and hurled against the land, over and over, day after day, can far exceed the combined energy budget for both insolation (solar energy) and wind power in localized areas. Captured and turned into useful forms, that energy could make a very substantial contribution to meeting local energy needs.

Numerous attempts have been made to use wave energy to drive electrical generators. Generally these take the form of a floating bar that moves up and down as the wave passes. When coupled to a dynamo, this mechanical energy can be converted to electricity in the same way that a waterwheel or steam turbine works. England, with a long coastline facing the stormy North Sea, plans to build an extensive system of wave-energy platforms. Unfortunately for developers of this energy source, the stormy coasts where waves are strongest are usually far from major population centers that need the power. In addition, the storms that bring this energy destroy the equipment intended to exploit it.

An intriguing proposal has been developed for small rotating cylinders called ducks that would sit just offshore, bobbing in the

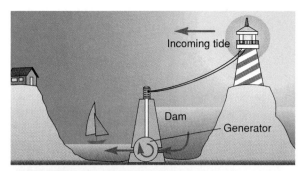

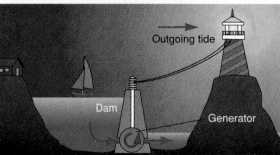

Figure 22.26

Tidal power station. Both incoming and outgoing tides are held back by a dam. The difference in water levels generates electricity in both directions as water runs through reversible turbogenerators.

surf and generating electricity. There are claims that British energy officials deliberately suppressed this low-tech approach because it might compete with conventional sources.

Ocean Thermal Electric Conversion

Temperature differentials between upper and lower layers of the ocean's water also are a potential source of renewable energy. Energy can be extracted in either a closed-cycle or open-cycle system. In a closed-cycle **ocean thermal electric conversion (OTEC)** system, heat from sun-warmed upper ocean layers is used to evaporate a working fluid, such as ammonia or Freon, which has a low boiling point. The pressure of the gas produced is high enough to spin turbines to generate electricity. Cold water then is pumped from the depths of the ocean to condense the gas.

An open-cycle system uses seawater itself as the working fluid (fig. 22.27). Warm seawater is sprayed into a chamber, where low air pressure causes some water to evaporate. The resulting water vapor drives a massive, low-pressure turbogenerator and then passes through two condensers cooled by deep ocean water. The first one, a closed surface condenser, produces distilled (desalinized) water. The other is a direct contact condenser in which cold water is sprayed directly into the steam flow, condensing the remaining water vapor and creating the vacuum that drives the system. The outflow from this condenser is discharged into the sea. This process produces fresh water, as well as electricity, but it requires very large turbines and inlet pipes, making it expensive to install.

As long as a temperature difference of about 20°C (36°F) exists between the warm upper layers and cooling water, useful

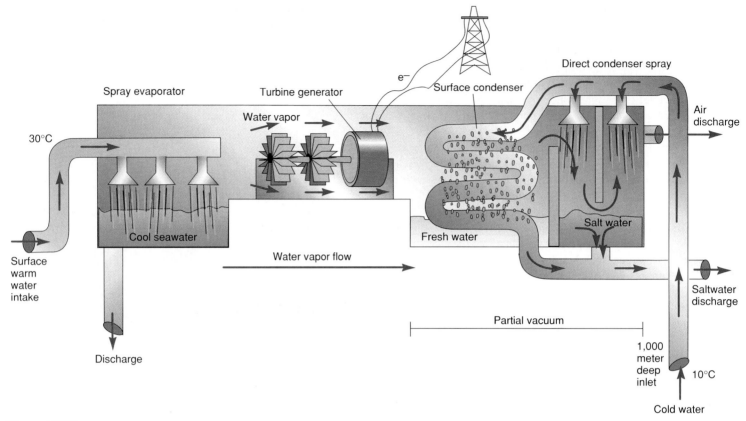

Figure 22.27

Open-cycle ocean thermal electric generator produces both fresh water and electricity as long as the water temperature differential is at least 20°C (36°F).

amounts of net power can, in principle, be generated with one of these systems. This differential corresponds, generally, to a depth of about 1000 m in tropical seas. The places where this much temperature difference is likely to be found close to shore are islands that are the tops of volcanic seamounts, such as Hawaii, or the edges of continental plates along subduction zones (chapter 16) where deep trenches lie just offshore. The west coast of Africa, the south coast of Java, and a number of South Pacific islands, such as Tahiti, have usable temperature differentials for OTEC power.

Disadvantages of OTEC include the energy cost of pumping up deep waters (which could eliminate most of the energy advantage of the system), saltwater corrosion of pipes and equipment, vulnerability to storm damage, and ecological destabilization from the upwelling of deep, nutrient-rich water and local alterations of ocean temperatures.

Research in Renewables

Public interest and funding for renewable energy has been fickle over the past 20 years. Immediately after the oil price shocks of the 1970s, government spending for alternative energy research and development in the United States grew to nearly $750 million per year. When Ronald Reagan became president in 1980, unfortu-

nately, support for renewables was slashed by almost 90 percent (fig. 22.28). Budding industries in solar, wind, biofuels, and cogeneration that were making terrific progress were cut off at the knees. Research funds were gradually being restored in the 1990s,

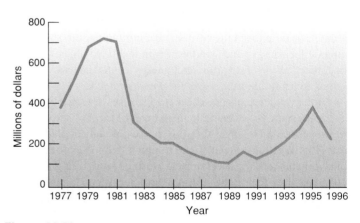

Figure 22.28

Renewable energy research expenditures in the United States (*1996 = proposed).

Source: U.S. Department of Energy.

but as this is being written in 1995, Congress once more is debating a 50 percent cut in sustainable energy. If this budget passes, renewable energy will be about 1 percent of all energy department expenditures. Nuclear power and fossil fuels will get the other 99

percent. Write to your congressional representatives to find out what has happened and to express your opinion on what our priorities ought to be in this area.

Summary

Several sustainable energy sources could reduce or eliminate our dependence on fossil fuels and nuclear energy. Some of these sources have been used for centuries but have been neglected since fossil fuels came into widespread use. Passive solar heat, fuelwood, windmills, and waterwheels, for instance, once supplied a major part of the external energy for human activities. With increased concern about the dangers and costs associated with conventional commercial energy, these ancient energy sources are being reexamined as part of a more sustainable future for humankind.

Exciting new technologies have been invented to use renewable energy sources. Active solar air and water heating, for instance, require less material and function more quickly than passive solar collection. Parabolic mirrors can produce temperatures high enough to be used as process heat in manufacturing. Ocean thermal electric conversion, tidal and wave power stations, and geothermal steam sources can produce useful amounts of energy in some localities. One of the most promising technologies is direct electricity generation by photovoltaic cells. Since solar energy is available everywhere, photovoltaic collectors could provide clean, inexpensive, nonpolluting, renewable energy, independent of central power grids or fuel-supply systems.

Biomass also may have some modern applications. In addition to direct combustion, biomass can be converted into methane or ethanol, which are clean-burning, easily storable, and transportable fuels. These alternative uses of biomass also allow nutrients to be returned to the soil and help reduce our reliance on expensive, energy-consuming artificial fertilizers.

Many of these sustainable energy sources depend on technology that is still experimental and too expensive to compete well with established energy industries. If the economies of mass production and marketing were applied, these new technologies could be made available more cheaply and more dependably. It may take special funding and other governmental incentives to make sustainable energy competitive. The subsidies for renewable energy sources have been especially meager in comparison to the billions of dollars spent on nuclear energy, large-scale hydropower, and fossil fuel extraction and utilization.

Although conventional and alternative energy sources offer many attractive possibilities, conservation often is the least expensive and easiest solution to energy shortages. Even basic conservation efforts, such as turning off lights, can save large amounts of energy when practiced by many people. More major conservation methods, such as home insulation and energy-efficient appliances and transportation, can drastically reduce energy consumption and similarly reduce energy expenses. Our natural resources, our environment, and our pocketbooks all benefit from careful and efficient energy consumption.

▼ Questions for Review

1. What is cogeneration and how does it save energy?

2. Explain the principle of net energy yield. Give some examples.

3. What is the difference between active and passive solar energy?

4. How do photovoltaic cells work?

5. Describe the advantages and disadvantages of multiple-bladed, two-bladed, and Darrieus windmills.

6. Describe some problems with wood burning in both industrialized nations and Third World nations.

7. How is methane made?

8. What are some advantages and disadvantages of large hydroelectric dams?

9. What are some examples of biomass fuel other than wood?

10. Describe how tidal power and ocean thermal conversion work.

▼ Questions for Critical Thinking

1. What alternative energy sources are most useful in your region and climate?

2. What can you do to conserve energy where you live? In personal habits? In your home, dormitory, or workplace?

3. What massive heat storage materials can you think of that could be attractively incorporated into a home?

4. Do you think building wind farms in remote places, parks, or scenic wilderness areas would be damaging or unsightly?

5. If you were a government energy administrator or planner, how would you distribute energy research funds?

6. What are the advantages and disadvantages of being disconnected from central utility power?

7. Can you think of environmental consequences associated with tidal or geothermal energy? If so, how can they be mitigated?

8. You are offered a home solar energy system that costs $10,000 but saves you $1000 a year. Will you take it at this rate? If the cost were higher and the payoff time longer, what is the threshold at which you would not buy the system?

▼ Key Terms

active solar systems   491
cogeneration   491
energy efficiency   489
eutectic chemicals   493
gasohol   502
geothermal energy   508
low-head hydropower   504
micro-hydro generators   504
net energy yield   490

ocean thermal electric
  conversion (OTEC)   509
parabolic mirrors   493
passive heat absorption   491
photovoltaic cell   495
run-of-the-river flow   504
tidal station   509
wind farms   506

▼ Additional Information on the Internet

Alternative Fuels Data Center
http://afdc.nrel.gov/

American Hydrogen Association
http://www.getnet.com/charity/aha/

American Solar Energy Society
http://www.engr.wisc.edu/centers/sel/ases/ases2.html/

American Wind Energy Association
http://www.igc.apc.org/awea/

Biofuels Information Network
http://www.esd.ornl.gov/BFDP/BFDPMOSAIC/binmenu.html/

California Energy Commission
http://www.energy.ca.gov/energy/

Electric Vehicle Sites on the Web
http://northshore.shore.net/~kester/websites.html/

EV Sites on the Web
http://northshore.shore.net/~kester/websites.html#alt.energy/

Joint European Torus
http://www.jet.uk/

The Journal of Solar Energy
http://www.rt66.com/rbahm/sejx.htm/

Mr. Solar
http://www.netins.net/showcase/solarcatalog/

National Ignition Facility
http://www-phys.llnl.gov/X_Div/nif.html/

National Renewable Energy Laboratory
http://nrelinfo.nrel.gov/

Passive Solar Energy Project
http://www.tuns.ca/~banderrj/solar.html/

Sandia National Laboratories Renewable Energy Technologies
http://www.sandia.gov/Renewable_Energy/renewable.html/

Solstice, Center for Renewable Energy and Sustainable Technology
http://solstice.crest.org/

SunSite—Alternative Energy
ftp://sunsite.unc.edu/pub/academic/environment/alternative-energy/

US Fusion Energy Sciences Program
http://wwwofe.er.doe.gov/

Note: Further readings appropriate to this chapter are listed on p. 606.

Part 5

Society and the Environment

At least 1.5 billion people lack life's basic necessities. Improving their lives without impoverishing the environment remains a challenge.

And the people asked him, "What then shall we do?" And he said, "Let him who has two tunics share with him who has none; and let him who has food do likewise."

Luke 3:10–11

Bulldozers pack down refuse at the Fresh Kills landfill on Staten Island, NY.

CHAPTER 23: Solid, Toxic, and Hazardous Waste

Objectives

After studying this chapter, you should be able to:

- identify the major components of the waste stream and describe how wastes have been—and are being—disposed of in North America and around the world.

- explain how incinerators work, as well as the advantages and disadvantages they offer.

- summarize the benefits, problems, and potential of recycling and reusing wastes.

- analyze some alternatives for reducing the waste we generate.

- understand what hazardous and toxic wastes are and how we dispose of them.

- evaluate the options for hazardous waste management.

- outline some ways we can destroy or permanently store hazardous wastes.

What a Long, Strange Journey It Has Been

On August 31, 1986, the cargo ship Khian Sea *loaded one month's production of ash (14,000 tons) from the Philadelphia municipal incinerator and set off on an odyssey that symbolizes a predicament we all share. The ship's first port of call in search of a dumping place for its noxious cargo was the poor Caribbean nation of Haiti. Four thousand tons of ash had been carried ashore by the time representatives of the environmental group Greenpeace alerted local residents to the potentially dangerous levels of arsenic, mercury, dioxins, and other toxins in the wastes. Authorities ordered the ship to take its objectionable goods elsewhere.*

For twenty-four months, this pariah ship wandered from port to port in the Caribbean, across to West Africa, around the Mediterranean, through the Suez Canal, past India, and over to Singapore looking for a place to dump its toxic load. Its name was changed from Khian Sea *to* Felicia *to* Pelacano. *Its registration was transferred from Liberia to the Bahamas to Honduras in an attempt to hide its true identity, but nobody*

wanted it or its contents. Like Coleridge's ancient mariner, it seemed cursed to wander the oceans forever. Two years, three names, four continents, and eleven countries later, the onerous cargo was still aboard.

Then, somewhere on the Indian Ocean between Columbo, Sri Lanka, and Singapore, 14,000 tons of toxic ash disappeared. When questioned about this remarkable occurrence, the crew had no comment except that it was all gone. Everyone assumes, of course, that once the ship was out of sight of land, the ash was dumped into the ocean.

If this were just an isolated incident, perhaps it wouldn't matter too much. However, some 3 million tons of toxic and hazardous waste goes to sea every year. How much ends up in the ocean and how much is deposited in poor countries is unknown.

The problem is that we generate vast amounts of unwanted stuff every year. Places to put these wastes are becoming more and more scarce as their contents becoming increasingly unpleasant and dangerous. None of us want it in our backyards. Vandals dump it in out-of-the-way places. Rich communities and nations send it to their impoverished neighbors. How can we stop these perilous practices? In this chapter, we will look at the kinds

of waste we produce, who makes them, what problems their disposal causes, or how we might reduce our waste production and dispose of our wastes in environmentally safe ways. ▶

Solid Waste

Waste is everyone's business. We all produce wastes in nearly everything we do. According to the Environmental Protection Agency, the United States produces 11 billion tons of solid waste each year. Nearly half of that amount consists of agricultural waste, such as crop residues and animal manure, which are generally recycled into the soil on the farms where they are produced. They represent a valuable resource as ground cover to reduce erosion and fertilizer to nourish new crops, but they also constitute the single largest source of nonpoint air and water pollution in the country. About one-third of all solid wastes are mine tailings, overburden from strip mines, smelter slag, and other residues produced by mining and primary metal processing. Most of this material is stored in or near its source of production and isn't mixed with other kinds of wastes. Improper disposal practices, however, can result in serious and widespread pollution.

Industrial waste—other than mining and mineral production—amounts to some 400 million metric tons per year in the United States. Most of this material is recycled, converted to other forms, destroyed, or disposed of in private landfills or deep injection wells. About 60 million metric tons of industrial waste falls in a special category of hazardous and toxic waste, which we will discuss later in this chapter.

Municipal waste—a combination of household and commercial refuse—amounts to about 300 million metric tons per year in the United States (fig. 23.1). That's approximately 1.24 tons for each man, woman, and child every year—twice as much per capita as Europe or Japan, and five to ten times as much as most developing countries.

The Waste Stream

Does it surprise you to learn that you generate that much garbage? Think for a moment about how much we discard every year. There are organic materials, such as yard and garden wastes, food wastes, and sewage sludge from treatment plants; junked cars; worn out furniture; and consumer products of all types. Newspapers, magazines, advertisements, and office refuse make paper one of our major wastes (fig. 23.2). In spite of recent progress in recycling, many of the 200 *billion* metal, glass, and plastic food and beverage containers used every year in the United States end up in the trash. Wood, concrete, bricks, and glass come from construction and demolition sites, dust and rubble from landscaping and road building. All of this varied and voluminous waste has to arrive at a final resting place somewhere.

The **waste stream** is a term that describes the steady flow of varied wastes that we all produce, from domestic garbage and yard

Figure 23.1

Bulldozers pack down trash at a municipal landfill near Niagara Falls, Canada. Rainwater percolating through unsealed landfills carries toxic pollutants into the groundwater and contaminates drinking supplies. Landfills are being closed all over the United States, but other forms of waste disposal are not being developed, or are equally controversial.

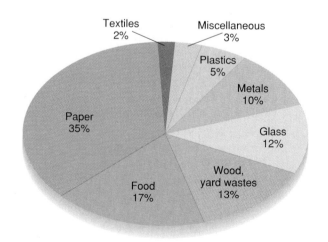

Figure 23.2

Composition of domestic waste in the United States, by weight. Plastics make up only 5 percent of the weight but up to 15 percent of the volume of our waste.

wastes to industrial, commercial, and construction refuse. Many of the materials in our waste stream would be valuable resources if they were not mixed with other garbage. Unfortunately, our collecting and dumping processes mix and crush everything together, making separation an expensive and sometimes impossible task. In a dump or incinerator, much of the value of recyclable materials is lost.

Another problem with refuse mixing is that hazardous materials in the waste stream get dispersed through thousands of tons of miscellaneous garbage. This mixing makes the disposal or burning of what might have been rather innocuous stuff a difficult, expensive, and risky business. Spray paint cans, pesticides, batteries (zinc, lead, or mercury), cleaning solvents, smoke detectors containing radioactive material, and plastics that produce dioxins and PCBs when burned are mixed willy-nilly with paper, table scraps, and other nontoxic materials. The best thing to do with household toxic and hazardous materials is to separate them for safe disposal or recycling, as we will see later in this chapter.

Waste Disposal Methods

Where are our wastes going now? In this section, we will examine some historic methods of waste disposal as well as some future options. Notice that our presentation begins with the least desirable—but most commonly used—measures and proceeds to discuss some preferable options. Keep in mind as you read this that modern waste management reverses this order and stresses the "three R's" of reduction, reuse, and recycling before destruction and, finally, secure storage of wastes.

Open Dumps

For many people, the way to dispose of waste is to simply drop it someplace. Open, unregulated dumps are still the predominant method of waste disposal in most developing countries. The giant Third World megacities have enormous garbage problems. Mexico City, the largest city in the world, generates some 10,000 tons of trash *each day*. Until recently, most of this torrent of waste was left in giant piles, exposed to the wind and rain, as well as rats, flies, and other vermin. Manila, in the Philippines, has at least ten huge open dumps. The most notorious is called Smoky Mountain because of its constant smoldering fires (fig. 23.3). Thousands of people live and work on this 30-meter-high heap of refuse. They spend their days sorting through the garbage for edible or recyclable materials. Health conditions are abysmal, but these people have nowhere else to go. The government would like to close these dumps, but how will the residents be housed and fed? Where else will the city put its garbage?

Most developed countries forbid open dumping, at least in metropolitan areas, but illegal dumping is still a problem. You have undoubtedly seen trash accumulating along roadsides and in vacant, weedy lots in the poorer sections of cities. Is this just a question of aesthetics? Consider the problem of waste oil and solvents. An estimated 200 million liters of waste motor oil are poured into the sewers or allowed to soak into the ground every year in the United States. This is about five times as much as was spilled by

Figure 23.3

Scavengers sort through the trash at "Smoky Mountain," one of the huge metropolitan dumps in Manila, Philippines. Some 20,000 people live and work on these enormous garbage dumps. The health effects are tragic.

the *Exxon Valdez* in Alaska in 1989! No one knows the volume of other solvents disposed of by similar methods.

Increasingly, these toxic chemicals are showing up in the groundwater supplies on which nearly half the people in America depend for drinking (chapter 20). An alarmingly small amount of oil or other solvents can pollute large quantities of drinking or irrigation water. One liter of gasoline, for instance, can make a million liters of water undrinkable. The problem of illegal dumping is likely to become worse as acceptable sites for waste disposal become more scarce and costs for legal dumping escalate. We clearly need better enforcement of antilittering laws as well as a change in our attitudes and behavior.

Ocean Dumping

The oceans are vast, but not so large that we can continue to treat them as carelessly as has been our habit. Every year some 25,000 metric tons (55 million lbs) of packaging, including half a million bottles, cans, and plastic containers, are dumped at sea. Beaches, even in remote regions, are littered with the nondegradable flotsam and jetsam of industrial society (see fig. 20.14). About 150,000 tons (330 million lbs) of fishing gear—including more than 1000 km (660 mi) of nets—are lost or discarded at sea each year. Environmental groups estimate that 50,000 northern fur seals are entangled in this refuse and drown or starve to death every year in the North Pacific alone.

Until 1988, many cities in the United States dumped municipal refuse, industrial waste, sewage, and sewage sludge in the ocean. Federal legislation now prohibits this dumping. New York City, the last to stop offshore sewage sludge disposal, finally ended this practice in 1992. Still, 60 million to 80 million cubic meters of dredge spoil are disposed of at sea. Some people claim that the deep abyssal plain is the most remote, stable, and innocuous place to dump our wastes. Others argue that we know too little about the values of these remote places or the rare species that live there to smother them with sludge and debris.

Landfills

Over the past fifty years most American and European cities have recognized the health and environmental hazards of open dumps. Increasingly, cities have turned to **landfills,** where solid waste disposal is regulated and controlled. To decrease smells and litter and to discourage insect and rodent populations, landfill operators are required to compact the refuse and cover it every day with a layer of dirt (fig. 23.4). This method helps control pollution, but the dirt fill also takes up as much as 20 percent of landfill space. Since 1994, all operating landfills in the United States have been required to control such hazardous substances as oil, chemical compounds, toxic metals, and contaminated rainwater that seeps through piles of waste. An impermeable clay and/or plastic lining underlies and encloses the storage area. Drainage systems are installed in and around the liner to catch drainage and to help monitor any chemicals that may be leaking. Modern municipal solid-waste landfills now have many of the safeguards of hazardous waste repositories described later in this chapter.

More careful attention is now paid to the siting of new landfills. Sites located on highly permeable or faulted rock formations are passed over in favor of sites with less leaky geologic foundations. Landfills are being built away from rivers, lakes, floodplains, and aquifer recharge zones rather than near them, as was often done in the past. More care is being given to a landfill's long-term effects so that costly cleanups and rehabilitation can be avoided.

Historically, landfills have been a convenient and relatively inexpensive waste-disposal option in most places, but this situation is changing rapidly. Rising land prices and shipping costs, as well as increasingly demanding landfill construction and maintenance requirements, are making this a more expensive disposal method. The cost of disposing a ton of solid waste in Philadelphia went from $20 in 1980 to more than $100 in 1990. Union County, New York, experienced an even steeper price rise. In 1987, it paid $70 to get rid of a ton of waste; a year later, that same ton cost $420, or about

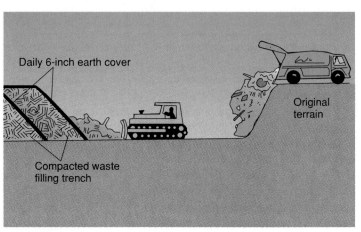

Figure 23.4

In a sanitary landfill, trash and garbage are crushed and covered each day to prevent accumulation of vermin and spread of disease. A waterproof lining is now required to prevent leaching of chemicals into underground aquifers.

$10 for a typical garbage bag. The United States now spends about $10 billion per year to dispose of trash. A decade from now, it may cost us $100 billion per year to dispose of our trash and garbage.

Suitable places for waste disposal are becoming scarce in many areas. Other uses compete for open space. Citizens have become more concerned and vocal about health hazards, as well as aesthetics. It is difficult to find a neighborhood or community willing to accept a new landfill. More than half of all U.S. cities will exhaust their landfills by 1995. Since 1984, when stricter financial and environmental protection requirements for landfills took effect, more than 1200 of the 1500 existing landfills in the United States have closed. Many major cities already have no more dumping space. They export their trash, at enormous expense, to neighboring communities and even other states. More than half the solid waste from New Jersey goes out of state, some of it up to 800 km (500 mi) away.

Exporting Waste

As disposal costs rise and restrictions on what can be dumped become more stringent, many European and American cities and industries are sending their wastes abroad to less-developed countries, as the story of the *Khian Sea* illustrates. In 1988, the environmental organization Greenpeace identified more than fifty plans to send wastes from the United States and Europe to Africa, Latin America, and the Middle East. This is becoming a touchy political issue. Local people usually aren't told what is in the waste being dumped on their land. Provincial officials don't have the resources or the knowledge to test or regulate toxic materials. In the West African nation of Guinea, children were found playing on a huge pile of incinerator ash from Philadelphia. Nigerian officials discovered 8000 leaking drums of radioactive waste stored in an open yard in the port city of Koko.

In 1985, two Americans were jailed in the United States for exporting 6000 l (1500 gal) of toxic waste to Zimbabwe under the guise of cleaning fluid. West African countries arrested sixty-two corrupt local officials and foreigners for illegally dumping danger-ous materials.

It's not surprising that exports are attractive. It commonly costs $2000 per barrel to dispose of toxic waste in the United States, but some African countries will take that barrel for $50. Toxic trash doesn't only move from north to south. In the 1970s and 1980s West Germany paid East Germany some $600 million to accept 5.5 million tons of household rubbish and 800,000 tons of toxic waste. After reunification, ironically, the Bonn govern-ment is now having to pay billions of Deutsch marks to properly dispose of those wastes. In 1994, despite opposition from the United States, Japan, and Germany, the United Nations passed a ban on international transport of toxic wastes. Waste exports for re-cycling, however, were exempted until 1997.

As we will discuss later in this chapter, garbage imperialism also operates within the richer countries as well. Poor neighbor-hoods and minority populations are much more likely than richer ones to be recipients of dumps, waste incinerators, and other lo-cally unwanted land uses (LULUs). In recent years, attention has turned to Indian reservations, which are exempt from some state and federal regulations concerning waste disposal. Virtually every tribe in America has been approached with schemes to store wastes on reservations.

Incineration and Resource Recovery

Landfilling is still the disposal method for the vast majority of mu-nicipal waste in the United States (fig. 23.5). Faced with growing piles of garbage and a lack of available landfills at any price, how-ever, public officials are investigating other disposal methods. The method to which they frequently turn is burning. Another term com-monly used for this technology is **energy recovery,** or waste-to-energy, because the heat derived from incinerated refuse is a useful resource. Burning garbage can produce steam used directly for heat-ing buildings or generating electricity. Internationally, well over 1000 waste-to-energy plants in Brazil, Japan, and Western Europe generate much-needed energy while also reducing the amount that needs to be landfilled. In the United States, more than 110 waste in-cinerators burn 45,000 tons of garbage daily. Some of these are sim-ple incinerators; others produce steam and/or electricity.

Types of Incinerators

Municipal incinerators are specially designed burning plants capa-ble of burning thousands of tons of waste per day. In some plants, refuse is sorted as it comes in to remove unburnable or recyclable materials before combustion. This is called **refuse-derived fuel** because the enriched burnable fraction has a higher energy content than the raw trash. Another approach, called **mass burn,** is to dump everything smaller than sofas and refrigerators into a giant furnace and burn as much as possible (fig. 23.6). This technique avoids the expensive and unpleasant job of sorting through the garbage for nonburnable materials, but it often causes greater problems with air pollution and corrosion of burner grates and chimneys.

In either case, residual ash and unburnable residues represent-ing 10 to 20 percent of the original volume are taken to a landfill

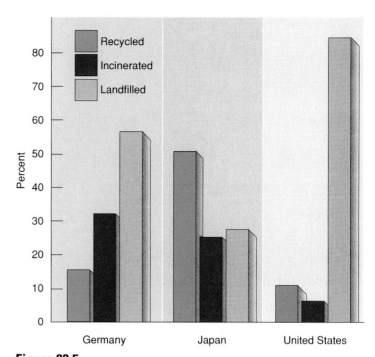

Figure 23.5

Percentages of municipal solid waste recycled, incinerated, and buried in Germany, Japan, and the United States.

Source: United Nations Environment Programme.

for disposal. Because the volume of burned garbage is reduced by 80 to 90 percent, disposal is a smaller task. However, the residual ash usually contains a variety of toxic components that make it an environmental hazard if not disposed of properly. Ironically, one worry about incinerators is whether enough garbage will be avail-able to feed them. Some communities in which recycling has been really successful have had to buy garbage from neighbors to meet contractual obligations to waste-to-energy facilities. In other places, fears that this might happen have discouraged recycling efforts.

Incinerator Cost and Safety

The cost-effectiveness of garbage incinerators is the subject of heated debates. Initial construction costs are high—usually be-tween $100 million and $300 million for a typical municipal facil-ity. Tipping fees at an incinerator, the fee charged to haulers for each ton of garbage dumped, are often much higher than those at a landfill. As landfill space near metropolitan areas becomes more scarce and more expensive, however, landfill rates are certain to rise. It may pay in the long run to incinerate refuse so that the life-time of existing landfills will be extended.

Environmental safety of incinerators is another point of con-cern. The EPA has found alarmingly high levels of dioxins, furans, lead, and cadmium in incinerator ash. These toxic materials were more concentrated in the fly ash (lighter, airborne particles capable of penetrating deep into the lungs) than in heavy bottom ash. Dioxin levels can be as high as 780 parts per billion. One part per billion of TCDD, the most toxic dioxin, is considered a health concern. All

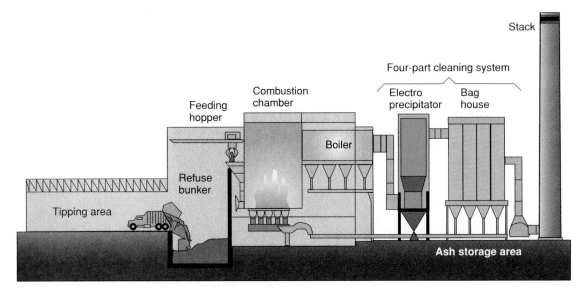

Stack

Four-part cleaning system

Electro precipitator

Bag house

Combustion chamber

Feeding hopper

Boiler

Refuse bunker

Tipping area

Ash storage area

Figure 23.6

A diagram of a municipal "mass-burn" garbage incinerator. Steam produced in the boiler can be used to generate electricity or to heat nearby buildings.

of the incinerators studied exceeded cadmium standards, and 80 percent exceeded lead standards. Proponents of incineration argue that if they are run properly and equipped with appropriate pollution-control devices, incinerators are safe to the general public. Opponents counter that neither public officials nor pollution control equipment can be trusted to keep the air clean. They argue that recycling and source reduction efforts are better ways to deal with waste problems.

The EPA, which generally supports incineration, acknowledges the health threat of incinerator emissions but holds that the danger is very slight. The EPA estimates that dioxin emissions from a typical municipal incinerator may cause one death per million people in seventy years of operation. Critics of incineration claim that a more accurate estimate is 250 deaths per million in seventy years.

One way to reduce these dangerous emissions is to remove batteries containing heavy metals and plastics containing chlorine before wastes are burned. Bremen, West Germany, is one of several European cities now trying to control dioxin and PCB emissions by keeping all plastics out of incinerator waste. Bremen is requiring households to separate plastics from other garbage. This is expected to eliminate nearly all dioxins and PCBs and prevent the expense of installing costly pollution-control equipment that otherwise would be necessary to keep the burners operating. Minneapolis has initiated a recycling program for the small "button" batteries used in hearing aids, watches, and calculators in an attempt to lower mercury emissions from its incinerator.

Shrinking the Waste Stream

Having less waste to discard is obviously better than struggling with disposal methods, all of which have disadvantages and drawbacks. In this section we will explore some of our options for recycling, reuse, and reduction of the wastes we produce.

Recycling

The term *recycling* has two meanings in common usage. Sometimes we say we are *recycling* when we really are *reusing* something, such as refillable beverage containers. In terms of solid waste management, however, **recycling** is the reprocessing of discarded materials into new, useful products (fig. 23.7). Some recycling processes reuse materials for the same purposes; for instance, old aluminum cans and glass bottles are usually melted and recast into new cans and bottles. Other recycling processes turn old materials into entirely new products. Old tires, for instance, are shredded and turned into rubberized road surfacing. Newspapers become cellulose insulation, kitchen wastes become a valuable soil

Figure 23.7

Trucks with multiple compartments pick up residential recyclables at curbside, greatly reducing the amount of waste that needs to be buried or burned. For many materials, however, collection costs are too high and markets are lacking for recycling to be profitable.

amendment, and steel cans become new automobiles and construction materials.

There have been some dramatic successes in recycling in recent years. Minnesota, for instance, now claims a 40-percent recycling rate, something thought unattainable a decade ago. The high value of aluminum scrap ($650 per ton in 1995) has spurred a large percentage of aluminum recycling nearly everywhere. About two-thirds of all aluminum cans are now recycled; up from only 15 percent 20 years ago. This recycling is so rapid that half of all the aluminum cans now on grocery shelves will be made into another can within two months. In 1995, a combination of shortages of virgin wood pulp and a sudden demand for recycled paper drove up the price of used paper and cardboard from $30 a ton to $160 a ton in 12 months. Thieves and scavengers were stealing used paper from curbside containers and collection centers.

Benefits of Recycling

Recycling is usually a better alternative to either dumping or burning wastes. It saves money, energy, raw materials, and land space, while also reducing pollution. Recycling also encourages individual awareness and responsibility for the refuse produced. Curbside pickup of recyclables costs around $35 per ton, as opposed to the $80 paid to dispose of them at an average metropolitan landfill. Many recycling programs cover their own expenses with materials sales and may even bring revenue to the community.

Another benefit of recycling is that it could cut our waste volumes drastically and reduce the pressure on disposal systems. Philadelphia is investing in neighborhood collection centers that will recycle 600 tons a day, enough to eliminate the need for a previously planned, high-priced incinerator. New York City, down to one available landfill but still producing 27,000 tons of garbage a day, has set a target of 50 percent waste reduction to be accomplished by recycling office paper and household and commercial waste. New York's curbside collection service, projected to be the nation's largest, should more than pay for itself simply in avoided disposal costs.

Japan probably has the most successful recycling program in the world (see fig. 23.5). Half of all household and commercial wastes in Japan are recycled while the rest is about equally incinerated or landfilled. By comparison, the United States landfills more than 80 percent of all solid waste. Japanese families diligently separate wastes into as many as seven categories, each picked up on a different day (fig. 23.8). Would we do the same? Some authors say that Americans are too lazy to recycle. North Stonington, Connecticut, however, faced with escalating disposal costs, reduced its waste volume by two-thirds in just two years. Overall, we already are making progress in removing material from the waste stream (fig. 23.9).

Recycling lowers our demands for raw resources. In the United States, we cut down 2 million trees every day to produce newsprint and paper products, a heavy drain on our forests. Recycling the print run of a single Sunday issue of the New York Times would spare 75,000 trees. Every piece of plastic we make reduces the reserves supply of petroleum and makes us more de-

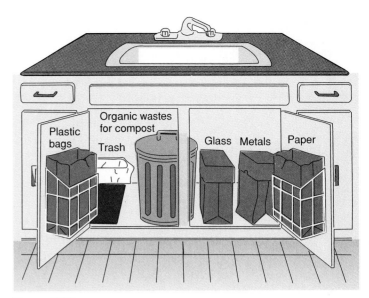

Figure 23.8

Source separation in the kitchen—the first step in a strong recycling program.

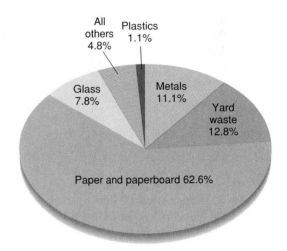

Figure 23.9

Percentage composition (by weight) of recycled materials in the United States in 1990.

Source: Environmental Protection Agency, 1992.

pendent on foreign oil. Recycling 1 ton of aluminum saves 4 tons of bauxite (aluminum ore) and 700 kg (1540 lb) of petroleum coke and pitch, as well as keeping 35 kg (77 lb) of aluminum fluoride out of the air.

Recycling also reduces energy consumption and air pollution. Plastic bottle recycling could save 50 to 60 percent of the energy needed to make new ones. Making new steel from old scrap offers up to 75 percent energy savings. Producing aluminum from scrap instead of bauxite ore cuts energy use by 95 percent, yet we still throw away more than a million tons of aluminum every year. If

Solid, Toxic, and Hazardous Waste

In Depth: Building a Home Compost Pile

Home composting is easy, beneficial, and educational. It takes very little effort or attention because millions of tiny microorganisms do almost all the work for you. All you have to do is supply them with a food supply, a little water, and oxygen.

To get started, you need a supply of organic material. Grass clippings, leaves, wood chips, vegetable peelings, sawdust, seaweed, manure, squash and tomato vines, watermelon rinds, coffee grounds, even hair, paper, and old rags are all good compost material. You can simply pile everything on the ground in an out-of-the-way place, or if you prefer a tidier operation, you can build a container or bin (fig. 1). Use boards, bricks, screen, an old barrel, or any other material that will hold the compost. Leave some spaces or holes in the side of your bin so that air can reach the lower layers—or turn the pile over frequently.

Layer material in the pile, alternating nitrogen-rich stuff such as

Figure 1

Composting is a good way to convert yard waste, vegetable scraps, and other organic materials into useful garden mulch. Mix everything together, keep it moist and well aerated, and in a few weeks, you will have a rich, odor-free mulch. This three-bin system allows you to have batches in different stages.

grass, manure, and food scraps with carbon-rich substances such as cornstalks, dry leaves, straw, wood chips, and paper. A pile that is too high in carbon will take a long time to work. Too much nitrogen will produce an odor like urine or ammonia gas; it also will make the pile slimy and putrid. The ideal ratio of carbon to nitrogen is 30:1. Since you proba-

---

aluminum recovery were doubled worldwide, more than a million tons of air pollutants would be eliminated every year.

Reducing litter is an important benefit of recycling. Ever since disposable paper, glass, metal, foam, and plastic packaging began to accompany nearly everything we buy, these discarded wrappings have collected on our roadsides and in our lakes, rivers, and oceans. Without incentives to properly dispose of beverage cans, bottles, and papers, it often seems easier to just toss them aside when we have finished using them. Litter is a costly as well as unsightly problem. We pay an estimated thirty-two cents for each piece of litter picked up by crews along state highways, which adds up to $500 million every year. "Bottle-bills" requiring deposits on bottles and cans have reduced littering in many states.

Creating Incentives for Recycling

In many communities, citizens have done such a good job of collecting recyclables that a glut has developed. Mountains of waste materials accumulate in warehouses (fig. 23.10) because there are no markets for them. Too often wastes that we carefully separate for recycling end up being mixed together and dumped in a landfill or incinerator.

Our present public policies often tend to favor extraction of new raw materials. Energy, water, and raw materials are often sold to industries below their real cost to create jobs and stimulate the economy. Setting the prices of natural resources at their real cost would tend to encourage efficiency and recycling. State, local, and national statutes requiring government agencies to purchase a minimum amount of recycled material have helped create a market for used materials. Each of us can play a role in creating markets, as well. If we buy things made from recycled materials—or ask for them if they aren't available—we will help make it possible for recycling programs to succeed.

Composting

Pressed for landfill space, many cities have banned yard waste from municipal garbage. Rather than bury this valuable organic material, they are turning it into a useful product through **composting:** biological degradation or breakdown of organic matter under aerobic (oxygen-rich) conditions. The organic compost resulting from this process makes a nutrient-rich soil amendment that aids water retention, slows soil erosion, and improves crop yields.

bly don't have the means to determine the composition of what you put in your pile, just use a variety of materials and don't overload with any one thing. If you have a very large amount of dry material—say leaves in the fall—you might want to add some extra manure or commercial fertilizer to give the compost a shot of nitrogen.

To function well, a compost pile shouldn't be either too big or too little. If the pile is too large, oxygen doesn't get into the middle. A pile that is too small doesn't retain enough heat for the microorganisms to grow optimally. One to two meters wide and a meter (3 ft) deep is about right. Given a good nutrient supply and plenty of air, bacteria and fungi growing in the compost pile will produce a temperature of about 70° C (160° F), enough to kill most pathogens and weed seeds. Turning the pile over frequently (every week or two) will mix the components and provide enough fresh air to keep the pile working well and will prevent the sour smell of anaerobic (oxygen-starved) fermentation.

The other essential ingredient for the microorganisms is water. If you live in a rainy climate or put lots of fresh vegetables and grass clippings in your compost pile, you won't need supplemental moisture, but if you use lots of dry leaves or live in a very dry place, you may need to add some water from time to time. The compost should be moist but not saturated. Too much water blocks oxygen penetration (that's why bogs are anaerobic and organic material doesn't decompose).

How long will your compost pile take to complete its work? That depends on temperature, moisture, texture, and contents. In the summer, a few weeks should produce a dark, soft, crumbly material that smells earthy and looks like rich potting soil. Composting will even work in the winter—but slowly if you live in a cold climate. Branches, stone fruit, and large chunks of material decay very slowly. Shredding, chipping, or chopping the starting material into small pieces will speed up the process.

There are some things that you shouldn't put into your compost pile. Don't add meat, fish, cheese, butter, milk, peanut butter, oil, grease, bones, skin or other animal products. They don't compost well, they tend to smell bad, and they attract pests and vermin. Don't add plants infected with disease or insects that could survive the compost pile's heat (such as scale insects, rusts, smuts, etc.). Avoid plants that take too long to break down such as pine needles, eucalyptus leaves or bark, or oak leaves. The tannins and acids they contain are natural bactericides and fungicides that inhibit composting. Cat and dog feces may contain harmful pathogens that can survive composting—especially if your pile doesn't reach optimal temperature. The compost produced could be infectious.

If you avoid these materials, keep pests and vermin out, and maintain a neat, well-aerated, odor-free compost pile, you shouldn't have any complaints from your neighbors. And after a few weeks, you will have a valuable soil amendment to spread on your yard or in your garden, or to use in potting house plants.

Figure 23.10
Mountains of paper, plastic, and glass are accumulating in warehouses because collection programs have been more successful than material marketing efforts. Federal and state governments need to encourage use of recycled materials. Each of us can help by buying recycled products.

A home compost pile is an easy and inexpensive way to dispose of organic waste in an interesting and environmentally friendly way (In Depth, above).

Large-scale municipal or industrial compost facilities have some of the same public relations and siting problems as incinerators and landfills. Neighbors complain of noise, odors, vermin, and increased traffic from poorly run facilities. These problems don't have to occur, but when you are dealing with huge amounts of garbage it's easy for the situation to get out of hand.

Energy from Waste

Every year we throw away the energy equivalent of 80 million barrels of oil in organic waste in the United States. In developing countries, up to 85 percent of the waste stream is food, textiles, vegetable matter, and other biodegradable materials. Worldwide, at least one-fifth of municipal waste is organic kitchen and garden refuse. In a landfill, much of this matter is decomposed by microorganisms generating billions of cubic meters of methane ("natural gas"), which contributes to global warming if allowed to escape into the atmosphere (chapter 17). Many cities are drilling methane wells in their landfills to capture this valuable resource.

This valuable organic material can be burned in an incinerator rather than being buried in landfills, but there are worries about air pollution from incineration. Organic wastes also can be decomposed in large, oxygen-free digesters to produce methane under more controlled conditions than in a landfill and with less air pollution than mass garbage burning.

Anaerobic digestion also can be done on a small scale. Millions of household methane generators provide fuel for cooking and lighting for homes in China and India (chapter 22). In the United States, some farmers produce all the fuel they need to run their farms—both for heating and to run trucks and tractors—by generating methane from animal manure.

Reuse

Even better than recycling or composting is cleaning and reusing materials in their present form, thus saving the cost and energy of remaking them into something else. We do this already with some specialized items. Auto parts are regularly sold from junkyards, especially for older car models. In some areas, stained glass windows, brass fittings, fine woodwork, and bricks salvaged from old houses bring high prices. Some communities sort and reuse a variety of materials received in their dumps (fig. 23.11).

In many cities, glass and plastic bottles are routinely returned to beverage producers for washing and refilling. The reusable, refillable bottle is the most efficient beverage container we have. This is better for the environment than remelting and more profitable for local communities. A reusable glass container makes an

Figure 23.11

Reusing discarded products is a creative and efficient way to reduce wastes. This dump in San Francisco has become a recycling center, a valuable source of secondary materials, and a money saver for the whole community.

average of fifteen round-trips between factory and customer before it becomes so scratched and chipped that it has to be recycled. Reusable containers also favor local bottling companies and help preserve regional differences.

Since the advent of cheap, lightweight, disposable food and beverage containers, many small, local breweries, canneries, and bottling companies have been forced out of business by huge national conglomerates. These big companies can afford to ship food and beverages great distances as long as it is a one-way trip. If they had to collect their containers and reuse them, canning and bottling factories serving large regions would be uneconomical. Consequently, the national companies favor recycling rather than refilling because they prefer fewer, larger plants and don't want to be responsible for collecting and reusing containers.

In less affluent nations, reuse of all sorts of manufactured goods is an established tradition. Where most manufactured products are expensive and labor is cheap, it pays to salvage, clean, and repair products. Cairo, Manila, Mexico City, and many other cities have large populations of poor people who make a living by scavenging. Entire ethnic populations may survive on scavenging, sorting, and reprocessing scraps from city dumps.

Producing Less Waste

What is even better than reusing materials? Generating less waste in the first place. Table 23.1 describes some contributions you can make to reducing the volume of our waste stream. Industry also can play an important role in source reduction. In Minnesota, 3M Company saved over $500 million since 1975 by changing manufacturing processes, finding uses for waste products, and listening to employees' suggestions. What is waste to one division is a treasure to another.

Excess packaging of food and consumer products is one of our greatest sources of unnecessary waste. Paper, plastic, glass, and metal packaging material make up 50 percent of our domestic trash by volume. Much of that packaging is primarily for marketing and has little to do with product protection (fig. 23.12). Manufacturers and retailers might be persuaded to reduce these wasteful practices if consumers ask for products without excess packaging. Canada's National Packaging Protocol (NPP) recommends that packaging minimize depletion of virgin resources and production of toxins in manufacturing. The preferred hierarchy is (1) no packaging, (2) minimal packaging, (3) reusable packaging, and (4) recyclable packaging. This plan sets a target of 50 percent reduction in excess packaging by December 31, 2000.

Communities also can have an impact. Berkeley, California; Portland, Oregon; Minneapolis and St. Paul, Minnesota; and about twenty-five other cities have passed ordinances requiring that fast-food restaurants package food in paper or other biodegradable wrappings, both to reduce litter and to protect the atmosphere. In 1990, responding to nationwide pressure, Burger King and McDonald's announced they would stop using plastic foam hamburger boxes.

Where disposable packaging is necessary, we still can reduce the volume of waste in our landfills by using biodegradable mate-

Table 23.1

What Can You Do to Reduce Waste?

1. Buy foods that come with less packaging; shop at farmers' markets or co-ops, using your own containers.

2. Take your own washable refillable beverage container to meetings or convenience stores.

3. When you have a choice at the grocery store between plastic, glass, or metal containers for the same food, buy the reusable or easier-to-recycle glass or metal.

4. When buying plastic products, pay a few cents extra for environmentally degradable varieties.

5. Separate your cans, bottles, papers, and plastics for recycling.

6. Wash and reuse bottles, aluminum foil, plastic bags, etc., for your personal use.

7. Compost yard and garden wastes, leaves, and grass clippings.

8. Write to your senators and representatives and urge them to vote for container deposits, recycling, and safe incinerators or landfills.

Source: Minnesota Pollution Control Agency.

Figure 23.12

How much more do we need? Where will we put what we already have?

Reprinted with special permission of King Features Syndicate.

rials. Usually this means no plastics. Recently, however, plastics have become available that do break down in the environment under ideal circumstances. **Photodegradable plastics** break down when exposed to ultraviolet radiation. **Biodegradable plastics** incorporate such materials as cornstarch that can be decomposed by microorganisms. Several states have introduced legislation requiring biodegradable or photodegradable six-pack beverage yokes, fast-food packaging, and disposable diapers. However, these degradable plastics don't decompose completely; they only break down to small particles that remain in the environment. In doing so, they can release toxic chemicals into the environment. Furthermore, they make recycling less feasible and may lead people to believe that littering is OK.

Some environmental groups are beginning to think that we have put too much emphasis on recycling. Many people think that if they recycle aluminum cans and newspapers they are doing everything they can for the environment. In surveys conducted by the New Jersey-based Environmental Research Associates, only two in ten people surveyed could define waste reduction. Companies that make "throw away" or heavily packaged products generally favor recycling because it allows them to continue business-as-usual. While recycling is an important part of waste management, we have to remember that it is actually the third "R" in the waste hierarchy. The two preferred methods—reduction and reuse—get lost in our enthusiasm for recycling.

Hazardous and Toxic Wastes

The most dangerous aspect of the waste stream we have described is that it often contains highly toxic and hazardous materials that are injurious to both human health and environmental quality. We now produce and use a vast array of flammable, explosive, caustic, acidic, and highly toxic chemical substances for industrial, agricultural, and domestic purposes (fig. 23.13). According to the EPA, industries in the United States generate about 265 million metric tons of *officially* classified hazardous wastes each year, slightly more than 1 ton for each person in the country. In addition, considerably more toxic and hazardous waste material is generated by industries or processes not regulated by the EPA. Shockingly, at least 40 million metric tons (22 billion lbs) of toxic and hazardous wastes are released into the air, water, and land in the United States each year. The biggest source of these toxins are the chemical and petroleum industries (fig. 23.14).

What Is Hazardous Waste?

Legally, a **hazardous waste** is any discarded material, liquid or solid, that contains substances known to be (1) fatal to humans or laboratory animals in low doses, (2) toxic, carcinogenic, mutagenic, or teratogenic to humans or other life-forms, (3) ignitable with a flash point less than 60° C, (4) corrosive, or (5) explosive or highly reactive (undergoes violent chemical reactions either by itself or when mixed with other materials). Notice that this definition includes both toxic and hazardous materials as defined in chapter 9. Certain compounds are exempt from classification as hazardous waste if they are accumulated in less than 1 kg (2.2 lb) of commercial chemicals or 100 kg of contaminated soil, water, or debris. Even larger amounts (up to 1000 kg) are exempt when stored at an approved waste treatment facility for the purpose of being beneficially used, recycled, reclaimed, detoxified, or destroyed.

Figure 23.13

According to the U.S. Environmental Protection Agency, industries produce about one ton of hazardous waste per year for every person in the United States. Much of this waste is dumped illegally or disposed of in environmentally dangerous ways.

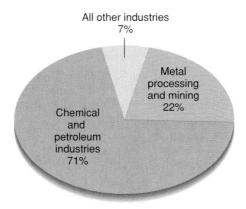

All other industries 7%

Metal processing and mining 22%

Chemical and petroleum industries 71%

Figure 23.14

Producers of hazardous wastes in the United States.

Hazardous Waste Disposal

Most hazardous waste is recycled, converted to nonhazardous forms, stored, or otherwise disposed of on-site by the generators—chemical companies, petroleum refiners, and other large industrial facilities—so that it doesn't become a public problem. The 60 million tons per year that does enter the waste stream or the environment represent one of our most serious environmental problems. Much of this material can be a serious threat to both environmental quality and human health.

Finding ways to safely ship, handle, and dispose of this enormous volume of dangerous material is a great challenge. For years, little attention was paid to this material. Wastes stored on private property, buried, or allowed to soak into the ground were considered of little concern to the public. An estimated 5 billion metric tons of highly poisonous chemicals were improperly disposed of in the United States between 1950 and 1975 before regulatory controls became more stringent.

Federal Legislation

Two important federal laws regulate hazardous waste management and disposal in the United States. The Resource Conservation and Recovery Act (RCRA, pronounced "rickra") of 1976 is a comprehensive program that requires "cradle to grave" recordkeeping and management of toxic and hazardous substances. A complex set of rules require generators, shippers, users, and disposers of these materials to keep meticulous account of everything they handle and what happens to it.

The Comprehensive Environmental Response, Compensation and Liability Act (CERCLA or Superfund Act), passed in 1980 is aimed at rapid containment, cleanup, or remediation of abandoned toxic waste sites. This statute authorizes the Environmental Protection Agency to undertake emergency actions when a threat exists that toxic material will leak into the environment. The agency is empowered to bring suit for the recovery of its costs from the responsible parties such as site owners, operators, waste generators, or transporters.

The government does not have to prove that anyone violated a law or what role they played in a superfund site. Rather, liability under CERCLA is "strict, joint, and several," meaning that anyone associated with a site can be held responsible for the entire cost of cleaning it up no matter how much of the mess they made. In some cases, property owners have been assessed millions of dollars for removal of wastes left there years earlier by previous owners. This strict liability has been a headache for the real estate and insurance businesses.

CERCLA was amended in 1995 to make some of its provisions less onerous. In cases where treatment is unavailable or too costly and it is likely that a less-costly remedy will become available within a reasonable time, interim containment is now allowed. The EPA also now has the discretion to set site-specific cleanup levels rather than adhere to rigid national standards. Congress agreed with industry that not every site needs to be cleaned up so you could safely eat the dirt. If the most likely future use of a site is as a parking lot or industrial site, a less stringent level of cleanup might be judged adequate.

Superfund Sites

The EPA estimates that there are at least 36,000 abandoned hazardous waste disposal sites in the United States. The General Accounting Office (GAO) places the number much higher, perhaps 425,000 sites when all are identified. By 1994, some 1290 sites had been placed on the National Priority List (NPL) for cleanup with financing from the 1980 federal Superfund Act (fig. 23.15). The EPA expects that at least 4500 sites eventually will be included on the NPL and that the total cost for cleanup will be more than $100 billion. The GAO estimates the cost may be as high as $500 billion.

The Superfund was first established as a $1.6 billion pool, but has been increased more than tenfold and will probably go much higher in the future. Cleanup of individual sites has been more expensive than anticipated, and the work has progressed slowly. After 15 years of work and $20 billion in expenditures, the EPA has cleaned up, or contained, only 200 out of 1290 sites on the NPL.

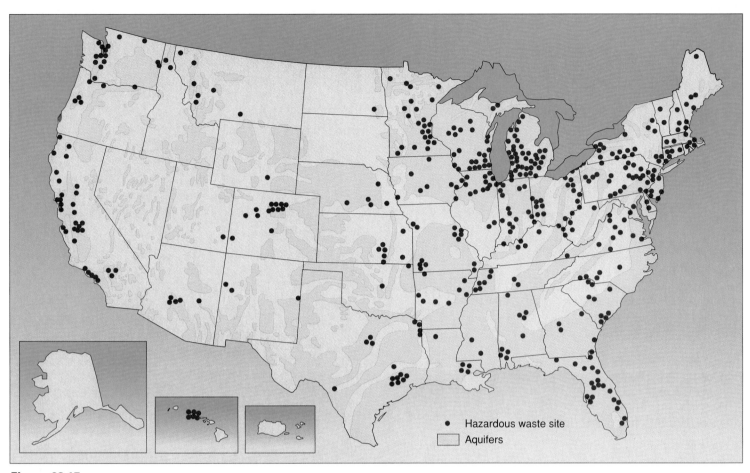

Figure 23.15

Some of the hazardous waste sites on the EPA priority cleanup list. Sites located on aquifer recharge zones represent an especially serious threat. Once groundwater is contaminated, cleanup is difficult and expensive. In some cases, it may not be possible.

Source: Data from the Environmental Protection Agency.

Critics charge that 80 percent of the money has been used for legal fees and high-priced consultants while only a pittance has been available for actual cleanup.

What qualifies a site for placement on the NPL? These sites are considered to be especially hazardous to human health and environmental quality because they are known to be leaking or have a potential for leaking supertoxic, carcinogenic, teratogenic, or mutagenic materials (chapter 9). So far, 444 toxic pollutants have been identified at these sites. The ten most commonly found substances are lead, trichloroethylene, toluene, benzene, PCBs, chloroform, phenol, arsenic, cadmium, and chromium. These or other hazardous materials are known to have contaminated groundwater at 75 percent of the sites now on the NPL. In addition, 56 percent of these sites have contaminated surface waters, and airborne materials are found at 20 percent of the sites.

Where are these thousands of hazardous waste sites, and how did they get contaminated? Wastes come from countless sources and are disposed of in just as many ways (fig. 23.16). Minority neighborhoods are disproportionately likely to be the site of a toxic waste dump or other pollution source (In Depth, p. 530). Some are large

and well known; others are long forgotten—perhaps only a few barrels buried in the ground or stored in a warehouse somewhere. In this section, we will review a few of the major categories of waste sites.

Hazardous Landfills and Dumps

Many of the worst Superfund sites are landfills and old dumps where industries buried metal drums full of a variety of toxic chemicals on their own or leased property (fig. 23.17). Sometimes the volumes were truly staggering. The Velsicol Chemical Company, for instance, buried at least 250,000 barrels containing about 60 million l (16 million gal) of toxic pesticide residue in Hardeman County, Tennessee, in the 1960s. These chemicals have leaked from the barrels and contaminated groundwater aquifers used for drinking water by local residents.

Toxic landfills are frequently in or near urban areas. The revelation that homes and a school in Niagara Falls, New York, were built over a waste dump containing 20,000 metric tons of toxic chemical waste awakened many people to the dangers of improper waste disposal and played a catalytic role in passage of the Superfund Act. There are thousands of other abandoned dumps,

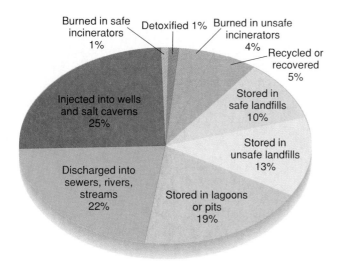

Figure 23.16

Fate of hazardous wastes in the United States. Estimated percentages of volumes of waste disposal in various legal and illegal manners.

Source: Data from the Congressional Budget Office, 1989.

some reported to be much larger and more dangerous than the one at Love Canal.

Waste Lagoons and Injection Wells

In many areas, liquid wastes weren't even put in barrels before they were dumped; they were simply pumped into lagoons and holding ponds or pumped into deep wells. All these techniques threaten to contaminate groundwater aquifers. The Stringfellow Acid Pit in California is a notorious example. Between 1956 and 1972, more than 120 million l (32 million gal) of toxic chemicals were poured into shallow ponds at this site in Riverside County on the eastern edge of Los Angeles. A chemical plume is now creeping through the aquifer that supplies drinking water to 500,000 people in the Los Angeles Basin.

Warehousing and Illegal Dumping

Large quantities of hazardous wastes have simply been stacked in old warehouses and abandoned. The high concentrations of flammable materials and highly poisonous chemicals at some of these sites create a lethal combination. On April 21, 1980, a fire broke out in an old warehouse at an inactive waste treatment facility in Elizabeth, New Jersey. More than 20,000 leaking and corroded drums containing pesticides, explosives, radioactive wastes, acids, and other hazardous substances were piled in and around the building. A cloud of toxic gas drifted over heavily populated areas close to the site. Contaminated water from fighting the fire ran off into Newark Bay.

The site had been licensed as an incinerator facility for hazardous waste, but, in fact, the operators—some of whom have been linked with organized crime in New York and New Jersey—were simply collecting disposal fees and stacking the barrels in warehouses and on parking lots. A former employee of the company

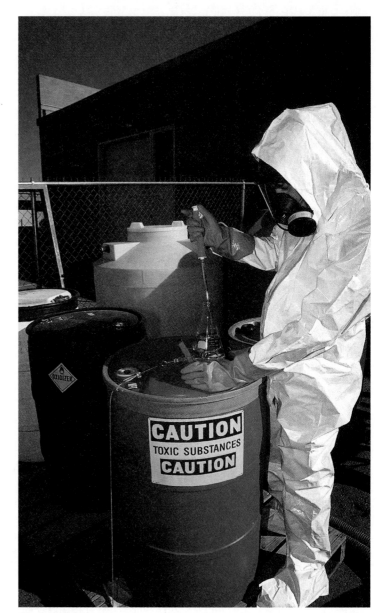

Figure 23.17

In many places, large quantities of hazardous wastes have been dumped in old warehouses and simply abandoned. Combinations of flammable materials and poisonous chemicals can make a lethal mix.

was quoted as saying: "There's millions to be made in this business. The overhead is nothing. You find a place—don't buy it, don't buy anything; just rent it and fill it up. When they catch up to you, just declare bankruptcy and get out."

The EPA estimates that in the past as much as 90 percent of the hazardous wastes handled by private contractors were disposed of by unsafe methods. So-called midnight dumpers use a variety of illegal and highly dangerous methods to get rid of unwanted toxic wastes. Sometimes liquid wastes are emptied at night into a storm sewer or river. Some drivers pick up a tanker full of wastes and simply drive down a deserted country road with the discharge valve open. "About 60 miles is all it takes to get rid of a load," boasted one driver, "and the only way I can get caught is if the windshield wipers or the tires of the car behind me start melting."

Options for Hazardous Waste Management

What shall we do with toxic and hazardous wastes? In our homes, we can reduce waste generation and choose less toxic materials. Buy only what you need for the job at hand. Use up the last little bit or share leftovers with a friend or neighbor. Many common materials that you probably already have make excellent alternatives to commercial products (table 23.2). Dispose of unneeded materials responsibly (fig. 23.18).

Produce Less Waste

As with other wastes, the safest and least expensive way to avoid hazardous waste problems is to avoid creating the wastes in the first place. Manufacturing processes can be modified to reduce or eliminate waste production. In Minnesota, the 3M Company reformulated products and redesigned manufacturing processes to eliminate more than 140,000 metric tons of solid and hazardous wastes, 4 billion l (1 billion gal) of wastewater, and 80,000 metric tons of air pollution each year. They frequently found that these new processes not only spared the environment but also saved money by using less energy and fewer raw materials.

Recycling and reusing materials also eliminates hazardous wastes and pollution. Many waste products of one process or industry are valuable commodities in another. Already, about 10 percent of the wastes that would otherwise enter the waste stream in the United States are sent to surplus material exchanges where they are sold as raw materials for use by other industries. This figure could probably be raised substantially with better waste management. In Europe, at least one-third of all industrial wastes are exchanged through clearinghouses where beneficial uses are found. This represents a double savings: the generator doesn't have to pay for disposal, and the recipient pays little, if anything, for raw materials.

Convert to Less Hazardous Substances

Several processes are available to make hazardous materials less toxic. *Physical treatments* tie up or isolate substances. Charcoal or resin filters absorb toxins. Distillation separates hazardous components from aqueous solutions. Precipitation and immobilization in ceramics, glass, or cement isolate toxins from the environment so that they become essentially nonhazardous. One of the few ways to dispose of metals and radioactive substances is to fuse them in silica at high temperatures to make a stable, impermeable glass that is suitable for long-term storage.

Incineration is applicable to mixtures of wastes. A permanent solution to many problems, it is quick and relatively easy but not necessarily cheap—nor always clean—unless it is done correctly. Wastes must be heated to over 1000° C (2000° F) for a sufficient period of time to complete destruction. The ash resulting from thorough incineration is reduced in volume up to 90 percent and often is safer to store in a landfill or other disposal site than the original wastes. Nevertheless, incineration remains a highly controversial topic (What Do *You* Think?, p. 533).

Table 23.2

Alternatives to Hazardous Household Chemicals

Chrome cleaner: Use vinegar and non-metallic scouring pad.

Copper cleaner: Rub with lemon juice and salt mixture.

Floor cleaner: Mop linoleum floors with 1 cup vinegar mixed with 2 gallons of water. Polish with club soda.

Brass polish: Use Worcestershire sauce.

Silver polish: Rub with toothpaste on a soft cloth.

Furniture polish: Rub in olive, almond, or lemon oil.

Ceramic tile cleaner: Mix 1/4 cup baking soda, 1/2 cup white vinegar, and 1 cup ammonia in 1 gallon warm water (good general purpose cleaner).

Drain opener: Use plunger or plumber's snake, pour boiling water down drain.

Upholstery cleaner: Clean stains with club soda.

Carpet shampoo: Mix 1/2 cup liquid detergent in 1 pint hot water. Whip into stiff foam with mixer. Apply to carpet with damp sponge. Rinse with 1 cup vinegar in 1 gal water. Don't soak carpet—it may mildew.

Window cleaner: Mix 1/3 cup ammonia, 1/4 cup white vinegar in 1 quart warm water. Spray on window. Wipe with soft cloth.

Spot remover: For butter, coffee, gravy, or chocolate stains: Sponge up or scrape off as much as possible immediately. Dab with cloth dampened with a solution of 1 teaspoon white vinegar in 1 quart cold water.

Toilet cleaner: Pour 1/2 cup liquid chlorine bleach into toilet bowl. Let stand for 30 minutes, scrub with brush, flush.

Pest control: Spray plants with soap-and-water solution (3 tablespoons soap per gallon water) for aphids, mealybugs, mites, and whiteflies. Interplant with pest repellent plants such as marigolds, coriander, thyme, yarrow, rue, and tansy. Introduce natural predators such as ladybugs or lacewings.

Indoor pests: Grind or blend 1 garlic clove and 1 onion. Add 1 tablespoon cayenne pepper and 1 quart water. Add 1 tablespoon liquid soap.

Moths: Use cedar chips or bay leaves.

Ants: Find where they are entering house, spread cream of tartar, cinnamon, red chili pepper, or perfume to block trail.

Fleas: Vacuum area, mix brewer's yeast with pet food.

Mosquitoes: Brewer's yeast tablets taken daily repel mosquitoes.

Note: test cleaners in small, inconspicuous area before using.

Several sophisticated features of modern incinerators improve their effectiveness. Liquid injection nozzles atomize liquids and mix air into the wastes so they burn thoroughly. Fluidized bed burners pump air from the bottom up through burning solid waste as it travels on a metal chain grate through the furnace. The air velocity

*I*n Depth: Environmental Justice

Who do you suppose lives closest to toxic waste dumps, Superfund sites, or other polluted areas in your city or county? If you answered poor people and minorities, you are probably right. Everyday experiences tell us that minority neighborhoods are much more likely to have high pollution levels and unpopular industrial facilities such as toxic waste dumps, landfills, smelters, refineries and incinerators than are white, middle, or upper class neighborhoods.

One of the first systematic studies showing this inequitable distribution of environmental hazards based on race in the United States was conducted by Robert D. Bullard in 1978. Asked for help by a predominantly black community in Houston that was slated for a waste incinerator, Bullard discovered that all five of the city's existing landfills and six of eight incinerators were located in African-American neighborhoods. In a book entitled *Dumping on Dixie,* Bullard showed that this pattern of risk exposure in minority communities is common throughout the United States.

In 1987, the Commission for Racial Justice of the United Church of Christ published an extensive study of environmental racism. Its conclusion was that race is the most significant variable in determining the location of toxic waste sites in the United States. Among the findings of this study are:

- 3 of the 5 largest commercial hazardous waste landfills accounting for about 40 percent of all hazardous waste disposal in the United States are located in predominantly Black or Hispanic communities.

- 60 percent of African-Americans and Latinos and nearly half of all Asians, Pacific Islanders, and Native Americans live in communities with uncontrolled toxic waste sites (fig. 1).

- The average percentage of the population made up by minorities in communities without a hazardous waste facility is 12 percent. By contrast, communities with one hazardous waste facility have, on average, twice as high (24 percent) a minority population, while those with two or more such facilities

Figure 1

Native Americans march in protest of toxic waste dumping on tribal lands.

is sufficient to keep the burning waste partially suspended. Plenty of oxygen is available, and burning is quick and complete. Afterburners add to the completeness of burning by igniting gaseous hydrocarbons not consumed in the incinerator. Scrubbers and precipitators remove minerals, particulates, and other pollutants from the stack gases.

Chemical processing can transform materials so they become nontoxic. Included in this category are neutralization, removal of metals or halogens (chlorine, bromine, etc.), and oxidation. The Sunohio Corporation of Canton, Ohio, for instance, has developed a process called PCBx in which chlorine in such molecules as PCBs is replaced with other ions that render the

compounds less toxic. A portable unit can be moved to the location of the hazardous waste, eliminating the need for shipping them.

Biological waste treatment taps the great capacity of biological organisms to absorb, accumulate, and detoxify a variety of toxic compounds. Bacteria in activated sludge ponds, aquatic plants (such as water hyacinths or cattails), soil microorganisms, and other species remove toxic materials and purify effluents. Biotechnology offers exciting possibilities for finding or creating organisms to eliminate specific kinds of hazardous or toxic wastes. By using a combination of classic genetic selection techniques and high-technology gene-transfer techniques, for instance, scientists

average three times as high a minority population (38 percent) as those without one.

- The "dirtiest" or most polluted zip codes in California are in riot-torn South Central Los Angeles where the population is predominantly African-American or Latino. Three-quarters of all blacks and half of all Hispanics in Los Angeles live in these polluted areas, while only one-third of all whites live there.

Race, not class or income, is the strongest determinant of who is exposed to environmental hazards. While poor people in general are more likely to live in polluted neighborhoods than rich people, the discrepancy between the pollution exposure of middle class blacks and middle class whites is even greater than the difference between poorer whites and blacks. Where upper class whites can "vote with their feet" and move out of polluted and dangerous neighborhoods, minorities are restricted by color barriers and prejudice to less desirable locations.

Racial inequities also are revealed in the way the government cleans up toxic waste sites and punishes polluters (fig.2). White communities see faster responses and get better results once toxic wastes are discovered than do minority communities. For instance, the federal Environmental Protection Agency takes 20 percent longer to

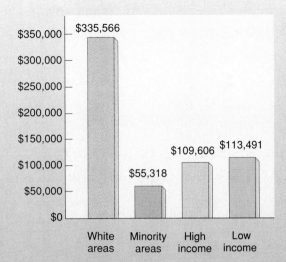

Figure 2

Hazardous waste law enforcement. The average fines or penalties imposed *per site* for violation of the Resource Conservation and Recovery Act vary dramatically with racial composition—but not wealth—of the communities where waste was dumped.

Source: M. LaVelle and M. Coyle, *The National Law Journal*, Vol. 15:52–56, No. 3, Sept. 21, 1992.

place a hazardous waste site in a minority community on the Superfund National Priority List than it does for one in a white community. Penalties assessed against polluters of white communities average six times higher than those against polluters of minority communities. Clean-up is more thorough in white communities as well. Most toxic waste in white communities are treated, that is, removed or destroyed. By contrast, waste sites in minority neighborhoods are generally only "contained" by putting a cap over them, leaving contaminants in place to potentially resurface or leak into groundwater at a later date. The growing environmental justice movement works to combine civil rights and social justice with environmental concerns to call for a decent, livable environment and equal environmental protection for everyone.

ETHICAL CONSIDERATIONS

Where do you stand on this issue? Suppose that a minority family is planning to move into your neighborhood. Would you welcome or shun them? Or suppose that the minority community claims they have been subjected to more than their fair share of pollution and toxic waste. Would you support a proposal to build the next toxic waste facility or incinerator in your neighborhood as a matter of fairness?

---

have recently been able to generate bacterial strains that are highly successful at metabolizing PCBs. There are concerns about releasing such exotic organisms into the environment, however (chapter 13). It may be better to keep these organisms contained in enclosed reaction vessels and feed contaminated material to them under controlled conditions.

Store Permanently

Inevitably, there will be some materials that we can't destroy, make into something else, or otherwise cause to vanish. We will have to store them out of harm's way. There are differing opinions about how best to do this.

Retrievable Storage. Dumping wastes in the ocean or burying them in the ground generally means that we have lost control of them. If we learn later that our disposal technique was a mistake, it is difficult, if not impossible, to go back and recover the wastes. For many supertoxic materials, the best way to store them may be in **permanent retrievable storage.** This means placing waste storage containers in a secure building, salt mine, or bedrock cavern where they can be inspected periodically and retrieved, if necessary, for repacking or for transfer if a better means of disposal is developed. This technique is more expensive than burial in a landfill because the storage area must be guarded and monitored continuously to prevent leakage, vandalism, or other dispersal of toxic

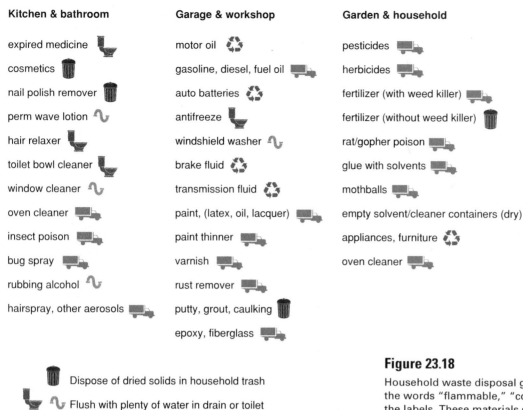

Kitchen & bathroom

expired medicine

cosmetics

nail polish remover

perm wave lotion

hair relaxer

toilet bowl cleaner

window cleaner

oven cleaner

insect poison

bug spray

rubbing alcohol

hairspray, other aerosols

Garage & workshop

motor oil

gasoline, diesel, fuel oil

auto batteries

antifreeze

windshield washer

brake fluid

transmission fluid

paint, (latex, oil, lacquer)

paint thinner

varnish

rust remover

putty, grout, caulking

epoxy, fiberglass

Garden & household

pesticides

herbicides

fertilizer (with weed killer)

fertilizer (without weed killer)

rat/gopher poison

glue with solvents

mothballs

empty solvent/cleaner containers (dry)

appliances, furniture

oven cleaner

Dispose of dried solids in household trash

Flush with plenty of water in drain or toilet

Save and bring to a hazardous waste collection site

Recycle

Figure 23.18

Household waste disposal guide. Solvent-containing products have the words "flammable," "combustible," or "petroleum distillates" on the labels. These materials should *not* be disposed of in a drain or toilet. Never mix products containing bleach with those containing ammonia. A toxic gas can form!

materials. Remedial measures are much cheaper with this technique, however, and it may be the best system in the long run.

Secure Landfills. One of the most popular solutions for hazardous waste disposal has been landfilling. Although, as we saw earlier in this chapter, many such landfills have been environmental disasters, newer techniques make it possible to create safe, modern **secure landfills** that are acceptable for disposing of many hazardous wastes. The first line of defense in a secure landfill is a thick bottom cushion of compacted clay that surrounds the pit like a bathtub (fig. 23.19). Moist clay is flexible and resists cracking if the ground shifts. It is impermeable to groundwater and will safely contain wastes. A layer of gravel is spread over the clay liner and perforated drain pipes are laid in a grid to collect any seepage that escapes from the stored material. A thick polyethylene liner, protected from punctures by soft padding materials, covers the gravel bed. A layer of soil or absorbent sand cushions the inner liner and the wastes are packed in drums, which then are placed into the pit, separated into small units by thick berms of soil or packing material.

When the landfill has reached its maximum capacity, a cover much like the bottom sandwich of clay, plastic, and soil—in that order—caps the site. Vegetation stabilizes the surface and improves its appearance. Sump pumps collect any liquids that filter through the landfill, either from rainwater or leaking drums. This water is treated and purified before being released. Monitoring wells check groundwater around the site to ensure that no toxins have escaped.

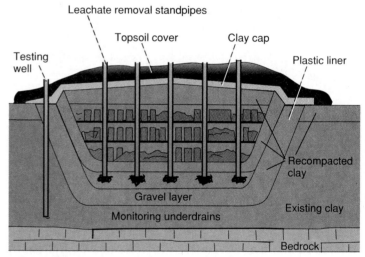

Figure 23.19

A secure landfill for toxic waste. Contents are enclosed by a thick plastic liner and two or more layers of impervious compacted clay. A gravel bed between the clay layers collects any leachate, which can then be pumped out and treated. Testing wells sample for escaping contaminants.

What Do You Think?
Incinerating Hazardous Wastes

Managing toxic chemical wastes is among the most serious environmental challenges facing industrial countries. In the United States alone, more than a ton of such wastes are produced per person each year. How to manage this material has been the subject of intense debate across the country.

Recently there has been a move toward greater use of incinerators to dispose of these dangerous chemicals. This has drawn heated opposition from many environmentalists. The recent battle over Waste Technologies Industries' new facility in East Livermore, Ohio has been perhaps the most widely publicized of these fights (fig. 1). All agree that these substances pose risks to human health and the environment, and that past disposal practices have contaminated groundwater and exposed unacceptably large numbers of people to toxins. So what is the controversy all about?

Disagreements often involve disputes over facts, disputes over interpretations, and simply differences of opinion. As you read the arguments below, try to identify the specific matters at issue.

Environmentalists point out that incineration fails to destroy all the toxic materials and that the residues escape into the environment through stack gases. They also point out that burning certain kinds of wastes will unavoidably create dioxins as by-products. Because dioxins are highly carcinogenic, environmentalists feel we should take considerable effort to keep them from being released. Incineration opponents are also concerned about the significant number of other totally new products of incomplete combustion (PICs) that will inevitably be formed and released. Opponents are concerned that these PICs are not even considered when the effectiveness of incinerators is determined.

Proponents of incineration counter that burning physically destroys a much higher percentage of the toxic material than any other treatment technology, and that this is highly preferable to having to store the material somewhere. They point to EPA regulations that require a reduction of 99.99 percent of the toxic material before an incineration facility can be permitted. Further, they point out that the dioxin removal requirement is 99.9999 percent. As for PICs, proponents argue that when incineration facilities are well run, release of these substances is at such low levels that health risks are minimal. Incineration opponents feel that the facilities do not actually work as effectively as EPA assumes they do.

Incineration opponents fear that widespread adoption of incineration will destroy incentives for industry to reuse and recycle toxic wastes. Incineration advocates agree that reuse and recycling are preferable to disposal, but contend that because incineration costs so much, industry will still be motivated to seek out these alternatives. They also claim that there will always be certain kinds of wastes that simply are not reusable or recyclable.

Finally, environmentalists contend that EPA underestimates the total health risk posed by incinerators. They point out that once released, the toxic materials enter the food chain and, when humans eat food and drink milk produced in the surrounding area, they add to the exposure from the air alone. In order to reduce this risk, opponents argue that allowable stack emission rates must be reduced to lower levels than currently in the rules.

When people disagree on factual matters, it is easy to understand how they can formulate different premises and come to different conclusions. Issues of fact ought to be resolvable. On which issues of fact do the two sides disagree? What procedures could be used to clarify these facts?

Opinions are positions taken that are not based on facts. Disputes involving differences of opinion can be less easily resolved. What aspects of the incineration debate seem opinion-centered instead of fact-centered? How do you think opinion-based disagreements are best resolved?

Figure 1

Actor Martin Sheen joins local activists in a protest in East Liverpool, OH, site of the largest hazardous waste incinerator in the United States. About 1000 people marched to the plant to pray, sing, and express their opposition. Involving celebrities draws attention to your cause. A peaceful, well-planned rally builds support and acceptance in the broader community.

Solid, Toxic, and Hazardous Waste

Most landfills are buried below ground level to be less conspicuous; however, in areas where the groundwater table is close to the surface, it is safer to build above-ground storage. The same protective construction techniques are used as in a buried pit. An advantage to such a facility is that leakage is easier to monitor because the bottom is at ground level.

Transportation of hazardous wastes to disposal sites is of concern because of the risk of accidents. Emergency preparedness officials conclude that the greatest risk in most urban areas is not nuclear war or natural disaster but crashes involving trucks or trains carrying hazardous chemicals through densely packed urban corridors. Another worry is who will bear financial responsibility for abandoned waste sites. The material remains toxic long after the businesses that created it are gone. As is the case with nuclear wastes (chapter 21), we may need new institutions for perpetual care of these wastes.

Summary

We produce enormous volumes of solid waste in industrialized societies, and there is an increasing problem of how to dispose of those wastes in an environmentally safe manner. In this chapter, we have looked at the character of our solid and hazardous wastes. We have surveyed the ways we dispose of our wastes and the environmental problems associated with waste disposal.

Solid wastes are domestic, commercial, industrial, agricultural, and mining wastes that are primarily nontoxic. About 80 percent of our domestic and industrial wastes are deposited in landfills; most of the rest is incinerated or recycled. Landfills are often messy and leaky, but they can be improved with impermeable clay or plastic linings, drainage, and careful siting. Incineration can destroy organic compounds, but whether incinerators can or will be operated satisfactorily is a matter of debate. Recycling is growing nationwide, encouraged by the economic and environmental benefits it brings. City leaders tend to doubt the viability of recycling programs, but successful programs have been sustained in other countries and in some American cities.

Near major urban centers, land suitable for waste disposal is becoming increasingly scarce and expensive. Costs for solid waste disposal totaled nearly $10 billion in the United States in 1990 and are expected to climb to $100 billion per year by the end of this century. A few cities now ship their refuse to other states or even other countries, but worries about toxic and hazardous material in the waste are leading to increasing resistance to shipping or storing it.

Hazardous and toxic wastes, when released into the environment, cause such health problems as birth defects, neurological disorders, reduced resistance to infection, and cancer. Environmental losses include contamination of water supplies, poisoning of the soil, and destruction of habitat. The major categories of hazardous wastes are ignitable, corrosive, reactive, explosive, and toxic. About 60 million metric tons of hazardous waste enter the waste stream each year in the United States, more than 90 percent of which is disposed of by environmentally unsound practices. Some materials that cause the most concern are heavy metals, solvents, and synthetic organic chemicals such as halogenated hydrocarbons, organophosphates, and phenoxy herbicides.

Disposal practices for solid and hazardous wastes have often been unsatisfactory. Thousands of abandoned, often unknown waste disposal sites are leaking toxic materials into the environment. Some alternative techniques for treating or disposing of hazardous wastes include not making the material in the first place, incineration, secure landfill, and physical, chemical, or biological treatment to detoxify or immobilize wastes. People are often unwilling to have transfer facilities, storage sites, disposal operations, or transportation of hazardous or toxic materials in or through their cities. Questions of safety and liability remain unanswered in solid and hazardous waste disposal.

▼ Questions for Review

1. What are solid wastes and hazardous wastes? What is the difference between them?

2. How much solid and hazardous waste do we produce each year in the United States? How do we dispose of the waste?

3. Why are landfill sites becoming rare around most major urban centers in the United States? What steps are being taken to solve this problem?

4. Describe some concerns about waste incineration.

5. List some benefits and drawbacks of recycling wastes. What are the major types of materials recycled from municipal waste and how are they used?

6. What is composting, and how does it fit into solid waste disposal?

7. Describe some ways that we can reduce the waste stream to avoid or reduce disposal problems.

8. List ten toxic substances in your home and how you would dispose of them.

9. What are some illegal hazardous waste disposal methods, and why are they dangerous?

10. What societal problems are associated with waste disposal? Why do people object to waste handling in their neighborhoods?

▼ Questions for Critical Thinking

1. A toxic waste disposal site has been proposed for the Pine Ridge Indian Reservation in South Dakota. Many tribal members oppose this plan, but some favor it because of the jobs and income it will bring to an area with 70 percent unemployment. If local people choose immediate survival over long-term health, should we object or intervene?

2. There is often a tension between getting your personal life in order and working for larger structural changes in society. Evaluate the trade-offs between spending time and energy sorting recyclables at home compared to working in the public arena on a bill to ban excess packaging.

3. Should industry officials be held responsible for dumping chemicals that were legal when they did it but are now known to be extremely dangerous? At what point can we argue that they *should* have known about the hazards involved?

4. Look at the discussion of recycling or incineration presented in this chapter. List the premises (implicit or explicit) that underlie the presentation as well as the conclusions (stated or not) that seem to be drawn from them. Do the conclusions necessarily follow from these premises?

5. Suppose that your brother or sister has decided to buy a house in Love Canal because it costs $20,000 less than a comparable house elsewhere. What do you say to him or her?

6. Is there an overall conceptual framework or point of view in this chapter? If you were presenting a discussion of solid or hazardous waste to your class, what would be your conceptual framework?

7. Is there a fundamental difference between incinerating municipal, medical, or toxic industrial waste? Would you oppose an incinerator for one type of waste in your neighborhood but not others? Why, or why not?

8. The Netherlands incinerates much of its toxic waste at sea by a shipborne incinerator. Would you support this as a way to dispose of our wastes as well? What are the critical considerations for or against this approach?

▼ Key Terms

biodegradable plastics   525
composting   522
energy recovery   519
hazardous waste   525
landfills   518
mass burn   519

permanent retrievable storage   531
photodegradable plastics   525
recycling   520
refuse-derived fuel   519
secure landfills   532
waste stream   516

▼ Additional Information on the Internet

Basel Convention on the Control of Transboundary Movements of Hazardous Wastes and Their Disposal
http://www.unep.ch/sbc.html/

Center for Nuclear and Toxic Waste Management
http://cnwm.berkeley.edu/cnwm/cnwm.html/

Cornell Composting
http://www.cals.cornell.edu/dept/compost/

EcoWatch
http://www.cadvision.com/Home_Pages/accounts/vut/ecowatch.htm

European Recycling and the Environment
http://www.tecweb.com/recycle/eurorec.htm

Global Recycling Network
http://grn.com/grn/

Great Plains/Rocky Mountain Hazardous Substance Research Center
http://www.engg.ksu.edu/HSRC/

Hazardous Materials Management
http://www.io.org/~hzmatmg/

Lawrence Livermore National Laboratory Environmental Technologies Program
http://www-ep.es.llnl.gov/www-ep/aet.html/

Preserving Resources through Integrated Sustainable Management of Wastes
http://www.wrfound.org.uk/

Recycler's World
http://granite.sentex.net:80/recycle/

Waste Management Education and Research Consortium
http://www.nmsu.edu:80/~werc/

Note: Further readings appropriate to this chapter are listed on p. 606.

The approach to downtown Miami, FL, is dominated by a maze of freeways.

CHAPTER 24: Urbanization and Sustainable Cities

Objectives

After studying this chapter, you should be able to:

- distinguish between a rural village, a city, and a megacity.

- recognize the push and pull factors that lead to urban growth.

- appreciate the growth rate of giant metropolitan urban areas such as Mexico City, as well as the problems this growth engenders.

- visualize living conditions for ordinary citizens in the megacities of the developing world.

- understand the causes and consequences of noise and crowding in cities.

- critique options for suburban design.

- see the connection between sustainable economic development, social justice, and the solution of urban problems.

Pneumonic Plague in Surat

In September, 1994, panicky residents of the Indian city of Surat fought to get on trains going through town, while porters and passengers already on board barricaded doors and windows and fought back the stampede with fists and knives. What were the people of Surat trying so desperately to escape from? And why were those on the trains so dead set against them? The answer is that a dreaded disease had appeared in Surat that hadn't been seen for centuries in most places. Pneumonic plague, a more virulent relative of bubonic plague, was sweeping through the slums lining the banks of the muddy Tapti River. Victims usually would die within hours of showing the first symptoms. The official death toll from this epidemic was 120 people, but the real number was probably several times higher.

Why had this terrible killer suddenly emerged? We may never know exactly, but poverty, rapid urbanization, and the horrible environmental conditions in the slums of Surat all played a role. A short distance north of Bombay, Surat had undergone an

industrial boom during the previous 20 years. The urban population surged to more than 2 million, more than half of whom live in shantytowns made of cardboard, plastic, flattened oil drums, or palm branches. Others crowd together in shabby hostels, sleeping 15 to 20 in a room. The shantytowns have no sewers or running water, and the teeming hostels are not much better. Nearly a million people living in extremely high densities have no place to defecate except in filthy alleys or along the river banks where many also bathe and get their drinking water. Until the plague struck, there was no garbage removal in the slums. Rats, fleas, and flies swarmed around rotting refuse helping to spread diseases. Poor nutrition and clouds of air pollution from factories, smoky buses, and burning garbage weakened people's natural immunity.

Embarrassed by sudden world publicity, city officials removed more than 6000 tons of garbage from Surat's slums and dispatched an army of rat catchers in an attempt to bring the

epidemic under control. They were less able to provide safe housing, food, sanitation, or better jobs for the residents. Far from being an exception, unfortunately, the terrible urban conditions in Surat are becoming common experiences for an increasing number of people in developing countries (fig. 24.1).

Nearly half the people in the world now live in urban areas. Demographers predict that by the end of the twenty-first century, 80 or 90 percent of all humans will live in cities and that some giant interconnecting metropolitan areas could have hundreds of millions of residents. In this chapter, we will look at how cities came into existence, why people live there, and what the environmental conditions of cities have been, are now, and might be in the future. Some of the most severe urban problems in the world are found in the giant megacities of the developing countries. Far more lives may be threatened by the desperate environmental conditions in these cities than by any other issue we have studied. We also will look at a few of those problems and possible solutions that could improve the urban environment. ▶

Figure 24.1

More than half of all city dwellers in Indonesia live in fetid slums and shantytowns like this one in Jakarta. Appalling conditions like this now affect more than one billion people worldwide.

Urbanization

Since their earliest origins, cities have been centers of education, religion, commerce, record keeping, communication, and political power. As cradles of civilization, cities have influenced culture and society far beyond their proportion of the total population (fig. 24.2). Until recently, however, only a small percentage of the world's people lived permanently in urban areas and even the greatest cities of antiquity were small by modern standards. The vast majority of humanity has always lived in rural areas where farming, fishing, hunting, timber harvesting, animal herding, mining, or other natural resource-based occupations provided support.

Since the beginning of the Industrial Revolution some three hundred years ago, however, cities have grown rapidly in both size and power. In every developing country, the transition from an agrarian society to an industrial one has been accompanied by **urbanization,** an increasing concentration of the population in cities and a transformation of land use and society to a metropolitan pattern of organization. Industrialization and urbanization bring many benefits—especially to the top members of society—but they also cause many problems, as we will detail in this chapter.

What Is a City?

Just what makes up an urban area or a city? Definitions differ. The U.S. Census Bureau considers any incorporated community to be

Figure 24.2

Since their earliest origins, cities have been centers of education, religion, commerce, politics, and culture. Unfortunately, they have also been sources of pollution, crowding, disease, and misery. The Parthenon, one of the most beautiful buildings in the world, is being eaten away by air pollution, vibrations, and the other urban ills of Athens, Greece.

a city, regardless of size, and defines any city with more than 2500 residents as urban. More meaningful definitions are based on *functions*. In a **rural area,** most residents depend on agriculture or other ways of harvesting natural resources for their livelihood. In an **urban area,** by contrast, a majority of the people are not directly dependent on natural resource-based occupations.

A **village** is a collection of rural households linked by culture, custom, family ties, and association with the land (fig. 24.3). A **city,** by contrast, is a differentiated community with a population and resource base large enough to allow residents to specialize in arts, crafts, services, or professions rather than natural resource-based occupations. While the rural village often has a sense of security and connection, it also can be stifling. A city offers more freedom to experiment, to be upwardly mobile, and to break from restrictive traditions, but it can be harsh and impersonal (fig. 24.4).

Beyond about 10 million inhabitants, an urban area is considered a supercity or **megacity.** Megacities in many parts of the world have grown to enormous size. In the United States, urban areas between Boston and Washington, DC, have merged into a nearly continuous megacity (sometimes called Bos-Wash) containing about 35 million people. The Tokyo-Yokohama-Osaka-Kobe corridor contains nearly 50 million people. The Paris Basin (or Ille de France) encompasses nearly half of that country. Because these agglomerations have expanded beyond what we normally think of as a city, some geographers prefer to think of urbanized **core regions** that dominate the social, political, and economic life of most countries (fig. 24.5). Architect and city planner C. A. Doxiadis predicts

Figure 24.3

A traditional hill-tribe village in northern Thailand is closely tied to the land through culture, economics, and family relationships. While the timeless pattern of life here gives a great sense of identity, it can also be stifling and repressive.

that as core regions of adjacent countries merge, the sea coasts and major river valleys of most continents will be covered with continuous strip cities, which he calls ecumenopolises, each containing billions of people.

Figure 24.4

A city is a differentiated community with a large enough population and resource base to allow specialization in arts, crafts, services, and professions. Although there are many disadvantages in living so closely together, there are also many advantages. Cities are growing rapidly, and most of the world's population will live in urban areas if present trends continue.

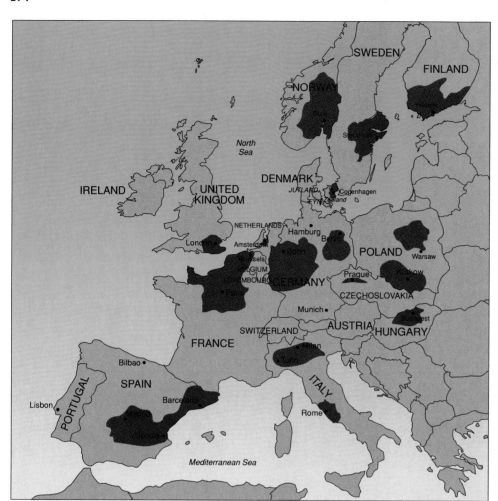

Figure 24.5

Map of core regions or major metropolitan conglomerations of Europe. Europe is so highly urbanized and densely populated that the Paris-Brussels-Amsterdam-Berlin-Prague-Kraków-Warsaw axis is on the verge of becoming a single giant megacity.

World Urbanization

The United States underwent a dramatic rural-to-urban shift in the nineteenth and early twentieth centuries (fig. 24.6). Now many developing countries are experiencing a similar demographic movement. In 1850, only about 2 percent of the world's population lived in cities. By 1995, 45 percent of the people in the world were urban. By the end of this century, for the first time in history, more people will live in cities than in the country. Only Africa and South Asia remain predominantly rural, but people there are swarming into cities in ever-increasing numbers. About three-fourths of the people in Europe, North America and Latin America already live in cities (table 24.1). Some urbanologists predict that by 2100, the whole world will be urbanized to the levels now seen in developed countries.

As figure 24.7 shows, 90 percent of the population growth over the next fifty years is expected to occur in the less-developed countries of the world. Three-quarters of that growth will be in the already overcrowded cities of the least affluent countries, such as India, China, Mexico, and Brazil. The combined population of these cities is expected to jump from its present 1 billion to nearly 4 billion by the year 2025. Meanwhile, rural populations in these countries are expected to decline somewhat as rural people migrate into the cities.

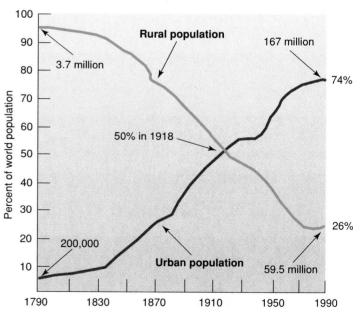

Figure 24.6

United States rural-to-urban shift. Percentages of rural and urban populations, 1790 to 1990.

Source: U.S. Bureau of the Census.

Table 24.1
Urban share of total population (percent)

Region	1950	1994	2025
Asia	29	34	55
Africa	15	34	54
Europe	56	73	83
Former USSR	39	69	78
Latin America	41	74	85
North America	64	76	85
Oceania	61	70	75
World	29	45	61

Source: United Nations, *Prospects of World Urbanization,* 1994.

Table 24.2
Top ten metropolitan regions (populations in millions)

1900		1950		2000 (estimated)	
London	6.6	New York	12.3	Mexico City	25.6
New York	4.2	London	8.7	São Paulo	22.1
Paris	3.3	Tokyo	6.7	Tokyo	19.0
Berlin	2.4	Paris	5.4	Shanghai	17.0
Chicago	1.7	Shanghai	5.3	New York	16.8
Vienna	1.6	Buenos Aires	5.0	Calcutta	15.7
Tokyo	1.5	Chicago	4.9	Bombay	15.4
St. Petersburg	1.4	Moscow	4.8	Beijing	14.0
Philadelphia	1.4	Calcutta	4.4	Los Angeles	13.9
Manchester	1.3	Los Angeles	4.0	Jakarta	13.7

Source: United Nations, *State of the World Population 1993.*

Recent urban growth has been particularly dramatic in the largest cities, especially those of the developing world. In 1900, thirteen cities had populations over 1 million; all except Tokyo were in Europe or North America (table 24.2). By 2000, only three of the ten largest cities will be in developed countries. By 1995, there were 235 metropolitan areas of more than 1 million people. A century ago, London was the only city with more than 5 million people; now nineteen cities have populations larger than that. Some futurists predict that by 2025 at least four hundred cities will have populations of 1 million or more, and ninety-three supercities each will have 5 million or more residents. Three-fourths of those cities will be in the developing nations (fig. 24.8).

The growth rate of the most rapidly expanding cities and the sizes they are predicted to reach are truly astounding (table 24.2). Many cities are growing at rates above 5 percent per year, which means they will double in less than fifteen years. Mexico City, for instance, with a population of about 2.45 million in 1950, was the largest city in the world in 1995, with more than 20 million residents. Now growing at a rate of about two thousand people per day (fig. 24.9), it is expected to exceed 30 million by the year 2000, an elevenfold increase in fifty years. If that growth trend continues, Mexico City could have 100 million inhabitants by the middle of the next century.

Can a city function with that many people? Can it supply food, water, sanitation, transportation, jobs, housing, police protection, fire control, electricity, education, and other public services necessary to sustain a civilized way of life? Adding 750,000 new people annually to Mexico City amounts to building a new city the size of Baltimore or San Francisco every year. This growth is occurring, as is most urban growth in the world, in a country with a sagging economy, an unstable government, and a high foreign debt load. The environmental costs (air and water pollution, soil erosion,

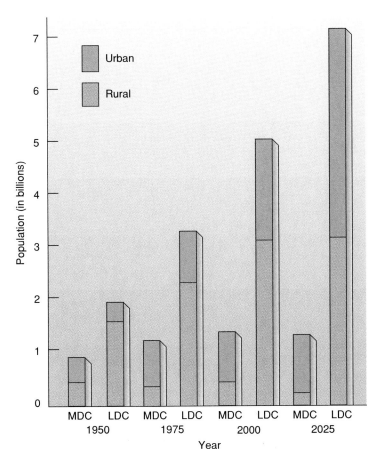

Figure 24.7

Distribution of urban and rural population in more-developed (MDC) and less-developed (LDC) countries, 1950–2025. Over the next 30 years, more than 90% of all urban growth will take place in the less developed countries.

Source: Data from *Prospects of World Urbanization,* 1994, United Nations.

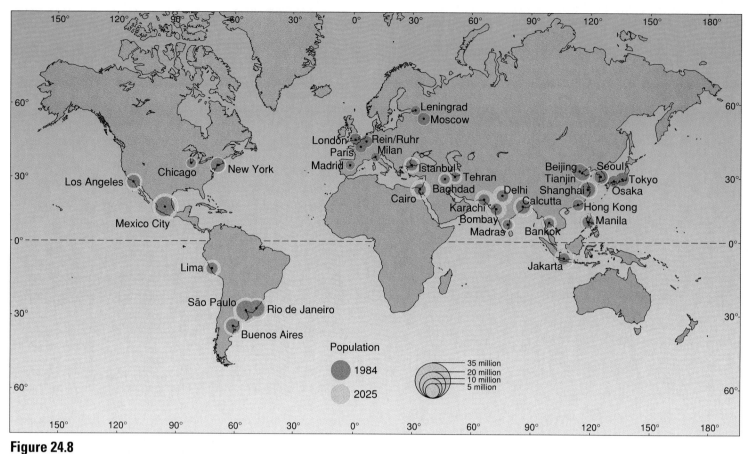

Figure 24.8

By 2025, at least 400 cities will have populations of 1 million or more, and 93 supercities will have populations above 5 million. Three-fourths of the world's largest cities will be in developing countries that already have trouble housing, feeding, and employing their people.

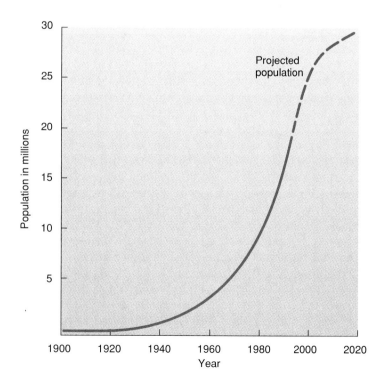

solid and hazardous waste accumulation, noise and congestion) as well as the human costs (lack of housing, jobs, transportation, health care, social services, and living space) are tragic.

Causes of Urban Growth

Urban populations grow in two ways: by natural increase (more births than deaths) and by immigration. Natural increase is fueled by improved food supplies, better sanitation, and advances in medical care that reduce death rates and cause populations to grow both within cities and in the rural areas around them (chapter 7). In Latin America and East Asia, natural increase is responsible for two-thirds of urban population growth. In Africa and West Asia, immigration is the largest source of urban growth. Immigration to cities can be caused both by **push factors** that force people out of the country and by **pull factors** that draw them into the city.

Figure 24.9

Growth of Mexico City metropolitan area, 1900–2020.

Source: U.N. Population Council.

Immigration Push Factors

People migrate to cities for many reasons. In some areas, the countryside is overpopulated and simply can't support more people. The "surplus" population is forced to migrate to cities in search of jobs, food, and housing. Not all rural-to-urban shifts are caused by overcrowding in the country, however. In some places, economic forces or political, racial, or religious conflicts drive people out of their homes. The countryside may actually be depopulated by such demographic shifts. The United Nations estimated that in 1992 at least 10 million people fled their native country and that another 30 or 40 million were internal refugees within their own country, displaced by political, economic, or social instability. Many of these refugees end up in the already overcrowded megacities of the developing world.

Land tenure patterns and changes in agriculture also play a role in pushing people into cities. The same pattern of agricultural mechanization that made farm labor obsolete in the United States early in this century is spreading now to developing countries. Furthermore, where land ownership is concentrated in the hands of a wealthy elite, subsistence farmers are often forced off the land so it can be converted to grazing lands or monoculture cash crops (chapter 10). Speculators and absentee landlords let good farmland sit idle that otherwise might house and feed rural families.

Immigration Pull Factors

Even in the largest and most hectic cities, many people are there by choice, attracted by the excitement, vitality, and opportunity to meet others like themselves. Cities offer jobs, housing, entertainment, and freedom from the constraints of village traditions. Possibilities exist in the city for upward social mobility, prestige, and power not ordinarily available in the country. Cities support specialization in arts, crafts, and professions for which markets don't exist elsewhere.

Modern communications also draw people to cities by broadcasting images of luxury and opportunity. An estimated 90 percent of the people in Egypt, for instance, have access to a television set. The immediacy of television makes city life seem more familiar and attainable than ever before. We see pictures of beggars and homeless people on the streets of teeming Third World cities and generally assume they have no other choice, but many of these people want to be in the city. In spite of what appears to be dismal conditions, living in the city may be preferable to what the country had to offer.

Government Policies

Government policies often favor urban over rural areas in ways that both push and pull people into the cities. Developing countries commonly spend most of their budgets on improving urban areas (especially around the capital city where leaders live), even though only a small percentage of the population lives there or benefits directly from the investment. This gives the major cities a virtual monopoly on new jobs, housing, education, and opportunities, all of which bring in rural people searching for a better life. In Peru, for example, Lima accounts for 20 percent of the country's population, but has 50 percent of the national wealth, 60 percent of the manufacturing, 65 percent of the retail trade, 73 percent of the industrial

wages, and 90 percent of all banking in the country. Similar statistics pertain to São Paulo, Mexico City, Manila, Cairo, Lagos, Bogota, and a host of other cities.

Governments often manipulate exchange rates and food prices for the benefit of more politically powerful urban populations but at the expense of rural people. Importing lower-priced food pleases city residents, but local farmers then find it uneconomical to grow crops. As a result, an increased number of people leave rural areas to become part of a large urban workforce, keeping wages down and industrial production high. Zambia, for instance, sets maize prices below the cost of local production to discourage farming and to maintain a large pool of workers for the mines. Keeping the currency exchange rate high stimulates export trade but makes it difficult for small farmers to buy the fuels, machinery, fertilizers, and seeds that they need. This depresses rural employment and rural income while stimulating the urban economy. The effect is to transfer wealth from the country to the city.

Current Urban Problems

Large cities in both developed and developing countries face similar challenges in accommodating the needs and by-products of dense populations. The problems are most intense, however, in rapidly growing cities of developing nations.

The Developing World

As figure 24.7 shows, 90 percent of the human population growth in the next century is expected to occur in the developing world, mainly in Africa, Asia, and South America. Almost all of that growth will occur in cities—especially the largest cities—which already have trouble supplying food, water, housing, jobs, and basic services for their residents. The unplanned and uncontrollable growth of those cities causes tragic urban environmental problems. You may have noticed that nearly every chapter in this book has a boxed reading presenting an ethical dilemma. This whole chapter is a dilemma of how people can survive under the conditions found in the megacities of the world. Consider as you read this what responsibilities we in the richer countries have to help others who are less fortunate and how we might do so.

Traffic and Congestion

A first-time visitor to a supercity—particularly in a less-developed country—is often overwhelmed by the immense crush of pedestrians and vehicles of all sorts that clog the streets. The noise, congestion, and confusion of traffic make it seem suicidal to venture onto the street. Jakarta, for instance, is one of the most densely populated cities in the world (fig. 24.10). Traffic is chaotic almost all the time. People commonly spend three or four hours each way commuting to work from outlying areas. Calcutta also has monumental traffic problems. The Howrah Bridge over the Hooghly River carries 30,000 motor vehicles and 500,000 pedestrians daily. Gigantic traffic jams occur day and night and there is almost always a wait to cross over.

Figure 24.10
Motorized rickshaws, motor scooters, bicycles, street vendors, and pedestrians all vie for space on the crowded streets of Jakarta. The heat, noise, smells, and sights are overpowering. In spite of the difficulties of living here, people work hard and have hope for the future.

Air Pollution

The dense traffic (commonly old, poorly maintained vehicles), smoky factories, and use of wood or coal fires for cooking and heating often create a thick pall of air pollution in the world's supercities. Lenient pollution laws, corrupt officials, inadequate testing equipment, ignorance about the sources and effects of pollution, and lack of funds to correct dangerous situations usually exacerbate the problem. What is its human toll? An estimated 60 percent of Calcutta's residents are thought to suffer from respiratory diseases linked to air pollution. Lung cancer mortality in Shanghai is reported to be four to seven times higher than rates in the countryside. Mexico City, which sits in a high mountain bowl with abundant sunshine, little rain, high traffic levels, and frequent air stagnation, has one of the highest levels of photochemical smog (chapter 18) in the world.

Sewer Systems and Water Pollution

Few cities in developing countries can afford to build modern waste treatment systems for their rapidly growing populations. The World Bank estimates that only 35 percent of urban residents in developing countries have satisfactory sanitation services. The situation is especially desperate in Latin America, where only 2 percent of urban sewage receives any treatment. In Egypt, Cairo's sewer system was built about fifty years ago to serve a population of 2 million people. It is now being overwhelmed by more than 11 million people. Less than one-tenth of India's 3000 towns and cities have even partial sewage systems and water treatment facilities. Some 150 million of India's urban residents lack access to sanitary sewer systems. In Colombia, the Bogota River, 200 km (125 mi) downstream from

Bogota's 5 million residents, still has an average fecal bacteria count of 7.3 million cells per liter, more than 700,000 times the safe drinking level and 3500 times higher than the limit for swimming.

Some 400 million people, or about one-third of the population in developing world cities do not have safe drinking water, according to the World Bank. Although city dwellers are somewhat more likely than rural people to have clean water, this still represents a large problem. Where people have to buy water from merchants, it often costs one hundred times as much as piped city water and may not be any safer to drink. Many rivers and streams in Third World countries are little more than open sewers, and yet they are all that poor people have for washing clothes, bathing, cooking, and—in the worst cases—for drinking. Diarrhea, dysentery, typhoid, and cholera are widespread diseases in these countries, and infant mortality is tragically high (chapter 20).

Housing

The United Nations estimates that at least 1 billion people—20 percent of the world's population—live in crowded, unsanitary slums of the central cities and in the vast shantytowns and squatter settlements that ring the outskirts of most Third World cities. Around 100 million people have no home at all. In Bombay, for example, it is thought that half a million people sleep on the streets, sidewalks, and traffic circles because they can find no other place to live (fig. 24.11). In São Paulo, perhaps 1 million "street kids" who have run away from home or been abandoned by their parents live however and wherever they can. This is surely a symptom of a tragic failure of social systems.

Slums are generally legal but inadequate multifamily tenements or rooming houses, either custom built to rent to poor people or converted from some other use. The chals of Bombay, for example, are high-rise tenements built in the 1950s to house immigrant workers. Never very safe or sturdy, these dingy, airless buildings are already crumbling and often collapse without warning. Eighty-four percent of the families in these tenements live in a single room; half of those families consist of six or more people. Typically, they have less than 2 square meters of floor space per person and only one or two beds for the whole family. They may share kitchen and bathroom facilities down the hall with fifty to seventy-five other people. Even more crowded are the rooming houses for mill workers where up to twenty-five men sleep in a single room only 7 meters square. Because of this crowding, household accidents are a common cause of injuries and deaths in Third World cities, especially to children. Charcoal braziers or kerosene stoves used in crowded homes are a routine source of fires and injuries. With no place to store dangerous objects beyond the reach of children, accidental poisonings and other mishaps are a constant hazard.

Shantytowns are settlements created when people move onto undeveloped lands and build their own houses. Shacks are built of corrugated metal, discarded packing crates, brush, plastic sheets, or whatever building materials people can scavenge. Some shantytowns are simply illegal subdivisions where the landowner rents land without city approval. Others are spontaneous or popular settlements or **squatter towns** where people occupy land without the owner's permission. Sometimes this occupation involves thou-

Figure 24.11

In Bombay, as many as half a million people sleep on the streets because they have no other place to live. Ten times as many live in crowded, dangerous slums and shantytowns throughout the city.

Figure 24.12

A shantytown or spontaneous settlement on the outskirts of Mexico City. Millions of people live with no water, power, or sanitation in *colonias* such as this.

sands of people who move onto unused land in a highly organized, overnight land invasion, building huts and laying out streets, markets, and schools before authorities can root them out (fig. 24.12). In other cases, shantytowns just gradually "happen."

Called *barriads, barrios, favelas,* or *turgios* in Latin America, *bidonvillas* in Africa, or *bustees* in India, shantytowns surround every megacity in the developing world. They are not an exclusive feature of poor countries, however. Some 250,000 immigrants and impoverished citizens live in the *colonias* along the southern Rio Grande in Texas. Only 2 percent have access to adequate sanitation. Most live in conditions as awful as you would see in any Third World city (What Do *You* Think?, p. 546).

About three-quarters of the residents of Addis Ababa, Ethiopia, or Luanda, Angola, live in squalid refugee camps. Two-thirds of the population of Calcutta live in unplanned squatter settlements and nearly half of the 20 million people in Mexico City live in uncontrolled, unauthorized shantytowns. Many governments try to clean out illegal settlements by bulldozing the huts and sending riot police to drive out the settlers, but the people either move back in or relocate in another shantytown elsewhere.

These popular but unauthorized settlements usually lack sewers, clean water supplies, electricity, and roads. Often the land on which they are built was not previously used because it is unsafe or unsuitable for habitation. In Bhopal, India, and Mexico City, for example, squatter settlements were built next to deadly industrial sites. In Rio de Janeiro, La Paz (Bolivia), Guatemala City, and Caracas (Venezuela), they are perched on landslide-prone hills. In Bangkok, thousands of people live in shacks built over a fetid tidal swamp. In Lima (Peru), Khartoum (Sudan), and Nouakchott, (Mauritania), shantytowns have spread onto sandy deserts. In Manila, 20,000 people live in huts built on towering mounds of garbage and burning industrial waste in city dumps (see fig. 23.3).

As desperate and inhumane as conditions are in these slums and shantytowns, many people do more than merely survive there. They keep themselves clean, raise families, educate their children, find jobs, and save a little money to send home to their parents. They learn to live in a dangerous, confusing, and rapidly changing world and have hope for the future. The people have parties; they sing and laugh and cry. They are amazingly adaptable and resilient. In many ways, their lives are no worse than those in the early industrial cities of Europe and America a century ago. Perhaps continuing development will bring better conditions to cities of the Third World as it has for many in the First World.

The Developed World

For the most part, the rapid growth of central cities that accompanied industrialization in nineteenth- and early twentieth-century

What Do You Think? Texas Colonias

For more than 1600 km (1000 mi), the Rio Grande River forms the border between Texas and Mexico. As the river approaches the Gulf of Mexico, it meanders across a broad, fertile valley with a subtropical climate perfect for citrus groves, sugar cane fields, and winter vegetables. A large population of farm workers, largely of Mexican ancestry, lives in the four southern-most border counties of Texas (Cameron, Hidalgo, Starr, and Willacy). Most of these workers are migrant laborers, picking winter crops in Texas from November to March, then traveling north for summer field jobs. More than 250,000 residents of these four counties live in *colonias:* unplanned, unregulated housing developments ranging from half a dozen houses plunked down in a field to neighborhoods of 10,000 people adjacent to an existing city.

Conditions in the *colonias* often are abysmal (fig. 1). Residents typically earn less than half the national average. More than half of all children live in poverty. Sewers, clean drinking water, electricity, paved roads, and garbage service often are inadequate or totally lacking. Diseases rarely seen in the rest of the United States—malaria, encephalitis, cholera, typhoid, typhus, dengue fever, even leprosy—afflict *colonias* residents. With a total population over 1 million people, these four counties have no public hospital and little public health service. There are fewer than half as many doctors per 1000 residents in this part of Texas as in adjacent regions of Mexico.

Figure 1

Substandard housing, inadequate sanitation, and a lack of public services often characterize *colonias* along the Texas/Mexico border.

How did the *colonias* get established? Texas is unique in that its counties have no zoning authority. Landowners can sell or rent land for any purpose and under any conditions the market will bear. When a farmer decides to sell land for a housing development, there is no oversight. Lots commonly are sold for little or nothing down on a contract-for-deed. People who can't qualify for a conventional loan find this very attractive; if they miss even a single monthly payment, however, the deed holder can repossess without paying them anything for previous payments or building investments.

Lack of building codes or strict county inspections can be helpful to *colonia* residents. They can build their houses a little at a time, adding rooms as they get some money and doing the work themselves. And in a region where racial and class discrimination is common, living among people similar to yourself can be more comfortable than trying to integrate neighborhoods where you really aren't welcome.

When pressed to provide services to *colonias,* local officials usually protest that they can't afford to do so. New upper class housing developments built next door to existing *colonias* have modern services, but commissioners claim this is justified because owners of high-priced houses pay more taxes. Besides, they argue, *colonia* residents really don't want, or won't pay for, complete hook-ups and other services.

What do you think? Is this an example of environmental racism? What evidence would you look for to determine if this situation has arisen from deliberate discrimination or benign neglect? Given that *colonias* exist and that environmental conditions and health problems don't stop at their borders, what policies would you adopt if you were a local official?

Europe and North America has now slowed or even reversed. London, for instance, once the most populous city in the world, has lost nearly 2 million people, dropping from its high of 8.6 million in 1939 to about 6.7 million now. While the greater metropolitan area surrounding London has been expanding to about 10 million inhabitants, it is now only the twelfth largest city in the world.

Areas of Progress

Many of the worst urban environmental problems of industrialized countries have been substantially reduced in recent years. Air and water quality have improved greatly. Working conditions and housing are better for most people. Improved sanitation and medical care have reduced or totally eliminated many of the worst communicable diseases that once were the scourge of cities. Dispersal of the people and businesses to the suburbs has significantly reduced crowding and congestion. Still, city life is generally regarded as being more stressful than life in the country or suburbs.

The major problems now facing cities of the United States tend to be associated with decay and blight. Businesses and jobs have fled to the suburbs along with the middle class. Tax revenues are down. The city infrastructure—streets, sewers, schools, and housing—is getting old and dilapidated. Drug dealers, pawn shops, bars, sleazy rooming houses, graffiti, broken windows, trash, and abandoned cars have become symbols of the worst parts of inner cities. Freeways make it much easier for suburbanites to get into and out of the city, but they also destroy housing, cause noise, increase air pollution, and create vast concrete landscapes that divide neighborhoods and make travel difficult for people without cars. The United States Agency for International Development (USAID) has recently begun turning its attention from the developing world to inner cities in this country, where many people live in more desperate and degrading conditions than do residents of Third World countries.

Many of the social problems of central cities are a result of a concentration of racial minorities, new immigrants, and the poor, elderly, handicapped, or others at the bottom of the social scale who were left behind when the middle and upper classes moved to the suburbs. Unemployment in some inner-city areas is routinely 50 percent or higher, and jobs that are available generally are menial, minimum-wage work with little prospect for advancement. Homelessness is an increasing problem of cities. Welfare agencies estimate that 3.5 million people are homeless in the United States and ten to twenty times that number live in substandard housing. About one-third of the homeless are now families. Of the single females living on the street, perhaps three-fourths are mentally disturbed and would formerly have been cared for in hospitals or institutions. Of the single, homeless men, two-thirds are elderly, poor, disabled, drug addicts, or alcoholics (fig. 24.13).

Urban Renewal

In the 1950s, the United States began a massive program of **urban renewal.** Billions of dollars were spent to revitalize blighted sections of inner cities. Old neighborhoods and decaying industrial areas were cleared or renovated, but often to create luxury housing, high-rise office towers, freeways, sports arenas, concert halls, or other projects that benefit the upper classes but do little for the people who once lived there. Many cohesive neighborhoods with strong traditions, cultural identity, and affordable housing were torn down to make way for "festival centers." People displaced from these neighborhoods can't afford to move to the suburbs and don't have transportation to commute to work in distant sites.

Much of the public housing built in the 1950s and 1960s consisted of massive blocks of high-rise apartment buildings that tend to concentrate and aggravate the problems that brought people there in the first place. Bleak, industrial architecture tends to isolate residents rather than promote community identity. There are few jobs nearby and few opportunities to escape from the culture of poverty and violence. Vandalism, drug problems, gang activity, and crime rates are high. Police often refuse to respond to calls

Figure 24.13

A homeless man sleeps on a park bench in Santa Monica, CA, with all his belongings in a shopping cart nearby. Welfare agencies estimate that 3.5 million people in the United States are homeless and ten to twenty times that number live in substandard housing.

from these projects because they are so dangerous. Among some infamous examples are the Cabrini-Green project in Chicago and the Pruitt-Igoe project in St. Louis, Missouri. Built in 1958 to house about 10,000 people, Pruitt-Igoe won architectural awards for modern design. It was poorly built and badly maintained, however, and lacked space in which residents could feel secure. By 1970, it was so run down and dangerous that most residents had left. The city concluded that all thirty-three buildings were beyond repair and demolished the entire project in 1976 (fig. 24.14). Ironically, in 1994, the United States Agency for International Development announced that it would begin assistance programs for American cities. Parts of Los Angeles, Detroit, Baltimore, and other large cities have conditions as bad as any developing country. It's fortunate that we have an agency with experience in disaster relief, but sad that conditions at home have gotten so desperate.

Not all public housing has been so disastrous. In the Cochran project, also in St. Louis and not far from Pruitt-Igoe, residents have formed an association to clean up and manage the buildings. They hire their own security guards, set standards for cleanliness and orderliness, and evict incorrigible troublemakers. The association also mediates disputes, organizes day care, and helps residents find jobs. By taking charge of their own lives and building a sense of community, these people have created a better place to live.

Noise

Every year since 1973, the U.S. Department of Housing and Urban Development has conducted a survey to find out what city residents dislike about their environment. And every year the same factor has been named most objectionable. It is not crime, pollution,

Figure 24.14

The Pruitt-Igoe public housing project in St. Louis was a dismal failure in urban planning. All 33 high-rise buildings were demolished in 1976.

or congestion; it is noise—something that reaches every part of the city every day.

We have known for a long time that prolonged exposure to noises, such as loud music or the roar of machinery, can result in hearing loss. Evidence now suggests that noise-related stress also causes a wide range of psychological and physiological problems ranging from irritability to heart disease. An increasing number of people are affected by noise in their environment. By age 40, nearly everyone in America has suffered hearing deterioration in the higher frequencies. An estimated 10 percent of Americans (24 million people) suffer serious hearing loss, and the lives of another 80 million people are significantly disrupted by noise.

What is noise? There are many definitions, some technical and some philosophical. What is music to your ears might be noise to someone else. Simply defined, noise pollution is any unwanted sound or any sound that interferes with hearing, causes stress, or disrupts our lives. Sound is measured either in dynes, watts, or decibels (table 24.3). Note that decibels (db) are logarithmic; that is, a 10-db increase represents a tenfold increase in sound energy.

City noises come from many sources. Traffic is generally the most omnipresent noise. Cars, trucks, and buses create a roar that permeates nearly everywhere in the city. Near airports, jets thunder overhead, stopping conversation, rattling dishes, sometimes even cracking walls. Jackhammers rattle in the streets; sirens pierce the air; motorcycles, lawnmowers, snowblowers, and chain saws cre-

ate an infernal din; and music from radios, TVs, and loudspeakers fills the air everywhere.

The sensitivity and discrimination of our hearing is remarkable. Normally, humans can hear sounds from 16 hertz to 20,000 hertz (cycles per second). A young child whose hearing has not yet been damaged by excess noise can hear the whine of a mosquito's wings at the window when less than one quadrillionth (1×10^{-15}) of a watt per cm^2 is reaching the eardrum.

Prolonged exposure to sounds above about 90 decibels can permanently damage the sensitive mechanism of the inner ear. By age 30, most Americans have lost 5 db of sensitivity and can't hear anything above 16,000 Hertz (Hz); by age 65, the sensitivity reduction is 40 db for most people, and all sounds above 8000 Hz are lost. By contrast, in the Sudan, where the environment is very quiet, even 70-year-olds have no significant hearing loss.

Extremely loud sounds—above 130 db, the level of a loud rock band or music heard through earphones at a high setting—actually can destroy sensory nerve endings, causing aberrant nerve signals that the brain interprets as a high-pitched whine or whistle. You may have experienced ringing ears after exposure to very loud noises. Coffee, aspirin, certain antibiotics, and fever also can cause ringing sensations, but they usually are temporary.

A persistent ringing is called tinnitus. It has been estimated that 94 percent of the people in the United States suffer some degree of tinnitus. For most people, the ringing is noticeable only in

Table 24.3
Sources and effects of noise

Source	Sound Pressure (dynes/cm^2)	Decibels (db)	Power at Ear	Effects
Shotgun blast (1 m)		150	10^{-1}	Instantaneous damage
Stereo headphones (full volume)	2000	140		Hearing damage in 30 sec
50-hp siren (at 100 m)		130	10^{-3}	Pain threshold
Jet takeoff (at 200 m)	200	120		Hearing damage in 7.5 min
Heavy metal rock band		110	10^{-5}	Hearing damage in 30 min
Power mower, motorcycle	20	100		Damage in 2 hr
Heavy city traffic		90	10^{-7}	Damage in 8 hr
Loud classical music	2	80		OSHA 8-hour standard
Vacuum cleaner		70	10^{-9}	Concentration disrupted
Normal conversation	0.2	60		Speech disrupted
Background music		50	10^{-11}	
Bedroom	0.02	40		Quiet
Library		30	10^{-13}	
Soft whisper	0.002	20		Very quiet
Leaves rustling in the wind		10	10^{-15}	Barely audible
Mosquito wings at 4 m	0.0002	0		Hearing threshold youth 1000–4000 Hz

a very quiet environment, and we rarely are in a place that is quiet enough to hear it. About thirty-five out of one thousand people have tinnitus severely enough to interfere with their lives. Sometimes the ringing becomes so loud that it is unendurable, like shrieking brakes on a subway train. Unfortunately, there is not yet a treatment for this distressing disorder.

One of the first charges to the EPA when it was founded in 1970 was to study noise pollution and to recommend ways to reduce the noise in our environment. Standards have since been promulgated for noise reduction in automobiles, trucks, buses, motorcycles, mopeds, refrigeration units, power lawnmowers, construction equipment, and airplanes. The EPA is considering ordering that warnings be placed on power tools, radios, chain saws, and other household equipment. The Occupational Safety and Health Agency also has set standards for noise in the workplace that have considerably reduced noise-related hearing losses.

Noise is still all around us. In many cases, the most dangerous noise is that to which we voluntarily subject ourselves. Perhaps if people understood the dangers of noise and the permanence of hearing loss, we would have a quieter environment.

Transportation and City Growth

Transportation has always played an essential role in city development. Most cities are located at crossroads, river fords, seaports, junctions of major rivers, agricultural centers, or other sites where travelers were brought together and merchandise could be traded, bought, or sold. Transportation is needed to bring building materials, energy sources, and food into the city and to remove wastes. Most older cities have been remodeled and reshaped repeatedly by the changing transportation systems that provide their life-giving circulation.

Before the Industrial Revolution, most cities were compact and densely populated (fig. 24.15a). Except for a few broad ceremonial boulevards, streets were narrow, twisting, and often barely passable for wheeled traffic. Transportation was by sailboat, horse-drawn vehicles, or on foot. The wealthiest citizens usually lived in the middle of the city close to the centers of power and prestige, while the poorer people lived on the outskirts. Housing was mainly multistoried apartment complexes over street-level shops.

Automobiles brought many changes to cities. The middle class moved to single-family housing developments that filled in open spaces between transit lines (fig. 24.15c). The wealthy moved out of town to rural estates and satellite cities. Business offices remained downtown, but shopping centers sprang up at the nodes where major streets intersected transit lines. In the country, small towns and crossroads stores that had been built at about 5-mile intervals to accommodate horse-and-buggy travelers dwindled when people were able to travel to the county seat or the nearest big city to do their shopping. The development of chain stores and brand-name merchandise was largely due to this new mobility of shoppers.

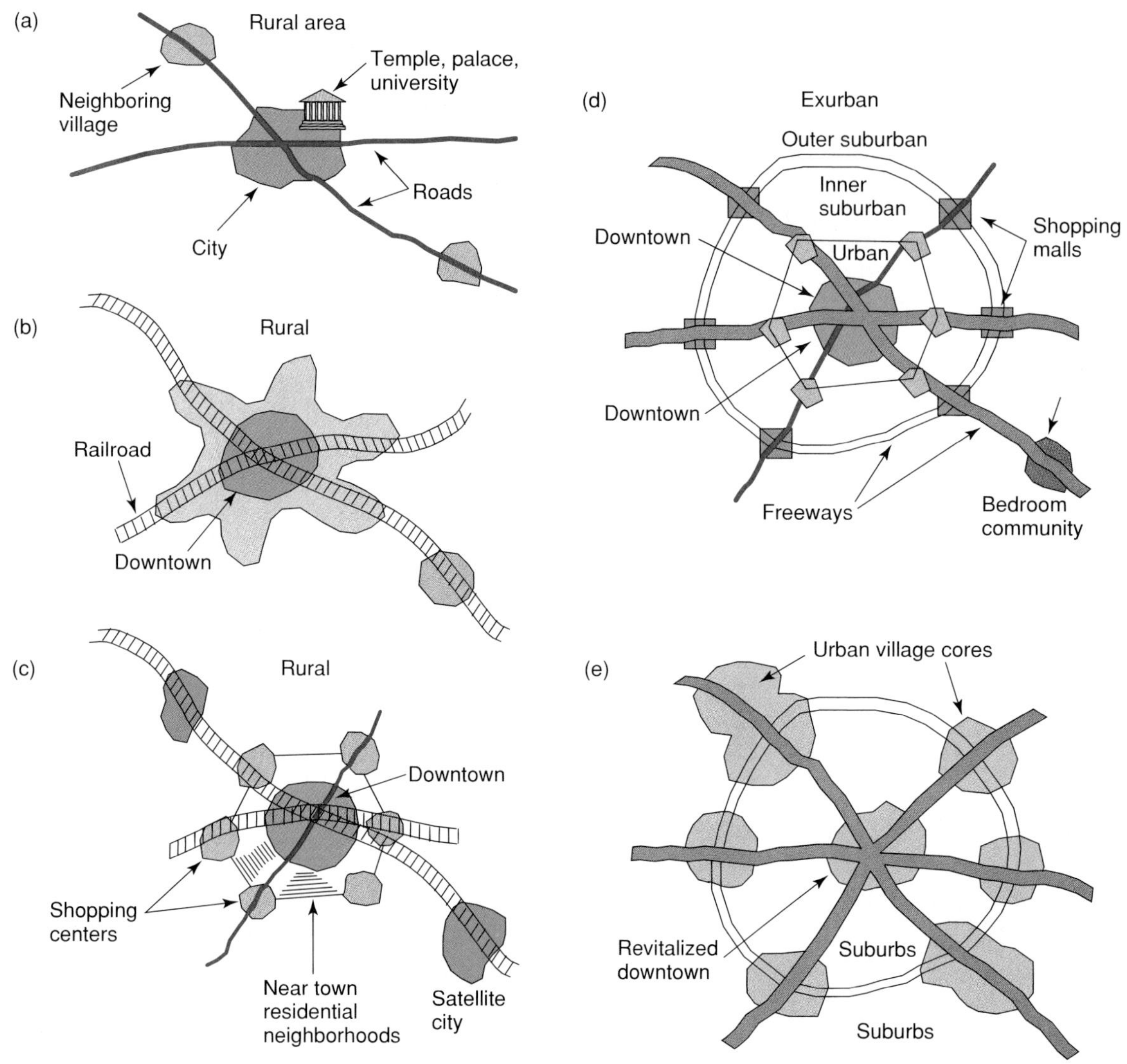

(a)

Neighboring village

Rural area

Temple, palace, university

Roads

City

(b)

Rural

Railroad

Downtown

(c)

Rural

Downtown

Shopping centers

Near town residential neighborhoods

Satellite city

(d) Exurban

Outer suburban

Inner suburban

Downtown

Urban

Shopping malls

Downtown

Freeways

Bedroom community

(e)

Urban village cores

Revitalized downtown

Suburbs

Suburbs

Figure 24.15

Patterns of city development from preindustrial (*a*) to polycentric postindustrial (*e*). Note how housing areas follow rail lines in early industrial stage (*b*). Introduction of the automobile allows people to move away from the rail lines (*c*). Freeways (*d*) divert business to shopping malls and encourage growth of suburban rings beyond the central city. Finally, (*e*) these urban village cores develop a full range of services and amenities that rival the central core. Removal of heavy manufacturing from downtown allows gentrification and revitalization.

The decision to build freeways has been called the most important land-use decision ever made in the United States. Freeways have profoundly reshaped where we live, work, and shop, and how we get from place to place. Shopping malls located at freeway interchanges have largely replaced downtown department stores and have become the urban centers of expanding rings of suburbs around major cities (fig. 24.15*d*). Freeways allow us to travel with greater privacy, freedom, convenience, and speed (usually) than a mass transit system.

But freeways also bring many problems. They have torn through neighborhoods, choked cities with traffic, enormously increased energy consumption, and caused pollution, noise, and urban sprawl (fig. 24.16). Some of the most contentious U.S. environmental battles of the last two decades concerned freeways planned to cut across residential neighborhoods, parks, scenic and historic areas, or farmlands. Many people were initiated into environmental activism through their opposition to a particular stretch of freeway that threatened an area they cared about.

Los Angeles epitomizes the modern postindustrial city. Its multiple suburban centers are linked by a 600-mile-long network of

Figure 24.16

The freeway makes it easy for suburban residents to get into and out of the city, but it consumes land, brings pollution into the city, creates noise and visual blight, and destroys neighborhoods.

freeways. It has been estimated that two-thirds of downtown Los Angeles and one-third of the total metropolitan area is devoted to roads, parking lots, service stations, and other automobile-related uses. Five million vehicles crowd onto the roads and highways each day, causing about 85 percent of both the air pollution and urban noise in the metropolitan area. Most Los Angeles freeways no longer have morning rush in one direction and evening rush in the other; traffic now comes to a standstill morning and evening in both directions on many sections, while busy intersections have slowdowns throughout the day and into the night. It can take three hours to drive 20 miles at peak traffic periods. Between commuting to work and driving to beaches, the mountains, or for shopping, many people spend a substantial portion of their lives on the freeways.

What's next now that we seem to have saturated our urban areas with automobiles? Some cities are using modern mass transit systems to redesign where people live and work. Shopping centers are developing into urban villages with a complete set of services and facilities (see fig. 24.15e). The old downtown, freed of the congestion and pollution of heavy industry, is undergoing revitalization as a cultural center. In the next section we will look at some ideas for how cities might be reconceived and rebuilt as better places to live.

City Planning

City planning has a long history. Many of the earliest cities in Mesopotamia, India, and Egypt were built in an orderly and civilized fashion. The Greek city-states of the first millennium B.C. reached a level of architectural beauty and rational city planning that has rarely been surpassed (see fig. 24.2). Miletus and Pergamum on the Ionian coast of Turkey, for instance, and Alexandria on the Nile delta in Egypt, were splendid examples of Greek city planning. They were beautifully situated, with an orderly, unified design, magnificent buildings, and a pleasant mix of public and private spaces.

Residents of these cities regarded them as oases of comfort, convenience, safety, and delight in an otherwise cruel and dangerous world (see fig. 2.10). Cities were seen as elysian refuges protected by a sacred wall from the chaos of evil spirits, wild animals, and dangerous people in the profane world outside. City temples were regarded as homes for the gods; their gardens were sanctuaries of beauty and sacred order. As centers of education, religion, political power, commerce, communication, and technology, cities offered many advantages over life in the country.

The Greeks considered optimum city size to be 30,000 to 50,000 people. This size was large enough to support a theater, university, library, temple, lively marketplace, and an active political and intellectual life, yet small enough to draw food from the surrounding region and compact enough so that a person could walk out into the country within a few minutes from anywhere in the city. These planned cities resemble efficient ecosystems in which inhabitants participate in exchanges of matter and energy with their environment.

In the sixteenth and seventeenth centuries, Rome, London, Paris, Berlin, Vienna, and Brussels all were rebuilt with spacious boulevards, parks, and imposing public buildings following the re-

Figure 24.17

Pierre L'Enfant's plan for Washington, DC, is considered one of the best urban designs in North America. Spacious boulevards, gracious traffic circles, and grand ceremonial malls create open space and interesting patterns.

Reproduced by the U.S. Geological Survey for the Library of Congress.

discovery of Greek and Roman principles of city planning. The design of these European capitals is reflected in Pierre L'Enfant's plan for Washington, DC, which is one of the most notable city plans in the United States (fig. 24.17). Washington's rectangular street grid is cut by diagonal avenues converging on the Capitol and Executive Mansion, forming numerous squares, circles, and triangles that make interesting spaces for pocket parks. The broad, tree-shaded streets and impressive vistas created by this plan make Washington a grand setting for the national capital.

Garden Cities and New Towns

The twentieth century has seen numerous experiments in building **new towns** for society at large that try to combine the best features of the rural village and the modern city. One of the most influential of all urban planners was Ebenezer Howard (1850–1929), who not only wrote about ideal urban environments but also built real cities to test his theories. In his *Garden Cities of Tomorrow,* written in 1898, Howard proposed that the congestion of London could be relieved by moving whole neighborhoods to **garden cities** separated from the central city by a greenbelt of forests and fields. Howard was the first modern urban planner to advocate comprehensive land-use planning and to reintroduce the Greek concept of organic growth and human scale to the city.

In the early 1900s, Howard worked with architect Raymond Unwin to build Letchworth and Welwyn Garden just outside of London. Interurban rail transportation provided access to these cities. Houses were clustered in "superblocks" surrounded by parks, gardens, and sports grounds. Streets were curved. Safe and convenient walking paths and overpasses protected pedestrians from traffic. Businesses and industries were screened from housing areas by vegetation. Each city was limited to about 30,000 people to facilitate social interaction. Housing and jobs were designed to create a mix of different kinds of people and to integrate work, social activities, and civic life. Trees and natural amenities were carefully preserved and the towns were laid out to maximize social interactions and healthful living. Care was taken to meet residents' psychological needs for security, identity, and stimulation.

Letchworth and Welwyn Garden each have seventy to one hundred people per acre. This is a true urban density, about the same as New York City in the early 1800s and five times as many people as most suburbs today. By planning the ultimate size in advance and choosing the optimum locations for housing, shopping centers, industry, transportation, and recreation, Howard believed he could create a hospitable and satisfying urban setting while protecting open space and the natural environment. He intended to create parklike surroundings that would preserve small-town values and encourage community spirit in neighborhoods.

Letchworth and Welwyn Garden were the first of thirty-two new towns established in Great Britain, which now house about 1 million people. The Scandinavian countries have been especially successful in building garden cities. The former Soviet Union also built about two thousand new towns. Some are satellites of existing cities, and others are entirely new communities far removed from existing urban areas.

Planned communities also have been built in the United States following the theories of Ebenezer Howard, but most plans have been based on personal automobiles rather than public transit. In the 1920s, Lewis Mumford, Clarence Stein, and Henry Wright drew up plans that led to the establishment of Radburn, New Jersey, and Chatham Village (near Pittsburgh). Reston, Virginia, and Columbia, Maryland, both were founded in the early 1960s and are widely regarded as the most successful attempts to build new towns of their era (fig. 24.18). Another movement to build new towns according to Howard's principles has sprung up in the 1990s. Towns such as Seaside in northern Florida and Laguna West near Sacramento cluster houses to save open space and create a sense of community. Commercial centers are located within a few minutes walk of most houses, and streets are designed to encourage pedestrians and to provide places to gather and visit.

Cities of the Future

Is the expansion of cities—especially onto agricultural land—a waste of land and resources? Some people feel that growing cities decrease local agricultural resources and corrupt once pristine countryside and quaint rural towns. Others see this as a healthy opportunity to decongest central cities and build smaller, more pleasant, more livable communities that have the benefits of both technological cities and rural villages. Perhaps the demographic shifts of the last forty years have merely been a transition between

Figure 24.18

By clustering buildings together, Reston, VA, has maintained a high percentage of open space as public commons. Pedestrian walkways are separated from roads to provide a safe, pleasant way to get anywhere in the village.

nineteenth-century industrial cities and garden cities of tomorrow. The polycentric network of urban villages in surrounding metropolitan areas may be a step toward more human-scale communities.

An alternative to spreading the population across a wide area of the countryside is to build upward. This model, which depends strongly on technology, has been called the **technopolis,** vertical city, or city of the future. The central hub of most big American cities now is dominated by skyscrapers and a highly technological environment (fig. 24.19). The emerging supercities of the Third World are also moving toward this style, in part because of its association with wealth, power, and progress.

There have been many proposals for gigantic, vertical supercities based on skyscrapers and completely artificial environments—what the Swiss architect Le Corbusier called "a machine for living." An example was Buckminster Fuller's proposed mile-high geodesic dome to cover most of Manhattan. Fuller also designed a hollow, tetrahedronal building two hundred stories high that would float on the ocean and could house 1 million people. It could be built in stages, he suggested, like a beehive, with trailer-like modules inserted in successive layers into a mountainous, open-truss framework. Some architects have talked seriously about the eventual possibility of erecting 500-story, mile-high buildings! The psychological and physiological effects of living 5000 feet above street level can only be imagined.

Japan is now building eighteen new high-technology cities intended to be centers for economic and scientific growth in the next century. With names like Teletopoia, Agripolis, and New Media City, these regional research, education, and marketing centers will have innovative housing, enclosed shopping malls, and high-technology communication and transportation systems. They will concentrate on leading-edge research and industries, such as fifth-generation supercomputers, biotechnology, lasers, ceramics, and bioelectronics. Some such high-tech cities may be giant, floating

Figure 24.19

For many people, the gleaming, climate-controlled, artificial environment of the modern high-rise city is the crowning achievement of modern civilization. To others, it seems brutal and dehumanizing. What do you think?

structures similar to Buckminster Fuller's visionary proposals. Others may consist of a maze of tunnels and chambers entirely underground, not unlike a giant ant nest, opening onto huge twenty-story-deep air wells. This plan would conserve energy and preserve scarce land surface; however, many technical and psychological problems will need to be overcome.

Urban Redesign

Most people in the United States now live in suburbs and will probably continue to do so in the foreseeable future. What can be done to make them more humane and environmentally sustainable?

Although suburbs have many advantages, they also have disadvantages. Most suburbs are too spread out for people to meet and interact, except with their immediate neighbors. Residential streets are empty during much of the day. Suburbs often have no sense of community and can be places of alienation and loneliness, just as cities often are (fig. 24.20). Although many urban problems have been eliminated, so have many of the finer aspects of city living. Suburbs, for instance, lack the activities, energy, and diversity that make cities exciting and dynamic. They have very limited artistic, cultural, and educational opportunities, compared to cities.

The low population density of the suburb makes public transport prohibitively expensive and private automobiles essential. Car pools for work, school, and extracurricular activities help fill group transportation needs. The uniform, single-family, detached houses of the suburbs offer few options to those who don't fit into the traditional, middle-class, nuclear family. As children grow up and leave home, parents often find the house bigger than they need. Single-parent families, households of single adults, the elderly, and the poor tend to have difficulty finding affordable suburban housing that fits their needs.

What can or should be done to create ideal suburban environments? Obviously not everyone in America will want to leave established suburban communities for new garden communities or wilderness utopias. Nor should they. Abandoning existing buildings and civic infrastructures would be a terrible waste, and if everyone moves to the country, natural areas will be obliterated. We need, instead, to find ways to remodel and revitalize existing suburban cities, reduce their problems, and adapt to the changing needs of their residents.

How can cities be redesigned to make them more diverse, flexible, and energy efficient? The following ten proposals might give them some of the better aspects of both the rural village and the big city.

- Limit city size or organize them in modules of 30,000 to 50,000 people, large enough to be a complete city but small enough to be a community. A greenbelt of agricultural and recreational land around the city limits growth while promoting efficient land use. By careful planning and cooperation with neighboring regions, a city of 50,000 people can have real urban amenities such as museums, performing arts centers, schools, hospitals, etc.

- Determine in advance where development will take place. This protects property values and prevents chaotic development in which the lowest uses drive out the better ones. It also recognizes historical and cultural values, agricultural resources, and such ecological factors as impact on wetlands, soil types, groundwater replenishment and protection, and preservation of aesthetically and ecologically valuable sites.

- Turn shopping malls into real city centers that invite people to stroll, meet friends, or listen to a debate or a street musician (fig. 24.21). If there aren't one hundred places for an impromptu celebration, a place isn't a real city. Another test of a city is a vital nightlife. Design city spaces with sidewalk cafes, pocket parks, courtyards, balconies, and porticoes that shelter pedestrians,

Urbanization and Sustainable Cities

Figure 24.20
This housing development south of San Francisco inspired Malvina Reynolds to write a song about little houses on a hillside that look just the same.

bring people together, and add life and security to the street. Restaurants, theaters, shopping areas, and public entertainment that draw people to the streets generate a sense of spontaneity, excitement, energy, and fun.

- Locate everyday shopping and services so people can meet daily needs with greater convenience, less stress, less automobile dependency, and less use of time and energy. This might be accomplished by encouraging small-scale commercial develop- ment in or close to residential areas. Perhaps we should once again have "mom and pop" stores on street corners or in homes.

- Increase jobs in the community by locating offices, light industry, and commercial centers in or near suburbs, or by enabling work at home via computer terminals. These alternatives save commut- ing time and energy and provide local jobs. There are also concerns, however, about work-at-home employees being exploited in low-paying "sweatshop" conditions by unscrupulous employers. Some safeguards may be needed.

- Encourage walking or the use of small, low-speed, energy- efficient vehicles (microcars, motorized tricycles, bicycles, etc.) for many local trips now performed by full-size automobiles. Creating special traffic lanes, reducing the number or size of parking spaces, or closing shopping streets to big cars might encourage such alternatives.

- Promote more diverse, flexible housing as alternatives to conventional, detached single-family houses. "In-fill" building between existing houses saves energy, reduces land costs, and might help provide a variety of living arrangements. Allowing owners to turn unused rooms into rental units provides space for those who can't afford a house and brings income to retired people who don't need a whole house themselves. Allowing single-parent families or groups of unrelated adults to share housing and to use facilities cooperatively also provides alternatives to those not living in a traditional nuclear family. One of the great "discoveries" of urban planning is that mixing various types of housing—individual

Figure 24.21
Many cities have redesigned core shopping areas to be more "user friendly." Pedestrian shopping malls, such as this one in Boulder, CO, create space for entertainment, dining, and chance encounters and provide opportunities for building a sense of community.

Figure 24.22
Communal gardens in Seattle, WA, use once-vacant land to grow food and give local residents a chance to experience nature in a highly urban setting.

homes, townhouses, and high-rise apartments—can be attractive if buildings are aesthetically arranged in relation to one another.

- Create housing "superblocks" that use space more efficiently and foster a sense of security and community. Widen peripheral arterial streets and provide pedestrian overpasses so traffic flows smoothly around residential areas; then reduce interior streets within blocks to narrow access lanes with speed bumps and barriers to through traffic so children can play more safely. The land released from streets can be used for gardens, linear parks, playgrounds, and other public areas that will foster community spirit and encourage people to get out and walk. Cars can be parked in remote lots or parking ramps, especially where people have access to public transit and can walk to work or shopping.

- Make cities more self-sustainable by growing food locally, recycling wastes and water, using renewable energy sources, reducing noise and pollution, and creating a cleaner, safer environment (fig. 24.22). Reclaimed inner-city space or a greenbelt of agricultural and forestland around the city provides food and open space as well as such valuable ecological services as

purifying air, supplying clean water, and protecting wildlife habitat and recreation land.

- Invite public participation in decisionmaking. Emphasize local history, culture, and environment to create a sense of community and identity. Create local networks in which residents take responsibility for crime prevention, fire protection, and home care of children, the elderly, sick, and disabled. Coordinate regional planning through metropolitan boards that cooperate with but do not supplant local governments.

Sustainable Development in the Third World

What can be done to improve conditions in Third World cities? Curitiba, Brazil, is an outstanding example of what can be done, even

CASE STUDY
Curitiba: An Environmental Showcase

Curitiba, a Brazilian city of about 2 million people located on the Atlantic coast about 650 km (400 mi) southwest of Rio de Janeiro, has acquired a worldwide reputation for its innovative urban planning and environmental protection policies. Tree-lined streets, clean air, smoothly-flowing traffic, and absence of litter and garbage make this one of Latin America's most liveable cities.

The architect of this remarkable program is mayor Jaime Lerner, who began in 1962 as a student leader protesting the proposed destruction of Curitiba's historic downtown center. Elected mayor nine years later, he has worked for more than two decades to preserve the city and make it a more beautiful and habitable place. Now Governor of the State of Parana, Lerner's success has thrust the city into the international spotlight as a shining example of what conservation and good citizenship can do to improve urban environments, even in Third World cities.

The heart of Curitiba's environmental plan is education for both children and adults. Signs posted along roadways proclaim "50 kg of paper equals one tree" and "recycle; it pays." School children study ecology along with Portuguese and math. With the assistance of children who encourage their parents, the city has successfully instituted a complex recycling plan that requires careful separation of different kinds of materials. The city calculates that 1200 trees per day are saved by paper recycled in this program. "Imagine if the whole of Brazil did this with an urban population 60 times greater than Curitiba," Lerner exclaims. "We could save 26 million trees per year!" More than 70 percent of the residents now recycle. Food and bus coupons are given out in exchange for recyclables, providing an alternative to welfare while also protecting the environment.

Another area in which Curitiba is setting an example is transportation. Faced with a population that tripled in two decades, bringing increasing levels of traffic congestion and air pollution, Curitiba had a transportation dilemma. The choices were either to bulldoze freeways through the historic heart of the city or to institute mass transit. The city chose mass transit. Now more than three quarters of the city's population leave their automobiles at home every day and take special express buses to work. The system is so successful that ridership has increased from 25,000 passengers a day twenty years ago to more than 1.25 million per day now. The result is not only less congestion and pollution but major energy savings.

Other measures to clear the air and reduce congestion include a limit on building height and construction of an industrial park outside city boundaries. Maximum use is made of all materials and buildings. Worn-out buses become city training centers, an old military fort is a cultural center, and a gunpowder depot is now a theater. Water and energy conservation are practiced widely. Even litter—a ubiquitous component of most Brazilian cities—is absent in Curitiba. People are so imbued with city pride that they keep their surroundings spotless. To further beautify their city civic volunteers have planted 1.5 million trees—more than any other place in Brazil.

Although many residents initially were skeptical of this environmental plan, now a remarkable 99 percent of the city's inhabitants would not want to live anywhere else. The World Bank uses Curitiba as an example of what can be done through civic leadership and public participation to clean up the urban environment. Some people claim that Curitiba, with its cool climate and high percentage of European immigrants, may be a special case among Brazilian cities. Lerner claims that Curitiba has no special features except concern, creativity, and communal efforts to care for its environment. Could you start a similar program in your hometown?

---

in relatively undeveloped countries to improve transportation, protect central cities, and create a sense of civic pride (see Case Study, p. 556). Other cities have far to go, however, before they reach this standard. Among the immediate needs are housing, clean water, sanitation, food, education, health care, and basic transportation for their residents. The World Bank estimates that interventions to improve living conditions in urban households in the developing world could average the annual loss of almost 80 million "disability-free" years of life. This is about twice the feasible benefit estimated from all other environmental programs studied by the bank.

Some countries, recognizing the need to use vacant urban land, are redistributing unproductive land or closing their eyes to illegal land invasions. Indonesia, Peru, Tanzania, Zambia, and Pakistan have learned that squatter settlements make a valuable contribution to meeting national housing needs. Squatters' rights are being upheld in some cases, and such services as water, sewers, schools, and electricity are being provided to the settlements (fig. 24.23). Some countries intervene directly in land distribution and land prices. Tunisia, for instance, has a "rolling land bank" to buy and sell land. This strong and effective program controls urban land prices and reduces speculation and unproductive land ownership.

Many planners argue that social justice and sustainable economic development are answers to the urban problems we have discussed in this chapter. If people have the opportunity and money to buy better housing, adequate food, clean water, sanitation, and other things they need for a decent life they will do so. Democracy, security, and improved economic conditions help in slowing population growth and reducing rural-to-city movement. An even

Figure 24.23

In this *colonia* on the outskirts of Mexico City, residents work with the government to bring in electricity, water, and sewers to shantytowns and squatter settlements. Like many Third World countries, Mexico recognizes that helping people help themselves is the best way to improve urban living.

more important measure of progress may be institution of a social welfare safety net guaranteeing that old or sick people will not be abandoned and alone.

Some countries have accomplished these goals even without industrialization and high incomes. Sri Lanka, for instance, has lessened the disparity between the core and periphery of the country. Giving all people equal access to food, shelter, education, and health care eliminates many incentives for interregional migration. Both population growth and city growth have been stabilized, even though the per capita income is only $800 per year. China has done something similar on a per capita income of around $300 per year.

Whether sustained, environmentally sound economic development is possible for a majority of the world's population remains one of the most important and most difficult questions in environmental science. The unequal relationship between the richer "Northern" countries and their impoverished "Southern" neighbors is a major part of this dilemma. Some people argue that the best hope for developing countries may be to "delink" themselves from the established international economic systems and develop direct south-south trade based on local self-sufficiency, regional cooperation, barter, and other forms of nontraditional exchange that are not biased in favor of the richer countries.

Summary

A rural area is one in which a majority of residents are supported by methods of harvesting natural resources. An urban area is one in which a majority of residents are supported by manufacturing, commerce, or services. A village is a rural community. A city is an urban community with sufficient size and complexity to support economic specialization and to require a higher level of organization and opportunity than is found in a village.

Urbanization in the United States over the past two hundred years has caused a dramatic demographic change. A similar shift is occurring in most parts of the world. Only Africa and South Asia remain predominantly rural, and cities are growing rapidly there as well. By the end of this century, we expect that more than half the world's people will live in urban areas. Most of that urban growth will be in the supercities of the Third World. A century ago only thirteen cities had populations above 1 million; now there are 235 such cities. In the next century, that number will probably double again, and three-fourths of those cities will be in the Third World.

Cities grow by natural increase (births) and migration. People move into the city because they are "pushed" out of rural areas or because they are "pulled" in by the advantages and opportunities of the city. Huge, rapidly growing cities in the developing world often have appalling environmental conditions.

Among the worst problems faced in these cities are traffic congestion, air pollution, inadequate or nonexistent sewers and waste disposal systems, water pollution, and housing shortages. Millions of people live in slums and shantytowns where conditions would crush any but the strongest spirit, yet these people raise families, educate their children, learn new jobs and new ways of living, and have hope for the future.

The problems of developed world cities tend to be associated with decay and blight. Over the past fifty years, many in the urban middle and upper classes moved to the suburbs, leaving the old, very poor, handicapped, and economically marginal people in the inner city. Increasing joblessness and poverty in the inner city create a cycle of poverty from which it is difficult to escape. Still, there are ways that we can improve cities in both the developed and the developing world to make them healthier, safer, and more environmentally sound, socially just, and culturally fulfilling than they are now.

▼ Questions for Review

1. What is the difference between a city and a village and between rural and urban?

2. How many people now live in cities, and how many live in rural areas worldwide?

3. What changes in urbanization are predicted to occur in the next fifty years, and where will that change occur?

4. Identify the ten largest cities in the world. Has the list changed in the past fifty years? Why?

5. When did the United States pass the point at which more people live in the city than the country? When will the rest of the world reach this point?

6. Describe the current conditions in Mexico City. What forces contribute to its growth?

7. Describe the difference between slums and shantytowns.

8. Why are urban areas in U.S. cities decaying?

9. How has transportation affected the development of cities? What have been the benefits and disadvantages of freeways?

10. Describe some ways that American cities and suburbs could be redesigned to be more ecologically sound, socially just, and culturally amenable.

▼ Questions for Critical Thinking

1. Picture yourself living in a rural village or a Third World city. What aspects of life there would you enjoy? What would be the most difficult for you to accept?

2. Are there fundamental differences between the lives of homeless people in First World and Third World cities? Where would you rather be?

3. A city could be considered an ecosystem. Using what you learned in chapters 2 and 3, describe the structure and function of a city in ecological terms.

4. Extrapolating from laboratory animals to humans is always a difficult task. How would you interpret the results of laboratory experiments on stress and crowding in terms of human behavior?

5. Weigh the costs and benefits of automobiles in modern American life. Would we have been better off if the internal combustion engine had never been invented?

6. Boulder, Colorado, has been a leader in controlling urban growth. One consequence is that housing costs have skyrocketed and poor people have been driven out. If you lived in Boulder, would you vote for additional population limits? What do you think is an optimum city size?

7. Ten proposals are presented in this chapter for suburban redesign. Which of them would be appropriate or useful for your community? Try drawing up a plan for the ideal design of your neighborhood.

8. How much do you think the richer countries are responsible for conditions in the developing countries? How much have people there brought on themselves? What role should, or could, we play in remedying their problems?

▼ Key Terms

city 539	shantytowns 544
core regions 539	slums 544
garden cities 551	squatter towns 544
megacity 539	technopolis 552
new towns 551	urban area 539
pull factors 542	urbanization 538
push factors 542	urban renewal 547
rural area 539	village 539

▼ Additional Information on the Internet

Earth Wave Directory
http://www.aloha.net/epf/dsg/directory.html

Environments By Design
http://www.zone.org/ebd/

European Sustainable Cities and Towns Campaign
http://www.iclei.org/europe/suscam.html/

Housing and Urban Development
http: //www.dworks.co.jp/hudmx/E01.html/

Public Technology, Inc.
http://pti.nw.dc.us/

Sustainable Cities
http://www.oneworld.org/overviews/cities/girardet_urbanage.html/

United Nations Development Programme
http://www.undp.org/

Urban Ecology
http://www.best.com/~schmitty/ueindex.htm

Urban Habitat Project
gopher://gopher.igc.apc.org:70/11/orgs/urbanhab/

US Department of Housing and Urban Development
http://www.hud.gov/

US Green Building Council
http://gwis.circ.gwu.edu/~greenu/usgbc.html/

Note: Further readings appropriate to this chapter are listed on p. 607.

Environmental Engineering

*A*ndrew Graham

Andrew Graham is a senior project manager at Peer Environmental Engineering Resources, a small consulting company in suburban Minneapolis, Minnesota. Choosing a rapidly growing field and hunting carefully and systematically for a job were important aspects of Andrew's career path. His example illustrates the importance of planning and networking in identifying and pursuing the job you want.

Growing up in the hill country of central Texas, Andrew developed a love for hiking, camping, and other outdoor activities. In college he majored in geology to combine his interests in science with field work, hoping to become a researcher or natural resource manager. After graduating from college, however, he discovered that research work generally requires an advanced degree and that field-work positions often are low-paying and hard to find. Most jobs for geologists are in extractive industries such as oil or hard-rock mining, areas in which he was not particularly interested.

Finding the right situation was a two-year process for Andrew. He supported himself during this long exploration by working as a gardener at the Minnesota Arboretum, a job that allowed him to be outdoors and to learn more about the region. The first year of his job search Andrew spent exploring the job market and assessing his options. One of the most useful approaches he found was talking to recent geology graduates from his college who were already employed, asking them where and how they had found jobs. He learned from these conversations that environmental engineering was a rapidly expanding field with many opportunities for work he found appealing.

By carrying out a series of informational interviews, Andrew obtained insights into how the system works as well as good interviewing practice. When he got to real job interviews, he knew what to wear, what questions to ask, and how to present himself in the best light.

After sending out many resumes and interviewing for a number of openings, Andrew finally landed a job with a mid-sized consulting firm that specializes in environmental assessment for real estate companies, clean-up design and monitoring for industry, and regulatory compliance issues for firms that create or handle toxic or hazardous chemicals. Initially, this job required a fair amount of field work inspecting properties, supervising drilling or sampling crews, and overseeing clean-up operations. Now, as a project manager, he spends most of his time writing reports, dealing with clients, and supervising the work of others.

In his present position, Andrew has a great deal of independence and responsibility to set his own schedule and determine how he carries out a particular job but also a lot of pressure to work efficiently. In industry, the bottom line of profit is ever present.

To students planning environmental careers, Andrew suggests that a strong technical education is important but that a broad liberal education is useful as well. Technical knowledge will get you a job, but your ability to communicate, work well with others, and manage projects will get you promoted. When job hunting, be persistent and choose an area that you really want. Careful analysis and preparation of the type Andrew employed can help you avoid dead-end jobs and ensure that you will find a place where you want to be.

Members of the Kenyan Greenbelt movement led by Wanari Maathai plant trees to build community and improve their environment. Grassroots nongovernmental organizations such as this have emerged as powerful voices for social justice and cleaner, more sustainable environments worldwide.

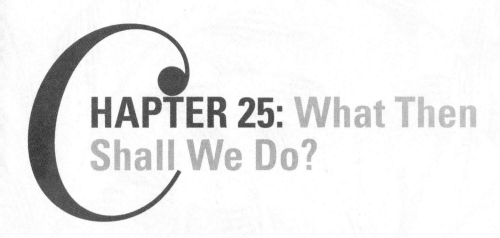

CHAPTER 25: What Then Shall We Do?

Objectives

After studying this chapter, you should be able to:

- be aware of the goals and opportunities in environmental education and environmental careers.

- recognize opportunities for making a difference through the goods and services we choose as well as the limits of green consumerism.

- compare the differences between radical and mainline environmental groups and the tactics they employ to bring about social change.

- summarize some practical suggestions for how to reduce consumption, organize an environmental campaign,

- communicate effectively with the media, and write to elected officials.

- appreciate the need for sustainable development and explain how nongovernmental groups work toward this goal.

- evaluate how green politics and governments function nationally and internationally to help protect the earth.

- formulate your own philosophy and action plan for what you can and should do to create a better world and a sustainable environment.

Making a Difference: Wangari Maathai

"When you talk only about problems, you disempower people. You make them feel there is nothing they can do, that there is no hope," says Wangari Maathai, founder of the Kenyan Greenbelt Movement (fig. 25.1). Looking for something positive to empower poor people and help restore self-reliance and community spirit, Maathai began her grassroots organizing with the simple step of planting seven trees in her backyard on World Environment Day, June 5, 1977. Since then, the Greenbelt Movement has grown to include more than 50,000 women and over a million school-children who have planted more than 10 million trees. And more than 80 percent of those trees have survived, a far greater proportion than in most reforestation projects.

As you have learned through your study of environmental science, most African nations are desperately poor and suffer

Figure 25.1

Wangari Maathai, founder of the Kenyan Greenbelt Movement, has received many awards for her work for social justice and environmental restoration.

tragic environmental degradation. Deforestation plays a big role in this cycle of poverty and loss. Forest destruction results in soil erosion, fuelwood shortages, biodiversity reduction, and the drying up of springs and streams. Most people think there is nothing they can do about these problems, but Wangari Maathai has shown that planting a tree is a simple, easy way by which we can help restore our environment.

A shelterbelt of trees around a farm field or African village has immediate practical benefits providing firewood, building materials, fruits, and fodder. It also beautifies the environment, provides shade, shelter from the wind and blowing dust, and creates a home for birds and small animals. Furthermore, planting trees brings people together to work for the common good. The motto of the Greenbelt Movement is Harambee, Swahili for "everyone pull together." Once established, this attitude can benefit many aspects of community life.

Originally dismissed as "just a bunch of women," the Greenbelt Movement has become an important political power in Kenya and a model for how ordinary people can make a difference. It may seem incongruous for Maathai, who was born to an upper class family and was the first woman in her country to earn a Ph.D. and become a professor at the University of Nairobi, to be working with illiterate rural women and barefoot foresters. She says, however, "when you are educated and have been exposed to knowledge, experience, and information, one of the things you should do is to be a useful person, to be a stimulant, a catalyst in your own society."

Maathai's involvement has not been without personal cost. Her husband divorced her because he disliked her activism and independence. In 1992 she was arrested and jailed briefly for her opposition to the corrupt, dictatorial rule of President Daniel arap Moi. After her release, Maathai joined a group of mothers and sisters protesting illegal detention of political prisoners. When armed policemen attacked and brutally beat up women and children at a rally in a Nairobi park, Maathai was seriously injured and was taken to a hospital in critical condition. Although now recovered, she still carries physical and emotional scars from her treatment.

There also have been rewards from this work. Maathai received the Right Livelihood Prize in 1984, the United Nations Environment Programme Global 500 Award in 1987, and the Goldman Environmental Prize in 1991, among others. She has been invited to many conferences and conventions. She has hobnobbed with world leaders. Maybe most satisfying, she has seen positive results in the environment and in the lives and status of ordinary women in her community as a result of her work.

Throughout this book, we have examined many urgent environmental dilemmas, as well as some ways that these problems could be overcome. As we pointed out in chapter 1, there are good reasons to be both pessimistic and optimistic about our prospects for the future. Clearly, many of the problems that face us seem to be getting worse. Still, we have made some encouraging progress in many areas. Although population growth continues at an unsustainable pace in many countries, most industrialized nations—and

some poorer countries—have dramatically reduced birth rates in the past century. We seem poised on the brink of breakthroughs in renewable energy supplies, environmentally friendly agricultural techniques, and better pollution control technology.

Perhaps most importantly, knowledge about and concern for our environment worldwide is at an all-time high. It may not be unreasonable to hope that we can continue to eliminate poverty and resource waste, and to create a healthier, happier, more just, and more meaningful life for everyone. In many ways, the choices we make now in managing our resources and our relationships with others determine what our lives—and those of our children—will be in the future (fig. 25.2). In this chapter, we will examine some things we can do individually and collectively to protect nature and to create sustainable societies and a better future for everyone. ▶

Figure 25.2

What will our environmental future be? In many ways, the choices we make now in managing resources and working with others to protect and restore our environment will determine what our lives and those of our children will be like in years to come.

Environmental Education

In 1990 Congress passed the National Environmental Education Act establishing two broad goals: (1) to improve understanding among the general public of the natural and built environment and the relationships between humans and their environment, including global aspects of environmental problems, and (2) to encourage postsecondary students to pursue careers related to the environment. Specific objectives proposed to meet these goals include developing an awareness and appreciation of our natural and social/cultural environment, knowledge of basic ecological concepts, acquaintance with a broad range of current environmental issues, and experience in using investigative, critical-thinking, and problem-solving skills in solving environmental problems. Several states, including Arizona, Florida, Maryland, Minnesota, Pennsylvania, and Wisconsin, have successfully incorporated these goals and objectives into their curricula.

Environmental Literacy

Speaking to the first of these broad goals, former Environmental Protection Agency administrator William K. Reilly called for broad **environmental literacy** in which every citizen is fluent in the principles of ecology and has a "working knowledge of the basic grammar and underlying syntax of environmental wisdom." Environmental literacy, according to Reilly can help create a stewardship ethic—a sense of duty to care for and manage wisely our natural endowment and our productive resources for the long haul. "Environmental education," he says, "boils down to one profoundly important imperative: preparing ourselves for life in the next century. When the 21st century rolls around, it will not be enough for a few specialists to know what is going on while the rest of us wander about in ignorance."

Environmental Careers

The need for both environmental educators and environmental professionals opens up many job opportunities in environmental fields. The EPA estimates, for example, that 100,000 new professionals will be required in the United States in the next five years alone to deal with hazardous waste problems. Scientists are needed to understand the natural world and the effects of human activity on the environment. Lawyers and other specialists are needed to develop government and industry policy, laws, and regulations to protect the environment. Engineers are needed to develop technologies and products to clean up pollution and to prevent its production in the first place. Economists, geographers, and social scientists are needed to evaluate the costs of pollution and resource depletion and to develop solutions that are socially, culturally, politically, and economically appropriate for different parts of the world. In addition, business will be looking for a new class of environmentally literate and responsible leaders who appreciate how products sold and services rendered affect our environment.

Trained people are essential in these professions at every level, from technical and clerical support staff to top managers. Unfortunately, while the need for environmental professionals is growing rapidly, the number of students enrolling in courses leading to these jobs is declining. Perhaps the biggest national demand over the next few years will be for environmental educators to help train an environmentally literate populace and to encourage the next generation to prepare themselves for jobs in this area. We urgently need many more teachers at every level who are trained in environmental education. Outdoor activities and natural sciences are important components of this mission (fig. 25.3), but environmental topics such as responsible consumerism, waste disposal, and respect for nature can and should be incorporated into reading, writing, arithmetic, and every other part of education. Table 25.1 presents some guidelines for an environmental education program.

Can environmental protection and resource conservation—a so-called green perspective—be a strategic advantage in business? Many companies think so. An increasing number are jumping on the environmental bandwagon, and most large corporations now have an environmental department. A few are beginning to explore integrated programs to design products and manufacturing processes to minimize environmental impacts. Often called "design for the environment," this approach is intended to avoid problems at the beginning rather than deal with them later on a case-by-case basis. In the long run, executives believe this will save money and make their businesses more competitive in future markets. The

Figure 25.3

Outdoor activities are an essential part of environmental education, which strives to introduce environmental literacy to the general population and to encourage students to prepare for environmental careers.

What Then Shall We Do?

Table 25.1
Outcomes from environmental education

The natural context: An environmentally educated person understands the scientific concepts and facts that underlie environmental issues and the interrelationships that shape nature.

The social context: An environmentally educated person understands how human society is influencing the environment, as well as the economic, legal, and political mechanisms that provide avenues for addressing issues and situations.

The valuing context: An environmentally educated person explores his or her values in relation to environmental issues; from an understanding of the natural and social contexts, the person decides whether to keep or change those values.

The action context: An environmentally educated person becomes involved in activities to improve, maintain, or restore natural resources and environmental quality for all.

Source: *A Greenprint for Minnesota,* Minnesota Office of Environmental Education, 1993.

Figure 25.4
Many interesting, well-paid jobs are opening up in environmental fields. Here an environmental technician uses a video probe to inspect a residential sewer line.

alternative is to face increasing pollution control and waste disposal costs—now estimated to be more than $100 billion per year for all American businesses—as well as to be tied up in expensive litigation and administrative proceedings.

The market for pollution-control technology and know-how is also expected to be huge. Cleaning up the former East Germany alone is expected to cost some $200 billion. Many companies are positioning themselves to cash in on this enormous market. Germany and Japan appear to be ahead of America in the pollution control field because they have had more stringent laws for many years, giving them more experience in reducing effluents.

The rush to "green up" business is good news for those looking for jobs in environmentally related fields, which are predicted to be among the fastest growing areas of employment during the next few years. The federal government alone projects a need to hire some 10,000 people per year in a variety of environmental disciplines (fig. 25.4). How can you prepare yourself to enter this market? The best bet is to get some technical training: environmental engineering, analytical chemistry, microbiology, ecology, limnology, groundwater hydrology, or computer science all have great potential. Currently, a chemical engineer with a graduate degree and some experience in an environmental field can practically name his or her salary. Some other very good possibilities are environmental law and business administration, both rapidly expanding fields.

For those who aren't inclined toward technical fields there are still opportunities for environmental careers. A good liberal arts education will help you develop skills such as communication, critical thinking, balance, vision, flexibility, and caring that should serve you well. Large companies need a wide variety of people; small companies need a few people who can do many things well.

We hope you have noticed the profiles throughout this book describing a variety of young people and the environmental careers they are pursuing. The tips they offer for how you might prepare yourself for similar jobs should help in your job search. There are many opportunities if you think creatively and are persistent.

Individual Accountability

Some prime reasons for our destructive impacts on the earth are our consumption of resources and disposal of wastes. Technology has made consumer goods and services cheap and readily available in the richer countries of the world. As you already know, we in the industrialized world use resources at a rate out of proportion to our percentage of the population. If everyone in the world were to attempt to live at our level of consumption, given current methods of production, the results would surely be disastrous. In this section we will look at some options for consuming less and reducing our environmental impacts.

How Much Is Enough?

The first questions we should ask ourselves is whether we really need to consume so much (fig. 25.5). How much ought we leave for other people and future generations? As chapter 2 points out, in-

Figure 25.5
Is this our highest purpose?

© 1990 Bruce von Alten.

tergenerational justice is an important component of environmental ethics. Although advertising urges us to buy and discard more, are our lives really better with more stuff, or simply more complicated and stressful? Shopping has become a major pastime in many industrialized countries. The average American spends more than two hours every week shopping for goods other than food. We own twice as many cars, drive two-and-a-half times as far, use twenty times as much plastic, and travel twenty-five times as far by air as did our parents in 1950.

To avoid waste production, we can practice **precycling,** making environmentally sound decisions at the store and reducing waste before we buy. We are already making good progress in this area, but we could do more. Table 25.2 offers some suggestions for reducing our consumption levels by buying only what we need.

Shopping for Green Products

Obviously, we can never reduce our consumption levels to zero. We can, however, make sound, informed decisions about the products we buy to select those that are least environmentally damaging in production, use, and disposal. As consumers demand environmentally friendly products, manufacturers, food producers, and merchants are moving to safer, more humane, and more sustainable consumer items. Although each of our individual choices makes a small impact, collectively they can be important.

With surveys showing nine out of ten of us willing to pay more for products and packaging that don't harm the environment, many merchandisers are moving quickly to cash in on green consumerism. Green sales and services are projected to nearly quintuple, from $1.8 billion in 1990 to $8.8 billion in 1995. Some claims made by green marketers are of questionable validity, however. Consumers must look closely to avoid "green scams." Many terms used in advertising are vague and have little meaning. For example:

- "Nontoxic" suggests that a product has no harmful effects on humans. Since there is no legal definition of the term, however, it can have many meanings. How nontoxic is the product? And to whom? Substances not poisonous to humans can be harmful to other organisms.

- "Biodegradable," "recyclable," "reusable," or "compostable" may be technically correct but not signify much. Almost everything will biodegrade *eventually,* but it may take thousands of years. Similarly, almost anything is potentially recyclable or reusable; the real question is whether there are programs to do so in your community. If the only recycling or composting program for a particular material is half a continent away, this claim has little value.

- "Natural" is another vague and often misused term. Many natural ingredients—lead or arsenic, for instance—are highly toxic. Synthetic materials are not necessarily more dangerous or environmentally damaging than those created by nature.

- "Organic" can connote different things in different places. Some states have standards for organic food, but others do not. On items such as shampoos and skin-care products, "organic" may have no significance at all. Most detergents and oils are organic chemicals, whether they are synthesized in a laboratory or found in nature. Few of these products are likely to have pesticide residues anyway.

- "Environmentally friendly," "environmentally safe," and "won't harm the ozone layer" are often empty claims. Since there are no standards to define these terms, anyone can use them. How much energy and nonrenewable material are used in manufacture, shipping, or use of the product? How much waste is generated, and how will the item be disposed of when it is no longer functional? One product may well be more environmentally benign than another, but be careful who makes this claim.

Blue Angels and Green Seals

Products that claim to be environmentally friendly are being introduced at twenty times the normal rate for consumer goods. To help consumers make informed choices, several national programs have

    What Then Shall We Do?

Table 25.2
Reducing consumption

Purchase Less

Ask yourself whether you really need more stuff.

Avoid buying things you don't need or won't use.

Use items as long as possible (and don't replace them just because a new product becomes available).

Use the library instead of purchasing books you read.

Make gifts from materials already on hand, or give nonmaterial gifts.

Reduce Excess Packaging

Carry reusable bags when shopping and refuse bags for small purchases.

Buy items in bulk or with minimal packaging; avoid single-serving foods.

Choose packaging that can be recycled or reused.

Avoid Disposable Items

Use cloth napkins, handkerchiefs, and towels.

Bring a washable cup to meetings; use washable plates and utensils rather than single-use items.

Buy pens, razors, flashlights, and cameras with replaceable parts.

Chose items built to last and have them repaired; you will save materials and energy while providing jobs in your community.

Conserve Energy

Walk, bicycle, or use public transportation.

Turn off (or avoid turning on) lights, water, heat, and air conditioning when possible.

Put up clotheslines or racks in the backyard, carport, or basement to avoid using a clothes dryer.

Carpool and combine trips to reduce car mileage.

Save Water

Water lawns and gardens only when necessary.

Use water-saving devices and fewer flushes with toilets.

Don't leave water running when washing hands, food, dishes, and teeth.

Based on material by Karen Oberhauser, Bell Museum Imprint, University of Minnesota, 1992. Used by permission.

been set up to independently and scientifically analyze environmental impacts of major products. Germany's Blue Angel, begun in 1978, is the oldest of these programs. Endorsement is highly sought after by producers since environmentally conscious shoppers have shown that they are willing to pay more for products they know have minimum environmental impacts. To date, more than 2000 products display the Blue Angel symbol. They range from re-

cycled paper products, energy-efficient appliances, and phosphate-free detergents to refillable dispensers.

Similar programs are being proposed in every Western European country as well as in Japan and North America. Some are autonomous, nongovernmental efforts like the United States' new Green Seal program (managed by the Alliance for Social Responsibility in New York). Others are quasigovernmental institutions such as the Canadian Environmental Choice programs (fig. 25.6).

The best of these organizations attempt "cradle-to-grave" **life-cycle analysis** (fig. 25.7) that evaluates material and energy inputs and outputs at each stage of manufacture, use, and disposal of the product. While you need to consider your own situation in making choices, the information supplied by these independent agencies is generally more reliable than self-made claims from merchandisers.

Limits of Green Consumerism

To quote Kermit the Frog, "It's not easy being green." Even with the help of endorsement programs, doing the right thing from an environmental perspective may not be obvious. Often we are faced with complicated choices. Do the social benefits of buying rainforest nuts justify the energy expended in transporting them here, or would it be better to eat only locally grown products? In switching from Freon propellants to hydrocarbons, we spare the stratospheric ozone but increase hydrocarbon-caused smog. By choosing reusable diapers over disposable ones, we decrease the amount of material going to the landfill, but we also increase water pollution, energy consumption, and pesticide use (cotton is one of the most pesticide-intensive crops grown in the United States).

When the grocery store clerk asks you, "Paper or plastic?" you probably choose paper and feel environmentally virtuous, right? Everyone knows that plastic is made by noxious chemical processes from nonrenewable petroleum or natural gas. Paper from naturally growing trees is a better environmental choice, isn't it? Well, not

(a)   (b)

Figure 25.6

American (a) and Canadian (b) symbols that will be used to indicate products that are "environmentally superior."

(a) Reproduced with permission, Green Seal, Inc., Washington, DC. (b) M—Official mark of Environment Canada.

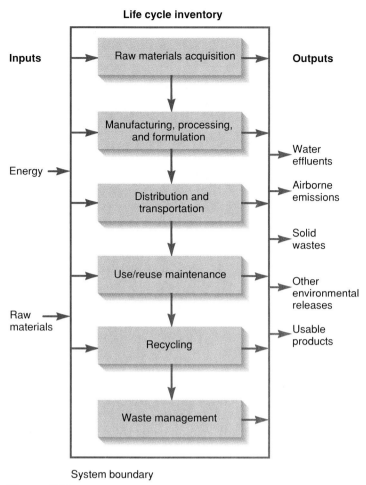

Life cycle inventory

Inputs

Raw materials acquisition → Outputs

Energy →

Manufacturing, processing, and formulation → Water effluents

Distribution and transportation → Airborne emissions

Solid wastes

Use/reuse maintenance → Other environmental releases

Raw materials →

Recycling → Usable products

Waste management →

System boundary

Figure 25.7

At each stage in its life cycle, a product receives inputs of materials and energy and produces outputs of materials or energy that move to subsequent phases and wastes that are released into the environment. The relationship between inputs and outputs and the six stages of a life-cycle inventory is shown.

Source: After Society of Environmental Toxicology and Chemistry, 1991.

necessarily. In the first place, paper making consumes water and causes much more water pollution than does plastic manufacturing. Paper mills also release air pollutants, including foul-smelling sulfides and captans as well as highly toxic dioxins.

Furthermore, the brown paper bags used in most supermarkets are made primarily from virgin paper. Recycled fibers aren't strong enough for the weight they must carry. This paper comes mainly from intensively managed, heavily fertilized, clearcut agroforestry plantations. It takes a great deal of energy to pulp wood and dry newly made paper. Paper is also heavier and bulkier to ship than plastic. Although the polyethylene used to make a plastic bag contains many calories, in the end, paper bags are generally more energy-intensive to produce and market than plastic ones.

If both paper and plastic go to a landfill in your community, the plastic bag takes up less space. It doesn't decompose in the landfill, but neither does the paper in an air-tight, water-tight landfill. If paper is recycled but plastic is not, then the paper bag may be the better choice. If you are lucky enough to have both paper and plastic recycling, the plastic bag is probably a better choice since it recycles more easily and produces less pollution in the process. The best choice of all is to bring your own reusable cloth bag.

Complicated, isn't it? We often must make decisions without complete information, but it's important to make the best choices we can. Don't assume that your neighbors are wrong if they reach conclusions different from yours. They may have valid considerations of which you are unaware. The truth is that simple black and white answers often don't exist.

Paying Attention to What's Important

Taking personal responsibility for our environmental impact can have many benefits. Recycling, buying "green" products, and other environmental good works not only set good examples for your friends and neighbors, they also strengthen your sense of involvement and commitment in valuable ways. There are limits, however, to how much we can do individually through our buying habits and personal actions to bring about the fundamental changes needed to save the earth. Green consumerism generally can do little about larger issues of global equity, chronic poverty, oppression, and the suffering of millions of people in the Third World. There is a danger that exclusive focus on such problems as whether to choose paper or plastic bags will divert our attention from the greater need to change basic institutions.

During the 1970s, many people joined the back-to-the-land movement. Abandoning mass society, they moved to places like the mountains of North Carolina or remote valleys in Vermont where they experimented with self-sufficient lifestyles, living in converted school buses, teepees, or geodesic domes; growing their own food; cutting firewood; carrying water; and performing other aspects of alternative, utopian living. This lifestyle can be satisfying and morally uplifting but may contribute little toward helping other people live better lives.

Remaining a part of mass society, living an ordinary life, and devoting time and energy to bringing about important structural changes may ultimately be more valuable than withdrawing to "do your own thing." Rather than sort recyclables for which there is no market, for instance, your time might be better spent working for legislation to prevent excess packaging and to create markets for recycled materials. As in many other cases, there is no simple choice between these alternatives. Some of us will focus primarily on living sustainably, others will work to bring about change in the system, and still others will find some balance between these two options.

Collective Actions

While a few exceptional individuals can be effective working alone to bring about change, most of us find it more productive and more satisfying to work with others (Case Study, p. 570).

Collective action multiplies individual power (fig. 25.8). You get encouragement and useful information from meeting regularly with others who share your interests. It's easy to get discouraged by the slow pace of change; having a support group helps maintain your enthusiasm. You should realize, however, that there is a broad spectrum of environmental and social action groups. Some will suit your particular interests, preferences, or beliefs more than others. In this section, we will look at some environmental organizations as well as options for getting involved.

Student Environmental Groups

A number of organizations have been established to teach ecology and environmental ethics to elementary and secondary school students, as well as to get them involved in active projects to clean up their local community. Groups such as Kids Saving the Earth or Eco-Kids Corps are an important way to reach this vital audience. Family education results from these efforts as well. In a World Wildlife Fund survey, 63 percent of young people said they "lobby" their parents about recycling and buying environmentally responsible products.

Organizations for secondary and college students often are among our most active and effective groups for environmental change. The largest student environmental group in North America is the Student Environmental Action Coalition (SEAC). Formed in 1988 by students at the University of North Carolina at Chapel Hill, SEAC has grown rapidly to more than 30,000 members in some 500 campus environmental groups. SEAC is both an umbrella organization and grassroots network that functions as an information clearinghouse and a training center for student leaders. Member groups undertake a diverse spectrum of activities ranging from politically neutral recycling promotion to confrontational protests of government or industrial projects (fig. 25.9). National conferences bring together thousands of activists who share tactics and inspiration while also having fun. If there isn't a group on your campus, why not look into organizing one?

Another important student organizing group is the network of Public Interest Research Groups active on most campuses in the United States. While not focused exclusively on the environment, the PIRGs usually include environmental issues in their priorities for research. By becoming active, you could probably introduce environmental concerns to your local group if they are not already working on problems of importance to you.

One of the most important skills that you are likely to learn in SEAC or other groups committed to social change is how to organize. Organizing is a dynamic process in which you must constantly adapt to changing conditions. Some basic principles apply in most situations, however (table 25.3). Remember that you are not alone. Others share your concerns and want to work with you to bring about change; you just have to find them. There is power in numbers. As Margaret Mead once said, "Never doubt that a small, highly committed group of individuals can change the world; indeed, it is the only thing that ever has."

Using the communication media to get your message out is an important part of the modern environmental movement. Table 25.4 suggests some important considerations in planning a media campaign.

Figure 25.8

Students cleaning up the beach in Santa Cruz, CA, after an oil spill. Ordinary people can make a contribution to improving the environment. Doing so can be personally rewarding and can foster public spirit.

Figure 25.9

Imaginative costumes and street theater can get your message across in a humorous, nonthreatening way that is attractive to the media if carefully executed. Here, a talking trash can tells students the benefits of recycling.

Table 25.3
Organizing an Environmental Campaign

1. What do you want to change? Are your goals realistic, given the time and resources you have available?

2. What and who will be needed to get the job done? What resources do you have now and how can you get more?

3. Who are the stakeholders in this issue? Who are your allies and constituents? How can you make contact with them?

4. How will your group make decisions and set priorities? Will you operate by consensus, majority vote, or informal agreement?

5. Have others already worked on this issue? What successes or failures did they have? Can you learn from their experience?

6. Who has the power to give you what you want or to solve the problem? Which individuals, organizations, corporations, or elected officials should be targeted by your campaign?

7. What tactics will be effective? Using the wrong tactics can alienate people and be worse than taking no action at all.

8. Are there social, cultural, or economic factors that should be recognized in this situation? Will the way you dress, talk, or behave offend or alienate your intended audience? Is it important to change your appearance or tactics to gain support?

9. How will you know when you have succeeded? How will you evaluate the possible outcomes?

10. What will you do when the battle is over? Is yours a single-issue organization, or will you want to maintain the interest, momentum, and network you have established?

Source: Based on material from "Grassroots Organizing for Everyone" by Claire Greensfelder and Mike Roselle from *Call to Action,* 1990 Sierra Book Club Books.

Mainline Environmental Organizations

Among the oldest, largest, and most influential environmental groups in the United States are the National Wildlife Federation, the World Wildlife Fund, the Audubon Society, the Sierra Club, the Izaak Walton League, Friends of the Earth, Greenpeace, Ducks Unlimited, the Natural Resources Defense Council, and The Wilderness Society. Sometimes known as the "group of 10," these organizations are criticized by radical environmentalists for their tendency to compromise and cooperate with the establishment. Although many of these groups were militant—even extremist—in their formative stages, they now tend to be more staid and conservative. Members are mostly passive and know little about the inner workings of the organization, joining as much for publications or social aspects as for their stands on environmental issues.

Table 25.4
Using the Media to Influence Public Opinion

Shaping opinion, reaching consensus, electing public officials, and mobilizing action are accomplished primarily through the use of the communications media. To have an impact in public affairs, it is essential to know how to use these resources. Here are some suggestions:

1. *Assemble a press list.* Learn to write a good press release by studying books from your public library on press relations techniques. Get to know reporters from your local newspaper and TV stations.

2. *Appear on local radio and TV talk shows.* Get experts from local universities and organizations to appear.

3. *Write letters to the editor, feature stories, and news releases.* You may include black and white photographs. Submit them to local newspapers and magazines. Don't overlook weekly community shoppers and other "freebie" newspapers, which usually are looking for newsworthy material.

4. *Try to get editorial support from local newspapers, radio, and TV stations.* Ask them to take a stand supporting your viewpoint. If you are successful, send a copy to your legislator and to other media managers.

5. *Put together a public service announcement and ask local radio and TV stations to run it* (preferably not at 2 A.M.). Your library or community college may well have audio-visual equipment that you can use. Cable TV stations usually have a public access channel and will help with production.

6. *If there are public figures in your area who have useful expertise, ask them to give a speech or make a statement.* A press conference, especially in a dramatic setting, often is a very effective way of attracting attention.

7. *Find celebrities or media personalities to support your position.* Ask them to give a concert or performance, both to raise money for your organization and to attract attention to the issue. They might like to be associated with your cause (fig. 25.10).

8. *Hold a media event that is photogenic and newsworthy.* Clean up your local river and invite photographers to accompany you. Picket the corporate offices of a polluter, wearing eye-catching costumes and carrying humorous signs. Don't be violent, abusive, or obnoxious; it will backfire on you. Good humor usually will go farther than threats.

9. *If you hear negative remarks about your issue on TV or radio, ask for free time under the Fairness Doctrine to respond.* Stations have to do a certain amount of public service to justify relicensing and may be happy to accommodate you.

10. *Ask your local TV or newspaper to do a documentary or feature story about your issue or about your organization and what it is trying to do.* You will not only get valuable free publicity, but you may inspire others to follow your example.

CASE STUDY
Citizens Against Toxic Waste

In the mid-1970s, Lois Gibbs was an ordinary housewife living happily with her husband and two children in a suburban neighborhood of Niagara Falls, New York (fig. 1). When her son became ill with an unusual combination of epilepsy and low white blood cell count—neither of which had ever been known in her family—she was puzzled and concerned. In talking with neighbors, Lois discovered that many other families were suffering from a variety of unexplained illnesses including cancers, childhood leukemias, allergies, and birth defects. Some remembered that years before the neighborhood had been the site of a smelly, industrial dump called Love Canal into which the Hooker Chemical Company had dumped more than 20,000 tons of chemical waste.

Although the city bought the dump from Hooker in 1953 and covered the pools of oily chemicals and leaking drums with dirt, residents had continued to complain that nasty-smelling liquids oozed out of low areas and that black, rancid residue leaked into their basements after every rain. An engineering firm found that solvents, oils, pesticides, herbicides, and other toxic chemicals were migrating through the soil. The school that the Gibbs children attended had been built directly over the dump.

Lois began circulating a petition among her neighbors demanding that the dump be cleaned up and that the school be protected or closed. Going door-to-door to talk about health problems and pollution took a great deal of courage on Lois' part. With only a high-school education, she felt very uneasy challenging company experts and governmental authorities who scoffed at "housewife research" and suggested that she ought to stay at home and pay attention to her cooking and sewing.

Lois and her neighbors persisted, however. They read about Love Canal and toxic wastes. They attended numerous meetings, organized a citizen's group, prepared press releases, met with city, state, and federal officials.

Figure 1

Lois Gibbs, leader of the Love Canal Homeowners' Association, went on to found the Citizen's Clearinghouse for Hazardous Waste, a national leader in organizing opposition to toxic dumps.

Finally they hired a lawyer and filed a lawsuit against the city, the Board of Education, and Hooker Chemical. Not everyone thought this activism and agitation was a good idea. For most people in this working class neighborhood, their home equity was their only form of savings. With all the controversy about Love Canal, those homes were worthless. Residents couldn't sell, and most couldn't afford to just walk away. Some would rather have not known what was in their basements.

By 1979, the protests were getting militant. An EPA study found chromosome abnormalities in 11 of 36 people who lived near Love Canal. Toxicologists recommended that pregnant women and young children be moved immediately from the neighborhood. But what about everyone else? If it wasn't safe for some to live there, how could it be safe for others? At one point an angry crowd threatened to hold EPA officials hostage until the government offered some relief.

Finally, in 1981, 810 families were evacuated, the school was torn down, and remediation work began. Underground streams have been diverted and a containment system of thick clay walls around the dump and two impervious caps over the top has been installed to prevent further migration of toxic wastes. The 239 houses immediately adjacent to the dump have been demolished. Some 236 previously abandoned houses in the next ring out have been declared habitable again and are now being sold at bargain prices to people who don't know or don't care about their previous history.

In 1988, Occidental Petroleum (the parent company of Hooker Chemical) agreed to pay $250 million in damages to Love Canal residents. Although they were repaid for property losses, the stresses and disruptions were traumatic for many families. This incident has awakened many of us to the dangers of toxic wastes and the careless way they have been discarded. Love Canal was the prime incentive for passage of the Superfund, which was intended to provide funds for immediate relocation and cleanup of other toxic waste sites without having to spend years in research and litigation to determine who was responsible.

In 1981, Lois moved to Washington, DC, and founded the Citizens Clearinghouse for Hazardous Wastes. She has become one of the leading citizen advocates in the United States on problems of toxic wastes and chemical spills. With a full-time staff of 13, and donations from individuals and foundations, the Clearinghouse has become a major force for environmental protection. Lois Gibbs says, "Average people *can* change the world. All you need is common sense, determination, persistence, and patience. It's a false premise that one must be an expert to make a difference. Major changes have come from local people being angry and speaking out." Could you follow her lead?

Figure 25.10
Celebrities can help raise money and bring visibility to your cause. Here Whitney Houston, Sting, Luciano Pavarotti, and Elton John perform at a Rainforest Foundation Benefit at Carnegie Hall.

Still, these groups are powerful and important forces in environmental protection. Their mass membership, large professional staffs, and long history give them a degree of respectability and influence not found in newer, smaller groups. The Sierra Club, for instance, with nearly half a million members and chapters in almost every state, has a national staff of about four hundred, an annual budget over $20 million, and twenty full-time professional lobbyists in Washington. These national groups have become a potent force in Congress, especially when they band together to pass specific legislation such as the Alaska National Interest Lands Act or the Clean Air Act.

In a survey that asked congressional staff and officials of government agencies to rate the effectiveness of groups that attempt to influence federal policy on pollution control, the top five were national environmental organizations. In spite of their large budgets and important connections, the American Petroleum Institute, the Chemical Manufacturers Association, and the Edison Electric Institute ranked far behind these environmental groups in terms of influence.

Some environmental groups such as the Environmental Defense Fund (EDF), The Nature Conservancy (TNC), National Resources Defense Council (NRDC), and the Wilderness Society (WS) have limited contact with ordinary members except through their publications. They depend on a professional staff to carry out the goals of the organization through litigation (EDF and NRDC), land acquisition (TNC), or lobbying (WS). Although not often in the public eye, these groups can be very effective because of their unique focus. The Nature Conservancy buys land of high ecological value that is threatened by development. With more than $50 million in cash purchases and $44 million in donated lands, it now owns some nine hundred parcels of land, the largest privately

owned system of nature sanctuaries in the world. Altogether, it has saved more than 1 million ha (2.47 million ac) of land in its forty-year history.

Broadening the Environmental Agenda

The environmental movement in the United States tends to be overwhelmingly white, middle-class, and suburban. Few blue-collar workers and even fewer minorities are involved, especially in leadership or professional positions. This is unfortunate in several ways. The movement will probably never be truly successful until it has a broad-based coalition of support. Furthermore, as we saw in chapter 2, poor people are often most seriously affected by toxic waste dumps, noise, urban blight, and air and water pollution. They should be included in discussions of how to remedy these problems. Unfortunately, most environmental groups have acquired a reputation for caring about plants, animals, and wilderness areas more than about people. Recently, a number of grass-roots citizen's groups have begun to demand cleaner local environments. Often organized around tenant councils or other social justice issues, these groups could be a powerful force for general environmental protection.

Deep or Shallow Environmentalism?

Norwegian philosopher Arne Naess criticizes the superficial commitment of environmentalists who claim to be green but are quick to compromise and who do little to bring about fundamental change. He characterizes this as **shallow ecology** and contrasts it with what he calls **deep ecology,** a profoundly biocentric worldview that calls for a fundamental shift in our attitudes and behavior. An equally radical but more humanist philosophy called **social ecology** is proposed by Murray Bookchin, who draws on the communitarian anarchism of the Russian geographer Peter Kropotkin.

Among the tenants of both these "dark green" philosophies are voluntary simplicity; rejection of anthropocentric attitudes; intimate contact with nature; decentralization of power; support for cultural and biological diversity; a belief in the sacredness of nature; and direct personal action to protect nature, improve the environment, and bring about essential societal change. They differ mainly on the necessity for personal freedom and communal interaction.

Both Naess and Bookchin criticize mainstream environmental groups that favor pragmatic work within the broadly agreed-upon social agenda and the established political system to bring about incremental, progressive reform rather than radical revolution. Most progressives naturally object to being called shallow, light, or superficial; they prefer instead to characterize their position as moderate rather than radical, or reformist rather than revolutionary. They tend to view the deep ecologists and anarchists as unrealistic zealots who would rather be righteous than effective.

Aside from name calling, there are some important philosophical questions in this debate. One of the questions is whether we face an environmental crisis that requires radical action or only "problems" that can best be overcome by working within established channels. Another important question is whether it is better to overturn society to explore new ways of living or to work for progressive change within existing political, economic, and social systems. Finally, is it more important to work for personal perfection or collective improvement? There may be no single answer to any of these questions; it's good to have people working in many different ways to find solutions.

Radical Environmental Groups

A striking contrast to the mainline conservation organizations are the **direct action** groups, such as Earth First!, Sea Shepherd, and a few other groups that form either the "cutting edge" or the "radical fringe" of the environmental movement, depending on your outlook. Often associated with the deep ecology philosophy and bioregional ecological perspective, the strongest concerns of these militant environmentals tend to be animal rights and protection of wild nature. Their main tactics are civil disobedience and attention-grabbing actions, such as guerrilla street theater, picketing, protest marches, road blockades, and other demonstrations. Many of these techniques are borrowed from the civil rights movement and Mahatma Gandhi's nonviolent civil disobedience. While often more innovative than the mainstream organizations, pioneering new issues and new approaches, the tactics of these groups can be controversial.

Members of Earth First! perched in Douglas firs marked for felling in Oregon, chained themselves to a giant tree-smashing bulldozer to prevent forest clearing in Texas, and blockaded roads being built into wilderness areas. In some cases, the protests can be humorous and lighthearted, such as skits or street theater that get a point across in a non-threatening way (fig. 25.11).

A more problematic tactic is **monkey wrenching,** or environmental sabotage, a concept made popular by author Edward Abbey's book *The Monkey Wrench Gang.* Among the actions advocated by some Earth First!ers are driving large spikes in trees to protect them from loggers, vandalizing construction equipment, pulling up survey stakes for unwanted developments, and destroying billboards.

This raises some difficult ethical questions. While each of us has a duty to resist immoral authority and destruction of nature, what kinds of action are justified? Suppose that in spiking trees you endanger loggers or mill workers; what moral responsibility do you bear? On a pragmatic level, does violent action engender or validate counterviolence?

Sea Shepherd, a radical offshoot from Greenpeace, is determined to stop the killing of marine mammals. Convinced that peaceful protests were not working, the members took direct action by sinking whaling vessels and destroying machinery at whale processing stations in Iceland. Greenpeace criticized this destructive approach, perhaps because of memories of the sinking of its ship *Rainbow Warrior* by the French secret service in New Zealand in 1985. They and other critics of ecotage (ecological sabotage) see these acts as environmental terrorism that give all environmentalists a bad name. Sea Shepherd's Paul Watson says, "Our aggressive nonviolence is just doing what must be done, according to the dictates of our own conscience." Earth First!'s Dave Foreman (sometimes described as a kamikaze environmentalist) says, "Extremism

Figure 25.11

Street theater can be a humorous, yet effective, way to convey a point in a non-threatening way. Confrontational tactics get attention, but they may alienate those who might be your allies and harden your opposition.

in defense of Mother Earth is no vice." What do you think? Do ends justify means?

Antienvironmental Backlash

As the world becomes increasingly urbanized and interconnected, social controls become necessary for us to live together safely and peacefully. Some people, however, find these restrictions confining. Especially in the western United States and Alaska, where people cling to traditions of independence, self-reliance, and personal freedom, a widespread feeling prevails that outsiders and the government represent a threat to local customs and values. Those whose livelihood is based on mining, logging, ranching, fishing, farming, and other methods of resource-extraction see jobs, communities, and ways of life coming to an end (fig. 25.12).

To many who live in sparsely populated areas—like much of the West—big government seems to be alien yet all-invasive, interfering constantly and unnecessarily in their lives. Environmental rules and regulations are an important part of this cultural clash. Restrictions on resource access, recreational opportunities, development rights, and other ways that people have become accustomed to using the land seem an infringement. Often the rhetoric on both sides of this argument rises to shrill extremes (What Do *You* Think?, p. 574).

Drawing on the anger, suspicion, and resentment felt by many rural residents, a number of antienvironmental groups have sprung up in recent years. Grouped together under the banner of the **wise use movement,** these organizations generally stand for little or no government, unlimited access to resources, less pollution control, and reduced protection for wild nature and endangered species. The Center for the Defense of Free Enterprise, founded by conser-

Figure 25.12
To families dependent on jobs in logging, mining, ranching, or other resource-based occupations, environmental regulations may seem threatening. How do we balance human needs against those of nature?

link between some of these groups and armed survivalists, tax protesters, and antigovernment militias is a stark contrast to radical environmental groups, whose activism generally is directed against property and ideas, not people.

These antienvironmental groups have had little national political power until recently. The Republican Congressional victory in 1994, however, has given them an influential voice. As this text is being written late in 1995, hundreds of bills have been introduced in Congress to roll back 50 years of social progress and environmental protection. Among their aims are to: (1) severely weaken the Clean Water Act; (2) effectively repeal the Clean Air Act—especially its 1990 amendments; (3) eliminate the Endangered Species Act; (4) sell federal lands including parks, wildlife refuges, grazing lands, and forests; (5) force government, under so-called "takings bills," to pay polluters not to pollute and developers not to destroy wetlands; (6) rescind funds for upgrading and monitoring water treatment systems under the Safe Drinking Water Act; (7) mandate increased logging on federal lands at bargain rates, especially the now-protected old-growth forests in the Northwest and Alaska; (8) remove citizens' right-to-know about toxic chemicals; (9) block reform of the 123 year-old mining law that gives away billions of dollars every year in public minerals and land to foreign and U.S. conglomerates; and (10) open all public lands to motorized recreation.

Current Interior Secretary Bruce Babbitt calls this a "war on the environment." Perhaps, however, it will only prove to be a skirmish. Ask your instructor what has happened to this agenda.

Global Issues

At the first United Nations Conference on the Environment in 1972 in Stockholm, Prime Minister Indira Gandhi of India declared that "poverty is the greatest danger to the environment." Twenty years of research and subsequent international conferences have shown the truth of that statement. In 1992, the United Nations Conference on Environment and Development focused on the plight of the poor and the need for sustainable development to alleviate poverty and provide ways to avoid environmental destruction.

The needs are tremendous. The United Nations estimates that about 1 billion people worldwide live in acute poverty and lack access to secure food supplies, safe drinking water, education, health services, infrastructure (roads, markets, etc.), land, credit, and jobs (fig. 25.13). Even more—around 1.5 billion—are exposed to dangerous air pollution, especially in urban areas and in homes where smoky fires provide the only means of cooking. About 1.8 billion people—nearly one-third of the world population—have inadequate sanitation, leading to the spread of infectious diseases that debilitate the poor and prevent them from working effectively to improve their conditions.

The connection between poverty and the environment has been a major focus of this book. Having to meet urgent short-term needs for survival, the poor are forced to "mine" their natural capital through excessive tree cutting, poor farming and grazing practices that lead to erosion and nutrient depletion, and overharvesting of fisheries and wildlife. The consequent degradation of the

vative fund raiser Alan Gottlieb and former Sierra Clubber Ron Arnold, leads the way in setting an agenda for the wise use movement. Sending out 30 million pieces of direct mail each year, Gottlieb and Arnold raise $5 million annually for themselves and for antienvironmental causes.

The National Cattlemen's Association and the National Farm Bureau, with hundreds of thousands of members and annual budgets approaching $10 million, compare in size and influence to some of the largest environmental groups. Their main concerns are with farming and ranching issues such as water projects and grazing on federal lands. People for the West, with 15,000 members in more than 200 local chapters, claims to be a grassroots, populist organization. However, 12 of its 13 original board members and almost all of its $1 million budget come from big timber or mining corporations. The Mountain States Legal Foundation, founded by former Interior Secretary James Watt and funded by donations from Coors Beer and other corporations, provides legal services to antienvironmental groups such as ranchers who tried to stop wolf reintroduction in Yellowstone National Park.

The California Desert Coalition is an unlikely association of off-road bikers, miners, ranchers, and oilmen who resent passage of the 1995 California Desert Protection Act and the new national parks it established. An offshoot of this group is the "Sahara Club," which advocates intimidation and violence—including assassination—to terrorize opponents. As James Watt, said in 1993, "If the troubles from environmentalists cannot be solved in the jury box or at the ballot box, perhaps the cartridge box should be used." The

What Then Shall We Do?

What Do *You* Think?
Evaluating Extremist Claims

Attaining a sustainable future requires us to change some of the ways we impact the earth. Some problems will likely turn out less serious and others more so, than we currently believe. Undoubtedly new concerns have yet to be uncovered. The issues will continue to stimulate strong feelings on both sides for several reasons. The problems are potentially very serious. Uncertainty is likely to continue to surround many of the issues. And, most importantly, needed changes often will challenge our world views.

Sometimes advocates on both sides get carried away and attempt to persuade through the use of biased language instead of relying on objective evidence to influence the thought of others. Critical thinking requires that conclusions be based on the objective evaluation of objective evidence. Bias, a person's inclination to favor or oppose a particular conclusion, is a formidable barrier to critical thinking.

Being able to detect bias enables a person to determine whether a claim or position is being presented objectively. There are a number of strategies to help identify bias. One clue as to whether or not a claim is being presented fairly is the person's choice of language. Many words have positive or negative connotations that can reflect emotion rather than objectivity.

The viewpoints presented below are positions that have appeared in print. Notice that bias can range from obvious to subtle.

Writer A. *Despite the hysterics of a few pseudo-scientists, there is no reason to believe in global warming.*

This writer seeks to persuade by using language to ridicule the view that the global warming threat is real. Would you believe a "*pseudo-scientist,*" especially one using "*hysterics?*" Such emotionally-charged language attacks the credibility of the opposing view. The implication is that only kooks think global warming likely. Would the sentence have lost its emotional punch if it had said: "Despite the claims of researchers, there is no reason to believe in global warming."

Writer B. *With the collapse of Marxism, environmentalism has become the new refuge of socialist thinking. The environment is a great way to advance a political agenda that favors central planning and an intrusive government. What better way to control someone's property than to subordinate one's private property rights to environmental concerns?*

This writer attempts to discredit environmental concerns not with evidence, but by linking them to an unpopular political ideology. *Marxism* and *socialist* are highly negatively charged terms in the minds of most Americans. The writer seeks to influence the reader's opinion by claiming environmental concern is akin to accepting a distasteful political belief.

Writer C. *The battle to feed humanity is over. In the 1970s and 1980s hundreds of millions of people will starve to death in spite of any crash programs embarked upon now. At this late date nothing can prevent a substantial increase in the world death rate, although many lives could be saved through dramatic programs to 'stretch' the carrying capacity of the earth. . . . But these programs will only provide a stay of execution unless they are accompanied by determined and successful efforts at population control.*

Words and phrases conveying bias include: "*The battle. . . is over.*" "Hundreds of millions *will* starve to death . . . . *Crash programs. . .* will only provide a *stay of execution. . . .*" The writer hopes to persuade the reader through an emotional appeal rather than convince the reader through reason.

All people have biases. They are a part of our conceptual frameworks, which in turn, are rooted in our level of knowledge, past experiences, purposes, and vested interests. Identifying bias in ourselves and others is a necessary skill in the critical thinking process. What are some other strategies one could use to detect bias in addition to looking at language?

(Writer of A and B: Rush Limbaugh; C: Paul Ehrlich)

---

land results in further impoverishment of the people who depend on that land for survival (fig. 25.14). As the World Bank *World Development Report* says, "Without adequate environmental protection, human development is undermined; without human development, environmental protection will fail."

Public Opinions and Environmental Protection

In spite of the recent gains in political power by antienvironmental forces in the United States mentioned earlier in this chapter, people around the world are becoming increasingly aware of the damaging effects of pollution on their own health and prosperity. These attitudes are reflected in public demands for environmental cleanup and protection. In a 1990–1993 World Values survey carried out in 43 countries representing 70 percent of the world population, almost everyone polled—96 percent of all respondents—said they are in favor of environmental protection. Perhaps even more important, 65 percent said that they would pay more taxes if they were assured that the money would go toward pollution control.

In general, public concern about environmental protection tends to be highest where pollution is most severe. More than 75 percent of urban residents of China, South Korea, and Chile, for instance, would pay higher taxes for clean air. In places where urban air reaches hazardous levels more than 200 days per year, it is not surprising that pollution control is a popular topic. But our attitudes about our environment also are shaped by cultural values. Interestingly, some of the highest levels of public support for environmental protection are observed in countries such as Norway, Sweden, Denmark, and the Netherlands where environments are among the world's cleanest.

Figure 25.13

More than 1 billion people—like this Indonesian family—live in acute poverty, lacking access to secure food supplies, safe drinking water, housing, education, health services, and decent jobs.

These countries also have high levels of what are called " **postmaterialist" values** that emphasize quality of life over acquisition of material goods. Postmaterialists place democracy, freedom, creativity, and humane society above making money or accumulating power. Countries with high levels of postmaterialist values are significantly more likely to take action to protect their environment than are those where the public is preoccupied with materialism (fig. 25.15). An encouraging trend is that young people in many societies tend to be much less materialistic than their parents or grandparents. In a survey of West European countries, people over 65 years were ten times more likely to agree with materialistic than postmaterialistic goals. Among people under 25 years in those same countries, on the other hand, postmaterialistic values predominate.

A majority of United States residents also strongly support environmental protection. A 1994 Times-Mirror poll of a random sample of Americans found the following:

- 82 percent wanted stricter laws to protect the environment.

- 76 percent said that water pollution laws have not gone far enough.

Poorest people in all developing countries
(millions)

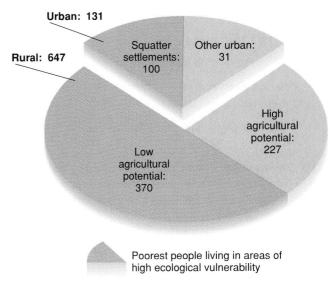

Poorest people living in areas of high ecological vulnerability

Figure 25.14

A total of some 470 million people, or 60 percent of the developing world's 780 million *poorest* people, live in rural or urban areas of high ecological vulnerability—areas where ecological destruction or severe environmental hazards threaten their well-being.

Source: Data from H. Jeffrey Leonard in *Environment and the Poor*, by Richard E. Feinberg and Valeriana Kallan. Copyright © 1989 Transaction Publishers.

Action taken to protect the environment
(by prevailing values in 12 nations)

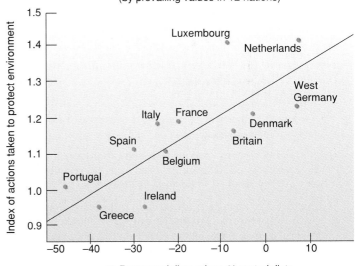

Figure 25.15

Countries where postmateralist (quality of life) values outweigh materialistic values tend to be significantly more likely to act in favor of environmental protection than are countries where people are primarily concerned with acquiring money and material goods.

Source: R. Inglehart, "Public Support for Environmental Protection," *Political Science and Politics*, March 1995.

- 67 percent would pay higher prices for environmental defense.

- 66 percent believe that a clean environment and economic growth go hand in hand.

- 66 percent said that air pollution laws should be stronger.

- 63 percent support more federal spending for the environment.

- 60 percent would choose a clean environment over economic growth if compromise is impossible.

- 54 percent think we need more protection for wild and natural areas.

These results show that politicians who claim a mandate from voters to eliminate environmental regulations are not really representing majority views. Perhaps as a new generation takes control, postmaterialist values will infuse the decision-making process in this country as it already has in many Nordic countries.

Sustainable Development

How can we accomplish the twin goals of improving human well-being and protecting our common environment? The developing countries of the world need access to more efficient, less-polluting technologies and to learn from the successes and failures of the developed countries (table 25.5). Technology transfer is essential. Environmental and resource protection benefits both the richer countries of the world and the poorer ones. It is in our best interests to help finance protection of our common future in some equitable way. Maurice Strong, chair of the 1992 Earth Summit, estimates that development aid from the richer countries should be some $150 billion per year, while internal investments in environ-

Table 25.5
An agenda for sustainable development

1. Remove subsidies that encourage excessive use of fossil fuels, water, pesticides, minerals, and logging.

2. Clarify rights to own and manage lands, forests, and fisheries to give people access to the means of production and security to invest in long-range improvements.

3. Improve sanitation and provide clean water, education (especially for girls), family planning, agricultural extension, credit, and research that helps the poor.

4. Empower, educate, and involve local people in their own long-term interests.

5. Provide rural jobs, access to markets, housing, and other reforms that encourage people to remain in villages rather than to crowd into already overstressed urban areas.

6. Protect natural habitat and biodiversity along with critical watersheds, wetlands, forests, and other fragile ecological areas.

mental protection by developing countries will need to be about twice that amount.

International Nongovernmental Organizations

The rise in international environmental organizations in recent years has been phenomenal. At the Stockholm Conference in 1972, only a handful of environmental groups attended, almost all from First World countries. Twenty years later, at the Rio Earth Summit, more than 30,000 individuals representing several thousand environmental groups, many from the Third World, held a global Ecoforum to debate issues and form alliances for a better world. We call these groups working for social change **nongovernmental organizations (NGOs).** They have become a powerful aspect of environmental protection.

Some NGOs are located primarily in the more highly developed countries of the North and work mainly on local issues. Others are headquartered in the North but focus their attention on the problems of developing countries in the South. Still others are truly global, with active groups in many different countries. A few are highly professional, combining private individuals with representatives of government agencies on quasi governmental boards or standing committees with considerable power. Others are at the fringes of society, sometimes literally voices crying in the wilderness. Many work for political change, more specialize in gathering and disseminating information, and some undertake direct action to protect a specific resource.

Public education and consciousness-raising using protest marches, demonstrations, civil disobedience, and other participatory public actions and media events are generally important tactics for these groups. Greenpeace, for instance, carries out well-publicized confrontations with whalers, seal hunters, toxic-waste dumpers, and others who threaten very specific and visible resources (fig. 25.16). Greenpeace may well be the largest environmental organization in the world, with some 2.5 million contributing members.

In contrast to these highly visible groups, others choose to work behind the scenes, but their impact may be equally important. Conservation International has been a leader in debt-for-nature swaps to protect areas particularly rich in biodiversity. It also has some interesting initiatives in economic development, seeking products made by local people that will provide income along with environmental protection.

Green Government and Politics

While winning the hearts and minds of the public and influencing policymakers are important and gratifying work, it is sometimes necessary to dive into the nitty-gritty of politics or administration to bring about meaningful change. In this section, we will explore how these political systems function, as well as some ways in which you can get involved.

Green Politics

In many countries with a parliamentary form of government, **"green" political parties** based on environmental interests have become a

Figure 25.16
As whale blood pours from a Japanese whaling factory ship Greenpeace protesters in a rubber dinghy draw attention to the killing of endangered species.

Figure 25.17
Delegates to the Green Party convention in Germany debate their political, social, economic, and environmental platform. Although the Greens lost almost all their seats in the National Parliament in 1990, they remain a force in local politics.

political force in recent years. The largest and most powerful green party is Germany's *Die Grünen,* which has controlled as much as 10 percent of the seats in the *Bundestag,* or parliament (fig. 25.17). It is a grassroots, egalitarian, council-style movement, committed to participatory democracy, environmental protection, and a fundamental transformation of society. An essential premise is that ceaseless industrial growth destroys both people and the environment and must, therefore, be stopped. The Greens have formed coalitions with antinuclear weapons protesters, feminists, human rights advocates, and other public interest groups, but have struggled continuously over whether to work with or oppose the ruling centrist party. Political realists argue that they could be more effective within a larger coalition; idealists vow never to compromise their principles.

The majority-rule political system in the United States makes it extremely difficult to start a new political party at the national level. Green candidates have been successful at the local and state levels, however. In 1990, Alaska became the first state to give a green party official standing. Jim Sykes, the Green gubernatorial candidate, received 3.2 percent of the total vote, enough to guarantee the party a spot on future ballots. National groups such as the League of Conservation Voters, and "green committees of correspondents," work to introduce environmental issues into party politics and support candidates who share their environmental concerns. Campus Greens are strong at many colleges and universities. The four key values they espouse are (1) ecological wisdom, (2) peace and social justice, (3) grassroots democracy, and (4) freedom from violence.

Green Plans

Several national governments have undertaken integrated environmental planning for a sustainable future. Canada, New Zealand, Sweden, and Denmark all have so-called **green plans** or comprehensive, long-range national environmental strategies. The best of these plans weave together complex systems, such as water, air, soil, and energy, and mesh them with human factors such as economics, health, and carrying capacity. Perhaps the most thorough and well-thought-out green plan in the world is that of the Netherlands.

Developed in the 1980s through a complex process involving the public, industry, and government, the 400-page Dutch plan contains 223 policy changes aimed at reducing pollution and establishing economic stability. Three important mechanisms have been adopted for achieving these goals: integrated life-cycle management, energy conservation, and improved product quality. These measures should make consumer goods last longer and be more easily recycled or safely disposed of when no longer needed. For example, auto manufacturers are now required to design cars so they can be repaired or recycled rather than being discarded.

Among the guiding principles of the Dutch Green Plan are: (1) the "stand-still" principle that says environmental quality will not deteriorate, (2) abatement at the source rather cleaning up afterwards, (3) the "polluter pays" principle says that the users of a resource pay for negative effects of that use, (4) prevention of unnecessary pollution, (5) application of the best practicable means of pollution control, (6) carefully controlled waste disposal, and (7) motivating people to behave responsibly.

The Netherlands have invested billions of guilders in implementing this comprehensive plan. Some striking successes already have been accomplished. Between 1980 and 1990, emissions of sulfur dioxide, nitrogen oxides, ammonia, and volatile organic compounds were reduced 30 percent. By 1992, pesticide use had been reduced 20 percent from 1988 levels, and chlorofluorocarbon use

What Then Shall We Do?

had been virtually eliminated. As of 1995, industrial wastewater discharges into the Rhine River were 50 to 70 percent less than a decade earlier. Some 250,000 ha (more than 600,000 ac) of former wetlands currently are being restored as nature preserves and 40,000 ha (99,000 ac) of forest are being replanted (fig. 25.18). This is remarkably generous and foresighted in such a small, densely populated country, but the Dutch have come to realize they cannot live without nature.

Not all goals have been met so far. Planned reductions in CO_2 emissions failed to materialize when cheap fuel prices encouraged fuel-inefficient cars. Currently, a carbon tax is being considered. A sudden population increase caused by immigration from developing

Figure 25.18

Under the Dutch Green Plan, 250,000 ha (600,000 ac) of drained agricultural land is being restored to wetland and 40,000 ha (99,000 ac) are being replanted as woodland.

countries and Eastern Europe also complicates plan implementation, but the basic framework of the Dutch plan has much to recommend it, nevertheless. Other countries would be more sustainable and less environmentally destructive if they were to adopt similar plans.

National Legislation

It is important for individuals to express their opinions about national programs and policies. Figure 25.19 shows the pathway that a piece of legislation follows in the United States from inception to being signed into law by the president. The idea for a law can originate from an ordinary citizen as well as from an elected official. After introduction, the bill is sent to the appropriate committee for hearings. Sometimes the hearing process is extensive, and may include field hearings in which ordinary citizens have a chance to express their opinions (fig. 25.20).

Contact your legislator to find out if there will be field hearings in your city. If they are scheduled, go even if you don't intend to speak. It is an educational experience to see how the process works. If you intend to testify, try to coordinate your presentation with others who share your position. Well-organized, factual testimony can make a good impression and have a positive impact both on elected officials and the public.

Almost all of the advance work in getting a bill ready for introduction and in guiding it through the hearings process is done by the legislative staff. These people are key in the success or failure of bills in which you are interested. It is well worth getting to know them and working with them. You could become a legislative aide. Being in the "corridors of power" is an exciting and educational experience.

Few of us get directly involved in the legislative process, but you can make your wishes and opinions known by communicating with your elected representatives. Table 25.6 gives some suggestions for how to write effectively. Most representatives get little mail, so a single, well-written letter from a constituent can make a difference. Even on controversial issues, a representative might get only 50–100 letters. Since each congressional district includes about half a million people, the legislator tends to assume that each letter represents the views of five thousand to ten thousand constituents. Your voice has great impact! If you don't have time to write, you can call the local or Washington office. You probably will not talk directly to your representative or senator, but your opinion will be registered by an aide. Another option is to send a telegram or e-mail message. Most legislators now have internet addresses (see end of this chapter).

The Courts

In the 1960s and 70s, the judicial branch of government was highly responsive to environmental and social concerns. Much of this activist philosophy has been lost through recent court appointments, but citizens still can seek relief in the courts from unjust laws or actions in two ways. One way is to bring a civil suit, asking for payment of damages that were caused by a private individual, corporation, or governmental agency and that injured you or your property. The other way is to ask for a judgment by the court about the constitutionality of laws passed by Congress or the adequacy and legality of regulations established by an administrative agency.

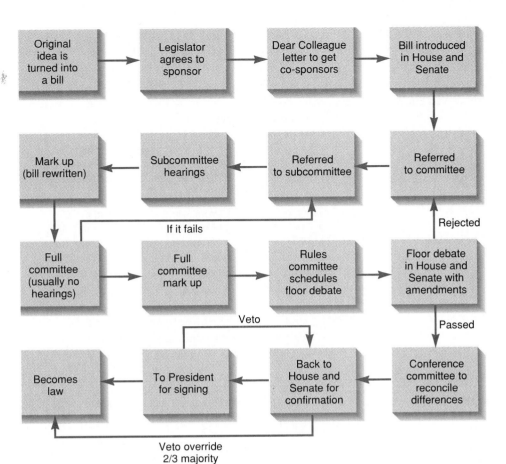

Figure 25.19
The path of a bill from inception to actual law.

Figure 25.20
Citizens line up to testify at a legislative hearing. Speaking about public issues is a right that few people in the world enjoy. By getting involved in the legislative process, you can be informed and have an impact on governmental policy.

If the court finds a law or regulation to be improper, it can issue an injunction to stop implementation or application of that law or regulation. It also can order an agency to rewrite its regulations and may even direct what the new regulations should be.

There are several problems that make lawsuits difficult for ordinary citizens who want to protect the environment or some other public resource:

- You have to show that the action or law you oppose is illegal.

- You have to establish that you have standing in court (a right to be heard). To do so, you must prove that you are directly affected.

- Suits are expensive; a major suit might cost hundreds of thousands of dollars in legal fees, court fees, witnesses, etc.

- It may take years before a suit is finally settled, and by that time it may be too late to save the resource.

- You have to prove that the defendant (an agency, corporation, or individual) is responsible for the harm that you allege. A corporation can admit that they produce a toxic chemical, but you have to show, beyond reasonable doubt, that it was their chemical that caused the problem.

- Counter lawsuits are often brought against environmental activists by businesses, developers, and government agencies whose projects they oppose. Called strategic lawsuits against political participation (or SLAPPs), these punitive suits are usually thrown out of court eventually but can have a chilling effect because of the time and expense required for defense.

What Then Shall We Do?

Table 25.6

How to write to your elected officials

1. Address your letter properly.

   a. *Your representative:*
   The Honorable _____,
   House Office Building
   Washington, DC 20515
   Dear Representative _____,

   b. *Your senators:*
   The Honorable _____,
   Senate Office Building
   Washington, DC 20510
   Dear Senator _____,

   c. *The president:*
   The President
   The White House
   1600 Pennsylvania Avenue N.W.
   Washington, DC 20500
   Dear Mr. President,

2. Tell who you are and why you are interested in this subject. Be sure to give your return address.

3. Always be courteous and reasonable. You can disagree with a particular position, but be respectful in doing so. You will gain little by being shrill, hostile, or abusive.

4. Be brief. Keep letters to one page or less. Cover only one subject, and come to the point quickly. Trying to cover several issues confuses the subject and dilutes your impact.

5. Write in your own words. It is more important to be authentic than polished. Don't use form letters or stock phrases provided by others. Speak or write from your own personal experiences and interests. Try to show how the issue affects the legislator's own district and constituents.

6. If you are writing about a specific bill, identify it by number (for instance, H.R. 321 or S. 123). You can get a free copy of any bill or committee report by writing to the House Document Room, U.S. House of Representatives, Washington, DC 20515 or the Senate Document Room, U.S. Senate, Washington, DC 20510.

7. Ask your legislator to vote a specific way, support a specific amendment, or take a specific action. Otherwise you will get a form response that says: "Thank you for your concern. Of course I support clean air, pure water, apple pie, and motherhood."

8. If you have expert knowledge or specifically relevant experience, share it. But don't try to intimidate, threaten, or dazzle your representative. Don't pretend to have vast political influence or power. Legislators quickly see through artifice and posturing; they are professionals in this field.

9. If possible, include some reference to the legislator's past action on this or related issues. Show that you are aware of his or her past record and are following the issue closely.

10. Follow up with a short note of thanks after a vote on an issue that you support. Show your appreciation by making campaign contributions or working for candidates who support issues important to you.

11. Try to meet your senators and representatives when they come home to campaign, or visit their office in Washington if you are able. If they know who you are personally, you will have more influence when you call or write.

12. Join with others to exert your combined influence. An organization is usually more effective than isolated individuals.

In spite of these difficulties, the courts often have been the most successful place for environmental organizations to change how we manage our environment. More than one hundred public-interest law firms in the United States specialize in social and environmental issues. Several environmental organizations, such as the Environmental Defense Fund, the National Resources Defense Council, and the Sierra Club Legal Foundation, act primarily or exclusively through litigation (bringing suits in court).

The Executive Branch

Of the roughly 14 million federal employees in the United States, the vast majority work in the executive branch. Many of these civil servants are in the administrative agencies that carry out and enforce the laws passed by the legislature. They monitor, manage, control, and protect our resources and our environment. Figure 25.21 shows the major agencies and branches of government that have environmental responsibilities. The Environmental Protection Agency is often regarded as the main guardian of environmental quality. It has responsibility for regulating air and water pollution, solid and hazardous wastes, toxic substances, noise, radiation, and certain pesticides.

The Department of the Interior manages the national parks, wildlife refuges, wild rivers, historic sites, Bureau of Land Management lands, the Bureau of Reclamation, and the Bureau of Mining. It is by far the largest land manager in the country. The Department of Agriculture administers the national forests, the Agricultural Research Service, the Soil Conservation Service, and the Food Safety Inspection Service. The Department of Health administers the Public Health Service and the Food and Drug

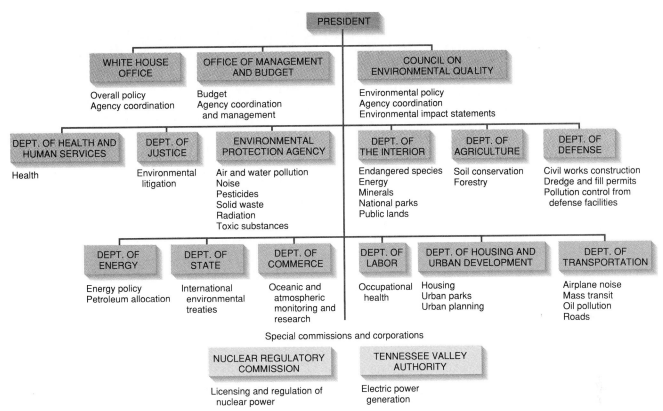

Figure 25.21

Major agencies of the Executive Branch of the U.S. federal government with responsibility for resource management and environmental protection.

Source: Data from U.S. General Accounting Office.

Administration. The Department of Labor is responsible for occupational safety and health. The Department of Energy is responsible for nuclear energy and fossil fuels.

These administrative agencies represent a tremendous bureaucracy, which often is ponderous and unresponsive, but then, we have a big country, and running it is a big job. For the most part, our civil service is made up of dedicated, professional people. If you ever have traveled in a less-developed country, you will appreciate how open and honest our government is. Our system has been criticized for responding incrementally to problems. It may make improvements more slowly than we would like in some situations, but it also makes mistakes more slowly. If it had the power to make sudden leaps, it might make them in the wrong direction.

Environmental Impact Statements

One of the most useful tools for those concerned about environmental protection is a provision in the National Environmental Policy Act of 1970 requiring all federal agencies to prepare an **environmental impact statement (EIS)** analyzing the effects of any major program that it plans to undertake. As we discussed earlier in this chapter, it often is difficult for citizens to prove they are personally injured by a pollutant or an action of the government. It

usually is not difficult, however, to show that an agency has prepared an inadequate EIS or has failed to consider important environmental consequences of their actions.

Intervening in administrative proceedings has become much easier since the Freedom of Information Act (FOIA) was passed requiring federal agencies to make public all minutes of meetings, correspondence, and other official documents. This means that you can find out what an agency considered when preparing an EIS, who they talked to, and why they decided as they did. It often is fairly easy to show that undue consideration was given to commercial interests or that environmental values were overlooked or deliberately ignored. In some cases, simply asking for an EIS will prompt an agency to reconsider a harmful project. In other cases, you may have to ask the courts to order one. Often the delay caused by litigation over an EIS gives time to organize other opposition. Sometimes just generating a discussion of the adverse effects of a project is enough to kill it. Letters to your legislators are considered private correspondence and cannot be made public under the FOIA.

International Treaties and Conventions

No international body has authority to pass binding environmental laws, nor are there international agencies with power to regulate resources on a global scale. An international court sits at the Hague in The Netherlands, but it has little power to enforce its decisions. Powerful nations can simply ignore the court, as the United States did when it was found in violation of international law after it mined the harbors of Nicaragua in 1983. Still, there is movement toward environmental protection on a worldwide scale through special covenants, treaties, and multilateral agreements.

Several successful international conventions and treaties have been passed. The 1979 Convention on Long-Range Transboundary Air Pollution was the first multilateral agreement on air pollution and the first environmental accord involving all the nations of Eastern and Western Europe and North America. The 1989 Accord on Chlorofluorocarbon Emissions (chapter 18) was achieved remarkably quickly to address the threat to the stratospheric ozone shield. It was the first treaty between the industrialized North and the developing South. Recognizing the need for economic expansion by the less-developed countries, this treaty requires greater reductions from the bigger polluters than the smaller ones and a sharing of new technology to ease the effects of change on poorer nations.

The Antarctic Treaty of 1961 reserves the Antarctic continent for peaceful scientific research and bans all military activities in the region. Proposed exploitation of Antarctic mineral and energy resources raises important questions about regulating the global commons. In 1992, the United States proposed dividing up Antarctic resources on a "first come, first served" basis, arguing that those who invest in developing and harvesting the resource deserve to enjoy the profits. The less-developed nations resisted, claiming that these resources are a common heritage of all humankind that should benefit everyone, not just those wealthy enough to afford the technology to harvest them. They see this as the first application of a new economic order, in which we share more equally in resource benefits. A compromise was finally reached in which it was agreed that no commercial activities will be allowed in Antarctica for fifty years. Many environmentalists believe that this last pristine continent should be set aside as an international wilderness area.

The U.N. Conference on Environment and Development in 1992 considered a wide range of international environmental issues. Among the most important of these are the convention on biodiversity, which calls for the conservation and sustainable use of biological diversity together with a fair sharing of the wealth produced from these resources. Only the United States, among all the world's nations, refused to sign because the administration claimed that benefit-sharing provisions were incompatible with existing intellectual property rights.

A convention on global climate change was also expected to be a centerpiece of the Earth Summit. Due to objections of the United States over timetables and specific commitments to limit emissions, only a vague and general document passed. Still, many important issues were discussed and the contacts made in this process, both government-to-government and person-to-person networking, may turn out to have been its most important contribution. The "Earth Charter" or "Rio Declaration" adopted by the delegates contains some important principles on both environmental protection and sustainable development (table 25.7).

Table 25.7 Key principles of the Earth Charter

1. Every state has a sovereign right to exploit its own resources in accordance with its own policies, providing it doesn't harm the environment elsewhere.

2. All peoples have a right to sustainable development.

3. Environmental protection must be an integral part of development.

4. Unsustainable patterns of production and consumption must be reduced and appropriate demographic policies must be promoted.

5. People have a right to information and the opportunity to participate in political processes.

6. Polluters should pay for their environmental impacts through internalization of costs and the use of appropriate taxes, charges, and incentives for environmental protection.

Source: Adopted by the United Nations Conference on Environment and Development, Rio de Janeiro, Brazil 1992.

Figure 25.22

Schoolchildren collect litter and trash along the roadside in Queensland, Australia. Each of us can do something to improve our common environment.

In the end, although a global perspective is essential for both environmental protection and social justice, we need to, as stated in E.F. Schumacher's famous dictum, "think globally but act locally" (fig. 25.22).

Summary

Current trends are not predictions of what must be, but rather warnings of what may be, depending on the course we follow. Knowledge about, and concern for, our environment is at an all-time worldwide high. Perhaps it is not unreasonable to hope that we will find ways to achieve social goals and environmental protection. The main goals of environmental education are to expand environmental literacy among the general public and to prepare students for environmental careers. Environmental literacy is a basic understanding of the natural and built environments and the relationships between humans and their environment. There are many opportunities for careers at a variety of different levels in the environmental field.

Businesses are rushing to invent and market environmental technology and "green" consumer goods. We have to look carefully at the claims made for "environmentally friendly" products. Some are true but others are not. We also must ask ourselves how much we need or have a right to consume.

Deep ecology is a philosophy that calls for voluntary simplicity, rejection of anthropocentrism, decentralization, and direct personal action. Progressive environmentalists, on the other hand, advocate working within existing systems to bring about gradual reform. There is much we can do through our individual choices of goods and services, but we must be careful to avoid green scams and fraudulent marketing. Independent, scientific, lifetime analysis can help us make informed decisions.

There are many opportunities for collective action to bring about social change. Student environmental groups offer chances to network with others, learn useful organizing techniques, do good work, and have fun. The large national "mainline" environmental organizations have a degree of respectability, power, and influence unmatched by smaller, independent groups. Monkey wrenching or direct action is a highly controversial tactic espoused by some radical environmentalists. A majority of the world's population supports environmental safeguards. Many people are willing to pay more for cleaner air and water, but antienvironmental forces also are becoming more vocal.

Sustainable development promises to alleviate acute poverty while also protecting the environment. Many nongovernmental organizations (NGOs) as well as intergovernmental treaties and conventions work toward these twin goals. In many countries, green politics espouses four key values of ecological wisdom, peace and social justice, grassroots democracy, and freedom from violence. While it is difficult to introduce new national parties in the United States, there are opportunities to work within the legislative, judicial, and administrative agencies to bring about change.

▼ Questions for Review

1. Summarize the goals of environmental education.

2. What are deep and shallow ecology? Why do progressives object to being called shallow ecologists?

3. List some things you can do to reduce resource consumption.

4. Discuss some of the limits to green consumerism and how you can spot fraudulent or meaningless green marketing claims.

5. Review the trade-offs between paper and plastic bags. Which is better in your view?

6. List ten key issues in organizing an environmental campaign or using media to influence public opinion.

7. What are the ten largest, oldest mainline environmental organizations in the United States?

8. Describe and evaluate the tactics used by radical environmental groups. Do you subscribe to monkey wrenching?

9. Describe the goals and tactics of the antienvironmental wise use groups. In what way are their claims extremist?

10. List the four key values of the Green party.

11. Explain how the legislative, judicial, and administrative branches of government work to protect the environment.

12. Compare the intent and effectiveness of three international environmental conventions or treaties.

▼ Questions for Critical Thinking

1. An article in a major environmental journal criticized a group of kayakers who paddled down a river in the Arctic National Wildlife Refuge to protest proposed oil drilling there. As they stood on the beach in their polypropylene-neoprene-gortex paddling outfits, next to their plastic kayaks and nylon tents, waiting for a charter plane to pick them up on the first leg of a trip of several thousand miles back home, they deplored our excessive use of petroleum. Was this hypocritical? When do the ends justify the means?

2. The wise use movement appeals to people who fear that their livelihood and way of life are threatened by those who want to "lock up" resources in parks and nature preserves. What would you say to convince them that their loss is worthwhile?

3. People in many cultures traditionally regard wild nature as disagreeable and dangerous. How would you approach environmental education if you were assigned to such a country?

4. Do you support the conservative, mainstream approach to conservation followed by the mainline environmental groups or do you prefer the more challenging, innovative approaches of more radical groups? What are the advantages and disadvantages of each?

5. Extremists on both sides of the environmental debate sometimes use violence against property—and sometimes against people—to advance their agendas. Would you ever condone taking the law into

What Then Shall We Do?

your own hands for an environmental cause? Under what circumstances and with what limits might you do so?

6. Is sustainable development possible? When will we come up against the limits to further growth in population, resource consumption, and conversion of natural ecosystems to suit human needs?

7. Do you agree that sustainable development has the potential to simultaneously reduce poverty *and* protect the environment? What responsibility do those of us in rich nations bear toward the rest of the world?

8. Choose one of the ambiguous dilemmas presented in this chapter such as whether it is better to use paper or plastic bags. What additional information would you need to choose between alternatives? How should competing claims or considerations be weighed in an issue such as this?

9. There are good reasons to be pessimistic about what we are doing to our environment, but there also are signs of progress and reasons for optimism. Where do you stand on this spectrum?

10. The Talmud says, "If not now, when?" How might this apply to environmental science?

▼ Key Terms

deep ecology 571
direct action 572
environmental impact
 statement (EIS) 581
environmental literacy 563
green plans 577
green political parties 576
life-cycle analysis 566

monkey wrenching 572
nongovernmental organizations
 (NGOs) 576
postmaterialist values 575
precycling 565
shallow ecology 571
social ecology 571
wise use movement 572

▼ Additional Information on the Internet

Blueprint for a Green Campus
 http://netspace.org/environ/earthnet/

Campus Green Vote
 http://www.cgv.org/cgv

CIEL, Center for International Environmental Law
 http://www.econet.org/ciel/

Communications for a Sustainable Future
 http://csf.colorado.edu/

The Earth Council
 http://terra.ecouncil.ac.cr/

Environmental Activism Online
 http://www.envirolink.org/elib/action/

Environmental Education Network
 http://www.envirolink.org/enviroed/

Environmentally Active Colleges, Universities, and Schools
 http://gwis.circ.gwu.edu/~greenu/ecu.html/

Environmental Organizations
 http://www.econet.apc.org/econet/en.orgs.html/

Friends of the Earth
 http://www.foe.co.uk/

Global Action and Information Network
 http://www.igc.apc.org/gain/

Global Resource Information Database
 http://www.grid.umep.ch/gridhome.html/

Governmental World Wide Web Resources Relating to the Environment
 http://www.envirolink.org/envirogov.html

Greenpeace
 http://www.greenpeace.org/

The Handbook for a Better Future
 http://www.mit.edu:8001/people/howes/environ.html

Infoterra
 http://pan.cedar.unil /ie.ac.at/gopher_ref/unep/

International Institute for Sustainable Development
 http://iisd1.iisd.ca/

Internet Green Marketplace
 http://www.envirolink.org/greenmarket/

League of Conservation Voters
 http://www.lcv.org/

'Lectric Law
 http://www.lectlaw.com

NCEET Gopher
 gopher://nceet.snre.umich.edu

One World Online
 http://www.oneworld.org/

Sierra Club
 http://www.sierraclub.org/

Student Environmental Action Coalition
 http://www.seac.org/

World Wide Web Virtual Library: Law: International and Environmental Law
 http://www.law.indiana.edu/law/intenvlaw.html/

Note: Further readings appropriate to this chapter are listed on p. 608.

Units of Measurement Metric/English Conversions

Length

1 meter = 39.4 inches = 3.28 feet = 1.09 yard

1 foot = 0.305 meters = 12 inches = 0.33 yard

1 inch = 2.54 centimeters

1 centimeter = 10 millimeters = 0.394 inch

1 millimeter = 0.001 meter = 0.01 centimeter = 0.039 inch

1 fathom = 6 feet = 1.83 meters

1 rod = 16.5 feet = 5 meters

1 chain = 4 rods = 66 feet = 20 meters

1 furlong = 10 chains = 40 rods = 660 feet = 200 meters

1 kilometer = 1000 meters = 0.621 miles = 0.54 nautical miles

1 mile = 5280 feet = 8 furlongs = 1.61 kilometers

1 nautical mile = 1.15 mile

Area

1 square centimeter = 0.155 square inch

1 square foot = 144 square inches = 929 square centimeters

1 square yard = 9 square feet = 0.836 square meters

1 square meter = 10.76 square feet = 1.196 square yards = 1 million square millimeters

1 hectare = 10,000 square meters = 0.01 square kilometers = 2.47 acres

1 acre = 43,560 square feet = 0.405 hectares

1 square kilometer = 100 hectares = 1 million square meters = 0.386 square miles = 247 acres

1 square mile = 640 acres = 2.59 square kilometers

Volume

1 cubic centimeter = 1 milliliter = 0.001 liter

1 cubic meter = 1 million cubic centimeters = 1000 liters

1 cubic meter = 35.3 cubic feet = 1.307 cubic yards = 264 US gallons

1 cubic yard = 27 cubic feet = 0.765 cubic meters = 202 US gallons

1 cubic kilometer = 1 million cubic meters = 0.24 cubic mile = 264 billion gallons

1 cubic mile = 4.166 cubic kilometers

1 liter = 1000 milliliters = 1.06 quarts = 0.265 US gallons = 0.035 cubic feet

1 US gallon = 4 quarts = 3.79 liters = 231 cubic inches = 0.83 imperial (British) gallons

1 quart = 2 pints = 4 cups = 0.94 liters

1 acre foot = 325,851 US gallons = 1,234,975 liters = 1234 cubic meters

1 barrel (of oil) = 42 US gallons = 159 liters

Mass

1 microgram = 0.001 milligram = 0.000001 gram

1 gram = 1000 milligrams = 0.035 ounce

1 kilogram = 1000 grams = 2.205 pounds

1 pound = 16 ounces = 454 grams

1 short ton = 2000 pounds = 909 kilograms

1 metric ton = 1,000 kilograms = 2200 pounds

Temperature

Celsius to Fahrenheit $°F = (°C \times 1.8) + 32$

Fahrenheit to Celsius $°C = (°F - 32) \div 1.8$

Energy and Power

1 erg = 1 dyne per square centimeter

1 joule = 10 million ergs

1 calorie = 4.184 joules

1 kilojoule = 1000 joules = 0.949 British Thermal Units (BTU)

1 megajoule = MJ = 1,000,000 joules

1 kilocalorie = 1000 calories = 3.97 BTU = 0.00116 kilowatt-hour

1 BTU = 0.293 watt-hour

1 kilowatt-hour = 1000 watt-hours = 860 kilocalories = 3400 BTU

1 horsepower = 640 kilocalories

1 quad = 1 quadrillion kilojoules = 2.93 trillion kilowatt-hours

Appendix *B*

Environmental Work, Study, and Travel Opportunities

Access
50 Beacon Street, Boston, MA 02108
(617) 720-JOBS

Gathers information on public interest and nonprofit internships and career opportunities.

American Friends Service Committee (AFSC)
1501 Cherry Street, Philadelphia, PA 19102
(215) 421-7000

AFSC is a Quaker organization that provides material assistance to grassroots groups in Asia, Africa, Latin America, the Middle East, and poor communities in the United States. AFSC also coordinates cross-cultural education programs, foreign policy programs, and human rights campaigns.

Amigos de las Americas
5618 Star Lane, Houston, TX 77057
(800) 231-7796 or (713) 782-5290

Volunteers lead public health projects in Latin America and the Caribbean. Each volunteer receives training and raises the funds needed for his or her project during the school year and then works with a partner in a village for six to eight weeks during the summer. Projects include teaching dental hygiene, distributing eyeglasses, helping to build latrines, and providing immunization against disease.

Canadian Organization for Development through Education
321 Chapel Street
Ottawa, Ontario K1N 7Z2
Canada
(613) 232-3569

Supports sister school projects between Canada and Uganda, Kenya, the Caribbean, and the Philippines.

Center for Global Education
c/o Augsburg College
731 21st Avenue South
Minneapolis, MN 55454
(612) 330-1159

Coordinates travel seminars designed to introduce participants to the realities of life in Mexico, Central America, the Caribbean, and the Philippines. Costs range between $900 and $2000. Some scholarships are available.

Cool It!
National Wildlife Federation
1400 16th Street NW
Washington, DC 20036
(202) 797-5435

Regional Cool It! coordinators work with students over the phone and conduct on-site workshops to help establish models of environmentally sound practices. Other services include a newsletter, job bank, and speakers' bureau.

Earth Day Resources
116 New Montgomery Street, Suite 530
San Francisco, CA 94105
(800) 727-8619 or (415) 495-5987

Maintains a network of grassroots organizations committed to environmental solutions at the local level. Offers materials such as lesson plans, fact sheets, and guidance for recycling and rideshare programs.

Earthwatch
P.O. Box 403
Watertown, MA 02272
(617) 926-8200

Earthwatch links volunteers with scientists and scholars on academic research expeditions in the United States and abroad. You might help to study rock art in Italy or help to make a documentary movie on Brazilian festivals. Costs range between $600 and $2500.

Eco-Net
18 DeBoom Street, Dept. GM
San Francisco, CA 94107
(415) 442-0220

An online computer network of more than 100 electronic bulletin boards that deal with environmental issues and job opportunities.

The Environmental Careers Organization, Inc.
286 Congress Street, Dept. GM
Boston, MA 02210
(617) 426-4375

Places college students and recent graduates in short-term paid internships with environmental groups and publishes a guide to environmental careers.

Environmental Project on Central America (EPOCA)
Earth Island Institute
300 Broadway, Suite 28
San Francisco, CA 94133-3312
(415) 788-3666

Sponsors reforestation brigades in Nicaragua. Brigadistas plant tens of thousands of trees to provide wind blocks and to prevent erosion and flooding on Nicaragua's farmland.

Foundation for International Training
200-1262 Don Mills Road, Don Mills, Ontario M3B 2W7, Canada
(416) 449-8838

The Foundation sponsors projects in community development, industry administration, and management in Third World countries. Participants should have some overseas experience as well as experience teaching a technical skill. Assignments are for six to eight weeks. All expenses are paid and a small honorarium is offered.

Gemquest
Global Exchange Motivators, Inc.
Montgomery County Intermediate Unit Building
Montgomery Avenue and Paper Mill Road
Erdenheim, PA 19118
(215) 233-9558

Organizes study and travel programs that promote cultural understanding through intensive contact between travelers and their hosts. Includes homestays and the possibility for university study abroad.

Global Action Plan
84 Ferry Hill Road
Woodstock, NY 12498
(914) 679-4830; fax (914) 679-4834

Provides workbook and support for the formation of small groups, or "Eco Teams," that meet to learn about simple and effective ways to reduce their impact on the environment. Each group quantifies their results, such as water saved each month, and adds the data to the results of other Eco Teams around the world.

Global Exchange
2940 16th Street, Room 307
San Francisco, CA 94103
(415) 255-7296

Creates tours designed to help people become more involved in Third World development efforts. Tours go to Central and South America, the Caribbean, and southern Africa.

Green Corps
1724 Gilpin Street
Denver, CO 80218
(303) 355-1881

An intensive training program in grassroots organizing, fund-raising, media, and campaign skills, followed by a year using your skills in a real life situation.

Habitat for Humanity
419 West Church Street
Americus, GA 31709
(912) 924-6935

Volunteers donate labor, money, and materials in this nonprofit Christian housing ministry to build and renovate homes in the United States and abroad.

Highlander Research and Education Center
Route 3, Box 370
New Market, TN 37820
(615) 933-3443

The Highlander Center organizes exchanges between community activists in Appalachia and the Third World, focusing on education, labor rights, health care, land use, toxics, and economic development.

International Bicycle Fund (IBF)
4247 135th Place SE
Bellevue, WA 98006-1319

Sponsors bicycle tours to Africa to encourage people-to-people contact and to learn about cultures, histories, and economies of the peoples visited. Costs are about $1000 plus airfare.

International Development Exchange (IDEX)
777 Valencia Street
San Francisco, CA 94110
(415) 621-1494

IDEX brings together school and church groups in the United States with grassroots development projects in the Third World.

MADRE
121 West 27th Street, Room 301
New York, NY 10001
(212) 627-0444

Madre works to build friendships between women in Central America, the Caribbean, and the United States.

Mennonite Central Committee
21 South 12th Street
Akron, PA 17501
(717) 859-1151

Volunteers work on health, education, social services, and community development projects in over fifty countries.

Overseas Development Network (ODN)
P.O. Box 1430
Cambridge, MA 02238
(617) 868-3002

A network of college activists, ODN organizes educational and fund-raising events dealing with Third World development issues.

Partners for Global Justice
4920 Piney Branch Road NW
Washington, DC 20011
(202) 723-8273

Partners for Global Justice empowers U.S. citizens to influence public policy effectively. Volunteers spend one year in a Third World country and one year in the United States working with cooperatives and community organizations.

Pax World Service
1111 16th Street NW, Suite 120
Washington, DC 20036
(202) 293-7290

A nonprofit organization that initiates and supports projects that encourage international understanding, reconciliation, and sustainable development. Sponsors hands-on development trips that work on community projects and educational, fact-finding trips.

Plowshares Institute
P.O. Box 243
Simsbury, CT 06070
(203) 651-9675, (203) 658-6645

Travel seminars to Africa, Asia, and Australia. Participants are asked to do advanced reading and preparation and to share experiences with others after the trip.

Public Interest Research Groups
The Fund for Public Interest Research,
29 Temple Place
Boston, MA 02111
(617) 292-4800

The PIRGs organize student groups, campaign on a variety of environmental and social justice issues, and provide numerous canvassing and fundraising jobs to recent college graduates.

SERVAS International
11 John Street, Suite 706
New York, NY 10038
(212) 267-0252

SERVAS is an international network of people interested in peace issues. SERVAS does not organize or lead trips. Instead, the organization compiles national directories of people in over seventy countries who are interested in meeting travelers for a short homestay, or just for coffee.

Student Conservation Association
1800 North Kent Street, Suite 1260
Arlington, VA 22209
(703) 524-2441

Volunteers work with stewardship and conservation programs on public land and in national parks. For example, you might help with a botany research project in Yellowstone National Park.

Student Environmental Action Coalition (SEAC)
P.O. Box 1168
Chapel Hill, NC 27514
(919) 967-4600

SEAC publishes a monthly newsletter for its members and sponsors over seventy gatherings across the country for students interested in environmental justice. Through the Campus Ecology Project, students can assess environmental quality and create strategies for change on their own campuses.

Youth Ambassadors of America
P.O. Box 5273
Bellingham, WA 98227

Exchange programs, conferences, and summits for young people.

Volunteers for Peace International Workcamps
Tiffany Road
Belmont, VT 05730
(802) 259-2759

Opportunities for young people of all nationalities to work together on two- and three-week hands-on projects in over 800 camps in thirty-six countries.

Witness for Peace
P.O. Box 567
Durham, NC 27702-0567
(919) 688-5049

A grassroots, faith-based organization committed to changing U.S. policy toward Central America through nonviolent action. WFP sponsors both long-term and short-term work delegations.

Appendix C

Environmental Organizations

This is a sampling of nongovernmental national and international environmental organizations. For the more complete, annually updated Conservation Directory, send $25 and $5.25 S&H to the National Wildlife Federation, 1400 Sixteenth Street NW, Washington, DC 20036. The directory includes many professional associations and state or local organizations not listed here. Your local library may have a copy. Most of the organizations publish a magazine or newsletter.

• Acid Rain Foundation, 1410 Varsity Drive, Raleigh, NC 27606. A nonprofit, nonpartisan organization that educates the public on air quality issues, including acid rain, air pollutants, and global climate changes.

• African Wildlife Foundation, 1717 Massachusetts Avenue NW, Washington, DC 20036 (202-265-8393). Finances and operates wildlife conservation projects in Africa.

• American Farmland Trust, 1920 N Street NW, Suite 400, Washington, DC 20036 (202-659-5170). Works for preservation of family farms and for soil conservation.

• American Museum of Natural History, Central Park West at 79th Street, New York, NY 10024 (212-769-5100). Conducts natural history research and publishes educational material.

• American Rivers, Inc., 801 Pennsylvania Avenue SE, Suite 400, Washington, DC 20003 (202-547-6900). Works for preservation of American rivers.

• Appalachian Mountain Club, 5 Joy Street, Boston, MA 02108 (617-523-0636). Sponsors trail maintenance, outdoor education, and recreational hikes and climbs. Operates a mountain hut system on the Appalachian trail.

• Bat Conservation International, P.O. Box 162603, Austin, TX 78716 (512-327-9721). Works to preserve bats and to change people's perceptions of bats.

• CARE, 660 First Avenue, New York, NY 10016 (212-686-3110). Offers technical and material assistance to developing countries focusing on health, nutrition, agriculture, environment, small business support, and emergency aid.

• Center for Environmental Education, 1725 DeSales Street NW, Suite 500, Washington, DC 20036 (202-429-5609). A nongovernmental, nonprofit organization that sponsors programs for protection of endangered species.

• Center for Science in the Public Interest, 1875 Connecticut Avenue NW, Suite 300, Washington, DC 20009 (202-332-9110). National consumer advocacy organization that focuses on health and nutrition.

• Cenozoic Society, P.O. Box 455, Richmond, VT 05477 (802-434-4077). Founded in 1991 as a splinter group from Earth First!, the Cenozoic Society publishes *Wild Earth,* a journal of conservation biology and wildlands activism.

• Citizen's Clearinghouse for Hazardous Wastes, P.O. Box 6806, Falls Church, VA 22040 (703-237-2249). Collects information on hazardous waste effects, disposal, and cleanup.

• Clean Water Fund, 1320 18th Street NW, Washington, DC 20036 (202-457-0336). Works for clean water and for protection of natural resources. Conducts voter education and public awareness projects through fourteen state groups.

• Conservation International, 1015 18th Street NW, Suite 1000, Washington, DC 20036 (202-429-5660). Buys land or trades foreign debt for land set aside for nature preserves in developing countries.

• Cousteau Society, Inc., 870 Greenbriar Cir., Suite 402, Chesapeake, VA 23320 (804-523-9335). Produces television films, lectures, books, and research on ocean quality and other resource issues.

• Cultural Survival, 53-A Church Street, Cambridge, MA 02138 (617-495-2562). Helps small, vulnerable societies survive the encroachments of governments and industrialized society.

• Defenders of Wildlife, 1101 14th St., NW, Suite 1400, Washington, DC 20005 (202-682-9400). Seeks to protect and restore native species, habitats, and ecosystems.

• Ducks Unlimited, Inc., 1 Waterfowl Way, Memphis, TN 38120 (901-758-3825). Perpetuates waterfowl by purchasing and protecting wetland habitat.

• Earth Island Institute, 300 Broadway, Suite 28, San Francisco, CA 94133 (415-788-3666). A clearinghouse for international information on environmental and resource issues. Founded by David Brower to bring together sources of conservation action and news.

• Earthwatch, P.O. Box 403N Mt. Auburn St., Watertown, MA 02272 (800-776-0188). Sponsors scientific field research worldwide. Recruits volunteers for a wide variety of research expeditions.

• Environmental Careers Organization, 286 Congress St., 3rd Floor, Boston, MA 02210 (617-426-4375). Promotes environmental careers through consulting and publications.

• Environmental Defense Fund, Inc., 257 Park Avenue South, New York, NY 10010 (212-505-2100). Protects environmental quality and public health through litigation and administrative appeals.

• Environmental Law Institute, 1616 P Street NW, Suite 200, Washington, DC 20036 (202-328-5150). Sponsors research and education on environmental law and policy.

• Food and Agriculture Organization of the United Nations, Via delle Terme di Caracalla, Rome 00100 Italy (Tele: 57971). A special agency of the United Nations established to improve food production, nutrition, and the health of all people.

• Friends of the Earth, 1025 Vermont Ave, NW, Suite 300, Washington, DC 20005 (202-783-7400). Affiliated with groups in thirty-two countries around the world. Works to form public opinion and to influence government policies to protect nature.

- Fund for Animals, Inc., 200 West 57th Street, New York, NY 10019 (212-246-2096). Advocacy group for humane treatment of all animals.

- Greater Yellowstone Coalition, P.O. Box 1874, 13 S. Willson, Bozeman, MT 59715 (406-586-1593). A small but influential organization dedicated to the preservation and protection of the Greater Yellowstone ecosystem.

- Green Party USA/The Greens, P.O. Box 30208, Kansas City, MO 64112 (816-931-9366). Headquarters of the green political movement in the U.S.

- Greenpeace USA, 1436 U Street NW, Washington, DC 20009 (202-462-1177). Worldwide organization that works to halt nuclear weapons testing, to protect marine animals, and to stop pollution and environmental degradation.

- Humane Society of the United States, 2100 L Street NW, Washington, DC 20037 (202-452-1100). Dedicated to the protection of both domestic and wild animals.

- Institute for Food and Development Policy, 398 60th St., Oakland, CA, 94618 (570-654-4400). An international research and education center working for social justice and environmental protection.

- Institute for Social Ecology, P.O. Box 89, Plainsfield, VT 85667. Offers courses for credit through Goddard College and workshops on social ecology, ecofeminism, urban design, and ecological planning.

- International Alliance for Sustainable Agriculture, 1701 University Avenue SE, Minneapolis, MN 55414 (612-331-1099). An alliance of organic farmers, researchers, consumers, and international organizations dedicated to sustainable agriculture.

- International Society for Ecological Economics, P.O. Box 1589, Solomons, MD 20688 (410-326-0794). Publishes the journal *Ecological Economics.*

- International Union for Conservation of Nature and Natural Resources (IUCN), US, 1400 16th St., NW, Washington, DC 20036 (202-797-5454). Promotes scientifically based action for the conservation of wild plants, animals, and resources.

- International Wolf Center, 1396 Hwy 169, Ely, MN 55731 (218-365-4695). Offers classes, tours, maintains displays, and a captive wolf pack for conservation education and research.

- Izaak Walton League of America, Inc., Conservation Center, 707 Conservation Lane, Gaithersburg, MD 20878. Educates the public on land, water, air, wildlife, and other conservation issues.

- Land Institute, 2440 E. Water Well Road, Salina, KS 67401 (913-823-5376). Carries out research on perennial species, prairie polycultures, and sustainable agriculture. Has training programs and conferences.

- League of Conservation Voters, 1707 L Street NW, Suite 750, Washington, DC 20036 (202-785-8683). A nonpartisan national political campaign committee that strives to elect environmentally responsible public officials. Publishes an annual evaluation of voting records of Congress.

- National Audubon Society, 700 Broadway, New York, NY 10003 (212-979-3000). One of the oldest and largest conservation organizations, Audubon has many educational and recreational programs as well as an active lobbying and litigation staff.

- National Parks and Conservation Association, 1776 Massachusetts Ave. NW, Suite 200, Washington, DC 20036 (202-223-6722). A private nonprofit organization dedicated to preservation, promotion, and improvement of our national parks.

- National Wildlife Federation, 1400 Sixteenth Street NW, Washington, DC 20036 (202-797-6800). Specializes in wildlife conservation but recognizes the importance of habitat and other resources to all living things. More than 5 million members.

- Natural Resources Defense Council, Inc., 40 W. 20th Street, New York, NY 10011 (212-727-2700). An environmental organization that monitors government agencies and brings legal action to protect the environment.

- (The) Nature Conservancy, 1815 North Lynn Street, Arlington, VA 22209 (703-841-5300). Works with state and federal agencies to identify ecologically significant natural areas. Manages a system of over 1000 nature sanctuaries nationwide.

- Planet Drum Foundation, P.O. Box 31251, San Francisco, CA 94131 (415-285-6556). Promotes bioregionalism, wise use of resources, and new attitudes toward nature.

- Population Reference Bureau, 1875 Connecticut Ave NW, Suite 540, Washington, DC 20009 (202-483-1100). Gathers, interprets, and publishes information on social, economic, and environmental implications of world population dynamics. Excellent data source.

- Rainforest Action Network, 450 Sansome St., Suite 700, San Francisco, CA 94111 (415-398-4404). Focuses on actions designed to save rainforest and to defend the rights of indigenous people around the world.

- Resources for the Future, 1616 P Street NW, Washington, DC 20036 (202-328-5000). Conducts research and provides education about natural resource conservation issues.

- Rodale Institute, 222 Main Street, Emmaus, PA 18098 (610-967-8509). A leading research institute for organic farming and alternative crops. Publishes magazines, books, and reports on regenerative farming.

- Save-the-Redwoods League, 114 Sansome Street, Room 605, San Francisco, CA 94104 (415-362-2352). Buys land, plants trees, and works with state and federal agencies to save redwood trees.

- Scientists' Institute for Public Information, 355 Lexington Avenue, New York, NY 10017 (212-661-9110). Enlists scientists and other experts in public information programs and public policy forums on a variety of environmental issues.

- Sea Shepherd Conservation Society, P.O. Box 628, Venice, CA 90294 (310-301-7325). An international marine conservation action program. Carries out field campaigns to call attention to and stop wildlife destruction and resource misuse.

- Sierra Club, 730 Polk Street, San Francisco, CA 94109 (415-776-2211). Founded in 1892 by John Muir and others to explore, enjoy, and protect the wild places of the earth. Conducts outings, educational programs, volunteer work projects, litigation, political action, and administrative appeals. Has one of the most comprehensive programs of any conservation organization.

- Society for Conservation Biology, Stanford University, Department of Biological Sciences, Stanford, CA 94305 (415-725-1852). A professional society for research and information. Publishes *Conservation Biology.*

- Soil and Water Conservation Society, 7515 NE Ankeny Rd., Ankeny, IA 50021 (515-289-2331). Promotes soil and water conservation.

- Student Conservation Association, Inc., Route 12A, River Road, Charlestown, NH 03603 (603-543-1700). Coordinates environmental internships and volunteer jobs with state and federal agencies and private organizations for students and adults.

- United Nations Environment Programme, P.O. Box 30552, Nairobi, Kenya; and United Nations, Rm. DC2-0803, New York, NY 10017 (212-963-8138). Coordinates global environmental efforts with United Nations agencies, national governments, and nongovernmental organizations.

- Waldebridge Ecological Centre, Worthyvale Manor Farm, Camelford, Cornwall P132 9TT England (Tele: 0840 212711). Conducts research and education on a variety of environmental issues. Publishes *The Ecologist,* an excellent global environmental journal.

- Wilderness Society, 900 17th Street NW, Washington, DC 20006 (202-833-2300). Dedicated to preserving wilderness and wildlife in America.

- World Resources Institute, 1709 New York Avenue NW, Washington, DC 20006 (202-638-6300). A policy research center that publishes excellent annual reports on world resources.

- Worldwatch Institute, 1776 Massachusetts Avenue NW, Washington, DC 20036 (202-452-1999). A nonprofit research organization concerned with global trends and problems. Publishes excellent periodic reports and annual summaries.

- World Wildlife Fund, 1250 24th Street NW, Washington, DC 20037 (202-293-4800). Seeks to protect the world's endangered wildlife and the habitat they need to survive.

- Zero Population Growth, Inc., 1400 16th Street NW, Suite 320, Washington, DC 20036 (202-332-2200). Advocates of worldwide population stabilization.

Appendix

United States Government Agencies

- Bureau of Land Management, 1620 L St. NW, Washington, DC 20240 (202-205-3801). Administers about half of all public lands, mainly in the western United States. Follows policy of multiple use for maximum public benefit.

- Bureau of Reclamation, Room 7654, Department of the Interior, Washington, DC 20240. Builds and operates federal water projects, mostly in the western states with the Army Corps of Engineers.

- Bureau of Sports Fisheries and Wildlife (formerly U.S. Fish and Wildlife Service), Washington, DC 20240 (202-208-4717). Carries out wildlife research and management. Enforces game, fish, and endangered species laws. Administers the national wildlife refuges.

- Council on Environmental Quality, 722 Jackson Place NW, Washington, DC 20503 (202-395-5750). Advises the president on environmental matters.

- Department of Agriculture, 14th Street and Independence Avenue SW, Washington, DC 20250 (202-720-8732). Manages national forests and grasslands and oversees farm prices, farm policies, and soil conservation. (http://www.usda.gov)

- Department of Health and Human Services: Food and Drug Administration, 5600 Fishers Lane, Rockville, MD 20857. Enforces laws requiring that foods and drugs be pure, safe, and wholesome.

- Department of the Interior, C Street between 18th and 19th NW, Washington, DC 20240 (202-208-3100). Administers national parks, monuments, wildlife refuges, and public lands. (http://www.usgs.gov.doi)

- Environmental Protection Agency, 401 M Street SW, Washington, DC 20460 (202-260-2090). Enforces clean air and clean water laws. Identifies, regulates, and purifies toxic and hazardous materials. (http://wwn.epa.gov)

- Forest Service, P.O. Box 96090, Washington, DC 20013 (202-250-0951). Administers national forests and grasslands.

- National Oceanic and Atmospheric Administration, Herbert C. Hoover Bldg., Rm 5128, 14th and Constitution Ave. NW, Washington, DC 20230 (202-482-3384). Part of the Department of Commerce. Promotes global environmental stewardship and conducts atmospheric and oceanic research. Good source of information on atmosphere and climate. (http://www.gov.doc)

- National Park Service, Interior Bldg, P.O. Box 37127, Washington, DC 20013 (202-208-4717). Administers the national park system, monuments, and wild and scenic rivers. Possible source of seasonal jobs.

- National Resource Conservation Service (formerly Soil Conservation Service), 14th and Independence Ave. SW, P.O. Box 2890, Washington, DC 20013 (202-447-4543). Provides technical and educational assistance for soil conservation and watershed protection.

urther Readings

Chapter 1

Barkham, J. P. 1995. "Environmental Needs and Social-Justice," *Biodiversity and Conservation* 4 (8): 857–868. An argument that the roots of environmental destruction lie in the lifestyle demands for increased material consumption and net flow of wealth from the South (less developed) to North (more developed) countries.

Brush, Stephen B. and Doren Stabinsky. 1995. *Valuing Local Knowledge: Indigenous People and Intellectual Property Rights.* Covelo, CA: Island Press. An innovative proposal to promote both cultural survival and biological resources by recognizing local systems of knowledge.

Bryant, Bunyan, ed. 1995. *Environmental Justice: Issues, Policies, and Solutions.* Covelo, CA: Island Press. An anthology of essays on social justice and environmental protection.

Crocker, D. A. 1995. "Consumption and Well-Being," *Philosophy and Public Policy* 15:12–17. How much do we really need?

Cunningham, W. P., et al. (eds). 1994. *Environmental Encyclopedia.* Detroit, MI: Gale Research Inc. A good general source of environmental information with nearly 1000 signed articles on a variety of topics.

Daily, G. C., Anne H. Ehrlich, and Paul R. Erlich. 1995. "Socioeconomic Equity—A Critical Element in Sustainability," *Ambio* 24(1): 58–59. Recognizes the importance of social justice in our quest for sustainability.

Easterbrook, Greg. 1995. *A Moment on the Earth: The Coming Age of Optimism.* New York: Penguin Books. A highly optimistic view of our future. Strongly criticized by many environmentalists.

Grove, Richard H. 1995. *Green Imperialism: Colonial Expansion, Tropical Island Edens, and the Origins of Environmentalism, 1600–1860.* Cambridge: Cambridge University Press. A wide-ranging survey of the environmental consequences of colonialism and exploitation.

Johnston, Barbara R., ed. 1994. *Who Pays the Price? The Sociocultural Context of Environmental Crisis.* Covelo, CA: Island Press. A statement from the Society for Applied Anthropology on human rights and the environment.

Long, L. E., et al. 1995. "A Ph/Uv-B Synergism in Amphibians," *Conservation Biology* 9(5):1301–1303. Does air pollution play a role in the disappearance of amphibians from the wild?

McKibben, Bill. (April) 1995. "An Explosion of Green," *Atlantic Monthly* 275(4):61–105. The forest cover of the eastern United States is as extensive today as it was 250 years ago. Can we count on the resilience and fecundity of nature to repair our damage?

Nagpal, Tanvi, and Camilla Foltz, eds. 1995. *A World of Difference: Giving Voices to Visions of Sustainability.* Washington, DC: World Resources Inc. A survey of the hopes and dreams for the future of people all over the world.

Phillips, Kathryn. 1994. *Tracking the Vanishing Frogs: An Ecological Mystery.* New York: Penguin Books. Why are frogs disappearing all over the world? A scientist surveys the evidence.

Redclift, M., and C. Sage, eds. 1994. *Strategies for Sustainable Development: Local Agendas for the Southern Hemisphere.* New York: John Wiley & Sons. A critique of usual plans for development.

Reid, David. 1995. *Sustainable Development: An Introductory Guide.* Washington, DC: EarthScan. An introduction to development and environmental protection.

Sachs, Aaron. 1996. "Upholding Human Rights and Environmental Justice," *State of the World.* Washington, DC: Worldwatch Institute. An exploration of human rights and environmental protection. Also published as *Worldwatch Paper #127.*

Sen, G. February 1995. "Creating Common Ground Between Environmentalists and Women—Thinking Locally, Acting Globally," *Ambio* 24(1):64–65. Seeks an accommodation between social justice and environmental protection.

Shiva, Vandana, ed. 1994. *Close to Home: Women Reconnect Ecology, Health and Development Worldwide.* Philadelphia: New Society Publishers. An ecofeminist view of sustainability.

United Nations. 1995. *The World's Women 1995: Trends and Statistics.* New York: The United Nations. Data gathered in preparation for the U.N. Women's Conference in Beijing.

Worster, Donald. 1993. "The Shaky Ground of Sustainability," in Wolfgang Sachs, ed. *Global Ecology: A New Arena of Political Conflict.* London: Zed Books. Claims that sustainable development is a misleading oxymoron.

Chapter 2

Atchia, Michael, and Shawna Troop, eds. 1995. *Environmental Management, Issues and Solutions.* New York: Wiley. Many specialists discuss a wide range of environmental issues with examples from many countries.

Brennan, A. 1995. "Ethics, Ecology and Economics," *Biodiversity and Conservation.* 4(8):798–811. A critique of utilitarian economics but argues that economic preference theory may be useful in environmental philosophy.

Bryant, Bunyan. 1995. *Environmental Justice: Issues, Policies, and Solutions.* Covelo, CA: Island Press. A good history of the social justice movement and methods of dealing with environmental problems.

Buege, Douglas J. (Spring) 1996. "The Ecologically Noble Savage Revisited," *Environmental Ethics* 18(1):71–88. Argues that we shouldn't romanticize all indigenous people as living in perfect harmony with their environment.

Callicott, J. Baird. 1994. *Earth's Insights: A Survey of Ecological Ethics from the Mediterranean Basin to the Australian Outback.* Traces environmental attitudes of all the major world religions as well as many traditional native cultures.

Carson, Rachel. 1962. *Silent Spring.* Boston: Houghton Mifflin. This landmark book awakened us to the dangers of indiscriminate pesticide use and started a new era of environmental activism.

Ebenreck, Sara. (Spring) 1996. "Opening Pandora's Box: The Role of Imagination in Environmental Ethics," *Environmental Ethics* 18(1):3–18. Dominant Western traditions rank abstract reasoning above imagination, but this article argues that imagination could play an important role in environmental philosophy.

Hartley, Troy W. (Fall) 1995. "Environmental Justice: An Environmental Civil Rights Value Acceptable to All World Views," *Environmental Ethics* 17(3):277–290. Calls for equitable distribution of environmental protection exists in most major religions and provide a philosophical basis for environmental justice.

Istock, Conrad A., and Robert S. Hoffmann. 1996. *Storm Over A Mountain Island: Conservation Biology and the Mt. Graham Affair.* Tucson: University of Arizona Press. A history of the conflict over building an astronomical observatory on top of Mt. Graham.

Kellert, Stephen R. 1995. *The Value of Life: Biological Diversity and Human Society.* An examination of the values of living diversity and our attitudes toward nature.

Mies, Maria, and Vandana Shiva. 1993. *Ecofeminism.* London: Zed Books. A presentation of ecofeminism from the viewpoint of women in developing countries.

Oelschlaeger, Max. 1994. *Caring For Creation.* New Haven: Yale University Press. An ecumenical approach to solving our environmental crisis.

O'Riordan, Timothy. (October) 1995. "Frameworks for Choice: Core Beliefs and the Environment," *Environment* 37(8):4–10. Environmental attitudes and actions often reflect deeper beliefs about the world.

Orr, David W. 1994. *Earth In Mind: On Education, Environment, and the Human Prospect.* Covelo, CA: Island Press. A critique of modern education and a plea for more environmental sensitivity.

Plumwood, Val. 1993. *Feminism and the Mastery of Nature.* London: Routledge. An ecofeminist critique of Western philosophy.

Rolston, Holmes, III. 1988. *Environmental Ethics.* Philadelphia: Temple University Press. Not easy reading, but one of the most rigorous and authoritative environmental ethics books available.

Ruether, Rosemary Radford. 1992. *Gaia and God: An Ecofeminist Theology of Earth Healing.* San Francisco: Harper Collins. An ecofeminist view of religion.

Sepanmaa, Yrjo. 1996. *The Beauty of Environment,* Second Edition. An analysis of the role of aesthetics in environmental valuation.

Sessions, George, ed. 1995. *Deep Ecology for the 21st Century: Readings on Philosophy and Practice of the New Environmentalism.* London: Shambhala Press. How can we live in harmony with nature?

Soule, Michael E. and Gary Lease. 1994. *Reinventing Nature? Responses to Postmodern Deconstruction.* A discussion of postmodernism and the meaning of nature.

Thomashow, Mitchell. 1995. *Ecological Identity: Becoming a Reflective Environmentalist.* Cambridge: MIT Press. A new approach to environmental education emphasizes hands-on learning as well as critical thinking about theory.

Weston, Anthony. 1994. *Back to Earth: Tomorrow's Environmentalism.* Philadelphia: Temple University Press. Goes beyond environmental ethics to generate a new approach to understanding our environment.

Chapter 3

Begon, M., et al. 1995. *Ecology: Individuals, Populations, and Communities,* 3d ed. London: Blackwell Science. A good basic ecology textbook.

Codispoti, Louis A. 1995. "Is the Ocean Losing Nitrate?" *Nature* 376:724. Application of geochemical cycles to the important problem of global climate regulation.

Freedman, Bill. 1995. *Environmental Ecology,* 2d ed. New York: Academic Press. A well-written basic ecology textbook.

Frosch, R. A. 1995. "Industrial Ecology: Adapting Technology for a Sustainable World," *Environment.* 37(10):16–24. How can we apply the science of ecology to industry?

Harper, David M., and Alastair J. D. Ferguson, eds. 1995. *The Ecological Basis For River Management.* Addresses the underlying ecological principles to practical problems of river management.

Keeling, C. D., et al. 1995. "Dynamics of Carbon Dioxide in the Atmosphere: Observations at Hawaii and the South Pole," *Nature* 375:666–670. A comparison of the 35-year history of measurements at Mauna Loa and the South Pole with respect to anthropogenic carbon emissions.

Maehr, David S., and James A. Cox. 1995. "Landscape Features and Panthers in Florida," *Conservation Biology* 9(5):1008–1015. A geographic information system shows landcover types of home ranges of these endangered top predators.

Risser, Paul G. 1995. "Biodiversity and Ecosystem Function," *Conservation Biology* 9(4):742–746. In our attempts to understand the relationship between biodiversity and the ways in which ecosystems respond to disturbance, we should look at structuring processes and ecological conditions rather than keystone species.

Wackernagel, Mathis, and William Rees. 1995. Our Ecological *Footprint.* Philadelphia: New Society Publishers. Applies ecological principles to issues of human resource use, waste disposal, and population growth.

Wilson, Mary F., and Karl C. Halupka. 1995. "Anadromous Fish as Keystone Species in Vertebrate Communities," *Conservation Biology* 9(3):489–497. A study of the role of anadromous fish as key food sources, habitat modifiers, and transporters of nutrients between ecosystems.

Chapter 4

Adams, Lowell W. 1994. *Urban Wildlife Habitats: A Landscape Perspective.* Minneapolis: University of Minnesota Press. Application of landscape ecology to wildlife management.

Bancroft, G. T., et al., 1995. "Deforestation and Its Effects on Forest-Nesting Birds in the Florida Keys," *Conservation Biology* 9(4):835–844. A good example of the effects of forest fragmentation on nesting success of woodland birds.

Belsky, A. Joy. 1995. "Spatial and Temporal Landscape Patterns in Arid and Semi-Arid African Savannas," in L. Hansson, et al., eds. *Mosaic Landscapes and Ecological Processes*. A good example of community structure in a real-world situation.

Hahn, D. Caldwell, and Jeff S. Hatfield. 1995. "Parasitism at the Landscape Scale: Cowbirds Prefer Forests," *Conservation Biology* 9(6):1415–1424. Contrary to earlier studies, this research presents evidence that cowbirds parasitize nests equally at the edge and forest interior.

Schiebinger, Londa. 1996. "The Loves of the Plants," *Scientific American* 274(2): 110–115. Our attitudes about what matters most in interactions between organisms colors our views of how the world works.

Schieck, J., et al. 1995. "Effects of Patch Size on Birds in Old-Growth Montane Forests," *Conservation Biology* 9(5):1072–1084. Shows that nonspecialists are more common than old-growth species in disturbed forests. For some species, however, small patches of remnant forest have larger populations than large old-growth patches.

Walker, Brian. 1995. "Conserving Biological Diversity Through Ecosystem Resilience," *Conservation Biology* 9(4):747–752. Explores the relationship between biodiversity, ecological redundancy, resilience, and ecosystem functions.

Wille, Chris. 1994. "The Birds and the Beans," *Audubon* 96(6):58–65. Coffee plantations have provided winter habitats for neotropical migratory birds, but sun-tolerant coffee varieties grown on massive plantations are causing a loss of cover trees that once sheltered winter residents.

Wilson, Edward O. 1994. *Naturalist*. Covelo, CA: Island Press. An excellent biography of one of the foremost naturalists of our time. Gives a good view of a life of field work.

Youth, Howard. 1994. "Flying Into Trouble," *Worldwatch* 7(1):10–19. A summary of the evidence for declining bird populations and what that shows about our environment.

Chapter 5

Baldwin, A. Dwight, Jr., et al. 1994. *Beyond Preservation: Restoring and Inventing Landscapes*. Minneapolis: University of Minnesota Press. See introductory articles by William Jordan and Frederic Turner arguing for human invention of landscapes.

Brown, Valerie A., et al. 1995. *Risks and Opportunities: Integrated Management of Environmental Conflict and Change*. Washington, DC: EarthScan. A project management framework for addressing environmental change that identifies stakeholders and strategic approaches.

Cairns, John, Jr. 1995. *Rehabilitating Damaged Ecosystems*, 2d ed. Covelo, CA: Island Press. An important handbook for practitioners, academics, and managers who want to restore degraded ecosystems.

Caughley, Graeme, and Anne Gunn. 1995. *Conservation Biology in Theory and Practice*. London: Blackwell Scientific. A text on management, restoration, and recovery of endangered species.

Dobson, Andrew. 1995. *Conservation and Biodiversity*. New York: W. H. Freeman. A broad overview of endangered species protection, conservation of biological communities, and ecosystem management.

Falk, Donald A., et al. 1996. *Restoring Diversity: Strategies for Reintroduction of Endangered Plants*. Covelo, CA: Island Press. Articles from a conference on restoring endangered plants.

Frenay, Robert. 1995. "Biorealism: Reading Nature's Blueprints," *Audubon* 97(5):70–79. Using nature as a model, a new science-based movement is redesigning many industrial processes.

Hannah, Lee, et al. 1995. "Human disturbance and Natural Habitat: A Biome-Level Analysis of a Global Data Set," *Biodiversity and Conservation* 4:128–155. Using a human disturbance and habitat index, researchers find that four of the top five disturbed biomes worldwide are in the temperate zone.

Holling, C. S. 1995. "What Barriers? What Bridges?" in Gunderson, C. S., et al. *Barriers and Bridges to the Renewal of Ecosystems and Institutions*. New York: Columbia University Press. The introductory essay in a landmark publication on ecosystem management and institutional change.

Hunter, Malcolm L., Jr. 1995. *Fundamentals of Conservation Biology*. London: Blackwell Scientific. An important marriage of ecosystem management and conservation biology in a single textbook.

Lapin, M., and B. V. Barnes. 1995. "Using the Landscape Ecosystem Approach to Assess Species and Ecosystem Diversity," *Conservation Biology* 9(5):1148–1158. Utilizes landscape ecology for endangered species management.

Lee, Kai N. 1993. *Compass and Gyroscope*. Covelo, CA: Island Press. An important and elegantly written account of adaptive management for environmental problems.

McManus, R. (May/June) 1996. "For Love of a Swamp," *Sierra* 81(3):42–3. Lead article in a special issue on wetlands. Great photos.

Meffe, Gary K., and C. Ronald Carroll. 1994. *Principles of Conservation Biology*. Sunderland, MA: Sinauer Associates. An excellent overview of the new science of conservation biology.

Mills, Stephanie. 1995. "Reforesting the Wastelands," *Earth Island* 10(4):23–24. How India's Auroville settlement is restoring a denuded landscape.

Nassauer, Joan I. 1995. "Culture and Changing Landscape Structure," *Landscape Ecology* 10(4):229–237. An integration of landscape dynamics and culture in an excellent example of landscape ecology.

Porter, Douglas R., and David A. Salvesen. 1995. *Collaborative Planning for Wetlands and Wildlife: Issues and Examples*. Covelo, CA: Island Press. Numerous case studies that demonstrate issues of creative wetland preservation and replacement.

Rutzler, Kalus, and Ilka C. Feller. 1996. "Caribbean Mangrove Swamps," *Scientific American* 274(3):94–99. A comprehensive study of a coastal mangrove swamp in Belize, one of the most important but endangered biological communities in the world.

Stanley, Thomas R., Jr. 1995. "Ecosystem Management and the Arrogance of Humanism," *Conservation Biology* 9(2):255–262. A detailed but vitriolic criticism of ecosystem management and restoration ecology.

Chapter 6

Cappuccino, Naomi, and Peter W. Price, eds. 1995. *Population Dynamics: New Approaches and Synthesis*. New York: Academic Press. A good blend of population biology and economic applications.

Hanski, I., and M. E. Gilpin. 1991. "Metapopulation Dynamics: Brief History and Conceptual Domain," *Biological Journal of the Linnean Society* 42:3–16. A history of the metapopulation concept and its application to conservation biology.

Miller, P. S. 1995. "Selective Breeding Programs for Rare Alleles: Examples from the Przewalski's Horse and California Condor Pedigrees," *Conservation Biology* 9(5):1262–1273. Application of population genetics to the conservation of rare species.

Peterson, Rolf O. 1995. *Wolves of Isle Royale: A Broken Balance.* Minocqua, WI: Willow Creek Press. The fascinating story of the best known wolf population in the world. Excellent photos.

Chapter 7

Abernethy, V. 1993. *Population politics: The Choices That Shape Our Future.* New York: Plenum Press. Both intensely personal and broadly public beliefs and choices determine how our population grows. Politics plays an important role in shaping population policies.

Bongaarts, J. 1994. "Can The Growing Human Population Feed Itself?" *Scientific American* 270, no. 3:36. An analysis of world population growth and food production potential.

Dale, A. (Winter/Spring) 1993. "Women and Population." *Women-Environments* 13, no. 3–4:26. Describes the emancipation of women as a key factor in birth control and environmental protection.

Dasgupta, Partha S. 1995. "Population, Poverty and the Local Environment," *Scientific American* 272(2):40–45. An economic study of the interrelationship between poverty, society, and the environment.

Goldscheider, E., ed. 1992. *Fertility Transitions, Family Structure, and Population Policy.* Boulder, CO: Westview Press. Articles by leading experts in demographics, sociology, and policy analysis explore the underlying forces in demographic transitions.

Hall, Charles A., et al. 1995. "The Environmental Consequences of Having a Baby in the United States," *Wildearth* 5(2):78–87. Most population growth is in developing countries but because of our inordinate consumption rates, a single baby born in the United States can have more environmental impact than a whole village in some parts of the world. See also, *Earth Island*

Journal 10(4):19–20 (Fall 1995) for a shorter version of this article.

Hardin, Garrett. 1993. *Living Within Limits: Ecology, Economics, and Population Taboos.* New York: Oxford University Press. A trenchant warning from our favorite iconoclast of the fallacy of thinking that humans are exempt from the laws of nature.

Hartman, Betsy. 1995. "Questioning the Population Consensus," *Earth Island Journal* 10(2):34–36. A feminist critique of the U.N. Population Conference in Cairo and claims that population growth is the cause of poverty and environmental degradation.

Jiggins, Janice. 1994. *Changing the Boundaries: Women-Centered Perspectives on Population and the Environment.* Covelo, CA: Island Press. An exploration of gender relations with respect to education, reproductive health, and access to resources as they apply to population growth.

Kane, Hal. 1995. *The Hour of Departure.* Washington, DC: Worldwatch Institute Paper 125. A study of the forces that create refugees.

Leisinger, Klaus M., and Karin Schmitt. 1994. *All Our People: Population Policy with a Human Face.* Covelo, CA: Island Press. Strategies for socially acceptable reductions in birth rates.

Livernash, Robert. 1995. "The Future of Populous Economies: China and India Shape Their Destinies." *Environment* 37(6):6–11. Unparalleled economic and population growth in these enormous countries pose serious environmental challenges for the region and the rest of the world.

Mazur, Laurie Ann, ed. 1994. *Beyond Numbers: A Reader on Population, Consumption, and the Environment.* Covelo, CA: Island Press. An interesting anthology of essays on the topics considered at the Conference on Population and Development held in Cairo, Egypt, in 1994.

Meyers, Norman, and Julian Simon. 1994. *Scarcity or Abundance: A Debate on the Environment.* New York: W. W. Norton. A prize-winning ecologist and cornucopian economist debate population, resources, and the future of our environment.

Chapter 8

Arrow, Ken, et al. 1995. "Economic Growth, Carrying Capacity, and the Environment,"

Ecological Economics 15(2):91–97. A consensus statement from a group of economists and ecologists on population growth and resource consumption. See subsequent articles in this same issue for responses.

Brandt, Barbara. 1995. *Whole Life Economics: Revaluing Daily Life.* Philadelphia: New Society Publishers. A prescription for personal and societal recovery from consumerism.

Cairncross, Francis. 1994. *Green, Inc.: A Guide to Business and the Environment.* Covelo, CA: Island Press. Resource economics including environmental cost-benefit analysis, natural resource accounting, and economic regulation.

Cohn, Susan. 1995. *Green at Work: Finding a Business Career That Works for the Environment,* 2d ed. Covelo, CA: Island Press. How to find a "green" job.

Duchin, F., and G. M. Lange. 1994. *Ecological Economics, Technological Change, and the Future of the Environment.* New York: Oxford University Press. A persuasive argument that sustainable development will require a significant change in technology and economics.

George, S. 1990. *A Fate Worse than Debt: The World Financial Crisis and the Poor.* New York: Grove Weidenfeld. An award-winning examination of the link between national debt, poverty, and environmental destruction.

Goudy, John, and Sabine O'Hara. 1995. *Economic Theory for Environmentalists.* St. Lucie Press. An easy-to-read explanation of basic economic theories for noneconomists.

Holm, Hans-Henrik, and Georg Sorenson, eds. 1995. *Whose World Order? Uneven Globalization and the End of the Cold War.* Boulder, CO: Westview Press. A discussion of the effects of globalization on society and the environment.

Jansson, AnnMari, et al. 1994. *Investing in Natural Capital: The Ecological Economics Approach to Sustainability.* Covelo, CA: Island Press. An important compilation of articles in the new field of ecological economics.

Jordan, A. (July/August) 1994. "Paying the Incremental Costs of Global Environmental Protection: The Evolving Role of GEF." *Environment* 36, no. 6:12. An analysis of the UN Global Environment Facility as the main conduit for development funds for poorer countries.

Krishnan, Rajaram, et al. 1995. *A Survey of Ecological Economics.* Covelo, CA: Island Press. A comprehensive summary of the history, theoretical frameworks, and applications of ecological economics.

MacNeill, J., et al. 1991. *Beyond Interdependence: The Meshing of the World's Economy and the Earth's Ecology.* New York: Oxford University Press. Builds on the work of the Brundtland Commission to argue the critical relationships between the global environment, the world economy, and international order.

Magraw, D. March 1994. "NAFTA's Repercussions: Is Green Trade Possible?" *Environment* 36, no. 2:14. The North American Free Trade Agreement illustrates the conflicts between free-trade and environmental protection.

Organization for Economic Cooperation and Development. 1994. *Managing the Environment: The Role of Economic Instruments.* Washington, DC: OECD. A view of environment and sustainable development from the wealthier nations.

Pearce, David, et al. 1995. "Debt and the Environment," *Scientific American* 272(6):52–57. Most social activists claim that national debt has contributed to environmental degradation in the developing world. These authors argue that debt may have been good for the environment and that market-based approaches are superior to command and control.

Power, Thomas M. 1996. *Extraction and the Environment: The Economic Battle to Control Our Natural Landscape.* An argument that our environment and natural resource base should not be sacrificed in short-term efforts to shore up industries that ultimately are not sustainable.

Prugh, Thomas. 1995. *Natural Capital and Human Economic Survival.* Solomons, MD: International Society for Ecological Economics. A brief but helpful overview of ecological economics.

Repetto, R. June 1992. "Accounting for Environmental Assets." *Scientific American* 266, no. 6:94. The national system of accounts used by the United Nations badly needs overhaul to consider depletion of natural resource stocks.

Rich, B. (January/February) 1994. "The Cuckoo in the Nest: 50 Years of Political Meddling by the World Bank," *The Ecologist* 24, no. 1:8. An indictment of the policies and practices of the world's largest and most powerful development bank.

Roodman, David M. 1996. "Harnessing the Market for the Environment," *State of the World* 1996. Washington, DC: Worldwatch Institute. Using market mechanisms to protect the environment.

Stavins, R. N., and B. W. Whitehead. September 1992. "Market-Based Incentives for Environmental Protection," *Environment* 34, no. 7:7. A good discussion of using market mechanisms for pollution control with many examples and case studies.

West, Karen. 1995. "Ecolabels: The Industrialization of Environmental Standards," *The Ecologist* 25(2):16–20. Claims that ecolabeling is little more than a marketing gimmick and that under GATT, even this weak instrument could be a barrier to trade.

World Bank. 1994. *World Development Report 1994.* Oxford: Oxford University Press. A good overview of world resources and natural resource economics with a special emphasis on the relationship between sustainable development, poverty, and the environment.

Chapter 9

Adler, Tina. 1994. "The Return of Thalidomide," *Science News* 146:424–425. A disturbing report of the misuse of this drug in Third World countries.

Bartecchi, Carl E., et al. 1995. "The Global Tobacco Epidemic," *Scientific American* 272(5):44–51. Despite the well-known medical effects of smoking, tobacco use is climbing rapidly worldwide.

Béland, P. (May) 1996. "The Beluga Whales of the St. Lawrence River," *Scientific American* 274(5):74–81. Great pods of whales once thronged the St. Lawrence estuary. Pollution appears to be their greatest current threat.

Blumenthal, Daniel S., and James Ruttenber. 1995. *Introduction to Environmental Health,* 2d ed. New York: Springer Publishing. A short but popular introductory text for health disciplines.

Cothern, C. Richard, ed. 1996. *Handbook for Environmental Risk Decision Making.* Covelo, CA: Island Press. A detailed reference on risk assessment and decision making. Aimed at managers but useful for upper division students as well.

Cutter, Susan L., ed. 1994. *Environmental Risks and Hazards.* New York: Prentice Hall. A broad collection of readings in environmental health.

Dana, Amy, and Tom Turner. 1993. "Currents of Controversy," *The Amicus Journal* 15(2):29–32. New studies reignite the debate on the cancer risks of power lines.

Davis, Devra Lee. 1995. "The Perplexing War on Cancer," *Public Health Reports* 110(3):364–369. Argues that we should be paying more attention to environmental risks in our war on cancer.

Francis, B. M. 1994. *Toxic Substances in the Environment.* Somerset, NJ: John Wiley & Sons. A detailed textbook of environmental toxicology.

Garrett, Laurie. 1994. *The Coming Plague: Newly Emerging Diseases in a World Out of Balance.* New York: Farrar, Straus and Giroux. Is the emergence of new strains of pathogens related to disruption and degradation of the environment?

Gerard, Michael B. 1994. *Whose Backyard, Whose Risk? Fear and Fairness in Toxic and Nuclear Waste Siting.* Cambridge, MA: MIT Press. Asks, "risks acceptable to whom?" in the siting of locally unwanted land uses.

Gibbs, Lois M. 1995. *Dying From Dioxin: A Citizen's Guide to Reclaiming Our Health and Rebuilding Democracy.* A rallying call for a public revolt against industrial pollution, especially from chlorinated hydrocarbons.

Hoffman, David J., et al., eds. 1995. *Handbook of Ecotoxicology.* Covelo, CA: Island Press. A group of eminent ecologists and wildlife biologists address the effects of toxic chemicals on natural ecosystems.

Koren, Herman, and Michael Bissesi. 1996. *Handbook of Environmental Health and Safety,* 3rd ed. Washington, DC: The National Environmental Health Association. A two volume compendium of everything you might want to know about environmental health and safety.

Mukerjee, Madhusree. 1995. "Toxins Abounding," *Scientific American* 273(1):22–23. Despite the lessons of Bhopal, chemical accidents are on the rise.

Newman, Jack. 1995. "How Breast Milk Protects Newborns," *Scientific American* 273(6):76–82. Mother's milk provides a special substances to boost and supplement the newborn's immune system.

Newman, Michael J., and Charles H. Jagoe. 1996. *Ecotoxicology.* Covelo, CA: Island Press. An extensive overview of the ecological effects of pollutants.

Platt, Anne E. 1996. "Confronting Infectious Diseases," *State of the World 1996,* pp. 114–132. Washington, DC: Worldwatch Institute. A good overview of the risks of infectious diseases.

Raloff, J. August 21, 1993. "EMFs Run Aground," *Science News* 144:124. A fairly accessible review of the evidence for and against human health effects of electromagnetic radiation.

Rea, William J. 1996. *Chemical Sensitivity,* vols. 1–4. An exhaustive treatment of the causes, effects, and recommended treatments for sensitivities to environmental toxins (also known as sick house syndrome) and occupational exposures.

Ross, J. F. 1995. "Risk, Part 1: The More We Learn, the Harder Our Choices," *Smithsonian* 26(8):42–55. A popular, readable account of risk assessment and the dangers we face.

Stone, Richard. 1995. "If the Mercury Soars, So May Health Hazards," *Science* 267:957–958. Predicts that rising temperatures could spread the vectors of disease.

Chapter 10

Anderson, J. R., et al. 1993. *Sustaining Growth in Agriculture: A Quantitative Review of Agricultural Research Investments.* The Hague, The Netherlands: International Service for National Agricultural Research. New research will be needed to increase future food production.

Bongaarts, J. (March) 1994. "Can the Growing Human Population Feed Itself?" *Scientific American* 270, no. 3:36. What are the prospects for extending and expanding world food production? There are signs of hope as well as reasons for concern.

Brown, J. Larry, and Ernesto Pollitt. 1996. *Malnutrition, Poverty and Intellectual Development.* 274(2):38–43. Lack of essential nutrients during a child's early development can stunt mental achievement for a lifetime.

Brush, S., et al. 1995. "Potato Diversity in the Andean Center of Crop Domestication," *Conservation Biology* 9(5): 1189–1198. Strategies for *in situ* conservation of crop genetics.

Doos, B. R. (March) 1994. "Environmental Degration, Global Food Production, and Risk for Large-Scale Migrations," *Ambio* 23, no. 2:124. Suggests that environmental degration will reduce food supplies and trigger mass migrations.

Fairlie, Simon, et al. 1995. "The Politics of Overfishing," *The Ecologist* 25(2/3):46–73. The lead article in a special double issue on the causes and consequences of overfishing.

Keen, David. 1995. "The Benefits of Famine: A Case Study of the Sudan," *The Ecologist* 25(6):214–220. Shows that famines may be deliberately caused by elites who benefit from the suffering of others.

Little, Peter, and Michael Watts, eds. 1994. *Living Under Contract: Contract Farming and Agrarian Transformation in Sub-Saharan Africa.* Madison: University of Wisconsin Press. Subsistence farmers increasingly are being brought into the commercial network with consequences both for society and the environment.

Minkin, Stephen F., and James K. Boyce. 1994. "Net Losses: Development Drains the Fisheries of Bangladesh," *The Amicus Journal* 16(3):36–40. An excellent article about the effects of economic development projects and commercial fish farms on fishing by poor people.

Parfit, Michael. 1995. "Diminishing Returns: Exploiting the Ocean's Bounty," *National Geographic* 188(5):2–37. Describes declining catches of ocean fish worldwide. Great photos.

Raeburn, Paul. 1995. *The Last Harvest: The Genetic Gamble that Threatens to Destroy American Agriculture.* New York: Simon and Schuster. Concentrating on only a few strains of commercial crops while letting wild ancestors go extinct could be a recipe for disaster.

Russell, Dick. 1995. "Fishing Down the Food Chain," *The Amicus Journal.* 17(3):16–23. Overexploitation is depleting fish populations and now the commercial fleet is turning to squid.

Safina, Carl. 1995. "The World's Imperiled Fish," *Scientific American* 273(5):46–53. During the 1950s and 60s, commercial fishing grew at three times the rate of the human population. Not surprisingly, many fish populations are in severe decline.

Tansey, Geoff, and Tony Worsley. 1995. *The Food System.* Washington, DC: EarthScan. Explains how agriculture, food technology, retailing, consumer preferences, politics, and economics combine to produce the food we eat.

Warrick, Joby, and Pat Smith. 1996. "Boss Hog," *The Amicus Journal* 18(1):36–40. Large-scale commercial agriculture is industrializing food production with serious environmental consequences.

Weindruch, Richard. 1996. "Caloric Restriction and Aging," *Scientific American* 274(1):46–53. Eating fewer calories can prolong your life, even if you are not overweight.

Wittwer, Sylvan H. 1995. *Food, Climate, and Carbon Dioxide.* What effects on food production might result from increasing atmospheric CO_2 levels and subsequent global climate change? Some areas and some crops may benefit while others could suffer. A timely warning about the consequences of tampering with nature.

Chapter 11

Bray, F. (July) 1994. "Agriculture for Developing Nations," *Scientific American* 271, no. 1:30. The capital-intensive, highly mechanized Western model of agriculture is inappropriate in many regions. Systems of intensive polyculture, exemplified by wet-paddy rice cultivation, may be better.

Brookfield, H., and C. Padoch. (June) 1994. "Appreciating Agrodiversity: A Look at the Dynamism and Diversity of Indigenous Farming Practices," *Environment* 36, no. 5:6. Evidence suggests that indigenous farming practices may be more suitable for certain ecosystems than modern, industrial agriculture.

Carroll, C. R., J. H. Vandermeer, and P. M. Rosset. 1990. *Agroecology.* New York: McGraw-Hill. Advocates a new synthesis of ecology and agriculture for a sustainable future.

Coleman, David C., and D. A. Crossley, Jr. 1996. *Fundamentals of Soil Ecology.* New York: Academic Press. A good soil science textbook.

Decker, A. M., et al. (January/February) 1994. "Legume Cover Crop Contributions to No-Tillage Corn Production." *Agronomy*

Journal 86, no. 1:126. Winter cover crops can supply nitrogen, reduce erosion, conserve water, and prevent weed growth.

Faeth, P. (January/February) 1994. "Building the Case for Sustainable Agriculture: Policy Lessons from India, Chile, and the Philippines." *Environment* 36, no. 1:16. Discusses ways to make sustainable practices profitable for small-scale farmers.

Gardner, Gary. 1996. "Preserving Agricultural Resources," *State of the World 1996.* Washington, DC: Worldwatch Institute, 78–94. Will the world be able to feed more people? What is happening to agricultural resources?

Hussain, M. J., and D. L. Doane. 1995. "Socioecological Determinants of Land Degradation and Rural Poverty in Northeast Thailand," *Environmental Conservation* 22(1):44–50. Top-down development models degrade the environment and fail to help poor people. More equitable approaches are suggested.

Kurlansky, Mark. 1995. "On Haitian Soil," *Audubon* 97(1):50–57. Haiti is one of the most environmentally degraded countries in the world. Social and political strife is a consequence.

Malhi, S. S., et al. (January/February) 1994. "Influence of Topsoil Removal on Soil Fertility and Barley Growth," *Soil Conservation Research* 49, no. 1:96. Not surprisingly, topsoil loss reduces fertility and productivity.

Plucknett, Donald L., and Donald L. Winkelmann. 1995. "Technology for Sustainable Agriculture," *Scientific American*: 273(3):182–187. Technology means more than machines, but we can't hope to feed the world by abandoning all technology.

Rissler, Jane F., and Margaret G. Mellon. 1996. *The Ecological Risks of Engineered Crops.* Cambridge, MA: MIT Press. What is the potential for transgenic crops to become weeds or to transfer their genes to wild weedy plants? A timely account of genetic engineering in agriculture.

Senft, D. (March) 1994. "Building Terraces Naturally," *Agricultural Research* 42, no. 3:11. Native grasses can trap sediment and build terraces naturally on eroding hillsides.

Thrupp, Lori Ann, et al. 1995. *Bitter-Sweet Harvests for Global Supermarkets.* Washington, DC: World Resources Institute.

Addresses the sustainability and equity of cash crop production in Latin America.

Wood, R. (January) 1993. "Soiling the Planet," *Discover,* 74–75. Publication of the land degradation map from the Soil Information Center in Wageningen, the Netherlands.

Worster, Donald. 1993. *The Wealth of Nature.* Oxford: Oxford University Press. One of our preeminent environmental historians looks at how farming, water appropriations, and land use have affected our environment.

Chapter 12

Antle, J. M., and P. L. Pingali. 1994. "Pesticides, Productivity, and Farmer Health: A Philippine Case Study," *American Journal of Agricultural Economics* 76:418–430. Details the risks to farmers in developing countries from dangerous agrochemicals.

Davis, Devra Lee, and H. Leon Bradlow. 1995. "Can Environmental Estrogens Cause Breast Cancer?" *Scientific American* 273(4):166–173. Suggests that estrogenic compounds in pesticides and other industrial chemicals may contribute to the incidence of breast cancer.

Dibb, Sue. 1995. "Swimming in a Sea of Oestrogens: Chemical Hormone Disrupters." *The Ecologist* 25(1):27–31. Endocrine hormone disrupting chemicals have become ubiquitous in our environment. How dangerous are they?

Holloway, Marguerite. 1994. "An Epidemic Ignored: Endometriosis Linked to Dioxin and Immunologic Dysfunction," *Scientific American* 270(4):24–26. This report links the rapid rise in endometriosis to dioxin exposure.

Kelce, William R., et al. 1995. "Persistent DDT metabolite p,p-DDE is a potent androgen receptor antagonist," *Nature* 375:581–586. Abnormalities in male sex development in wildlife and humans may be caused by disruption of androgen receptors by DDT and its metabolites.

Kendall, Ronald J. 1994. *The Population Ecology and Wildlife Toxicology of Agricultural Pesticide Use: A Modeling Initiative for Avian Species.* Chelsea, MI: Lewis Publishers. A modeling approach to population ecology and toxicology.

Luoma, Jon R. 1995. "Havoc in the Hormones," *Audubon* 97(4):60–67. A well-written article about the hazards of chemical pollutants to reproductive health and development.

Natural Resources Defence Council. 1993. *After Silent Spring: the Unsolved Problem of Pesticide Use in the United States.* New York: Natural Resources Defence Council. A critique of federal pesticide policies in the United States.

Prakash, Anand, and Jagadiswari Rao. 1996. *Botanical Pesticides in Agriculture.* Chelsea, MI: Lewis Publishers. Describes over 1000 plants that naturally repel agricultural pests as alternatives to synthetic pesticides.

Purdey, M. (May/June) 1994. "Degenerative Nervous Diseases and Chemical Pollution," *The Ecologist* 23, no. 3:100. Anecdotal evidence links nervous diseases such as multiple sclerosis and Parkinson's to pollution but the connection is difficult to prove. Should we worry?

Raloff, J. (22 January) 1994. "That Feminine Touch," *Science News* 145, no. 3:56. An up-to-date review of the evidence of male feminization in humans and other species due to "environmental hormones" such as pesticide residues.

Renner, Rebecca. 1995. "Dioxin Risk Reassessment Stalls; EPA To Create New Review Panel," *Environmental Science and Technology* 29(11):492–494. An interesting insight into the science and politics of evaluating environmental health risks.

Rola, Agnes C., and Prabhu L. Pingali. 1993. *Pesticides, Rice Productivity, and Farmers' Health.* Washington, DC: World Resources Institute. A short but useful summary of the state of farmers' health from the International Rice Research Institute in the Philippines.

Selcraig, B. (July/August) 1993. "Greens Fees," *Sierra* 78, no. 4:70. A look at pesticide use on golf courses and the price of a perfect lawn.

Sharpe, Richard M. 1995. "Another DDT Connection," *Nature* 375:538–539. Summarizes recent findings about the reproductive health effects of DDT and its metabolites.

Tolba, M. K., et al. 1992. *The World Environment 1972–1992.* London: Chapman & Hall. A report from the United Nations Environment Program on world environmental conditions.

Winters, D. (February) 1993. "Fly Wars," *Discover,* 42–53. An interesting account of public policy formation.

Chapter 13

Abramovitz, J. N. 1994. *Trends in Biodiversity Investments*. Washington, DC: World Resources Institute. Analyzes over 1000 projects in 100 developing countries designed to protect and restore biodiversity.

Bright, Chris. 1996. "Understanding the Threat of Bioinvasions," *State of the World 1996:* 95–114. An overview of the ecology and social effects of the spread of alien species.

Boucher, Norman. 1995. "Species of the Sprawl," *Wilderness* 58(209):11–24. Uses the California Gnatcatcher as an example of new approaches for saving habitat and endangered species.

Byrnes, Patricia. 1995. "Wild Medicine," *Wilderness* 59(210):28–33. What undiscovered valuable products are being lost as we destroy biodiversity?

Caicco, S. L., et al. 1995. "A Gap Analysis of the Management Status of the Vegetation of Idaho (USA)," *Conservation Biology* 9(3):498–511. A landmark GIS study of landcover and management.

Ceballos, Gerardo, and James H. Brown. 1995. "Global Patterns of Mammalian Diversity, Endemism, and Endangerment," *Conservation Biology* 9(3):559–568. Island countries such as Australia, Madagascar, Indonesia, and the Philippines tend to have the greatest number of endemic and endangered species.

Chapman, Lauren J., et al. 1995. "Hypoxia Tolerance in Twelve Species of East African Cichlids: Potential for Low Oxygen Refugia in Lake Victoria," *Conservation Biology* 9(5):1274–1288. Areas of anoxic water may provide a refuge for cichlids against predation by voracious Nile perch.

Clark, Tim W., et al. 1994. *Endangered Species Recovery*. Covelo, CA: Island Press. Analyzes the successes and failures of recovery programs and provides insights into both the biology and policy aspects of endangered species recovery.

Fernside, P. M., and J. Ferraz. 1995. "A Conservation Gap Analysis of Brazil Amazonian Vegetation," *Conservation Biology* 9(5):1134–1147. An important application of GIS to landscape analysis.

Goodrich, J. M., and S. W. Buskirk. 1995. "Control of Abundant Native Vertebrates for Conservation of Endangered Species," *Conservation Biology* 9(6):1357–1364. Species that adapt well to human-induced changes often become a threat to rare and endangered organisms. Recovery programs may require control of these unnaturally abundant species.

Grumbine, R. Edward, ed. 1994. *Environmental Policy and Biodiversity*. Covelo, CA: Island Press. An excellent overview of wildlife conservation and policymaking with interviews and articles by many leading experts in the field.

Jones, H. L. and J. Diamond. 1976. "Short-term-base studies of turnover in breeding bird populations on the California Channel Islands," *Condor* 78:526–549. One of the best empirical studies of island biogeology.

Kaufman, L. S. 1992. "The Lessons of Lake Victoria: Catastrophic Change in Species-Rich Freshwater Ecosystems," *BioScience* 42: 846–858. See introduction to this chapter.

Kellert, Stephen R. 1995. *The Value of Life: Biological Diversity and Human Society*. Covelo, CA: Island Press. Applies principles of environmental ethics to questions of conservation biology.

Lesica, Peter, and Fred W. Allendorf. 1995. "When Are Peripheral Populations Valuable for Conservation?" *Conservation Biology* 9(4):753–760. Local races, ecotypes, subspecies and isolated populations may represent a large proportion of the genetic diversity of a species.

Mann, Charles C., and Mark I. Plummer. 1995. "California vs. Gnatcatcher?" *Audubon* 97(1):38–49. Another article about new approaches to saving wildlife habitat.

Mech, L. David. 1995. "The Challenge and Opportunity of Recovering Wolf Populations," *Conservation Biology* 9(2):270–278. An assessment of recovery programs by one of the preeminent wolf researchers in the U.S.

Miller, Kenton R. 1995. *Balancing the Scales: Managing Biodiversity at the Bioregional Level*. Washington, DC: World Resources Institute. Case studies of biodiversity protection that includes institutional change and equitable use of resources.

Noss, Reid, and Allen Coperrider. 1994. *Saving Nature's Legacy*. Covelo, CA: Island Press. A thorough introduction to issues of land management and conservation biology.

Rabinowitz, A. 1995. "Helping A Species Go Extinct—the Sumatran Rhino in Borneo," *Conservation Biology* 9(3):482–488. Much of the money and effort put toward rhino conservation has focused on new technologies or political strategies that have done little to protect these rare mammals.

Reaka-Kudla, Marjorie L., et al. 1995. *Biodiversity II: Understanding and Protecting Our Biological Resources*. Covelo, CA: Island Press. Discusses new technology in biodiversity protection including molecular taxonomy and gap analysis.

Reid, Walter V. 1995. "Biodiversity and Health: Prescription for Progress," *Environment* 37(6):12–15. Collaboration between conservationists and medical researchers could help preserve biodiversity as well as aid in the search for new drugs.

Reid, W. V., et al. 1993. *Biodiversity Prospecting: Using Genetic Resources for Sustainable Development*. Washington, DC: World Resources Institute. Outlines policies needed to ensure sustainable and equitable use of biological resources.

Safina, C. 1995. "The World's Imperiled Fish," *Scientific American* 273(5)46–53. A chilling analysis of the precipitous declines of oceanic fish stocks.

Shiva, V., and R. Holla-Bhar. (November/December) 1993. "Intellectual Piracy and the Neem Tree," *The Ecologist* 23, no. 6:223. A Third World view of how valuable local knowledge and biological resources are expropriated by wealthy nations and transnational companies. Argues for alternatives to GATT.

Tarpy, C. (July) 1993, "New Zoos—Taking Down the Bars," *National Geographic* 184(1):2–37. Good discussion of the progress and problems of modern zoos. Great photos.

Watkins, T. H. 1996. "What's Wrong with the Endangered Species Act?" *Audubon* 98(1):36–42. Enemies in Congress are sharpening their knives in preparation for dismembering endangered species protection. Will they be successful?

Wayne, Robert I., and John L. Gittleman. 1995. "The Problematic Red Wolf," *Scientific American* 273(1):36–39. Is this rare animal a species or only a hybrid? Should we use resources to breed and reintroduce it?

Williams, T. (March/April) 1994. "Beyond Traps and Poison," *Audubon* 96 no. 2:28.

Criticisms and defenses of the Federal Animal Damage Control Program for predator and pest erradication.

Chapter 14

Acharya, Anjali. 1995. "Plundering the Boreal Forests," *Worldwatch* May/June 1995:21–29. Boreal forests in Canada and Russia are being liquidated at an alarming rate. Release of stored carbon could have serious global climatic effects.

Alback, A. J. 1995. "Mobilizing Rural People in Tanzania to Tree Planting—Why and How," *Ambio* 24(5):304–310. Argues that compulsory programs that allow people to choose where and how to plant trees will motivate reforestation.

Ascher, William. 1995. *Communities and Sustainable Forestry in Developing Countries.* San Francisco, CA: ICS Press, Institute for Contemporary Studies. In contrast to Alback (above) this author argues that benefits for local people are essential for success in community forestry programs.

Boot, R. G. A., and R. E. Gullison. 1995. "Approaches to Developing Sustainable Extraction Systems for Tropical Forest Products," *Ecological Applications* 5(4):869–903. Discusses criteria for sustainable harvest of forest products.

Callenbach, Ernest. 1995. *Bring Back the Buffalo! A Sustainable Future for America's Great Plains.* Covelo, CA: Island Press. The author of *Ecotopia* explores the history, politics, and potential for reestablishing a Buffalo Commons on the high plains.

Chatterjee, N. 1995. "Social Forestry in Environmentally Degraded Regions of India— Case Study of the Mayurakshi Basin," *Environmental Conservation* 22(1):20–30. An interesting case study of community agroforestry in a highly degraded area.

Dagget, Dan. 1994. *Beyond the Rangeland Conflict: Toward a West that Works.* Layton, UT: Gibbs Smith Publisher. Presents ways in which westerners can work together to reduce regulations, enforcement, and conflict over resource use.

Dudley, Nigel, et al. 1995. *Bad Harvest: The Timber Trade and the Degradation of Global Forests.* Washington, DC: Earthscan. Analyzes the effects of timber trade on world forests.

Durbin, Kathie. 1995. "The timber salvage scam," *The Amicus Journal* 17(3):29–31.

Acting at the behest of timber companies, Congress has attached riders to budget bills that essentially allow "logging without laws" of the last remaining old-growth timber in the U.S.

Dwyer, Lynn E., et al. 1995. "Property Rights Case Law and the Challenge to the Endangered Species Act," *Conservation Biology* 9(4):725–741. Warns that takings bills may pose a great threat to habitats and endangered species.

Faeth, P., C. Court, and R. Livernash. 1994. *Evaluating the Carbon Sequestration Benefits of Sustainable Forestry Projects in Developing Countries.* Washington, DC: World Resources Institute. Discusses the role of forest destruction and replanting in global warming.

Flynn, J. 1994. "The Falling Forest." *The Amicus Journal* 15, no. 4:34. A mysterious decline is affecting Appalachian forests. This article discusses some of the possible causes.

Goodstein, Carol. 1995. "Buffalo Comeback," *The Amicus Journal* 17(1):34–37. Native Americans work to help restore wild bison.

Jepma, C. J. 1995. *Tropical Deforestation: A Socio-Economic Approach.* Washington, DC: Earthscan. Investigates the underlying socioeconomic causes of deforestation.

Klinkenborg, Verlyn. 1995. "Crossing Borders," *Audubon* 97(5):34–47. An inspiring example of how ranchers, environmentalists, and federal agencies are working together in the badlands of New Mexico to preserve fragile rangelands.

Kolchugina, T. P., and T. S. Vinson. 1995. "Role of Russian Forests in the Global Carbon Balance," *Ambio* 24(5):258–264. Temperate forests account for a large sequestration of carbon. Vast logging operations in the Russian Far East and Siberia may upset the global carbon balance.

Passoff, M. 1994. "Clayoquot Protests Put British Columbia on Trial," *Earth Island Journal* 9, no. 1:30. Activists risk jail sentences to protest logging on Vancouver Island.

Ralston, J. 1994. "Paper Dreams: Can Kenaf Pay Off?" *Audubon* 96, no. 2:20. Kenaf can produce more fiber per acre than trees, but we need a market for kenaf paper.

Schelhas, John, and Russell Greenberg, eds. 1996. *Forest Patches in Tropical*

Landscapes. Brings together scientists and conservationists to address the biological and socioeconomic value of forest remnants. An outgrowth of policy studies at the Smithsonian Migratory Bird Center.

Schmidt, Michael J. 1996. "Working Elephants," *Scientific American* 274(1):82–87. Elephants can serve as an ecologically benign alternative to mechanical logging equipment in dense tropical forests.

Wald, Johanna H., and Sami Yassa. 1995. *Selling Our Heritage: Congressional Plans for America's Public Lands.* New York: Natural Resources Defense Council. An angry account of Congressional plans to exploit public lands.

Wuerthner, George. 1995. "Why Healthy Forests Need Dead Trees," *Earth Island Journal* 10(4):22–23. Explains why "salvage logging" poses an ecological disaster.

Chapter 15

Burke, Vincent J., and J. Whitfield Gibbons. 1995. "Terrestrial Buffer Zones and Wetland Conservation: A Case Study of Freshwater Turtles in a Carolina Bay," *Conservation Biology* 9(6):1365–1369. Official wetland boundaries often fail to protect critical habitat for endangered species.

Burks, David C., ed. 1995. *Place of the Wild: A Wildland Anthology.* A distinguished array of writers and activists reflect on the relationship between humans and wilderness.

Callicott, J. Baird. 1995. "Deep Grammar," *Wild Earth* 5(1):64–67. A defensive broadside in an ongoing debate with Dave Foreman, Reed Noss, and others about the ultimate meaning of wilderness and preservation as a religious ideology.

Conlin, Michael, and Tom Baum, eds. 1995. *Island Tourism: Management Principles and Practices.* New York: Wiley. A holistic look at tourism and small islands.

Coram, Robert. 1995. "National Marine Sanctuaries," *Audubon* 97(3):38–45. These twelve refuges for underwater treasures may be the best idea the U.S. has ever had.

Craighead, John J., et al. 1995. *The Grizzly Bears of Yellowstone: Their Ecology in the Yellowstone Ecosystem 1959–1992.* Covelo, CA: Island Press. The results of thirty-one years of groundbreaking, but controversial wildlife research.

Dunning, J. B., et al. 1995. "Patch Isolation, Corridor Effects, and Colonization by A Resident Sparrow in A Managed Pine Woodland," *Conservation Biology* 9(3):542–550. Evidence that corridors and patch size, shape, and isolation have demonstrable effects on colonization of complex landscapes.

Eckstrom, Christine K. 1996. "A Wilderness of Water: Pantanal," *Audubon* 98(2):54–62. This inland delta is the world's largest and richest wetland, but a huge development project threatens it.

Herliczek, Jeremy. 1996. "Where Is Ecotourism Going?" *The Amicus Journal* 18(1):31–35. Ecotourism is popular and profitable, but is it sustainable?

Johns, David. 1995. "The Wildlands Project," *Wild Earth* 5(1):4. An update on an ambitious proposal to set aside a large area of North America as wilderness.

Klein, Mary L., et al. 1995. "Effects of Ecotourism on Distribution of Waterbirds in a Wildlife Refuge," *Conservation Biology* 9(6):1454–1465. Tourist pressures can cause wildlife to avoid critical habitat.

Line, Les. 1995. "A System Under Siege," *Wilderness* 59(210):10–27. The National Wildlife Refuge system is threatened by a wide variety of human actions.

McKibben, B. (March/April) 1994. "The People and the Park," *Sierra* 79, no. 2:102. A history of the Adirondack Park and the problems of mixing people with wilderness.

Nepal, S. K., and K. E. Weber. 1995. "Prospects for Coexistence—Wildlife and Local People," *Ambio* 24(4):238–245. Active involvement of local people in protection and conservation of Royal Chitwan National Park in Nepal is dependent on satisfying their needs.

Newmark, W. D. (December) 1993. "The Role and Design of Wildlife Corridors with Examples from Tanzania." *Ambio* 22, no. 8:500. A practical discussion of the benefits of wildlife corridors.

———. 1995. "Extinction of Mammal Populations in Western North American National Parks," *Conservation Biology* 9(3):512–526. Most parks are too small to preserve top carnivore species for long times unless some mechanism is developed to facilitate gene flow between populations.

Noss, R. F. and L. D. Harris. 1986. "Nodes, Networks, and MUMs: Preserving Diversity At All Levels," *Environmental Management* 10:299–309. A good discussion of the role of corridors in maintaining minimum viable populations.

Peres, Carlos A., and John W. Terborgh. 1995. "Amazonian Nature Reserves: An Analysis of the Defensibility Status of Existing Conservation Units and Design Criteria for the Future," *Conservation Biology* 9(1):34–46. Many tropical nature reserves are woefully understaffed and are accessible to poachers by road or navigable rivers. Defensibility criteria are suggested for preserve design.

Porter, Douglas, R., and David A. Salvesen. 1994. *Collaborative Planning for Wetlands and Wildlife*. Covelo, CA: Island Press. Proposes guidelines for planning and design of wetlands.

Rauber, P. (September/October) 1993. "Look Who's Taking," *Sierra* 78, no. 5:43. Compensation bills are flooding state legislatures to prevent "taking" of private property through land-use planning and zoning bills. See also "Wishful Thinking," *Sierra* 79, no. 1:39, January 1994, concerning the county movement.

Romme, W. H., et al. 1995. "Aspen, Elk, and Fire in Northern Yellowstone-National Park," *Ecology* 76(7):2097–2106. Fire prevention and predator removal have created an artificial condition in Yellowstone. So many elk are now present that they browsed essentially all accessible aspen shoots in spite of dense sprouting after the 1988 fires.

Rowell, Galen. 1989. "Annapurna: Sanctuary for the Himalaya," *National Geographic* 176(3):391–405. An interesting account of a trek around the Annapurna Conservation Area Project with beautiful photographs.

Shafer, Craig. 1995. "Values and Shortcomings of Small Reserves," *BioScience* 45(2):80–88. Sometimes small habitat fragments are all that's left.

Snyder, Gary. 1995. "Cultivating Wildness," *Audubon* 97(3):64–70. A poet and philosopher describes how he and his family are attempting to live in harmony with nature and restore the Sierra ecosystem in which they live.

Stoltzenburg, William. 1996. "Building A Better Refuge," *Nature Conservancy* 46(1):18–23. Using new information and ideas, land-use planners are designing nature reserves to last.

Strittholt, James R., and Ralph E. Boerner. 1995. "Applying Biodiversity Gap Analysis in a Regional Nature Reserve Design for the Edge of Appalachia, Ohio (U.S.A.)," *Conservation Biology* 9(5):1492–1505. GIS can be helpful in assessing the best areas for a nature reserve.

Stutz, B., et al. (May/June) 1994. "Our National Seashores." *Audubon* 96, no. 3:38. A celebration of 10 of America's best beaches and efforts to protect them.

Wagner, Fredrick H., et al. 1995. *Wildlife Policies in the U.S. National Parks*. Covelo, CA: Island Press. A detailed examination of the history, values, goals, policies, and problems associated with wildlife management in national parks.

Watkins, T. H. 1995. "Beyond Mile Zero," *Wilderness* 58(208):9–14. An update on the political controversy about proposed oil drilling in the Alaska National Wildlife Refuge.

Williams, T. (May/June) 1996. "Seeking Refuge," *Audubon* 98(3):34–45. A critique of the U.S. Wildlife Refuge System.

Zaslowsky, Dyan, and T. H. Watkins. 1994. *These American Lands: Parks, Wilderness, and the Public Lands,* 2nd ed. Covelo, CA: Island Press. A sweeping overview of federal lands in the United States.

Chapter 16

Bonatti, E. (March) 1994. "The Earth's Mantle Below the Oceans," *Scientific American* 270, no. 3:44. Deep undersea exploration is challenging our understanding of plate tectonic movements and the mechanisms by which the mantle and crust exchange energy.

Clark, George R. 1995. "The Quake That Swallowed a City," *Earth* 4(2):34–39. A 17th century earthquake in Port Royal, Jamaica destroyed the pirate capital of the Caribbean. A second look at this famous disaster.

Davidson, Keay. 1995. "Kobe in California," *Earth* 4(3):20–21, 70–71. How likely is another big earthquake in California?

Grosser, J. R., et al. (March) 1994. "Heavy Metals in Stream Sediments in a Gold-Mining area near Los Andes, Colombia," *Ambio* 23, (2):146. Small-scale gold mining results in the release of a variety of toxic heavy metals into streams.

Hyndman, Roy D. 1995. "Giant Earthquakes of the Pacific Northwest," *Scientific American* 273(6):68–75. New studies of the geologic record show that massive quakes and tsunamis have struck the Northwest coast of North America.

Maxwell, Jessica. 1995. "True Nature," *Audubon* 97(5):82–87. A giant strip mine high in the Beartooth mountains adjacent to Yellowstone Park threatens water, wildlife, and wilderness.

Mileti, D. S., et al. (April) 1992. "Fostering Public Preparations for Natural Hazards: Lessons from the Parkfield Earthquake Predictions," *Environment* 34, no. 3:16. Interesting insights into how people react to risk analysis of natural hazards.

Milius, S. (August/September) 1992. "The Law with the Big Loophole," *National Wildlife,* 41. A critique of the 1872 Mining Law.

Montgomery, C. 1993. *Physical Geology,* 3rd ed. Dubuque, IA: Wm. C. Brown Publishers. A good basic geology text to add to this chapter's discussion.

Pendick, Daniel. 1995. "Return to Mount St. Helens," *Earth* 4(2):24–33. Life is returning to the slopes of Mt. St. Helens, but more eruptions are likely.

Sengupta, M. 1993. *Environmental Impacts of Mining: Monitoring, Restoration and Control.* Covelo, CA: Island Press. Aimed at professionals, this book has a motherlode of information about mining.

Shurkin, Joel, and Tom Yulsman. 1995. "Assembling Asia," *Earth* 4(3):52–59. A grand theory that the heart of Eurasia was assembled from chains of migrating volcanic islands.

Vogel, Shawna. 1995. "Inner Earth Revealed," *Earth* 4(4):42–49. A new view of the boundary between the core and mantle of the earth.

Witze, Alexandra. 1996. "Earth Online," *Earth Online* 5(2):52–59. A great collection of internet addresses for geological images and information.

Wuerthner, G. (Summer) 1992. "Hard Rock and Heap Leach," *Wilderness* 55, (197):14. One of several excellent articles on the effects of mining on the western landscape and abuses of the 1872 Mining Law.

Chapter 17

Broecker, Wallace S. 1995. "Chaotic Climate," *Scientific American* 273(5):62–69. Geologic records show that weather patterns have sometimes changed dramatically in a decade or less. Ocean currents may play a critical role.

Craig, P. P., and W. Kempton. April 1993. "European Perspectives on Global Climate Change." *Environment* 35, no. 3:16. European leaders are much more concerned about global warming than their American counterparts. This article explores how different histories, cultures, economics, and geography influence public policies and attitudes.

Davidson, Keay. 1995. "El Niño Strikes Again," *Earth* 4(3):24–33. The most recent El Niño has been the longest on record. Are weather patterns changing due to global warming?

Davis-Jones, Robert. 1995. "Tornadoes," *Scientific American* 273(2):48–57. Using an arsenal of ground- and air-based instruments, tornado hunters are learning more about these destructive storms.

Denning, A. Scott, et al. 1995. "Latitudinal Gradient of Atmospheric CO_2 Due to Seasonal Exchange With Land Biota," *Nature* 376:240–244. The concentration of carbon dioxide in the atmosphere is increasing because of human activities, but the rate of increase is only about half of the total emission rate. An advanced general circulation model suggests that CO_2 sinks must be stronger than previously suggested.

Flavin, Christopher. 1996. "Facing Up to the Risks of Climate Change," *State of the World 1996.* Washington, DC: Worldwatch Institute. What should we do in the face of potentially catastrophic global climate change?

Gore, R. April 1993. "Andrew Aftermath." *National Geographic* 183, no. 4:2. Good photos and description of the costliest natural disaster in U.S. history.

Hannien, H. 1995. "Assessing Ecological Implications of Climatic-Change: Can We Rely on Our Simulation Models?" *Climatic Change* 31(1):1–4. Current global climate models are relatively crude. Can we depend on them to make critical decisions?

Hodell, David, et al. 1995. "Possible Role of Climate in the Collapse of Classic Maya Civilization," *Nature* 375:391–394. Paleoecological records suggest that a severe drought around AD 800 may have triggered the sudden collapse of the Mayan civilization. See also A. Witze, *Earth* 4(5):22–23 for a popular version of this story.

Idso, Sherwood B. 1995. "CO_2 and the Biosphere: The Incredible Legacy of the Industrial Revolution," A Special Publication of the Department of Soil, Water and Climate, University of Minnesota. Idso, of the USDA's Water Conservation Laboratory in Phoenix, claims that plants grow better under elevated CO_2 levels. Rather than try to decrease emissions, he argues, we should increase them.

Keeling, C. D., et al. 1995. "Interannual Extremes in the Rate of Rise of Atmospheric Carbon Dioxide Since 1980," *Nature* 375:666–670. In recent years, the growth rate of atmospheric CO_2 has failed to match industrial emissions. Models suggest that this anomaly may be caused by interannual variations in global air temperature which alter both the terrestrial biospheric and oceanic carbon sinks.

Kerr, Richard A. 1995. "It's Official: First Glimmer of Greenhouse Warming Seen," *Science* 270:1566–1567. Working Group I of the Intergovernmental Panel on Climate Change suggests that there is already a discernible human influence on global climate.

Lee, Henry, ed. 1995. *Shaping National Responses to Climate Change.* Covelo, CA: Island Press. Presents a strategic framework by which alternative actions can be assessed and compared.

Lynch, Colum. 1996. "Global Warming," *The Amicus Journal.* Global warming is here. What are we going to do about it?

Moore, B. and B. H. Braswell. February 1994. "Understanding the Carbon Cycle." *Ambio* 23, no. 1:4. A good overview of the role of carbon cycle in planetary climate.

Peck, S. C., and T. J. Teisberg. 1995. "Optimal CO_2-Control Policy with Stochastic Losses from Temperature Rise," *Climatic Change* 31(1):19–34. Attempts to estimate costs and benefits of specific CO_2 reduction policies.

Pendick, D. 1996. "Hurricane Mean Season," *Earth* 5(3)24–33. 1995 was an unusually active hurricane season in the Caribbean. Is this a harbinger of global climate change?

Tans, P. P., and P. S. Bakwin. 1995. "Climate Change and Carbon Dioxide Forever." *Ambio* 24(6):376–378. Even if we reduce CO_2 emissions, the levels now in the atmosphere will take centuries to return to preindustrial levels. A case is made that we already know enough to start taking steps to reduce emissions.

Wigley, Tom. 1995. "A Successful Prediction?" *Nature* 376:463–464. Reviews the power of general circulation climate models to simulate past climate variations. The importance of correctly modeling cloud processes is a critical factor. See Wolf and Ploegh, p. 464 in this same issue.

Wiscombe, Warren J. 1995. "An Absorbing Mystery," *Nature* 376: 466–467. Reviews evidence that clouds may play a large but unappreciated role in regulating global climate.

Chapter 18

Albritton, D. and R. Watson, eds. 1995. *Scientific Assessment of Stratospheric Ozone: 1994.* World Meteorological Organization Global Ozone Research and Monitoring Project Report 37. Geneva: World Meteorological Organization. UV radiation reaching the Earth's surface appears to be increasing in polar regions. See also Lubin and Jensen *Nature* 377:710–713 (1995).

Charlson, R. J. (February) 1994. "Sulfate Aerosol and Climatic Change," *Scientific American* 270, no. 2:48. Sulfate particles scatter sunlight back into space, cooling the atmosphere and reducing global warming. Reducing sulfur pollution could greatly accelerate urban warming by greenhouse gases.

Easterbrook, Greg. 1994. "Forget PCB's, Radon, Alar: The World's Greatest Environmental Dangers Are Dung Smoke and Dirty Water." *The New York Times Magazine* September 11, 1994, p. 60–63. We spend billions of dollars to protect ourselves from exotic pollutants while not lifting a finger to help the 7.8 million poor children who die each year mainly from what they drink and breathe.

Gibbs, W. Wayt. 1995. "The Treaty That Worked—Almost," *Scientific American* 273(3):18–19. Will the black market for CFCs short-circuit the Montreal Protocol?

Godish, Thad. 1995. *Sick Buildings: Definition, Diagnosis and Mitigation.* Covelo, CA: Island Press. Discusses causal factors, risks, diagnostic measurements, and mitigation techniques for health effects related to poor indoor air quality.

Jones, A. E., and J. D. Shanklin. 1995. "Continued Decline of Total Ozone Over

Halley, Antarctica, since 1985," *Nature* 376:409–411. Reviews ten years of stratospheric ozone data.

Keith, Lawrence H., and Mary M. Walker. 1995. *Handbook of Air Toxics: Sampling, Analysis, and Properties.* Chelsea, MI: Lewis Publishers. Comprehensive coverage for regulators, consultants, engineers and others who need detailed information on toxics and the laws that apply to them.

Lents, J. M., and W. J. Kelly. (October) 1993. "Clearing the Air in Los Angeles," *Scientific American* 269, no. 4:32. Describes the history of air pollution in the Los Angeles Basin and how authorities plan to improve air quality.

Longstreth, J. D., et al. 1995. "Effects of Increased Solar Ultraviolet Radiation on Human Health," *Ambio* 24(3):153–165. Risks of eye diseases, cancer, and immune dysfunction from increased UV exposure are calculated.

Lubin, Dan, and Elsa H. Jensen. 1995. "Effects of Clouds and Stratospheric Ozone Depletion on Ultraviolet Radiation Trends," *Nature* 377:710–714. By the end of this century, trends in summer average local-noon UV dose rates should be significantly greater than natural cloud variability.

Parson, Edward A., and Owen Greene. 1995. "The Complex Chemistry of International Ozone Agreements," *Environment* 37(2):16–20. So far, the Montreal Protocol has been fairly successful but developing countries will need substantial financial assistance to meet their obligations.

Raloff, J. 1995. "Outdoor Carbon Monoxide: Risk to Millions," *Science News* 148:247. Even federally permissible levels of carbon monoxide can aggravate congestive heart failure, a life-threatening condition suffered by 3 million mostly older Americans.

Spix, C., et al. (November) 1993. "Air Pollution and Daily Mortality in Erfurt, East Germany, 1980–1989," *Environmental Health Perspectives* 101, no. 6:518. Air pollution from burning soft coal is shown to correlate to increased incidence of respiratory diseases and death in an East German city.

United Nations Environment Program. (March) 1994. "Air Pollution in the World's Megacities," *Environment* 36, no. 2:4. A special report from the UN on air quality in twenty of the world's largest cities. Although data is often incomplete, urban

air pollution presents a serious threat to human health and well-being.

Upgren, Arthur R. 1996. "Night Blindness," *The Amicus Journal* 17(4):22–24. Light pollution doesn't harm us directly but it interferes with astronomy and changes our experience of nature.

Vanderleun, J. C., et al. 1995. "Environmental Effects of Ozone Depletion," *Ambio* 24(3):138. The lead editorial of a special issue on this subject.

Chapter 19

Chau, K. C. 1995. "The Three Gorges Project of China: Resettlement Prospects and Problems," *Ambio* 24(2):98–102. Nineteen cities will be flooded and more than one million people dislocated by the largest dam project in the world.

Gardner, Gary. 1995. "From Oasis to Mirage: The Aquifers that Won't Be Replenished," *Worldwatch* 8(3):32–36. Underground aquifers of fossil water are being drained for short term gain but the water is irreplaceable.

Gleick, P. H. (April) 1994. "Water, War, and Peace in the Middle East," *Environment* 36, no. 3:6. Since antiquity water has been both a prize and a weapon of war. Growing scarcity makes this precious resource a continuing cause of conflict.

Hinrichsen, Don. 1995. "Waterworld," *The Amicus Journal* 17(2):23–27. A hundred years of plumbing, plantations, and politics in the Florida Everglades.

———. 1995. "Sea Change," *The Amicus Journal* 17(1):38–41. Claims that hope is drying up for the Aral Sea.

Horta, Korinna. 1995. "The Mountain Kingdom's White Oil: The Lesotho Highlands Water Project," *The Ecologist* 25(6):227–231. A massive scheme to ship water to South Africa already has cost 20,000 people part or all of their livelihood. Another case of World Bank financed environmental disaster.

Lansing, J. S. 1991. *Priests and Programmers: Technologies of Power in the Engineered Landscape of Bali.* Princeton, NJ: Princeton Univ. Press. A fascinating account of how water temples organized communal irrigation projects on this remarkable island.

Mascarenhas, Ophelia, and Peter G. Veit. 1995. *Indigenous Knowledge in Resource Management: Irrigation in Msanzi,*

Tanzania. Washington, DC: World Resources Institute. Villagers have successfully managed their water supply for irrigation despite a variety of serious, long-term ecological problems. Perhaps others could learn from their experience.

Mitchell, J. (November) 1993. "James Bay: Where Two Worlds Collide," *National Geographic* 184, no. 5:66. An excellent analysis of Quebec's James Bay hydropower project as part of a special issue devoted to water supply, development, pollution, and restoration problems.

Myers, M. F., and G. F. White. (December) 1993. "The Challenge of the Mississippi Flood," *Environment* 35, no. 10:6. A 500-year flood in the summer of 1993 has caused us to rethink wetland drainage, floodplain development, and dependence on channelization and levees for flood protection.

Postel, Sandra. 1996. "Forging a Sustainable Water Strategy," *State of the World 1996:* 40–59. Summarizes the state of our water supply and outlines options for the future.

Qing, Dai, ed. 1994. *Yangtze! Yangtze! Debate Over the Three Gorges Project.* Washington, DC: Earthscan. Two dozen Chinese engineers, scholars, scientists, and government officials state their case against the dam in one of the boldest critiques of Chinese government policy ever published. Qing, winner of the prestigious Goldman Environmental Prize for her opposition to the dam, was arrested and held without trial for nearly a year. The book is now banned in China.

Schneider, David. 1995. "On the Level: Central Asia's Inland Seas Curiously Rise and Fall," *Scientific American* 274(1):14. While the Aral Sea is falling, the nearby Caspian Sea is rising. Could these be natural fluctuations?

Schwarzbach, David A. 1995. "Promised Land (But What About the Water?)," *The Amicus Journal* 17(2):35–41. Heated politics over Israel's most precious commodity.

Strong-Aufhauser, Lisa. 1995. "The Mono Lake Water War," *Earth* 4(5):50–57. Mono Lake has been shrinking since Los Angeles began diverting inflowing streams. How much natural variation occurs in lake levels and how serious are the ecological effects of reduced volume? A nonconventional assessment.

Stutz, Bruce. 1994. "Water and Peace," *Audubon* 96(5):64–77. The struggle for water in the arid Middle East could bring war or it could foster peace.

Watson, Ian, and Alister Burnett. 1994. *Hydrology: An Environmental Approach.* Covelo, CA: Island Press. A good general textbook of hydrology.

Worster, D. 1985. *Rivers of Empire.* New York: Pantheon Press. A fascinating history of water use and abuse in the western United States.

Chapter 20

Anderson, D. M. (August) 1994. "Red Tides," *Scientific American* 271, no. 2:52. Blooms of toxic algae seem to have increased frequently in near-shore areas where pollution and run-off from disturbed land have changed ocean chemistry and disrupted normal sea life.

Briscoe, J. (May) 1993. "When the Cup Is Half Full: Improving Water and Sanitation Services in the Developing World," *Environment* 35, no. 4:6. Encouraging case studies of progress in cleaning water supplies for developing countries.

Davison, W., et al. 1995. "Controlled Reversal of Lake Acidification by Treatment with Phosphate Fertilizer," *Nature* 377:650–655. Artificial eutrophication can help neutralize acidic lake water without drastically altering community structure in some cases.

DePalma, Anthony. (August 1) 1995. "Treaty Partners Study Fate of Birds at Polluted Mexican Lake," *The New York Times:* B6. More than 40,000 migrant water birds died at a polluted agricultural reservoir near Leon, Mexico. Assessing the causes and solutions is one of the first tests of NAFTA.

Jones, Keith. 1994. "Waterborne Diseases," *New Scientist* 73:1–4. Waterborne diseases are a major cause of illness in developing countries and localized disease outbreaks in the developed world. What can we do about this?

Knopman, D. S., and R. A. Smith. (January/February) 1993. "Twenty Years of the Clean Water Act," *Environment* 35, no. 1:16. Reviews issues in the reauthorization debate. A lack of good data on water quality hinders decision making.

Lathem, Alexis. 1995. "Hydro Quebec Is Back," *Earth Island Journal* 10(3):25. The latest scheme in this gargantuan project is to dam and divert the Sainte Marguerite River, which flows through Innu territory on Quebec's Atlantic shore.

Malle, Karl-Geert. 1996. "Cleaning Up the River Rhine," *Scientific American* 274(1):70–79. Twenty years ago, pollution threatened to ruin both the Rhine's beauty and utility. International cooperation has brought many troublesome sources of chemical contamination under control.

Marison, Alan. 1994. "Great Flood of '93," *National Geographic* 185(1):42–81. A good picture of the social and environmental costs of a century of wetland drainage and dredging and diking of the great river.

Olsen, Erik D. 1995. *Trouble on Tap: Arsenic, Radioactive Radon, and Trihalomethanes in Our Drinking Water.* New York: The Natural Resources Defense Council. A useful summary of hazards in drinking water.

Rhodes, S. L., and K. B. Wiley. (September) 1993. "Great Lakes Toxic Sediments and Climate Change: Implications for Environmental Remediation," *Global Environmental Change* 3, no. 3:292. Climate change may complicate attempts to clean up the Great Lakes.

Sampson, Martin W. 1995. "Black Sea Environmental Cooperation: Three Tracks and the First Year of Action," *International Studies Association.* An excellent study of the politics of environmental protection in a turbulent region.

Skaare, J. U., et al. (February) 1994. "Mercury and Selenium in Arctic and Coastal Seals Off the Coast of Norway," *Environmental Pollution* 85, no. 2:153. High levels of toxic metals are found in marine mammals even in pristine areas.

White, K. (November) 1993. "No More Muddy Waters: Cleaning Up the Clean Water Act," *Environmental Health Perspectives* 101, no. 6:490. A good summary of the conflicting views about reauthorization of the U.S. Clean Water Act.

Chapter 21

Brown, L., et al. 1993. *Vital Signs.* Washington, DC: Worldwatch Institute. A comprehensive if gloomy analysis of environmental trends including energy supplies and consumption rates.

Davis, G. R. 1991. *Energy for Planet Earth.* New York: W. H. Freeman. Reading on energy from the *Scientific American.* Highly recommended.

Grossman, D., and S. Shulman. (March) 1994. "Verdict at Yucca Mountain," *Earth* 3, no. 2:54. Will the nuclear repository be safe? Despite $1.5 billion worth of studies, geologists continue to debate the merits of this complex project.

International Energy Agency. 1993. *World Energy Outlook*. Washington, DC: Organization for Economic Cooperation and Development. An analysis of world energy supplies and markets from a Eurocentric perspective.

Lenssen, N. 1993. "Providing Energy in Developing Countries," in *State of the World 1993*. Washington, DC: Worldwatch Institute. What options do developing countries have for power? An exploration of conventional and sustainable energy sources.

McCafferty, N. (May/June) 1996. "Life After Chernobyl," *Audubon* 98(3):66–75. A sad commentary on the aftermath of a nuclear disaster.

MacKenzie, J. J., et al. 1992. *The Going Rate: What It Really Costs To Drive*. Washington DC: World Resources Institute. A brief, but sobering, account of our dependence on automobiles.

Morgan, G. M. (March) 1993. "What Would It Take To Revitalize Nuclear Power in the United States?" *Environment* 35, no. 2:6. A balanced survey of nuclear power problems and possible solutions to them.

Scherr, S. Jacob, and David Schwarzbach. 1995. "Turning Points: The Future of Nuclear Power in Eastern Europe Is Being Decided In Two Small Towns," *The Amicus Journal* 13(4):13–16. The former Soviet Union and its allies are locked in a debate about nuclear power. Will plants now under construction be completed and allowed to operate?

Taubes, Gary. 1995. "Blowup at Yucca Mountain," *Science* 268 (3 June 1995):1836–1839. A theory raising the possibility of atomic explosions in a nuclear waste dump causes further doubts about the safety of underground storage of Yucca Mountain.

Watkins, T. H. 1995. "Beyond Mile Zero," *Wilderness* 58(208):9–14. An update on the political controversy about proposed oil drilling in the Alaska National Wildlife Refuge.

Whipple, C. G. (June) 1996. "Can Nuclear Waste Be Stored Safely at Yucca Mountain?" *Scientific American* 274 no. 6:72–79. A review of the controversy over nuclear waste storage.

Williams, Phil, and Paul N. Woessner. 1996. "The Real Threat of Nuclear Smuggling," *Scientific American*. 274(1):40–45. Poorly supervised nuclear stockpiles in Russia and elsewhere make the danger of a black market in radioactive materials all too real.

World Energy Council. 1992. *Survey of Energy Resources*. London: World Energy Council. A good source of up-to-date world statistics.

Chapter 22

Asmus, Peter. 1994. "Hot Air, Hot Tempers, and Cold Cash," *The Amicus Journal* 16(3): 30–35. Opposition to wind power brings together a strange coalition of activists.

Burmeister, G., and F. Kreith. 1993. *Energy Management and Conservation*. Washington, DC: National Conference of State Legislatures. Details state and regional programs for reducing energy consumption and improving energy efficiency.

Cavallo, A. J., et al. 1993. "Wind Energy: Technology and Economics," in *Renewable Energy: Sources for Fuels and Electricity*, edited by T. B. Johansson, et al. Washington, DC: Island Press. See also excellent articles on hydropower, biomass, and photovoltaic solar energy.

Cole, Nancy, and P. J. Skerrett. 1995. *Renewables Are Ready: People Creating Renewable Energy Solutions*. Washington, DC: The Union of Concerned Scientists. Inspirational stories of Americans who have made renewable energy a reality in their own communities.

Flavin, Christopher, and Nicholas Lenssen. 1994. *Power Surge*. New York: W.W. Norton. A strong case is made for renewable energy.

Gipe, Paul. 1993. *Wind Power For Home and Business*. Washington, DC: American Wind Energy Association. A comprehensive reference for those who want to know how wind energy works and how to use it.

Hoagland, William. 1995. "Solar Energy," *Scientific American* 273(3):170–173. Explores the potential for solar energy. See also article by Furth on fusion in same issue.

Lehman, Peter, and Christine Parra. 1994. "Hydrogen Fuel from the Sun." *Solar Today* 8 (Sept/Oct 1994):20–25. Hydrogen gas produced electrolytically with photovoltaic energy could substitute for natural gas in many applications.

Löfstedt, R. E. (March) 1993. "Hard Habits to Break: Energy Conservation Patterns in Sweden," *Environment* 35, no. 2:10. Although Swedes use much less energy per person on average than Americans, they will have to reduce their consumption if they want to eliminate nuclear power plants. Perhaps we will be able to learn something from them.

MacKenzie, J. J. 1994. *Electric and Hydrogen Vehicles: Transportation Technologies for the Twenty-first Century*. Washington, DC: World Resources Institute. A rigorous study of the potential of electric and hydrogen vehicles.

Patterson, W. 1994. *Power from Plants: The Global Implications of New Technologies for Electricity from Biomass*. London: Earthscan Publications. A good review of biomass potential.

Reynolds, Michael. 1993. *Earthship III*. Taos, NM: Earthship Publications. Detailed step-by-step manual for building rammed earth and tire construction houses.

Sastry, Anjali M., and Ashok J. Gadgil. 1996. "Bombay Efficient Lighting Large-scale Experiment (BELLE): Blueprint for Improving Energy Efficiency and Reducing Peak Electric Demand in a Developing Country," *Atmospheric Environment* 30(5):803–808. A plan for energy conservation in India.

Sperling, Daniel. 1994. *Future Drive: Electric Vehicles and Sustainable Transportation*. Covelo, CA: Island Press. Addresses strategies for more environmentally benign transportation system.

Steen, Athena S., et al. 1994. *The Straw Bale House*. Hopewell, CA: Real Goods Publishers. Practical advice and inspiring case studies of straw bale construction.

Sterrett, Frances S., ed. 1995. *Alternative Fuels and the Environment*. New York: CRC Press. Experts from many fields survey potential for solar, wind, geothermal, and biomass energy.

Chapter 23

Aquino, John T., ed. 1995. *Waste Age/Recycling Times' Recycling HandBook*. Washington, DC: Waste Age Publications. A definitive collection from the largest association in the field with chapters written by leading international authorities on waste management and recycling.

Boerner, C., and K. Chilton. (January/February) 1994. "False Economy: The Folly of Demand-Side Recycling," *Environment* 36, no. 1:6. Popular curbside recycling programs are producing a glut of materials for which there are no markets. Examines Germany's innovative packaging laws and realities of creating markets for unwanted materials.

Connett, P., and E. Connett. (January/February) 1994. "Municipal Waste Incineration: Wrong Question, Wrong Answer," *The Ecologist* 24, no. 1:14. An eloquent critique of incineration and arguments for waste reduction, reuse, and recycling.

Fairlie, S. (November/December) 1992. "Long Distance, Short Life: Why Big Business Favors Recycling," *The Ecologist,* 22 no. 6:276. Industry promotes recycling rather than reuse for two main reasons: recycling is better suited to centralized, long-distance markets; and it can be used as an environmental justification for short-life products.

Harker, Donald F., and Elizabeth U. Natter. 1994. *Where We Live: A Citizen's Guide to Conducting a Community Environmental Inventory.* Covelo, CA: Island Press. A practical workbook to help citizens find information concerning their local environment and to use that information in furthering environmental goals.

Hinchee, Robert E., et al. 1994. *Applied Biotechnology for Site Remediation.* Chelsea, MI: Lewis Publishers. Practical techniques for site remediation with biotechnology.

Hoffman, Andrew J. 1995. "An Uneasy Rebirth At Love Canal," *Environment* 37(2):4–9. After nearly 10 years of cleanup, families and businesses are moving back to Love Canal. How safe is it to live there?

Poore, P. (October/November) 1992. "Disposable Diapers Are OK," *Garbage* 4, no. 5:26. An interesting study both of the life cycle costs of disposable and reusable diapers and the politics and theology of recycling.

Rathje, William R., and Cullen Murphy. 1992. *Rubbish! The Archeology of Garbage.* New York: HarperCollins. A good introduction to garbology, the science of analyzing garbage with amazing stories about the longevity of ordinary materials in landfills.

Renner, Rebecca, and Jeff Johnson. 1995. "Renewed Superfund Debate May Turn On Retroactive Liability," *Environmental Science and Technology* 29(11):498–499. Who will pay for Superfund cleanups remains a hotly contested issue in Congress.

Russell, Dick. 1995. "Moving Mountains," *The Amicus Journal* 6(4):39–42. Describes how a Native American tribe organized to close a dump on their land.

Stigliani, W. M., et al. 1991. "Chemical Time Bombs: Predicting the Unpredictable," *Environment* 33, no. 4:4. We have treated our environment as a "sink" for waste disposal. Now some areas have become poisonous wastelands. How can these chemical time bombs be detected and defused?

Stoner, Daphne L. 1994. *Biotechnology for the Treatment of Hazardous Waste.* New York: CRC Press. Living organisms can provide a cost-effective means of detoxifying hazardous substances.

Waite, Richard. 1995. *Household Waste Recycling.* New York: EarthScan. A comprehensive look at recycling of household waste and how it could develop.

Chapter 24

Adams, Lowell W. 1994. *Urban Wildlife Habitats: A Landscape Perspective.* Minneapolis: University of Minnesota Press. Reveals how cities and towns can be made more accommodating for a wide diversity of species, including our own.

Aberly, Doug, ed. 1994. *Futures By Design: The Practice of Ecological Planning.* Philadelphia: New Society Publishers. An anthology of the history and theory of ecologically sound urban planning.

Clunies-Ross, T. (May/June) 1994. "Urban Renewal in Denmark: Fighting for Control," *The Ecologist* 24, no. 3:110. Residents of a Copenhagen slum are planning revitalization projects such as communal housing and permaculture gardens. Their plans are opposed by official urban renewal programs.

Cortes, Ernesto, Jr. 1994. "Reweaving the Fabric: The Iron Rule and the IAF Strategy for Power and Politics." in *Interwoven Destinies,* Henry G. Cisneros, ed. New York: W. W. Norton and Company. An interesting description of organizing principles for bringing about political change.

Dwyer, Augusta. 1994. *On the Line: Life on the US-Mexican Border.* London: Latin American Bureau. A disturbing look at conditions in the colonias and maquilladoras along the border.

Frenay, Robert. 1996. "Chattanooga Turnaround," *Audubon* 98(1): 82–92. A prime example of ecologically friendly urban renewal.

Goldstein, Eric A., and Mark A. Izeman. 1995. *New York City's Environmental Vital Signs: A Twenty-Five Year Check-Up.* New York: The Natural Resources Defense Council. An environmental check list for urban areas.

Lerner, Steve. 1996. "Brownfields of Dreams," *The Amicus Journal* 17(4):15–21. Can old industrial sites be turned into new economic hopes?

Mage, David, et al. 1996. "Urban Air Pollution in Megacities of the World," *Atmospheric Environment* 30(5):681–686. Details the findings of the Global Environment Monitoring System on air pollution in megacities of the developing world.

McGranahan, G. and J. Songsore. (July/August) 1994. "Wealth, Health, and the Urban Household," *Environment* 36, no. 6:4. A comparison of environmental conditions and household income in Accra, Ghana; Jakarta, Indonesia; and Sao Paulo, Brazil.

Nijkamp, Peter, and Adriaan Perrels. 1994. *Sustainable Cities in Europe.* New York: EarthScan. A report of the CITES program of the Commission of European Communities in urban redesign.

Nixon, Will. 1995. "Trashed Urban Rivers and the People Who Love Them," *The Amicus Journal* 17(3):25–28. Activists are reclaiming urban rivers in many cities.

Platt, Rutherford H., et al. 1994. *The Ecological City: Preserving and Restoring Urban Biodiversity.* Minneapolis: The University of Minnesota Press. An interdisciplinary look at the ecology of urban communities.

Rabinovitch, Jonas, and Josef Leitman. 1996. "Urban Planning in Curitiba," *Scientific American* 274(3):46–53. One of the best examples of city planning and sustainable urban development is in southern Brazil.

Rees, W. E. and W. Wackernagel. 1994. "Ecological Footprints and Appropriated Carrying Capacity," in *Investing in Natural Capacity: The Ecological Economics Approach to Sustainability,* edited by A. M. Jannson, et al. Washington, DC: Island

Press. Modern urban societies depend on resources from the surrounding countryside.

Sassen, S. 1994. *Cities in a World Economy.* Thousand Oaks, CA: Pine Forge Press. Shows how major urban areas have become interlinked in a system of world cities.

U.N. Environment Program. (March) 1994. "Air Pollution in the World's Megacities," *Environment* 36, no. 2:4. Although data are often fragmentary, the air quality in the world's megacities threatens the health of millions of urban residents.

Van der Ryn, Sim, and Stuart Cowan. 1995. *Ecological Design.* Covelo, CA: Island Press. A vision of how we could use ecological principles as the basis for urban design.

White, Rodney R. 1994. *Urban Environmental Management: Environmental Change and Urban Design.* New York: John Wiley & Sons. Offers a detailed examination of the problems of urban environments.

World Resources Institute. 1996. *World Resources 1996–1997.* Oxford, UK: Oxford University Press. A special issue focusing on the urban environment and sustainability prepared with the UN Environment Programme, the UN Development Program, and the World Bank in preparation for the 1996 Habitat II Conference in Istanbul, Turkey. Excellent data and analysis.

Yeang, Ken. 1995. *Designing with Nature: The Ecological Basis for Architectural Design.* New York: McGraw-Hill. Features examples of some ecologically comprehensive building designs.

Zaslowsky, Dyan. 1995. "The Battle of Boulder," *Wilderness* 58(209):25–29. Boulder, CO, has been struggling to preserve open space and wildlife habitat. A success story in urban design.

Chapter 25

Andrews, Richard. 1995. "Reform or Reaction: EPA at a Crossroads," *Environmental Science and Technology* 29(11):505–514. After 25 years, it's time to reassess the effectiveness and purpose of the EPA.

Bloom, David E. 1995. "International Opinion on the Environment," *Science* 269 (21 July 1995):354–358. Concern about the state of our environment is growing in both developed and developing countries.

Bolling, David M. 1994. *How to Save a River: A Handbook for Citizen Action.* Covelo, CA: Island Press. Detailed explanations of how to form an organization, raise money, plan a campaign, and cultivate the media for environmental protection.

Brick, Phil. 1995. "Determined Opposition: The Wise Use Movement Challenges Environmentalism," *Environment* 37(8):16–20. The success of the Wise Use Movement shows the need for environmental activists to reassess their strategies as the nation's political and social strategies change.

Cable, Sherry, and Charles Cable. 1995. *Environmental Problems/Grassroots Solutions: The Politics of Grassroots Environmental Conflict.* New York: St. Martin's Press. A short but interesting survey of environmental sociology.

Dovers, S. R., and J. W. Handmer. 1995. "Ignorance, the Precautionary Principle, and Sustainability," *Ambio* 24(2):92–97. In the face of uncertainty, ignorance auditing may help us plan for the future.

Dowie, Mark. 1995. *Losing Ground: American Environmentalism at the Close of the Twentieth Century.* Cambridge: MIT Press. A provocative critique of the mainstream environmental movement.

French, Hilary F. 1995. *Partnership for the Planet: An Environmental Agenda for the United Nations.* Washington, DC: World Watch Paper 126. Suggestions for global environmental policy.

Helvarg, David. 1994. *The War Against the Greens: The Wise-Use Movement, the New Right, and Anti-Environmental Violence.* San Francisco: Sierra Club Books. A shrill but chilling book about violence against environmentalists.

Hempel, Lamont C. 1995. *Environmental Governance: The Global Challenge.* Covelo, CA: Island Press. Considers institutional responses to global environmental challenges.

Inglehart, Ronald. 1995. "Public Support for Environmental Protection: Objective Problems and Subjective Values in 43 Societies," *PS: Political Science and Politics* March 1995:57–71. A 3-year values survey carried out in countries containing 70 percent of the world population reveals widespread support for environmental protection.

Johnson, Huey D. 1995. *Green Plans: Greenprint for Sustainability.* Lincoln, NE: University of Nebraska Press. Inspiring examples of countries with national plans for sustainable development.

Kane, Hal. 1996. "Shifting to Sustainable Industries," *State of the World 1996.* Washington, DC: Worldwatch Institute. How could renewable energy, recycling, and reengineering industry lead to a sustainable future?

Meadows, Donella. 1995. "Taking the Low Road: Property Rights Hustlers Make A Mockery of Community," *The Amicus Journal* 17(2):12–15. Social conservatives are using the 5th amendment to further their own interests at the expense of the common good.

Meyers, Norman, and Julian L. Simon. 1994. *Scarcity or Abundance? A Debate on the Environment.* New York: W.W. Norton. An interesting debate between an ecologist and a cornucopian economist over the state of the world.

Morris, David B. 1995. *Earth Warrior: Overboard with Paul Watson and the Sea Shepherd Conservation Society.* Golden, CO: Fulcrum Publishing. Watson undertakes dangerous confrontations with those he regards as a threat to our environment. Do ends justify means?

Natural Resources Defense Council. 1995. *The Year of Living Dangerously: Congress and the Environment in 1995.* New York: The Natural Resources Defense Council. An environmentalist view of the Republican Contract with America.

Orr, David, and David Ehrenfeld. 1995. "None So Blind: The Problem of Ecological Denial," *Conservation Biology* 9(5):985–987. Two leading educators address questions of denial of environmental problems.

Osborne, P. L. 1995. "Biological and Cultural Diversity in Papua New Guinea: Conservation Conflicts, Constraints, and Compromise," *Ambio* 24(4):231–237. Suggests that protecting cultural diversity goes hand-in-hand with conserving biodiversity.

Robbins, Jim. 1995. "Target Green," *Audubon* 97(4):82–85. In the West, extremists are threatening government agents who attempt to manage resources for the common good.

Russell, Dick. 1996. "Massachusetts Takes A Stand," *The Amicus Journal* 17(4):26–30.

State environment secretary Trudy Coxe seeks a green Republican path.

Russell, Milton. 1995. "Environmental Policy's Great Dilemma," *Environment* 37(2):13–15. One article in a series questioning the strict, command-and-control standards in current environmental laws. Makes an argument for more reasonable standards.

Schwab, Jim. 1995. *Deeper Shades of Green: The Rise of Blue-Collar and Minority Environmentalism in America.* San Francisco: Sierra Club Books. A history of the environmental justice and social justice movements.

Skelton, Renee. 1994. "Harlem Green-aissance," *The Amicus Journal* 16(3):14–17. Activists build community and rebuild their urban environment.

Taylor, Bron R., ed. 1995. *Ecological Resistance Movements: The Global Emergence of Radical and Popular Environmentalism.* Albany, NY: State University of New York Press. Readings on activism and theory.

Tolkar, Brian. 1995. "The Wise Use Backlash: Responding to Militant Anti-Environmentalism," *The Ecologist* 25(4):150–156. Proposes a return to grassroots community activism as the best opposition to radical anti-environmentalism.

Vollers, Maryanne. 1995. "Everyone Has Got to Breathe," *Audubon* 97(2):64–73. What happened when a Pennsylvania community demanded regulation of polluting industries

Wann, David. 1995. *Deep Design: Pathways to a Livable Future.* Covelo, CA: Island Press. Explores new ways of thinking about design for renewability, recyclability and nontoxicity.

Glossary

A

albedo A description of a surface's reflective properties.

allergens Substances that activate the immune system and cause an allergic response; may not be directly antigenic themselves but may make other materials antigenic.

alpine The high, treeless biogeographic zone of mountains that consists of slopes above the timberline.

altruistic preservation A philosophy of preserving nature for its own sake.

ambient air The air that surrounds us.

amino acid An organic compound containing an amino group and a carboxyl group; amino acids are the units or building blocks that make peptide and protein molecules.

anaerobic respiration The incomplete intracellular breakdown of sugar or other organic compounds in the absence of oxygen that releases some energy and produces organic acids and/or alcohol.

anemia Low levels of hemoglobin due to iron deficiency or lack of red blood cells.

annual A plant that lives for a single growing season.

anthropocentric The belief that humans hold a special place in nature; being centered primarily on humans and human affairs.

antigens Substances that stimulate the production of, and react with, specific antibodies.

aquifers Porous, water-bearing layers of sand, gravel, and rock below the earth's surface; reservoirs for groundwater.

arithmetic growth A pattern of growth that increases at a constant amount per unit time, such as 1, 2, 3, 4 or 1, 3, 5, 7.

artesian well The result of a pressurized aquifer intersecting the surface or being penetrated by a pipe or conduit, from which water gushes without being pumped; also called a spring.

asphyxiants Chemicals that exclude oxygen or actively interfere with oxygen uptake and distribution; includes inert chemicals, such as nitrogen gas or halothane, that can displace oxygen and fill enclosed spaces.

asthma A distressing disease characterized by shortness of breath, wheezing, and bronchial muscle spasms.

atmospheric deposition Sedimentation of solids, liquids, or gaseous materials from the air. Snow, rain, and dust are the most familiar examples, but acids, metals, and toxic organic chemicals also can be transported by winds and deposited far from the source of origin.

atom The smallest unit of matter that has the characteristics of an element; consists of three main types of subatomic particles: protons, neutrons, and electrons.

autotroph An organism that synthesizes food molecules from inorganic molecules by using an external energy source, such as light energy.

B

barrier islands Low, narrow, sandy islands that form offshore from a coastline.

BAT *See* best available, economically achievable technology.

Batesian mimicry Evolution by one species to resemble the coloration, body shape, or behavior of a related species that is protected from predators by a venomous stinger, bad taste, or some other adaptation.

bedrock Unbroken, solid rock overlain by sand, gravel, or soil.

benthos The bottom of a sea or lake.

best available, economically achievable technology (BAT) The best pollution control available; the Clean Water Act effectively negates this category by stipulating that equipment must be economically feasible.

best practical control technology (BPT) The best technology for pollution control available at reasonable cost and operable under normal conditions.

beta particles High-energy electrons released by radioactive decay.

bill A piece of legislation introduced in Congress and intended to become law.

bioaccumulation The selective absorption and concentration of molecules by cells.

biocentric preservation A philosophy that emphasizes the fundamental right of living organisms to exist and to pursue their own goods.

biocentrism The belief that all creatures have rights and values; being centered on nature rather than humans.

biocide A broad-spectrum poison that kills a wide range of organisms.

biodegradable plastics Plastics that can be decomposed by microorganisms.

biodiversity The genetic, species, and ecological diversity of the organisms in a given area.

biogeochemical cycles Movement of matter within or between ecosystems; caused by living organisms, geological forces, or chemical reactions. The cycling of nitrogen, carbon, sulfur, oxygen, phosphorus, and water are examples.

biogeographical area An entire self-contained natural ecosystem and its associated land, water, air, and wildlife resources.

biological or biotic factors Organisms and products of organisms that are part of the environment and potentially affect the life of other organisms.

biological community The populations of plants, animals, and microorganisms living and interacting in a certain area at a given time.

biological controls Use of natural predators, pathogens, or competitors to regulate pest populations.

biological oxygen demand (BOD) A standard test for measuring the amount of dissolved oxygen utilized by aquatic microorganisms over a five-day period.

biological pests Organisms that reduce the availability, quality, or value of resources useful to humans.

biological resources The earth's organisms.

biomagnification Increase in concentration of certain stable chemicals (for example, heavy metals or fat-soluble pesticides) in successively higher trophic levels of a food chain or web.

biomass The total mass or weight of all the living organisms in a given population or area.

biomass fuel Organic material produced by plants, animals, or microorganisms that can be burned directly as a heat source or converted into a gaseous or liquid fuel.

biomass pyramid A metaphor or diagram that explains the relationship between the amounts of biomass at different trophic levels.

biome A broad, regional type of ecosystem characterized by distinctive climate and soil conditions and a distinctive kind of biological community adapted to those conditions.

bioregionalism Organization of human activities according to natural geographic or ecological boundaries and associations. This philosophy emphasizes a sense of place and living within the resources of one's local ecosystem.

biosphere The zone of air, land, and water at the surface of the earth that is occupied by organisms.

biosphere reserves World heritage sites identified by the IUCN as worthy for national park or wildlife refuge status because of high biological diversity or unique ecological features.

biota All organisms in a given area.

biotic Pertaining to life; environmental factors created by living organisms.

biotic potential The maximum reproductive rate of an organism, given unlimited resources and ideal environmental conditions. Compare with environmental resistance.

birth control Any method used to reduce births, including celibacy, delayed marriage, contraception; devices or medication that prevent implantation of fertilized zygotes, and induced abortions.

black lung disease Inflammation and fibrosis caused by accumulation of coal dust in the lungs or airways. *See* respiratory fibrotic agents.

blue revolution New techniques of fish farming that may contribute as much to human nutrition as miracle cereal grains but also may create social and environmental problems. Commercial fish farming already is a $5 billion per year industry, and domestic fish ponds supply as much as two-thirds of the protein consumed by subsistence farmers in many countries.

bog An area of waterlogged soil that tends to be peaty; fed mainly by precipitation; low productivity; some bogs are acidic.

boreal forest A broad band of mixed coniferous and deciduous trees that stretches across northern North America (and also Europe and Asia); its northernmost edge, the **taiga,** intergrades with the arctic **tundra.**

BPT *See* best practical control technology.

breeder reactor A nuclear reactor that produces fuel by bombarding isotopes of uranium and thorium with high-energy neutrons that convert inert atoms to fissionable ones. These reactors are inherently less stable and more dangerous than normal nuclear reactors.

bronchitis An inflammation of bronchial linings that causes persistent cough, copious production of sputum, and involuntary muscle spasms that constrict airways.

C

cancer Invasive, out-of-control cell growth that results in malignant tumors.

captive breeding Raising plants or animals in zoos or other controlled conditions to produce stock for subsequent release into the wild.

carbohydrate An organic compound consisting of a ring or chain of carbon atoms with hydrogen and oxygen attached; examples are sugars, starches, cellulose, and glycogen.

carbon cycle The circulation and reutilization of carbon atoms, especially via the processes of photosynthesis and respiration.

carbon monoxide (CO) Colorless, odorless, nonirritating but highly toxic gas produced by incomplete combustion of fuel, incineration of biomass or solid waste, or partially anaerobic decomposition of organic material.

carbon sink Places of carbon accumulation, such as in large forests (organic compounds) or ocean sediments (calcium carbonate); carbon is thus removed from the carbon cycle for moderately long to very long periods of time.

carbon source Originating point of carbon that reenters the carbon cycle; cellular respiration and combustion.

carcinogens Substances that cause cancer.

carnivores Organisms that mainly prey upon animals.

carrying capacity The maximum number of individuals of any species that can be supported by a particular ecosystem on a long-term basis.

cash crops Crops that are sold rather than consumed or bartered.

catastrophic systems Dynamic systems that jump abruptly from one seemingly steady state to another without any intermediate stages.

cell Minute biological compartments within which the processes of life are carried out.

cellular respiration The process in which a cell breaks down sugar or other organic compounds to release energy used for cellular work; may be **anaerobic** or **aerobic,** depending on the availability of oxygen.

chain reaction A self-sustaining reaction in which the fission of nuclei produces subatomic particles that cause the fission of other nuclei.

chaotic systems Systems that exhibit variability, which may not be necessarily random, yet whose complex patterns are not discernible over a normal human timescale.

chemical bond The force that holds molecules together.

chemosynthesis Autotrophic synthesis of organic compounds by certain bacteria; uses energy from inorganic compounds.

chloroplasts Chlorophyll-containing organelles in eukaryotic organisms; sites of photosynthesis.

chronic effects Long-lasting results of exposure to a toxin; can be a permanent change caused by a single, acute exposure or a continuous, low-level exposure.

chronic food shortages Long-term undernutrition and malnutrition; usually caused by people's lack of money to buy food or lack of opportunity to grow it themselves.

circle of poisons Importation of food contaminated with pesticides banned for use in this country but made here and sold abroad.

city A differentiated community with a sufficient population and resource base to allow residents to specialize in arts, crafts, services, and professional occupations.

clear-cut Cutting every tree in a given area, regardless of species or size; an appropriate harvest method for some species; can be destructive if not carefully controlled.

climate A description of the long-term pattern of weather in a particular area.

climax community A relatively stable, long-lasting community reached in a successional series; usually determined by climate and soil type.

closed canopy A forest where tree crowns spread over 20 percent of the ground; has the potential for commercial timber harvests.

cloud forests High mountain forests where temperatures are uniformly cool and fog or mist keeps vegetation wet all the time.

coal gasification The heating and partial combustion of coal to release volatile gases, such as methane and carbon monoxide; after pollutants are washed out, these gases become efficient, clean-burning fuel.

coal washing Coal technology that involves crushing coal and washing out soluble sulfur compounds with water or other solvents.

Coastal Zone Management Act Legislation of 1972 that gave federal money to thirty seacoast and Great Lakes states for development and restoration projects.

co-composting Microbial decomposition of organic materials in solid waste into useful soil additives and fertilizer; often, extra organic material in the form of sewer sludge, animal manure, leaves, and grass clippings are added to solid waste to speed the process and make the product more useful.

coevolution Mutually reinforcing evolutionary changes in interacting partners in an association.

cofactor Nonprotein components needed by enzymes in order to function; often minerals or vitamins.

cogeneration The simultaneous production of electricity and steam or hot water in the same plant.

cold front A moving boundary of cooler air displacing warmer air.

coliform bacteria Bacteria that live in the intestines (including the colon) of humans and other animals; used as a measure of the presence of feces in water or soil.

commensalism A symbiotic relationship in which one member is benefited and the second is neither harmed nor benefited.

communal resource management systems Resources managed by a community for long-term sustainability.

community ecology The study of interactions of all populations living in the ecosystem of a given area.

competitive exclusion A theory that no two populations of different species will occupy the same niche and compete for exactly the same resources in the same habitat for very long; disputed by some ecologists who see biological communities as highly individualistic and variable.

complexity (ecological)   The number of species at each trophic level and the number of trophic levels in a community.

composting   The biological degradation of organic material under aerobic (oxygen-rich) conditions to produce compost, a nutrient-rich soil amendment and conditioner.

compound   A molecule made up of two or more kinds of atoms held together by chemical bonds.

condensation   The aggregation of water molecules from vapor to liquid or solid when the saturation concentration is exceeded.

condensation nuclei   Tiny particles that float in the air and facilitate the condensation process.

conifers   Needle-bearing trees that produce seeds in cones.

conservation of matter   In any chemical reaction, matter changes form; it is neither created nor destroyed.

Consolidated Metropolitan Statistical Area (CMSA) An area with more than one million people formed by the merger of two or more population centers meeting other specified requirements of the Census Bureau.

consumer   An organism that obtains energy and nutrients by feeding on other organisms or their remains. *See also* heterotroph.

consumption   The fraction of withdrawn water that is lost in transmission or that is evaporated, absorbed, chemically transformed, or otherwise made unavailable for other purposes as a result of human use.

contour plowing   Plowing along hill contours; reduces erosion.

control rods   Neutron-absorbing material inserted into spaces between fuel assemblies in nuclear reactors to regulate fission reaction.

conurbation   An area with more than one million people that is formed by the merger of two or more population centers and that meets other specified requirements.

convection currents   Rising or sinking air currents that stir the atmosphere and transport heat from one area to another. Convection currents also occur in water; *see* spring overturn.

conventional pollutants   The seven substances (sulfur dioxide, carbon monoxide, particulates, hydrocarbons, nitrogen oxides, photochemical oxidants, and lead) that make up the largest volume of air quality degradation; identified by the Clean Air Act as the most serious threat of all pollutants to human health and welfare; also called criteria pollutants.

cool deserts   Deserts such as the American Great Basin characterized by cold winters and sagebrush.

coral reefs   Prominent oceanic features composed of hard, limy skeletons produced by coral animals; usually formed along edges of shallow, submerged ocean banks or along shelves in warm, shallow, tropical seas.

core   The dense, intensely hot mass of molten metal, mostly iron and nickel, thousands of kilometers in diameter at the earth's center.

core region   The primary industrial region of a country; usually located around the capital or largest port; has both the greatest population density and the greatest economic activity of the country.

Coriolis effect   The influence of friction and drag on air layers near the earth; deflects air currents to the direction of the earth's rotation.

cornucopian fallacy   The belief that nature is limitless in its abundance and that perpetual growth is not only possible but essential.

corridor   A strip of natural habitat that connects two adjacent nature preserves to allow migration of organisms from one place to another.

cost/benefit analysis   An evaluation of large-scale public projects by comparing the costs and benefits that accrue from them.

cover crops   Plants, such as rye, alfalfa, or clover, that can be planted immediately after harvest to hold and protect the soil.

criteria pollutants   *See* conventional pollutants.

critical factor   The single environmental factor closest to a tolerance limit for a given species at a given time. *See* limiting factors.

croplands   Lands used to grow crops.

crude birth rate   The number of births in a year divided by the midyear population.

crude death rate   The number of deaths per thousand persons in a given year; also called crude mortality rate.

crust   The cool, lightweight, outermost layer of the earth's surface that floats on the soft, pliable underlying layers; similar to the "skin" on a bowl of warm pudding.

cultural eutrophication   An increase in biological productivity and ecosystem succession caused by human activities.

D

debt-for-nature swap   Forgiveness of international debt in exchange for nature protection in developing countries. Environmental groups and nongovernmental organizations often pay banks to write off uncollectable debts of developing countries at a steep discount in exchange for a promise by the debtor country to establish nature preserves.

deciduous   Trees and shrubs that shed their leaves at the end of the growing season.

decline spiral   A catastrophic deterioration of a species, community, or whole ecosystem; accelerates as functions are disrupted or lost in a downward cascade.

decomposers   Fungi and bacteria that break complex organic material into smaller molecules.

deep ecology   A philosophy that calls for a profound shift in our attitudes and behavior based on voluntary simplicity; rejection of anthropocentric attitudes; intimate contact with nature; decentralization of power; support for cultural and biological diversity; a belief in the sacredness of nature; and direct personal action to protect nature, improve the environment, and bring about fundamental societal change.

degradation (of water resource)   Deterioration in water quality due to contamination or pollution; makes water unsuitable for other desirable purposes.

Delaney Clause   A controversial amendment to the Federal Food, Drug, and Cosmetic Act, added in 1958, prohibiting the addition of any known cancer-causing agent to processed foods, drugs, or cosmetics.

delta   Fan-shaped sediment deposit found at the mouth of a river.

demand   The amount of a product that consumers are willing and able to buy at various possible prices, assuming they are free to express their preferences.

demographic transition   A pattern of falling death and birth rates in response to improved living conditions; could be reversed in deteriorating conditions.

demography   Vital statistics about people: births, marriages, deaths, etc.; the statistical study of human populations relating to growth rate, age structure, geographic distribution, etc., and their effects on social, economic, and environmental conditions.

denitrifying bacteria   Free-living soil bacteria that convert nitrates to gaseous nitrogen and nitrous oxide.

dependency ratio   The number of nonworking members compared to working members for a given population.

desalinization (or desalination)   Removal of salt from water by distillation, freezing, or ultrafiltration.

desert   A type of biome characterized by low moisture levels and infrequent and unpredictable precipitation. Daily and seasonal temperatures fluctuate widely.

desertification   Denuding and degrading a once-fertile land, initiating a desert-producing cycle that feeds on itself and causes long-term changes in soil, climate, and biota of an area.

detritivore   Organisms that consume organic litter, debris, and dung.

dew point   The temperature at which condensation occurs for a given concentration of water vapor in the air.

dieback   A sudden population decline; also called a population crash.

diminishing returns   A condition in which unrestrained population growth causes the standard of living to decrease to a subsistence level where poverty, misery, vice, and starvation makes life permanently drab and miserable. This dreary prophecy has led economics to be called "the dismal science."

direct action   Civil disobedience, guerrilla street theater, picketing, protest marches, road blockades, demonstrations, and other techniques borrowed from the civil rights movement and applied to environmental protection.

discharge The amount of water that passes a fixed point in a given amount of time; usually expressed as liters or cubic feet of water per second.

disclimax community *See* equilibrium community.

discount rate The amount we discount or reduce the value of a future payment. When you borrow money from the bank at 10 percent annual interest, you are in effect saying that having the money now is worth 10 percent more to you than having the same amount one year from now.

disease A deleterious change in the body's condition in response to destabilizing factors, such as nutrition, chemicals, or biological agents.

dissolved oxygen (DO) content Amount of oxygen dissolved in a given volume of water at a given temperature and atmospheric pressure; usually expressed in parts per million (ppm).

diversity (species diversity, biological diversity) The number of species present in a community (species *richness*), as well as the relative *abundance* of each species.

DNA Deoxyribonucleic acid; the long, double-helix molecule in the nucleus of cells that contains the genetic code and directs the development and functioning of all cells.

dominant plants Those plant species in a community that provide a food base for most of the community; they usually take up the most space and have the largest biomass.

drip irrigation Uses pipe or tubing perforated with very small holes to deliver water one drop at a time directly to the soil around each plant. This conserves water and reduces soil waterlogging and salinization.

dry alkali injection Spraying dry sodium bicarbonate into flue gas to absorb and neutralize acidic sulfur compounds.

E

ecocentric (ecologically centered) A philosophy that claims moral values and rights for both organisms and ecological systems and processes.

ecofeminism A pluralistic, nonhierarchical, relationship-oriented philosophy that suggests how humans could reconceive themselves and their relationships to nature in nondominating ways as an alternative to patriarchal systems of domination.

ecojustice Justice in the social order and integrity in the natural order.

ecological development A gradual process of environmental modification by organisms.

ecological equivalents Different species that occupy similar ecological niches in similar ecosystems in different parts of the world.

ecological niche The functional role and position of a species (population) within a community or ecosystem, including what resources it uses, how and when it uses the resources, and how it interacts with other populations.

ecological succession The process by which organisms occupy a site and gradually change

environmental conditions so that other species can replace the original inhabitants.

ecology The scientific study of relationships between organisms and their environment. It is concerned with the life histories, distribution, and behavior of individual species as well as the structure and function of natural systems at the level of populations, communities, and ecosystems.

economic development A rise in real income *per person;* usually associated with new technology that increases productivity or resources.

economic growth An increase in the total wealth of a nation; if population grows faster than the economy, there may be real economic growth, but the share per person may decline.

ecosystem A specific biological community and its physical environment interacting in an exchange of matter and energy.

ecosystem management An integration of ecological, economic, and social goals in a unified systems approach to resource management.

ecosystem restoration To reinstate an entire community of organisms to as near its natural condition as possible.

ecotage Direct action (guerrilla warfare) or sabotage in defense of nature; also called monkey wrenching.

ecotone A boundary between two types of ecological communities.

ecotourism A combination of adventure travel, cultural exploration, and nature appreciation in wild settings.

edge effects A change in species composition, physical conditions, or other ecological factors at the boundary between two ecosystems. Some organisms flourish at this boundary and benefit from processes such as habitat fragmentation that increase edge area. Other organisms are harmed by increasing edge effects.

effluent sewerage A low-cost alternative sewage treatment for cities in poor countries that combines some features of septic systems and centralized municipal treatment systems. A septic tank near each residence collects and processes solid wastes—and must be pumped periodically just like a septic tank—while liquid effluent is collected and piped to a central treatment plant. This avoids drainfields in urban areas and yet allows sewer pipes and treatment plants to be downsized, saving money.

electron A negatively-charged subatomic particle that orbits around the nucleus of an atom.

electrostatic precipitators The most common particulate controls in power plants; fly ash particles pick up an electrostatic surface charge as they pass between large electrodes in the effluent stream, causing particles to migrate to the oppositely charged plate.

element A molecule composed of one kind of atom; cannot be broken into simpler units by chemical reactions.

El Niño A climatic change marked by shifting of a large warm water pool from the western Pacific Ocean towards the east. Nutrient-rich upwelling currents along the coast of South America are blocked by this sea change and fisheries fail catastrophically. An El Niño event normally is accompanied by droughts in Australia and Southeast Asia together with heavy rain and snow in western North America.

emigration The movement of members from a population.

emission standards Regulations for restricting the amounts of air pollutants that can be released from specific point sources.

emphysema An irreversible, obstructive lung disease in which airways become permanently constricted and alveoli are damaged or destroyed.

endangered species A species considered to be in imminent danger of extinction.

endemism A state in which species are restricted to a single region.

energy The capacity to do work (that is, to change the physical state or motion of an object).

energy efficiency A measure of energy produced compared to energy consumed.

energy pyramid A representation of the loss of useful energy at each step in a food chain.

energy recovery Incineration of solid waste to produce useful energy.

environment The circumstances or conditions that surround an organism or group of organisms as well as the complex of social or cultural conditions that affect an individual or community.

environmental ethics A search for moral values and ethical principles in human relations with the natural world.

environmental hormones Chemical pollutants that substitute for, or interfere with, naturally-occurring hormones in our bodies; these chemicals may trigger reproductive failure, developmental abnormalities, or tumor promotion.

environmental impact statement (EIS) An analysis, required by provisions in the National Environmental Policy Act of 1970, of the effects of any major program a federal agency plans to undertake.

environmentalism Active participation in attempts to solve environmental pollution and resource problems.

environmental justice Combines civil rights with environmental protection to demand a safe, healthy, life-giving environment for everyone.

environmental literacy Fluency in the principles of ecology that gives us a working knowledge of the basic grammar and underlying syntax of environmental wisdom.

environmental resistance All the limiting factors that tend to reduce population growth rates and set the maximum allowable population size or carrying capacity of an ecosystem.

environmental resources Anything an organism needs that can be taken from the environment.

environmental science The systematic, scientific study of our environment as well as our role in it.

enzymes Molecules, usually proteins or nucleic acids, that act as catalysts in biochemical reactions.

epidemiology The study of the distribution and causes of disease and injuries in human populations.

epiphyte A plant that grows on a substrate other than the soil, such as the surface of another organism.

equilibrium community Also called a **disclimax community;** a community subject to periodic disruptions, usually by fire, that prevent it from reaching a climax stage.

estuary A bay or drowned valley where a river empties into the sea. Fresh water mingling with salt water brings in sediment and nutrients and creates a gradient of salinity that makes estuaries among the most diverse and biologically productive ecosystems on earth.

eukaryotic cell A cell containing a membrane-bounded nucleus and membrane-bounded organelles.

eutectic chemicals Phase-changing chemicals used in heat storage systems to store a large amount of energy in a small volume.

eutrophic Rivers and lakes rich in organisms and organic material (*eu* = truly; *trophic* = nutritious).

evaporation The process in which a liquid is changed to vapor (gas phase).

evergreens Coniferous trees and broad-leaved plants that retain their leaves year-round.

evolution A theory that explains how random changes in genetic material and competition for scarce resources cause species to change gradually.

exhaustible resources Generally considered the earth's geologic endowment: minerals, nonmineral resources, fossil fuels, and other materials present in fixed amounts in the environment.

existence value The importance we place on just knowing that a particular species or a specific organism exists.

exponential growth Growth at a constant rate of increase per unit of time; can be expressed as a constant fraction or exponent. *See* geometric growth.

external costs Expenses, monetary or otherwise, borne by someone other than the individuals or groups who use a resource.

extinction The irrevocable elimination of species; can be a normal process of the natural world as species out-compete or kill off others or as environmental conditions change.

extirpate To destroy totally; extinction caused by direct human action, such as hunting, trapping, etc.

F

family planning Controlling reproduction; planning the timing of birth and having as many babies as are wanted and can be supported.

famines Acute food shortages characterized by large-scale loss of life, social disruption, and economic chaos.

fauna All of the animals present in a given region.

fecundity The physical ability to reproduce.

fen An area of waterlogged soil that tends to be peaty; fed mainly by upwelling water; low productivity.

feral A domestic animal that has taken up a wild existence.

fermentation (alcoholic) A type of anaerobic respiration that yields carbon dioxide and alcohol; used in commercial fermentation processes, including production of raised bakery dough products and alcoholic beverages.

fertility Measurement of actual number of offspring produced through sexual reproduction; usually described in terms of number of offspring of females, since paternity can be difficult to determine.

fibrosis The general name for accumulation of scar tissue in the lung.

fidelity A principle that forbids misleading or deceiving any creature capable of being mislead or deceived. We are to be truthful in our dealings with others.

filters A porous mesh of cotton cloth, spun glass fibers, or asbestos-cellulose that allows air or liquid to pass through but holds back solid particles.

fire-climax community An equilibrium community maintained by periodic fires; examples include grasslands, chaparral shrubland, and some pine forests.

first law of thermodynamics States that energy is *conserved;* that is, it is neither created nor destroyed under normal conditions.

First World The industrialized capitalist or market-economy countries of Western Europe, North America, Japan, Australia, and New Zealand.

flood An overflow of water onto land that normally is dry.

floodplains Low lands along riverbanks, lakes, and coastlines subjected to periodic inundation.

flora All of the plants present in a given region.

flue-gas scrubbing Treating combustion exhaust gases with chemical agents to remove pollutants. Spraying crushed limestone and water into the exhaust gas stream to remove sulfur is a common scrubbing technique.

fluidized bed combustion High pressure air is forced through a mixture of crushed coal and limestone particles, lifting the burning fuel and causing it to move like a boiling fluid. Fresh coal and limestone are added continuously to the top of the combustion bed while ash and slag are drawn off below.

food aid Financial assistance intended to boost less-developed countries' standards of living.

food chain A linked feeding series; in an ecosystem, the sequence of organisms through which energy and materials are transferred, in the form of food, from one trophic level to another.

food security The ability of individuals to obtain sufficient food on a day-to-day basis.

food surpluses Excess food supplies.

food web A complex, interlocking series of individual food chains in an ecosystem.

forest management Scientific planning and administration of forest resources for sustainable harvest, multiple use, regeneration, and maintenance of a healthy biological community.

fossil fuels Petroleum, natural gas, and coal created by geological forces from organic wastes and dead bodies of formerly living biological organisms.

Fourth World A political/economic category describing very poor nations that have neither market economies nor central planning and are either not developing or are developing very slowly. Also used to describe indigenous communities within wealthier nations.

freezing condensation A process that occurs in the clouds when ice crystals trap water vapor. As the ice crystals become larger and heavier, they begin to fall as rain or snow.

fresh water Water other than seawater; covers only about 2 percent of earth's surface, including streams, rivers, lakes, ponds, and water associated with several kinds of wetlands.

freshwater ecosystems Ecosystems in which the fresh (nonsalty) water of streams, rivers, ponds, or lakes plays a defining role.

front The boundary between two air masses of different temperature and density.

frontier An unexploited natural area at the leading edge of human settlement.

frontier mentality The idea that the world has an unlimited supply of resources for human use regardless of the consequences to natural ecosystems and the biosphere.

fuel assembly A bundle of hollow metal rods containing uranium oxide pellets; used to fuel a nuclear reactor.

fuel-switching A change from one fuel to another.

fuelwood Branches, twigs, logs, wood chips, and other wood products harvested for use as fuel.

fugitive emissions Substances that enter the air without going through a smokestack, such as dust from soil erosion, strip mining, rock crushing, construction, and building demolition.

fungi One of the five kingdom classifications; consists of nonphotosynthetic, eukaryotic organisms with cell walls, filamentous bodies, and absorptive nutrition.

fungicide A chemical that kills fungi.

G

Gaia hypothesis A theory that the living organisms of the biosphere form a single, complex interacting system that creates and maintains a habitable Earth; named after Gaia, the Greek Earth mother goddess.

gamma rays   Very short wavelength forms of the electromagnetic spectrum.

gap analysis   A biogeographical technique of mapping biological diversity and endemic species to find gaps between protected areas that leave endangered habitats vulnerable to disruption.

garden city   A new town with special emphasis on landscaping and rural ambience.

gasohol   A mixture of gasoline and ethanol.

gene   A unit of heredity; a segment of DNA nucleus of the cell that contains information for the synthesis of a specific protein, such as an enzyme.

gene banks   Storage for seed varieties for future breeding experiments.

general fertility rate   Representation of population age structure and fecundity; crude birth rate multiplied by the percentage of fecund women (between approximately fifteen and forty-four years of age) by 1000.

genetic assimilation   The disappearance of a species as its genes are diluted through crossbreeding with a closely related species.

geometric growth   Growth that follows a geometric pattern of increase, such as 2, 4, 8, 16, etc. *See* exponential growth.

geothermal energy   Energy drawn from the internal heat of the earth, either through geysers, fumaroles, hot springs, or other natural geothermal features, or through deep wells that pump heated groundwater.

germ plasm   Genetic material that may be preserved for future agricultural, commercial, and ecological values (plant seeds or parts or animal eggs, sperm, and embryos).

global environmental citizenship   A shift in our attention from pollution in a specific place to a concern about the life-support systems of the whole planet.

grasslands   A biome dominated by grasses and associated herbaceous plants.

green plans   Integrated national environmental plans for reducing pollution and resource consumption while achieving sustainable development and environmental restoration.

green political parties   Political organizations based on environmental protection, participatory democracy, grassroots organization, and sustainable development.

green revolution   Dramatically increased agricultural production brought about by "miracle" strains of grain; usually requires high inputs of water, plant nutrients, and pesticides.

gross national product (GNP)   The sum total of all goods and services produced in a national economy. Gross domestic product (GDP) is used to distinguish economic activity within a country from that of off-shore corporations.

groundwater   Water held in gravel deposits or porous rock below the earth's surface; does not include water or crystallization held by chemical bonds in rocks or moisture in upper soil layers.

gully erosion   Removal of layers of soil, creating channels or ravines too large to be removed by normal tillage operations.

H

habitat   The place or set of environmental conditions in which a particular organism lives.

Hadley cells   Circulation patterns of atmospheric convection currents as they sink and rise in several intermediate bands.

half-life (of radioisotopes)   The length of time required for half the nuclei in a sample to change into another isotope.

hazardous   Describes chemicals that are dangerous, including flammables, explosives, irritants, sensitizers, acids, and caustics; may be relatively harmless in diluted concentrations.

hazardous waste   Any discarded material containing substances known to be toxic, mutagenic, carcinogenic, or teratogenic to humans or other life-forms; ignitable, corrosive, explosive, or highly reactive alone or with other materials.

health   A state of physical and emotional well-being; the absence of disease or ailment.

heap-leach extraction   A technique for separating gold from extremely low-grade ores. Crushed ore is piled in huge heaps and sprayed with a dilute alkaline-cyanide solution, which percolates through the pile to extract the gold, which is separated from the effluent in a processing plant. This process has a high potential for water pollution.

heat   A form of energy transferred from one body to another because of a difference in temperatures.

heat capacity   The amount of heat energy that must be added or subtracted to change the temperature of a body; water has a high heat capacity.

heat of vaporization   The amount of heat energy required to convert water from a liquid to a gas.

herbicide   A chemical that kills plants.

herbivore   An organism that eats only plants.

heterotroph   An organism that is incapable of synthesizing its own food and, therefore, must feed upon organic compounds produced by other organisms.

high-level waste repository   A place where intensely radioactive wastes can be buried deep in the ground and remain unexposed to groundwater and earthquakes for tens of thousands of years. The first such site chosen in the United States is Yucca Mountain, Nevada, although it is not certain that a waste repository will ever be built there.

high-quality energy   Intense, concentrated, and high-temperature energy that is considered high-quality because of its usefulness in carrying out work.

homeostasis   Maintaining a dynamic, steady state in a living system through opposing, compensating adjustments.

Homestead Act   Legislation passed in 1862 allowing any citizen or applicant for citizenship over twenty-one years old and head of a family to acquire 160 acres of public land by living on it and cultivating it for five years.

host organism   An organism that provides lodging for a parasite.

hot desert   Deserts of the American Southwest and Mexico; characterized by extreme summer heat and cacti.

human ecology   The study of the interactions of humans with the environment.

human resources   Human wisdom, experience, skill, labor, and enterprise.

humus   Sticky, brown, insoluble residue from the bodies of dead plants and animals; gives soil its structure, coating mineral particles and holding them together; serves as a major source of plant nutrients.

hypothyroidism   Listlessness and other metabolic symptoms caused by low thyroid hormone levels.

I

igneous rocks   Crystalline minerals solidified from molten magma from deep in the earth's interior; basalt, rhyolite, andesite, lava, and granite are examples.

inbreeding depression   In a small population, an accumulation of harmful genetic traits (through random mutations and natural selection) that lowers viability and reproductive success of enough individuals to affect the whole population.

industrial revolution   Advances in science and technology that have given us power to understand and change our world.

industrial timber   Trees used for lumber, plywood, veneer, particleboard, chipboard, and paper; also called roundwood.

inertial confinement   A nuclear fusion process in which a small pellet of nuclear fuel is bombarded with extremely high-intensity laser light.

infiltration   The process of water percolation into the soil and pores and hollows of permeable rocks.

inflammatory response   A complex series of interactions between fragments of damaged cells, surrounding tissues, circulating blood cells, and specific antibodies; typical of infections.

informal economy   Small-scale family businesses in temporary locations outside the control of normal regulatory agencies.

insecticide   A chemical that kills insects.

insolation   Incoming solar radiation.

instrumental value   Value or worth of objects that satisfy the needs and wants of moral agents. Objects that can be used as a means to some desirable end.

intangible resources   Abstract commodities, such as open space, beauty, serenity, genius, information, diversity, and satisfaction.

integrated pest management (IPM)   IPM is an ecologically based pest-control strategy that relies on natural mortality factors, such as natural enemies, weather, cultural control methods, and carefully applied doses of pesticides.

internal costs The expenses, monetary or otherwise, borne by those who use a resource.

internalizing costs Planning so that those who reap the benefits of resource use also bear all the external costs.

interplanting The system of planting two or more crops, either mixed together or in alternating rows, in the same field; protects the soil and makes more efficient use of the land.

interspecific competition In a community, competition for resources between members of *different* species.

intraspecific competition In a community, competition for resources among members of the *same* species.

ionizing radiation High-energy electromagnetic radiation or energetic subatomic particles released by nuclear decay.

ionosphere The lower part of the thermosphere.

ions Electrically-charged atoms that have gained or lost electrons.

irritants Corrosives (strong acids), caustics (alkaline reagents), and other substances that damage biological tissues on contact.

irruptive growth *See* Malthusian growth.

island biogeography The study of rates of colonization and extinction of species on islands or other isolated areas based on size, shape, and distance from other inhabited regions.

island effects Reductions in species diversity caused by reduction in ecosystem area.

isotopes Forms of a single element that differ in atomic mass due to a different number of neutrons in the nucleus.

J

J curve A growth curve that depicts exponential growth; called a J curve because of its shape.

jet streams Powerful winds or currents of air that circulate in shifting flows; similar to oceanic currents in extent and effect on climate.

joule A unit of energy. One joule is the energy expended in 1 second by a current of 1 amp flowing through a resistance of 1 ohm.

K

keystone species A species whose impacts on its community or ecosystem are much larger and more influential than would be expected from mere abundance. This could be a top predator, a plant that shelters or feeds other organisms, or an organism that plays a critical ecological role.

kinetic energy Energy contained in moving objects such as a rock rolling down a hill, the wind blowing through the trees, or water flowing over a dam.

known resources Minerals or other useful environmental materials or services that are identified and partially mapped; may not be environmentally or socially acceptable or economically feasible to exploit.

kwashiorkor A widespread human protein deficiency disease resulting from a starchy diet low in protein and essential amino acids.

L

landfills Land disposal sites for solid waste; operators compact refuse and cover it with a layer of dirt to minimize rodent and insect infestation, wind-blown debris, and leaching by rain.

land reform Democratic redistribution of landownership to recognize the rights of those who actually work the land to a fair share of the products of their labor.

land rehabilitation A utilitarian program to repair damage and make land useful to humans.

landscape ecology The study of the reciprocal effects of spatial pattern on ecological processes. A study of the ways in which landscape history shapes the features of the land and the organisms that inhabit it as well as our reaction to, and interpretation of, the land.

landslide The sudden fall of rock and earth from a hill or cliff. Often triggered by an earthquake or heavy rain.

LD50 A chemical dose lethal to 50 percent of a test population.

less-developed countries (LDC) Nonindustrialized nations characterized by low per capita income, high birth and death rates, high population growth rates, and low levels of technological development.

life-cycle analysis Evaluation of material and energy inputs and outputs at each stage of manufacture, use, and disposal of a product.

life expectancy The average number of years lived by a group of individuals after reaching a given age; the probable number of years of survival for an individual of a given age.

life span The longest period of life reached by a type of organism.

lignite Soft, brown to black coal of low BTU value with original plant components still discernible.

limiting factors Chemical or physical factors that limit the existence, growth, abundance, or distribution of an organism. The **principle of limiting factors** states that for each physical factor in the environment, there are both minimum and maximum limits (**tolerance limits**) beyond which a particular species cannot survive. The single factor closest to a tolerance limit for a given species at a given time is the **critical factor**.

lipid A nonpolar organic compound that is insoluble in water but soluble in solvents, such as alcohol and ether; includes fats, oils, steroids, phospholipids, and carotenoids.

liquid metal fast breeder A nuclear power plant that converts uranium 238 to plutonium 239; thus, it creates more nuclear fuel than it consumes. Because of the extreme heat and density of its core, the breeder uses liquid sodium as a coolant.

logistic growth Growth rates regulated by internal and external factors that establish an equilibrium with environmental resources; species may grow exponentially when resources are unlimited but slowly as the carrying capacity is reached. *See* S curve.

longevity The length or duration of life; compare to survivorship.

low-head hydropower Small-scale hydro technology that can extract energy from small headwater dams; causes much less ecological damage.

low-quality energy Diffuse, dispersed energy at a low temperature that is difficult to gather and use for productive purposes.

LULUs *L*ocally *u*nwanted *l*and *u*ses such as toxic waste dumps, incinerators, smelters, airports, freeways, and other sources of environmental, economic, or social degradation.

M

magma Molten rock from deep in the earth's interior; called lava when it spews from volcanic vents.

magnetic confinement A technique for enclosing a nuclear fusion reaction in a powerful magnetic field inside a vacuum chamber.

malignant tumor A mass of cancerous cells that have left their site of origin, migrated through the body, invaded normal tissues, and are growing out of control.

malnourishment A nutritional imbalance caused by lack of specific dietary components or inability to absorb or utilize essential nutrients.

Malthusian growth A population explosion followed by a population crash; also called irruptive growth.

Man and Biosphere (MAB) program A design for nature preserves that divides protected areas into zones with different purposes. Critical ecosystem functions and endangered wildlife are protected in a central core region where limited scientific study is the only human access allowed. Ecotourism and research facilities are located in a relatively pristine buffer zone around the core, while sustainable resource harvesting and permanent habitation are allowed in multiple-use peripheral regions.

mantle A hot, pliable layer of rock that surrounds the earth's core and underlies the cool, outer crust.

marasmus A widespread human protein deficiency disease caused by a diet low in calories and protein or imbalanced in essential amino acids.

marine Living in or pertaining to the sea.

market equilibrium The dynamic balance between supply and demand under a given set of conditions in a "free" market (one with no monopolies or government interventions).

marsh Wetland without trees; in North America, this type of land is characterized by cattails and rushes.

mass burn Incineration of unsorted solid waste.

matter Something that occupies space and has mass.

Mediterranean climate areas Specialized landscapes with warm, dry summers; cool, wet winters; many unique plant and animal adaptations; and many levels of endemism.

megacity *See* megalopolis.

megalopolis Also known as a megacity or supercity; megalopolis indicates an urban area with more than 10 million inhabitants.

megawatt (MW) Unit of electrical power equal to one thousand kilowatts or one million watts.

mesosphere The atmospheric layer above the stratosphere and below the thermosphere; the middle layer; temperatures are usually very low.

metabolism All the energy and matter exchanges that occur within a living cell or organism; collectively, the life processes.

metamorphic rock Igneous and sedimentary rocks modified by heat, pressure, and chemical reactions.

micro-hydro generators Small power generators that can be used in low-level rivers to provide economical power for four to six homes, freeing them from dependence on large utilities and foreign energy supplies.

Milankovitch cycles Periodic variations in tilt, eccentricity, and wobble in the earth's orbit; Milutin Milankovitch suggested that it is responsible for cyclic weather changes.

milpa agriculture An ancient farming system in which small patches of tropical forests are cleared and perennial polyculture agriculture practiced and is then followed by many years of fallow to restore the soil; also called swidden agriculture.

mineral A naturally occurring, inorganic, crystalline solid with definite chemical composition and characteristic physical properties.

mitigation Repairing or rehabilitating a damaged ecosystem or compensating for damage by providing a substitute or replacement area.

mixed perennial polyculture Growing a mixture of different perennial crop species (where the same plant persists for more than one year) together in the same plot; imitates the diversity of a natural system and is often more stable and more suitable for sustainable agriculture than monoculture of annual plants.

molecule A combination of two or more atoms.

monitored, retrievable storage Holding wastes in underground mines or secure surface facilities such as dry casks where they can be watched and repackaged, if necessary.

monkey wrenching Environmental sabotage such as driving large spikes in trees to protect them from loggers, vandalizing construction equipment, pulling up survey stakes for unwanted developments, and destroying billboards.

monoculture agroforestry Intensive planting of a single species; an efficient wood production approach, but one that encourages pests and disease infestations and conflicts with wildlife habitat or recreation uses.

monsoon A seasonal reversal of wind patterns caused by the different heating and cooling rates of the oceans and continents.

montane coniferous forests Coniferous forests of the mountains consisting of belts of different forest communities along an altitudinal gradient.

moral agents Beings capable of making distinctions between right and wrong and acting accordingly. Those whom we hold responsible for their actions.

moral subjects Beings that are not capable of distinguishing between right or wrong or that are not able to act on moral principles and yet are susceptible of being wronged by others. This category assumes some rights or inherent values in moral subjects that gives us duties or obligations towards them.

morbidity Illness or disease.

more-developed countries (MDC) Industrialized nations characterized by high per capita incomes, low birth and death rates, low population growth rates, and high levels of industrialization and urbanization.

mortality Death rate in a population; the probability of dying.

Muellerian mimicry Evolution by one species to resemble the coloration, body shape, or behavior of an unrelated species that is protected from predators by a venomous stinger, bad taste, or some other adaptation.

mulch Protective ground cover, including manure, wood chips, straw, seaweed, leaves, and other natural products, or synthetic materials, such as heavy paper or plastic, that protect the soil, save water, and prevent weed growth.

multiple use Many uses that occur simultaneously; used in forest management; limited to mutually compatible uses.

mutagens Agents, such as chemicals or radiation, that damage or alter genetic material (DNA) in cells.

mutation A change, either spontaneous or by external factors, in the genetic material of a cell; mutations in the gametes (sex cells) can be inherited by future generations of organisms.

mutualism A symbiotic relationship between individuals of two different species in which both species benefit from the association.

N

NAAQS *N*ational *A*mbient *A*ir *Q*uality *S*tandard; federal standards specifying the maximum allowable levels (averaged over specific time periods) for regulated pollutants in ambient (outdoor) air.

natality The production of new individuals by birth, hatching, germination, or cloning.

natural history The study of where and how organisms carry out their life cycles.

natural increase Crude death rate subtracted from crude birth rate.

natural landscaping A landscaping design that uses carefully contrived meadows, forests, and natural-looking ponds and marshes to create an illusion of wild nature.

natural resources Goods and services supplied by the environment.

natural selection The mechanism for evolutionary change in which environmental pressures cause certain genetic combinations in a population to become more abundant; genetic combinations best adapted for present environmental conditions tend to become predominant.

neo-Malthusian A belief that the world is characterized by scarcity and competition in which too many people fight for too few resources. Named for Thomas Malthus, who predicted a dismal cycle of misery, vice, and starvation as a result of human overpopulation.

net energy yield Total useful energy produced during the lifetime of an entire energy system minus the energy used, lost, or wasted in making useful energy available.

neurotoxins Toxic substances, such as lead or mercury, that specifically poison nerve cells.

neutron A subatomic particle, found in the nucleus of the atom, that has no electromagnetic charge.

new towns Experimental urban environments that seek to combine the best features of the rural village and the modern city.

nihilists Those who believe the world has no meaning or purpose other than a dark, cruel, unceasing struggle for power and existence.

NIMBY *N*ot *I*n *M*y *B*ackyard: the rallying cry of those opposed to LULUs.

nitrate-forming bacteria Bacteria that convert nitrites into compounds that can be used by green plants to build proteins.

nitrite-forming bacteria Bacteria that combine ammonia with oxygen to form nitrites.

nitrogen cycle The circulation and reutilization of nitrogen in both inorganic and organic phases.

nitrogen-fixing bacteria Bacteria that convert nitrogen from the atmosphere or soil solution into ammonia that can then be converted to plant nutrients by nitrite- and nitrate-forming bacteria.

nitrogen oxides Highly reactive gases formed when nitrogen in fuel or combustion air is heated to over 650° C (1200° F) in the presence of oxygen or when bacteria in soil or water oxidize nitrogen-containing compounds.

noncriteria pollutants *See* unconventional pollutants.

nongovernmental organizations (NGOs) A term referring collectively to pressure and research groups, advisory agencies, political parties, professional societies, and other groups concerned about environmental quality, resource use, and many other issues.

noninterference A principle that requires us to refrain from interfering with the lives or destinies of other moral agents or moral subjects except to satisfy essential needs.

nonmalfeasance Doing no harm to any being that has goods (values, interests, moral values) of their own.

nonpoint sources   Scattered, diffuse sources of pollutants, such as runoff from farm fields, golf courses, construction sites, etc.

nonrenewable resources   Materials or services from the environment that are not replaced or replenished by natural processes at a rate comparable to our use of the resource; a resource depleted or exhausted by use.

North/South division   A description of the fact that most of the world's wealthier countries tend to be in North America, Europe, and Japan while the poorer countries tend to be located closer to the equator.

nuclear fission   The radioactive decay process in which isotopes split apart to create two smaller atoms.

nuclear fusion   A process in which two smaller atomic nuclei fuse into one larger nucleus and release energy; the source of power in a hydrogen bomb.

nucleic acids   Large organic molecules made of nucleotides that function in the transmission of hereditary traits, in protein synthesis, and in control of cellular activities.

nucleus   The center of the atom; occupied by protons and neutrons. In cells, the organelle that contains the chromosomes (DNA).

nuées ardentes   Deadly, denser-than-air mixtures of hot gases and ash ejected from volcanoes.

numbers pyramid   A diagram showing the relative population sizes at each trophic level in an ecosystem; usually corresponds to the biomass pyramid.

O

oceanic islands   Islands in the ocean; formed by breaking away from a continental landmass, volcanic action, coral formation, or a combination of sources; support distinctive communities.

ocean shorelines   Rocky coasts and sandy beaches along the oceans; support rich, stratified communities.

ocean thermal electric conversion (OTEC)   Energy derived from temperature differentials between warm ocean surface waters and cold deep waters. This differential can be used to drive turbines attached to electric generators.

offset allowances   A controversial component of air quality regulations that allows a polluter to avoid installation of control equipment on one source with an "offsetting" pollution reduction at another source.

oligotrophic   Condition of rivers and lakes that have clear water and low biological productivity (*oligo* = little; *trophic* = nutrition); are usually clear, cold, infertile headwater lakes and streams.

omnivore   An organism that eats both plants and animals.

open access system   A commonly-held resource for which there are no management rules.

open canopy   A forest where tree crowns cover less than 20 percent of the ground; also called woodland.

open range   Unfenced, natural grazing lands; includes woodland as well as grassland.

open system   A system that exchanges energy and matter with its environment.

optimum   The most favorable condition in regard to an environmental factor.

orbital   The space or path in which an electron orbits the nucleus of an atom.

organic compounds   Complex molecules organized around skeletons of carbon atoms arranged in rings or chains; includes biomolecules, molecules synthesized by living organisms.

overburden   Overlying layers of noncommercial sediments that must be removed to reach a mineral or coal deposit.

overnutrition   Receiving too many calories.

overshoot   The extent to which a population exceeds the carrying capacity of its environment.

oxygen cycle   The circulation and reutilization of oxygen in the biosphere.

oxygen sag   Oxygen decline downstream from a pollution source that introduces materials with high biological oxygen demands.

ozone   A highly reactive molecule containing three oxygen atoms; a dangerous pollutant in ambient air. In the stratosphere, however, ozone forms an ultraviolet absorbing shield that protects us from mutagenic radiation.

P

parabolic mirrors   Curved mirrors that focus light from a large area onto a single, central point, thereby concentrating solar energy and producing high temperatures.

parasite   An organism that lives in or on another organism, deriving nourishment at the expense of its host, usually without killing it.

parent material   Undecomposed mineral particles and unweathered rock fragments beneath the subsoil; weathering of this layer produces new soil particles for the layers above.

particulate material   Atmospheric aerosols, such as dust, ash, soot, lint, smoke, pollen, spores, algal cells, and other suspended materials; originally applied only to solid particles but now extended to droplets of liquid.

parts per billion (ppb)   Number of parts of a chemical found in one billion parts of a particular gas, liquid, or solid mixture.

parts per million (ppm)   Number of parts of a chemical found in one million parts of a particular gas, liquid, or solid mixture.

parts per trillion (ppt)   Number of parts of a chemical found in one trillion (10^{12}) parts of a particular gas, liquid, or solid mixture.

passive heat absorption   The use of natural materials or absorptive structures without moving parts to gather and hold heat; the simplest and oldest use of solar energy.

pasture   Enclosed domestic meadows or managed grazing lands.

patchiness   Within a larger ecosystem, the presence of smaller areas that differ in some physical conditions and thus support somewhat different communities; a diversity-promoting phenomenon.

pathogen   An organism that produces disease in a host organism, disease being an alteration of one or more metabolic functions in response to the presence of the organism.

peat   Deposits of moist, acidic, semidecayed organic matter.

pellagra   Lassitude, torpor, dermatitis, diarrhea, dementia, and death brought about by a diet deficient in tryptophan and niacin.

peptides   Two or more amino acids linked by a peptide bond.

perennial species   Plants that grow for more than two years.

permafrost   A permanently frozen layer of soil that underlies the arctic tundra.

permanent retrievable storage   Placing waste storage containers in a secure building, salt mine, or bedrock cavern where they can be inspected periodically and retrieved, if necessary, for repacking or for transfer if a better means of disposal or reuse is developed.

pest   Any organism that reduces the availability, quality, or value of a useful resource.

pest resurgence   Rebound of pest populations due to acquired resistance to chemicals and nonspecific destruction of natural predators and competitors by broadscale pesticides.

pesticide   Any chemical that kills, controls, drives away, or modifies the behavior of a pest.

pesticide rain   Long-range transport of pesticides by air currents and deposition through precipitation in sites far from its origin; analogous to acid rain.

pesticide treadmill   A need for constantly increasing doses or new pesticides to prevent pest resurgence.

pH   A value that indicates the acidity or alkalinity of a solution on a scale of 0 to 14, based on the proportion of H^+ ions present.

phosphorus cycle   The movement of phosphorus atoms from rocks through the biosphere and hydrosphere and back to rocks.

photochemical oxidants   Products of secondary atmospheric reactions. *See* smog.

photodegradable plastics   Plastics that break down when exposed to sunlight or to a specific wavelength of light.

photosynthesis   The biochemical process by which green plants and some bacteria capture light energy and use it to produce chemical bonds. Carbon dioxide and water are consumed while oxygen and simple sugars are produced.

photosynthetic efficiency   The percentage of available light captured by plants and used to make useful products.

photovoltaic cell   An energy-conversion device that captures solar energy and directly converts it to electrical current.

physical or abiotic factors   Nonliving factors, such as temperature, light, water, minerals, and climate, that influence an organism.

phytoplankton   Microscopic, free-floating, autotrophic organisms that function as producers in aquatic ecosystems.

pioneer species   In primary succession on a terrestrial site, the plants, lichens, and microbes that first colonize the site.

plankton   Primarily microscopic organisms that occupy the upper water layers in both freshwater and marine ecosystems.

plasma   A hot, electrically neutral gas of ions and free electrons.

poachers   Those who hunt wildlife illegally.

point sources   Specific locations of highly concentrated pollution discharge, such as factories, power plants, sewage treatment plants, underground coal mines, and oil wells.

pollution   To make foul, unclean, dirty; any physical, chemical, or biological change that adversely affects the health, survival, or activities of living organisms or that alters the environment in undesirable ways.

pollution charges   Fees assessed per unit of pollution based on the "polluter pays" principle.

polycentric complex   Cities with several urban cores surrounding a once dominant central core.

population   A group of individuals of the same species occupying a given area.

population crash   A sudden population decline caused by predation, waste accumulation, or resource depletion; also called a dieback.

population explosion   Growth of a population at exponential rates to a size that exceeds environmental carrying capacity; usually followed by a population crash.

population hurdle   The need for investment in infrastructure and social services in a rapidly growing population that prevents the capital investment necessary for real economic development.

population momentum   A potential for increased population growth as young members reach reproductive age.

postmaterialist values   A philosophy that emphasizes quality of life over acquisition of material goods.

potential energy   Stored energy that is latent but available for use. A rock poised at the top of a hill or water stored behind a dam are examples of potential energy.

power   The rate of energy delivery; measured in horsepower or watts.

precycling   Making environmentally sound decisions at the store and reducing waste before we buy.

predation   The act of feeding by a predator.

predator   An organism that feeds directly on other organisms in order to survive; live- feeders, such as herbivores and carnivores.

prevention of significant deterioration   A clause of the Clean Air Act that prevents degradation of existing clean air; opposed by industry as an unnecessary barrier to development.

price elasticity   A situation in which supply and demand of a commodity respond to price.

primary pollutants   Chemicals released directly into the air in a harmful form.

primary productivity   Synthesis of organic materials (biomass) by green plants using the energy captured in photosynthesis.

primary standards   Regulations of the 1970 Clean Air Act; intended to protect human health.

primary succession   An ecological succession that begins in an area where no biotic community previously existed.

primary treatment   A process that removes solids from sewage before it is discharged or treated further.

principle of competitive exclusion   A result of natural selection whereby two similar species in a community occupy different ecological niches, thereby reducing competition for food.

producer   An organism that synthesizes food molecules from inorganic compounds by using an external energy source; most producers are photosynthetic.

production frontier   The maximum output of two competing commodities at different levels of production.

productivity   The amount of biological matter or biomass produced in a given area during a given unit of time.

prokaryotic   Cells that do not have a membrane-bounded nucleus or membrane-bounded organelles.

Promethean environmentalism   A form of technological optimism that predicts that human ingenuity and enterprise will find cures for all our problems.

promoters   Agents that are not carcinogenic but that assist in the progression and spread of tumors; sometimes called cocarcinogens.

pronatalist pressures   Influences that encourage people to have children.

proteins   Chains of amino acids linked by peptide bonds.

proton   A positively-charged subatomic particle found in the nucleus of an atom.

public trust   A doctrine obligating the government to maintain public lands in a natural state as guardians of the public interest.

pull factors (in urbanization)   Conditions that draw people from the country into the city.

push factors (in urbanization)   Conditions that force people out of the country and into the city.

R

radioactive   An unstable isotope that decays spontaneously and releases subatomic particles or units of energy.

radioactive decay   A change in the nuclei of radioactive isotopes that spontaneously emit high-energy electromagnetic radiation and/or subatomic particles while gradually changing into another isotope or different element.

radioactive isotope   An unstable form of an element that spontaneously emits either high- energy electromagnetic radiation or subatomic particles (or both). The decay rate describes the time necessary for a radioactive isotope to change to a stable element.

radionucleides   Isotopes that exhibit radioactive decay.

rain forest   A forest with high humidity, constant temperature, and abundant rainfall (generally over 380 cm [150 in] per year); can be tropical or temperate.

rain shadow   Dry area on the downwind side of a mountain.

rangeland   Grasslands and open woodlands suitable for livestock grazing.

raprenox   A process to reduce nitrogen oxide emissions by mixing combustion gases with isocyanic acid, which converts nitrogen oxides to dimolecular nitrogen gas.

recharge zone   Area where water infiltrates into an aquifer.

recycling   Reprocessing of discarded materials into new, useful products; not the same as reuse of materials for their original purpose, but the terms are often used interchangeably.

red tide   A population explosion or bloom of minute, single-celled marine organisms called dinoflagellates. Billions of these cells can accumulate in protected bays where the toxins they contain can poison other marine life.

reduced tillage systems   Systems, such as minimum till, conserve-till, and no-till, that preserve soil, save energy and water, and increase crop yields.

refuse-derived fuel   Processing of solid waste to remove metal, glass, and other unburnable materials; organic residue is shredded, formed into pellets, and dried to make fuel for power plants.

regenerative farming   Farming techniques and land stewardship that restore the health and productivity of the soil by rotating crops, planting ground cover, protecting the surface with crop residue, and reducing synthetic chemical inputs and mechanical compaction.

regulations   Rules established by administrative agencies; regulations can be more important than statutory law in the day-to-day management of resources.

rehabilitate land   A utilitarian program to make an area useful to humans.

relative humidity   At any given temperature, a comparison of the actual water content of the air with the amount of water that could be held at saturation.

relativists   Those who believe moral principles are always dependent on the particular situation.

renewable resources   Resources normally replaced or replenished by natural processes; resources not depleted by moderate use; examples include solar energy, biological resources such as forests and fisheries, biological organisms, and some biogeochemical cycles.

residence time The length of time a component, such as an individual water molecule, spends in a particular compartment or location before it moves on through a particular process or cycle.

resilience The ability of a community or ecosystem to recover from disturbances.

resistance (inertia) The ability of a community to resist being changed by potentially disruptive events.

resource partitioning In a biological community, various populations sharing environmental resources through specialization, thereby reducing direct competition. *See also* ecological niche.

resource scarcity A shortage or deficit in some resource.

respiratory fibrotic agents Special class of irritants, including chemical reagents and particulate materials, that damages the lungs, causing scar tissue formation that lowers respiratory capacity.

restoration ecology Seeks to repair or reconstruct ecosystems damaged by human actions.

restore ecosystems To reinstate an entire community of organisms to naturally occurring association.

rill erosion The removing of thin layers of soil as little rivulets of running water gather and cut small channels in the soil.

risk Probability that something undesirable will happen as a consequence of exposure to a hazard.

risk assessment Evaluation of the short-term and long-term risks associated with a particular activity or hazard; usually compared to benefits in a cost-benefit analysis.

RNA Ribonucleic acid; nucleic acid used for transcription and translation of the genetic code found on DNA molecules.

rock cycle The process whereby rocks are broken down by chemical and physical forces; sediments are moved by wind, water, and gravity, sedimented and reformed into rock, and then crushed, folded, melted, and recrystallized into new forms.

ruminant animals Cud-chewing animals, such as cattle, sheep, goats, and buffalo, with multichambered stomachs in which cellulose is digested with the aid of bacteria.

runoff The excess of precipitation over evaporation; the main source of surface water and, in broad terms, the water available for human use.

run-of-the-river flow Ordinary river flow not accelerated by dams, flumes, etc. Some small, modern, high-efficiency turbines can generate useful power with run-of-the-river flow or with a current of only a few kilometers per hour.

rural area An area in which most residents depend on agriculture or the harvesting of natural resources for their livelihood.

S

S curve A curve that depicts logistic growth; called an S curve because of its shape.

salinity Amount of dissolved salts (especially sodium chloride) in a given volume of water.

salinization A process in which mineral salts accumulate in the soil, killing plants; occurs when soils in dry climates are irrigated profusely.

saltwater intrusion Movement of saltwater into freshwater aquifers in coastal areas where groundwater is withdrawn faster than it is replenished.

saturation point The maximum concentration of water vapor the air can hold at a given temperature.

scavenger An organism that feeds on the dead bodies of other organisms.

science A methodical, precise, objective way to study the natural world.

scientific method A systematic, precise, objective study of a problem. Generally this requires observation, hypothesis development and testing, data gathering, and interpretation.

secondary pollutants Chemicals modified to a hazardous form after entering the air or that are formed by chemical reactions as components of the air mix and interact.

secondary recovery technique Pumping pressurized gas, steam, or chemical- containing water into a well to squeeze more oil from a reservoir.

secondary standards Regulations of the 1970 Clean Air Act intended to protect materials, crops, visibility, climate, and personal comfort.

secondary succession Succession on a site where an existing community has been disrupted.

secondary treatment Bacterial decomposition of suspended particulates and dissolved organic compounds that remain after primary sewage treatment.

second law of thermodynamics States that, with each successive energy transfer or transformation in a system, less energy is available to do work.

Second World The industrialized, socialist, centrally planned economy nations of Eastern Europe and the former Soviet Union and its allies.

secure landfill A solid waste disposal site lined and capped with an impermeable barrier to prevent leakage or leaching. Drain tiles, sampling wells, and vent systems provide monitoring and pollution control.

sedimentary rock Deposited material that remains in place long enough or is covered with enough material to compact into stone; examples include shale, sandstone, breccia, and conglomerates.

sedimentation The deposition of organic materials or minerals by chemical, physical, or biological processes. Sediments can be transported from their source to their place of deposition by gravity, wind, water, or ice. If subjected to sufficient heat, pressure, or chemical reactions, sediments can solidify into sedimentary rock.

selective cutting Harvesting only mature trees of certain species and size; usually more expensive than clear-cutting, but it is less disruptive for wildlife and often better for forest regeneration.

seriously undernourished Those who receive less than 80 percent of their minimum daily caloric requirements; are likely to suffer permanently stunted growth, mental retardation, and other social and developmental disorders.

shallow ecology A critical term applied to superficial environmentalists who claim to be green but are quick to compromise and who do little to bring about fundamental change. Many of those accused of being shallow ecologists counter that they are merely pragmatic, progressive reformers who prefer to work within the established social contract rather than try to overthrow it.

shantytowns Settlements created when people move onto undeveloped lands and build their own shelter with cheap or discarded materials; some are simply illegal subdivisions where a landowner rents land without city approval; others are land invasions.

sheet erosion Peeling off thin layers of soil from the land surface; accomplished primarily by wind and water.

sinkholes A large surface crater caused by the collapse of an underground channel or cavern; often triggered by groundwater withdrawal.

sludge Semisolid mixture of organic and inorganic materials that settles out of wastewater at a sewage treatment plant.

slums Legal but inadequate multifamily tenements or rooming houses; some are custom built for rent to poor people, others are converted from some other use.

smog The term used to describe the combination of smoke and fog in the stagnant air of London; now often applied to photochemical pollution products or urban air pollution of any kind.

social ecology A socialist/humanist philosophy based on the communitarian anarchism of the Russian geographer Peter Kropotkin. It shares much with deep ecology except that it is more humanist in its outlook.

social justice Equitable access to resources and the benefits derived from them; a system that recognizes inalienable rights and adheres to what is fair, honest, and moral.

soil A complex mixture of weathered mineral materials from rocks, partially decomposed organic molecules, and a host of living organisms.

soil horizons Horizontal layers that reveal a soil's history, characteristics, and usefulness.

southern pine forest United States coniferous forest ecosystem characterized by a warm, moist climate.

species A population of morphologically similar organisms that can reproduce sexually among themselves but that cannot produce fertile offspring when mated with other organisms.

species diversity   The number and relative abundance of species present in a community.

species recovery plan   A plan for restoration of an endangered species through protection, habitat management, captive breeding, disease control, or other techniques that increase populations and encourage survival.

spring overturn   Springtime lake phenomenon that occurs when the surface ice melts and the surface water temperature warms to its greatest density at 4° C and then sinks, creating a convection current that displaces nutrient-rich bottom waters.

squatter towns   Shantytowns that occupy land without owner's permission; some are highly organized movements in defiance of authorities; others grow gradually.

stability   In ecological terms, a dynamic equilibrium among the physical and biological factors in an ecosystem or a community; relative homeostasis.

stable runoff   The fraction of water available year-round; usually more important than total runoff when determining human uses.

Standard Metropolitan Statistical Area (SMSA)   An urbanized region with at least 100,000 inhabitants with strong economic and social ties to a central city of at least 50,000 people.

statutory law   Rules passed by a state or national legislature.

steady-state economy   Characterized by low birth and death rates, use of renewable energy sources, recycling of materials, and emphasis on durability, efficiency, and stability.

stewardship   A philosophy that holds that humans have a unique responsibility to manage, care for, and improve nature.

stomates   The small openings in leaves, herbaceous stems, and fruits through which gases and water vapor pass.

strategic minerals   Materials a country cannot produce itself but that it uses for essential materials or processes.

stratosphere   The zone in the atmosphere extending from the tropopause to about 50 km (30 mi) above the earth's surface; temperatures are stable or rise slightly with altitude; has very little water vapor but is rich in ozone.

streambank erosion   Washing away of soil from banks of established streams, creeks, or rivers, often as a result of the removal of trees and brush along streambanks or cattle damage to the banks.

stress   Physical, chemical, or emotional factors that place a strain on an animal. Plants also experience physiological stress under adverse environmental conditions.

stress-related disease   *See* stress shock.

stress shock   A loose set of physical, psychological, and/or behavioral changes thought to result from the stress of excess competition and extreme closeness to other members of the same species.

strip-farming   Planting different kinds of crops in alternating strips along land contours; when one crop is harvested, the other crop remains to protect the soil and prevent water from running straight down a hill.

strip-mining   Removing surface layers over coal seams using giant, earth-moving equipment; creates a huge open-pit from which coal is scooped by enormous surface-operated machines and transported by trucks; an alternative to deep mines.

structure (in ecological terms)   Patterns of organization, both spatial and functional, in a community.

sublimation   The process by which water can move between solid and gaseous states without ever becoming liquid.

subsidence   A settling of the ground surface caused by the collapse of porous formations that result from withdrawal of large amounts of groundwater, oil, or other underground materials.

subsoil   A layer of soil beneath the topsoil that has lower organic content and higher concentrations of fine mineral particles; often contains soluble compounds and clay particles carried down by percolating water.

sulfur dioxide   A colorless, corrosive gas directly damaging to both plants and animals.

supply   The quantity of that product being offered for sale at various prices, other things being equal.

surface mining   Some minerals are also mined from surface pits. *See* strip-mining.

surface tension   A condition in which the water surface meets the air and acts like an elastic skin.

survivorship   The percentage of a population reaching a given age or the proportion of the maximum life span of the species reached by any individual.

sustainable agriculture   An ecologically sound, economically viable, socially just, and humane agricultural system. Stewardship, soil conservation, and integrated pest management are essential for sustainability.

sustainable development   A real increase in well-being and standard of life for the average person that can be maintained over the long-term without degrading the environment or compromising the ability of future generations to meet their own needs.

sustained yield   Utilization of a renewable resource at a rate that does not impair or damage its ability to be fully renewed on a long-term basis.

swamp   Wetland with trees, such as the extensive swamp forests of the southern United States.

swidden agriculture   *See* milpa agriculture.

symbiosis   The intimate living together of members of two different species; includes **mutualism, commensalism,** and, in some classifications, **parasitism.**

synergistic effects   When an injury caused by exposure to two environmental factors together is greater than the sum of exposure to each factor individually.

systemic   A condition or process that affects the whole body; many metabolic poisons are systemic.

T

taiga   The northernmost edge of the boreal forest, including species-poor woodland and peat deposits; intergrading with the arctic tundra.

tailings   Mining waste left after mechanical or chemical separation of minerals from crushed ore.

taking   Unconstitutional confiscation of private property.

technological optimists   Those who believe that technology and human enterprise will find cures for all our problems. Also called Promethean environmentalism.

technopolis   Also called a vertical city; this model of city development proposes that cities grow vertically instead of horizontally.

tectonic plates   Huge blocks of the earth's crust that slide around slowly, pulling apart to open new ocean basins or crashing ponderously into each other to create new, larger landmasses.

temperate rainforest   The cool, dense, rainy forest of the northern Pacific coast; enshrouded in fog much of the time; dominated by large conifers.

temperature   A measure of the speed of motion of a typical atom or molecule in a substance.

teratogens   Chemicals or other factors that specifically cause abnormalities during embryonic growth and development.

terracing   Shaping the land to create level shelves of earth to hold water and soil; requires extensive hand labor or expensive machinery, but it enables farmers to farm very steep hillsides.

territoriality   An intense form of intraspecific competition in which organisms define an area surrounding their home site or nesting site and defend it, primarily against other members of their own species.

tertiary treatment   The removal of inorganic minerals and plant nutrients after primary and secondary treatment of sewage.

thermal plume   A plume of hot water discharged into a stream or lake by a heat source, such as a power plant.

thermocline   In water, a distinctive temperature transition zone that separates an upper layer that is mixed by the wind (the epilimnion) and a colder, deep layer that is not mixed (the hypolimnion).

thermodynamics   A branch of physics that deals with transfers and conversions of energy.

thermodynamics, first law   Energy can be transformed and transferred, but cannot be destroyed or created.

thermodynamics, second law   With each successive energy transfer or transformation, less energy is available to do work.

thermoplastics   Soft plastics composed of single-chain, unlinked polymers, such as polyethylene, polypropylene, polyvinylchloride, polystyrene, and polyester, that can be remelted and reformed to make useful products.

thermoset polymers Hard plastics composed of cross-linked molecular networks, such as acrylic, phenolic, or epoxy resins, that cannot be remelted or recycled.

thermosphere The highest atmospheric zone; a region of hot, dilute gases above the mesosphere extending out to about 1600 km (1000 mi) from the earth's surface.

Third World Less-developed countries that are not capitalistic and industrialized (First World) or centrally-planned socialist economies (Second World); not intended to be derogatory.

threatened species While still abundant in parts of its territorial range, this species has declined significantly in total numbers and may be on the verge of extinction in certain regions or localities.

tidal station A dam built across a narrow bay or estuary traps tide water flowing both in and out of the bay. Water flowing through the dam spins turbines attached to electric generators.

tied ridges Series of ridges running at right angles to each other so that runoff is blocked in all directions and water is allowed to soak into the soil.

timberline In mountains, the highest-altitude edge of forest that marks the beginning of the treeless alpine tundra.

tolerance limits *See* limiting factors.

tool-making revolution The invention of tools—probably first of stone, wood, or bone— sometime in the early Paleolithic period. Knowledge of tool-making and tool-using technologies may have predated modern humans, as evidenced by tool utilization by other primates.

topsoil The first true layer of soil; layer in which organic material is mixed with mineral particles; thickness ranges from a meter or more under virgin prairie to zero in some deserts.

tornado A violent storm characterized by strong swirling winds and updrafts; tornadoes form when a strong cold front pushes under a warm, moist air mass over the land.

total fertility rate The number of children born to an average woman in a population during her entire reproductive life.

total growth rate The net rate of population growth resulting from births, deaths, immigration, and emigration.

toxic colonialism Shipping toxic wastes to a weaker or poorer nation.

toxins Poisonous chemicals that react with specific cellular components to kill cells or to alter growth or development in undesirable ways; often harmful, even in dilute concentrations.

tradable permits Pollution quotas or variances that can be bought or sold. If you have a permit to emit a certain amount of some pollutant, I might buy a portion of that amount from you rather than put pollution controls on my equipment. You, in turn, would have to reduce your emissions accordingly.

tragedy of the commons An inexorable process of degradation of communal resources due to selfish self-interest of "free riders" who use or destroy more than their fair share of common property. *See* open access systems.

transitional zone A zone in which populations from two or more adjacent communities meet and overlap.

transpiration The evaporation of water from plant surfaces, especially through stomates.

trauma Injury caused by accident or violence.

trombe wall An interior, heat-absorbing wall; may be water-filled glass tubes that absorb heat rays and let light into interior rooms.

trophic level Step in the movement of energy through an ecosystem; an organism's feeding status in an ecosystem.

tropical rainforests Forests in which rainfall is abundant—more than 200 cm (80 in) per year—and temperatures are warm to hot year-round.

tropical seasonal forest Semievergreen or partly deciduous forests tending toward open woodlands and grassy savannas dotted with scattered, drought-resistant tree species; distinct wet and dry seasons, hot year-round.

tropopause The boundary between the troposphere and the stratosphere.

troposphere The layer of air nearest to the earth's surface; both temperature and pressure usually decrease with increasing altitude.

tsunami Giant seismic sea swells that move rapidly from the center of an earthquake; they can be 10 to 20 meters high when they reach shorelines hundreds or even thousands of kilometers from their source.

tundra Treeless arctic or alpine biome characterized by cold, harsh winters, a short growing season, and potential for frost any month of the year; vegetation includes low-growing perennial plants, mosses, and lichens.

U

unconventional air pollutants Toxic or hazardous substances, such as asbestos, benzene, beryllium, mercury, polychlorinated biphenyls, and vinyl chloride, not listed in the original Clean Air Act because they were not released in large quantities; also called noncriteria pollutants.

unconventional oil Resources such as shale oil and tar sands that can be liquefied and used like oil.

undernourished Those who receive less than 90 percent of the minimum dietary intake over a long-term time period; they lack energy for an active, productive life and are more susceptible to infectious diseases.

undiscovered resources Potential supplies of a mineral or other useful material believed to exist based on history, scientific theory, or general knowledge of geology, biology, or geography of an unexplored area.

universalists Those who believe that some fundamental ethical principles are universal and unchanging. In this vision, these principles are valid regardless of the context or situation.

upwelling Convection currents within a body of water that carry nutrients from bottom sediments toward the surface.

urban area An area in which a majority of the people are not directly dependent on natural resource-based occupations.

urbanization An increasing concentration of the population in cities and a transformation of land use to an urban pattern of organization.

urban renewal Programs to revitalize old, blighted sections of inner cities.

utilitarian conservation A philosophy that resources should be used for the greatest good for the greatest number for the longest time.

utilitarianism *See* utilitarian conservation.

V

vertical city *See* technopolis.

vertical stratification The vertical distribution of specific subcommunities within a community.

village A collection of rural households linked by culture, custom, and association with the land.

visible light A portion of the electromagnetic spectrum that includes the wavelengths used for photosynthesis.

vitamins Organic molecules essential for life that we cannot make for ourselves; we must get them from our diet; they act as enzyme cofactors.

volatile organic compounds Organic chemicals that evaporate readily and exist as gases in the air.

vulnerable species Naturally rare organisms or species whose numbers have been so reduced by human activities that they are susceptible to actions that could push them into threatened or endangered status.

W

warm front A long, wedge-shaped boundary caused when a warmer advancing air mass slides over neighboring cooler air parcels.

waste stream The steady flow of varied wastes, from domestic garbage and yard wastes to industrial, commercial, and construction refuse.

water cycle The recycling and reutilization of water on Earth, including atmospheric, surface, and underground phases and biological and nonbiological components.

water droplet coalescence A mechanism of condensation that occurs in clouds too warm for ice crystal formation.

waterlogging Water saturation of soil that fills all air spaces and causes plant roots to die from lack of oxygen; a result of overirrigation.

water table The top layer of the zone of saturation; undulates according to the surface topography and subsurface structure.

weather Description of the physical conditions of the atmosphere (moisture, temperature, pressure, and wind).

weathering Changes in rocks brought about by exposure to air, water, changing temperatures, and reactive chemical agents.

wetlands Ecosystems of several types in which rooted vegetation is surrounded by standing water during part of the year. *See also* swamp, marsh, bog, fen.

wilderness An area of undeveloped land affected primarily by the forces of nature; an area where humans are visitors who do not remain.

Wilderness Act Legislation of 1964 recognizing that leaving land in its natural state may be the highest and best use of some areas.

wildlife Plants, animals, and microbes that live independently of humans; plants, animals, and microbes that are not domesticated.

wildlife refuges Areas set aside to shelter, feed, and protect wildlife; due to political and economic pressures, refuges often allow hunting, trapping, mineral exploitation, and other activities that threaten wildlife.

windbreak Rows of trees or shrubs planted to block wind flow, reduce soil erosion, and protect sensitive crops from high winds.

wind farms Large numbers of windmills concentrated in a single area; usually owned by a utility or large-scale energy producer.

Wise Use Movement A coalition of ranchers, loggers, miners, industrialists, hunters, off-road vehicle users, land developers, and others who call for unrestricted access to natural resources and public lands.

withdrawal A description of the total amount of water taken from a lake, river, or aquifer.

woodland A forest where tree crowns cover less than 20 percent of the ground; also called open canopy.

work The application of force through a distance; requires energy input.

world conservation strategy A system of maintaining essential ecological processes, preserving genetic diversity, and ensuring that utilization of species and ecosystems is sustainable.

World Ocean The interconnected world seas and oceans.

X ray Very short wavelength in the electromagnetic spectrum; can penetrate soft tissue; although it is useful in medical diagnosis, it also damages tissue and causes mutations.

Y

yellowcake The concentrate of 70 to 90 percent uranium oxide extracted from crushed ore.

Z

zero population growth (ZPG) The number of births at which people are just replacing themselves; also called the replacement level of fertility.

zone of aeration Upper soil layers that hold both air and water.

zone of leaching The layer of soil just beneath the topsoil where water percolates, removing soluble nutrients that accumulate in the subsoil; may be very different in appearance and composition from the layers above and below it.

zone of saturation Lower soil layers where all spaces are filled with water.

Credits

Photographs

Atlas Map:

Map from the *Goode's World Atlas* © 1996 by Rand McNally, R.L. 96-S-31; Scale icon © Times Mirror Higher Education Group, Inc./Bob Coyle, photographer

Part Openers

1: © Bob Clemenz Photography; 2: © Kevin Morris/Tony Stone Images; 3: © Carl Frank/Photo Researchers, Inc.; 4: © Peter May/Peter Arnold, Inc.; 5: © Jeremy Horner/Tony Stone Images

Profiles:

1: Ann Wendland; 2: Courtesy of Micaela Martinez; 3: Courtesy of Marian Taylor; 4: Courtesy of Jim Bartlett; 5: Courtesy of Jeremy Sterk; 6: Courtesy of Lee Lebbon; 7: Courtesy of Tracy Williams; 8: Courtesy of Lisa Vernegaard; 9: Courtesy of Sarah Lloyd; 10: Courtesy of Andrew Graham

Chapter 1

Opener: © John Cancalosi/Peter Arnold, Inc.; 1.1: © Michael Fogden/Animals Animals/Earth Scenes; 1.3: © Kurgan-Lisnet/Gamma Liaison; 1.4: © Bettmann Archive; 1.5: Photo by David Swanlund, Courtesy of Save-The-Redwoods League; 1.6: AP/Wide World Photos; page 17: © John Chiasson/Gamma Liaison International; 1.7: © Earth Imaging/Tony Stone Images; 1.8: © Craig Tuttle/Stock Market; 1.9: © Norbert Schiller/The Image Works; 1.12 © Gary Milburn/Tom Stack & Associates; 1.13 © Kit King/Gamma Liaison; 1.18: W.P. Cunningham; 1.19: © Tony Stone Images; 1.20: Photo by N.R. Farbman, Life Magazine © Time Warner; page 21: © Laurie Rubin/Tony Stone Images; 1.22: *The Slaughter of the Jacques on the Bridge at Meaux*. From Froissart's Chronicles. Louis de Bruges copy, c. 1460. Courtesy of Bibliotheque Nationale

Chapter 2

Opener: © Nicholas DeVore/Tony Stone Images; 2.1: © Hollyman/Gamma Liaison; 2.2: Courtesy of Vatican Museums; 2.3: © B. Hemphill/Photo Researchers, Inc.; 2.4: © Grant Heilman; 2.5: © Carr Clifton; 2.6: © Inga Spence/Tom Stack & Associates; 2.7: W.P. Cunningham; 2.8: *Nataraja: Siva as King of Dance*. Bronze, h. 111.5 cm. South India, Chola

Period, 11th c. © The Cleveland Museum of Art, 1996 Purchase from the J.H. Wade Fund, 1930.331; 2.9: © Sam Kittner; 2.10: W.P. Cunningham; 2.11: © Sam Kittner/Greenpeace Photo; 2.13, 2.14: W.P. Cunningham; 2.16: The Bettmann Archive; 2.17: © Martha Cooper/Peter Arnold, Inc.; 2.19: © Shahn Kermani/Gamma Liaison

Chapter 3

Opener: © Tom Tietz/Tony Stone Images; 3.1: © Thomas Kitchin/Tom Stack & Associates; 3.4: © E. H. Newcomb & W.P. Wergin, University of Wisconsin/Biological Photo Service; 3.5: © Carl Purcell/Photo Researchers, Inc.; page 61: © George J. Bernard/Animals Animals/Earth Scenes; 3.17: © David Dennis/Tom Stack & Associates

Chapter 4

Opener: © Larry Ulrich/Tony Stone Images; 4.1: © Leonard L. Rue/Visuals Unlimited; 4.3: © S. J. Krasemann/Photo Researchers, Inc.; 4.4 a,b: © Breck Kent/Animals Animals/Earth Scenes; 4.6: © Robert C. Simpson/Valan Photos; 4.9: © Ray Coleman/Photo Researchers, Inc.; 4.10: © D.P. Wilson/Photo Researchers, Inc,; page 76: © John Gerlach/Tony Stone Images; 4.11: © Randy Morse/Tom Stack & Associates; 4.12: © Billy Beatty/Visuals Unlimited; page 79: © Milton Tierney/Visuals Unlimited; 4.13, 4.14: W.P. Cunningham; 4.15: © Brian Parker/Tom Stack & Associates; 4.16 a,b: Dr. Paul A. Opler; 4.17 a,b: © Edward S. Ross; 4.21: © Fred Bavendam/Peter Arnold, Inc.; 4.22: W.P. Cunningham; 4.25: Dr. Carl W. Bollwinkel; 4.26: W.P. Cunningham; 4.27: © Gerard Lacz/Peter Arnold, Inc.

Chapter 5

Opener: © L.W. Richardson/Photo Researchers, Inc.; 5.1: Courtesy University of Wisconsin - Madison Archives; 5.4: W.P. Cunningham; 5.5: © Joe McDonald/Visuals Unlimited; 5.6, 5.7, 5.8, 5.9, 5.10: W.P. Cunningham; 5.13: © John D. Cunningham/Visuals Unlimited; 5.14: © Andrew Martinez/Photo Researchers, Inc; 5.16: © Mike Bacon/Tom Stack & Associates; 5.19: Courtesy J. Morgan Grove, School of Forestry Environmental Sciences, Yale University; 5.20: © Larry Kolvoord/The Image Works; 5.21: Courtesy South Florida Water Management District; 5.22:

Courtesy University of Wisconsin - Madison Archives; 5.23: Courtesy University of Wisconsin Arboretum; 5.24: © State of Minnesota, Department of Natural Resources. Reprinted with Permission; 5.25: © Peter Menzel Photography; 5.26: © R.J. Goldstein/Visuals Unlimited

Chapter 6

Opener: © Mark Peterson/Tony Stone Images; 6.1: W. P. Cunningham; 6.3: © Tom McHugh/Photo Researchers, Inc.; 6.5: Courtesy of National Museum of Natural History, Smithsonian Institution

Chapter 7

Opener: © Louisa Preston/Photo Researchers, Inc.; 7.1: John W. Cunningham; page 144: © Forrest Anderson/Gamma Liaison; 7.18 a,b,d,f: © Times Mirror Higher Education Group, Inc./Bob Coyle, photographer; 7.18e: Courtesy of Ortho-McNeil

Chapter 8

Opener: © Glen Allison/Tony Stone Images; 8.1: W.P. Cunningham; 8.3: © Lowell Georgia/Photo Researchers, Inc.; 8.4: © Audrey Lang/Valan Photos; 8.5: W.P. Cunningham; page 169: © Frank J. Miller/Photo Researchers, Inc.; page 173: © Zed Nelson/Panos Pictures

Chapter 9

Opener: © David J. Cross/Peter Arnold, Inc.; 9.1: © Raghu Rai/Magnum Photos; 9.2: Courtesy of Stanley Erlandsen, University of Minnesota; 9.3: WHO/World Bank/R. Witlin; 9.4: Courtesy Donald R. Hopkins; 9.5: © John Maier, Jr./JB Pictures; page 189: © John Barr/Gamma Liaison

Chapter 10

Opener: © Jim Nilsen/Tom Stack & Associates; 10.1: © Audrey Topping Cohen/Photo Researchers, Inc.; 10.2: © Allen Tannenbaum/Sygma; 10.3: © Scott Daniel Peterson/Gamma Liaison; 10.4: © Paul Souders Photography; 10.5: © Lester Bergman/Corbis Media; page 213: © Pete Saloutos/Tony Stone Images; 10.11: W.P. Cunningham; 10.12: Courtesy Agricultural Research Service, USDA; 10.14: W.P. Cunningham; 10.15: © Link/Visuals Unlimited; 10.17: © Victor Englebert/Photo Researchers, Inc.; 10.18: © Norbert Schiller/The Image Works; 10.19: © Thierry A. Brun

Chapter 11

Opener: © Tony Craddock/Tony Stone Images; 11.1: Courtesy Artur Rothstein/Library of Congress; 11.2 © Arthur Rothstein/National Geographic; 11.8: © Jacques Jangoux/Tony Stone Images; 11.13: Courtesy Soil Conservation Service, USDA; 11.14: © Larry Miller/Photo Researchers Inc.; 11.16: © Dennis Budd Gray/Stock Boston; page 242: © Mike Maloney/*San Francisco Chronicle*; 11.19: © Joe Munroe/Photo Researchers, Inc.; 11.20: W.P. Cunningham; 11.21: Courtesy Dave Hanson, College of Agriculture, University of Minnesota; 11.2: © Tom Sweeny/*Minneapolis Star Tribune*

Chapter 12

Opener: © Inga Spence/Tom Stack & Associates; 12.1: © Tom J. Ulrich/Visuals Unlimited; 12.2: © Rod Planck/Photo Researchers, Inc.; 12.3: © Steve Wilson/Entheos; 12.4: UPI Bettmann; 12.6: © Yoram Lehmann/Peter Arnold, Inc.; 12.7: © Stephen Dalton/Photo Researchers, Inc.; 12.8: © SPL/Photo Researchers, Inc.; 12.9: © Joe Munroe/Photo Researchers, Inc.; page 257: Dan Winters © 1993 The Walt Disney Co. Reprinted with permission of *Discover Magazine*; 12.12: © Norm Thomas/Photo Researchers, Inc.; 12.13: AP/Wide World; 12.14: © Brian Parker/Tom Stack & Associates; page 260: © Howard Suzuki, Aquatic Life Sculptures/*Discover Magazine*; 12.16: © Cary Wolinsky/Stock Boston; 12.17: Courtesy Florida Department of Agriculture and Consumer Service's 12.18: Photo by the late J.K. Holloway, from *Biological Control*, C.B. Huffaker, ed. © Plenum Press, NY, 1971; 12.20: © W.C. Wood/Tanimura & Antle, Inc.; 12.23: W.P. Cunningham

Chapter 13

Opener: © Tim Davis/Tony Stone Images; 13.1: © Patrice Ceisel/Visuals Unlimited; 13.2: © Tim Davis/Photo Researchers, Inc.; 13.4: W.P. Cunningham; 13.6: Courtesy of INBio, San Jose, Costa Rica; 13.7: © Paul Chesley/Tony Stone Image; 13.10: © Bill Beatty/Visuals Unlimited; 13.13: © John Cancalosi/Tom Stack & Associates; page 280 © Mike Bacon/Tom Stack & Associates; 13.14: © Patrick Forestier/Gamma Liaison; 13.15: © Lynn Funkhouser/Peter Arnold, Inc.; 13.16: Courtesy U.S. Department of Agriculture Animal & Plant Health Inspection Service; 13.18: © Tom McHugh/Photo Researchers, Inc.; 13.19: © T. Leeson/Photo Researchers, Inc.; 13.24: W.P. Cunningham; 13.25: © Tom Myers Photography; 13.26: Courtesy of Dr. Ronald Tilson, Minnesota Zoo

Chapter 14

Opener: © Larry Ulrich/Tony Stone Images; 14.1: W.P. Cunningham; 14.5: Courtesy Daniel Kammen, Princeton University; 14.7: © G. Prance/Visuals Unlimited; 14.9: W.P. Cunningham; 14.10: © Chip & Jull Isenhart/Tom Stack & Associates; 14.11: Courtesy Andy White, College of Agriculture, University of Minnesota; page 301: Courtesy Dan Janzen; 14.12, 14.13: © Greg Vaughn/Tom Stack & Associates; 14.15: © Gary Braasch/Tony Stone Images; 14.16: © Earl Dotter; 14.17: © *Seattle Times*/Gamma Liaison; 14.18: © John W. Bova/Photo Researchers, Inc.; 14.21: © Eastcott/Momatiuk/The Image Works; 14.23: © William and Marcy Levy/Photo Researchers, Inc.; 14.24: © Francois Gohier/Photo Researchers, Inc.; 14.25: © Jay Pasachoff/Visuals Unlimited; 14.26: AP/Wide World Photos; 14.27: © G. Prance/Visuals Unlimited

Chapter 15

Opener: © Greg Probst/Tony Stone Images; 15.1: © Galen Rowell, 1987, Mountain Light Photography; 15.2: © A. Louis Goldman/Photo Researchers, Inc.; 15.3: W.P. Cunningham; 15.4: © John Gerlach/Visuals Unlimited; 15.5: © Lee Battaglia, 1973/Photo Researchers, Inc.; 15.6: American Heritage Center, University of Wyoming; 15.7: © Gary Milburn/Tom Stack & Associates; page 323: Courtesy Michigan State University Com Tech Lab, Perciles Gomes Photographer; 15.10: © Carl and Ann Purcell/Words & Pictures; 15.13: Courtesy R.O. Bierregaard; 15.15: W. P. Cunningham; page 331: Courtesy Robert O. Bierregaard; 15.16: © Francois Gohier/Photo Researchers, Inc.; 15.17: © Jeff Foott Productions; 15.18: Steve Raymer © National Geographic Society; 15.19: © Fred Whitehead/Animals Animals/Earth Scenes; 15.20: © Chris Harris/Gamma Liaison; 15.21: © James L. Shaffer; 15.22: © Times Mirror Higher Education Group, Inc./Bob Coyle, photographer; 15.23: © Stephen Rose/Gamma Liaison

Chapter 16

Opener: © Times Mirror Higher Education Group, Inc./Doug Sherman, photographer; 16.1: © Reuter/Sankei Shimbun; 16.7: Courtesy E.B. Hardin, U.S. Geological Survey; page 350: © Bryan F. Peterson; 16.10: W.P. Cunningham; 16.11: Courtesy of the Library of Congress; 16.13: © Joseph Netts Photo Researchers, Inc.; 16.15: © Herman J. Kokojah/Black Star; 16.16: Courtesy K. Jackson/USAF; 16.17: © Al Seib/*LA Times*

Chapter 17

Opener: © Wm.L. Wantland/Tom Stack & Associates; 17.1.: © John J. Lopinot/*Palm Beach Post*; 17.2: W.P. Cunningham; 17.9: © Science VU/Visuals Unlimited; 17.11: © Victor Englebert/Photo Researchers, Inc.; 17.14: © Peter Gilday

Chapter 18

Opener: © Martin Rogers/Tony Stone Images; 18.1a © Leonard L. T. Rhodes/Animals Animals/Earth Scenes; 18.1b: © Joseph Sohm/Photo Researchers, Inc.; 18.8 © 1988 Ted Spiegel/Black Star; 18.11: Courtesy NOAA; 18.14, 18.15: W.P. Cunningham; 18.17: © Simon Fraser/SPL/Photo Researchers, Inc.; 18.18: © John D. Cunningham/Visuals Unlimited; 18.23: © Klaus Reisinger/Black Star

Chapter 19

Opener: © Times Mirror Higher Education Group, Inc./Doug Sherman, photographer; 19.1: © Ray Ellis/Photo Researchers, Inc.; 19.2, 19.6, 19.7, 19.12, page 422: W.P. Cunningham; 19.13: © Penny Tweedie/Tony Stone Images; 19.15: © M. Timothy O'Keefe/Tom Stack & Associates; 19.17: Courtesy Los Angeles Department of Water & Power; 19.18: © John Gerlach/Visuals Unlimited; page 428: © Sylvan H. Wittwer/Visuals Unlimited; 19.19: Courtesy Tim McCabe, Soil Conservation Service, USDA; 19.20: Whitewater Photography; 19.23: © Sylvan H. Wittwer/Visuals Unlimited

Chapter 20

Opener: © Tom Stack/Tom Stack & Associates; 20.1: © James Nachtwey/Magnum Photos; 20.2: © Simon Fraser/Photo Researchers, Inc.; 20.3: © Joe McDonald/Animals Animals/Earth Scenes; 20.5: © Roger A. Clark, Jr./Photo Researchers, Inc.; 20.7: © W.A. Banaszewski/Visuals Unlimited; 20.8: © The Heirs of W. Eugene Smith/Black Star; 20.9: © Rob & Melissa Simpson/Valan Photos; 20.10: © Lawrence Lowry/Photo Researchers, Inc.; page 445: © Hank Andrews/Visuals Unlimited; 20.13: © Les Stone/Sygma; 20.16: © Ken Sakamoto/Black Star; 20.17: © Frans Lanting/Photo Researchers, Inc.; 20.20: © Times Mirror Higher Education Group, Inc./Bob Coyle, photographer

Chapter 21

Opener: © Noel Quidu/Gamma Liaison; 21.1: © W. Ormerod/Visuals Unlimited; 21.2: © Owen Franken/Stock Boston; 21.8: © Charles E. Rotkin/Corbis Media; 21.11: Courtesy American Petroleum Institute; page 472: Mary Ann Cunningham; 21.13: © Jerry Irwin/Photo Researchers, Inc.; page 478: © TASS/Sovfoto/Eastfoto; 21.23: © Arthur Grace/Sygma

Chapter 22

Opener: © Times Mirror Higher Education Group, Inc./Doug Sherman, photographer; 22.1: Courtesy Paul Hoeke, Province North-Holland; 22.3: Courtesy Volvo; 22.9: © David Wells/The Image Works; 22.10, 22.13: W.P. Cunningham; 22.14: © G. Turner/Sacramento Municipal Utility District; 22.15: Dr. Pratt; 22.16: © Robert E. Ford/Biological Photo Service; 22.18: © Wolfgang Kaehler; 22.20: © Robert Frerck/Odyssey Productions; 22.21: Courtesy Burkhardt Turbines; 22.2: © W. Shields/Valan Photos; 22.24: © George Steinmetz Photography

Chapter 23

Opener: © Ray Pfortner/Peter Arnold, Inc.; 23.1: © Thomas Kitchin/Tom Stack & Associates; 23.3: © Fred McConnaughey/Photo Researchers, Inc.; 23.7: W.P. Cunningham; 23.10: © Mike Brisson; 23.11: Urban Ore, Inc.; 23.13: © Michael Greenlar/The Image Works, page 532: © Barbara Gauntt/*The Clarion-Ledger*, Jackson, MS; 23.17: © J. Cancalosi/Valan Photos; page 535: © Piet Van Lier/Impact Visuals

Chapter 24

Opener: © Malcolm S. Kirk/Peter Arnold, Inc.; 24.1: W.P. Cunningham; 24.2: © Robert Frerck/Odyssey Productions; 24.3: W.P. Cunningham; 24.4: © Walter Frerck/Odyssey Productions; 24.10: W.P. Cunningham; 24.11: © Louis Psihoyos/Matrix International, Inc.; 24.12, page 548: W.P. Cunningham; 24.13: © Joseph Sohm/Photo Researchers, Inc.; 24.14: AP/Wide World Photos; 24.16: © Tom Myers Photography; 24.18, 24.19, 24.20: W.P. Cunningham; 24.21: © Tom Myers/Tom Stack & Associates; 24.22: © Kevin R. Morris/Tony Stone Images; 24.23: W. P. Cunningham

Chapter 25

Opener: © William Campbell/Peter Arnold, Inc.; 25.1: © Wendy Stone/Gamma Liaison; 25.2: © Sam Kittner/Greenpeace Photo; 25.3: © Robert Queen, Wisconsin Department of Natural Resources; 25.4: © Frank Pedrick/The Image Works; page 572: Courtesy Lois Gibbs, Citizens Clearinghouse for Hazardous Waste; 25.8: © Esther Henderson/Photo Researchers, Inc.; 25.9: W. P. Cunningham; 25.10: © Mark Greenberg/Gamma Liaison; 25.11: W. P. Cunningham; 25.12: © Bryan F. Peterson; 25.13: W. P. Cunningham; 25.16: © Culley/Greenpeace Photo; 25.17: © Photoreporters, Inc.; 25.18: © David L. Brown/Tom Stack & Associates; 25.20: © Bob Daemmrich/The Image Works; 25.22: © Bill Bachman/Photo Researchers, Inc.

Illustrators

Bensen Studios:

1.2, Case Study 2.1-fig 1, 2.18, 3.2, 3.9, 3.10, 3.11, 3.20, 4.23, 5.15, Box 5.2-fig 1, 6.8, 7.7, 7.8, 7.16, 7.21, 7.22, Box 7.1-fig 1, Box 7.1-fig 2, 8.18, 8.19, In Depth 9.1-fig 1, 9.6, 9.8, 10.8b, 10.10, 12.11, 13.8, 13.9, 13.12, 13.21, 13.22, 14.2, 14.3, 14.6, 14.19, 14.20, 15.9, 15.11, 15.12, 17.13, 17.15b, 17.18, 17.20, Box 17.1-fig 1, 18.5, Box 18.1-fig 1, 18.10, 18.12, 18.13, Box 19.1-fig 1, 20.11, 21.3, 21.4, 21.6, 21.12, 21.21, Box 22.1-fig 1, 22.28, 23.18, 24.7, 25.15

Line Art

Chapter 1

14: From Jerome Fellmann, et al., *Human Geography* 4th ed., with data from World Bank and United Nations. Copyright © 1995 Times Mirror Higher Education Group, Inc., Dubuque, Iowa. All Rights Reserved. Reprinted by permission.

Chapter 3

18: From Michael D. Morgan, et al., *Environmental Science.* Copyright © 1991 Times Mirror Higher Education Group, Inc., Dubuque, Iowa. All Rights Reserved. Reprinted by permission.

Chapter 4

From Peter H. Raven and George B. Johnson, *Biology,* 4th ed. Copyright © 1996 Times Mirror Higher Education Group, Inc., Dubuque, Iowa. All Rights Reserved. Reprinted with permission. 23: From Temple, Stanley A. "Predicting Impacts of Habitat Fragmentation on Forest Birds: A Comparison of Two Models" in Verner, Jared, Michael L. Morrison, and C. John Ralph. *Wildlife Two Thousand: Modeling Habitat Relationships of Terrestrial Vertebrates* © 1986. (Madison: The University of Wisconsin Press.) Reprinted by permission of the University of Wisconsin Press.

Chapter 7

2: Figure redrawn with permission from *Population Bulletin*, vol. 18, no. 1, 1985, Population Reference Bureau. 5: Thomas W. Merrick, with PRB staff, "World Population in Transition," *Population Bulletin*, Vol. 41, No. 2 (reprinted 1991). 6: Map from *The Nystrom Desk Atlas* © 1994 NYSTROM. 14: Reprinted from *Beyond the Limits*, copyright 1992 by Meadows, Meadows and Randers. With permission from Chelsea Green Publishing Co., White River Junction, Vermont. Used by permission of the Canadian Publishers, McClelland & Stewart, Toronto; by permission of Earthscan Publications Ltd., London, UK.

Chapter 8

13, 14: Reprinted from *Beyond the Limits,* copyright 1992 by Meadows, Meadows and Randers. With permission from Chelsea Green Publishing Co., White River Junction, Vermont. Used by permission of the Canadian Publishers, McClelland & Stewart, Toronto; by permission of Earthscan Publications Ltd., London, UK. 15: *Environment,* June 1985. Reprinted with permission of the Helen Dwight Reid Educational Foundation. Published by Heldref Publications, 1319 Eighteenth St., N.W., Washington, D.C. 20036-1802. Copyright © 1985. 18: Source: E.S. Goodstein, Economic Policy Institute, Washington, DC. 19: Modified from Raymond Guzzle and Christopher Barrett, personal communication. page 422: Source UNICEF, *State of the World's Children 1993,* p. 6.

Chapter 10

6: Based on data from USDA as illustrated for Peter H. Raven and George B. Johnson, *Understanding Biology,* 3d ed. Copyright © 1995 Times Mirror Higher Education Group, Dubuque, Iowa. Reprinted by permission. All Rights Reserved. 13: From Jerome Fellmann, et al, *Human Geography,* 4th ed. Copyright © 1995 Times Mirror Higher Education Group, Inc., Dubuque, Iowa. All Rights Reserved. Reprinted by permission.

Chapter 13

9: Reprinted with permission from J. Curtis in William L. Thomas (ed.), *Man's Role in Changing the Face of the Earth* © 1956 by The University of Chicago Press. All rights reserved.

Chapter 14

22: From Jerome Fellmann, et al., *Human Geography,* 4th ed., from data in H.E. Dregne, *Desertification of Arid Lands,* Harwood Academic Press, 1982; and *World Map of Desertification,* UNESCO/FAO. Copyright © 1995 Times Mirror Higher Education Group, Inc., Dubuque, Iowa. Reprinted by permission.

Chapter 16

2: Redrawn from Carla W. Montgomery, *Physical Geography,* 3d ed. Copyright © 1993 Times Mirror Higher Education Group, Inc., Dubuque, Iowa. All Rights Reserved. Reprinted by permission. 12: Source: George Laycock, *Audubon Magazine,* vol. 91(7), July 1989, p. 77.

Chapter 17

12: Government of Canada, 1991, *The State of Canada's Environment.* Reproduced with the permission of the Minister of Supply and Services Canada, 1996.

Chapter 18

10: Government of Canada, 1991, *The State of Canada'a Environment.* Reproduced with the permission of the Minister of Supply and Services Canada, 1996. 21: Used with permission of General Motors Corporation.

Chapter 21

17: Courtesy of Northern States Power Company, Minneapolis. 20: Reprinted with permission from *Science News,* the weekly newsmagazine of science, copyright 1993 by Science Service, Inc. 21: Redrawn with permission of Atomic Energy of Canada Limited. 24: Reprinted from *Worldwatch Paper #75,* "Reassessing Nuclear Power" by Christopher Flavin (1987) with permission of Worldwatch Institute, Washington, D.C.

Index

8195

DATE DUE

JA 22 '99			
JY 25 '99			
MAY 0 1 2001			
1/16/04			
GAYLORD			PRINTED IN U.S.A